Springer Tracts in Modern Physics

Volume 265

Springer Tracts in Modern Physics

Springer Tracts in Modern Physics provides comprehensive and critical reviews of topics of current interest in physics. The following fields are emphasized: Elementary Particle Physics, Condensed Matter Physics, Light Matter Interaction, Atomic and Molecular Physics, Complex Systems, Fundamental Astrophysics.

Suitable reviews of other fields can also be accepted. The editors encourage prospective authors to correspond with them in advance of submitting a manuscript. For reviews of topics belonging to the above mentioned fields, they should address the responsible editor as listed below.

Special offer: For all clients with a print standing order we offer free access to the electronic volumes of the Series published in the current year.

More information about this series at http://www.springer.com/series/426

Kamakhya Prasad Ghatak

Dispersion Relations in Heavily-Doped Nanostructures

Springer

Kamakhya Prasad Ghatak
Department of Basic Science and Humanities
Institute of Engineering and Management
Salt Lake, Kolkata
India

ISSN 0081-3869 ISSN 1615-0430 (electronic)
Springer Tracts in Modern Physics
ISBN 978-3-319-20999-9 ISBN 978-3-319-21000-1 (eBook)
DOI 10.1007/978-3-319-21000-1

Library of Congress Control Number: 2015943034

Springer Cham Heidelberg New York Dordrecht London

Printed on acid-free paper

Springer International Publishing AG Switzerland is part of Springer Science+Business Media (www.springer.com)

Courage in adversity, patience in prosperity,
modesty in fame, indifference in own glory,
deep love for foes, staying in solitude,
all these are naturally found in wise persons

Knowledge is proud, he knows too much, but
the wise is humble, he knows no more

What lies before us and what lies behind us
are tiny matters as compared to what lies
WITHIN US

Smile and passion are real debit cards of life.
Pay now and use later. But ego and vindictive
attitude are like credit cards of life.
Use now and pay later.

Iron does not become steel without through
fire. I must be prepared to undergo the test of
fire, if I wish to enjoy the success in life.

To be the best, I must extract the best from myself

This book is dedicated to Dr. Amarnath Chakravarti, Ex. Professor and the Head of the Department of the Institute of Radio Physics and Electronics of the University of Calcutta, the Ph.D. Supervisor and Philosopher of the present author.

Sir, your teaching such creativity bring,
I must not change my state with a King

Preface

The creation of Nano Electronics, the subset of the generalized set Physics, is based on the following two important concepts:

- The symmetry of the wave vector space of charge carriers in electronic materials having various band structures is being reduced from a 3D closed surface to a quantized 2D closed surface, quantized non-parabolas and fully quantized wave vector space leading to the formation of 0D systems such as ultrathin films (UFs), doping superlattices, inversion and accumulation layers, quantum wells (QWs), quantum well superlattices, carbon nanotubes, nanowires (NWs), quantum wire superlattices, magnetic quantization, magneto size quantization, quantum dots (QDs), magneto inversion and accumulation layers, magneto quantum well superlattices, magneto NIPIs, quantum dot superlattices and other field aided nanostructures.
- The advent of modern experimental methods, namely Fine Line Lithography (FLL), Metallo-Organic Chemical Vapor Deposition (MOCVD), Molecular Beam Epitaxy (MBE), etc., for fabricating low-dimensional nanostructured systems.

Quantum confined materials have gained much interest in modern physics because of their importance to unlock both new scientific revelations and multi-dimensional altogether unheard of technological applications. In UFs, quantization of the motion of carriers in the direction perpendicular to the surface exhibits the two-dimensional carrier motion of charge carriers, and the third direction is being quantized. Another one-dimensional structure known as NWs has been proposed to investigate the physical properties in these materials, where the carrier gas is quantized in two transverse directions and they can move only in the longitudinal direction. As the concept of quantization increases from 1D to 3D, the degree of freedom of the free carriers decreases drastically and the total density-of-states (DOS) function changes from Heaviside step function to the Dirac's delta function forming QDs which, in turn, depend on the carrier energy spectra in different materials. An enormous range of important applications of such low-dimensional structures for modern physics in the quantum regime, together with a rapid increase in computing power, have generated considerable interest in the study of the optical properties of quantum effect devices based on various new materials of reduced dimensionality. Examples of such new applications include quantum switches, quantum registers, quantum sensors, heterojunction field-effect, quantum logic gates, quantum well and quantum wire transistors, quantum cascade lasers, high-frequency microwave circuits, high-speed digital networks, high-resolution terahertz spectroscopy, advanced integrated circuits, superlattice photo-oscillator, superlattice photo-cathodes, resonant tunneling diodes and transistors, superlattice coolers, thermoelectric devices, thin film transistors, micro-optical systems, intermediate-band solar cells, high performance infrared imaging systems, optical modulators, optical switching systems, single electron/molecule electronics, nanotube-based diodes, and other nanoelectronic devices [1–14].

Although many new effects in quantized structures have already been reported, the interest in further research of different other aspects of such quantum-confined materials is becoming increasingly important. One such significant concept is the Dispersion Relations (DRs) of carriers in semiconductors and their nanostructures, which occupies a singular position in the arena of Modern Physics and related disciplines in general and whose importance [15–36] has already been established since the inception of the theory of band structure of Solid State Physics. The concept of DRs is of fundamental importance for not only the characterization of semiconductor nanostructures, but also for the study of carrier transport in semiconductors and their quantized counterparts through proper formulation of the Boltzmann Transport equation which, in turn, needs the corresponding carrier energy spectra of the heavily doped materials and is still one of the open research problems. **It is important to note that six important transport quantities, namely the effective carrier mass (ECM), density-of-states (DOS) function, the sub-band energy and the measurement of band gap in the presence of strong light waves, intense electric field and heavy doping are in disguise in the very important concept of DR. Besides, the acoustic mobility limited momentum relaxation time is inversely proportional to the respective DOS function of a particular semiconductor and integral over the DOS function leads to carrier statistics under the condition of extreme carrier degeneracy which, in turn, is connected to the 25 important transport topics of quantum effect devices, namely** *the Landau Dia and Pauli's Para Magnetic Susceptibilities* [37], *the Einstein's Photoemission* [38], *the Einstein Relation* [39], *the Debye Screening Length* [40], *the Generalized Raman gain* [41], *the Normalized Hall coefficient* [42], *the Fowler-Nordheim Field Emission* [43], *the Gate Capacitance* [44], *the Thermoelectric Power* [45], *the Plasma Frequency* [46], *the Magneto-Thermal effect in Quantized Structures* [47], *the Activity coefficient* [48], *the Reflection coefficient* [49], *the Heat Capacity* [50], *the Faraday rotation* [51], *the Optical Effective Mass* [52], *the Carrier contribution to the elastic constants* [53], *the Diffusion coefficient of the minority carriers* [54], *the Nonlinear optical response* [55], *the Third order nonlinear optical susceptibility* [56], *the Righi-Leduc coefficient* [57], *the Electric Susceptibility* [58], *the Electric Susceptibility Mass* [59], *the*

Electron Diffusion Thermo-power [60] *and the Hydrostatic Piezo-resistance Coefficient* [61] *respectively.*

The present monograph solely investigates DRs in heavily doped nanostructures of nonlinear optical, III–V, II–VI, gallium phosphide, germanium, platinum antimonide, stressed, IV–VI, lead germanium telluride, tellurium, II–V, zinc and cadmium diphosphides, bismuth telluride, III–V, II–VI, IV–VI and HgTe/CdTe quantum well HD superlattices with graded interfaces under magnetic quantization, III–V, II–VI, IV–VI and HgTe/CdTe HD effective mass superlattices under magnetic quantization, quantum confined effective mass superlattices and superlattices of HD optoelectronic materials with graded interfaces in addition to other quantized systems. Incidentally, even after 40 years of continuous effort, we see that complete investigation of the DR comprising the whole set of HD materials and allied sciences is really a sea and permanently enjoys the domain of impossibility theorems. DRs have different forms for different materials and change under one-, two- and three-dimensional quantum confinement of charge carriers. *It is rather curious to note that for the* 31 *important concepts, only 6 monographs have been written* [62–67] *and the remaining* 25 *books will appear in the future, hopefully from the readers of this book.* In this context, it may be mentioned that the available reports on the said areas cannot afford to cover even an entire chapter containing detailed investigations on DRs in semiconductors and their quantized structures.

It is worth remarking that the effects of crossed electric and quantizing magnetic fields on the transport properties of semiconductors having various band structures have been relatively less investigated compared to the corresponding magnetic quantization, although the study of cross-fields is of fundamental importance with respect to the addition of new physics and related experimental findings in modern quantum effect devices. It is well known that in the presence of electric field (E_0) along x-axis and the quantizing magnetic field (B) along z-axis, the DRs of carriers in semiconductors become modified, for which the carrier moves in both the z and y directions respectively. The motion along y direction is purely due to the presence of E_0 along x-axis and in the absence of an electric field, the effective electron mass along y-axis tends to infinity indicating the fact that the electron motion along y-axis is forbidden. The effective electron mass of the isotropic, bulk semiconductors having parabolic energy bands exhibit mass anisotropy in the presence of cross-fields and this anisotropy depends on the electron energy, the magnetic quantum number, the electric and the magnetic fields respectively, although the effective electron mass along z-axis is a constant quantity. In 1966, Zawadzki and Lax [68] derived the expression of DR for III–V semiconductors in accordance with the two-band model of Kane under cross-fields configuration, which generates the interest to study this particular topic of solid-state science in general [69].

It is well known that heavy doping and carrier degeneracy are the keys to unlock the important properties of semiconductors; they are especially instrumental in dictating the characteristics of Ohmic and Schottky contacts respectively [70]. It is an amazing fact that although heavily doped semiconductors (HDS) have been investigated in the literature, the study of the corresponding DRs of HDS ***is still one of the open research problems***. We have obtained the ***exact E-k dispersion***

relations in HD nonlinear optical, III–V, II–VI, gallium phosphide, germanium, platinum antimonide, stressed, IV–VI, lead germanium telluride, tellurium, II–V, zinc and cadmium di-phosphides, bismuth telluride, III–V, II–VI, IV–VI and HgTe/CdTe quantum well HD superlattices with graded interfaces under magnetic quantization, III–V, II–VI, IV–VI and HgTe/CdTe HD effective mass superlattices under magnetic quantization, quantum confined effective mass superlattices and superlattices of HD optoelectronic materials with graded interfaces respectively. Our method is not related to the DOS technique as used in the aforementioned works. From the electron energy spectrum, one can obtain the DOS but the DOS technique, as used in the literature, cannot generate the DRs. ***Therefore, our study is more fundamental than those in the existing literature, because the Boltzmann transport equation, which controls the study of the charge transport properties of semiconductor devices, can be solved if and only if the DR is known.***

This book is divided into five parts, each containing 1, 11, 5, 1 and 3 chapters, respectively; it is partially based on our ongoing research on the DR in HDS from 1974 and an attempt has been made to present a cross section of the DR for a wide range of HDS and their quantized-structures with varying carrier energy spectra under various physical conditions. **It may be noted that among the 21 chapters of this book, two-third of the portion is new, whereas the remaining one-third is based on our previous eight books from Springer with necessary modifications in a condensed way not only for the larger cross section of readers and potential buyers to enjoy, but also to satisfy the self-consistent and sufficient conditions of Mathematics**. The single chapter in Part I deals with DRs in HD Quantum Wells (QWs), NWs and QDs, respectively, in the presence of surface magnetic field. In Chap. 1 we study the DR in heavily doped QWs, NWs and QDs of HD III–V, ternary, quaternary materials and IV–VI semiconductors in the presence of surface magnetic field, respectively, on the basis of **newly formulated electron energy spectra**. We also investigate the same in the presence of cross-fields. **This chapter explores the study of DR in cylindrical QD of HD III–V semiconductors in the presence of crossed electric and magnetic fields and in the presence of arbitrarily oriented magnetic field in QWs of HD III–V materials respectively. It is important to note that the surface magnetic field applied parallel to the surface makes effective carrier mass quantum number dependent, whose contribution to the oscillatory mobility would be important.**

Part II deals with the influence of quantum confinement on the DR of non-parabolic HDS and in Chap. 2 we study the DR in UFs of HD nonlinear optical materials on the basis of a generalized electron dispersion law introducing the anisotropies of the effective masses and the spin orbit splitting constants, respectively, together with the inclusion of the crystal field splitting within the framework of the ***k.p*** formalism. ***We observe that the complex electron dispersion law in HDS instead of the real one occurs from the existence of the essential poles in the corresponding electron energy spectrum in the absence of band tails.*** The physical picture behind the existence of the complex energy spectrum in heavily doped nonlinear optical semiconductors is the interaction of the impurity atoms in the tails with the splitting constants of the valance bands. The more the interaction, the more the prominence of the complex part than the other case. In the absence of band tails, there is no interaction of impurity atoms in the tails with the spin orbit constants and, consequently, the complex part vanishes. One important consequence of the HDS forming band tails is that ***the effective mass exists in the forbidden zone, which is impossible without the effect of band tailing. In the absence of band tails, the effective mass in the band gap of semiconductors is infinity. Besides, depending on the type of unperturbed carrier energy spectrum, the new forbidden zone will appear within the normal energy band gap for HDS.***

The results of HD III–V (e.g. InAs, InSb, GaAs, etc.), ternary (e.g. $Hg_{1-x}Cd_xTe$), quaternary (e.g. $In_{1-x}Ga_xAs_{1-y}P_y$ lattice matched to InP) compounds form a special case of our generalized analysis under certain limiting conditions. The DR in HD UFs of II–VI, IV–VI, stressed Kane type semiconductors, Te, GaP, $PtSb_2$, Bi_2Te_3, Ge, GaSb, II–V, lead germanium telluride, zinc and cadmium diphosphides has also been investigated in the appropriate sections. The importance of the aforementioned semiconductors has also been described in the same chapter. In the absence of band tails together with certain limiting conditions, all the results for all the DRs for all the HD UFs of Chap. 1 get simplified into the form of isotropic parabolic energy bands exhibiting the necessary mathematical compatibility test. In Chaps. 3 and 4, the DR for nanowires (NWs) and quantum dots (QDs) of all the materials of Chap. 2 have respectively been investigated. As a collateral study, we observe that the EEM in such QWs and NWs become a function of size quantum number, the Fermi energy, the scattering potential and other constants of the system, which is the intrinsic property of such 2D and 1D electrons.

With the advent of modern experimental techniques of fabricating nanomaterials as already noted, it is also possible to grow semiconductor superlattices (SLs) composed of alternative layers of two different degenerate layers with controlled thickness [71]. These structures have found wide applications in many new devices such as photodiodes [56], photoresistors [72], transistors [73], light emitters [74], tunneling devices [75], and others [76–88]. The investigations of the physical properties of narrow gap SLs have increased extensively since they are important for optoelectronic devices and because the quality of heterostructures involving narrow gap materials has been improved. It may be written in this context that doping superlattices are crystals with a periodic sequence of ultrathin film layers [89, 90] of the same semiconductor with the intrinsic layer in between, together with the opposite sign of doping. All the donors will be positively charged and all the acceptors negatively. This periodic space charge causes a periodic space charge potential which quantizes the motions of the carriers in the z direction together with the formation of sub-band energies. *In* Chap. *5, the DR in doping superlattices of HD nonlinear optical, III–V, II–VI, IV–VI, and stressed Kane type semiconductors has been investigated. In this case we note that the EEM in such doping supper lattices becomes a function of nipi sub-band index, surface electron concentration, Fermi energy, the scattering potential and other constants of the system, which is the intrinsic property of such 2D quantized systems.*

It is well known that the electrons in bulk semiconductors in general have three-dimensional freedom of motion. When these electrons are confined in a one-dimensional potential well, whose width is of the order of the carrier wavelength, the motion in that particular direction gets quantized while that along the other two directions remains free. Thus, the energy spectrum appears in the shape of discrete levels for one-dimensional quantization, each of which has a continuum for two-dimensional free motion. The transport phenomena of such one-dimensional confined carriers have recently been studied [7.1–7.20] with great interest. For the metal-oxide-semiconductor (MOS) structures, the work functions of the metal and the semiconductor substrate are different and the application of an external voltage at the metal-gate causes the change in charge density at the oxide semiconductor interface leading to a bending of the energy bands of the semiconductor near the surface. As a result, a one-dimensional potential well is formed at the semiconductor interface. The spatial variation of the potential profile is so sharp that for considerably large values of the electric field, the width of the potential well becomes of the order of the de Broglie wavelength of the carriers. The Fermi energy, which is near the edge of the conduction band in the bulk, becomes nearer to the edge of the valance band at the surface creating inversion layers. The energy levels of the carriers bound within the potential well get quantized and form electric sub bands. Each sub band corresponds to a quantized level in a plane perpendicular to the surface leading to a quasi two-dimensional electron gas. Thus, the extreme band bending at low temperature allows us to observe the quantum effects at the surface [91]. Although considerable work has already been done regarding the various physical properties of different types of inversion layers having various band structures, nevertheless it appears from the literature that there lies scope in the investigations made while the interest for studying different other features of accumulation layers is becoming increasingly important. In Chap. 6, the DR in accumulation layers of HD nonlinear optical, III–V, II–VI, IV–VI, stressed Kane type semiconductors and Ge have been investigated. For the purpose of relative comparisons, we have also studied the DR in inversion layers of the aforementioned materials. *It is interesting to note that the EEM in such layers is a function of electric sub-band index, surface electric field, Fermi energy, scattering potential and other constants of the system, which is the intrinsic property of such 2D electrons.*

It is worth remarking that the effects of quantizing magnetic field (B) on the band structures of compound semiconductors are more striking than those of the parabolic one and are easily observed in experiments. A number of interesting physical features originate from the significant changes in the basic energy wave vector relation of the carriers caused by the magnetic field. Valuable information could also be obtained from experiments under magnetic quantization regarding important physical properties such as Fermi energy and effective masses of the carriers, which affect almost all the transport properties of the electron devices [92] of various materials having different carrier dispersion relations [93, 94].

Specifically, in Chap. 7 we study the DR in HD nonlinear optical materials in the presence of magnetic quantization. The results of HD III–V (e.g. InAs, InSb, GaAs etc.), ternary (e.g. $Hg_{1-x}Cd_xTe$), quaternary (e.g. $In_{1-x}Ga_xAs_{1-y}P_y$ lattice matched to InP) compounds form a special case of our generalized analysis under certain limiting conditions. The magneto DR for HD II–VI, IV–VI, stressed Kane type semiconductors, Te, GaP, $PtSb_2$, Bi_2Te_3, Ge, GaSb, II–V and lead germanium telluride has also been investigated by formulating the respective appropriate HD energy band structure. In the absence of band tails, together with certain limiting conditions, all the results for all the DRs for all the HD compounds as considered in this chapter get simplified into the well-known parabolic energy bands under magnetic quantization exhibiting the necessary mathematical compatibility test.

Chapter 8 ***investigates the DR under cross-field configuration in HD nonlinear optical, III–V, II–VI, IV–VI and stressed Kane type semiconductors respectively. This chapter also tells us that the EEM in all the cases is a function of the finite scattering potential, the magnetic quantum number, the electric field, the quantizing magnetic field and the Fermi energy even for HD semiconductors, whose bulk electrons in the absence of band tails are defined by the parabolic energy bands.*** In Chap. 9 the DR in HDs of non-parabolic semiconductors under magneto-size quantization has been studied for all the materials of Chap. 7. In Chap. 10 the DR in HD ultrathin films under cross-fields configuration has been investigated for all the materials of Chap. 8. In Chap. 11 the magneto DR in doping superlattices has been investigated for all the cases of Chap. 5. The magneto DR in accumulation and inversion layers has been explored in Chap. 12 for all the case of Chap. 6.

In Part III we study the DRs in quantum confined superlattices (SLs). It is well known that Keldysh [95] first suggested the fundamental concept of a SL, although it was successfully experimentally realized by Esaki and Tsu [96]. The importance of SLs in the field of nanoelectronics has already been described in [97–99]. The most extensively studied III–V SL is that consisting of alternate layers of GaAs and $Ga_{1-x}Al_xAs$ owing to the relative ease of fabrication. The GaAs layer forms quantum wells and $Ga_{1-x}Al_xAs$ form potential barriers. The III–V SLs are attractive for the realization of high speed electronic and optoelectronic devices [100]. In addition to SLs of the usual structure, SLs with more complex structures such as II–VI [101], IV–VI [102] and HgTe/CdTe [103] SLs have also been proposed. The IV–VI SLs exhibit quite different properties compared to the III–V SL due to the peculiar band structure of the constituent materials [104]. The epitaxial growth of II–VI SL is a relatively recent development and the primary motivation for studying the mentioned SLs made of materials with large band gap is in their potential for optoelectronic operation in the blue [104]. HgTe/CdTe SLs have raised a great deal of attention since 1979 as a promising new material for long wavelength infrared detectors and other electro-optical applications [105]. Interest in Hg-based SLs further increased as new properties with potential device applications were revealed [106]. These features arise from the unique zero band gap material HgTe [107] and the direct band gap semiconductor CdTe, which can be described by the three-band mode of Kane [108]. The combination of the aforementioned materials with specified dispersion relation makes HgTe/CdTe SL very attractive, especially because of the possibility to tailor the material properties for various applications by varying the energy band constants of the SLs. In addition, for effective mass SLs, the electronic sub-bands appear continually in real space [109].

We note that all the aforementioned SLs have been proposed with the assumption that the interfaces between the layers are sharply defined, of zero thickness, i.e., devoid of any interface effects. The SL potential distribution may be then considered as a one-dimensional array of rectangular potential wells. The aforementioned advanced experimental techniques may produce SLs with physical interfaces between the two materials crystallo-graphically abrupt; adjoining their interface will change at least on an atomic scale. As the potential form changes from a well (barrier) to a barrier (well), an intermediate potential region exists for the electrons. The influence of finite thickness of the interfaces on the electron dispersion law is very important, since the electron energy spectrum governs the electron transport in SLs.

In Chap. 13 the DR in III–V, II–VI, IV–VI, HgTe/CdTe and strained layer quantum well heavily doped superlattices (QWHDSLs) with graded interfaces is studied. Besides, the DR in III–V, II–VI, IV–VI, HgTe/CdTe and strained layer quantum well HD effective mass superlattices, respectively, has also been explored. In Chaps. 14 and 15 the DRs in quantum wire HD superlattices and quantum dot HD superlattices have, respectively, been investigated for all the cases of Chap. 13. Chapter 16 contains the study of the DR in HD SLs under magnetic quantization for all the cases of Chap. 13. In the last Chap. 17 of Part III, we have studied the DR in quantum well HD superlattices under magnetic quantization for all the cases of Chap. 13.

With the advent of nanophotonics, there has been considerable interest in studying the optical processes in semiconductors and their nanostructures in the presence of intense light waves [110]. ***It appears from the literature that the investigations in the presence of external intense photo-excitation have been carried out on the assumption that the carrier energy spectra are invariant quantities under strong external light waves, which is not fundamentally true.*** The physical properties of semiconductors in the presence of strong light waves which alter the basic dispersion relations have relatively been much less investigated in [111, 112] as compared with the cases of other external fields and in opto-electronics the influence of strong light waves is needed for the characterization of low-dimensional opto-electronic devices. The solo Chap. 18 of Part IV investigates the DR in bulk specimens HD Kane type semiconductors under intense light waves. The same chapter studies DR in the presence of magnetic quantization, cross-fields configuration, QWs, NWs, QDs, magneto size quantization, inversion and accumulation layers, magneto inversion and magneto accumulation layers, doping superlattices, magneto doping superlattices, QWHD, NWHD and QDHD effective mass superlattices, magneto QWHD effective mass superlattices, magneto HD effective mass superlattices, QWHD, NWHD and QDHD superlattices with graded interfaces, magneto QWHD superlattices with graded interfaces and magneto HD superlattices with graded interfaces respectively.

With the advent of nanodevices, the built-in electric field becomes so large that the electron energy spectrum changes fundamentally [113–115] ***instead of being invariant*** and Chap. 19 of Part V of this book investigates the DR under intense electric field in bulk specimens of HD III–V, ternary and quaternary semiconductors. The same chapter also explores the influence of electric field on the DR in the presence of magnetic quantization, cross-fields configuration, QWs, NWs, QDs, magneto size quantum effect, inversion and accumulation layers, magneto inversion and magneto accumulation layers, doping superlattices, magneto doping superlattices, QWHD, NWHD and QDHD effective mass superlattices, magneto QWHD effective mass superlattices, magneto HD effective mass superlattices, QWHD, NWHD and QDHD superlattices with graded interfaces, magneto QWHD superlattices with graded interfaces and magneto HD superlattices with graded interfaces and, respectively, magnetic quantization, size quantization, accumulation layers, HD doping superlattices and effective mass HD superlattices under magnetic quantization respectively. **It is interesting to note that the EEM depends on the strong electric field** (which is not observed elsewhere) together with the fact that the EEM in the said systems depends on the respective quantum numbers in addition to the Fermi energy, the scattering potential and others system constants which are the characteristics features of such heterostructures.

Chapter 20 explores **28 different applications, namely***Carrier Statistics, Thermoelectric Power, Debye Screening Length, Carrier contribution to the elastic constants, Diffusivity-mobility ratio, Measurement of Band-gap in the presence of Light Waves, Diffusion coefficient of the minority carriers, Nonlinear optical response, Third order nonlinear optical susceptibility, Generalized Raman gain, The plasma frequency, The activity coefficient, Magneto-Thermal effect in Quantized Structures, Normalized Hall coefficient, Reflection coefficient, Heat Capacity, Magnetic Susceptibilities, Faraday rotation, Fowler-NordheimFied Emission, Optical Effective Mass, Einstein's Photoemission, Righi-Leduc coefficient, Electric Susceptibility, Electric Susceptibility Mass, Electron Diffusion Thermo-power, Hydrostatic Piezo-resistance Coefficient, Relaxation time for Acoustic Mode Scattering and Gate Capacitance* of the content of this book and Chap. 21 contains the conclusions and future research. It is needless to say that this monograph is based on the '***iceberg principle***' [116] and the rest of it will be explored by researchers from different appropriate fields. *Since there is no existing report devoted solely to the study of DR for HD quantized structures to the best of our knowledge, we earnestly hope that the present book will a useful reference source for the present and the next generation of readers and researchers of materials and allied sciences in general. We have discussed enough regarding DRs in different quantized HD materials although a number of new computer-oriented numerical analyses are being left for the purpose of being computed by the readers, to generate the new graphs and the inferences from them which altogether is a sea by itself.* Since the production of an error-free first edition of any book from every point of view is a permanent member of impossibility theorems, therefore in spite of our joint concentrated efforts for a couple of years together with the seasoned team of Springer, the same stands very true for this monograph also. ***Various expressions and a few chapters of this book are appearing for the first time in printed form.*** Suggestions from the readers for the improvement of the book will be highly appreciated for the purpose of inclusion in future editions, if any. ***In this book, from***

the first chapter till the end, we have presented 200 open research problems for graduate students, PhD aspirants, researchers and engineers in this pinpointed research topic. We strongly hope that alert readers of this monograph will not only solve the said problems by removing all the mathematical approximations and establishing the appropriate uniqueness conditions, but will also generate new research problems both theoretical and experimental and, thereby, transform this monograph into a solid book. Incidentally, our readers after reading this book will easily understand how little is presented and how much more is yet to be investigated on this exciting topic, which is the signature of coexistence of new physics and advanced mathematics combined with the inner fire for performing creative researches in this context by young scientists, as like Kikoin [117] we feel that "***A young scientist is no good if his teacher learns nothing from him and gives his teacher nothing to be proud of***". We emphatically assert that the problems presented here form an integral part of this book and will be useful for readers to initiate their own contributions on the DR for HDS and their quantized counterparts. Like Sakurai [118], we firmly believe *"The reader who has read the book but cannot do the exercise has learned nothing"*. It is nice to note that if we assign the alphabets A–Z, the positive integers from 1 to 26 chronologically, then the word "*ATTITUDE* " receives the perfect score 100 and is the vital quality needed from the readers since *attitude* is the ladder on which all other virtues mount.

In this monograph, we have investigated ***the expressions of effective electron mass and the sub-band energy has been formulated throughout this monograph as a collateral study for the purpose of in-depth investigations of the mentioned important pinpointed research topics***. Thus, in this book, the readers will get much information regarding the influence of quantization in HD low-dimensional materials having different band structures. Although the name of the book is an example of extremely high Q-factor, from the content one can easily infer that it should be useful for graduate courses on materials science, condensed matter physics, solid-state electronics, nanoscience and technology and solid-state sciences and devices in many universities and the institutions in addition to both Ph.D. students and researchers in the aforementioned fields. *Last but not the least, the author hopes that his humble effort will kindle the desire to delve deeper into this fascinating and deep topic by anyone engaged in materials research and device development either in academics or in industries.*

References

1. M. Rieth, W. Schommers (eds.), *Handbook of Computational and Theoretical Nanoscience*, vol. 1–10 (American Scientific Publishers, Los Angeles, 2006); M. Rieth, W. Schommers, J. Nanosci. Nanotechnol. **2,** 679 (2002)
2. H.S. Nalwa (ed.), *Encyclopedia of Nanoscience and Nanotechnology*, vol. 1–25 (American Scientific Publishers, Los Angeles, 2004/2011); H.S. Nalwa (ed.), *Soft Nanomaterials*, vols. 1–2 (American Scientific Publishers, Los Angeles, 2009); H.S. Nalwa (ed.), *Magnetic Nanostructures* (American Scientific Publishers, Los Angeles, 2002)

3. H.S. Nalwa (ed.), *Nanomaterials for Energy Storage Applications* (American Scientific Publishers, Los Angeles, 2009); H.S. Nalwa (ed.), *Handbook of Nanostructured Biomaterials and Their Applications in Nanobiotechnology*, vol. 2 (American Scientific Publishers, Los Angeles, 2005); S. Bandyopadhyay, H.S. Nalwa (eds.) *Quantum Dots and Nanowires,* vol. 5 (American Scientific Publishers, Los Angeles, 2003)
4. T.Y. Tseng, H.S. Nalwa (eds.), *Handbook of Nanoceramics and Their Based Nanodevices*, vol. 1–5 (American Scientific Publishers, Los Angeles, 2009); K. Ariga, H.S. Nalwa (eds.), *Bottom-up Nanofabrication*, vol. 1–6 (American Scientific Publishers, Los Angeles, 2009); K.P. Ghatak, S.N. Biswas, *Acta. Phys. Slovaca.* **43**, 425 (1993)
5. Y. Lin, H.S. Nalwa (eds.), *Handbook of Electrochemical Nanotechnology*, vol. 1–2 (American Scientific Publishers, Los Angeles, 2009); A.K. Mohanty, M. Misra, H.S. Nalwa (eds.), *Packaging Nanotechnology* (American Scientific Publishers, Los Angeles, 2009)
6. S. Yan, Y. Shi, Z. Xiao, M. Zhou, W. Yan, H. Shen, D. Hu, J. Nanosci. Nanotechnol. **12**, 6873 (2012); J.R.C. Smirnov, E. Jover, R. Amade, G. Gabriel, R. Villa, E. Bertran, J. Nanosci. Nanotechnol. **12**, 6941 (2012)
7. M.G. Mashapa, N. Chetty, S.S. Ray, J. Nanosci. Nanotechnol. **12**, 7030 (2012); D.H. Kim, J. H. You, J.H. Kim, K.H. Yoo, T.W. Kim, J. Nanosci. Nanotechnol. **12**, 5687 (2012); A. Ooba, Y. Fujimura, K. Takahashi, T. Komine, R. Sugita, J. Nanosci. Nanotechnol. **12**, 7411 (2012); C.W. Tan, B.K. Tay, J. Nanosci. Nanotechnol. **12**, 707 (2012)
8. S.J. Noh, Y. Miyamoto, M. Okuda, N. Hayashi, Y.K. Kim, J. Nanosci. Nanotechnol. **12**, 428 (2012); S. Padalkar, K. Schroeder, Y.H. Won, H.S. Jang, L. Stanciu, J. Nanosci. Nanotechnol. **12**, 227 (2012); K.Y. Lee, B. Kumar, H.K. Park, W.M. Choi, J.Y. Choi, S. W. Kim, J. Nanosci. Nanotechnol. **12**, 1551 (2012)
9. J.M. Wu, J. Nanosci. Nanotechnol. **12**, 1434 (2012); Y. Zhang, L. Su, K. Lubonja, C. Yu, Y. Liu, D. Huo, C. Hou, Y. Lei, Sci. Adv. Mater. **4**, 825 (2012)
10. R. Tamang, B. Varghese, E.S. Tok, S. Mhaisalkar, C.H. Sow, Nanosci. Nanotechnol. Lett. **4**, 716 (2012); Y. Yue, M. Chen, Y.J, Nanosci. Nanotechnol. Lett. **4,** 414 (2012)
11. S.M. Adhikari, A. Sakar, K.P. Ghatak, Quantum Matter **2**, 455 (2013); K.P. Ghatak, P.K. Bose, S. Bhattacharya, A. Bhattacharjee, D. De, S. Ghosh, Quantum Matter **2**, 83 (2013); K. P. Ghatak, S. Bhattacharya, A. Mondal, S. Debbarma, P. Ghorai, A. Bhattacharjee, Quantum Matter **2,** 25 (2013)
12. N. Paitya, S. Bhattacharya, D. De, K.P. Ghatak, Quantum Matter **1**, 63 (2012); S. Bhattacharya, N. Paitya, K.P. Ghatak, J. Comput. Theor. Nanosci. **10**, 1999 (2013)
13. S. Choudhury, D. De, S. Mukherjee, A. Neogi, A. Sinha, M. Pal, S.K. Biswas, S. Pahari, S. Bhattacharya, K.P. Ghatak, J. Comput. Theor. Nanosci. **5**, 375 (2008); K.P. Ghatak, S. Bhattacharya, D. De, R. Sarkar, S. Pahari, A. Dey, A.K. Dasgupta, S.N. Biswas, J. Comput. Theor. Nanosci. **5**, 1345 (2008); S. Bhattacharya, R. Sarkar, D. De, S. Mukherjee, S. Pahari, A. Saha, S. Roy, N.C. Paul, S. Ghosh, K.P. Ghatak, J. Comput. Theor. Nanosci. **6**, 112 (2009)
14. S. Bhattacharya, S. Chowdhury, S. Ghoshal, S.K. Biswas, D. De, K.P. Ghatak, J. Comput. Theor. Nanosci. **3**, 423 (2006); S. Bhattacharya, D. De, S. Chowdhury, S. Karmakar, D.K. Basu, S. Pahari, K.P. Ghatak, J. Comput. Theor. Nanosci. **3,** 280 (2006); K.P. Ghatak, S. Karmakar, D. De, S. Pahari, S.K. Charaborty, S.K. Biswas, S. Chowdhury, J. Comput. Theor. Nanosci. **3**, 153 (2006); K.P. Ghatak, P.K. Bose, A.R. Ghatak, D. De, S. Pahari, S.K. Chakraborty, S.K. Biswas, J. Comput. Theor. Nanosci. **2**, 423 (2005)
15. N. Paitya, K.P. Ghatak, J. Adv. Phys. **1**, 161 **(2012)**; S. Debbarma, A. Bhattacharjee, S. Bhattacharyya, A. Mondal, K. P. Ghatak, *J. Adv. Phys.* 1, 84 (2012); S. Bhattacharya, D. De, S.M. Adhikari, K.P. Ghatak, Superlattices Microstruct. **51**, 203 (2012); D. De, S. Bhattacharya, S.M. Adhikari, A. Kumar, P.K. Bose, K.P. Ghatak, Beilstein, J. Nanotechnol. **2**, 339 (2011)
16. S. Bhattacharya, S. Choudhury, K.P. Ghatak, Superlattices Microstruct. **48**, 257 (2010); S. Pahari, S. Bhattacharya, S. Roy, A. Saha, D. De, K.P. Ghatak, Superlattices Microstruct. **46**, 760 (2009); D. De, K.P. Ghatak, J. Mater. Sci.: Mater. Electron. **20**, 185 (2009)

17. S.M. Adhikari, D. De, J.K. Baruah, S. Chowdhury, K.P. Ghatak, Adv. Sci. Focus **1**, 54 (2013); S. Bhattacharya, S. Pahari, R. Sarkar, S. Ghosh, K.P. Ghatak, Physica B **403**, 3635 (2008)
18. L.J. Singh, S. Choudhury, D. Baruah, S.K. Biswas, S. Pahari, K.P. Ghatak, Physica B **368**, 188 (2005); P.K. Chakraborty, S. Bhattacharya, K.P. Ghatak, J. Appl. Phys. **98**, 053517 (2005)
19. P.K. Chakraborty, S. Ghoshal, K.P. Ghatak, Physica B **382**, 26 (2006); S. Chowdhary, L. J. Singh, K.P. Ghatak, Physica B **365**, 5 (2005); K.P. Ghatak, D.K. Basu, S. Bhattacharya, S. Ghosal, S.K. Biswas, A. Mallik and G. Majumdar, Nonlinear Opt. Quantum Opt. **33**, 221 (2005); L.J. Singh, S. Choudhury, S.S. Roy, K.P. Ghatak, Electr. Eng. **87**, 1 (2005)
20. K.P. Ghatak, S.K. Biswas, D. De, S. Ghosal, S. Chatterjee, Physica B **353**, 127 (2004); P.K. Chakraborty, S.K. Biswas, K.P. Ghatak, Physica B **352**, 111 (2004); P.K. Chakraborty, L. J. Singh, K.P. Ghatak, J. Appl. Phys. **95**, 5311 (2004); P.K Chakraborty, G.C Datta, K.P Ghatak, Physica B **339**, 198 (2003)
21. P.K. Chakraborty, K.P. Ghatak, J. Phys. Chem. Solids **62**, 1061 (2001); K.P. Ghatak, J.Y. Siddiqui, B. Nag, Phys. Lett. A **282**, 428 (2001); P.K. Chakraborty, K.P. Ghatak, J. Appl. Phys. **89**, 1075 (2001)
22. P.K. Chakraborty, K.P. Ghatak, J. Phys. D Appl. Phys. **32**, 2438 (1999); B. Nag, K. P. Ghatak, J. Phys. Chem. Solids **59**, 713 (1998); K.P. Ghatak, B. Nag, J. Phys. Chem. Solids **59**, 411 (1998), K.P. Ghatak, B. Nag, Phys. Status. Solidi. B **205**, 519 (1998)
23. K.P. Ghatak, J.P. Banerjee, B. Nag, J. Appl. Phys. **83**, 1420 (1998); B. Nag, K.P. Ghatak, Nonlinear Opt. **19**, 1 (1998); J.P. Banerjee, S. Ghosh, B. Nag, K.P. Ghatak, Int. J. Electron. **82**, 335 (1997)
24. K.P Ghatak, D.K Basu, B. Nag, J. Phys. Chem. Solids **58**, 133 (1997); B. Mitra, D.K. Basu, B. Nag, K.P. Ghatak, Nonlinear Opt. **17**, 171 (1997); K.P. Ghatak, B. Mitra, B. Nag, Phys. Status. Solidi. B. **199**, 95 (1997); B. Nag, K.P. Ghatak, Phys. Scr. **54**, 657 (1996)
25. K.P. Ghatak, D.K. Basu, D. Basu, B. Nag, II Nuovo Cimento D **18**, 947 (1996); K.P. Ghatak, S. Beta, A. Ali, B. Nag, Fizika A **5**, 111 (1996); K.P. Ghatak, B. De, B. Nag, P.K. Chakraborty, Nonlinear Opt. **16**, 221 (1996); K.P. Ghatak, J.P Banerjee, B. Goswami, B. Nag, Nonlinear Opt. **16**, 241 (1996)
26. K.P. Ghatak, J.P. Banerjee, M. Mitra, B. Nag, Nonlinear opt. **16**, 193 (1996); K.P. Ghatak, M. Mitra, B. Goswami, B. Nag, Nonlinear opt. **16**, 167 (1996); K.P. Ghatak, B. Nag, Nanostruct. Mater. **5**, 769 (1995); K.P. Ghatak, B. Nag, D. Bhattacharyya, J. Low Temp. Phys. **101**, 983 (1995)
27. K.P. Ghatak, D. Bhattacharya, D. Basu, B. Nag, Phys. Status. Solidi. B **191**, 141 (1995); K. P. Ghatak, D. Bhattacharyya, Phys. Scr. **52**, 343 (1995); D. Bhattacharyya, K.P. Ghatak, Phys. Status. Solidi. B **187**, 523 (1995); K.P. Ghatak, S.N. Banik, *Fizika* A **4**, 33 (1995)
28. K.P. Ghatak, M. Mondal, Phys. Status. Solidi. B **185**, K5 (1994); K.P. Ghatak, S.N. Biswas, Solid State Electron. **37**, 1437 (1994); K.P. Ghatak, Acta Phys. Hung. **74**, 167 (1994); K. P. Ghatak, D. Bhattacharyya, Phys. Lett. A **184**, 366 (1994)
29. S. Banik, K.P. Ghatak, S. Biswas, Fizika A **3**, 77 (1994); K.P. Ghatak, S.N. Banik, Fizika A **3**, 127 (1994); K.P. Ghatak, S.N. Banik, Fizika A **3**, 155 (1994); K.P. Ghatak, D. Bhattacharyya, Phys. Status. Solidi. B **181**, 127 (1994); K.P. Ghatak, S. Biswas, Fizika A **3**, 7 (1994); K.P. Ghatak, D. Bhattacharya, Phys. Status. Solidi. B **179**, 383 (1993); P.K. Chakraborty, K.P. Ghatak, J. Appl. Phys. **74**, 3246 (1993)
30. K.P. Ghatak, S.N. Biswas, Nanostruct. Mater. **2**, 91 (1993); K.P. Ghatak, Fizika A. **2**, 133 (1993); K.P. Ghatak, S.N. Biswas, II Nuovo Cimento D **15**, 87 (1993)
31. K.P. Ghatak, A. Ghoshal, S.N. Biswas, II Nuovo Cimento D **15**, 39 (1993); K.P. Ghatak, Fizika A **2**, 59 (1993); K.P. Ghatak, M. Mondal, Phys. Status. Solidi. B **175**, 113 (1993); K. P. Ghatak, B. Mitra, II Nuovo Cimento D **15**, 97 (1993)
32. K.P. Ghatak, Acta Phys. Hung. **72**, 147 (1992); K.P. Ghatak, Il Nuovo Cimento D **14**, 1187 (1992); P.K. Chakraborty, K.P. Ghatak, J. Appl. Phys. **72**, 2086 (1992); K.P. Ghatak, A. Ghoshal, B. Mitra, Il Nuovo Cimento D **14**, 903 (1992)

33. S.N. Biswas, K.P. Ghatak, Int. J. Electron. **73**, 287 (1992); K.P. Ghatak, B. Mitra, Phys. Scr. **46**, 182 (1993); K.P. Ghatak, B. Mitra, Int. J. Electron. **72**, 541 (1992); K.P. Ghatak, M. Mondal, Phys. Status. Solidi. B **170**, 57 (1992)
34. K.P. Ghatak, A. Ghoshal, Phys. Status. Solidi. B **170**, K27 (1992); K.P. Ghatak, M. Mondal, J. Appl. Phys. **71**, 1277 (1992); K.P. Ghatak, M. Mondal, Fizika A **1**, 113 (1992); K. P. Ghatak, Fizika A **1**, 197 (1992)
35. K.P. Ghatak, Fizika A **1**, 207 (1992); M. Mondal, A. Ghoshal, K.P. Ghatak, Il Nuovo Cimento D **14**, 63 (1992); K.P. Ghatak, S.N. Biswas, J. Appl. Phys. **70**, 4309 (1991); K. P. Ghatak, Il Nuovo Cimento D **13**, 1321 (1991)
36. C. Majumdar, A.B. Maity, A.N. Chakravarti, Phys. Stat. Sol. B **140**, K7 (1987); C. Majumdar, A.B. Maity, A.N. Chakravarti, Phys. Stat. Sol. B **141**, K35 (1987); N.R. Das, K. K. Ghosh, D. Ghoshal, Phys. Stat. Sol. B **197**, 97 (1996); C. Majumdar, A.B. Maity, A.N. Chakravarti, Phys. Stat. Sol. B, **144**, K13 (1987); N.R. Das, A.N. Chakravarti, Phys. Stat. Sol. B **176**, 335 (1993); S. Sen, N.R. Das and A.N. Chakravarti, Jour. of Phys: Conden. Mat. **19**, 186205 (2007); N.R. Das, S. Ghosh, A.N. Chakravarti, Phys. Stat. Sol. B **174**, 45 (1992); A.B. Maity, C. Majumdar, A.N. Chakravarti, Phys. Stat. Sol. B **144**, K93, (1987); A.B. Maity, C. Majumdar, A.N. Chakravarti, Phys. Stat. Sol. B **149**, 565 (1988)
37. M. Mondal, K.P. Ghatak, Phys. Status. Solidi. B **126**, K41 (1984); D.R. Choudhury, A.K. Chowdhury, K.P. Ghatak, A.N. Chakravarti, Phys. Status. Solidi. B **98**, K141 (1980); K. P. Ghatak, M. Mondal, Phys. Status. Solidi. B **135**, K119 (1986); K.P. Ghatak, S. Biswas, Society of Photo-Optical Instrumentation Engineers (SPIE) Conference Series, **818**, 115 (1987), E.V. Rozenfeld, Yu. P. Irkhin, P.G. Guletskii, Sov. Phys. Solid State **29**, 1134 (1987)
38. K.P. Ghatak, D. De, S. Bhattacharya, *Photoemission from Optoelectronic Materials and Their Nanostructures*. Springer Series in Nanostructure Science and Technology (Springer, Heidelberg, 2009); R.K. Pathria, *Statistical Mechanics*, 2nd edn. (Butterworth-Heinemann, Oxford, 1996); L. Torres, L. Lopez-Diaz, O. Alejos, J. Iniguez, J. Appl. Phys. **85**, 6208 (1999); S. Tiwari, S. Tiwari, Cryst. Res. Technol. **41**, 78 (2006); R. Houdré, C. Hermann, G. Lampel, P.M. Frijlink, A.C. Gossard, Phys. Rev. Lett. **55**, 734 (1985); K.P. Ghatak, S. Bhattacharya, K.M. Singh, S. Choudhury, S. Pahari, Physica B **403**, 2116 (2008); H. Shimoda, B. Gao, X.-P. Tang, A. Kleinhammes, L. Fleming, Y. Wu, O. Zhou, Phys. Rev. Lett. **88**, 015502 (2002); M. Maillard, P. Monchicourt, M.P. Pileni, Chem. Phys. Lett. **380**, 704 (2003)
39. A.N. Chakravarti, B.R. Nag, Phys. Status. Solidi. A **14**, K55 (1972); Int. J. Elect. **37** (281) (1974); Phys. Status. Solidi. A **14**, K23 (1972); A.N. Chakravarti, D.P. Parui, Can. J. Phys. **51**, 451 (1973). D. Mukherjee, A.N.Chakravarti, B.R. Nag, Phys. Phys. Status. Solidi. A **26**, K27 (1974)
40. K.P. Ghatak, S.Bhattacharya, J. Appl. Phys. **102**, 073704 (2007); M. Mondal, K.P. Ghatak, Phys. Lett. **102A**, 54 (1984); P.K. Chakraborty, G.C. Datta, K.P. Ghatak, Phys. Scr. **68**, 368 (2003). E.O. Kane, Solid State Electron. **28**, 3 (1985); W. Zawadzki, in *Handbook on Semiconductors*, ed. by W. Paul (North Holland, New York, 1982), 1, p. 715
41. B.S. Wherreff, W. Wolland, C.R. Pidgeon, R.B. Dennis, S.D. Smith, in *Proceedings of the 12th International Conference of the Physics of the Semiconductors*, ed. by M.H. Pilkahn, R. G. Tenbner (Staffgard, 1978), p. 793
42. F.J. Blatt, *J. Phys. Chem. Solids***17**, 177 (1961)
43. R.H. Fowler, L. Nordheim, Proc. Roy. Soc. Lond., Ser. A **119**,173 (1928); A.Van Der Ziel, *SolidState Physical Electronics* (Prentice-Hall, Inc., Englewood Cliffs, USA, 1957), p.176; B. Mitra, K.P. Ghatak, Phys. Lett. A **357**, 146 (1990); K.P. Ghatak, M. Mondal, J. Mag. Mag. Mater. **74**, 203 (1988); K.P. Ghatak, B. Mitra, Phys. Lett. **A156**, 233 (1991); K. P. Ghatak, A. Ghosal, S.N. Biswas, M. Mondal, Proc. of SPIE, USA, **1308**, 356 (1990)
44. K.P. Ghatak, S. Biswas, J. Vac. Sci. Technol. B **7**(1), 104 (1989); K.P. Ghatak, M. Mondal, Z. Phys. B: Condens. Matter **64**, 223 (1986); D.R. Choudhury, A.K. Chowdhury, A.N. Chakravarti, Phys. Scr. **22**, 656 (1981); D.R. Choudhury, A.K. Chowdhury, A.N. Chakravarti, Czech. J. Phys. B, **30** (1980); D.R. Choudhury, A.K. Chowdhury, A.N.

Chakravarti, B.R. Nag, Phys. Status Solidi A **58**, K51 (1980); B. Mitra, A. Ghoshal, K. P. Ghatak, Phys. Status Solidi B **153**, K209 (1989); M. Mondal, K.P. Ghatak, Phys. Status Solidi A **93**, 377 (1986); D.R. Choudhury, A.K. Chowdhury, A.N. Chakravarti, Appl. Phys. **22**, 145 (1980)

45. K.v. Klitzing, G. Dorda, M. Pepper, Phys. Rev. Lett. **45**, 494 (1980); K.v. Klitzing, Rev. Mod. Phys. **58**, 519 (1986); W. Zawadzki, in *Two-Dimensional Systems, Heterostructures and Superlattices*, ed. by G. Bauer, F. Kuchar, H. Heinrich Springer Series in Solid State Sciences, vol. 53, (Springer, Berlin, Heidelberg, 1984)
46. C.H. Peschke, Phys. Status. Solidi. B **191**, 325 (1995)
47. A. Isihara, Y. Shiwa, J. Phys. C **18**, 4703 (1985); V.G.I. Ravish, B.A. Efimova, I.A Smirnov, *Semiconducting Lead Chalocogenides* (Plenum Press, New York, 1970), p. 163
48. C.J. Hwang, J.R. Brews, J. Phys. Chem. Solids **32**, 837 (1971);P.T Landsberg, H.C Cheng, Phys. Rev. 32B, 8021 (1985); H. Reiss, C.S. Fuller, F.J. Morin, Bell Syst. Tech. J. **35**, 535 (1956); P.T. Landsberg, A.G. Guy, Phys. Rev. B, **28**, 1187 (1983)
49. D.K. Roy, *Quantum Mechanical Tunneling and its Applications*, (World Scientific Publication Co. Ptc. Ltd, Singapore, 1986), p. 309
50. W. Zawadzki, Physica **127**, 388 (1980)
51. H. Piller, in *Semiconductors and Semimetals, vol. 8*, ed. by R.K. Willardson, A.C. Beer (Academic Press, New York, 1972), p. 124
52. O. Madelung, *Physics of III–V compounds* (John Wiley and Sons, Inc. New York, 1966)
53. A.K. Sreedhar, S.C. Gupta, Phys. Rev. B **5**, 3160 (1972); R.W. Keyes, IBM. J. Res. Dev. **5**, 266 (1961); Solid State Phys. **20**, 37, (1967); B. Nag, K.P. Ghatak, Nonlinear. Opt. **19**, 1 (1998); K.P. Ghatak, B. Nag, Nanostruct. Mater. **10**, 923 (1998); B. Nag, K.P. Ghatak, J. Phys. Chem. Sol. **58**, 427, (1997)
54. K.P. Ghatak, S. Bhattacharya, D. De, *Einstein Relation in Compound Semiconductors and Their Nanostructures*, Springer Series in Materials Science, vol. 116 (Springer-Verlag, Germany, 2008)
55. A.S. Filipchenko, I.G. Lang, D.N. Nasledov, S.T. Pavlov, L.N. Radaikine, Phys. Status. Solidi. B **66**, 417 (1974)
56. M. Wegener, *Extreme Nonlinear Optics* (Springer-Verlag, Germany, 2005)
57. B.M. Askerov, N.F. Gushimzade, M.M. Panakhov, Sov. Phys. Solid State **29**, 465 (1987)
58. B. Mitra, A. Ghoshal, K.P. Ghatak, Phys. Status. Solidi. B **155**, K23 (1989)
59. S.N. Biswas, N. Chattopadhyay, K.P. Ghatak, Phys. Status. Solidi. B **141**, K47 (1987)
60. J.A. Woollam, Phys. Rev. **3**, 1148 (1971)
61. V.V. Kaminoskii, N.N. Stepanav, l.M. Smirnov, Sov. Phys. Solid State **27**, 1295 (1985)
62. K.P Ghatak, S Bhattacharya, *Thermo Electric Power in Nano structured Materials Strong Magnetic Fields,* Springer Series in Materials Science, vol. 137 (Springer, Heidelberg, 2010)
63. S. Bhattacharya, K.P. Ghatak, *Fowler-Nordheim Field Emission: Effects in Semiconductor Nanostructures,* Springer Series in Solid state Sciences, vol. 170 (Springer, Heidelberg, 2012)
64. S. Bhattacharya, K.P. Ghatak, *Effective Electron Mass in Low Dimensional Semiconductors,* Springer Series in Materials Sciences, vol. 167 (Springer, Heidelberg, 2013)
65. K.P. Ghatak, S. Bhattacharya, *Debye Screening Length: Effects of Nanostructured Materials*, Springer Tracts in Modern Physics, vol. 255 (Springer, Heidelberg, 2014)
66. K.P. Ghatak, S. Bhattacharya, *Heavily Doped 2D Quantized Structures and the Einstein Relation*, Springer Tracts in Modern Physics, vol. 260 (Springer, Heidelberg, 2015)
67. K.P. Ghatak, *Einstein's Photo-Emission: Emissions from Heavily Doped Nanostructures*, Springer Tracts in Modern Physics, Vol. 262 (Springer, Heidelberg, 2015)
68. W. Zawadzki, B. Lax, Phys. Rev. Lett. **16**, 1001 (1966)
69. B. Mitra, A. Ghoshal, K.P. Ghatak, Phys. Status. Solidi. B **154**, K147 (1989); K.P. Ghatak, B. Goswami, M. Mitra, B. Nag, Nonlinear Opt. Quantum Opt. **16**, 9 (1996); M. Mondal, K. P. Ghatak, Ann. Phys. **46**, 502 (1989); M.J. Harrison, Phys. Rev. A **29**, 2272 (1984); J. Zak, W. Zawadzki, Phys. Rev. **145**, 536 (1966); W. Zawadzki, Q.H. Vrehen, B. Lax, Phys. Rev.

148, 849 (1966); Q.H. Vrehen, W. Zawadzki, M. Reine, Phys. Rev. **158**, 702 (1967); M.H. Weiler, W. Zawadzki, B. Lax, Phys. Rev. **163**, 733 (1967); W. Zawadzki, J. Kowalski, Phys. Rev. Lett. **27**, 1713 (1971); C. Chu, M.-S Chu, T. Ohkawa, Phys. Rev. Lett. **41**, 653 (1978); P. Hu, C.S. Ting, Phys. Rev. B **36**, 9671 (1987); E.I. Butikov, A.S. Kondratev, A.E. Kuchma, Sov. Phys. Solid State **13**, 2594 (1972); M. Mondal, K.P. Ghatak, Physica Status Solidi (b), **133**, K67 (1986); M. Mondal, N. Chattopadhyay, K.P. Ghatak, J. Low Temp. Phys. **66**, 131 (1987); K.P. Ghatak, M. Mondal, Z. Phys. B **69**, 471 (1988); M. Mondal, K.P. Ghatak, Phys. Lett. A **131A**, 529 (1988); M. Mondal, K.P. Ghatak, Phys. Status. Solidi. B Germany, **147**, K179 (1988); B. Mitra, K.P. Ghatak, Phys. Lett. **137A**, 413 (1989); B. Mitra, K.P. Ghatak, Physica Status Solidi (b), **164**, K13 (1991); K.P. Ghatak, B. Mitra, Int. J. Electron. **70**, 345 (1991); Ghatak, N. Chattopadhyay, S.N. Biswas, *Proc. Society of Photo-Optical and Instrumentation Engineers (SPIE)*, **836**, Optoelectronic Materials, Devices, Packaging and Interconnects, 203 (1988); K.P. Ghatak, M. Mondal, S. Bhattacharyya, *SPIE***1284**, 113 (1990); K.P. Ghatak, *SPIE***1280**, Photonic materials and optical bistability 53 (1990); K.P. Ghatak, S.N. Biswas, *SPIE,* Growth and Characterization of Materials for Infrared Detectors and Nonlinear Optical Switches **1484**, 149 (1991); K. P. Ghatak, SPIE, Fiber Optic and Laser Sensors IX **1584**, 435 (1992).

70. S.N. Mohammad, J. Appl. Phys. **97**, 063703 (2005); K. Suzue, S.N. Mohammad, Z.F. Fan, W. Kim, O. Aktas, A.E. Botchkarev, H. Morkoç, J. Appl. Phys. **80**, 4467 (1996); S.N. Mohammad, Z.F. Fan, W. Kim, O. Aktas, A.E. Botchkarev, A. Salvador, H. Morkoç, Electron. Lett. **32**, 598 (1996); Z. Fan, S.N. Mohammad,W. Kim, O. Aktas, A.E. Botchkarev, K. Suzue, H. Morkoç, J. Electron. Mater. **25**, 1703 (1996); C. Lu, H. Chen, X. Lv, X. Xia, S.N. Mohammad, J. Appl. Phys. **91**, 9216 (2002)
71. N.G. Anderson, W.D. Laidig, R.M. Kolbas, Y.C. Lo, J. Appl. Phys. **60**, 2361 (1986); D. De, S. Bhattacharya, S.M. Adhikari, A. Kumar, P.K. Bose, K.P. Ghatak, Beilstein J. Nanotechnol. **2**, 339 (2012); D. De, A. Kumar, S.M. Adhikari, S. Pahari, N. Islam, P. Banerjee, S.K. Biswas, S. Bhattacharya, K.P. Ghatak, Superlattices Microstruct. **47,** 377 (2010); S. Pahari, S. Bhattacharya, K.P. Ghatak, J. Comput. Theor. Nanosci. **6**, 2088 (2009); K.P. Ghatak, J. Mukhopadhyay J.P. Banerjee, SPIE proceedings series **4746**, 1292 (2002); K.P. Ghatak, S. Dutta, D.K. Basu, B. Nag, Il Nuovo Cimento **D 20,** 227 (1998); K.P. Ghatak, B. De, Mat. Res. Soc. Proc. **300,** 513 (1993); K.P. Ghatak, B. Mitra, Il Nuovo Cimento. **D 15**, 97 (1993); K. P. Ghatak, Int. Soc.Opt. and Photon. Proc. Soc. Photo Opt. Instru. Eng., **1626**, 115 (1992); S. N. Biswas, K.P. Ghatak, Int. J. Electron. Theor. Exp. **70,** 125 (1991); B. Mitra, K.P. Ghatak, Phys. Lett. A. **142**, 401 (1989); K.P. Ghatak, B Mitra, A Ghoshal, Phys. Status. Solidi. B **154**, K121 (1989); B. Mitra, K.P. Ghatak, Phys. Status. Solidi. B. **149**, K117 (1988)
72. F. Capasso, *Semiconductors and Semimetals***22**, 2 (1985)
73. F. Capasso, K. Mohammed, A.Y. Cho, R. Hull, A.L. Hutchinson, Appl. Phys. Lett. **47,** 420 (1985)
74. F. Capasso, R.A. Kiehl, J. Appl. Phys. **58**, 1366 (1985)
75. K. Ploog, G.H. Doheler, Adv. Phys. **32**, 285 (1983)
76. F. Capasso, K. Mohammed, A.Y. Cho, Appl. Phys. Lett. 478 (1986)
77. R. Grill, C. Metzner, G.H. Döhler, Phys. Rev. B **63**, 235316 (2001); Phys. Rev. B **61**, 15614 (2000)
78. A.R. Kost, M.H. Jupina, T.C. Hasenberg, E.M. Garmire, J. Appl. Phys. **99**, 023501 (2006)
79. A.G. Smirnov, D.V. Ushakov, V.K. Kononenko, Proc. SPIE **4706**, 70 (2002)
80. D.V. Ushakov, V.K. Kononenko, I.S. Manak, Proc. SPIE **4358**, 171 (2001)
81. J.Z. Wang, Z.G. Wang, Z.M. Wang, S.L. Feng, Z. Yang, Phys. Rev. B **62**, 6956 (2000)
82. A.R. Kost, L. West, T.C. Hasenberg, J.O. White, M. Matloubian, G.C. Valley, Appl. Phys. Lett. **63**, 3494 (1993)
83. S. Bastola, S.J. Chua, S.J. Xu, J. Appl. Phys. **83**, 1476 (1998)
84. Z.J. Yang, E.M. Garmire, D. Doctor, J. Appl. Phys. **82**, 3874 (1997)
85. G.H. Avetisyan, V.B. Kulikov, I.D. Zalevsky, P.V. Bulaev, Proc. SPIE **2694,** 216 (1996)

86. U. Pfeiffer, M. Kneissl, B. Knüpfer, N. Müller, P. Kiesel, G.H. Döhler, J.S. Smith, Appl. Phys. Lett. **68**, 1838 (1996)
87. H.L. Vaghjiani, E.A. Johnson, M.J. Kane, R. Grey, C.C. Phillips, J. Appl. Phys. **76**, 4407 (1994)
88. P. Kiesel, K.H. Gulden, A. Hoefler, M. Kneissl, B. Knuepfer, S.U. Dankowski, P. Riel, X.X. Wu, J.S. Smith, G.H. Doehler, Proc. SPIE **85**, 278 (1993)
89. G.H. Doheler, Phys. Script. **24,** 430 (1981)
90. S. Mukherjee, S.N. Mitra, P.K. Bose, A.R. Ghatak, A. Neoigi, J.P. Banerjee, A. Sinha, M. Pal, S. Bhattacharya, K.P. Ghatak, J. Comput. Theor. Nanosci **4**, 550 (2007)
91. J.J. Quinn, P.J. Styles (eds.) *Electronic Properties of Quasi Two Dimensional Systems* (North Holland, Amsterdam, 1976); G.A. Antcliffe, R.T. Bate, R.A. Reynolds, in *Proceedings of the International Conference*, ed. by D.L. Carter, R.T. Bate Physics of Semi-metals and Narrow-Gap semiconductors vol. 499 (Pergamon Press, Oxford, 1971); G. Paasch, T. Fiedler, M. Kolar, I. Bartos, Phys. Status. Solidi. B **118**, 641 (1983); Th. Lindner, G. Paasch, J. Appl. Phys. **102,** 054514 (2007); S. Lamari, J. Appl. Phys. **91**, 1698 (2002); K.P. Ghatak, M. Mondal, J. Appl. Phys. **70**, 299 (1991); B. Mitra, K.P. Ghatak, Solid State Electron. **32**, 177 (1989); M. Mondal, K.P. Ghatak, Phys. Script. **31**, 613 (1985); M. Mondal, K.P. Ghatak, Acta Phys. Polon. A **67,** 983 (1985); M. Mondal, K.P. Ghatak, Phys. Status. Solidi. B **128**, K21 (1985); K.P. Ghatak, M. Mondal, Phys. Status. Solidi. B **135**, 819 (1986); M. Mondal, K.P. Ghatak, Phys. Status. Solidi. B **139**, 185 (1987); K.P. Ghatak, N. Chatterjee, M. Mondal, Phys. Status. Solidi. B **139,** K25 (1987); K.P. Ghatak, A. Ghosal, Phys. Status. Solidi. B **151,** K135 (1989); K.P. Ghatak, N. Chattopadhyay, M. Mondal, Appl. Phys. **A 48**, 365 (1989)
92. N. Miura, in *Physics of Semiconductors in High Magnetic Fields*, Series on Semiconductor Science and Technology (Oxford University Press, USA, 2007); K.H.J. Buschow, F.R. de Boer, *Physics of Magnetism and Magnetic Materials* (Springer, New York, 2003); D. Sellmyer, R. Skomski (eds.), *Advanced Magnetic Nanostructures* (Springer, New York, 2005); J.A.C. Bland, B. Heinrich (eds.), *Ultrathin Magnetic Structures III: Fundamentals of Nanomagnetism (Pt. 3)* (Springer-Verlag, Germany, 2005); B.K. Ridley, *Quantum Processes in Semiconductors*, 4th edn. (Oxford publications, Oxford, 1999); J.H. Davies, *Physics of Low Dimensional Semiconductors* (Cambridge University Press, UK, 1998); S. Blundell, *Magnetism in Condensed Matter*, Oxford Master Series in Condensed Matter Physics (Oxford University Press, USA, 2001); C. Weisbuch, B. Vinter, *Quantum Semiconductor Structures: Fundamentals and Applications* (Academic Publishers, USA, 1991); D. Ferry, *Semiconductor Transport* (CRC, USA, 2000); M. Reed (ed.), *Semiconductors and Semimetals: Nanostructured Systems* (Academic Press, USA, 1992); T. Dittrich, *Quantum Transport and Dissipation* (Wiley-VCH Verlag GmbH, Germany, 1998); A.Y. Shik, *Quantum Wells: Physics and Electronics of Two dimensional Systems* (World Scientific, USA, 1997)
93. K.P. Ghatak, M. Mondal, Z. Naturforsch., A **41a**, 881 (1986); K.P. Ghatak, M. Mondal, J. Appl. Phys. **62**, 922 (1987); K.P. Ghatak, S.N. Biswas, Phys. Status. Solidi. B **140**, K107 (1987); K.P. Ghatak, M. Mondal, Phys. Status. Solidi. B **139**, 195 (1987); K.P. Ghatak, M. Mondal, Phys. Status. Solidi. B **148**, 645 (1988); K.P. Ghatak, S.N. Biswas, J. Low Temp. Phys. **78**, 219 (1990); K.P. Ghatak, M. Mondal, Phys. Status. Solidi. B **160**, 673 (1990); K.P. Ghatak, A. Ghoshal, B. Mitra, Nouvo Cimento D **13D**, 867 (1991); K. P. Ghatak, M. Mondal, Phys. Status. Solidi. B **148**, 645 (1989); K.P. Ghatak, Nouvo Cimento D **13D**, 1321 (1992); K.P. Ghatak, S.N. Biswas, Nonlinear Opt. **4**, 347 (1993)
94. A.N. Chakravarti, K.P. Ghatak, A. Dhar, S. Ghosh, Phys. Status. Solidi. B **105**, K55 (1981); A.N. Chakravarti, A.K. Choudhury, K.P. Ghatak, Phys. Status. Solidi. B **63**, K97 (1981); A. N. Chakravarti, A.K. Choudhury, K.P. Ghatak, S. Ghosh, A. Dhar, Appl. Phys. **25**, 105 (1981); A.N. Chakravarti, K.P. Ghatak, G.B. Rao, K.K. Ghosh, Phys. Status. Solidi. B **112**, 75 (1982); A.N. Chakravarti, K.P. Ghatak, K.K. Ghosh, H.M. Mukherjee, Phys. Status. Solidi. B **116**, 17 (1983); M. Mondal, K.P. Ghatak, Phys. Status. Solidi. B **133**, K143 (1984);

M. Mondal, K.P. Ghatak, Phys. Status. Solidi. B **126**, K47 (1984); M. Mondal, K.P. Ghatak, Phys. Status. Solidi. B **129**, K745 (1985); M. Mondal, K.P. Ghatak, Phys. Scr. **31**, 615 (1985); M. Mondal, K.P. Ghatak, Phys. Status. Solidi. B **135**, 239 (1986); M. Mondal, K. P. Ghatak, Phys. Status. Solidi. B **135**, K21 (1986); M. Mondal, S. Bhattacharyya, K. P. Ghatak, Appl. Phys. A **42A**, 331 (1987); B. Mitra, A. Ghoshal, K.P. Ghatak, Phys. Status. Solidi. B **150**, K67 (1988); M. Mondal, K.P. Ghatak, Phys. Status. Solidi. B **146**, K97 (1988); B. Mitra, K.P. Ghatak, Solid State Electron. **32**, 515 (1989); B. Mitra, K.P. Ghatak, Phys. Lett. **135A**, 397 (1989)

95. L.V. Keldysh, Sov. Phys. Solid State **4**, 1658 (1962)
96. L. Esaki, R. Tsu, IBM J. Res. Dev. **14**, 61 (1970)
97. R. Tsu, *Superlattices to nanoelectronics* (Elsevier, Amsterdam, 2005); E.L. Ivchenko, G. Pikus, *Superlattices and other heterostructures* (Springer, Berlin 1995)
98. G. Bastard, *Wave mechanics applied to heterostructures, Editions de Physique* (Les Ulis, France, 1990)
99. K.P. Ghatak, B. De, *MRS Proceedings, vol.* 65 (Cambridge University Press, Cambridge, 1994), p. 299; K.P. Ghatak, B. De, *MRS Proceedings, vol.* 242 (Cambridge University Press, Cambridge, 1992), p. 377; K.P. Ghatak, San Diego,'91, San Diego, CA, Int. Soc. Opt. Photonics 282 (1991); K.P. Ghatak, Properties of II–VI Semiconductors: Bulk Crystals, Epitaxial Films, Quantum Well Structures, and Dilute Magnetic Systems, Materials Research Society, **161**, 371 (1990); K.P. Ghatak, S.N. Biswas, *MRS Proceedings, vol.* 161 (Cambridge University Press, Cambridge, 1989); S. Bhattacharyya, K.P. Ghatak, S. Biswas, OE/Fibers **87**, 73 (1987);
100. K.V. Vaidyanathan, R.A. Jullens, C.L. Anderson, H.L. Dunlap, Solid State Electron. **26**, 717 (1983)
101. B.A. Wilson, IEEE, J. Quantum Electron. **24**, 1763 (1988)
102. M. Krichbaum, P. Kocevar, H. Pascher, G. Bauer, IEEE, J. Quantum Electron. **24**, 717 (1988)
103. J.N. Schulman, T.C. McGill, Appl. Phys. Lett. **34,** 663 (1979)
104. H. Kinoshita, T. Sakashita, H. Fajiyasu, J. Appl. Phys. **52**, 2869 (1981)
105. L. Ghenin, R.G. Mani, J.R. Anderson, J.T. Cheung, Phys. Rev. B, **39**, 1419 (1989)
106. C.A. Hoffman, J.R. Mayer, F.J. Bartoli, J.W. Han, J.W. Cook, J.F. Schetzina, J.M. Schubman, Phys. Rev. B. **39**, 5208 (1989)
107. V.A. Yakovlev, Sov. Phys. Semicon. **13**, 692 (1979)
108. E.O. Kane, J. Phys. Chem. Solids **1**, 249 (1957)
109. H. Sasaki, Phys. Rev. B **30**, 7016 (1984)
110. P.K. Bose, N. Paitya, S. Bhattacharya, D. De, S Saha, K.M. Chatterjee, S. Pahari, K. P. Ghatak, Quant. Matter. **1**, 89 (2012)
111. S. Bhattarchya, D. De, S. Ghosh, P. Banerjee, S. Saha, M. Mitra, B. Nag, M. Pal, S.K. Biswas, K.P. Ghatak, J. Comput. Theor. Nanosci. **7**, 1066 (2010); K.P. Ghatak, S. Bhattacharya, S. Pahari, S.N. Mitra, P.K. Bose, D. De, J. Phys. Chem. Solids. **70**, 122 (2009)
112. S. Bhattacharya, N.C. Paul, D. De, K.P. Ghatak, Phys. B: Condens. Matter **403**, 4139 (2008)
113. P.K. Chakraborty, S. Choudhury, K.P. Ghatak, Phys. B **387**, 333 (2007)
114. S.M. Adhikari, K.P. Ghatak, Quant. Matter **2**, 296 (2013)
115. S.M. Adhikari, K.P. Ghatak, J. Nanoeng. Nanomanuf. **3**, 48 (2013)
116. A. Pais, J. Robert, *Oppenheimer* (Oxford University Press, UK, 2006), p. xviii
117. I.K. Kikoin, *Science for Everyone: Encounters with Physicists and Physics* (Mir Publishers, Russia, 1989), p. 154
118. J.J. Sakurai, Advanced Quantum Mechanics (Addison–Wesley Publishing Company, USA, 1967)

Acknowledgments

I am grateful to A.N. Chakravarti, former Head of the Department of the Institute of Radio Physics and Electronics of the University of Calcutta, my mentor and an *ardent advocator of the concept of theoretical minimum of Landau*, who convinced a 21-year–old Circuit theorist that Condensed Matter Physics, in general, is the hidden dual dance of quantum mechanics and statistical mechanics together with advanced mathematical techniques; even to appreciate the incredible beauty, he not only placed a stiff note for me to derive all the equations in the Monumental Course of Theoretical Physics, the Classics of Landau–Lifshitz together with the two-volume classics of Morse–Feshbach 40 years ago, but also forced me to stay in the *creative critical zone of research till date*. I express my gratitude to Late B.R. Nag, former Head of the Departments of Radio Physics and Electronics and Electronic Science of the University of Calcutta, to whom I am ever grateful as a student and research worker, the Semiconductor Grandmaster of India for his three books on semiconductor science in general and more than 200 research papers (many of them are absolutely in *honor's class*) which still fire my imagination. I consider myself to be rather fortunate to study Mathematics under the direct influence of Late S.C. Dasgupta, former Head of the Department of Mathematics of the then Bengal Engineering College, Shibpur (presently Indian Institute of Engineering Science and Technology) and M. Mitra (both of them were formidable Applied Mathematicians with deep physical insight and could solve any problem from the two-volume classics of Morse–Feshbach instantly on the blackboard) of the said Department of Mathematics for teaching me deeply the various methods of different branches of Applied Mathematics with special emphasis on ***analytic number theory*** when I was pursuing the bachelor's degree in the branch of Electronics and Telecommunication Engineering 45 years ago. I offer my thanks to Late S.S. Boral, former Founding Head of the Department of Electronics and Telecommunication Engineering of the then Bengal Engineering College, Shibpur, for teaching me the Theoretical Acoustics from the Classic of Morse and Ingard by urging me to solve the problems. I am grateful to S.K. Sen, former Head of the Department of Electrical Engineering of Bengal Engineering College, Shibpur,

Ex-Vice-Chancellor of Jadavpur University and Ex-Power Minister of West Bengal State for teaching me deeply nonlinear network analysis and synthesis and nonlinear control systems in general. I am indebted to Late C.K. Majumdar, Head of the Department of Physics of the University of Calcutta to light the fire for Theoretical Physics.

I am grateful to all my friends and colleagues of the last 40 years from my school days till date for forming my inner fire to solve independently the problems from five volumes of Berkley Physics Course, three volumes of Sakurai, ten volumes of Addison–Wesley Modular Series on Solid-State Devices, two volumes of Cohen-Tannoudji et.al., three volumes of Edwards, three volumes of Carslaw, six of volumes Guillemin, three of volumes Tuttle, two of volumes Scott, two volumes of Budak, five volumes of Chen, five volumes of Smirnov.... It is curious to note that they have been insisting me in the real sense of the term to verify all the entries of Gradshteyn–Ryzhik and three volumes of the Bateman manuscript project for the past 40 years. It may be noted that the academic output of a good scholar is the response of a good quality RC coupled amplifier in the mid-band zone, whereas the same for a fine research worker is the Dirac's delta function. Incidentally, I can safely say that I belong to neither as a consequence of the academic surroundings which this life presents to me. I express my warm gratitude to H.L. Hartnagel, D. Bimberg, W.L. Freeman and W. Schommers for various academic interactions spanning over the past three decades. I remember the sweet memories of P.N. Robson, I.M. Stephenson, V.S. Letokhov, J. Bodnar and K.F. Brennan with true reverence for inspiring me with the ebullient idea that the publications of the research monographs from internationally reputed publishers containing innovative concepts is of prime importance to excel in creative research activity.

On 1 January 2007, I wrote a letter to Late P.T. Landsberg (popularly known as P.T.L to his scientific friends) for writing a book on Dispersion Relations of nanostructured materials which can at least transform me into a better scientist, but P.T.L in turn requested me repeatedly to complete the proposed book at the earliest and often expressed his desire to write a foreword for this book. Incidentally, due to previous other heavy academic and related commitments I was unable to finish this present monograph and his wife Mrs. Sophia Landsberg in her letter of intense sadness dated 12 January 2009 informed me that from 2008 onward due to Alzheimer's disease her husband had to give up scientific works. The disappearance of P.T.L from my research scenario is a big shock to me which I have to bear till my own physical disappearance. I still remember his power-packed words, "The definition of Kamakhya Prasad Ghatak (K.P.G) = M.Tech. + Ph.D. + D. Engg. + more than 300 research papers in nanoscience and technology in SCI Journals + more than 60 research papers in the proceedings of the international conferences of SPIE and MRS of USA + Ph.D. guide of more than 30 Ph.D. students + many research monographs form Springer-Verlag, Germany + HOD + Dean… is a **big zero to P.T.L. if K.P.G. cannot live in a world full of new creative concepts and not merely repeating the past successes, since past success is a dead concept at least to P.T.L.**

On 21 December 1974 A.N. Chakravarti (an internationally recognized expert on Einstein Relation in general), my mentor, in my first interaction with him emphatically energized me by making myself acquainted with the ***famous Fermi–Dirac Integrals*** and told me that ***I must master*** "Semiconductor Statistics" (Pergamon Press, London, 1962) by J.S. Blakemore for my initiation in semiconductor physics. Later, I came into deep touch with K. Seeger, the well-known author of the book "Semiconductor Physics" (Springer Series in Solid State Sciences, vol. 40, Springer-Verlag, Germany, 1982). The solid Mathematical Physicist Late P.N. Butcher has been a steady hidden force since 1983, before his sad demise, with respect to our scripting the series in band structure-dependent properties of nanostructured materials. He urged me repeatedly regarding it and to tune with his high rigorous academic standard, myself with the prominent members of my research group wrote [1] as the first one, [2] as the second one, [3] as the third one, [4] as the fourth one, [5] as the fifth one, [6] as the sixth one, [7] as the seventh one, [8] as the eighth one, and the present monograph as the ninth one. Both P.T.L and P.N. Butcher visited the Institute of Radio Physics and Electronics of the University of Calcutta, my ALMA MATER where I started my research as an M.Tech. student and later as a faculty member. I formed permanent covalent bonds with J.S. Blakemore, K. Seeger, P.N. Butcher and P.T.L through letters (Pre email era) and these four first-class semiconductor physicists, in turn, shared with pleasure their vast creative knowledge of semiconductors and related sciences with a novice like me.

I offer special thanks to Late N. Guhachoudhury of Jadavpur University for instilling in me the thought that the *academic output = ((desire X determination X dedication) − (false enhanced self ego pretending like a true friend although a real unrecognizable foe))* although a thank you falls in the band gap regime for my beloved better-half, See, who really formed the backbone of my long unperturbed research career, since in accordance with "Sanatan" Hindu Dharma, the fusion of marriage has transformed us to form a single entity, where the individuality is being lost. I am grateful to all the members of my research group (from 1983 till date) for not only quantum confining me in the infinitely deep quantum wells of ***Ramanujan and Rabindranath*** but also inspiring me to teach quantum mechanics and related topics from the eight volumes classics of Greiner et al.

In this context, from the flashback of my memory I wish to offer my indebtedness to M. Mondal, the first member of my research team, who in 1983 joined with me to complete his Ph.D. work under my guidance when R.K. Poddar, the then Vice-chancellor of the University of Calcutta selected me as a Lecturer (presently Assistant Professor) of the famous Department of Radio Physics and Electronics. In 1987, S.K. Sen, the then Vice-chancellor of Jadavpur University accepted me as the Reader (presently Associate Professor) in the Department of Electronics and Telecommunication Engineering and since then a flood of young researchers (more than 12 in number consisting of B. Mitra, A. Ghoshal, D. Bhattacharya, A.R. Ghatak, …) around me created my research team and insisted me to work with them at the @ of 16 h per day, including holidays in different directions of research for the purpose of my creative conversion from an ordinary engineer to a 360° research scientist; consequently, I enjoyed the high rate of research publications in different reputed international journals in various aspects of band structure-dependent properties of quantized structures. It is nice to note that the said young talented researchers obtained their respective Ph.D. degrees under my direct supervision. Incidentally in 1994, R.K. Basu, the then Vice-chancellor of the University of Calcutta selected me as a Professor in the Department of Electronic Science and another flood of research overwhelmed me in a new direction of materials science. The persons responsible for this change include S. Datta, S. Sengupta, A. Ali…. The 11th and 12th names of this golden series are S. Bhattacharya and D. De respectively, who in turn formed permanent covalent bonds with me not only with respect to research (S. Bhattacharya and D. De are respectively the co-authors of seven and two monographs in different series of Springer) but also in all aspects of life in general.

It is curious to note that after serving 18 years as a Professor in the Department of Electronic Science, in 2012, P.K. Bose, the then Director of the National Institute of Technology, Agartala requested me to join as a Professor and Departmental Head in Electronics and Communication Engineering. Being my life-long friend, I accepted his offer (and later as a DEAN) and more than ten young scholars around me again directed my research in an altogether new direction. In 2015, the respected Director Professor S. Chakrabarti of Institute of Engineering and Management in Saltlake City Kolkata kindly offered me the position of Research Director and senior Professor in his famous Institute at the fag end of my academic life to complete my last run towards the creative knowledge temple with my new young research workers and I am grateful to him for his creative gesture. In my 40+ years of teaching life (I have the wide experience of teaching Physics, Mathematics, Applied Mechanics(from engineering statics up to nonlinear mechanics including indeterminate structures) and 70 % of the courses of Electronics and Telecommunication and Electrical Engineering respectively) and 40+ years of research life (mostly in Materials Science, Nanoscience and Number Theory), I have finally realized that teaching without research is body without brain and research without teaching is body without blood **although my all-time hero, creatively prolific number theorist Godfrey Harold Hardy in his classic book entitled "A Mathematician's Apology" (Cambridge University Press, 1990) tells us "I hate teaching".**

Incidentally, one young theoretician friend of mine often tells me that many works in theoretical semiconductor science are based on the following seven principles:

1. Principles of placing the necessary and sufficient conditions of a proof in the band gap regime.
2. Principles of interchange of the summation and integration processes and unconditioned convergences of the series and integrals.
3. Principles of random applications of one electron theory and superposition theorem in studying the properties of semiconductors, although the many body effects are very important together with the fact that the nature is fundamentally nonlinear in nature.

4. Principles of using the invariant band structure concept of semiconductors even in the presence of strong external fields (light, electric, heavy doping, etc.) and the random applications of perturbation theory, which is in a sense quantum mechanical Taylor series without considering the related consequences.
5. Principle of random applications of the binomial theorem without considering the important concept of branch cut.
6. Principle of little discussion regarding the whole set of materials science comprising different compounds having various band structures under different physical conditions compared with the simplified two-band model of Kane for III–V semiconductors.
7. Principle of using the Fermi's golden rule, the band structure, and the related features which are valid for non-degenerate semiconductors to materials having degenerate carrier concentrations directly.

Although my young friend is a purist in his conjecture, there are no doubt certain elements of truth inside his beautiful comments. I hope that our readers will present their intricate and advanced theories after paying due weightage of his aforementioned seven principles.

I am grateful to P.N. Singh, the Ex-Chairman of National Institute of Technology Agartala for his encouragement and enthusiastic support and the present Director, G. Mugeraya for his keen academic insight and the creation of a congenial atmosphere to promote research and higher learning among the faculty members. I offer special thanks to S.N. Pradhan, the Head of the Department Electronics and Communication Engineering of the said Institute, my previous working place for extending his helping hand for my research scholars in the real sense of the term. I must express my gratitude to Professor B. Chatterjee of the Computer Science and Engineering Department of my present working place, one of the strong members of my research group, for offering important suggestions for the condensed presentation of this monograph. I offer special thanks to T.N. Sen, T. Datta, K. Sarkar, M. Chakarborty, M. Debbarma, P. Bhardwaj, N. Debbarma, R. Das and other members of my research team for placing their combined efforts towards the development of this book in the DO-LOOP of a computer and for critically reading the manuscript in its last phase before sending it to C. Ascheron, Executive Editor Physics, Springer-Verlag. Last but not the least, I am grateful for ever to our life long time tested friend S. Sanyal, Principal, Lake School for Girls, Kolkata, for not only motivating me at rather turbulent moments of my academic career but for also instilling in me the concept that ***the ratio of total accumulated knowledge in my chosen field of research to my own contribution in my topic of research tends to infinity at any time and is the definition of non-removable pole in the transfer function of my life***.

As always, myself with the members of my research team are grateful to C. Ascheron, Executive Editor Physics, Springer Verlag, in the real sense of the term for his inspiration and priceless technical assistance from the very start of our first monograph from Springer. **C. Ascheron is the hidden force behind the publications of nine monographs, the collective output of myself and my**

research group for the last 40 years, from Springer. I am also grateful in the real sense of the term to Peter Wölfle, Editor Condensed matter theory of Springer Tract in Modern Physics for triggering me in the right direction at the right time so that my books can, in turn, be accepted by Springer. We owe a lot to A. Duhm, Associate Editor Physics, Springer, and E. Sauer, assistant to Ascheron. Naturally, the author is responsible for non-imaginative shortcomings. I firmly believe that our Mother Nature has propelled this Project in her own unseen way in spite of several insurmountable obstacles.

Kolkata, India Kamakhya Prasad Ghatak
February 2015

References

1. K.P. Ghatak, S. Bhattacharya, D. De, *Einstein Relation in Compound Semiconductors and their Heterostructures,* Springer Series in Materials Science, vol. 116 (Springer, Germany, 2009)
2. K.P. Ghatak, D. De, S. Bhattacharya, *Photoemission from Optoelectronic Materials and their Nanostructures*, Springer Series in Nanostructure Science and Technology (Springer, Heidelberg, 2009)
3. K.P. Ghatak, S. Bhattacharya, *Thermo Electric Power in Nano structured Materials Strong Magnetic Fields*, Springer Series in Materials Science, Vol. 137 (Springer, Heidelberg, 2010)
4. S. Bhattacharya, K.P. Ghatak, *Fowler-Nordheim Field Emission: Effects in Semiconductor Nanostructures*, Springer Series in Solid state Sciences, Vol. 170 (Springer, Heidelberg, 2012)
5. S. Bhattacharya, K.P. Ghatak, *Effective Electron Mass in Low Dimensional Semiconductors,* Springer Series in Materials Sciences, Vol. 167 (Springer, Heidelberg, 2013)
6. K.P. Ghatak, S. Bhattacharya, *Debye Screening Length: Effects of Nanostructured Materials*, Springer Tracts in Modern Physics, Vol. 255 (Springer, Heidelberg, 2014)
7. K.P. Ghatak, S. Bhattacharya, *Heavily Doped 2D Quantized Structures and the Einstein Relation*, Springer Tracts in Modern Physics, Vol. 260 (Springer, Heidelberg, 2015)
8. K.P. Ghatak, *Einstein's Photo-emission: Emission from Heavily Doped Quantized Structures*, Springer Tracts in Modern Physics, Vol. 262 (Springer, Heidelberg, 2015)

Contents

Part V Dispersion Relations in HD Kane Type Semiconductors in the Presence of Intense Electric Field

Symbols

α	Band nonparabolicity parameter
a	The lattice constant
a_0, b_0	The widths of the barrier and the well for superlattice structures
A_0	The amplitude of the light wave
$\vec{A}$	The vector potential
$A(E, n_z)$	The area of the constant energy 2D wave vector space for ultrathin films
B	Quantizing magnetic field
B_2	The momentum matrix element
b	Bandwidth
c	Velocity of light
C_1	Conduction band deformation potential
C_2	A constant which describes the strain interaction between the conduction and valance bands
ΔC_{44}	Second order elastic constant
ΔC_{456}	Third order elastic constant
δ	Crystal field splitting constant
Δ_0	Interface width
$\Delta\left(\frac{1}{B}\right)$	Period of SdH oscillation
d_0	Superlattice period
$D_0(E)$	Density-of-states (DOS) function
$D_B(E)$	DOS function in magnetic quantization
$D_B(E, \lambda)$	DOS function under the presence of light waves
d_x, d_y, d_z	Nano thickness along the x, y and z-directions
$\Delta_{\parallel}$	Spin-orbit splitting constants parallel
$\Delta_{\perp}$	Spin-orbit splitting constants perpendicular to the C-axis
Δ	Isotropic spin-orbit splitting constant
d^3k	Differential volume of the k space
ϵ	Energy as measured from the center of the band gap

ε	Trace of the strain tensor
ε_0	Permittivity of free space
ε_∞	Semiconductor permittivity in the high frequency limit
ε_{sc}	Semiconductor permittivity
ΔE_g	Increased band gap
$\|e\|$	Magnitude of electron charge
E	Total energy of the carrier
E_0, ζ_0	Electric field
E_g	Band gap
E_i	Energy of the carrier in the ith band
E_{ki}	Kinetic energy of the carrier in the ith band
E_F	Fermi energy
E_{FB}	Fermi energy in the presence of magnetic quantization
E_n	Landau sub-band energy
E_{Fs}	Fermi energy in the presence of size quantization
$\bar{E}_{Fn}$	Fermi energy for nipis
E_{FSL}	Fermi energy in superlattices
$\vec{\varepsilon}_s$	Polarization vector
E_{FQWSL}	Fermi energy in quantum wire superlattices with graded interfaces
E_{F_L}	Fermi energy in the presence of light waves
$E_{F_{BL}}$	Fermi energy under quantizing magnetic field in the presence of light waves
E_{F2DL}	2D Fermi energy in the presence of light waves
E_{F1DL}	1D Fermi energy in the presence of light waves
E_{g0}	Unperturbed band-gap
$Erfc$	Complementary error function
Erf	Error function
E_{F_h}	Fermi energy of HD materials
$\bar{E}_{hd}$	Electron energy within the band gap
F_s	Surface electric field
$F(V)$	Gaussian distribution of the impurity potential
$F_j(\eta)$	One parameter Fermi–Dirac integral of order j
f_o	Equilibrium Fermi–Dirac distribution function of the total carriers
f_{oi}	Equilibrium Fermi–Dirac distribution function of the carriers in the ith band
g_v	Valley degeneracy
G	Thermoelectric power under classically large magnetic field
G_0	Deformation potential constant
g^*	Magnitude of the band edge g-factor
h	Planck's constant
$\hat{H}$	Hamiltonian
$\hat{H}'$	Perturbed Hamiltonian

$H(E - E_n)$	Heaviside step function
$\hat{i}, \hat{j}$ and $\hat{k}$	Orthogonal triads
i	Imaginary unit
I	Light intensity
j_{ci}	Conduction current contributed by the carriers of the ith band
k	Magnitude of the wave vector of the carrier
k_B	Boltzmann's constant
λ	Wavelength of the light
$\bar{\lambda}_0$	Splitting of the two spin-states by the spin-orbit coupling and the crystalline field
$\bar{l}, \bar{m}, \bar{n}$	Matrix elements of the strain perturbation operator
L_x, L_z	Sample length along x and z directions
L_0	Superlattices period length
L_D	Debye screening length
m_1	Effective carrier masses at the band-edge along x direction
m_2	Effective carrier masses at the band-edge along y direction
m_3	The effective carrier masses at the band-edge along z direction
m_2'	Effective-mass tensor component at the top of the valence band (for electrons) or at the bottom of the conduction band (for holes)
m_i^*	Effective mass of the ith charge carrier in the ith band
$m_{\parallel}^*$	Longitudinal effective electron masses at the edge of the conduction band
$m_{\perp}^*$	Transverse effective electron masses at the edge of the conduction band
m_c	Isotropic effective electron masses at the edge of the conduction band
$m_{\perp,1}^*, m_{\parallel,1}^*$	Transverse and longitudinal effective electron masses at the edge of the conduction band for the first material in superlattice
m_r	Reduced mass
m_v	Effective mass of the heavy hole at the top of the valance band in the absence of any field
n	Landau quantum number
n_x, n_y, n_z	Size quantum numbers along the x, y and z directions
n_{1D}, n_{2D}	1D and 2D carrier concentration
n_{2Ds}, n_{2Dw}	2D surface electron concentration under strong and weak electric field
$\bar{n}_{2Ds}, \bar{n}_{2Dw}$	Surface electron concentration under the strong and weak electric field quantum limit
n_i	Mini-band index for nipi structures
$N_{nipi}(E)$	DOS function for nipi structures
$N_{2DT}(E)$	2D DOS function
$N_{2D}(E, \lambda)$	2D DOS function in the presence of light waves

$N_{1D}(E,\lambda)$	1D DOS function in the presence of light waves
n_0	Total electron concentration
$\bar{n}_0$	Electron concentration in the electric quantum limit
n_i	Carrier concentration in the ith band
P	Isotropic momentum matrix element
P_n	Available noise power
$P_{\parallel}$	Momentum matrix elements parallel to the direction of crystal axis
$P_{\perp}$	Momentum matrix elements perpendicular to the direction of crystal axis
$\vec{r}$	Position vector
S_i	Zeros of the Airy function
$\vec{S}_0$	Momentum vector of the incident photon
t	Timescale
t_c	Tight binding parameter
T	Absolute temperature
$\tau_i(E)$	Relaxation time of the carriers in the ith band
$u_1\left(\vec{k},\vec{r}\right), u_2\left(\vec{k},\vec{r}\right)$	Doubly degenerate wave functions
$V(E)$	Volume of k space
V_0	Potential barrier encountered by the electron
$V(\vec{r})$	Crystal potential
x, y	Alloy compositions
Z_t	Classical turning point
μ_i	Mobility of the carriers in the ith band
μ	Average mobility of the carriers
$\zeta(2r)$	Zeta function of order $2r$
$\Gamma(j+1)$	Complete Gamma function
η	Normalized Fermi energy
η_g	Impurity scattering potential
ω_0	Cyclotron resonance frequency
θ	Angle
μ_0	Bohr magnetron
ω	Angular frequency of light wave
$\uparrow', \downarrow'$	Spin up and down function

About the Author

Kamakhya Prasad Ghatak Born in India on 12 April 1953, Professor K. P. Ghatak obtained the B.E. degree in Electronics and Telecommunication Engineering from the then Bengal Engineering College Shibpur (presently Indian Institute of Engineering Science and Technology) of the University of Calcutta in 1974 and the M.Tech. degree from the Institute of Radio Physics and Electronics of the University of Calcutta in 1976. He obtained the Ph.D. (Tech) degree from the University of Calcutta in 1988 on the basis of 27 published research papers in international peer-reviewed SCI Journals, which is still a record in the said institute. He joined as Lecturer in the Institute of Radio Physics and Electronics of the University of Calcutta in 1983, Reader in the Department of Electronics and Telecommunication Engineering of Jadavpur University in 1987 and Professor in the Department of Electronic Science of the University of Calcutta in 1994 and was at the top of the merit list in all the cases respectively. From March 2012, he worked in the Department of Electronics and Communication Engineering of National Institute of Technology, Agartala, Tripura, as Professor and Departmental Head and later acted as Dean (Faculty Welfare) of the said institute. From January 2015 he has joined as Research Director and Senior Professor at Institute of Engineering and Management, Salt Lake Electronics Complex, Sector V, Kolkata, India. Professor K.P. Ghatak is the First Recipient of the Degree of Doctor of Engineering of Jadavpur University in 1991 since the university's inception in 1955 and in the same year he received the Indian National Science Academy Visiting Fellowship to IIT-Kharagpur. He is the principal co-author of more than 400 research papers on Semiconductor Nanoscience in different reputed foreign journals; among them are more than 300 research papers in eminent S.C.I. Journals, more than 60 research papers in the Proceedings of SPIE and MRS of USA (https://scholar.google.co.in/citations?user=q8z3_CMAAAAJ) and many of his papers are often cited. At present, the h-index, i10-index, total citations and the maximum citation of a research paper within two years of publication of Professor Ghatak are 23, 114, 3132 and 280 respectively. He is the invited Speaker of SPIE, MRS, etc., the referee and editor of different reputed journals, supervisor of more than two dozen Ph.D. candidates in various aspects of semiconductor nanoscience with many of them working as professors, associate professors and assistant professors in different universities and academic institutions respectively. He is the principal co-author of the EIGHT research monographs from Springer-Verlag, Germany entitled "Einstein Relation in Compound Semiconductors and Their Nanostructures" Springer Series in Materials Science, Vol. 116, 2009, which is the first monograph on Einstein relation containing more than 100 open research problems, "Photoemission from Optoelectronic materials and their nanostructures", Springer Series in Nanostructure Science and Technology, USA, in 2009 as the first monograph solely devoted to Photoemission from nanostructured compounds containing more than 100 open research problems, "Thermo electric power in nano structured materials: Strong magnetic fields" Springer Series in Materials Science Vol. 137, 2010 which is the first monograph on this topic containing 150 open research problems, "Fowler-Nordheim Field Emission: Effects in Semiconductor Nanostructures "Springer Series in Solid State Sciences, Vol. 170, 2012 which is the first monograph on this topic containing 200 open research problems, "Effective electron mass in low dimensional semiconductors" Springer Series in Materials Science Vol. 167, 2013 which is the first monograph on this pinpointed topic containing 250 open research problems and "Debye Screening Length: Effects of nano-structured materials" Springer Tracts in Modern Physics, Vol. 255, 2014 which is the first book solely devoted to this particular topic containing more than 100 open research problems, "Heavily-Doped 2D-Quantized Structures and the Einstein Relation" Springer Tracts in Modern Physics, Vol. 260, 2014 which is the first book solely devoted to this particular topic containing 200 open research problems and "Einstins photoemission: emission from Heavily-Doped Quantized Structures" Springer Tracts in Modern Physics, Vol. 262, 2015 which is the first book solely devoted in this context containing more than 200 open research problems respectively. He is the principal editor of the two edited monographs entitled "Bismuth:: Characteristics, Production and Applications: Series in Materials Science and Technologies" and "Quantum Dots and Quantum Cellular Automata: Recent Trends and Applications: Series in Nanotechnology Science" of NOVA, USA respectively. The All Indian Council for Technical Education selected the first Research and Development project in his life for the best project award in Electronics and the second best research project award considering all the branches of Engineering for the year 2006. In 2011, the University Grant Commission recommended a research project to him and placed him at the top in the list of awardees. His 40-year-long teaching interests include Nonlinear Network Analysis and Synthesis, Nonlinear Devices, Nonlinear Mechanics, Nonlinear Integral and Partial Differential equations and his present research interests include semiconductor nanoscience and number theory respectively. His brief CV has been enlisted in many biographical references in the USA and UK. Professor Ghatak is a firm believer of the principle of theoretical minimum of Landau and being an ardent follower of the GITA (well known as the theory of KRIYA YOGA), his vision, mission and passion is to stay deeply in solitude to become a permanent member of the endless world of self-realization.

Part I
Dispersion Relations in HD Quantum Wells, Nano Wires and Dots in the Presence of Magnetic Field

Respect is received when it is given.

Chapter 1
The DRs in Low Dimensional HD Systems in the Presence of Magnetic Field

Their is one mental light year distance between a "HUMAN BEING" and "BEING HUMAN".

1.1 Introduction

In recent years, with the advent of fine lithographical methods [1, 2] molecular beam epitaxy [3], organometallic vapor-phase epitaxy [4], and other experimental techniques, the restriction of the motion of the carriers of bulk materials in one (QWs, doping super-lattices, accumulation, and inversion layers), two (nanowires) and three (quantum dots, magneto-size quantized systems, magneto inversion layers, magneto accumulation layers, quantum dot super-lattices, magneto QW super-lattices, and magneto doping superlattices) dimensions have in the last few years, attracted much attention not only for their potential in uncovering new phenomena in nano-science but also for their interesting quantum device applications [5–8]. In QWs, the restriction of the motion of the carriers in the direction normal to the film (say, the z direction) may be viewed as carrier confinement in an infinitely deep 1D rectangular potential well, leading to quantization (known as quantum size effect (QSE)) of the wave vector of the carriers along the direction of the potential well, allowing 2D carrier transport parallel to the surface of the film representing new physical features not exhibited in bulk semiconductors [9–13]. The low-dimensional hetero-structures based on various materials are widely investigated because of the enhancement of carrier mobility [14]. These properties make such structures suitable for applications in QWs lasers [15], hetero-junction FETs [16, 17], high-speed digital networks [18–21], high-frequency microwave circuits [22], optical modulators [23], optical switching systems [24], and other devices. The constant energy 3D wave-vector space of bulk semiconductors becomes 2D wave-vector surface in QWs due to dimensional quantization. Thus, the concept of reduction of symmetry of the wave-vector space and its consequence can unlock the physics of low-dimensional structures.

K.P. Ghatak, *Dispersion Relations in Heavily-Doped Nanostructures*,
Springer Tracts in Modern Physics 265, DOI 10.1007/978-3-319-21000-1_1

It may be noted that in nano wires (NWs), the restriction of the motion of the carriers along two directions may be viewed as carrier confinement by two infinitely deep 1D rectangular potential wells, along any two orthogonal directions leading to quantization of the wave vectors along the said directions, allowing 1D carrier transport [25]. With the help of modern experimental techniques, such one dimensional quantized structures have been experimentally realized and enjoy an enormous range of important applications in the realm of nano-science in the quantum regime. They have generated much interest in the analysis of nano-structured devices for investigating their electronic, optical and allied properties [26–28]. Examples of such new applications are based on the different transport properties of ballistic charge carriers which include quantum resistors [29–31], resonant tunneling diodes and band filters [32, 33], quantum switches [34], quantum sensors [35–37], quantum logic gates [38, 39], quantum transistors and sub tuners [40–42], heterojunction FETs [43], high-speed digital networks [44], high-frequency microwave circuits [45], optical modulators [46], optical switching systems [47, 48], and other devices [49–51].

It is well known that as the dimension of the QWs increases from 1D to 3D, the degree of freedom of the free carriers decreases drastically and the density-of-states function changes from the Heaviside step function in QWs to the Dirac's delta function in Quantum Dot (QD) [52]. The QDs can be used for visualizing and tracking molecularprocesses in cells using standard fluorescence microscopy [53–55]. Theydisplay minimal photo-bleaching [56], thus allowing molecular tracking over prolonged periods and consequently, single molecule can be tracked by using optical fluorescencemicroscopy [57]. The salient features of quantum dot lasers [58] include low threshold currents, higher power, and great stability as compared with the conventional one and the QDs find extensive applications in nano-robotics [59], neural networks [60] and high density memory or storage media [61]. The QDs are also used in nano-photonics [62] because of their theoretically high quantum yield and have been suggested as implementations of Q-bits for quantum information processing [63]. The QDs also find applications in diode lasers [64], amplifiers [65], and optical sensors [66]. High-quality QDs are well suited for optical encoding [67] because of their broad excitation profiles and narrow emission spectra. The new generations of QDs have far-reaching potential for the accurate investigations of intracellular processes at the single-molecule level, high-resolution cellular imaging, long-term in vivo observation of cell trafficking, tumor targeting, and diagnostics [68]. The QD nanotechnology is one of the most promising candidates for use in solid-state quantum computation [69]. It may also be noted that the QDsare being used in single electron transistors [70], photovoltaic devices [71], photoelectrics [72], ultrafast all-optical switches and logic gates [73], organic dyes [74] and in other types of nano devices.

The III–V compounds find applications in infrared detectors [75], quantum dot light emitting diodes [76], quantum cascade lasers [77], QWs wires [78], opto-electronic sensors [79], high electron mobility transistors [80], etc. The electron energy spectrum of III–V semiconductors can be described by the three- and two-band models of Kane [81, 82], together with the models of Stillman et al. [83],

Newson and Kurobe [84] and, Palik et al. [85] respectively. In this context it may be noted that the ternary and quaternary compounds enjoy the singular position in the entire spectrum of optoelectronic materials. The ternary alloy $Hg_{1-x}Cd_xTe$ is a classic narrow gap compound. The band gap of this ternary alloy can be varied to cover the spectral range from 0.8 to over 30 μm [86] by adjusting the alloy composition. $Hg_{1-x}Cd_xTe$ finds extensive applications in infrared detector materials and photovoltaic detector arrays in the 8–12 μm wave bands [87]. The above uses have generated the $Hg_{1-x}Cd_xTe$ technology for the experimental realization of high mobility single crystal with specially prepared surfaces. The same compound has emerged to be the optimum choice for illuminating the narrow sub-band physics because the relevant material constants can easily be experimentally measured [88]. Besides, the quaternary alloy $In_{1-x}Ga_xAs_yP_{1-y}$ lattice matched to InP, also finds wide use in the fabrication of avalanche photo-detectors [89], hetero-junction lasers [90], light emitting diodes [91] and avalanche photodiodes [92], field effect transistors, detectors, switches, modulators, solar cells, filters, and new types of integrated optical devices are made from the quaternary systems [93].

The Lead Chalcogenides (PbTe, PbSe, and PbS) are IV–VI non-parabolic semiconductors whose studies over several decades have been motivated by their importance in infrared IR detectors, lasers, light-emitting devices, photo-voltaic, and high temperature thermo-electrics [94–98]. PbTe, in particular, is the end compound of several ternary and quaternary high performance high temperature thermoelectric materials [99–103]. It has been used not only as bulk but also as films [104–107], QWs [108] super-lattices [109, 110] nanowires [111] and colloidal and embedded nano-crystals [112–115], and PbTe films doped with various impurities have also been investigated [116–123] These studies revealed some of the interesting features that had been seen in bulk PbTe, such as Fermi level pinning and, in the case of superconductivity [124].

In recent years there is considerable interest in studying the Dispersion Relations (DRs) of the carriers of different technologically important compound semiconductors and their nano structures under different physical conditions [125–140] because of their importance in the characterization in semiconductor nano structures. The formulations of DRs lead to the Density-of-States Function (DOS) which in turn plays the key role in determining the electron statistics in semiconductors and their quantized counter parts. In this chapter in Sects. 1.2.1–1.2.3 of the theoretical background Sect. 1.2 we study the DR in QWs, NWs and QDs of HD III–V semiconductors in the presence of magnetic field respectively. In Sects. 1.2.4–1.2.6 of the theoretical background Sect. 1.2 we study the same for QWs, NWs and QDs of HD III–V semiconductors in the presence of crossed fields respectively. Besides the Sects. 1.2.7–1.2.9 explore the DR in QWs, NWs and QDs of HD IV–VI semiconductors in the presence of magnetic field. The Sect. 1.2.10 contains the study of the DR in cylindrical QD of HD III–V semiconductors in the presence of crossed electric and magnetic fields. The Sect. 1.2.11 contains the study of DR in QWs of HD III–V materials in the presence of an arbitrarily oriented magnetic field.

The Sect. 1.3 contains the results and discussion. The Sect. 1.4 contains sixteen open research problems in this context.

1.2 Theoretical Background

1.2.1 The DR in Quantum Wells of HD III–V, Ternary and Quaternary Materials in the Presence of Magnetic Field

The DR of the conduction electrons in bulk samples of III–V ternary and quaternary materials can be written as

$$E(1 + \alpha E) = \frac{\hbar^2 k^2}{2m_c} \tag{1.1a}$$

where E is the electron energy as measured from the edge of the conduction band in the vertically upward direction in the absence of any quantization, $\alpha = \frac{1}{E_g}$, E_g is the band gap and m_c is the effective electron mass at the edge of the conduction band.

Let us consider a thin layer III–V semiconductor of rectangular cross-section. In such a structure the electron motion is quantized along the z-direction, (a being the nano thickness along z-direction) resulting in formation of electric sub-bands corresponding to different quantum numbers. The electrons motion is free along in the x-y plane and the magnetic field B is applied along the y-direction.

In the absence of magnetic field, $E(k)$ dispersion relation of the electrons in quasi 2D structures can be expressed as

$$E(1 + \alpha E) = \frac{\hbar^2}{2m_c}\left(\frac{n\pi}{a}\right)^2 + \frac{\hbar^2 k_x^2}{2m_c} + \frac{\hbar^2 k_y^2}{2m_c} \tag{1.1b}$$

In this case, the potential function assumes the form

$$V(z) = 0, 0 < z < a$$

and $V(z) = \alpha, 0 < z$ and for $z > a$.

The corresponding Eigen function in such structure can be written as

$$\begin{aligned}\psi_n(x, y, z) = \psi_n^0 = u_n &= \sqrt{\frac{2}{aL_xL_y}}\sin\left(\frac{n\pi}{a}z\right)\exp(ik_xx + ik_yy)\\ &= \sqrt{\frac{2}{aL_xL_y}}\psi_n(z)\exp(ik_xx + ik_yy), \left[\psi_n(z) = \sin\left(\frac{n\pi}{a}z\right)\right]\end{aligned} \tag{1.2}$$

The (1.1a) can be expressed as

$$\hat{E} = a\hat{p}^2 - b\hat{p}^4$$

where, $a = \frac{1}{2m_c}$, $b = \alpha a^2$ and $\vec{p} = -i\hbar\vec{\nabla}$.

Therefore the Hamiltonian can be expressed as

$$\hat{H} = a(-i\hbar\vec{\nabla} + e\vec{A})^2 - b(-i\hbar\vec{\nabla} + e\vec{A})^4 \tag{1.3}$$

where $\vec{A}$ is the vector potential.

We know that $\vec{B} = \vec{\nabla} \times \vec{A}$ and let us choose Coulomb gauge here as $\vec{\nabla} \cdot \vec{A} = 0$.

Since magnetic field is along y-direction, so

$$\vec{B} = \vec{j}B, \quad \vec{B} = (0, B, 0)$$

Let $\vec{A} = \vec{i}Bz$ so that it satisfies the above two relations i.e.

$$\vec{\nabla} \cdot \vec{A} = \frac{\partial}{\partial x}(Bz) + 0 + 0 = 0$$

and

$$\vec{B} = \vec{\nabla} \cdot \vec{A} = \begin{vmatrix} \vec{i} & \vec{j} & \vec{k} \\ \frac{\partial}{\partial x} & \frac{\partial}{\partial y} & \frac{\partial}{\partial z} \\ Bz & 0 & 0 \end{vmatrix} = \vec{i} \cdot 0 + \vec{j}\frac{\partial}{\partial z}(Bz) + \vec{k}\left[-\frac{\partial}{\partial y}(Bz)\right] = \vec{j}B$$

Applying this in (1.3), we have the Hamiltonian for the system as

$$\begin{aligned}
\hat{H} &= a(-i\hbar\vec{\nabla} + \vec{i}eBz)^2 - b(-i\hbar\vec{\nabla} + \vec{i}eBz)^4 \\
&= a\left[\left(-i\hbar\frac{\partial}{\partial x} + eBz\right)\vec{i} + \vec{j}\left(-i\hbar\frac{\partial}{\partial y}\right) + \vec{k}\left(-i\hbar\frac{\partial}{\partial z}\right)\right]^2 \\
&\quad - b\left[\left(-i\hbar\frac{\partial}{\partial x} + eBz\right)\vec{i} + \vec{j}\left(-i\hbar\frac{\partial}{\partial y}\right) + \vec{k}\left(-i\hbar\frac{\partial}{\partial z}\right)\right]^4 \\
&= a(\hat{H}_1)^2 - b(\hat{H}_1)^4
\end{aligned} \tag{1.4}$$

Again,

$$\hat{H} = \hat{H}_0 + \hat{H}' \tag{1.5}$$

where $\hat{H}_0$ is referred to as unperturbed Hamiltonian.

$\hat{H}'$ is the perturbation which has to be calculated and the effect of H' is assumed to be small.

$$\hat{H}\psi_n = E_n\psi_n \text{ is the Schrodinger equation of the system.} \tag{1.6}$$

$\hat{H}_0 u_n = E_n^0 u_n$ is the Eigen value equation for the unperturbed Hamiltonian so that

$$E_n^0 = \frac{\hbar^2}{2m_c}\left(\frac{n\pi}{a}\right)^2 + \frac{\hbar^2 k_y^2}{2m_c} + \frac{\hbar^2 k_x^2}{2m_c} - \alpha\left[\frac{\hbar^2}{2m_c}\left(\frac{n\pi}{a}\right)^2 + \frac{\hbar^2 k_y^2}{2m_c} + \frac{\hbar^2 k_x^2}{2m_c}\right]^2 \tag{1.7}$$

$$\begin{aligned}\psi_n^0 = u_n &= \sqrt{\frac{2}{aL_xL_y}}\sin\left(\frac{n\pi}{a}z\right)\exp(ik_xx + ik_yy)\\ &= \psi_n(z)\exp(ik_xx + ik_yy)\end{aligned}$$

Now,

$$\begin{aligned}\hat{H}_1\psi_n^0 &= \left[\left(-i\hbar\frac{\partial}{\partial x} + eBz\right)\vec{i} + \vec{j}\left(-i\hbar\frac{\partial}{\partial y}\right) + \vec{k}\left(-i\hbar\frac{\partial}{\partial z}\right)\right]u_n\\ &= \vec{i}(-i\hbar\cdot ik_x + eBz)u_n + \vec{j}(-i\hbar ik_y)u_n\\ &\quad + \vec{k}\left(-i\hbar\frac{n\pi}{a}\right)\sqrt{\frac{2}{aL_xL_y}}\cos\left(\frac{n\pi}{a}z\right)\exp(ik_xx + ik_yy)\end{aligned} \tag{1.8}$$

The term $(\hat{H}_1)^2\psi_n^0$ is being calculated in the following way

$$\begin{aligned}(\hat{H}_1)^2\psi_n^0 &= \hat{H}_1(\hat{H}_1\psi_n^0) = \left[\left(-i\hbar\frac{\partial}{\partial x} + eBz\right)\vec{i} + \vec{j}\left(-i\hbar\frac{\partial}{\partial y}\right) + \vec{k}\left(-i\hbar\frac{\partial}{\partial z}\right)\right]\cdot(\hat{H}_1\psi_n^0)\\ &= (-i\hbar\cdot ik_x + eBz)(-i\hbar\cdot ik_x + eBz)u_n + (-i\hbar\cdot ik_y)(-i\hbar\cdot ik_y)u_n\\ &\quad + \left(-i\hbar\frac{n\pi}{a}\right)\left(-i\hbar\frac{n\pi}{a}\right)(-u_n)(\hat{H}_1)^2\psi_n^0 = (\hbar k_x + eBz)^2u_n\\ &\quad + \hbar^2k_y^2u_n + \hbar^2\left(\frac{n\pi}{a}\right)^2u_n\end{aligned} \tag{1.9}$$

Therefore

$$(\hat{H}_1)^2\psi_n^0 = \left[\left\{\hbar^2k_x^2 + \hbar^2\left(\frac{n\pi}{a}\right)^2 + \hbar^2k_y^2\right\} + e^2B^2z^2 + 2\hbar k_xeBz\right]u_n \tag{1.10}$$

$$(\hat{H}_1)^3\psi_n^0 = \left[\left(-i\hbar\frac{\partial}{\partial x} + eBz\right)\vec{i} + \vec{j}\left(-i\hbar\frac{\partial}{\partial y}\right) + \vec{k}\left(-i\hbar\frac{\partial}{\partial z}\right)\right](\hat{H}_1)^2\psi_n^0$$

Therefore

$$(\hat{H}_1)^3\psi_n^0 = \vec{i}\left[(-i\hbar\cdot ik_x + eBz)\left\{\hbar^2k_x^2 + \hbar^2k_y^2 + \hbar^2\left(\frac{n\pi}{a}\right)^2 + e^2B^2z^2 + 2\hbar k_x eBz\right\}\right]u_n$$
$$+ \vec{j}\left[(-i\hbar\cdot ik_y)\left\{\hbar^2k_x^2 + \hbar^2k_y^2 + \hbar^2\left(\frac{n\pi}{a}\right)^2 + e^2B^2z^2 + 2\hbar k_x eBz\right\}\right]u_n$$
$$+ \vec{k}\left[\left(-i\hbar\cdot\frac{n\pi}{z}\right)\left\{\hbar^2k_x^2 + \hbar^2k_y^2 + \hbar^2\left(\frac{n\pi}{a}\right)^2\right.\right.$$
$$\left.\left.+ e^2B^2z^2 + 2\hbar k_x eBz\right\}\right]\sqrt{\frac{2}{a}}\cos\left(\frac{n\pi}{a}z\right)\exp(ik_xx + ik_yy)$$
$$+ \vec{k}(-i\hbar e^2B^2 2z + -i\hbar 2_x eB)u_n \tag{1.11}$$

$$(\hat{H}_1)^3\psi_n^0 = \vec{i}\left[(\hbar k_x + eBz)\left\{\hbar^2k_x^2 + \hbar^2k_y^2 + \hbar^2\left(\frac{n\pi}{a}\right)^2 + e^2B^2z^2 + 2_x eBz\right\}\right]u_n$$
$$+ \vec{j}\left[(_y)\left\{\hbar^2k_x^2 + \hbar^2k_y^2 + \hbar^2\left(\frac{n\pi}{a}\right)^2 + e^2B^2z^2 + 2\hbar k_x eBz\right\}\right]u_n$$
$$+ \vec{k}\left[\left(-i\hbar\cdot\frac{n\pi}{a}\right)\left\{\hbar^2k_x^2 + \hbar^2k_y^2 + \hbar^2\left(\frac{n\pi}{a}\right)^2\right.\right.$$
$$\left.\left.+ e^2B^2z^2 + 2_x eBz\right\}\right]u_n' + \vec{k}[-2i\hbar e^2B^2z - 2i\hbar^2k_x eB]u_n \tag{1.12}$$

where $\left[u_n' = \sqrt{\frac{2}{a}}i_xi_y\cos\left(\frac{n\pi}{a}\right)z\exp(ik_xx + ik_yy)\right]$

Again

$$(\hat{H}_1)^4\psi_n^0 = \left[\left(-i\hbar\frac{\partial}{\partial x} + eBz\right)\vec{i} + \vec{j}\left(-i\hbar\frac{\partial}{\partial y}\right) + \vec{k}\left(-i\hbar\frac{\partial}{\partial z}\right)\right](\hat{H}_1)^3\psi_n^0$$
$$= (\hbar k_x + eBz)^2\left\{\hbar^2k_x^2 + \hbar^2k_y^2 + \hbar^2\left(\frac{n\pi}{a}\right)^2 + e^2B^2z^2 + 2\hbar k_x eBz\right\}u_n$$
$$+ (\hbar^2k_y^2)\left\{\hbar^2k_x^2 + \hbar^2k_y^2 + \hbar^2\left(\frac{n\pi}{a}\right)^2\right.$$
$$\left.+ e^2B^2z^2 + 2\hbar k_x eBz\right\}u_n + \left(-i\hbar\frac{n\pi}{a}\right)\left(-i\hbar\frac{n\pi}{a}\right)$$
$$\left\{\hbar^2k_x^2 + \hbar^2k_y^2 + \hbar^2\left(\frac{n\pi}{a}\right)^2 + e^2B^2z^2 + 2\hbar k_x eBz\right\}(-u_n)$$
$$+ \left(-i\hbar\frac{n\pi}{a}\right)(-i\hbar)(e^2B^2 2z + 2\hbar k_x eBz)u_n'$$
$$+ (-i\hbar)(-2i\hbar^2e^2B^2z - 2i\hbar^2k_x eB)\left(\frac{n\pi}{a}\right)u_n'$$
$$+ (-i\hbar)(-2i\hbar e^2B^2)u_n = (\hbar^2k_x^2 + 2\hbar k_x eBz + e^2B^2z^2)$$
$$\left\{\hbar^2k_x^2 + \hbar^2k_y^2 + \hbar^2\left(\frac{n\pi}{a}\right)^2 + e^2B^2z^2 + 2\hbar k_x eBz\right\}u_n$$
$$+ \hbar^2k_y^2\left\{\hbar^2k_x^2 + \hbar^2\left(\frac{n\pi}{a}\right)^2 e^2B^2z^2 + 2\hbar k_x eBz\right\}u_n$$
$$+ \hbar^2\left(\frac{n\pi}{a}\right)^2\left\{\hbar^2k_x^2 + \hbar^2k_y^2 + \hbar^2\left(\frac{n\pi}{a}\right)^2 + e^2B^2z^2 + 2\hbar k_x eBz\right\}u_n$$
$$- 2\hbar^2e^2B^2u_n - \hbar^2\left(\frac{n\pi}{a}\right)(4e^2B^2z + 4\hbar k_x eB)\sqrt{\frac{2}{a}}i_xi_y\cos\left(\frac{n\pi}{a}z\right)\exp(ik_xx + ik_yy) \tag{1.13}$$

Thus finally we get

$$(\hat{H}_1)^4\psi_n^0 = \left\{\hbar^2k_x^2 + \hbar^2k_y^2 + \hbar^2\left(\frac{n\pi}{a}\right)^2\right\}^2 u_n + 2\left\{\hbar^2k_x^2 + \hbar^2k_y^2 + \hbar^2\left(\frac{n\pi}{a}\right)^2\right\}(e^2B^2z^2 + 2\hbar k_x eBz)u_n + (e^2B^2z^2 + 2\hbar k_x eBz)^2u_n - 2\hbar^2e^2B^2u_n - \hbar^2\left(\frac{n\pi}{a}\right)(4e^2B^2z + 4\hbar k_x eB)u_n' \quad (1.14)$$

Thus the total Hamiltonian of the system is

$$\hat{H} = a(\hat{H}_1)^2 - b(\hat{H}_1)^4 = \frac{\hbar^2k_x^2}{2m_c} + \frac{\hbar^2k_y^2}{2m_c} + \frac{\hbar^2}{2m_c}\left(\frac{n\pi}{a}\right)^2 + \frac{e^2B^2z^2}{2m_c} + \frac{2\hbar k_x eBz}{2m_c} - \alpha\left[\frac{\hbar^2k_x^2}{2m_c} + \frac{\hbar^2k_y^2}{2m_c} + \frac{\hbar^2}{2m_c}\left(\frac{n\pi}{a}\right)^2\right]^2 - 2\alpha\left\{\frac{\hbar^2k_x^2}{2m_c} + \frac{\hbar^2k_y^2}{2m_c} + \frac{\hbar^2}{2m_c}\left(\frac{n\pi}{a}\right)^2\right\}\cdot\left(\frac{e^2B^2z^2}{2m_c} + \frac{2\hbar k_x eBz}{2m_c}\right) - \alpha\left(\frac{e^2B^2z^2}{2m_c} + \frac{2\hbar k_x eBz}{2m_c}\right)^2 + 2\alpha\left(\frac{\hbar eB}{2m_c}\right)^2 + \frac{\alpha\hbar^2}{2m_c}\left(\frac{n\pi}{a}\right)\left(\frac{4e^2B^2z}{2m_c} + \frac{4_x eB}{2m_c}\right)\cot\left(\frac{n\pi}{a}z\right) \quad (1.15)$$

Therefore the perturbed Hamiltonian is given by

$$\hat{H}' = \frac{e^2B^2z^2}{2m_c} + \frac{\hbar k_x eBz}{m_c} + 2\alpha\left(\frac{\hbar eB}{2m_c}\right)^2 + \frac{\alpha\hbar^2}{2m_c}\left(\frac{n\pi}{a}\right)\left(\frac{2e^2B^2z}{m_c} + \frac{2\hbar k_x eB}{m_c}\right)\cot\left(\frac{n\pi}{a}z\right) - 2\alpha\left\{\frac{\hbar^2k_x^2}{2m_c} + \frac{\hbar^2k_y^2}{2m_c} + \frac{\hbar^2}{2m_c}\left(\frac{n\pi}{a}\right)^2\right\}\cdot\left(\frac{e^2B^2z^2}{2m_c} + \frac{\hbar k_x eBz}{m_c}\right) - \alpha\left(\frac{e^2B^2z^2}{2m_c} + \frac{\hbar k_x eBz}{m_c}\right)^2 \quad (1.16)$$

So the first order perturbation to the energy is

$$E_n^{(1)} = \hat{H}'_{nn} = \int_\tau u_n^*\hat{H}'u_n d\tau = <u_n|\hat{H}'|u_n> = \frac{\hbar k_x eB}{m_c}<z> + \frac{e^2B^2}{2m_c}<z^2> + 2\alpha\left(\frac{\hbar eB}{2m_c}\right)^2 + \frac{\alpha\hbar^2}{2m_c}\left(\frac{n\pi}{a}\right)\frac{2e^2B^2}{m_c}\int_0^a\left(\frac{2}{a}\right)z\sin\left(\frac{n\pi}{a}z\right)\cos\left(\frac{n\pi}{a}z\right)dz + \frac{\alpha\hbar^2}{2m_c}\left(\frac{n\pi}{a}\right)\frac{2\hbar k_x eB}{m_c}\int_0^a\left(\frac{2}{a}\right)\sin\left(\frac{n\pi}{a}z\right)\cos\left(\frac{n\pi}{a}z\right)dz - 2\alpha\left\{\frac{\hbar^2k_x^2}{2m_c} + \frac{\hbar^2k_y^2}{2m_c} + \frac{\hbar^2}{2m_c}\left(\frac{n\pi}{a}\right)^2\right\}\frac{e^2B^2}{2m_c}<z^2> - 2\alpha\left\{\frac{\hbar^2k_x^2}{2m_c} + \frac{\hbar^2k_y^2}{2m_c} + \frac{\hbar^2}{2m_c}\left(\frac{n\pi}{a}\right)^2\right\}\frac{\hbar k_x eBz}{m_c}<z> - \alpha\frac{e^4B^4}{4m_c^2}<z^4> - 2\alpha\frac{e^2B^2}{2m_c}\frac{\hbar k_x eB}{m_c}<z^3> - \alpha\frac{\hbar^2k_x^2e^2B^2}{m_c^2}<z^2> \quad (1.17)$$

in which

$$<z> \;=\; \int_{\tau} u_n^* z u_n d\tau \;=\; \left(\frac{2}{a}\right)\int_0^a \sin^2\left(\frac{n\pi}{a}z\right)dz \tag{1.18}$$

$$<z^2> \;=\; \int_{\tau} u_n^* z^2 u_n d\tau \;=\; \left(\frac{2}{a}\right)\int_0^a z^2\sin^2\left(\frac{n\pi}{a}z\right)dz \tag{1.19}$$

$$<z^3> \;=\; \int_{\tau} u_n^* z^3 u_n d\tau \tag{1.20}$$

$$<z^4> \;=\; \int_{\tau} u_n^* z^4 u_n d\tau \tag{1.21}$$

Now $\int_0^a \mathrm{z}\sin^2\left(\frac{n\pi}{a}z\right)\mathrm{dz} = \frac{1}{2}\int_0^a \mathrm{z}\left(1 - \cos\frac{2n\pi}{a}z\right)\mathrm{dz} = \frac{1}{2}\int_0^a \mathrm{z\,dz} - \frac{1}{2}\int_0^a \mathrm{z}\cos\left(\frac{2n\pi}{a}\right)z\,\mathrm{dz}$

$$= \frac{a^2}{4} - \frac{1}{2}\left[\mathrm{z}\int \cos\left(\frac{2n\pi}{a}z\right)\mathrm{dz} - \int\left\{\cos\frac{2n\pi}{a}z\,\mathrm{dz}\right\}\right]_0^a$$

$$= \frac{a^2}{4} - \frac{1}{2}\left[\frac{az}{2n\pi}\sin\frac{2n\pi}{a}z - \frac{a}{2n\pi}\right]\int \sin\left(\frac{2n\pi}{a}\right)z\,dz\Bigg]_0^a$$

$$= \frac{a^2}{4} - \frac{1}{2}\left[\frac{az}{2n\pi}\sin\left(\frac{2n\pi z}{a}\right) + \left(\frac{a}{2n\pi}\right)^2\cos\left(\frac{2n\pi z}{a}\right)\right]_0^a$$

$$= \frac{a^2}{4} - \frac{1}{2}\left[\frac{a^2}{2n\pi}\sin 2n\pi + \frac{a^2}{4n^2\pi^2}\cos 2n\pi - 0 - \frac{a^2}{4n^2\pi^2}\right]_0^a$$

$$= \frac{a^2}{4} - \frac{1}{2}[0 + 1 - 1] = \frac{a^2}{4}$$

Therefore from (1.18) we can write,

$$<z> \;=\; \frac{2}{a}\frac{a^2}{4} = \frac{a}{2} \tag{1.22}$$

$$\text{Again} \int_0^a z^2 \sin^2\left(\frac{n\pi}{a}z\right)dz = \frac{1}{2}\int_0^a z^2\left(1 - \cos\frac{2n\pi}{a}z\right)dz$$

$$= \frac{1}{2}\int_0^a z^2 dz - \frac{1}{2}\int_0^a z^2 \cos\left(\frac{2n\pi}{a}\right)z\, dz$$

$$= \frac{a^3}{6} - \frac{1}{2}\left[z^2 \int \cos\left(\frac{2n\pi}{a}z\right)dz - \int\left\{2z\left(\cos\frac{2n\pi}{a}z\right)dz\right\}\right]_0^a$$

$$= \frac{a^3}{6} - \frac{1}{2}\left[\frac{az^2}{2n\pi}\sin\left(\frac{2n\pi}{a}z\right) - \frac{2a}{2n\pi}\int z\sin\left(\frac{2n\pi}{a}\right)z\, dz\right]_0^a$$

$$= \frac{a^3}{6} + \left(\frac{a}{2n\pi}\right)\int_0^a z\sin\left(\frac{2n\pi z}{a}\right)dz = \frac{a^3}{6} + \left(\frac{a}{2n\pi}\right)$$

$$\left[\int_0^a z\sin\left(\frac{2n\pi z}{a}\right)dz - \int_0^a 1\left\{\sin\left(\frac{2n\pi z}{a}\right)dz\right\}dz\right]$$

$$= \frac{a^3}{6} + \frac{a}{2n\pi}\left[\left\{\frac{az}{2n\pi} - \cos\left(\frac{2n\pi z}{a}\right)\right\}\right]_0^a + \frac{a}{2n\pi}\int_0^a \cos\left(\frac{2n\pi z}{a}\right)dz$$

$$= \frac{a^3}{6} + \frac{a}{2n\pi}\left[-\frac{a^2}{2n\pi}\cos 2n\pi + 0 + \left(\frac{a}{2n\pi}\right)^2 \sin\left(\frac{2n\pi z}{a}\right)\right]_0^a$$

$$= \frac{a^3}{6} + \left(\frac{a}{2n\pi}\right)\left[-\frac{a^2}{2n\pi} + \left(\frac{a}{2n\pi}\right)^2 \sin 2n\pi - 0\right] = \left(\frac{a^3}{6} - \frac{a^3}{4n^2\pi^2}\right)$$

Therefore, from (1.19) we can write,

$$<z^2> = \frac{2}{a}\left(\frac{a^3}{6} - \frac{a^3}{4n^2\pi^2}\right) = \left(\frac{a^2}{3} - \frac{a^2}{2n^2\pi^2}\right) \tag{1.23}$$

$$\text{Since} \int_0^a z^3 \sin^2\left(\frac{n\pi}{a}z\right)dz = \frac{1}{2}\int_0^a z^3\left(1 - \cos\frac{2n\pi}{a}z\right)dz$$

$$= \frac{1}{2}\int_0^a z^3 dz - \frac{1}{2}\int_0^a z^3 \cos\left(\frac{2n\pi}{a}z\right)dz$$

$$= \frac{a^4}{8} - \frac{1}{2}\left[z^3 \int_0^a \cos\left(\frac{2n\pi}{a}z\right)dz\right.$$

$$\left. - 3\int_0^a z^2\left\{\int \cos\left(\frac{2n\pi}{a}z\right)dz\right\}dz\right]$$

$$= \frac{a^4}{8} - \frac{1}{2}\left[\frac{z^3 a}{2n\pi}\sin\frac{2n\pi}{a}z\int_0^a -3\frac{a}{2n\pi}\int_0^a z^2\sin\left(\frac{2n\pi z}{a}\right)dz\right]$$

$$= \frac{a^4}{8} + \frac{3}{2}\left(\frac{a}{2n\pi}\right)\int_0^a z^2\sin\left(\frac{2n\pi z}{a}\right)dz$$

$$= \frac{a^4}{8} + \frac{3}{2}\left(\frac{a}{2n\pi}\right)\left[z^2\int_0^a \sin\left(\frac{2n\pi z}{a}\right)dz - 2\int_0^a z\int \sin\left(\frac{2n\pi}{a}z\right)dz\}dz\right]$$

$$= \frac{a^4}{8} + \frac{3}{2}\left(\frac{a}{2n\pi}\right)\left[-\frac{z^2 a}{2n\pi}\cos\left(\frac{2n\pi z}{a}\right)\int_0^a + \frac{a}{n\pi}\int_0^a z\cos\left(\frac{2n\pi z}{a}\right)dz\right]$$

$$= \frac{a^4}{8} + \frac{3}{2}\left(\frac{a}{2n\pi}\right)\left[-\frac{a^3}{2n\pi}\cos(2n\pi) - 0 + \left(\frac{a}{n\pi}\right)\int_0^a z\cos\left(\frac{2n\pi}{a}z\right)dz\right]$$

$$= \frac{a^4}{8} - \frac{3a^4}{8n^2\pi^2} + \frac{3a^2}{4n^2\pi^2}\int_0^a z\cos\left(\frac{2n\pi}{a}z\right)dz = \left(\frac{a^4}{8} - \frac{3a^4}{8n^2\pi^2}\right)$$

Therefore from (1.20) we can write,

$$<z^3> = \frac{2}{a}\left(\frac{a^4}{8} - \frac{3a^4}{8n^2\pi^2}\right) = \frac{a^3}{4} - \frac{3a^3}{4n^2\pi^2} \tag{1.24}$$

$$<z^4> = \frac{2}{a}\int_0^a z^4\sin^2\left(\frac{n\pi}{a}\right)zdz = \frac{1}{2}\int_0^a z^4\left(1 - \cos\frac{2n\pi}{a}z\right)dz$$

$$= \frac{1}{a}\int_0^a z^4 dz - \frac{1}{a}\int_0^a z^4\cos\frac{2n\pi}{a}zdz$$

$$= \frac{a^4}{5} - \frac{1}{a}\left[z^4\frac{a}{2n\pi}\sin\frac{2n\pi}{a}z\int_0^a - \frac{4a}{2n\pi}\int_0^a z^3\sin\left(\frac{2n\pi}{a}z\right)dz\right]$$

$$= \frac{a^4}{5} + \frac{2}{n\pi}\int_0^a z^3\sin\left(\frac{2n\pi}{a}z\right)dz$$

$$= \frac{a^4}{5} + \frac{2}{n\pi}\left[-z^3\frac{a}{2n\pi}\cos\left(\frac{2n\pi}{a}Z\right)\int_0^a + 3\frac{a}{2n\pi}\int_0^a z^2\cos\left(\frac{2n\pi}{a}z\right)dz\right]$$

$$= \frac{a^4}{5} - \frac{a^4}{n^2\pi^2} + \frac{3a}{n^2\pi^2}\int_0^a z^2 \cos\left(\frac{2n\pi}{a}z\right)dz$$

$$= \frac{a^4}{5} - \frac{a^4}{n^2\pi^2} + \frac{3a}{n^2\pi^2}\left[-\frac{a}{n\pi}\int_0^a z^2 \sin\left(\frac{2n\pi}{a}z\right)dz\right]$$

$$= \frac{a^4}{5} - \frac{a^4}{n^2\pi^2} + \frac{3a}{n^2\pi^2}\left[-\frac{a}{n\pi}(-\frac{a^2}{2n\pi})\right] = \frac{a^4}{5} - \frac{a^4}{n^2\pi^2} + \frac{3a}{n^2\pi^2}\left(\frac{a^3}{2n^2\pi^2}\right)$$

$$<z^4> = \frac{a^4}{5} - \frac{a^4}{n^2\pi^2} + \frac{3a^4}{2n^4\pi^4}$$

Since the basic wave function is normalized, the first order contribution of the Eigen energy value can be written as

$$\begin{aligned} E_n^{(1)} = {} & \frac{\hbar k_x eB}{m_c}\langle z\rangle + \frac{e^2B^2}{2m_c}\langle z^2\rangle + 2\alpha\left(\frac{\hbar eB}{2m_c}\right)^2 - \frac{\alpha\hbar^2}{2m_c}\frac{e^2B^2}{m_c} \\ & - 2\alpha\left\{\frac{\hbar^2k_x^2}{2m_c} + \frac{\hbar^2k_y^2}{2m_c} + \frac{\hbar^2}{2m_c}\left(\frac{n\pi}{a}\right)^2\right\}\frac{\hbar k_x eB}{m_c}\langle z\rangle \\ & - 2\alpha\left\{\frac{\hbar^2k_x^2}{2m_c} + \frac{\hbar^2k_y^2}{2m_c} + \frac{\hbar^2}{2m_c}\left(\frac{n\pi}{a}\right)^2\right\}\frac{e^2B^2}{2m_c}\langle z^2\rangle \\ & - \alpha\frac{\hbar^2k_x^2e^2B^2}{m_c^2}\langle z^2\rangle - \frac{\alpha\hbar k_x e^3B^3}{m_c^2}\langle z^3\rangle - \alpha\left(\frac{e^4B^4}{4m_c^2}\right)\langle z^4\rangle \end{aligned} \quad (1.25)$$

So the total energy is

$$\begin{aligned} E = E_n^0 + E_n^{(1)} = {} & \frac{\hbar^2k_x^2}{2m_c} + \frac{\hbar^2k_y^2}{2m_c} + \frac{\hbar^2}{2m_c}\left(\frac{n\pi}{a}\right)^2 - \alpha\left[\frac{\hbar^2k_x^2}{2m_c} + \frac{\hbar^2k_y^2}{2m_c} + \frac{\hbar^2}{2m_c}\left(\frac{n\pi}{a}\right)^2\right]^2 \\ & + \frac{\hbar k_x eB}{m_c}\langle z\rangle + \frac{e^2B^2}{2m_c}\langle z^2\rangle + 2\alpha\left(\frac{\hbar eB}{2m_c}\right)^2 \\ & - 2\alpha\left(\frac{\hbar eB}{2m_c}\right)^2 - \frac{2\alpha\hbar k_x eB}{m_c}\left\{\frac{\hbar^2k_x^2}{2m_c} + \frac{\hbar^2k_y^2}{2m_c} + \frac{\hbar^2}{2m_c}\left(\frac{n\pi}{a}\right)^2\right\}\langle z\rangle \\ & - \alpha\left(\frac{e^2B^2}{m_c}\right)\left\{\frac{\hbar^2k_x^2}{2m_c} + \frac{\hbar^2k_y^2}{2m_c} + \frac{\hbar^2}{2m_c}\left(\frac{n\pi}{a}\right)^2\right\}\langle z^2\rangle \\ & - \alpha\left(\frac{\hbar k_x eB}{m_c}\right)^2\langle z^2\rangle - \alpha\left(\frac{e^2B^2}{2m_c}\right)^2\langle z^4\rangle - \alpha\left(\frac{\hbar k_x e^3B^3}{m_c^2}\right)^2\langle z^3\rangle \\ = {} & \frac{\hbar^2}{2m_c}\left(\frac{n\pi}{a}\right)^2 + \frac{\hbar^2k_y^2}{2m_c} + \frac{1}{2m_c}[\hbar k_x + eB\langle z\rangle]^2 + \frac{e^2B^2}{2m_c}[\langle z^2\rangle - \langle z^2\rangle] - \alpha\left[\frac{\hbar^2k_x^2}{2m_c} + \frac{\hbar^2k_y^2}{2m_c}\right. \\ & \left. + \frac{\hbar^2}{2m_c}\left(\frac{n\pi}{a}\right)^2 + \frac{\hbar k_x eB}{m_c}\langle z\rangle\right]^2 + \alpha\left(\frac{\hbar k_x eB}{m_c}\right)^2\langle z^2\rangle - \alpha\left(\frac{e^2B^2}{2m_c}\right)\left\{\frac{\hbar^2k_x^2}{2m_c} + \frac{\hbar^2k_y^2}{2m_c} + \frac{\hbar^2}{2m_c}\left(\frac{n\pi}{a}\right)^2\right\} \\ & - \alpha\left(\frac{\hbar k_x eB}{m_c}\right)^2\langle z^2\rangle - \alpha\left(\frac{e^2B^2}{2m_c}\right)^2\langle z^4\rangle - \alpha\left(\frac{\hbar k_x e^3B^3}{m_c^2}\right)\langle z^3\rangle \end{aligned}$$

Therefore,

$$
\begin{aligned}
E = {} & \frac{\hbar^2}{2m_c}\left(\frac{n\pi}{a}\right)^2 + \frac{\hbar^2 k_y^2}{2m_c} + \frac{1}{2m_c}(\hbar k_x + eB\langle z\rangle)^2 + \frac{e^2B^2}{2m_c}\left[\langle z^2\rangle - \langle z\rangle^2\right] \\
& - \alpha\left(\frac{\hbar k_x eB}{m_c}\right)^2\left[\langle z^2\rangle - \langle z\rangle^2\right] - \alpha\left(\frac{e^2B^2}{2m_c}\right)^2\langle z\rangle^4 - \alpha\left(\frac{\hbar k_x e^3B^3}{m_c^2}\right)\langle z^3\rangle \\
& - \alpha\left(\frac{e^2B^2}{m_c}\right)\left\{\frac{\hbar^2k_x^2}{2m_c} + \frac{\hbar^2k_y^2}{2m_c} + \frac{\hbar^2}{2m_c}\left(\frac{n\pi}{a}\right)^2\right\}\langle z^2\rangle - \alpha\left[\frac{\hbar^2}{2m_c}\left(\frac{n\pi}{a}\right)^2 + \frac{\hbar^2k_y^2}{2m_c} + \frac{1}{2m_c}(\hbar k_x + eB\langle z\rangle)^2\right]^2 \\
& + \alpha\left(\frac{e^2B^2}{m_c}\right)\left[\frac{\hbar^2}{2m_c}\left(\frac{n\pi}{a}\right)^2 + \frac{\hbar^2k_y^2}{2m_c} + \frac{\hbar^2k_x^2}{2m_c}\right]\langle z\rangle^2 + \alpha\left(\frac{\hbar k_x e^3B^3}{m_c^2}\right)\langle z^3\rangle + \alpha\left(\frac{e^2B^2}{2m_c}\right)^2\langle z\rangle^4
\end{aligned}
\tag{1.26a}
$$

It is well known that the band tails are being formed in the forbidden zone of HDS and can be explained by the overlapping of the impurity band with the conduction and valence bands [141]. Kane [142] and Bonch Bruevich [143] have independently derived the theory of band tailing for semiconductors having unperturbed parabolic energy bands. Kane's model [142] was used to explain the experimental results on tunneling [144] and the optical absorption edges [145, 146] in this context. Halperin and Lax [147] developed a model for band tailing applicable only to the deep tailing states. Although Kane's concept is often used in the literature for the investigation of band tailing [148, 149], it may be noted that this model [142, 150] suffers from serious assumptions in the sense that the local impurity potential is assumed to be small and slowly varying in space coordinates [149]. In this respect, the local impurity potential may be assumed to be a constant. In order to avoid these approximations, we have developed in this book, the electron energy spectra for HDS for studying the EP based on the concept of the variation of the kinetic energy [141, 149] of the electron with the local point in space coordinates. This kinetic energy is then averaged over the entire region of variation using a Gaussian type potential energy. It may be noted that, a more general treatment of many-body theory for the DOS of HDS merges with one-electron theory under macroscopic conditions [141]. Also, the experimental results for the Fermi energy and others are the average effect of this macroscopic case. So, the present treatment of the one-electron system is more applicable to the experimental point of view and it is also easy to understand the overall effect in such a case [151]. In a HDS, each impurity atom is surrounded by the electrons, assuming a regular distribution of atoms, and it is screened independently [148, 150, 152]. The interaction energy between electrons and impurities is known as the impurity screening potential. This energy is determined by the inter-impurity distance and the screening radius, which is known as the screening length. The screening radius changes with the electron concentration and the effective mass.

Furthermore, these entities are important for HDS in characterizing the semiconductor properties [153, 154] and the modern electronic devices [148, 155]. The works on Fermi energy and the screening length in an n-type GaAs have already been initiated in the literature [156, 157], based on Kane's model. Incidentally, thelimitations of Kane's model [142, 149], as mentioned above, are also present in their studies.

The Gaussian distribution $F(V)$ of the impurity potential is given by [142, 149]

$$F(V) = \left(\pi\eta_g^2\right)^{-1/2} \exp\left(-V^2/\eta_g^2\right) \tag{1.26b}$$

where, η_g is the impurity screening potential. It appears from (1.26a) that the variance parameter η_g is not equal to zero, but the mean value is zero. Further, the impurities are assumed to be uncorrelated and the band mixing effect has been neglected in this simplified theoretical formalism.

We have to average the kinetic energy in the order to obtain the DR in this case in the presence of band tails. Using the (1.26a) and (1.26b), we can write that the DR in HDQW of III–V semiconductors (whose unperturbed conduction electrons obey the two band model of Kane) in the presence of a parallel magnetic field B along y-direction can be written

$$\begin{aligned}\gamma_3(E,\eta_g) = {} & \left[\frac{\hbar^2}{2m_c}\left(\frac{n\pi}{a}\right)^2 + \frac{\hbar^2 k_y^2}{2m_c} + \frac{1}{2m_c}(\hbar k_x + eB\langle z\rangle)^2\right] \\ & - \alpha\left[\frac{\hbar^2}{2m_c}\left(\frac{n\pi}{a}\right)^2 + \frac{\hbar^2 k_y^2}{2m_c} + \frac{1}{2m_c}(\hbar k_x + eB\langle z\rangle)^2\right]^2 \\ & + \frac{e^2B^2}{2m_c}\left[\langle z^2\rangle - \langle z\rangle^2\right] - \alpha\left(\frac{e^2B^2}{m_c}\right)\left\{\frac{\hbar^2 k_x^2}{2m_c} + \frac{\hbar^2 k_y^2}{2m_c} + \frac{\hbar^2}{2m_c}\left(\frac{n\pi}{a}\right)^2\right\}\left[\left\{\langle z^2\rangle - \langle z\rangle^2\right\}\right] \\ & - \alpha\left(\frac{\hbar k_x eB}{m_c}\right)^2\left[\langle z^2\rangle - \langle z\rangle^2\right] - \alpha\left(\frac{\hbar k_x e^3B^3}{m_c^2}\right)^2\left[\langle z^3\rangle - \langle z\rangle^3\right] - \alpha\left(\frac{e^2B^2}{2m_c}\right)^2\left[\langle z^4\rangle - \langle z\rangle^4\right]\end{aligned} \tag{1.27}$$

Putting $k_n = k_y = 0$ and $E = E_n$ in (1.27) we get

$$\begin{aligned}\gamma_3(E,\eta_g) = {} & \frac{\hbar^2}{2m_c}\left(\frac{n\pi}{a}\right)^2 + \frac{e^2B^2\langle z\rangle^2}{2m_c} - \alpha\left[\frac{\hbar^2}{2m_c}\left(\frac{n\pi}{a}\right)^2 + \frac{e^2B^2\langle z\rangle^2}{2m_c}\right]^2 + \frac{e^2B^2}{2m_c}\left[\langle z^2\rangle - \langle z\rangle^2\right] \\ & - \alpha\frac{e^2B^2}{m_c}\left[\frac{\hbar^2}{2m_c}\left(\frac{n\pi}{a}\right)^2\right]\left[\langle z^2\rangle - \langle z\rangle^2\right] - \alpha\left(\frac{e^2B^2}{2m_c}\right)^2\left[\langle z^4\rangle - \langle z\rangle^4\right]\end{aligned} \tag{1.28}$$

From (1.27) and (1.28) we get

$$\begin{aligned}
\gamma_3(E,\eta_g) &= \gamma_3(E_n,\eta_g) + \frac{\hbar^2 k_y^2}{2m_c} + \left(\frac{\hbar^2 k_x^2 + 2\hbar k_x eB\langle z\rangle}{2m_c}\right) \\
&\quad - \alpha\left[\left(\frac{\hbar^2 k_x^2}{2m_c} + \frac{\hbar^2 k_y^2}{2m_c} + \frac{{}_x eB\langle z\rangle}{m_c}\right) + \left\{\frac{\hbar^2}{2m_c}\left(\frac{n\pi}{a}\right)^2 + \frac{e^2B^2\rangle^2}{2m_c}\right\}\right]^2 \\
&\quad + \alpha\left[\frac{\hbar^2}{2m_c}\left(\frac{n\pi}{a}\right)^2 + \frac{e^2B^2\langle z\rangle^2}{m_c}\right]^2 - \alpha\frac{e^2B^2}{m_c}\left(\frac{\hbar^2 k_x^2}{2m_c} + \frac{\hbar^2 k_y^2}{2m_c}\right)\left[\langle z^2\rangle - \langle z\rangle^2\right] \\
&\quad - \alpha\frac{\hbar^2 k_x^2}{m_c}\cdot\frac{e^2B^2}{m_c}\left[\langle z^2\rangle - \langle z\rangle^2\right] - \alpha\left(\frac{\hbar k_x e^3B^3}{m_c^2}\right)\left[\langle z^3\rangle - \langle z\rangle^3\right]
\end{aligned} \tag{1.29}$$

Putting $\alpha = 0$ we get

$$\gamma_3(E,\eta_g) = f_n + \frac{\hbar^2 k_x^2}{2m_c} + \frac{\hbar^2 k_y^2}{2m_c} + \frac{\hbar k_x eB\langle z\rangle}{m_c} \tag{1.30}$$

where $f_n = \frac{\hbar^2}{2m_c}\left(\frac{n\pi}{a}\right)^2 + \frac{e^2B^2\langle z^2\rangle}{2m_c}$ putting the value $\frac{\hbar^2 k_x^2}{2m_c} + \frac{\hbar^2 k_y^2}{2m_c} + \frac{\hbar k_x eB\langle z\rangle}{m_c} = \gamma_3(E,\eta_g) - f_n$ in the perturbed term.

$\alpha\left[\left(\frac{\hbar^2 k_x^2}{2m_c} + \frac{\hbar^2 k_y^2}{2m_c} + \frac{\hbar k_x eB\langle z\rangle}{m_c}\right) + \left\{\frac{\hbar^2}{2m_c}\left(\frac{n\pi}{a}\right)^2 + \frac{e^2B^2\langle z^2\rangle}{2m_c}\right\}\right]^2$ of (1.29) we get

$$\bar{\gamma}(E,\eta_g) = \mathrm{A}(n)k_x^2 + 2b(n)k_x + c(n)k_y^2 \tag{1.31}$$

$$\begin{aligned}
\text{where } \bar{\gamma}(E,\eta_g) &= \Bigg[\Bigg[\gamma_3(E,\eta_g) - \gamma_3(E,\eta_g) + \alpha\Bigg[\gamma_3(E,\eta_g) - f_n \\
&\quad + \left\{\frac{\hbar^2}{2m_c}\left(\frac{n\pi}{a}\right)^2 + \frac{e^2B^2\langle z^2\rangle}{2m_c}\right\}\Bigg]^2 - \alpha\left[\frac{\hbar^2}{2m_c}\left(\frac{n\pi}{a}\right)^2 + \frac{e^2B^2\langle z^2\rangle}{m_c}\right]^2\Bigg],
\end{aligned}$$

$$A(n) = \left[\frac{\hbar^2}{2m_c} - \alpha\frac{e^2B^2}{m_c}\cdot\frac{\hbar^2}{2m_c}\left\{\langle z^2\rangle - \langle z\rangle^2\right\} - \alpha^2\frac{\hbar^2 e^2B^2}{2m_c^2}\left\{\langle z^2\rangle - \langle z\rangle^2\right\}\right],$$

$$2b(n) = \left[\frac{\hbar eB\langle z\rangle}{m_c} - \frac{\alpha\hbar e^3B^3}{m_c^2}\left\{\langle z^3\rangle - \langle z\rangle^3\right\}\right]$$

$$\text{and } \mathrm{c}(n) = \left[\frac{\hbar^2}{2m_c} - \alpha\frac{e^2B^2}{m_c}\cdot\frac{\hbar^2}{2m_c}\left[\langle z^2\rangle - \langle z\rangle^2\right]\right]$$

(1.31) can be written as

$$\frac{\left[k_x + \frac{b(n)}{A(n)}\right]^2}{\frac{1}{A(\mathrm{n})}\left[\bar{\gamma}(E,\eta_g) - \frac{b^2(n)}{A(n)}\right]} + \frac{k_y^2}{\frac{1}{c(n)}\left[\bar{\gamma}(E,\eta_g) - \frac{b^2(n)}{A(n)}\right]} = 1 \quad (1.32)$$

The (1.32) indicates that the DR in this case is quantized ellipses for constant electron energy and scattering potential in the $k_x - k_y$ plane. The area of the ellipse can be written as

$$\tau(E,n,\eta_g) = \pi\tau_1(E,n,\eta_g) \quad (1.33)$$

where $\tau_1(E,n,\eta_g) = \frac{1}{\sqrt{A(n)\mathrm{c}(\mathrm{n})}}\left[\bar{\gamma}(E,\eta_g) - \frac{b^2(n)}{A(n)}\right]$

It is interesting to note that the single important concept in the whole spectra of materials and allied sciences is the effective electron mass which is in disguise in the apparently simple (1.32), and can, briefly be described as follows:

Effective electron mass: The effective mass of the carriers in semiconductors, being connected with the mobility, is known to be one of the most important physical quantities, used for the analysis of electron devices under different operating conditions [158]. The carrier degeneracy in semiconductors influences the effective mass when it is energy dependent. Under degenerate conditions, only the electrons at the Fermi surface of n-type semiconductors participate in the conduction process and hence, the effective mass of the electrons corresponding to the Fermi level (EEM) would be of interest in electron transport under such conditions. The Fermi energy is again determined by the electron energy spectrum and the carrier statistics and therefore, these two features would determine the dependence of the effective electron mass in degenerate n-type semiconductors under the degree of carrier degeneracy. In recent years, various DRs have been proposed [159–161] which have created the interest in studying the effective mass in such materials under external conditions. It has, therefore, different values in different materials and varies with electron concentration, with the magnitude of the reciprocal quantizing magnetic field under magnetic quantization, with the quantizing electric field as in inversion layers, with the nano-thickness as in UFs and nano wires and with superlattice period as in the quantum confined superlattices of small gap semiconductors with graded interfaces having various carrier energy spectra [162–224].

The EEM in this case is given by

$$m^*(E,n,\eta_g) = \frac{\hbar^2}{2}\tau_1'(E,n,\eta_g) \quad (1.34)$$

Thus magnetic field makes the mass quantum number dependent.

It is important to note that in any semiconductor the EEM at the Fermi level is a function of electron concentration. In the case of HDS, the EEM at any energy is a function of electron concentration due to the presence of η_g, a concept impossible without band tailing.

The DOS function is given by

$$N_{QWHDB}(E,B,\eta_g) = \frac{g_v}{2\pi}\sum_{n=1}^{n_{\max}} \tau_1'(E,n,\eta_g)\mathrm{H}(\mathrm{E} - \mathrm{E}_{19,1}) \quad (1.35)$$

where $\mathrm{E}_{19,1}$ is the sub-band energy and can be expressed through the equation

$$\bar{\gamma}(E_{19,1},\eta_g) = 0 \quad (1.36)$$

1.2.2 The DR in Nano Wires of HD III–V Semiconductors in the Presence of Magnetic Field

The eigen function and the energy eigen values in this case are given by

$$\psi_{nl}^0 = \left(\frac{4}{ab}\right)^{\frac{1}{2}} \sin\left(\frac{n\pi z}{a}\right) \sin\left(\frac{l\pi x}{b}\right) \exp(iyk_y) = \mathrm{c}_1\psi_{nl}(x,z)\exp(iyk_y) \quad (1.37)$$

and

$$\begin{aligned} E_{nl}^0 = & \frac{\hbar^2}{2m_c}\left[\left(\frac{n\pi}{a}\right)^2 + \left(\frac{l\pi}{b}\right)^2\right] + \frac{\hbar^2 k_y^2}{2m_c} \\ & - \alpha\left[\frac{\hbar^2}{2m_c}\left\{\left(\frac{n\pi}{a}\right)^2 + \left(\frac{l\pi}{b}\right)^2\right\} + \frac{\hbar^2 k_y^2}{2m_c}\right]^2 \end{aligned} \quad (1.38)$$

where a and b are the nano thickness along z and y-directions respectively, $n(= 1,2,3,\ldots)$ is the size quantum number along z-direction and $l(= 1,2,3,\ldots)$ is the size quantum number along x-direction, $\mathrm{c}_1 = \left(\frac{4}{ab}\right)^{\frac{1}{2}}$ and $\psi_{nl}(x,z) = \sin\left(\frac{n\pi z}{a}\right)\sin\left(\frac{l\pi x}{b}\right)$.

In the presence of a magnetic field B along y-direction the term $\hat{H}_1\psi_{nl}^0$ assumes the from

$$\begin{aligned} \hat{H}_1\psi_{nl}^0 = & \left[\left(-i\hbar\frac{\partial}{\partial x} + eBz\right)\vec{i} + \vec{j}\left(-i\hbar\frac{\partial}{\partial y}\right) + \vec{k}\left(-i\hbar\frac{\partial}{\partial z}\right)\right]\psi_{nl}^0 \\ = & \left[-i\hbar c_1\left(\frac{l\pi}{b}\right)\cos\left(\frac{l\pi}{b}x\right)\sin\left(\frac{n\pi}{a}z\right)\exp(ik_y y) + eBz\psi_{nl}^0\right]\vec{i} \\ & + \vec{j}(-i\hbar\cdot ik_y)\psi_{nl}^0 + \vec{k}\left[-i\hbar c_1\left(\frac{n\pi}{b}\right)\sin\left(\frac{l\pi}{a}x\right)\cos\left(\frac{n\pi}{a}z\right)\exp(ik_y y)\right] \end{aligned} \quad (1.39)$$

Again

$$
\begin{aligned}
(\hat{H}_1)^2\psi_{nl}^0 &= \left(-i\hbar\frac{\partial}{\partial x}+eBz\right)\left[-i\hbar c_1\left(\frac{l\pi}{b}\right)\cos\left(\frac{l\pi}{b}x\right)\sin\left(\frac{n\pi}{a}z\right)\exp(ik_yy)+eBz\psi_{nl}^0\right] \\
&\quad +(-i\hbar\cdot ik_y)^2\psi_{nl}^0+\left(-i\hbar\frac{\partial}{\partial z}\right)\left[-i\hbar c_1\left(\frac{n\pi}{a}\right)\sin\left(\frac{l\pi}{b}x\right)\cos\left(\frac{n\pi}{a}z\right)\exp(ik_yy)\right] \\
&= \left[-i\hbar \;-\; i\hbar c_1\left(\frac{l\pi}{b}\right)^2 - \sin\left(\frac{l\pi}{b}x\right)\sin\left(\frac{n\pi}{a}z\right)\exp(ik_yy)\right. \\
&\quad \left.-2i_1\left(\frac{l\pi}{b}\right)\cos\left(\frac{l\pi}{b}x\right)\sin\left(\frac{n\pi}{a}z\right)\exp(ik_yy)+e^2B^2z^2\psi_{nl}^0\right]+\hbar^2k_y^2\psi_{nl}^0+\hbar^2\left(\frac{n\pi}{a}\right)^2\psi_{nl}^0 \\
&= \left[\hbar^2\left(\frac{l\pi}{b}\right)^2\psi_{nl}^0+e^2B^2z^2\psi_{nl}^0 - 2i\hbar eBzc_1\left(\frac{l\pi}{b}\right)\cos\left(\frac{l\pi}{b}x\right)\sin\left(\frac{n\pi}{a}z\right)\exp(ik_yy)\right] \\
&\quad +\hbar^2k_y^2\psi_{nl}^0+\hbar^2\left(\frac{n\pi}{a}\right)^2\psi_{nl}^0 = \left[\hbar^2\left(\frac{l\pi}{b}\right)^2+\hbar^2\left(\frac{n\pi}{a}\right)^2+\hbar^2k_y^2+e^2B^2z^2\right]\psi_{nl}^0 \\
&\quad -2i\hbar eBz\sqrt{\frac{4}{ab}}\left(\frac{l\pi}{b}\right)\cos\left(\frac{l\pi}{b}x\right)\sin\left(\frac{n\pi}{a}z\right)\exp(ik_yy)
\end{aligned}
$$

Therefore

$$
(\hat{H}_1)^2\psi_{nl}^0 = \left[\hbar^2\left(\frac{l\pi}{b}\right)^2+\hbar^2\left(\frac{n\pi}{a}\right)^2+\hbar^2k_y^2+e^2B^2z^2\right]\psi_{nl}^0 - 2i\hbar e\mathrm{B}z\left(\frac{l\pi}{b}\right)\cot\left(\frac{l\pi}{b}x\right)\psi_{nl}^0 \tag{1.40}
$$

$$
\begin{aligned}
(\hat{H}_1)^3\psi_{nl}^0 &= \left[\left(-i\hbar\frac{\partial}{\partial x}+eBz\right)\vec{i}+\vec{j}\left(-i\hbar\frac{\partial}{\partial y}\right)+\vec{k}\left(-i\hbar\frac{\partial}{\partial z}\right)\right]\cdot(\hat{H}_1^2)\psi_{nl}^0 \\
&= \vec{i}\left[\left\{\hbar^2\left(\frac{l\pi}{b}\right)^2+\hbar^2\left(\frac{n\pi}{a}\right)^2+\hbar^2k_y^2+e^2B^2z^2\right\}(-i\hbar)c_1\left(\frac{l\pi}{b}\right)\cos\left(\frac{l\pi}{b}x\right)\sin\left(\frac{n\pi}{a}z\right)\exp(ik_yy)\right. \\
&\quad +2(i\hbar eBz)(-i\hbar)c_1\left(\frac{l\pi}{b}\right)^2\sin\left(\frac{l\pi}{b}x\right)\sin\left(\frac{n\pi}{a}z\right)\exp(ik_yy) \\
&\quad +eBz\left\{\hbar^2\left(\frac{l\pi}{b}\right)^2+\hbar^2\left(\frac{n\pi}{a}\right)^2+\hbar^2k_y^2+e^2B^2z^2\right\}\psi_{nl}^0 \\
&\quad \left.-2i\hbar e^2B^2z^2c_1\left(\frac{l\pi}{b}\right)\cos\left(\frac{l\pi}{b}x\right)\sin\left(\frac{n\pi}{a}z\right)\exp(ik_yy)\right] \\
&\quad +\vec{j}\left[-i\hbar\left[\hbar^2\left(\frac{l\pi}{b}\right)^2+\hbar^2\left(\frac{n\pi}{a}\right)^2+\hbar^2k_y^2+e^2B^2z^2\right]\cdot(ik_y)\psi_{nl}^0\right. \\
&\quad \left.+(i\hbar)2i\hbar eBz(ik_y)c_1\left(\frac{l\pi}{b}\right)\cos\left(\frac{l\pi}{b}x\right)\cdot\sin\left(\frac{n\pi}{a}z\right)\exp(iyk_y)\right] \\
&\quad +\vec{k}\left[(-i\hbar)\left[\hbar^2\left(\frac{l\pi}{b}\right)^2+\hbar^2\left(\frac{n\pi}{a}\right)^2+\hbar^2k_y^2+e^2B^2z^2\right]\right. \\
&\quad c_1\left(\frac{n\pi}{a}\right)\cos\left(\frac{n\pi}{a}z\right)\sin\left(\frac{l\pi}{b}x\right)\exp(iyk_y)+(-i\hbar)2ze^2B^2\psi_{nl}^0 \\
&\quad +(-i\hbar)(-2i\hbar eB)c_1\left(\frac{l\pi}{b}\right)\cos\left(\frac{l\pi}{b}x\right)\sin\left(\frac{n\pi}{a}z\right)\exp(iyk_y) \\
&\quad \left.+(-i\hbar)(-2i\hbar eBz)c_1\left(\frac{l\pi}{b}\right)\left(\frac{n\pi}{a}\right)\cos\left(\frac{l\pi}{b}x\right)\cos\left(\frac{n\pi}{a}z\right)\exp(iyk_y)\right]
\end{aligned} \tag{1.41}
$$

$$
\begin{aligned}
(\hat{H}_1)^4\psi^0_{nl} = &\left[\left\{\hbar^2\left(\frac{n\pi}{a}\right)^2 + \hbar^2\left(\frac{l\pi}{b}\right)^2 + \hbar^2k_y^2 + e^2B^2z^2\right\}\right.\\
&\cdot(-i\hbar)^2c_1\left(\frac{l\pi}{b}\right)^2.\left(-\sin\left(\frac{l\pi}{b}x\right)\right)\sin\left(\frac{n\pi}{a}z\right)\exp(iyk_y)\\
&+ 2(-i\hbar)^2(i\hbar eBz)c_1\left(\frac{l\pi}{b}\right)^3\cos\left(\frac{l\pi}{b}x\right)\sin\left(\frac{n\pi}{a}z\right)\exp(iyk_y)\\
&+ eBz\left\{\hbar^2\left(\frac{l\pi}{b}\right)^2 + \hbar^2\left(\frac{n\pi}{a}\right)^2 + \hbar^2k_y^2 + e^2B^2z^2\right\}\\
&\cdot(-i\hbar)c_1\left(\frac{l\pi}{b}\right)\cos\left(\frac{l\pi}{b}x\right)\sin\left(\frac{n\pi}{a}z\right)\exp(iyk_y) - 2i\hbar e^2B^2z^2\\
&\cdot(-i\hbar)c_1\left(\frac{l\pi}{b}\right)^2.\left(-\sin\left(\frac{l\pi}{b}x\right)\right)\sin\left(\frac{n\pi}{a}z\right)\exp(iyk_y)\\
&+ eBz\left\{\hbar^2\left(\frac{l\pi}{b}\right)^2 + \hbar^2\left(\frac{n\pi}{a}\right)^2 + \hbar^2k_y^2 + e^2B^2z^2\right\}\\
&(-i\hbar)\left(\frac{l\pi}{b}\right)\cot\left(\frac{l\pi}{b}x\right)\psi^0_{nl} + 2\hbar^2e^2B^2z^2\left(\frac{l\pi}{b}\right)^2\psi^0_{nl}\\
&\left.+\left\{\hbar^2\left(\frac{l\pi}{b}\right)^2 + \hbar^2\left(\frac{n\pi}{a}\right)^2 + \hbar^2k_y^2 + e^2B^2z^2\right\}e^2B^2z^2\psi^0_{nl} - 2i^3B^3z^3\left(\frac{l\pi}{b}\right)\cot\left(\frac{l\pi}{b}x\right)\psi^0_{nl}\right]\\
&+\left[\hbar^2\left(\frac{l\pi}{b}\right)^2 + \hbar^2\left(\frac{n\pi}{a}\right)^2 + \hbar^2k_y^2 + e^2B^2z^2\right]\hbar^2k_y^2\psi^0_{nl}\\
&+ \hbar^2k_y^2(-2i\hbar eBz)\left(\frac{l\pi}{b}\right)\cot\left(\frac{l\pi}{b}x\right)\psi^0_{nl} + \hbar^2\left(\frac{n\pi}{a}\right)^2\\
&\left[\hbar^2\left(\frac{l\pi}{b}\right)^2 + \hbar^2\left(\frac{n\pi}{a}\right)^2 + \hbar^2k_y^2 + e^2B^2z^2\right]\psi^0_{nl} - 4\hbar^2e^2B^2z^2\left(\frac{n\pi}{a}\right)\cot\left(\frac{n\pi}{a}z\right)\psi^0_{nl} - 2\hbar^2e^2B^2\psi^0_{nl}\\
&+ 4i\hbar^3eB\left(\frac{l\pi}{b}\right)\left(\frac{n\pi}{a}\right)\cot\left(\frac{l\pi}{b}x\right)\cot\left(\frac{n\pi}{a}z\right)\psi^0_{nl} - 2i\hbar^3eBz\left(\frac{l\pi}{b}\right)\left(\frac{n\pi}{a}\right)^2\cot\left(\frac{l\pi}{b}x\right)\psi^0_{nl}
\end{aligned}
\tag{1.42}
$$

Thus the total Hamiltonian of the system is

$$\hat{H} = a(\hat{H}_1)^2 - b(\hat{H}_1)^4$$

Therefore

$$
\begin{aligned}
\hat{H} = &\left[\frac{\hbar^2}{2m_c}\left(\frac{l\pi}{b}\right)^2 + \frac{\hbar^2}{2m_c}\left(\frac{n\pi}{a}\right)^2 + \frac{\hbar^2k_y^2}{2m_c}\right] + \frac{e^2B^2z^2}{2m_c} - \frac{2i}{2m_c}\left(\frac{l\pi}{b}\right)\cot\left(\frac{l\pi}{b}x\right)\\
&- \alpha a^2\left[\hbar^2\left(\frac{l\pi}{b}\right)^2\left\{\hbar^2\left(\frac{l\pi}{b}\right)^2 + \hbar^2\left(\frac{n\pi}{a}\right)^2 + \hbar^2k_y^2 + e^2B^2z^2\right\}\right.
\end{aligned}
$$

$$
\begin{aligned}
&- 2i\hbar^3 eBz\left(\frac{l\pi}{b}\right)^3 \cot\left(\frac{l\pi}{b}x\right) - 2i\hbar eBz\left\{\hbar^2\left(\frac{l\pi}{b}\right)^2 + \hbar^2\left(\frac{n\pi}{a}\right)^2\right.\\
&\left.+\hbar^2 k_y^2 + e^2B^2z^2\right\}\left(\frac{l\pi}{b}\right)\cot\left(\frac{l\pi}{b}x\right) + 4\hbar^2 e^2B^2z^2\left(\frac{l\pi}{b}\right)^2\\
&+e^2B^2z^2\left\{\hbar^2\left(\frac{l\pi}{b}\right)^2 + \hbar^2\left(\frac{n\pi}{a}\right)^2 + \hbar^2k_y^2 + e^2B^2z^2\right\}\\
&- 2i\hbar e^3B^3z^3\left(\frac{l\pi}{b}\right)\cot\left(\frac{l\pi}{b}x\right) + \left\{\hbar^2\left(\frac{l\pi}{b}\right)^2 + \hbar^2\left(\frac{n\pi}{a}\right)^2 + \hbar^2k_y^2 + e^2B^2z^2\right\}\hbar^2k_y^2\\
&- 2i(\hbar^2k_y^2)\left(\frac{l\pi}{b}\right)\cot\left(\frac{l\pi}{b}x\right) + \hbar^2\left(\frac{n\pi}{a}\right)^2\\
&\left\{\hbar^2\left(\frac{l\pi}{b}\right)^2 + \hbar^2\left(\frac{n\pi}{a}\right)^2 + \hbar^2k_y^2 + e^2B^2z^2\right\}\\
&- 4\hbar^2e^2B^2z\left(\frac{n\pi}{a}\right)\cot\left(\frac{n\pi}{a}z\right) - 2\hbar^2e^2B^2\\
&\left.+ 4i\hbar^3eB\left(\frac{l\pi}{b}\right)\left(\frac{n\pi}{a}\right)\cot\left(\frac{l\pi}{b}x\right)\cot\left(\frac{n\pi}{a}z\right) - 2i\hbar^3eBz\left(\frac{l\pi}{b}\right)\left(\frac{n\pi}{a}\right)^2\cot\left(\frac{l\pi}{b}x\right)\right]
\end{aligned} \tag{1.43}
$$

Thus the perturbed Hamiltonian will be

$$
\begin{aligned}
\hat{H}' =\ &\frac{e^2B^2}{2m_c}(z^2) - \left(\frac{2i\hbar eB}{2m_c}\right)(z)\left(\frac{l\pi}{b}\right)\cot\left(\frac{l\pi}{b}x\right)\\
&- 2\alpha\left(\frac{e^2B^2}{2m_c}\right)\left[\frac{\hbar^2}{2m_c}\left(\frac{l\pi}{b}\right)^2 + \frac{\hbar^2}{2m_c}\left(\frac{n\pi}{a}\right)^2 + \frac{\hbar^2k_y^2}{2m_c}\right]\\
&(z^2) - \alpha\left(\frac{e^2B^2}{2m_c}\right)^2(z^4) + \alpha\left(\frac{2i\hbar eB}{2m_c}\right)(z)\\
&\left\{\frac{\hbar^2}{2m_c}\left(\frac{l\pi}{b}\right)^2 + \frac{\hbar^2}{2m_c}\left(\frac{n\pi}{a}\right)^2 + \frac{\hbar^2k_y^2}{2m_c}\right\}\left(\frac{l\pi}{b}\right)\cot\left(\frac{l\pi}{b}x\right)\\
&\alpha\left(\frac{4i\hbar e^3B^3}{4m_c^2}\right)(z^3)\left(\frac{l\pi}{b}\right)\cot\left(\frac{l\pi}{b}x\right) + \alpha\left(\frac{2i\hbar eB}{2m_c}\right)\left(\frac{\hbar^2k_y^2}{2m_c}\right)\\
&(z)\left(\frac{l\pi}{b}\right)\cot\left(\frac{l\pi}{b}x\right) + \alpha\left(\frac{2\hbar eB}{2m_c}\right)^2\left(\frac{n\pi}{a}z\right)\cot\left(\frac{n\pi}{a}z\right)\\
&+ \alpha\left(\frac{2i\hbar^3eB}{4m_c^2}\right)\left(\frac{l\pi}{b}\right)^3(z)\cot\left(\frac{l\pi}{b}x\right) + \alpha\left(\frac{2i\hbar^3eB}{4m_c^2}\right)\left(\frac{l\pi}{b}\right)\left(\frac{n\pi}{a}\right)^2\\
&(z)\cot\left(\frac{l\pi}{b}x\right) - \alpha\left(\frac{2\hbar eB}{2m_c}\right)^2(z^2)\left(\frac{l\pi}{b}\right)^2\\
&- 4\alpha\left(\frac{i\hbar^3eB}{4m_c^2}\right)^2\left(\frac{l\pi}{b}\right)\left(\frac{n\pi}{a}\right)\cot\left(\frac{l\pi}{b}x\right)\cot\left(\frac{n\pi}{a}z\right) + 2\alpha\left(\frac{\hbar eB}{2m_c}\right)^2
\end{aligned} \tag{1.44}
$$

So the first order perturbation to the energy is

$$E_{nl}^{(1)} = \hat{H}'_{nl} = \frac{\int_{-\infty}^{+\infty} \psi_{nl}^* \hat{H}' \psi_{nl} d\psi}{\int_{-\infty}^{+\infty} \psi_{nl}^* \psi_{nl} d\psi} \tag{1.45}$$

Now,

$$\int_{-\infty}^{+\infty} \psi_{nl}^* \psi_{nl} d\psi = \left[\frac{2}{a}\int_0^a \sin^2\left(\frac{n\pi}{a}z\right)\mathrm{dz}\right]\left[\frac{2}{b}\int_0^b \sin^2\left(\frac{l\pi}{b}\mathrm{x}\right)\mathrm{dx}\right] = 1 \tag{1.46}$$

$$\langle z^2\rangle = \left[\frac{2}{b}\int_0^b \sin^2\left(\frac{l\pi}{b}x\right)dx\right]\left[\frac{2}{a}\int_0^a z^2\sin^2\left(\frac{n\pi}{a}z\right)dz\right] \tag{1.47}$$

Now $\frac{2}{b}\int_0^b \sin^2\left(\frac{l\pi}{b}x\right)dx = 1.$
Therefore

$$\langle z^2\rangle_n = \frac{2}{a}\int_0^a z^2\sin^2\left(\frac{n\pi}{a}z\right)dz = \left(\frac{a^2}{3} - \frac{a^2}{2n^2\pi^2}\right) \tag{1.48}$$

$$\text{Again } \left\langle z\cot\left(\frac{l\pi}{b}x\right)\right\rangle = \left[\frac{2}{a}\int_0^a z^2\sin^2\left(\frac{n\pi}{a}z\right)dz\right]\cdot\left[\frac{2}{b}\int_0^b \cos\left(\frac{l\pi}{b}x\right)\sin\left(\frac{l\pi}{b}x\right)dx\right] = 0 \tag{1.49}$$

$$\text{Now } \frac{2}{b}\int_0^b \sin\left(\frac{l\pi}{b}x\right)\cos\left(\frac{l\pi}{b}x\right)dx = 0$$

We can write

$$\langle z^4\rangle_n = \frac{2}{a}\int_0^a z^4\sin^2\left(\frac{n\pi}{a}z\right)dz = \left(\frac{a^4}{5} - \frac{a^4}{n^2\pi^2} + \frac{3a^4}{2n^4\pi^4}\right) \tag{1.50}$$

$$\text{Besides } \left\langle z^3\cot\left(\frac{l\pi}{b}x\right)\right\rangle = \left[\frac{2}{a}\int_0^a z^3\sin^2\left(\frac{n\pi}{a}z\right)dz\right]\cdot\left[\frac{2}{b}\int_0^b \cos\left(\frac{l\pi}{b}x\right)\sin\left(\frac{l\pi}{b}x\right)dx\right] = 0 \tag{1.51}$$

$$\text{and } \left\langle z\cot\left(\frac{n\pi}{a}z\right)\right\rangle = \left[\frac{2}{a}\int_0^a z\cos\left(\frac{n\pi}{a}z\right)\sin\left(\frac{n\pi}{a}z\right)dz\right] = -\left(\frac{a}{2n\pi}\right) \quad (1.52)$$

We also obtain

$$\left\langle\cot\frac{l\pi}{b}x\cot\left(\frac{n\pi}{a}z\right)\right\rangle = \left[\frac{2}{a}\int_0^a \sin\left(\frac{n\pi}{a}z\right)\cos\left(\frac{n\pi}{a}z\right)dz\right]\cdot\left[\frac{2}{b}\int_0^b \sin\left(\frac{l\pi}{b}x\right)\cos\left(\frac{l\pi}{b}x\right)dz\right] = 0 \quad (1.53)$$

The first order correction to the energy is given by

$$\begin{aligned} E_{nl}^{(1)} = {} & \frac{e^2B^2}{2m_c}\langle z^2\rangle_n - 2\alpha\left(\frac{e^2B^2}{2m_c}\right)\left[\frac{\hbar^2}{2m_c}\left(\frac{l\pi}{b}\right)^2 + \frac{\hbar^2}{2m_c}\left(\frac{n\pi}{a}\right)^2 + \frac{\hbar^2k_y^2}{2m_c}\right]\langle z^2\rangle_n - \alpha\frac{e^2B^2}{2m_c}\langle z^4\rangle_n \\ & + \alpha\left(\frac{2\hbar eB}{2m_c}\right)^2\left(\frac{n\pi}{a}\right)\left(-\frac{a}{2n\pi}\right) - \alpha\left(\frac{2\hbar eB}{2m_c}\right)^2\left(\frac{l\pi}{b}\right)^2\langle z^2\rangle_n + 2\alpha\left(\frac{2\hbar eB}{2m_c}\right)^2 \end{aligned} \quad (1.54)$$

$$\begin{aligned} E_{nl}^{(1)} = {} & \frac{e^2B^2}{2m_c}\langle z^2\rangle_n - 2\alpha\left(\frac{e^2B^2}{2m_c}\right)\left[\frac{\hbar^2}{2m_c}\left(\frac{l\pi}{b}\right)^2 + \frac{\hbar^2}{2m_c}\left(\frac{n\pi}{a}\right)^2 + \frac{\hbar^2k_y^2}{2m_c}\right]\langle z^2\rangle_n \\ & - \alpha\left(\frac{2\hbar eB}{2m_c}\right)^2\left(\frac{l\pi}{b}\right)^2\langle z^2\rangle_n - \alpha\left(\frac{e^2B^2}{2m_c}\right)^2\langle z^4\rangle_n - \alpha\left(\frac{\hbar^2e^2B^2}{2m_c^2}\right)^2 + \alpha\left(\frac{\hbar^2e^2B^2}{2m_c^2}\right) \end{aligned} \quad (1.55)$$

The DR in NWs of HD III–V semiconductors in the presence of a parallel magnetic field B along y-direction whose unperturbed electrons obey the two band model of Kane can be written as

$$\begin{aligned} \gamma_3(E,\eta_g) = {} & \left[\frac{\hbar^2}{2m_c}\left(\frac{n\pi}{a}\right)^2 + \frac{\hbar^2}{2m_c}\left(\frac{l\pi}{b}\right)^2 + \frac{\hbar^2k_y^2}{2m_c}\right] - \alpha\left[\frac{\hbar^2}{2m_c}\left(\frac{n\pi}{a}\right)^2 + \frac{\hbar^2}{2m_c}\left(\frac{l\pi}{b}\right)^2 + \frac{\hbar^2k_y^2}{2m_c}\right]^2 \\ & + \frac{e^2B^2}{2m_c}\langle z^2\rangle_n - \alpha\left(\frac{2\hbar eB}{2m_c}\right)^2\left(\frac{l\pi}{b}\right)^2\langle z^2\rangle_n \\ & - 2\alpha\left(\frac{e^2B^2}{2m_c}\right)\left[\frac{\hbar^2}{2m_c}\left(\frac{l\pi}{b}\right)^2 + \frac{\hbar^2}{2m_c}\left(\frac{n\pi}{a}\right)^2 + \frac{\hbar^2k_y^2}{2m_c}\right]\langle z^2\rangle_n - \alpha\left(\frac{e^2B^2}{2m_c}\right)^2\langle z^4\rangle_n \end{aligned} \quad (1.56a)$$

(1.56a) can be written as

$$k_y^2 = \gamma_{100}(E, \eta_g, l, n, B) \tag{1.56b}$$

where

$$\gamma_{100}(E, \eta_g, l, n, B) = \frac{1}{2\alpha a^2}\left[a - 2\alpha\omega_1 a - \sqrt{(2\alpha\omega_1 a - a)^2 - 4\alpha a^2[\alpha\omega_1^2 - \omega_1 + \gamma_3(E, \eta_g)]}\right]$$

$$\omega_1 = \frac{\hbar^2}{2m_c}\left(\frac{n\pi}{a}\right)^2 + \frac{\hbar^2}{2m_c}\left(\frac{l\pi}{b}\right)^2,$$

$$\omega_2 = \frac{e^2B^2}{2m_c}\langle z^2\rangle_n - \alpha\left(\frac{\hbar eB}{2m_c}\right)^2\left(\frac{l\pi}{b}\right)^2\langle z^2\rangle_n - 2\alpha\left(\frac{e^2B^2}{2m_c}\right)[\omega_1]\langle z^2\rangle_n - \alpha\left(\frac{e^2B^2}{2m_c}\right)\langle z^4\rangle_n$$

and $\omega_3 = 2\alpha\left(\frac{e^2B^2}{2m_c}\right)a\langle z^4\rangle_n$

The EEM is given by

$$m^*(E, n, l, B) = \frac{\hbar^2}{2}\gamma'_{100}(E, \eta_g, l, n, B) \tag{1.56c}$$

The DOS function is given by

$$N_{1D}(E) = \frac{g_v}{\pi}\sum_{n=0}^{n_{\max}}\sum_{l=0}^{l_{\max}}\frac{\gamma'_{100}(E, \eta_g, l, n, B)H(E - E_{19,2})}{\sqrt{\gamma_{100}(E, \eta_g, l, n, B)}} \tag{1.56d}$$

where $E_{19,2}$ is the sub-band energy which can be expressed through the equation

$$\gamma_{100}(E_{19,2}, \eta_g, l, n, B) = 0 \tag{1.56e}$$

1.2.3 The DR in Quantum Dot of HD III–V Semiconductors in the Presence of Magnetic Field

The unperturbed energy eigen value and unperturbed wave function in this case can be written as

$$E^0_{nlr} = \left[\frac{\hbar^2}{2m_c}\left(\frac{n\pi}{a}\right)^2 + \frac{\hbar^2}{2m_c}\left(\frac{l\pi}{b}\right)^2 + \frac{\hbar^2}{2m_c}\left(\frac{r\pi}{c}\right)^2\right] - \alpha\left[\frac{\hbar^2}{2m_c}\left(\frac{n\pi}{a}\right)^2 + \frac{\hbar^2}{2m_c}\left(\frac{l\pi}{b}\right)^2 + \frac{\hbar^2}{2m_c}\left(\frac{r\pi}{c}\right)^2\right]^2 \tag{1.57}$$

and

$$\psi_{nlr} = c_{11}\psi(x, y, z) \tag{1.58}$$

where *l, r,* and *n* are the size quantum number along x, y, and z-directions, a, b, c are the nano thickness along z, x and y-directions respectively,

$$c_{11} = \left(\frac{8}{abc}\right)^{1/2} \quad \text{and} \quad \psi(x, y, z) = \sin\left(\frac{l\pi}{b}x\right)\sin\left(\frac{r\pi}{c}y\right)\sin\left(\frac{n\pi}{a}z\right)$$

In the presence of a magnetic field B parallel to the y-direction the total Hamiltonian $\hat{H}$ can be written as

$$\hat{H} = a\left(-i\hbar\vec{\nabla} + e\vec{A}\right)^2 - b\left(-i\hbar\vec{\nabla} + e\vec{A}\right)^4 \tag{1.59}$$

So the total Hamiltonian will be

$$\begin{aligned}\hat{H} &= a\left[\left(-i\hbar\frac{\partial}{\partial x} + eBz\right)\vec{i} + \vec{j}\left(-i\hbar\frac{\partial}{\partial y}\right) + \vec{k}\left(-i\hbar\frac{\partial}{\partial z}\right)\right]^2 \\ &\quad - b\left[\left(-i\hbar\frac{\partial}{\partial x} + eBz\right)\vec{i} + \vec{j}\left(-i\hbar\frac{\partial}{\partial y}\right) + \vec{k}\left(-i\hbar\frac{\partial}{\partial z}\right)\right]^4 \\ &= a(\hat{H}_1)^2 - b(\hat{H}_1)^4 = \hat{H}_0 + \hat{H}_1\end{aligned} \tag{1.60}$$

$\hat{H}_0$ corresponds to unperturbed Hamiltonian and $\hat{H}'$ is the perturbation to be calculated

$$\hat{H}_0\psi_{nlr} = E^0_{nlr}\psi_{nlr} \tag{1.61}$$

Now

$$\begin{aligned}\hat{H}_1\psi_{nlr} &= \left[\left(-i\hbar\frac{\partial}{\partial x} + eBz\right)\vec{i} + \vec{j}\left(-i\hbar\frac{\partial}{\partial y}\right) + \vec{k}\left(-i\hbar\frac{\partial}{\partial z}\right)\right]\psi_{nlr} \\ &= \left[-i\hbar\left(\frac{l\pi}{b}\right)c_{11}\cos\left(\frac{l\pi}{b}x\right)\sin\left(\frac{n\pi}{a}z\right)\sin\left(\frac{r\pi}{c}y\right) + eBz\psi_{nlr}\right] \\ &\quad \vec{i} + \vec{j}\left(-i\hbar\left(\frac{r\pi}{c}\right)c_{11}\right)\cos\left(\frac{r\pi}{c}y\right)\sin\left(\frac{n\pi}{a}z\right)\sin\left(\frac{l\pi}{b}x\right)\Big] \\ &\quad + \vec{k}\left(-i\hbar\left(\frac{n\pi}{a}\right)c_{11}\right)\cos\left(\frac{n\pi}{a}z\right)\sin\left(\frac{l\pi}{b}x\right)\sin\left(\frac{r\pi}{c}y\right)\Big]\end{aligned} \tag{1.62}$$

$$\hat{H}_1^2\psi_{nlr} = -(-i\hbar)^2\left(\frac{l\pi}{b}\right)^2\psi_{nlr} + 2eBz(-i\hbar)\left(\frac{l\pi}{b}\right)c_{11}\cos\left(\frac{l\pi}{b}x\right)\sin\left(\frac{n\pi}{a}z\right)\sin\left(\frac{r\pi}{c}y\right)$$
$$+ e^2B^2z^2\psi_{nlr}) - (-i\hbar)^2\left(\frac{r\pi}{c}\right)^2\psi_{nlr} - (-i\hbar)^2\left(\frac{n\pi}{a}\right)^2\psi_{nlr}$$
$$= \hbar^2\left[\left(\frac{n\pi}{a}\right)^2 + \left(\frac{l\pi}{b}\right)^2 + \left(\frac{rc}{c}\right)^2\right]\psi_{nlr} + e^2B^2z^2\psi_{nlr}$$
$$- 2i\hbar eBz\left(\frac{l\pi}{b}\right)c_{11}\cos\left(\frac{l\pi}{b}x\right)\sin\left(\frac{n\pi}{a}z\right)\sin\left(\frac{r\pi}{c}y\right)$$
$$= \left[\left(\frac{n\pi}{a}\right)^2\hbar^2 + \hbar^2\left(\frac{l\pi}{b}\right)^2 + \hbar^2\left(\frac{r\pi}{c}\right)^2 + e^2B^2z^2\right]\psi_{nlr}$$
$$- 2i\hbar eBz\left(\frac{l\pi}{b}\right)\cot\left(\frac{l\pi}{b}x\right)\psi_{nlr} \tag{1.63}$$

$$(\hat{H}_1)^3\psi^0_{nlr} = \left[\left(-i\hbar\frac{\partial}{\partial x} + eBz\right)\vec{i} + \vec{j}\left(-i\hbar\frac{\partial}{\partial y}\right) + \vec{k}\left(-i\hbar\frac{\partial}{\partial z}\right)\right](\hat{H}_1^2)\psi_{nlr}$$
$$= \vec{i}\left[(-i\hbar)\left\{\hbar^2\left(\frac{n\pi}{a}\right)^2 + \hbar^2\left(\frac{l\pi}{b}\right)^2 + \hbar^2\left(\frac{r\pi}{c}\right)^2 + e^2B^2z^2\right\}\right.$$
$$\left(\frac{l\pi}{b}\right)c_{11}\cos\left(\frac{l\pi}{b}x\right)\sin\left(\frac{n\pi}{a}z\right)\sin\left(\frac{r\pi}{c}y\right) + (-i\hbar)(-2i\hbar eBz)\left(\frac{l\pi}{b}\right)^2\cdot(-\psi_{nlr})$$
$$+ eBz\left\{\hbar^2\left(\frac{n\pi}{a}\right)^2 + \hbar^2\left(\frac{l\pi}{b}\right)^2 + \hbar^2\left(\frac{r\pi}{c}\right)^2 + e^2B^2z^2\right\}\psi_{nlr}$$
$$\left.- 2i\hbar e^2B^2z^2\left(\frac{l\pi}{b}\right)\cot\left(\frac{l\pi}{b}x\right)\psi_{nlr}\right]$$
$$+ \vec{j}\left[(-i\hbar)\left\{\hbar^2\left(\frac{n\pi}{a}\right)^2 + \hbar^2\left(\frac{l\pi}{b}\right)^2 + \hbar^2\left(\frac{r\pi}{c}\right)^2 + e^2B^2z^2\right\}\right.$$
$$\left(\frac{r\pi}{c}\right)c_{11}\cos\left(\frac{r\pi}{c}y\right)\sin\left(\frac{n\pi}{a}z\right)\sin\left(\frac{l\pi}{b}x\right)$$
$$+ (-i\hbar)(2i\hbar eBz)\left(\frac{l\pi}{b}\right)\left(\frac{r\pi}{c}\right)c_{11}\cos\left(\frac{l\pi}{b}x\right)\cos\left(\frac{r\pi}{c}y\right)$$
$$\left.\sin\left(\frac{n\pi}{a}z\right)\right] + \vec{k}\left[(-i\hbar)\left\{\left(\frac{n\pi}{a}\right)^2\hbar^2 + \hbar^2\left(\frac{l\pi}{b}\right)^2 + \hbar^2\left(\frac{r\pi}{c}\right)^2 + e^2B^2z^2\right\}\right.$$
$$\left(\frac{n\pi}{a}\right)c_{11}\cos\left(\frac{n\pi}{a}z\right)\sin\left(\frac{l\pi}{b}x\right)\sin\left(\frac{r\pi}{c}y\right) + (-i\hbar)\cdot 2e^2B^2z\psi_{nlr}$$
$$- 2i\hbar eB(-i\hbar)\left(\frac{l\pi}{b}\right)c_{11}\cos\left(\frac{l\pi}{b}x\right)\sin\left(\frac{n\pi}{a}z\right)\sin\left(\frac{r\pi}{c}y\right)$$
$$\left.- 2i\hbar eBz(-i\hbar)\left(\frac{l\pi}{b}\right)\left(\frac{n\pi}{a}\right)c_{11}\cos\left(\frac{l\pi}{b}x\right)\sin\left(\frac{n\pi}{a}z\right)\sin\left(\frac{r\pi}{c}y\right)\right] \tag{1.64}$$

$$
\begin{aligned}
(\hat{H}_1)^4\psi_{nlr} = & \left\{\hbar^2\left(\frac{n\pi}{a}\right)^2 + \hbar^2\left(\frac{l\pi}{b}\right)^2 + \hbar^2\left(\frac{r\pi}{c}\right)^2 + e^2B^2z^2\right\}(-i\hbar)\left(\frac{l\pi}{b}\right)^2 \\
& \cdot(-\psi_{nlr}) + 2i\hbar eBz(-i\hbar)^2\left(\frac{l\pi}{b}\right)^3 \\
& c_{11}\cos\left(\frac{l\pi}{b}x\right)\sin\left(\frac{n\pi}{a}z\right)\sin\left(\frac{r\pi}{c}y\right) \\
& + eBz\left\{\hbar^2\left(\frac{n\pi}{a}\right)^2 + \hbar^2\left(\frac{l\pi}{b}\right)^2 + \hbar^2\left(\frac{r\pi}{c}\right)^2 + e^2B^2z^2\right\}(-i\hbar)\left(\frac{l\pi}{b}\right) \\
& c_{11}\cos\left(\frac{l\pi}{b}x\right)\sin\left(\frac{n\pi}{a}z\right)\sin\left(\frac{r\pi}{c}y\right) - 2i\hbar e^2B^2z^2(-i\hbar)\left(\frac{l\pi}{b}\right)^2(-\psi_{nlr}) \\
& + (-i\hbar)eBz\left\{\hbar^2\left(\frac{n\pi}{a}\right)^2 + \hbar^2\left(\frac{l\pi}{b}\right)^2\right. \\
& \left. + \hbar^2\left(\frac{r\pi}{c}\right)^2 + e^2B^2z^2\right\}\left(\frac{l\pi}{b}\right)\cot\left(\frac{l\pi}{b}x\right)\psi_{nlr} + 2\hbar^2e^2B^2z^2\left(\frac{l\pi}{b}\right)^2\psi_{nlr} \\
& + e^2B^2z^2\left\{\hbar^2\left(\frac{n\pi}{a}\right)^2 + \hbar^2\left(\frac{l\pi}{b}\right)^2\right. \\
& \left. + \hbar^2\left(\frac{r\pi}{c}\right)^2 + e^2B^2z^2\right\}\psi_{nlr} - 2i\hbar e^3B^3z^3\left(\frac{l\pi}{b}\right)\cot\left(\frac{l\pi}{b}x\right)\psi_{nlr} \\
& + (-i\hbar)^2\left\{\hbar^2\left(\frac{n\pi}{a}\right)^2 + \hbar^2\left(\frac{l\pi}{b}\right)^2 + \hbar^2\left(\frac{r\pi}{c}\right)^2\right. \\
& \left. + e^2B^2z^2\right\}\left(\frac{r\pi}{c}\right)^2\cdot(-\psi_{nlr}) + 2i\hbar eBz\cdot(-i\hbar)^2\left(\frac{l\pi}{b}\right)\left(\frac{r\pi}{c}\right)^2\cot\left(\frac{l\pi}{b}x\right)\psi_{nlr} \\
& + \hbar^2\left(\frac{n\pi}{a}\right)^2\left[\hbar^2\left(\frac{n\pi}{a}\right)^2 + \hbar^2\left(\frac{l\pi}{b}\right)^2\right. \\
& \left. + \hbar^2\left(\frac{r\pi}{c}\right)^2 + e^2B^2z^2\right]\psi_{nlr} - 4\hbar^2e^2B^2z\left(\frac{n\pi}{a}\right)\cot\left(\frac{n\pi}{a}z\right)\psi_{nlr} \\
& - 2\hbar^2e^2B^2\psi_{nlr} + 4i\hbar^3eBz\left(\frac{l\pi}{b}\right)\left(\frac{n\pi}{a}\right) \\
& \cot\left(\frac{l\pi}{b}x\right)\cot\left(\frac{n\pi}{a}z\right)\psi_{nlr} - 2i\hbar^3eBz\left(\frac{l\pi}{b}\right)\left(\frac{n\pi}{a}\right)^2\cot\left(\frac{l\pi}{b}x\right)\psi_{nlr}
\end{aligned}
\tag{1.65}
$$

Thus the total Hamiltonian of the system is

$$
\begin{aligned}
\hat{H} &= a(\hat{H}_1)^2 - b(\hat{H}_1)^4 \\
&= \left[\frac{\hbar^2}{2m_c}\left(\frac{n\pi}{a}\right)^2 + \frac{\hbar^2}{2m_c}\left(\frac{l\pi}{b}\right)^2 + \frac{\hbar^2}{2m_c}\left(\frac{r\pi}{c}\right)^2\right] + \frac{e^2B^2z^2}{2m_c} - \frac{2i\hbar eBz}{2m_c}\left(\frac{l\pi}{b}\right)\cot\left(\frac{l\pi}{b}x\right) - \alpha a^2 \\
&\left[\hbar^2\left(\frac{l\pi}{b}\right)^2\left\{\hbar^2\left(\frac{l\pi}{b}\right)^2 + \hbar^2\left(\frac{n\pi}{a}\right)^2 + \hbar^2\left(\frac{r\pi}{c}\right)^2 + e^2B^2z^2\right\} - 2i\hbar^3 eBz\left(\frac{l\pi}{b}\right)^3\cot\left(\frac{l\pi}{b}x\right) - 2i\hbar eBz\right. \\
&\left\{\hbar^2\left(\frac{l\pi}{b}\right)^2 + \hbar^2\left(\frac{n\pi}{a}\right)^2 + \hbar^2\left(\frac{r\pi}{c}\right)^2 + e^2B^2z^2\right\}\left(\frac{l\pi}{b}\right)\cot\left(\frac{l\pi}{b}x\right) + 4\hbar^2e^2B^2z^2\left(\frac{l\pi}{b}\right)^2 + e^2B^2z^2 \\
&\left\{\hbar^2\left(\frac{l\pi}{b}\right)^2 + \hbar^2\left(\frac{n\pi}{a}\right)^2 + \hbar^2\left(\frac{r\pi}{c}\right)^2 + e^2B^2z^2\right\} - 2i\hbar e^3B^3z^3\left(\frac{l\pi}{b}\right)\cot\left(\frac{l\pi}{b}x\right) + \hbar^2\left(\frac{r\pi}{c}\right)^2 \\
&\left\{\hbar^2\left(\frac{l\pi}{a}\right)^2 + \hbar^2\left(\frac{n\pi}{b}\right)^2 + \hbar^2\left(\frac{r\pi}{c}\right)^2 + e^2B^2z^2\right\} - 2i\hbar eBz\left(\frac{\hbar^2r^2\pi^2}{c^2}\right)\left(\frac{l\pi}{b}\right)\cot\left(\frac{l\pi}{b}x\right) + \hbar^2\left(\frac{n\pi}{a}\right)^2 \\
&\left\{\hbar^2\left(\frac{l\pi}{b}\right)^2 + \hbar^2\left(\frac{n\pi}{a}\right)^2 + \hbar^2\left(\frac{r\pi}{c}\right)^2 + e^2B^2z^2\right\} - 4\hbar^2e^2B^2z\left(\frac{n\pi}{a}\right)\cot\left(\frac{n\pi}{a}z\right) - 2\hbar^2e^2B^2 + 4i\hbar^3 eB \\
&\left.\left(\frac{l\pi}{b}\right)\left(\frac{n\pi}{a}\right)\cot\left(\frac{l\pi}{b}x\right)\cot\left(\frac{n\pi}{a}z\right) - 2i\hbar^3 eBz\left(\frac{l\pi}{b}\right)\left(\frac{n\pi}{a}\right)^2\cot\left(\frac{l\pi}{b}x\right)\right]
\end{aligned}
\tag{1.66}
$$

The total Hamiltonian can be written as

$$
\begin{aligned}
\hat{H} &= \left[\frac{\hbar^2}{2m_c}\left(\frac{n\pi}{a}\right)^2 + \frac{\hbar^2}{2m_c}\left(\frac{l\pi}{b}\right)^2 + \frac{\hbar^2}{2m_c}\left(\frac{r\pi}{c}\right)^2\right] + \frac{e^2B^2}{2m_c}z^2 - \left(\frac{2i\hbar eB}{2m_c}\right)\cdot z\cdot\left(\frac{l\pi}{b}\right)\cot\left(\frac{l\pi}{b}x\right) - \alpha \\
&\left[\frac{\hbar^2}{2m_c}\left\{\left(\frac{n\pi}{a}\right)^2 + \left(\frac{l\pi}{b}\right)^2 + \left(\frac{r\pi}{c}\right)^2\right\} + \frac{e^2B^2z^2}{2m_c}\right]^2 + \alpha\frac{2i\hbar eBz}{4m_c^2}\left\{\hbar^2\left(\frac{l\pi}{b}\right)^2 + \hbar^2\left(\frac{n\pi}{a}\right)^2 + \hbar^2\left(\frac{r\pi}{c}\right)^2\right\} \\
&\left(\frac{l\pi}{b}\right)\cot\left(\frac{l\pi}{b}x\right) + \alpha\frac{4i\hbar e^3B^3z^3}{4m_c^2}\left(\frac{l\pi}{b}\right)\cot\left(\frac{l\pi}{b}x\right) \\
&+ \alpha\left(\frac{2i\hbar eBz}{4m_c^2}\right)\left(\frac{l\pi}{b}\right)\hbar^2\left(\frac{r\pi}{c}\right)^2\cot\left(\frac{l\pi}{b}x\right) + \alpha\left(\frac{2\hbar eB}{2m_c}\right)^2 \\
&z\left(\frac{n\pi}{a}\right)\cot\left(\frac{n\pi}{a}z\right) + \alpha\left(\frac{2i\hbar^3 eBz}{4m_c^2}\right)\left(\frac{l\pi}{b}\right)^3\cot\left(\frac{l\pi}{b}x\right) \\
&+ \alpha\left(\frac{2i\hbar e^3Bz}{4m_c}\right)\left(\frac{l\pi}{b}\right)\left(\frac{n\pi}{a}\right)^2\cot\left(\frac{l\pi}{b}x\right) - \alpha\left(\frac{2\hbar eB}{2m_c}\right)^2 \\
&z^2\left(\frac{l\pi}{b}\right)^2 - 4\alpha\left(\frac{i\hbar^3 eB}{4m_c^2}\right)\left(\frac{l\pi}{b}\right)\left(\frac{n\pi}{a}\right)\cot\left(\frac{l\pi}{b}x\right)\cot\left(\frac{n\pi}{a}z\right) + 2\alpha\left(\frac{2\hbar eB}{2m_c}\right)^2
\end{aligned}
\tag{1.67}
$$

Thus the perturbed Hamiltonian assumes the form

$$
\begin{aligned}
\hat{H}' = & \frac{e^2B^2}{2m_c}\langle z^2\rangle - \left(\frac{2i\hbar eB}{2m_c}\right)\langle z\rangle\left(\frac{l\pi}{b}\right)\cot\left(\frac{l\pi}{b}x\right) \\
& - 2\alpha\left(\frac{e^2B^2}{2m_c}\right)\left[\frac{\hbar^2}{2m_c}\left(\frac{n\pi}{a}\right)^2 + \frac{\hbar^2}{2m_c}\left(\frac{l\pi}{b}\right)^2 + \frac{\hbar^2}{2m_c}\left(\frac{r\pi}{c}\right)^2\right] \\
& \langle z^2\rangle - \alpha\left(\frac{e^2B^2}{2m_c}\right)z^4 + \alpha\left(\frac{2i}{2m_c}\right)\langle z\rangle\left\{\frac{\hbar^2}{2m_c}\left(\frac{l\pi}{b}\right)^2 + \frac{\hbar^2}{2m_c}\left(\frac{n\pi}{a}\right)^2 + \frac{\hbar^2}{2m_c}\left(\frac{r\pi}{c}\right)^2\right\}\left(\frac{l\pi}{b}\right) \\
& \cot\left(\frac{l\pi}{b}x\right) + \alpha\left(\frac{4i\hbar e^3B^3}{4m_c^2}\right)z^3\left(\frac{l\pi}{b}\right)\cot\left(\frac{l\pi}{b}x\right) \\
& + \alpha\left(\frac{2i\hbar eB}{2m_c}\right)\frac{\hbar^2}{2m_c}\left(\frac{r\pi}{c}\right)^2\left(\frac{l\pi}{b}\right)\langle z\rangle\cot\left(\frac{l\pi}{b}x\right) \\
& + \alpha\left(\frac{2i\hbar eB}{2m_c}\right)^2\left(\frac{n\pi}{a}\right)\langle z\rangle\cot\left(\frac{n\pi}{a}z\right) + \alpha\left(\frac{2i\hbar^3 eB}{4m_c^2}\right) \\
& \left(\frac{l\pi}{b}\right)^3\langle z\rangle\cot\left(\frac{l\pi}{b}x\right) + \alpha\left(\frac{2i\hbar^3 eB}{4m_c^2}\right)\left(\frac{l\pi}{b}\right)\left(\frac{n\pi}{a}\right)^2\langle z\rangle\cot\left(\frac{l\pi}{b}x\right) \\
& + \alpha\left(\frac{2i\hbar eB}{2m_c}\right)^2\langle z^2\rangle\left(\frac{l\pi}{b}\right)^2 - 4\alpha\left(\frac{i\hbar^3 eB}{b}\right)\left(\frac{l\pi}{b}\right)\left(\frac{n\pi}{a}\right) \\
& \cot\left(\frac{l\pi}{b}x\right)\cot\left(\frac{n\pi}{a}z\right) + 2\alpha\left(\frac{\hbar eB}{2m_c}\right)^2
\end{aligned}
\tag{1.68}
$$

So the first order perturbation is

$$E^{(1)}_{nlr} = \int_\tau \psi^*_{nlr}\hat{H}'\psi_{nlr}d\tau$$

Since $\int_\tau \psi^*_{nlr}\hat{H}'\psi_{nlr}d\tau = \left[\frac{2}{a}\int_0^a \sin^2\left(\frac{n\pi}{a}z\right)dz\right].\left[\frac{2}{b}\int_0^b \sin^2\left(\frac{l\pi}{b}x\right)dx\right].\left[\frac{2}{c}\int_0^c \sin^2\left(\frac{r\pi}{c}y\right)dy\right] = 1.$
Therefore

$$
\begin{aligned}
E^{(1)}_{nlr} = & \frac{e^2B^2}{2m_c}\langle z^2\rangle_n - 2\alpha\left(\frac{e^2B^2}{2m_c}\right)\left[\frac{\hbar^2}{2m_c}\left(\frac{n\pi}{a}\right)^2 + \frac{\hbar^2}{2m_c}\left(\frac{l\pi}{b}\right)^2 + \frac{\hbar^2}{2m_c}\left(\frac{r\pi}{c}\right)^2\right]\langle z^2\rangle_n \\
& - \alpha\left(\frac{e^2B^2}{2m_c}\right)^2\langle z^4\rangle_n + \alpha\left(\frac{2\hbar eB}{2m_c}\right)^2\left(\frac{n\pi}{a}\right)\left(-\frac{a}{2n\pi}\right) \\
& + 2\alpha\left(\frac{\hbar eB}{2m_c}\right)^2 - \alpha\left(\frac{2\hbar eB}{2m_c}\right)^2\langle z^2\rangle_n\left(\frac{l\pi}{b}\right)^2 \\
E^{(1)}_{nl} = & \frac{e^2B^2}{2m_c}\langle z^2\rangle_n - \alpha\left(\frac{2\hbar eB}{2m_c}\right)^2\left(\frac{l\pi}{b}\right)^2\langle z^2\rangle_n \\
& - 2\alpha\left(\frac{e^2B^2}{2m_c}\right)\left[\frac{\hbar^2}{2m_c}\left(\frac{n\pi}{a}\right)^2 + \frac{\hbar^2}{2m_c}\left(\frac{l\pi}{b}\right)^2 + \frac{\hbar^2}{2m_c}\left(\frac{r\pi}{c}\right)^2\right] \\
& \langle z^2\rangle_n - \alpha\left(\frac{e^2B^2}{2m_c}\right)^2\langle z^4\rangle_n - \frac{\alpha}{2}\left(\frac{\hbar eB}{2m_c}\right)^2 + \frac{\alpha}{2}\left(\frac{\hbar eB}{2m_c}\right)^2
\end{aligned}
\tag{1.69}
$$

Therefore the DR in QD of HD III–V semiconductors in the presence of a parallel magnetic field B is givenby

$$\gamma_3(E,\eta_g) = \left[\frac{\hbar^2}{2m_c}\left(\frac{n\pi}{a}\right)^2 + \frac{\hbar^2}{2m_c}\left(\frac{l\pi}{b}\right)^2 + \frac{\hbar^2}{2m_c}\left(\frac{r\pi}{c}\right)^2\right] - \alpha\left[\frac{\hbar^2}{2m_c}\left(\frac{n\pi}{a}\right)^2 + \frac{\hbar^2}{2m_c}\left(\frac{l\pi}{b}\right)^2 + \frac{\hbar^2}{2m_c}\left(\frac{r\pi}{c}\right)^2\right]^2 + \frac{e^2B^2}{2m_c}\langle z^2\rangle_n - \alpha\left(\frac{2\hbar eB}{2m_c}\right)^2\left(\frac{l\pi}{b}\right)^2\langle z^2\rangle_n - 2\alpha\left(\frac{e^2B^2}{2m_c}\right)\left[\frac{\hbar^2}{2m_c}\left(\frac{n\pi}{a}\right)^2 + \frac{\hbar^2}{2m_c}\left(\frac{l\pi}{b}\right)^2 + \frac{\hbar^2}{2m_c}\left(\frac{r\pi}{c}\right)^2\right]\langle z^2\rangle_n - \alpha\left(\frac{e^2B^2}{2m_c}\right)^2\langle z^4\rangle_n \quad (1.70a)$$

The DOS function in this case is given by

$$N(E) = \frac{2g_v}{d_x d_y d_z}\sum_{n=0}^{n_{\max}}\sum_{l=0}^{l_{\max}}\sum_{r=0}^{r_{\max}}\delta'(E - E_{19,3}) \quad (1.70b)$$

where $E_{19,3}$ is the totally quantized energy in this case.

1.2.4 The DR in Quantum Wells of HD III–V Semiconductors in the Presence of Cross Fields

The total Hamiltonian and the unperturbed wave function in the present case in the presence of the magnetic field B along y direction and the crossed electric field E_0 along z direction can, respectively, be written as

$$\hat{H} = a(-i\hbar\vec{\nabla} + e\vec{A})^2 - b(-i\hbar\vec{\nabla} + e\vec{A})^4 - eE_0 z \quad (1.71)$$

$$\psi_n^0 = \sqrt{\frac{2}{aL_xL_y}}\sin\left(\frac{n\pi}{a}z\right)\exp(ik_x x + ik_y y) \quad (1.72)$$

In this case the total Hamiltonian can be expressed as

$$\begin{aligned}\hat{H} &= \frac{\hbar^2 k_x^2}{2m_c} + \frac{\hbar^2 k_y^2}{2m_c} + \frac{\hbar^2}{2m_c}\left(\frac{n\pi}{a}\right)^2 + \frac{e^2B^2z^2}{2m_c} + \frac{2\hbar k_x eBz}{2m_c} \\ &\quad - \alpha\left[\frac{\hbar^2}{2m_c}\left(k_x^2 + k_y^2 + \left(\frac{n\pi}{a}\right)^2\right)\right]^2 - 2\alpha\left\{\frac{\hbar^2 k_x^2}{2m_c} + \frac{\hbar^2 k_y^2}{2m_c} + \frac{\hbar^2}{2m_c}\left(\frac{n\pi}{a}\right)^2\right\} \\ &\quad \left(\frac{e^2B^2z^2}{2m_c} + \frac{2\hbar k_x eBz}{2m_c}\right) - \alpha\left[\frac{e^2B^2z^2}{2m_c} + \frac{2\hbar k_x eBz}{2m_c}\right]^2 + 2\alpha\left(\frac{\hbar eB}{2m_c}\right)^2 \\ &\quad + \alpha\frac{\hbar^2}{2m_c}\left(\frac{n\pi}{a}\right)\left(\frac{4e^2B^2z}{2m_c} + \frac{4\hbar k_x eB}{2m_c}\right)^2 \cot\left(\frac{n\pi}{a}z\right) - eE_0 z\end{aligned} \tag{1.73}$$

Therefore, the DR can be written as

$$\begin{aligned}E &= \frac{\hbar^2}{2m_c}\left(\frac{n\pi}{a}\right)^2 + \frac{\hbar^2 k_y^2}{2m_c} + \frac{1}{2m_c}(\hbar k_x + eB<z>)^2 \\ &\quad - \alpha\left[\frac{\hbar^2}{2m_c}\left(\frac{n\pi}{a}\right)^2 + \frac{\hbar^2 k_y^2}{2m_c} + \frac{1}{2m_c}(\hbar k_x + eB<z>)^2\right]^2 \\ &\quad + \frac{e^2B^2}{2m_c}[<z^2> - <z>^2] - \alpha\left(\frac{e^2B^2}{2m_c}\right)\left\{\frac{\hbar^2 k_x^2}{2m_c} + \frac{\hbar^2 k_y^2}{2m_c} + \frac{\hbar^2}{2m_c}\left(\frac{n\pi}{a}\right)^2\right\} \\ &\quad [<z^2> - <z>^2] - \alpha\left(\frac{\hbar k_x eB}{m_c}\right)^2 \\ &\quad [<z^2> - <z>^2] - \alpha\left(\frac{\hbar k_x e^3B^3}{m_c^2}\right)^2[<z^3> - <z>^3] - \alpha\left(\frac{e^2B^2}{2m_c}\right)^2 \\ &\quad [<z^4> - <z>^4] - eE_0<z>\end{aligned} \tag{1.74}$$

If the electric field E_0 is along x direction, the total Hamiltonian can be written as

$$\hat{H} = a(-i\hbar\vec{\nabla} + e\vec{A})^2 - b(-i\hbar\vec{\nabla} + e\vec{A})^4 - eE_0x \tag{1.75}$$

(1.74) can be expressed as

$$\hat{H} = \frac{\hbar^2 k_x^2}{2m_c} + \frac{\hbar^2 k_y^2}{2m_c} + \frac{\hbar^2}{2m_c}\left(\frac{n\pi}{a}\right)^2 + \frac{e^2B^2z^2}{2m_c}$$
$$+ \left(\frac{2\hbar k_x eBz}{2m_c}\right) - \alpha\left[\frac{\hbar^2}{2m_c}\left(k_x^2 + k_y^2 + \left(\frac{n\pi}{a}\right)^2\right)\right]^2$$
$$- 2\alpha\left\{\frac{\hbar^2 k_x^2}{2m_c} + \frac{\hbar^2 k_y^2}{2m_c} + \frac{\hbar^2}{2m_c}\left(\frac{n\pi}{a}\right)^2\right\}\left(\frac{e^2B^2z^2}{2m_c} + \frac{2\hbar k_x eBz}{2m_c}\right)$$
$$- \alpha\left[\frac{e^2B^2z^2}{2m_c} + \frac{2\hbar k_x eBz}{2m_c}\right]^2 + 2\alpha\left(\frac{\hbar eB}{2m_c}\right)^2$$
$$+ \alpha\frac{\hbar^2}{2m_c}\left(\frac{n\pi}{a}\right)\left(\frac{4e^2B^2z}{2m_c} + \frac{4\hbar k_x eB}{2m_c}\right)^2 \cot\left(\frac{n\pi}{a}z\right) - eE_0x \tag{1.76}$$

Thus the perturbed part = $<eE_0x> = eE_0\frac{2}{aL_xL_y}\int_0^a \sin^2\left(\frac{n\pi}{a}z\right)dz\int_0^{L_x} xdx\int_0^{L_y} dy = eE_0\frac{2}{aL_xL_y}\left(\frac{a}{2}\right)\frac{L_x^2}{2}L_y = eE_0\left(\frac{L_x}{2}\right)$.

The DR in Quantum Wells of HD III–V Semiconductors in the Presence of Cross Fields is given by

$$\gamma_3(E,\eta_g) = \frac{\hbar^2}{2m_c}\left(\frac{n\pi}{a}\right)^2 + \frac{\hbar^2 k_y^2}{2m_c} + \frac{1}{2m_c}(\hbar k_x + \mathrm{eB}<z>)^2$$
$$- \alpha\left[\frac{\hbar^2}{2m_c}\left(\frac{n\pi}{a}\right)^2 + \frac{\hbar^2 k_y^2}{2m_c} + \frac{1}{2m_c}(\hbar k_x + \mathrm{eB}<z>)^2\right]^2$$
$$+ \frac{e^2B^2}{2m_c}[<z^2> - <z>^2] - \alpha\left(\frac{e^2B^2}{m_c}\right)\left\{\frac{\hbar^2 k_x^2}{2m_c} + \frac{\hbar^2 k_y^2}{2m_c} + \frac{\hbar^2}{2m_c}\left(\frac{n\pi}{a}\right)^2\right\}$$
$$[<z^2> - <z>^2] - \alpha\left(\frac{\hbar k_x eB}{m_c}\right)^2[<z^2> - <z>^2]$$
$$- \alpha\left(\frac{\hbar k_x e^3B^3}{m_c^2}\right)^2[<z^3> - <z>^3] - \alpha\left(\frac{e^2B^2}{2m_c}\right)^2[<z^4> - <z>^4] - eE_0\left(\frac{L_x}{2}\right) \tag{1.77}$$

The use of (1.77) leads to the expressions of EEM and DOS function, which is left to be done by the readers.

1.2.5 The DR in Nano-Wires of HD III–V Semiconductors in the Presence of Cross Fields

(a) The electric field E_0 is along z direction and the crossed magnetic field B is along y direction

The DR in this case is given by

$$\begin{aligned}\gamma_3(E,\eta_g) &= \left[\frac{\hbar^2}{2m_c}\left(\frac{n\pi}{a}\right)^2+\frac{\hbar^2}{2m_c}\left(\frac{l\pi}{b}\right)^2+\frac{\hbar^2k_y^2}{2m_c}\right]\\ &\quad-\alpha\left[\frac{\hbar^2}{2m_c}\left(\frac{n\pi}{a}\right)^2+\frac{\hbar^2}{2m_c}\left(\frac{l\pi}{b}\right)^2+\frac{\hbar^2k_y^2}{2m_c}\right]^2-eE_0<z>_n\\ &\quad+\frac{e^2B^2}{2m_c}\langle z^2\rangle_n-\alpha\left(\frac{2\hbar eB}{2m_c}\right)^2\left(\frac{l\pi}{b}\right)^2\langle z^2\rangle_n\\ &\quad-2\alpha\left(\frac{e^2B^2}{2m_c}\right)\left[\frac{\hbar^2}{2m_c}\left(\frac{n\pi}{a}\right)^2+\frac{\hbar^2}{2m_c}\left(\frac{l\pi}{b}\right)^2+\frac{\hbar^2k_y^2}{2m_c}\right]\langle z^2\rangle_n-\alpha\left(\frac{e^2B^2}{2m_c}\right)^2\langle z^4\rangle_n\end{aligned} \tag{1.78}$$

(b) The electric field E_0 is along x direction and the crossed magnetic field B is along y direction

In this case we can write

$$\begin{aligned}E &= \left[\frac{\hbar^2}{2m_c}\left(\frac{n\pi}{a}\right)^2+\frac{\hbar^2}{2m_c}\left(\frac{l\pi}{b}\right)^2+\frac{\hbar^2k_y^2}{2m_c}\right]\\ &\quad-\alpha\left[\frac{\hbar^2}{2m_c}\left(\frac{n\pi}{a}\right)^2+\frac{\hbar^2}{2m_c}\left(\frac{l\pi}{b}\right)^2+\frac{\hbar^2k_y^2}{2m_c}\right]^2\\ &\quad-eE_0<x>_n+\frac{e^2B^2}{2m_c}\langle z^2\rangle_n-\alpha\left(\frac{2\hbar eB}{2m_c}\right)^2\left(\frac{l\pi}{b}\right)^2\langle z^2\rangle_n\\ &\quad-2\alpha\left(\frac{e^2B^2}{2m_c}\right)\left[\frac{\hbar^2}{2m_c}\left(\frac{n\pi}{a}\right)^2+\frac{\hbar^2}{2m_c}\left(\frac{l\pi}{b}\right)^2+\frac{\hbar^2k_y^2}{2m_c}\right]\langle z^2\rangle_n\\ &\quad-\alpha\left(\frac{e^2B^2}{2m_c}\right)^2\langle z^4\rangle_n\end{aligned} \tag{1.79}$$

$$\begin{aligned}\text{The perturbed term} &= eE_0<x>_n=\frac{eE_0}{L_y}\frac{4}{ab}\int_0^a\sin^2\left(\frac{n\pi}{a}z\right)dz\int_0^b \mathrm{x}\sin^2\left(\frac{l\pi}{b}x\right)dx\int_0^{L_y}dy\\ &= \frac{eE_0}{L_y}\left(\frac{2}{a}\right)\frac{a}{2}L_y\left(\frac{2}{b}\right)\int_0^b x\sin^2\left(\frac{l\pi}{b}x\right)dx=\left(\frac{2}{b}\right)\frac{b^2}{4}\left(\frac{b}{2}\right)eE_0\end{aligned} \tag{1.80}$$

Therefore the DR in this case is given by

$$
\begin{aligned}
\gamma_3(E,\eta_g) = &\left[\frac{\hbar^2}{2m_c}\left(\frac{n\pi}{a}\right)^2 + \frac{\hbar^2}{2m_c}\left(\frac{l\pi}{b}\right)^2 + \frac{\hbar^2 k_y^2}{2m_c}\right] \\
&- \alpha\left[\frac{\hbar^2}{2m_c}\left(\frac{n\pi}{a}\right)^2 + \frac{\hbar^2}{2m_c}\left(\frac{l\pi}{b}\right)^2 + \frac{\hbar^2 k_y^2}{2m_c}\right]^2 \\
&- eE_0\left(\frac{b}{2}\right) + \frac{e^2B^2}{2m_c}\left\langle z^2\right\rangle_n - \alpha\left(\frac{2\hbar eB}{2m_c}\right)^2\left(\frac{l\pi}{b}\right)^2\left\langle z^2\right\rangle_n \\
&- 2\alpha\left(\frac{e^2B^2}{2m_c}\right)\left[\frac{\hbar^2}{2m_c}\left(\frac{n\pi}{a}\right)^2 + \frac{\hbar^2}{2m_c}\left(\frac{l\pi}{b}\right)^2 + \frac{\hbar^2 k_y^2}{2m_c}\right] \\
&\left\langle z^2\right\rangle_n - \alpha\left(\frac{e^2B^2}{2m_c}\right)^2\left\langle z^4\right\rangle_n
\end{aligned}
\tag{1.81}
$$

The use of (1.78) and (1.81) lead to the expressions of EEM and DOS function, which is left to be done by the readers.

1.2.6 The DR in Quantum Dot of HD III–V Semiconductors in the Presence of Cross Fields

(a) The electric field E_0 is along z direction and the crossed magnetic field B is along y direction

The DR in this case is given by

$$
\begin{aligned}
\gamma_3(E,\eta_g) = &\left[\frac{\hbar^2}{2m_c}\left(\frac{n\pi}{a}\right)^2 + \frac{\hbar^2}{2m_c}\left(\frac{l\pi}{b}\right)^2 + \frac{\hbar^2}{2m_c}\left(\frac{r\pi}{c}\right)^2\right] \\
&- \alpha\left[\frac{\hbar^2}{2m_c}\left(\frac{n\pi}{a}\right)^2 + \frac{\hbar^2}{2m_c}\left(\frac{l\pi}{b}\right)^2 + \frac{\hbar^2}{2m_c}\left(\frac{r\pi}{c}\right)^2\right]^2 \\
&- eE_0 <z>_n + \frac{e^2B^2}{2m_c}\left\langle z^2\right\rangle_n - \alpha\left(\frac{2\hbar eB}{2m_c}\right)^2\left\langle z^2\right\rangle_n \\
&- 2\alpha\left(\frac{e^2B^2}{2m_c}\right)^2\left[\frac{\hbar^2}{2m_c}\left(\frac{n\pi}{a}\right)^2 + \frac{\hbar^2}{2m_c}\left(\frac{l\pi}{b}\right)^2 + \frac{\hbar^2}{2m_c}\left(\frac{r\pi}{c}\right)^2\right]\left\langle z^2\right\rangle_n \\
&\alpha\left(\frac{e^2B^2}{2m_c}\right)^2\left\langle z^4\right\rangle_n
\end{aligned}
\tag{1.82a}
$$

The DOS function in this case can be written as

$$N(E) = \frac{2g_v}{abc}\sum_{n=1}^{n_{\max}}\sum_{l=1}^{l_{\max}}\sum_{r=1}^{r_{\max}}\delta'(E - E_{19,900}) \quad (1.82b)$$

where $E_{19,900}$ is the totally quantized energy in this case.

(b) The electric field E_0 is along x direction and the crossed magnetic field B is along y direction

In this case we can write

$$\begin{aligned}\gamma_3(E,\eta_g) = & \left[\frac{\hbar^2}{2m_c}\left(\frac{n\pi}{a}\right)^2 + \frac{\hbar^2}{2m_c}\left(\frac{l\pi}{b}\right)^2 + \frac{\hbar^2}{2m_c}\left(\frac{r\pi}{c}\right)^2\right] \\ & - \alpha\left[\frac{\hbar^2}{2m_c}\left(\frac{n\pi}{a}\right)^2 + \frac{\hbar^2}{2m_c}\left(\frac{l\pi}{b}\right)^2 + \frac{\hbar^2}{2m_c}\left(\frac{r\pi}{c}\right)^2\right]^2 \\ & + \frac{e^2B^2}{2m_c}\left\langle z^2\right\rangle_n - \alpha\left(\frac{2\hbar eB}{2m_c}\right)^2\left(\frac{l\pi}{b}\right)^2\left\langle z^2\right\rangle_n - 2\alpha\left(\frac{e^2B^2}{2m_c}\right)^2 \\ & \left[\frac{\hbar^2}{2m_c}\left(\frac{n\pi}{a}\right)^2 + \frac{\hbar^2}{2m_c}\left(\frac{l\pi}{b}\right)^2 + \frac{\hbar^2}{2m_c}\left(\frac{r\pi}{c}\right)^2\right]\left\langle z^2\right\rangle_n \\ & - \alpha\left(\frac{e^2B^2}{2m_c}\right)^2\left\langle z^4\right\rangle_n - eE_0\left(\frac{b}{2}\right)\end{aligned} \quad (1.83a)$$

The DOS function in this case can be written as

$$N(E) = \frac{2g_v}{abc}\sum_{n=1}^{n_{\max}}\sum_{l=1}^{l_{\max}}\sum_{r=1}^{r_{\max}}\delta'(E - E_{19,901}) \quad (1.83b)$$

where $E_{19,901}$ is the totally quantized energy in this case.

1.2.7 *The DR in Quantum Wells of HD IV–VI Semiconductors in the Presence of Magnetic Field*

(a) Cohen Model

In accordance with Cohen model, the DR in bulk specimens of IV–VI semiconductors can be written as

$$\left(\mathrm{E} - \frac{\hbar^2 k_y^2}{2m_2}\right)\left(1 + \alpha\mathrm{E} + \frac{\alpha\hbar^2 k_y^2}{2m_2'}\right) = \frac{\hbar^2 k_x^2}{2m_1} + \frac{\hbar^2 k_z^2}{2m_3} \tag{1.84}$$

Let us substitute $\frac{1}{2m_i} = a_i \quad i = 1, 2, 3 \quad \text{and} \quad \frac{1}{2m_2'} = a_2'$

Using the method of successive approximation we can write

$$\begin{aligned} E &= a_1 p_x^2 + a_3 p_z^2 + a_2 p_y^2 + \alpha(a_1 p_x^2 + a_3 p_z^2 + a_2 p_y^2)^2 \\ &\quad + \alpha a_2 p_y^2 (a_1 p_x^2 + a_3 p_z^2 + a_2 p_y^2) \\ &\quad - \alpha a_2' p_y^2 (a_1 p_x^2 + a_3 p_z^2 + a_2 p_y^2) + \alpha a_1 a_2' p_y^4 \end{aligned} \tag{1.85}$$

Let us choose the formation of QWs is along z direction in presence of magnetic field (B) along y direction.

Choosing the Coulomb gauge as

$\vec{A} = (0, 0, -Bx)$ and $\vec{\nabla} \cdot \vec{A} = 0$ together with the fact

$$\vec{B} = \vec{\nabla} \times \vec{A} = \begin{vmatrix} i & j & k \\ \partial/\partial x & \partial/\partial y & \partial/\partial z \\ 0 & 0 & -Bx \end{vmatrix}$$

In presence of perturbing magnetic field total Hamiltonian can be written as

$$
\begin{aligned}
\hat{H} =& a_1\left(-i\hbar\frac{\partial}{\partial x} + eA_x\right)^2 + a_2\left(-i\hbar\frac{\partial}{\partial y} + eA_y\right)^2 + a_3\left(-i\hbar\frac{\partial}{} + eA_z\right)^2 \\
& - \alpha\left[a_1\left(-i\hbar\frac{\partial}{\partial x} + eA_x\right)^2 + a_2\left(-i\hbar\frac{\partial}{\partial y} + eA_y\right)^2 + a_3\left(-i\hbar\frac{\partial}{\partial z} + eA_z\right)^2\right]^2 \\
& + \alpha a_2\left(-i\hbar\frac{\partial}{\partial y} + eA_y\right)^2\left[a_1\left(-i\hbar\frac{\partial}{\partial x} + eA_x\right)^2 + a_2\left(-i\hbar\frac{\partial}{\partial y} + eA_y\right)^2 + a_3\left(-i\hbar\frac{\partial}{\partial z} + eA_z\right)^2\right] \\
& - \alpha a_2'\left(-i\hbar\frac{\partial}{\partial y} + eA_y\right)^2\left[a_1\left(-i\hbar\frac{\partial}{\partial x} + eA_x\right)^2 + a_2\left(-i\hbar\frac{\partial}{\partial y} + eA_y\right)^2 + a_3\left(-i\hbar\frac{\partial}{\partial z} + eA_z\right)^2\right] \\
& + \alpha a_2 a_2'\left(-i\hbar\frac{\partial}{\partial y} + eA_y\right)^4
\end{aligned}
\tag{1.86}
$$

$$
\begin{aligned}
=& -a_1\hbar^2\frac{\partial^2}{\partial x^2} - a_2\hbar^2\frac{\partial^2}{\partial y^2} + a_3\left(-\hbar^2\frac{\partial^2}{\partial z^2} + \mathrm{e}^2B^2x^2 + 2\mathrm{i}\hbar\,\mathrm{e}\,Bx\frac{\partial}{\partial z}\right) \\
& - \alpha\left[-a_1\hbar^2\frac{\partial^2}{2} - a_2\hbar^2\frac{\partial^2}{2} + a_3\left(-\hbar^2\frac{\partial^2}{\partial z^2} + \mathrm{e}^2B^2x^2 + 2\hbar\,\mathrm{e}\,Bx\frac{\partial}{\partial z}\right)\right]^2 \\
& + \alpha a_2\left(-\hbar^2\frac{\partial^2}{\partial y^2}\right)\left[-a_1\hbar^2\frac{\partial^2}{\partial x^2} - a_2\hbar^2\frac{\partial^2}{\partial y^2} + a_3\left(-\hbar^2\frac{\partial^2}{\partial z^2} + \mathrm{e}^2B^2x^2 + 2\mathrm{i}\hbar\, e\,Bx\frac{\partial}{\partial z}\right)\right] \\
& + \alpha a_2'\left(-\hbar^2\frac{\partial^2}{\partial y^2}\right)\left[-a_1\hbar^2\frac{\partial^2}{\partial x^2} - a_2\hbar^2\frac{\partial^2}{\partial y^2} + a_3\left(-\hbar^2\frac{\partial^2}{\partial z^2} + \mathrm{e}^2B^2x^2 + 2\mathrm{i}\hbar\, e\,Bx\frac{\partial}{\partial z}\right)\right] \\
& - \alpha a_2'\left(-\hbar^2\frac{\partial^2}{\partial y^2}\right)\left[-a_1\hbar^2\frac{\partial^2}{\partial x^2} - a_2\hbar^2\frac{\partial^2}{\partial y^2} + a_3\left(-\hbar^2\frac{\partial^2}{\partial z^2} + e^2B^2x^2 + 2\mathrm{i}\hbar\, e\,Bx\frac{\partial}{\partial z}\right)\right] \\
& + \alpha a_2 a_2'\left(-\hbar^4\frac{\partial^4}{\partial y^4}\right)
\end{aligned}
\tag{1.87}
$$

Considering the effect of magnetic field and non parabolicity as perturbed parameters, the unperturbed wave equation can be written as

$$
-a_1\hbar^2\frac{\partial^2\psi_0}{\partial x^2} - a_2\hbar^2\frac{\partial^2\psi_0}{\partial y^2} - a_3\hbar^2\frac{\partial^2\psi_0}{\partial z^2} = E_0\psi_0 \tag{1.88}
$$

Therefore energy eigen value (E_n^0) and the eigen function assume the forms

$$
E_0 = E_n^0 = \frac{\hbar^2k_x^2}{2m_1} + \frac{\hbar^2k_y^2}{2m_2} + \frac{\hbar^2}{2m_3}\left(\frac{n\pi}{a}\right)^2 \tag{1.89}
$$

$$
\psi^0 = \psi_n^0 = \sqrt{\frac{2}{aL_xL_y}}\sin\left(\frac{n\pi}{a}z\right)\exp(ik_xx + ik_yy) \tag{1.90}
$$

The perturbed Hamiltonian can be calculated in the following way

$$\begin{aligned}
&\left[-a_1\hbar^2\frac{\partial^2}{\partial x^2} - a_2\hbar^2\frac{\partial^2}{\partial y^2} + a_3\left(-\hbar^2\frac{\partial^2}{\partial z^2} + \mathrm{e}^2B^2x^2 + 2\mathrm{i}\hbar\,\mathrm{e}\,Bx\frac{\partial}{\partial z}\right)\right]\psi_n \\
&= a_1\hbar^2k_x^2\psi_n + a_2\hbar^2k_y^2\psi_n + a_3\hbar^2\left(\frac{n\pi}{a}\right)^2\psi_n + a_3\mathrm{e}^2B^2x^2\psi_n \\
&\quad + 2a_3\mathrm{i}\hbar\mathrm{e}\,Bx\left(\frac{n\pi}{a}\right)\cot\left(\frac{n\pi}{a}z\right)\psi_n - a_1\hbar^2\frac{\partial^2}{\partial x^2} - a_2\hbar^2\frac{\partial^2}{\partial y^2} \\
&\quad + a_3\left(-\hbar^2\frac{\partial^2}{\partial z^2} + \mathrm{e}^2B^2x^2 + 2\mathrm{i}\hbar\mathrm{e}\,Bx\frac{\partial}{\partial z}\right).
\end{aligned}$$

$$\begin{aligned}
&\left[a_1\hbar^2k_x^2\psi_n + a_2\hbar^2k_y^2\psi_n + a_3\hbar^2\left(\frac{n\pi}{a}\right)^2\psi_n + a_3\mathrm{e}^2B^2x^2\psi_n + 2a_3\mathrm{i}\hbar\mathrm{e}\,B\left(\frac{n\pi}{a}\right)x\cot\left(\frac{n\pi}{a}z\right)\psi_n\right] \\
&= \left[a_1^2\hbar^4k_x^4\psi_n + a_1a_2\hbar^4k_x^2k_y^2\psi_n + a_1a_3\hbar^4k_x^2\left(\frac{n\pi}{a}\right)^2\psi_n\right. \\
&\quad + a_1a_3\hbar^2k_x^2e^2B^2x^2\psi_n - a_1a_3\hbar^2 2e^2B^2\psi_n \\
&\quad + 2a_1a_3\mathrm{i}\hbar^3\mathrm{e}\,B\left(\frac{n\pi}{a}\right)k_x^2x\cot\left(\frac{n\pi}{a}z\right)\psi_n + a_2a_1\hbar^4k_y^2k_x^2\psi_n \\
&\quad + a_2^2\hbar^4k_y^4\psi_n + a_2a_3\hbar^4\left(\frac{n\pi}{a}\right)^2k_y^2\psi_n + a_2a_3e^2B^2x^2\hbar^2k_y^2\psi_n \\
&\quad + 2a_2a_3i\hbar^3k_y^2eB\left(\frac{n\pi}{a}\right)x\cot\left(\frac{n\pi}{a}z\right)\psi_n + a_3a_1\hbar^4\left(\frac{n\pi}{a}\right)^2k_x^2\psi_n \\
&\quad + a_3a_2\hbar^4\left(\frac{n\pi}{a}\right)^2k_y^2\psi_n + a_3^2\hbar^4\left(\frac{n\pi}{a}\right)^4\psi_n + a_3^2\hbar^2\left(\frac{n\pi}{a}\right)^2e^2B^2x^2\psi_n \\
&\quad + 2a_3^2i\hbar^3eB\left(\frac{n\pi}{a}\right)^3x\cot\left(\frac{n\pi}{a}z\right)\psi_n \\
&\quad + \left[a_3a_1\hbar^2k_x^2e^2B^2 + a_3a_2\hbar^2k_y^2e^2B^2 + a_3^2\hbar^2\left(\frac{n\pi}{a}\right)^2e^2B^2\right]x^2\psi_n \\
&\quad + a_3^2e^4B^4x^4\psi_n + 2a_3^2i^3B^3\left(\frac{n\pi}{a}\right)x^3\cot\left(\frac{n\pi}{a}z\right)\psi_n \\
&\quad + 2a_3i\hbar eB\left[a_1\hbar^2k_x^2 + a_2\hbar^2k_y^2 + a_3\hbar^2\left(\frac{n\pi}{a}\right)^2\right]x\left(\frac{n\pi}{a}\right)\cot\left(\frac{n\pi}{a}z\right)\psi_n \\
&\quad + 2a_3^2i\hbar e^3B^3x^3\left(\frac{n\pi}{a}\right)\cot\left(\frac{n\pi}{a}z\right)\psi_n + 4a_3^2\hbar^2e^2B^2\left(\frac{n\pi}{a}\right)^2x^2\psi_n \\
&= \left[a_1\hbar^2k_x^2 + a_2\hbar^2k_y^2 + a_3\hbar^2\left(\frac{n\pi}{a}\right)^2\right]^2\psi_n - 2a_1a_3\hbar^2e^2B^2\psi_n \\
&\quad + 4a_3i\hbar eB\left[a_1\hbar^2k_x^2 + a_2\hbar^2k_y^2 + a_3\hbar^2\left(\frac{n\pi}{a}\right)^2\right]x\left(\frac{n\pi}{a}\right)\cot\left(\frac{n\pi}{a}z\right)\psi_n \\
&\quad + \left[2a_1a_3\hbar^2k_x^2e^2B^2 + 2a_2a_3\hbar^2k_y^2e^2B^2 + 6a_3^2\hbar^2\left(\frac{n\pi}{a}\right)^2e^2B^2\right]x^2\psi_n \\
&\quad + 4a_3^2i\hbar e^3B^3\left(\frac{n\pi}{a}\right)x^3\cot\left(\frac{n\pi}{a}z\right)\psi_n + a_3^2e^4B^4x^4\psi_n
\end{aligned} \tag{1.91}$$

$$
\begin{aligned}
\hat{H}\psi_n &= a_1\hbar^2k_x^2\psi_n + a_2\hbar^2k_y^2\psi_n + a_3\hbar^2\left(\frac{n\pi}{a}\right)^2\psi_n \\
&+ a_3e^2B^2x^2\psi_n + 2a_3i\hbar\, e\, B\left(\frac{n\pi}{a}\right)x\cot\left(\frac{n\pi}{a}z\right)\psi_n \\
&- \alpha\left[a_1\hbar^2k_x^2 + a_2\hbar^2k_y^2 + a_3\hbar^2\left(\frac{n\pi}{a}\right)^2\right]^2\psi_n + 2\alpha a_1a_3\hbar^2e^2B^2\psi_n \\
&- 4\alpha a_3i\hbar\, e\, B\left[a_1\hbar^2k_x^2 + a_2\hbar^2k_y^2 + a_3\hbar^2\left(\frac{n\pi}{a}\right)^2\right]\left(\frac{n\pi}{a}\right)x\cot\left(\frac{n\pi}{a}z\right)\psi_n \\
&- \alpha\left[2a_1a_3\hbar^2k_x^2e^2B^2 + 2a_2a_3\hbar^2k_y^2e^2B^2 + 6a_3^2\hbar^2\left(\frac{n\pi}{a}\right)^2e^2B^2\right]x^2\psi_n \\
&- 4\alpha i\hbar a_3^2e^3B^3\left(\frac{n\pi}{a}\right)x^3\cot\left(\frac{n\pi}{a}z\right)\psi_n - \alpha a_3^2e^4B^4x^4\psi_n \\
&+ \alpha a_2\left[a_1\hbar^4k_x^2k_y^2 + a_2\hbar^4k_y^4 + a_3\hbar^4k_y^2\left(\frac{n\pi}{a}\right)^2\right]\psi_n \\
&+ \alpha a_2a_3\hbar^2k_y^2e^2B^2x^2\psi_n + 2\alpha a_2a_0a_3i\hbar^3k_y^2e\, B\left(\frac{n\pi}{a}\right)x\cot\left(\frac{n\pi}{a}z\right)\psi_n \\
&- \alpha a_2'\left[a_1\hbar^4k_x^2k_y^2 + a_2\hbar^4k_y^4 + a_3\hbar^4k_y^2\left(\frac{n\pi}{a}\right)^2\right]\psi_n \\
&- \alpha a_2'a_3\hbar^2k_y^2e^2B^2x^2\psi_n - 2\alpha a_2'a_3i\hbar^3k_y^2e\, B\left(\frac{n\pi}{a}\right)x\cot\left(\frac{n\pi}{a}z\right)\psi_n \\
&+ \alpha a_2a_2'\hbar^4k_y^4\psi_n = \hat{H}_0\psi_n + \hat{H}'\psi_n
\end{aligned}
\tag{1.92}
$$

where $\hat{H}_0 = a_1\hbar^2k_x^2 + a_2\hbar^2k_y^2 + a_3\hbar^2\left(\frac{n\pi}{a}\right)^2$.

The remaining part correspond to perturbed Hamiltonian.

$$
\left\langle x\cot\left(\frac{n\pi}{a}z\right)\right\rangle = \frac{2}{a}\int_0^a \sin\left(\frac{n\pi}{a}z\right)\cos\left(\frac{n\pi}{a}z\right)dz\int_0^{L_x}\frac{1}{L_x}dx = 0,
$$

$$
\left\langle x^3\cot\left(\frac{n\pi}{a}z\right)\right\rangle = 0,\quad \langle x^2\rangle = \frac{1}{L_x}\int_0^{L_x}x^2dx = \frac{L_x^2}{3},\quad \langle x^4\rangle = \frac{1}{L_x}\int_0^{L_x}x^4dx = \frac{L_x^4}{5}
$$

The DR in this case is given by

$$\begin{aligned} E = & a_1\hbar^2k_x^2 + a_2\hbar^2k_y^2 + a_3\hbar^2\left(\frac{n\pi}{a}\right)^2 \\ & - \alpha\left[a_1\hbar^2k_x^2 + a_2\hbar^2k_y^2 + a_3\hbar^2\left(\frac{n\pi}{a}\right)^2\right]^2 \\ & + {}_2\hbar^2k_y^2\left[a_1\hbar^2k_x^2 + a_2\hbar^2k_y^2 + a_3\hbar^2\left(\frac{n\pi}{a}\right)^2\right] \\ & - \alpha a_2'\hbar^2k_y^2\left[a_1\hbar^2k_x^2 + a_2\hbar^2k_y^2 + a_3\hbar^2\left(\frac{n\pi}{a}\right)^2\right] \\ & + \alpha a_2a_2'\hbar^4k_y^4 + a_3e^2B^2\langle x^2\rangle + 2\alpha a_1a_3\hbar^2e^2B^2 \\ & - \alpha\left[2a_1a_3\hbar^2k_x^2 + a_2a_3\hbar^2k_y^2 + 6a_3^2\hbar^2\left(\frac{n\pi}{a}\right)^2\right]e^2B^2\langle x^2\rangle \\ & - \alpha a_3^2e^4B^4\langle x^4\rangle - \alpha a_2'a_3\hbar^2k_y^2e^2B^2\langle x^2\rangle \end{aligned} \quad (1.93)$$

Thus following the method of this chapter, the DR in HDQWs of IV–VI semiconductors can be written as

$$\begin{aligned} \gamma_3(E,\eta_g) = & a_1\hbar^2k_x^2 + a_2\hbar^2k_y^2 + a_3\hbar^2\left(\frac{n\pi}{a}\right)^2 - \alpha\left[a_1\hbar^2k_x^2 + a_2\hbar^2k_y^2 + a_3\hbar^2\left(\frac{n\pi}{a}\right)^2\right]^2 \\ & + \alpha a_2\hbar^2k_y^2\left[a_1\hbar^2k_x^2 + a_2\hbar^2k_y^2 + a_3\hbar^2\left(\frac{n\pi}{a}\right)^2\right] \\ & - \alpha a_2'\hbar^2k_y^2\left[a_1\hbar^2k_x^2 + a_2\hbar^2k_y^2 + a_3\hbar^2\left(\frac{n\pi}{a}\right)^2\right] \\ & + \alpha a_2a_2'\hbar^4k_y^4 + a_3e^2B^2\langle x^2\rangle + {}_2{}_1a_3\hbar^2e^2B^2 \\ & - \alpha\left[2a_1a_3\hbar^2k_x^2 + a_2a_3\hbar^2k_y^2 + 6a_3^2\hbar^2\left(\frac{n\pi}{a}\right)^2\right] \\ & e^2B^2\langle x^2\rangle - {}_3^2e^4B^4\langle x^4\rangle - \alpha a_2'a_3\hbar^2k_y^2e^2B^2\langle x^2\rangle \end{aligned} \quad (1.94)$$

Putting $\alpha = 0$, we get

$$\gamma_3(E,\eta_g) = a_1\hbar^2k_x^2 + a_2\hbar^2k_y^2 + a_3\hbar^2\left(\frac{n\pi}{a}\right)^2 + a_3\mathrm{e}^2B^2\langle x^2\rangle \quad (1.95)$$

(1.95) can be written as

$$\gamma_3(E,\eta_g) - a_3\mathrm{e}^2B^2\langle x^2\rangle - a_3\hbar^2\left(\frac{n\pi}{a}\right)^2 = a_1\hbar^2k_x^2 + a_2\hbar^2k_y^2 \quad (1.96)$$

Using the method of successive approximation we can write

$$\begin{aligned}\gamma_3(E,\eta_g) = {} & a_1\hbar^2k_x^2 + a_2\hbar^2k_y^2 + a_3\hbar^2\left(\frac{n\pi}{a}\right)^2 - \alpha(\gamma_3(E,\eta_g) - c_0)^2 \\ & + \alpha a_2(\gamma_3(E,\eta_g) - c_0)\hbar^2k_y^2 - \alpha a_2'(\gamma_3(E,\eta_g) - c_0)\hbar^2k_y^2 \\ & + \alpha a_2a_2'\hbar^4k_y^4 + c_0 + c_2 - 2\alpha a_1a_3e^2B^2\langle x^2\rangle\hbar^2k_x^2 \\ & - \alpha a_2a_3e^2B^2\langle x^2\rangle\hbar^2k_y^2 - 6\alpha a_{32}^2\hbar^2\left(\frac{n\pi}{a}\right)^2 e^2B^2\langle x^2\rangle \\ & - \alpha a_3^2e^4B^4\langle x^4\rangle - \alpha a_2'a_3e^2B^2\langle x^2\rangle\hbar^2k_y^2\end{aligned} \tag{1.97}$$

where $c_0 = a_3e^2B^2\langle x^2\rangle = \frac{a_3e^2B^2L_x^2}{3}$, $c_1 = a_3\hbar^2\left(\frac{n\pi}{a}\right)^2$.

The (1.97) can be written as

$$\begin{aligned}\gamma_3(E,\eta_g) - c_0 - c_1 - c_2 + \alpha(\gamma_3(E,\eta_g) - c_0)^2 + c_3 + c_4 = {} & \left[a_1\hbar^2 - 2\alpha\hbar^2a_1a_3e^2B^2\langle x^2\rangle\right]k_x^2 \\ & + \left[a_2\hbar^2 + \alpha a_2(\gamma_3(E,\eta_g) - c_0)\hbar^2 - \alpha a_2'(\gamma_3(E,\eta_g) - c_0)\hbar^2\right. \\ & \left.- \hbar^2\alpha a_2a_3e^2B^2\langle x^2\rangle - \hbar^2\alpha a_2'a_3e^2B^2\langle x^2\rangle\right]k_y^2 + \alpha a_2a_2'\hbar^4k_y^4\end{aligned} \tag{1.98}$$

The DR can be written as

$$\gamma_{10}(E,\eta_g) = \beta_1k_x^2 + \beta_2k_y^2 + \beta_3k_y^4 \tag{1.99}$$

where,

$$\begin{aligned}\gamma_{10}(E,\eta_g) &= \gamma_3(E,\eta_g) - c_0 - c_1 - c_2 + c_3 + c_4 + \alpha(\gamma_3(E,\eta_g) - c_0)^2 \\ c_2 &= 2\alpha a_1a_3\hbar^2e^2B^2,\ c_3 = 6\alpha a_3^2\hbar^2\left(\frac{n\pi}{a}\right)^2e^2B^2\left(\frac{L_x^2}{3}\right),\ c_4 = \alpha a_3^2e^4B^4\left(\frac{L_x^4}{5}\right),\end{aligned}$$

$$\begin{aligned}\beta_1 &= a_1\hbar^2 - 2\alpha\hbar^2a_1a_3e^2B^2\left(\frac{L_x^2}{3}\right) = a_1\hbar^2\left[1 - 2\alpha a_3e^2B^2\left(\frac{L_x^2}{3}\right)\right] = \text{positive quantity}, \\ \beta_2 &= a_2\hbar^2 + \alpha a_2(\gamma_3(E,\eta_g) - c_0)\hbar^2 - \alpha a_2'(\gamma_3(E,\eta_g) - c_0)\hbar^2 \\ & \quad - \hbar^2\alpha a_2a_3e^2B^2\left(\frac{L_x^2}{3}\right) - \hbar^2\alpha a_2'a_3e^2B^2\left(\frac{L_x^2}{3}\right) \\ & \simeq a_2\hbar^2 - 2\alpha\hbar^2a_2a_3e^2B^2\left(\frac{L_x^2}{3}\right) \simeq a_2\hbar^2\left(1 - 2\alpha a_2a_3e^2B^2\frac{L_x^2}{3}\right) \\ & \simeq \text{positive quantity} \quad \text{and} \quad \beta_3 = \alpha a_2a_2'\hbar^4 = \text{positive quantity}\end{aligned}$$

Therefore, from (1.99) we can write

$$k_x = \pm\frac{1}{\sqrt{\beta_1}}\sqrt{\gamma_{10}(E,\eta_g) - \beta_2k_y^2 - \beta_3k_y^4} \tag{1.100}$$

Since area is a positive quantity, therefore

$$k_x = \frac{1}{\sqrt{\beta_1}} \cdot \sqrt{\gamma_{10}(E, \eta_g) - \beta_2 k_y^2 - \beta_3 k_y^4} \tag{1.101}$$

Similarly we get $k_x = \pm\sqrt{\frac{-\beta_2}{2\beta_3} + \frac{\sqrt{\beta_2^2 + 4\beta_3\gamma_{10}(E,\eta_g)}}{2\beta_3}}$.

Since k_y is real, therefore $k_y = \pm\sqrt{-\frac{1}{2} \cdot \left(\frac{\beta_2}{\beta_3}\right) + \frac{1}{2}\sqrt{\left(\frac{\beta_2}{\beta_3}\right)^2 + \frac{4\gamma_{10}(E,\eta_g)}{\beta_3}}} = \pm u.$

So the area enclosed by the 2D surface as given by (1.99) can be written as

$$A = 2I = 2\int_{-u}^{+u} \frac{1}{\sqrt{\beta_1}}\sqrt{\gamma_{10}(E, \eta_g) - \beta_2 k_y^2 - \beta_3 k_y^4 dk_y} \tag{1.102}$$

where,

$$\begin{aligned} I &= \frac{1}{\sqrt{\beta_1}}\int_{-u}^{+u} \sqrt{\gamma_{10}(E, \eta_g) - \beta_2 k_y^2 - \beta_3 k_y^4 dk_y} \\ &= \frac{2}{\sqrt{\beta_1}}\int_{0}^{+u} \sqrt{\gamma_{10}(E, \eta_g) - \beta_2 k_y^2 - \beta_3 k_y^4 dk_y} \\ &= 2\sqrt{\frac{\beta_3}{\beta_1}}\int_{0}^{+u} \sqrt{\frac{\gamma_{10}(E, \eta_g)}{\beta_3} - \left(\frac{\beta_2}{\beta_3}\right)k_y^2 - k_y^4 dk_y} \\ &= 2\sqrt{\frac{\beta_3}{\beta_1}} \cdot I_1 \end{aligned}$$

Therefore

$$A = 2I = 4\sqrt{\frac{\beta_3}{\beta_1}} \cdot I_1 \tag{1.103}$$

where

$$I_1 = \int_{0}^{+u} \sqrt{\frac{\gamma_{10}(E, \eta_g)}{\beta_3} - \left(\frac{\beta_2}{\beta_3}\right)k_y^2 - k_y^4 dk_y} \tag{1.104}$$

Now

$$\frac{\gamma_{10}(E,\eta_g)}{\beta_3} - \left(\frac{\beta_2}{\beta_3}\right)k_y^2 - k_y^4 dk_y = (a^2 + x^2)(b^2 - x^2)$$
$$= a^2b^2 - a^2x^2 + b^2x^2 - x^4 = a^2b^2(a^2 - b^2)x^2 - x^4$$

So that

$$a^2b^2 = \frac{\gamma_{10}(E,\eta_g)}{\beta_3},\ a^2 - b^2 = \frac{\beta_2}{\beta_3} \quad \text{and} \quad x = k_y \text{ i.e, } dx = dk_y$$

$$I_1 = \int_0^u \sqrt{(a^2 + x^2)(b^2 - x^2)dx} \tag{1.105}$$

Now

$$(a^2 + b^2)^2 = (a^2 - b^2)^2 + 4a^2b^2 = \left(\frac{\beta_2}{\beta_3}\right)^2 + \frac{4\gamma_{10}(E,\eta_g)}{\beta_3} \tag{1.106}$$

$$a^2 + b^2 = \sqrt{\left(\frac{\beta_2}{\beta_3}\right)^2 + \frac{4\gamma_{10}(E,\eta_g)}{\beta_3}} \text{ for real } b.$$

$$\text{Therefore} \quad a^2 = \frac{1}{2}\left(\frac{\beta_2}{\beta_3}\right) + \frac{1}{2}\sqrt{\left(\frac{\beta_2}{\beta_3}\right)^2 + \frac{4\gamma_{10}(E,\eta_g)}{\beta_3}}$$

$$b^2 = \frac{1}{2}\sqrt{\left(\frac{\beta_2}{\beta_3}\right)^2 + \frac{4\gamma_{10}(E,\eta_g)}{\beta_3}} - \frac{1}{2}\left(\frac{\beta_2}{\beta_3}\right)$$

So all the coefficients of (1.105) are defined. From the expressions of u and b we note that $u = b$. Since b is real and positive, $u > 0$ and also $a > b$. Therefore $a > u$. Thus we observe that $a > u > 0$. Under this condition (1.105) can be written as

$$\begin{aligned} I_1 = {} & \frac{1}{3}\sqrt{a^2 + b^2}\cdot\left\{a^2F(\theta,r) - (a^2 - b^2)E(\theta,r)\right\} \\ & + \frac{u}{3}(u^2 + 2a^2 - b^2)\sqrt{\frac{b^2 - u^2}{a^2 + u^2}} \end{aligned} \tag{1.107}$$

where $F(\theta,r) = \int_0^\theta \frac{dt}{\sqrt{1 - r^2\sin^2 t}}$, $E(\theta,r) = \int_0^\theta \sqrt{1 - r^2\sin^2 t}dt$, $F(\theta,r)$ and $E(\theta,r)$ are known as incomplete Elliptic integral of first and second kind respectively in which $r^2 < 1$ and is known as modulus of the integral, $\theta = \sin^{-1}\left[\frac{u}{b}\sqrt{\frac{a^2 + b^2}{a^2 + u^2}}\right]$ and $r = \frac{b}{\sqrt{a^2 + b^2}}$.

In this case $b = u$, so that

$$I_1 = \frac{1}{3}\sqrt{a^2 + b^2} \cdot \left\{a^2 F(\theta, r) - (a^2 - b^2)E(\theta, r)\right\} \quad (1.108)$$

$$\text{When}\, b = u,\ \text{then}\, \theta = \frac{\pi}{2} \quad (1.109)$$

Therefore from (1.109) we get

$$\begin{aligned} I_1 &= \frac{1}{3}\sqrt{a^2 + b^2} \cdot \left\{a^2 F\left(\frac{\pi}{2}, r\right) - (a^2 - b^2)E\left(\frac{\pi}{2}, r\right)\right\} \\ &= \frac{1}{3}\sqrt{a^2 + b^2} \cdot \left\{a^2 k(r) - (a^2 - b^2)E(r)\right\} \end{aligned} \quad (1.110)$$

where $K(r)$ and $E(r)$ are known as the complete Elliptic integrals of first and second kind respectively.

In our case $r = \frac{b}{\sqrt{a^2 + b^2}}$ and since $a > b$ therefore $r^2 < 1$.

$$\text{Thus we write } r = \sqrt{\frac{b^2}{a^2 + b^2}} = \frac{1}{\sqrt{2}}\left[1 - \frac{\frac{\beta_2}{\beta_3}}{\sqrt{\left(\frac{\beta_2}{\beta_3}\right)^2 + \frac{4\gamma_{10}(\mathrm{E}, \eta_g)}{\beta_3}}}\right]^{1/2}.$$

Since $r \ll 1$.

Therefore

$$K(\mathrm{r}) = \frac{\pi}{2}\left[1 + (1/2)^2\mathrm{r}^2 + \left(\frac{1 \cdot 3}{2 \cdot 4}\right)^2 \mathrm{r}^4 + \cdots\cdots + \left\{\frac{(2\mathrm{n} - 1)!!}{2^n n!}\right\}^2 \mathrm{r}^{2n} + \cdots\cdots\right]$$

$$E(r) = \frac{\pi}{2}\left[1 - \frac{1}{2^2}\mathrm{r}^2 - \left(\frac{1^2 \cdot 3}{2^2 \cdot 4^2}\right)^2 \mathrm{r}^4 - \cdots\cdots - \left\{\frac{(2\mathrm{n} - 1)!!}{2^n n!}\right\}^2 \frac{\mathrm{r}^{2n}}{(2\mathrm{n} - 1)} - \cdots\cdots\right]$$

Since $r^2 \ll 1$, we take only r^2 terms as they provide maximum contribution in comparison with the higher order terms. Therefore

$$
\begin{aligned}
I_1 &= \frac{1}{3}(\sqrt{a^2+b^2})\{a^2K(\mathrm{r}) - (a^2 - b^2)E(\mathrm{r})\} \\
&= \frac{1}{3}(\sqrt{a^2+b^2})\left[a^2\frac{\pi}{2}\left\{1+\frac{1}{2^2}\mathrm{r}^2\right\} - (a^2 - b^2)\frac{\pi}{2}\left\{1 - \frac{1}{2^2}\mathrm{r}^2\right\}\right] \\
&= \frac{1}{3}(\sqrt{a^2+b^2})\left[\frac{a^2\pi}{2} + \frac{a^2r^2}{4}\left(\frac{\pi}{2}\right) - \frac{a^2\pi}{2} + \frac{a^2r^2}{4}\left(\frac{\pi}{2}\right) + b^2\frac{\pi}{2} - \frac{b^2r^2}{4}\left(\frac{\pi}{2}\right)\right] \\
&= \frac{1}{3}(\sqrt{a^2+b^2})\left(\frac{\pi}{2}\right)\frac{3b^2[2a^2+b^2]}{4(a^2+b^2)} \\
&= \left(\frac{\pi}{2}\right)\frac{b^2[2a^2+b^2]}{4\sqrt{(a^2+b^2)}} = \left(\frac{\pi}{2}\right)\frac{1}{4}\frac{(2a^2b^2+b^4)}{(a^2+b^2)^{1/2}} \\
&= \left(\frac{\pi}{8}\right)\frac{\frac{2\gamma_{10}(E,\eta_g)}{\beta_3} + \left[\frac{1}{2}\sqrt{\left(\frac{\beta_2}{\beta_3}\right)^2 + \frac{4\gamma_{10}(,\eta_g)}{\beta_3}} - \frac{1}{2}\left(\frac{\beta_2}{\beta_3}\right)\right]^2}{\left[\left(\frac{\beta_2}{\beta_3}\right)^2 + \frac{4\gamma_{10}(E,\eta_g)}{\beta_3}\right]^{1/2}} \\
&= \left(\frac{\pi}{8}\right)\frac{\left(\frac{\beta_2}{\beta_3}\right)^2\left[\frac{1}{2} + \frac{3\gamma_{10}(\mathrm{E},\eta_g)\beta_3}{\beta_2^2} - \frac{1}{2}\left\{1+\frac{4\gamma_{10}(\mathrm{E},\eta_g)\beta_3}{\beta_2^2}\right\}^{1/2}\right]}{\left(\frac{\beta_2}{\beta_3}\right)^{1/2}\left[1+\frac{4\gamma_{10}(,\eta_g)\beta_3}{\beta_2^2}\right]^{1/4}}
\end{aligned}
$$

So the general expression of Area (A) is given by

$$
\begin{aligned}
A &= 4\sqrt{\frac{\beta_3}{\beta_1}}\left(\frac{\pi}{8}\right)\left(\frac{\beta_2}{\beta_3}\right)^{3/2}\frac{\left[\frac{1}{2} + \frac{3\gamma_{10}(\mathrm{E},\eta_g)\beta_3}{\beta_2^2} - \frac{1}{2}\left\{1 + \frac{4\gamma_{10}(\mathrm{E},\eta_g)\beta_3}{\beta_2^2}\right\}^{1/2}\right]}{\left[1 + \frac{4\gamma_{10}(\mathrm{E},\eta_g)\beta_3}{\beta_2^2}\right]^{1/4}} \\
&= \left(\frac{\pi}{2}\right)\left(\frac{\beta_2}{\beta_3}\right)\left(\frac{\beta_2}{\beta_1}\right)^{1/2}\frac{\left[\frac{1}{2} + \frac{3\gamma_{10}(E,\eta_g)\beta_3}{\beta_2^2} - \frac{1}{2}\left\{1 + \frac{4\gamma_{10}(E,\eta_g)\beta_3}{\beta_2^2}\right\}^{1/2}\right]}{\left[1 + \frac{4\gamma_{10}(E,\eta_g)\beta_3}{\beta_2^2}\right]^{1/4}}
\end{aligned}
\quad (1.111)
$$

Since

$\frac{4\gamma_{10}(\mathrm{E},\eta_g)\beta_3}{\beta_2^2} < 1$ We can write

$\left[1 + \frac{4\gamma_{10}(\mathrm{E},\eta_g)\beta_3}{\beta_2^2}\right]^{1/4} \cong 1 + \frac{2\gamma_{10}(\mathrm{E},\eta_g)\beta_3}{\beta_2^2}$ and $\left[1 + \frac{4\gamma_{10}(\mathrm{E},\eta_g)\beta_3}{\beta_2^2}\right]^{1/4} \cong 1 + \frac{\gamma_{10}(\mathrm{E},\eta_g)\beta_3}{\beta_2^2}$.

Therefore we can write

$$A = \left(\frac{\pi}{2}\right)\left(\frac{\beta_2}{\beta_3}\right)\left(\frac{\beta_2}{\beta_1}\right)^{1/2}\left[\frac{3\gamma_{10}(\mathrm{E},\eta_g)\beta_3}{\beta_2^2} - \frac{\gamma_{10}(\mathrm{E},\eta_g)\beta_3}{\beta_2^2}\right]\left[1 + \frac{\gamma_{10}(\mathrm{E},\eta_g)\beta_3}{\beta_2^2}\right]^{-1}$$
$$= \pi\left(\frac{\beta_2}{\beta_3}\right)\left(\frac{\beta_2}{\beta_1}\right)^{1/2}\left[\frac{\gamma_{10}(\mathrm{E},\eta_g)\beta_3}{\beta_2^2}\right]\left[1 + \frac{\gamma_{10}(\mathrm{E},\eta_g)\beta_3}{\beta_2^2}\right]^{-1}$$
$$= \pi\left(\frac{\beta_2}{\beta_1}\right)^{1/2}\left(\frac{1}{\beta_2}\right)\gamma_{10}(\mathrm{E},\eta_g)\left[1 + \frac{\gamma_{10}(E,\eta_g)\beta_3}{\beta_2^2}\right]^{-1}$$

Therefore the simplified expression of area can be written as

$$A = \frac{\pi\gamma_{10}(\mathrm{E},\eta_g)}{\sqrt{\beta_1\beta_2}} \cdot \frac{1}{\left[1 + \frac{\gamma_{10}(\mathrm{E},\eta_g)\beta_3}{\beta_2^2}\right]} = \omega_{19,1}(\mathrm{E},\eta_g) \quad (1.112)$$

The EEM and the DOS function can respectively be written in this case as

$$m^*(\mathrm{E},\eta_g,n) = \frac{\hbar^2}{2\pi}\omega'_{19,1}(\mathrm{E},\eta_g) \quad (1.113)$$

$$N_{2DQWHD}(\mathrm{E},\eta_g) = \frac{gv}{(2\pi)^2}\sum_{n=1}^{n_{\max}}\omega'_{19,1}(\mathrm{E},\eta_g)H(\mathrm{E} - \mathrm{E}_{19,1}) \quad (1.114)$$

where $\mathrm{E}_{19,1}$ is the sub band energy and can be obtained from the equation

$$\gamma_{10}(E_{19,1},\eta_g) = 0 \quad (1.115)$$

(b) Model of McClure and Choi

In accordance with this model, the DR in bulk specimens of IV–VI materials can be written as

$$E(1 + \alpha E) = \frac{\hbar^2 k_x^2}{2m_1} + \frac{\hbar^2 k_y^2}{2m_2} + \frac{\hbar^2 k_z^2}{2m_3} + \frac{\alpha\hbar^4 k_y^4}{4m_1 m_2'} + \frac{\hbar^2 k_y^2}{2m_2}\alpha E\left(1 - \frac{m_2}{m_2'}\right) - \frac{\alpha\hbar^4 k_x^2 k_y^2}{4m_1 m_2} - \frac{\alpha\hbar^4 k_y^2 k_z^2}{4m_2 m_3} \quad (1.116)$$

Let us substitute

$$\frac{1}{2m_i} = a_i,\ i = 1,2,3,\quad \frac{1}{2m_2'} = a_2' \text{ and } \left(1 - \frac{m_2}{m_2'}\right) = a_0$$

Therefore (1.116) can be written as

$$E(1+\alpha E) = a_1p_x^2 + a_2p_y^2 + a_3p_z^2 + \alpha a_2a_2'p_y^4 + \alpha a_2a_0p_y^2E - \alpha a_1a_2p_x^2p_y^2 - \alpha a_2a_3p_y^2p_z^2 \tag{1.117}$$

Putting $\alpha = 0$, in (1.117) we get $E = a_1p_x^2 + a_2p_y^2 + a_3p_z^2$.

So, by the method of successive approximation we write

$$E = a_1p_x^2 + a_2p_y^2 + a_3p_z^2 - \alpha[a_1p_x^2 + a_2p_y^2 + a_3p_z^2]^2 + \alpha a_2a_2'p_y^4 + \alpha a_2a_0p_y^2(a_1p_x^2 + a_2p_y^2 + a_3p_z^2) - \alpha a_1a_2p_x^2p_y^2 - \alpha a_2a_3p_y^2p_z^2 \tag{1.118}$$

Choosing Coulomb Gange as

$$\vec{A} = (0,0,-Bx)\text{and}\vec{B}(0,B,0)$$

So that $\vec{\nabla}\vec{A} = 0$ and $\vec{B} = \vec{\nabla}X\vec{A}$.

In presence of perturbing magnetic field $(\vec{B})$ total Hamiltonian $(\hat{H})$ can be written as

$$\begin{aligned}\hat{H} &= a_1\left(-i\hbar\frac{\partial}{\partial x}\right)^2 + a_2\left(-i\hbar\frac{\partial}{\partial y}\right)^2 + a_3\left(-i\hbar\frac{\partial}{\partial z} - eBx\right)^2 \\ &- \alpha\left[a_1\left(-i\hbar\frac{\partial}{\partial x}\right)^2 + a_2\left(-i\hbar\frac{\partial}{\partial y}\right)^2 + a_3\left(-i\hbar\frac{\partial}{\partial z} - eBx\right)^2\right]^2 \\ &+ \alpha a_2a_2'\left(-i\hbar\frac{\partial}{\partial y}\right)^4 + \alpha a_2a_0\left(-i\hbar\frac{\partial}{\partial y}\right)^2\cdot\left[a_1\left(-i\hbar\frac{\partial}{\partial x}\right)^2 + a_2\left(-i\hbar\frac{\partial}{\partial y}\right)^2 + a_3\left(-i\hbar\frac{\partial}{\partial z} - eBx\right)^2\right] \\ &- \alpha a_1a_2\left(-i\hbar\frac{\partial}{}\right)^2\left(-i\hbar\frac{\partial}{}\right)^2 - \alpha a_2a_3\left(-i\hbar\frac{\partial}{\partial y}\right)^2\cdot\left(-i\hbar\frac{\partial}{\partial z} - eBx\right)^2\end{aligned} \tag{1.119}$$

Considering non-parabolicity and magnetic field $(\vec{B})$ as perturbations we can write

$$\hat{H}_0\psi_0 = E_0\psi_0 \tag{1.120}$$

Therefore from (1.120) we get

$$-a_1\hbar^2\frac{\partial^2\psi_0}{\partial x^2} - a_2\hbar^2\frac{\partial^2\psi_0}{\partial y^2} - a_3\hbar^2\frac{\partial^2\psi_0}{\partial z^2} = E_0\psi_0 \tag{1.121}$$

So that

$$E_0 = E_n^0 = \frac{\hbar^2 k_x^2}{2m_1} + \frac{\hbar^2 k_y^2}{2m_2} + \frac{\hbar^2}{2m_3}\left(\frac{n\pi}{a}\right)^2 \tag{1.122}$$

and

$$\psi_0 = \psi_n^0 = \sqrt{\frac{2}{aL_xL_y}}\sin\left(\frac{n\pi}{a}z\right)\exp(ik_xx + ik_yy) \tag{1.123}$$

Considering perturbing effects, the total Hamiltonian can be written as

$$\begin{aligned}\hat{H} = & -a_1\hbar^2\frac{\partial^2}{\partial x^2} - a_2\hbar^2\frac{\partial^2}{\partial y^2} + a_3\left(-\hbar^2\frac{\partial^2}{\partial z^2} + e^2B^2x^2 + 2i\hbar eBx\frac{\partial}{\partial z}\right)\\ & -\alpha\left[-a_1\hbar^2\frac{\partial^2}{\partial x^2} - a_2\hbar^2\frac{\partial^2}{\partial y^2} + a_3\left(-\hbar^2\frac{\partial^2}{\partial z^2} + e^2B^2x^2 + 2i\hbar eBx\frac{\partial}{\partial z}\right)\right]^2\\ & + \alpha a_2a_2'\left(\hbar^4\frac{\partial^4}{\partial x^4}\right) + \alpha a_2a_0\left(-\hbar^2\frac{\partial^2}{\partial y^2}\right)\\ & \left[-a_1\hbar^2\frac{\partial^2}{\partial x^2} - a_2\hbar^2\frac{\partial^2}{\partial y^2} + a_3\left(-\hbar^2\frac{\partial^2}{\partial z^2} + e^2B^2x^2 + 2i\hbar eBx\frac{\partial}{\partial z}\right)\right]\\ & - \alpha a_1a_2\hbar^4\left(\frac{\partial^2}{\partial x^2}\right)\left(\frac{\partial^2}{\partial y^2}\right) - \alpha a_2a_3\left(-\hbar^2\frac{\partial^2}{\partial y^2}\right)\left(-\hbar^2\frac{\partial^2}{\partial z^2} + e^2B^2x^2 + 2i\hbar eBx\frac{\partial}{\partial z}\right)\end{aligned} \tag{1.124}$$

Now

$$\begin{aligned}&\left[-a_1\hbar^2\frac{\partial^2}{\partial x^2} - a_2\hbar^2\frac{\partial^2}{\partial y^2} + a_3\left(-\hbar^2\frac{\partial^2}{\partial z^2} + e^2B^2x^2 + 2i\hbar eBx\frac{\partial}{\partial z}\right)\right]\psi_n = a_1\hbar^2k_x^2\psi_n\\ & + a_2\hbar^2k_y^2\psi_n + a_3\hbar^2\left(\frac{n\pi}{a}\right)^2\psi_n + a_3e^2B^2x^2\psi_n + 2a_3i\hbar eBx\left(\frac{n\pi}{a}\right)\cot\left(\frac{n\pi}{a}z\right)\psi_n\end{aligned}$$

$$\begin{aligned}&\left[-a_1\hbar^2\frac{\partial^2}{\partial x^2} - a_2\hbar^2\frac{\partial^2}{\partial y^2} + a_3\left(-\hbar^2\frac{\partial^2}{\partial z^2} + e^2B^2x^2 + 2i\hbar eBx\frac{\partial}{\partial z}\right)\right]^2\psi_n\\ = & \left[a_1\hbar^2k_x^2 + a_2\hbar^2k_y^2 + a_3\hbar^2\left(\frac{n\pi}{a}\right)^2\right]^2\psi_n - 2a_1a_3\hbar^2e^2B^2\psi_n\\ & + 4a_3i\left[a_1\hbar^2k_x^2 + a_2\hbar^2k_y^2 + a_3\hbar^2\left(\frac{n\pi}{a}\right)^2\right]\left(\frac{n\pi}{a}\right)x\cot\left(\frac{n\pi}{a}z\right)\psi_n\\ & + \left[2a_1a_3\hbar^2k_x^2 + 2a_2a_3\hbar^2k_y^2 + 6a_3^2\hbar^2\left(\frac{n\pi}{a}\right)^2\right]e^2B^2x^2\psi_n\\ & + 4a_3^2i\hbar e^3B^3\left(\frac{n\pi}{a}\right)x^3\cot\left(\frac{n\pi}{a}z\right)\psi_n + a_3^2e^4B^4x^4\psi_n\end{aligned}$$

$$
\begin{aligned}
&\hbar^4 \frac{\partial^4}{\partial y^4}(\psi_n) = \hbar^4 (ik_y)^4 \psi_n = \hbar^4 k_y^4 \psi_n \left(-\hbar^2 \frac{\partial^2}{\partial y^2}\right) \\
&\left[a_1 \hbar^2 k_x^2 \psi_n + a_2 \hbar^2 k_y^2 \psi_n + a_3 \hbar^2 \left(\frac{n\pi}{a}\right)^2 \psi_n + a_3 e^2 B^2 x^2 \psi_n + 2a_3 i\hbar eBx\left(\frac{n\pi}{a}\right)\cot\left(\frac{n\pi}{a}z\right)\psi_n\right] \\
&= \hbar^2 k_y^2 \left[a_1 \hbar^2 k_x^2 + a_2 \hbar^2 k_y^2 + a_3 \hbar^2 \left(\frac{n\pi}{a}\right)^2\right]\psi_n + a_3 \hbar^2 k_y^2 e^2 B^2 x^2 \psi_n \\
&\quad + 2a_3 i\hbar^2 k_y^2 eBx\left(\frac{n\pi}{a}\right)\cot\left(\frac{n\pi}{a}z\right)\psi_n \hbar^4 \left(\frac{\partial^2}{2}\right)\left(\frac{\partial^2}{\partial y^2}\right)\psi_n \\
&\quad = \hbar^4 \frac{\partial^2}{\partial x^2}\left[(ik_y)^2 \psi_n\right] = \hbar^4 (ik_y)^2 (ik_x)^2 \psi_n = \hbar^4 k_x^2 k_y^2 \psi_n \\
&\left(-\hbar^2 \frac{\partial^2}{\partial y^2}\right)\left[-\hbar^2 \frac{\partial^2}{\partial z^2} + e^2 B^2 x^2 + 2i\hbar eBx \frac{\partial}{\partial z}\right]\psi_n \\
&\quad = -\hbar^2 \frac{\partial^2}{\partial y^2}\left[\hbar^2 \left(\frac{n\pi}{a}\right)^2 \psi_n + e^2 B^2 x^2 \psi_n + 2i\left(\frac{n\pi}{a}\right)\cot\left(\frac{n\pi}{a}z\right)\psi_n\right] \\
&\quad = \hbar^2 k_y^2 \left[\hbar^2 \left(\frac{n\pi}{a}\right)^2 + e^2 B^2 x^2 + 2i\hbar eBx\left(\frac{n\pi}{a}\right)\cot\left(\frac{n\pi}{a}z\right)\right]\psi_n
\end{aligned}
\tag{1.125}
$$

Therefore

$$
\begin{aligned}
\hat{H}\psi_n &= a_1 \hbar^2 k_x^2 \psi_n + a_2 \hbar^2 k_y^2 \psi_n + a_3 \hbar^2 \left(\frac{n\pi}{a}\right)^2 \psi_n \\
&\quad + a_3 e^2 B^2 x^2 \psi_n + 2a_3 i\hbar eB\left(\frac{n\pi}{a}\right)x\cot\left(\frac{n\pi}{a}z\right)\psi_n \\
&\quad - \alpha\left[a_1 \hbar^2 k_x^2 + a_2 \hbar^2 k_y^2 + a_3 \hbar^2 \left(\frac{n\pi}{a}\right)^2\right]^2 \psi_n + 2_1 a_3 \hbar^2 e^2 B^2 \psi_n \\
&\quad - 4\alpha a_3 i\hbar eB\left[a_1 \hbar^2 k_x^2 + a_2 \hbar^2 k_y^2 + a_3 \hbar^2 \left(\frac{n\pi}{a}\right)^2\right] \\
&\quad \left(\frac{n\pi}{a}\right)x\cot\left(\frac{n\pi}{a}z\right)\psi_n - \alpha\left[2a_1 a_3 \hbar^2 k_x^2 + 2a_2 a_3 \hbar^2 k_y^2\right. \\
&\quad \left.+ 6a_3^2 \hbar^2 \left(\frac{n\pi}{a}\right)^2\right] e^2 B^2 x^2 \psi_n - 4\alpha i\hbar a_3^2 e^3 B^3 \left(\frac{n\pi}{a}\right)x^3 \cot\left(\frac{n\pi}{a}z\right)\psi_n \\
&\quad - \alpha a_3^2 e^4 B^4 x^4 \psi_n + \alpha a_2 a_2' \hbar^4 k_y^4 \psi_n + \alpha a_2 a_0 \hbar^2 k_y^2 \\
&\quad \left[a_1 \hbar^2 k_x^2 + a_2 \hbar^2 k_y^2 + a_3 \hbar^2 \left(\frac{n\pi}{a}\right)^2\right]\psi_n + {}_2\, a_0 a_3 \hbar^2 k_y^2 e^2 B^2 x^2 \psi_n \\
&\quad + 2\alpha a_2 a_0 a_3 i\hbar^3 k_y^2 eB\left(\frac{n\pi}{a}\right)x\cot\left(\frac{n\pi}{a}z\right)\psi_n - \alpha a_1 a_2 \hbar^4 k_x^2 k_y^2 \psi_n \\
&\quad - \alpha a_2 a_3 \hbar^4 k_y^2 \left(\frac{n\pi}{a}\right)^2 \psi_n - \alpha a_2 a_3 \hbar^2 k_y^2 e^2 B^2 x^2 \psi_n \\
&\quad - 2\alpha a_2 a_3 i\hbar^3 k_y^2 eB\left(\frac{n\pi}{a}\right)x\cot\left(\frac{n\pi}{a}z\right)\psi_n = \hat{H}_0 \psi_n + \hat{H}' \psi_n
\end{aligned}
\tag{1.126}
$$

where $\hat{H}_0$ is the unperturbed Hamiltonian $= a_1\hbar^2k_x^2 + a_2\hbar^2k_y^2 + a_3\hbar^2\left(\frac{n\pi}{a}\right)^2$.

Applying first order perturbation we get

$$\left\langle x\cot\left(\frac{n\pi}{a}z\right)\right\rangle = 0, \quad \left\langle x^3\cot\left(\frac{n\pi}{a}z\right)\right\rangle = 0, \quad \langle x^2\rangle = \frac{L_x^2}{3} \text{ and } \langle x^4\rangle = \frac{L_x^4}{5}$$

Now, $a_2a_0 = \frac{1}{2m_2}\left(1 - \frac{m_2}{m_2'}\right) = \frac{1}{2m_2} - \frac{1}{2m_2'} = (a_2 - a_2')$.

Therefore the DR in QWs of HD IV–VI semiconductors in the presence of magnetic field whose conduction electrons obey the model of McClure and Choican be written as

$$\begin{aligned}
\gamma_3(E,\eta_g) = {} & a_1\hbar^2k_x^2 + a_2\hbar^2k_y^2 + a_3\hbar^2\left(\frac{n\pi}{a}\right)^2 - \alpha\left[a_1\hbar^2k_x^2 + a_2\hbar^2k_y^2 + a_3\hbar^2\left(\frac{n\pi}{a}\right)^2\right]^2 \\
& + \alpha a_2a_2'\hbar^4k_y^4 + \alpha a_2\hbar^2k_y^2\left[a_1\hbar^2k_x^2 + a_2\hbar^2k_y^2 + a_3\hbar^2\left(\frac{n\pi}{a}\right)^2\right] \\
& - \alpha a_2'k_x^2k_y^2\left[a_1\hbar^2k_x^2 + a_2\hbar^2k_y^2 + a_3\hbar^2\left(\frac{n\pi}{a}\right)^2\right] \\
& - \alpha a_1a_2\hbar^4k_x^2k_y^2 - \alpha a_2a_3\hbar^4k_y^2\left(\frac{n\pi}{a}\right)^2 + a_3e^2B^2\langle x^2\rangle \\
& + 2\alpha a_1a_3\hbar^2e^2B^2 - \alpha a_3^2e^4B^4\langle x^4\rangle \\
& - \alpha[\left[2a_1a_3\hbar^2k_x^2 + 2a_2a_3\hbar^2k_y^2 + 6a_3^2\hbar^2\left(\frac{n\pi}{a}\right)^2\right]e^2B^2\langle x^2\rangle \\
& - \alpha a_2'a_3\hbar^2k_y^2e^2B^2\langle x^2\rangle
\end{aligned} \tag{1.127}$$

The use of (1.127) leads to the expressions of EEM and DOS, the derivations of which are left for the readers.

1.2.8 *The DR in Nano Wires of HD IV–VI Semiconductors in the Presence of Magnetic Field*

(a) McClure and Choi Model

In nano-wire let us assume the free electron motion is along y-direction and the magnetic field B is also applied along y-direction. Following (1.118) the Hamiltonian can be expressed as

$$\begin{aligned}\hat{H} &= a_1\left(-i\hbar\frac{\partial}{\partial x}\right)^2 + a_2\left(-i\hbar\frac{\partial}{\partial y}\right)^2 + a_3\left(-i\hbar\frac{\partial}{\partial z} - eBx\right)^2 \\ &\quad - \alpha\left[a_1\left(-i\hbar\frac{\partial}{\partial x}\right)^2 + a_2\left(-i\hbar\frac{\partial}{\partial y}\right)^2 + a_3\left(-i\hbar\frac{\partial}{\partial z} - eBx\right)^2\right]^2 \\ &\quad + \alpha a_2 a_2'\left(-i\hbar\frac{\partial}{\partial y}\right)^4 \alpha a_2 a_0\left(-i\hbar\frac{\partial}{\partial y}\right)^2 \cdot \\ &\quad \left[a_1\left(-i\hbar\frac{\partial}{\partial x}\right)^2 + a_2\left(-i\hbar\frac{\partial}{\partial y}\right)^2 + a_3\left(-i\hbar\frac{\partial}{\partial z} - eBx\right)^2\right] \\ &\quad - \alpha a_1 a_2'\left(-i\hbar\frac{\partial}{\partial y}\right)^2\left(-i\hbar\frac{\partial}{\partial y}\right)^2 - \alpha a_2 a_3\left(-i\hbar\frac{\partial}{\partial y}\right)^2\left[-i\hbar\frac{\partial}{\partial z} - eBx\right]^2\end{aligned} \tag{1.128}$$

Considering non-parabolicity and magnetic field as perturbed, unperturbed wave equation is given by

$$\hat{H}_0\psi_0 = E_0\psi_0 \tag{1.129}$$

The (1.129) can be written as

$$-a_1\hbar^2\frac{\partial^2\psi_0}{\partial x^2} - a_2\hbar^2\frac{\partial^2\psi_0}{\partial y^2} - a_3\hbar^2\frac{\partial^2\psi_0}{\partial z^2} = E_0\psi_0 \tag{1.130}$$

Therefore

$$E_0 = E_{nl} = \frac{\hbar^2}{2m_1}\left(\frac{l\pi}{b}\right)^2 + \frac{\hbar^2 k_y^2}{2m_2} + \frac{\hbar^2}{2m_3}\left(\frac{n\pi}{a}\right)^2 \tag{1.131}$$

and

$$\psi^0 = \psi_0 = \psi_{nl} = \sqrt{\frac{4}{abL_y}}\sin\left(\frac{l\pi}{b}\mathrm{x}\right)\sin\left(\frac{n\pi}{a}z\right)\exp(ik_y y) \tag{1.132}$$

Including perturbing effect, the total Hamiltonian can be written as

$$\hat{H} = \hat{H}_0 + \hat{H}' = -a_1\hbar^2\frac{\partial^2}{\partial x^2} - -a_2\hbar^2\frac{\partial^2}{\partial y^2} + a_3\left(-\hbar^2\frac{\partial^2}{\partial z^2} + e^2B^2x^2 + 2i\hbar eBx\frac{\partial}{\partial z}\right)$$
$$-\alpha\left[-a_1\hbar^2\frac{\partial^2}{\partial x^2} - a_2\hbar^2\frac{\partial^2}{\partial y^2} + a_3\left(-\hbar^2\frac{\partial^2}{\partial z^2} + e^2B^2x^2 + 2i\hbar eBx\frac{\partial}{\partial z}\right)\right]^2$$
$$+\alpha a_2 a_2'\left(-\hbar^4\frac{\partial^4}{\partial y^4}\right) + \alpha a_2 a_0\left(-\hbar^2\frac{\partial^2}{\partial y^2}\right)$$
$$\left[-a_1\hbar^2\frac{\partial^2}{\partial x^2} - a_2\hbar^2\frac{\partial^2}{\partial y^2} + a_3\left(-\hbar^2\frac{\partial^2}{\partial z^2} + e^2B^2x^2 + 2i\hbar eBx\frac{\partial}{\partial z}\right)\right]$$
$$-\alpha a_1a_2\hbar^4\left(\frac{\partial^2}{\partial x^2}\right)\left(\frac{\partial^2}{\partial y^2}\right) - \alpha a_2a_3\left(-\hbar^2\frac{\partial^2}{\partial y^2}\right)\left(-\hbar^2\frac{\partial^2}{\partial z^2} + e^2B^2x^2 + 2i\hbar eBx\frac{\partial}{\partial z}\right) \tag{1.133}$$

Now $\left[-a_1\hbar^2\frac{\partial^2}{\partial x^2} - a_2\hbar^2\frac{\partial^2}{\partial y^2} + a_3\left(-\hbar^2\frac{\partial^2}{\partial z^2} + e^2B^2x^2 + 2i\hbar eBx\frac{\partial}{\partial z}\right)\right]\psi_{nl}^0$

$$= a_1\hbar^2\left(\frac{l\pi}{b}\right)\psi_{nl}^0 + a_2\hbar^2k_y^2\psi^0 + a_3\hbar^2\left(\frac{n\pi}{a}\right)^2\psi^0$$
$$+ a_3e^2B^2x^2\psi^0 + 2a_3i\hbar eBx\left(\frac{n\pi}{a}\right)\cot\left(\frac{n\pi}{a}z\right)\psi^0 \tag{1.134}$$

Therefore

$$\left[-a_1\hbar^2\frac{\partial^2}{\partial x^2} - a_2\hbar^2\frac{\partial^2}{\partial y^2} + a_3\left(-\hbar^2\frac{\partial^2}{\partial z^2} + e^2B^2x^2 + 2i\hbar eBx\frac{\partial}{\partial z}\right)\right]^2\psi_{nl}^0$$
$$= \left[a_1\hbar^2\left(\frac{l\pi}{b}\right)^2 + a_2\hbar^2k_y^2 + a_3\hbar^2\left(\frac{n\pi}{a}\right)^2\right]^2\psi^0 - 2a_1a_3\hbar^2e^2B^2\psi^0$$
$$+4a_3i\hbar eB\left[a_1\hbar^2\left(\frac{l\pi}{b}\right)^2 + a_2\hbar^2k_y^2 + a_3\hbar^2\left(\frac{n\pi}{a}\right)^2\right]\left(\frac{n\pi}{a}\right)x\cot\left(\frac{n\pi}{a}z\right)\psi^0$$
$$+\left[2a_1a_3\hbar^2\left(\frac{l\pi}{b}\right)^2 + 2a_2a_3\hbar^2k_y^2 + 6a_3^2\hbar^2\left(\frac{n\pi}{a}\right)^2\right]e^2B^2x^2\psi^0$$
$$+4a_3^2i\hbar e^3B^3\left(\frac{n\pi}{a}\right)x^3\cot\left(\frac{n\pi}{a}z\right)\psi^0 + a_3^2e^4B^4x^4\psi^0 \tag{1.135}$$

$$\hbar^4 \frac{\partial^4}{\partial y^4}(\psi^0) = \hbar^4 (ik_y)^4 \psi^0 = \hbar^4 k_y^4 \psi^0$$

$$\hbar^4 \left(\frac{\partial^2}{\partial x^2}\right)\left(\frac{\partial^2}{\partial y^2}\right)\psi^0 = \hbar^4 \frac{\partial^2}{\partial x^2}\left[(ik_y)^2 \psi^0\right]$$

$$= \hbar^4 (-k_y^2)\left[-\left(\frac{l\pi}{b}\right)^2 \psi^0\right] = \hbar^4 k_y^2 \left(\frac{l\pi}{b}\right)^2 \psi^0$$

So

$$\hat{H}\psi_{nl}^0 = \hat{H}_0 \psi_{nl}^0 + \hat{H}' \psi_{nl}^0 \tag{1.136a}$$

Therefore

$$\begin{aligned}
\hat{H}\psi_{nl}^0 &= a_1 \hbar^2 \left(\frac{l\pi}{b}\right)^2 \psi^0 + a_2 \hbar^2 k_y^2 \psi^0 + a_3 \hbar^2 \left(\frac{n\pi}{a}\right)^2 \psi^0 + a_3 e^2 B^2 x^2 \psi^0 \\
&+ 2a_3 i\hbar eB\left(\frac{n\pi}{a}\right) x \cot\left(\frac{n\pi}{a} z\right)\psi^0 - \alpha\left[a_1 \hbar^2 \left(\frac{l\pi}{b}\right)^2\right. \\
&\left. + a_2 \hbar^2 k_y^2 + a_3 \hbar^2 \left(\frac{n\pi}{a}\right)^2\right]^2 \psi^0 + 2\alpha a_1 a_3 \hbar^2 e^2 B^2 \psi^0 - 4\alpha a_3 i\hbar eB \\
&\left[a_1 \hbar^2 \left(\frac{l\pi}{b}\right)^2 + a_2 \hbar^2 k_y^2 + a_3 \hbar^2 \left(\frac{n\pi}{a}\right)^2\right]\left(\frac{n\pi}{a}\right) x \cot\left(\frac{n\pi}{a} z\right)\psi^0 \\
&- \alpha\left[2a_1 a_3 \hbar^2 \left(\frac{l\pi}{b}\right)^2 + 2a_2 a_3 \hbar^2 k_y^2 + 6a_3^2 \hbar^2 \left(\frac{n\pi}{a}\right)^2\right] e^2 B^2 x^2 \psi^0 \\
&- 4\alpha a_3^2 i\hbar e^3 B^3 \left(\frac{n\pi}{a}\right) x^3 \cot\left(\frac{n\pi}{a} z\right)\psi^0 - \alpha a_3^2 e^4 B^4 x^4 \psi^0 \\
&+ \alpha a_2 a_2' \hbar^4 k_y^4 \psi^0 + \alpha a_2 a_0 \hbar^2 k_y^2 \left[a_1 \hbar^2 \left(\frac{l\pi}{b}\right)^2 + a_2 \hbar^2 k_y^2 + a_3 \hbar^2 \left(\frac{n\pi}{a}\right)^2\right]\psi^0 \\
&+ \alpha a_2 a_0 a_3 \hbar^2 k_y^2 e^2 B^2 x^2 \psi^0 + 2\alpha a_2 a_0 a_3 i\hbar^3 k_y^2 eB\left(\frac{n\pi}{a}\right) \\
&\cdot x \cot\left(\frac{n\pi}{a} z\right)\psi^0 - \alpha a_1 a_2 \hbar^4 k_y^2 \left(\frac{l\pi}{b}\right)^2 \psi^0 - \alpha a_2 a_3 \hbar^4 k_y^2 \left(\frac{n\pi}{a}\right)^2 \psi^0 \\
&- \alpha a_2 a_3 \hbar^2 k_y^2 e^2 B^2 x^2 \psi^0 - 2\alpha a_2 a_3 i\hbar^3 k_y^2 eB\left(\frac{n\pi}{a}\right) x \cot\left(\frac{n\pi}{a} z\right)\psi^0
\end{aligned} \tag{1.136b}$$

Applying first order perturbation we can write

$$\left\langle x\cot\left(\frac{n\pi}{a}z\right)\right\rangle = 0,\ \left\langle x^3\cot\left(\frac{n\pi}{a}z\right)\right\rangle = 0,\ \langle x^2\rangle = \frac{2}{b}\int_0^b x^2\sin^2\left(\frac{l\pi}{b}\right)x dx = \left(\frac{b^2}{3} - \frac{b^2}{2l^2\pi^2}\right) \text{ and}$$

$$\langle x^4\rangle = \frac{2}{b}\int_0^b x^4\sin^2\left(\frac{l\pi}{b}\right)x dx = \left(\frac{b^4}{5} - \frac{b^4}{l^2\pi^2} + \frac{3b^4}{2l^4\pi^4}\right)$$

The DR in this case is given by

$$\begin{aligned}
E = {} & a_1\hbar^2\left(\frac{l\pi}{b}\right)^2 + a_2\hbar^2k_y^2 + a_3\hbar^2\left(\frac{n\pi}{a}\right)^2 \\
& - \alpha\left[a_1\hbar^2\left(\frac{l\pi}{b}\right)^2 + a_2\hbar^2k_y^2 + a_3\hbar^2\left(\frac{n\pi}{a}\right)^2\right]^2 \\
& + \alpha a_2a_2'\hbar^4k_y^4 + \alpha a_2a_0\hbar^2k_y^2\left[a_1\hbar^2\left(\frac{l\pi}{b}\right)^2 + a_2\hbar^2k_y^2 + a_3\hbar^2\left(\frac{n\pi}{a}\right)^2\right] \\
& - \alpha a_1a_2\hbar^4\left(\frac{l\pi}{b}\right)^2k_y^2 - \alpha a_2a_3\hbar^4k_y^2\left(\frac{n\pi}{a}\right)^2 + a_3e^2B^2\langle x^2\rangle \\
& + 2\alpha a_1a_3\hbar^2e^2B^2 - \alpha a_3^2e^4B^4\langle x^4\rangle \\
& - \left[2a_1a_3\hbar^2\left(\frac{l\pi}{b}\right)^2 + 2a_2a_3\hbar^2k_y^2 + 6a_3^2\hbar^2\left(\frac{n\pi}{a}\right)^2\right]e^2B^2\langle x^2\rangle \\
& + \alpha a_2a_0a_3\hbar^2k_y^2e^2B^2\langle x^2\rangle - \alpha a_2a_3\hbar^2k_y^2e^2B^2\langle x^2\rangle
\end{aligned} \tag{1.137}$$

Now $a_2a_0 = (a_2 - a_2')$.

Therefore the DR in NWs of HD IV–VI semiconductors in the presence of magnetic field in the present case can be written as

$$\begin{aligned}
\gamma_3(E,\eta_g) = {} & a_1\hbar^2\left(\frac{l\pi}{b}\right)^2 + a_2\hbar^2k_y^2 + a_3\hbar^2\left(\frac{n\pi}{a}\right)^2 \\
& - \alpha\left[a_1\hbar^2\left(\frac{l\pi}{b}\right)^2 + a_2\hbar^2k_y^2 + a_3\hbar^2\left(\frac{n\pi}{a}\right)^2\right]^2 + 2a_2'\hbar^4k_y^4 \\
& + \alpha a_2a_0\hbar^2k_y^2\left[a_1\hbar^2\left(\frac{l\pi}{b}\right)^2 + a_2\hbar^2k_y^2 + a_3\hbar^2\left(\frac{n\pi}{a}\right)^2\right] \\
& - \alpha a_1a_2\hbar^4\left(\frac{l\pi}{b}\right)^2k_y^2 - \alpha a_2a_3\hbar^4k_y^2\left(\frac{n\pi}{a}\right)^2 + a_3e^2B^2\langle x^2\rangle \\
& + 2\alpha a_1a_3\hbar^2e^2B^2 - \alpha a_3^2e^4B^4\langle x^4\rangle \\
& - \left[2a_1a_3\hbar^2\left(\frac{l\pi}{b}\right)^2 + 2a_2a_3\hbar^2k_y^2 + 6a_3^2\hbar^2\left(\frac{n\pi}{a}\right)^2\right]e^2B^2\langle x^2\rangle \\
& + \alpha a_2a_0a_3\hbar^2k_y^2e^2B^2\langle x^2\rangle - \alpha a_2a_3\hbar^2k_y^2e^2B^2\langle x^2\rangle
\end{aligned} \tag{1.138}$$

The use of (1.138) leads to the expressions of EEM and DOS function, the derivations of which are left to the readers.

(b) Cohen model

The total Hamiltonian be

$$\begin{aligned}
\hat{H} = {} & a_1\left(-i\hbar\frac{\partial}{\partial x}\right)^2 + a_2\left(-i\hbar\frac{\partial}{\partial y}\right)^2 + a_3\left(-i\hbar\frac{\partial}{\partial z} - eBx\right)^2 \\
& - \alpha\left[a_1\left(-i\hbar\frac{\partial}{\partial x}\right)^2 + a_2\left(-i\hbar\frac{\partial}{\partial y}\right)^2 + a_3\left(-i\hbar\frac{\partial}{\partial z} - eBx\right)^2\right]^2 \\
& + \alpha a_2\left(-i\hbar\frac{\partial}{\partial y}\right)^2\left[a_1\left(-i\hbar\frac{\partial}{\partial x}\right)^2 + a_2\left(-i\hbar\frac{\partial}{\partial y}\right)^2 + a_3\left(-i\hbar\frac{\partial}{\partial z} - eBx\right)^2\right] \\
& - \alpha a_2'\left(-i\hbar\frac{\partial}{\partial y}\right)^2\left[a_1\left(-i\hbar\frac{\partial}{\partial x}\right)^2 + a_2\left(-i\hbar\frac{\partial}{\partial y}\right)^2 + a_3\left(-i\hbar\frac{\partial}{\partial z} - eBx\right)^2\right] + \alpha a_2a_2'\left(-i\hbar\frac{\partial}{\partial y}\right)^4
\end{aligned} \tag{1.139}$$

Considering non-parabolicity and magnetic field as perturbed, the unperturbed wave equation is

$$\hat{H}_0\psi_0 = E_0\psi_0 \tag{1.140}$$

$$-a_1\hbar^2\frac{\partial^2\psi_0}{\partial x^2} - a_2\hbar^2\frac{\partial^2\psi_0}{\partial y^2} - a_3\hbar^2\frac{\partial^2\psi_0}{\partial z^2} = E_0\psi_0$$

So that

$$E_0 = E_{nl}^{(0)} = \frac{\hbar^2}{2m_1}\left(\frac{l\pi}{b}\right)^2 + \frac{\hbar^2 k_y^2}{2m_2} + \frac{\hbar^2}{2m_3}\left(\frac{n\pi}{a}\right)^2 \tag{1.141}$$

$$\psi_0 = \psi_{nl}^0 = \sqrt{\frac{4}{abL_y}}\sin\left(\frac{l\pi}{b}x\right)\sin\left(\frac{n\pi}{a}z\right)\exp(ik_y y) \tag{1.142}$$

Including perturbing effect, the total Hamiltonian can be expressed as

$$\begin{aligned}\hat{H} = \hat{H}_0 + \hat{H}' = & -a_1\hbar^2\frac{\partial^2}{\partial x^2} - a_2\hbar^2\frac{\partial^2}{\partial y^2} + a_3\left(-\hbar^2\frac{\partial^2}{\partial z^2} + e^2B^2x^2 + 2i\hbar eBx\frac{\partial}{\partial z}\right)\\ & - \alpha\left[-a_1\hbar^2\frac{\partial^2}{2} - a_2\hbar^2\frac{\partial^2}{2} + a_3\left(-\hbar^2\frac{\partial^2}{\partial z^2} + e^2B^2x^2 + 2i\frac{\partial}{\partial z}\right)\right]^2\\ & + \alpha a_2\left(-\hbar^2\frac{\partial^2}{2}\right)\left[-a_1\hbar^2\frac{\partial^2}{\partial x^2} - a_2\hbar^2\frac{\partial^2}{\partial y^2} + a_3\left(-\hbar^2\frac{\partial^2}{\partial z^2} + e^2B^2x^2 + 2i\hbar eBx\frac{\partial}{\partial z}\right)\right]\\ & - \alpha a_2'\left(-\hbar^2\frac{\partial^2}{\partial y^2}\right)\left[-a_1\hbar^2\frac{\partial^2}{\partial x^2} - a_2\hbar^2\frac{\partial^2}{\partial y^2} + a_3\left(-\hbar^2\frac{\partial^2}{\partial z^2} + e^2B^2x^2 + 2i\hbar eBx\frac{\partial}{\partial z}\right)\right]\\ & + \alpha a_2 a_2'\left(\hbar^4\frac{\partial^4}{\partial y^4}\right)\end{aligned} \tag{1.143}$$

We can write

$$\begin{aligned}&\left[-a_1\hbar^2\frac{\partial^2}{\partial x^2} - a_2\hbar^2\frac{\partial^2}{\partial y^2} + a_3\left(-\hbar^2\frac{\partial^2}{\partial z^2} + e^2B^2x^2 + 2i\hbar eBx\frac{\partial}{\partial z}\right)\right]\psi_n^0\\ &= a_1\hbar^2\left(\frac{l\pi}{b}\right)^2\psi_n + a_2\hbar^2k_y^2\psi_n + a_3\hbar^2\left(\frac{n\pi}{a}\right)^2\psi_n + a_3e^2B^2x^2\psi_n\\ &\quad + 2a_3i\hbar eBx\left(\frac{n\pi}{a}\right)cot\left(\frac{n\pi}{a}z\right)\psi_n\end{aligned} \tag{1.144}$$

$$\left[-a_1\hbar^2\frac{\partial^2}{\partial x^2} - a_2\hbar^2\frac{\partial^2}{\partial y^2} + a_3\left(-\hbar^2\frac{\partial^2}{\partial z^2} + e^2B^2x^2 + 2i\hbar eBx\frac{\partial}{\partial z}\right)\right]^2\psi_n^0$$

$$= \left[a_1^4\hbar^4\left(\frac{l\pi}{b}\right)^4\psi_n + a_1a_2\hbar^4\left(\frac{l\pi}{b}\right)^2k_y^2\psi_n + a_1a_3\hbar^4\left(\frac{l\pi}{b}\right)^2\left(\frac{n\pi}{a}\right)^2\psi_n\right.$$

$$+ a_1a_3\hbar^2\left(\frac{l\pi}{b}\right)^2e^2B^2x^2\psi_n - 2a_1a_3\hbar^2e^2B^2\psi_n$$

$$+ 2a_1a_3i\hbar^3eBx\left(\frac{l\pi}{b}\right)^2\left(\frac{n\pi}{a}\right)\cot\left(\frac{n\pi}{a}z\right)\psi_n + a_2a_1\hbar^4\left(\frac{l\pi}{b}\right)^2k_y^2\psi_n + a_2^2\hbar^4k_y^4\psi_n$$

$$+ a_2a_3\hbar^4\left(\frac{n\pi}{a}\right)^2k_y^2\psi_n + a_2a_3e^2B^2x^2\hbar^2k_y^2\psi_n$$

$$+ 2a_2a_3i\hbar^3k_y^2eB\left(\frac{n\pi}{a}\right)x\cot\left(\frac{n\pi}{a}z\right)\psi_n + a_3a_1\hbar^4\left(\frac{l\pi}{b}\right)^2\left(\frac{n\pi}{a}\right)^2\psi_n$$

$$+ a_3a_2\hbar^4\left(\frac{n\pi}{a}\right)^2k_y^2\psi_n + a_3^2\hbar^4\left(\frac{n\pi}{a}\right)^4\psi_n + a_3^2\hbar^2\left(\frac{n\pi}{a}\right)^2e^2B^2x^2\psi_n$$

$$+ 2a_3^2i\hbar^3eB\left(\frac{n\pi}{a}\right)^3x\cot\left(\frac{n\pi}{a}z\right)\psi_n$$

$$+ \left[a_3a_1\hbar^2\left(\frac{l\pi}{b}\right)^2 + a_3a_2\hbar^2k_y^2 + a_3^2\hbar^2\left(\frac{n\pi}{a}\right)^2\right]e^2B^2x^2\psi_n$$

$$+ a_3^2e^4B^4x^4\psi_n + 2a_3^2i\hbar e^3B^3\left(\frac{n\pi}{a}\right)x^3\cot\left(\frac{n\pi}{a}z\right)\psi_n$$

$$+ 2a_3i\hbar eB\left[a_1\hbar^2\left(\frac{l\pi}{b}\right)^2 + a_2\hbar^2k_y^2 + a_3\hbar^2\left(\frac{n\pi}{a}\right)^2\right]\left(\frac{n\pi}{a}\right)x\cot\left(\frac{n\pi}{a}z\right)\psi_n$$

$$+ 2a_3^2i\hbar e^3B^3x^3\left(\frac{n\pi}{a}\right)\cot\left(\frac{n\pi}{a}z\right)\psi_n + 4a_3^2\hbar^2e^2B^2\left(\frac{n\pi}{a}\right)^2x^2\psi_n$$

$$= \left[a_1\hbar^2\left(\frac{l\pi}{b}\right)^2 + a_2\hbar^2k_y^2 + a_3\hbar^2\left(\frac{n\pi}{a}\right)^2\right]^2\psi_n - 2a_1a_3\hbar^2e^2B^2\psi_n$$

$$+ 4a_3i\hbar eB\left[a_1\hbar^2\left(\frac{l\pi}{b}\right)^2 + a_2\hbar^2k_y^2 + a_3\hbar^2\left(\frac{n\pi}{a}\right)^2\right]$$

$$\left(\frac{n\pi}{a}\right)x\cot\left(\frac{n\pi}{a}z\right)\psi_n + \left[2a_1a_3\hbar^2\left(\frac{l\pi}{b}\right)^2 + 2a_2a_3\hbar^2k_y^2 + 6a_3^2\hbar^2\left(\frac{n\pi}{a}\right)^2\right]e^2B^2x^2\psi_n$$

$$+ 4a_3^2i\hbar e^3B^3\left(\frac{n\pi}{a}\right)x^3\cot\left(\frac{n\pi}{a}z\right)\psi_n + a_3^2e^4B^4x^4\psi_n$$

$$-\left(-\hbar^2\frac{\partial^2}{\partial y^2}\right)\left[-a_1\hbar^2\frac{\partial^2}{2} - a_2\hbar^2\frac{\partial^2}{2} + a_3\left(-\hbar^2\frac{\partial^2}{\partial z^2} + e^2B^2x^2 + 2i\frac{\partial}{\partial z}\right)\right]\psi_n$$

$$= a_1\hbar^4\left(\frac{l\pi}{b}\right)^2k_y^2\psi_n + a_2\hbar^4k_y^4\psi_n + a_3\hbar^4\left(\frac{n\pi}{a}\right)^2k_y^2\psi_n + a_3\hbar^2e^2B^2k_y^2x^2\psi_n$$

$$+ 2a_3i\hbar^3k_y^2eBx\left(\frac{n\pi}{a}\right)\cot\left(\frac{n\pi}{a}z\right)\psi_n$$

Again

$$\alpha a_2a_2'\hbar^4\frac{\partial^4}{\partial y^4}\psi_n = \alpha a_2a_2'\hbar^4k_y^4\psi_n \tag{1.145}$$

So

$$\hat{H}\psi_n = \hat{H}_0\psi_n + \hat{H}'\psi_n \tag{1.146}$$

$$= a_1\hbar^2\left(\frac{l\pi}{b}\right)^2\psi_n + a_2\hbar^2k_y^2\psi_n + a_3\hbar^2\left(\frac{n\pi}{a}\right)^2\psi_n + a_3e^2B^2x^2\psi_n$$

$$+ 2a_3i\hbar eB\left(\frac{n\pi}{a}\right)x\cot\left(\frac{n\pi}{a}z\right)\psi_n - \alpha\left[a_1\hbar^2\left(\frac{l\pi}{b}\right)^2 + a_2\hbar^2k_y^2\right.$$

$$\left.+ a_3\hbar^2\left(\frac{n\pi}{a}\right)^2\right]^2\psi_n + 2\alpha a_1a_3\hbar^2e^2B^2\psi_n - 4\alpha a_3i\hbar eB$$

$$\left[a_1\hbar^2\left(\frac{l\pi}{b}\right)^2 + a_2\hbar^2k_y^2 + a_3\hbar^2\left(\frac{n\pi}{a}\right)^2\right]\left(\frac{n\pi}{a}\right)x\cot\left(\frac{n\pi}{a}z\right)\psi_n$$

$$-\alpha\left[2a_1a_3\hbar^2\left(\frac{l\pi}{b}\right)^2 + 2a_2a_3\hbar^2k_y^2 + 6a_3^2\hbar^2\left(\frac{n\pi}{a}\right)^2\right]e^2B^2x^2\psi_n$$

$$-4\alpha a_3^2i\hbar e^3B^3\left(\frac{n\pi}{a}\right)x^3\cot\left(\frac{n\pi}{a}z\right)\psi_n - \alpha a_3^2e^4B^4x^4\psi_n$$

$$+ \alpha a_2\hbar^2k_y^2\left[a_1\hbar^2\left(\frac{l\pi}{b}\right)^2 + a_2\hbar^2k_y^2 + a_3\hbar^2\left(\frac{n\pi}{a}\right)^2\right]\psi_n$$

$$+ \alpha a_2a_3\hbar^2e^2B^2k_y^2x^2\psi_n + 2\alpha a_2a_3i\hbar^3k_y^2eB\left(\frac{n\pi}{a}\right)x\cot\left(\frac{n\pi}{a}z\right)\psi_n$$

$$-\alpha a_2'\hbar^2k_y^2\left[a_1\hbar^2\left(\frac{l\pi}{b}\right)^2 + a_2\hbar^2k_y^2 + a_3\hbar^2\left(\frac{n\pi}{a}\right)^2\right]\psi_n - \alpha a_2'a_3\hbar^2k_y^2e^2B^2x^2\psi_n$$

$$-2\alpha a_2'a_3i\hbar^3k_y^2eB\left(\frac{n\pi}{a}\right)x\cot\left(\frac{n\pi}{a}z\right)\psi_n + \alpha a_2a_2'\hbar^4k_y^4\psi_n$$

The averages are calculated as follow:

$$\left\langle x\cot\left(\frac{n\pi}{a}z\right)\right\rangle = 0, \quad \left\langle x^3\cot\left(\frac{n\pi}{a}z\right)\right\rangle = 0,$$

$$\langle x^2\rangle = \frac{2}{b}\int_0^b x^2\sin^2\left(\frac{l\pi}{b}\right)x\,dx = \left(\frac{b^2}{3} - \frac{b^2}{2l^2\pi^2}\right) \text{ and}$$

$$\langle x^4\rangle = \frac{2}{b}\int_0^b x^4\sin^2\left(\frac{l\pi}{b}x\right)dx = \left(\frac{b^4}{5} - \frac{b^4}{l^2\pi^2} + \frac{3b^4}{2l^4\pi^4}\right)$$

The DR in this case is given by

$$
\begin{aligned}
\gamma_3(E,\eta_g) &= a_1\hbar^2\left(\frac{l\pi}{b}\right)^2 + a_2\hbar^2k_y^2 + a_3\hbar^2\left(\frac{n\pi}{a}\right)^2 \\
&\quad - \alpha\left[a_1\hbar^2\left(\frac{l\pi}{b}\right)^2 + a_2\hbar^2k_y^2 + a_3\hbar^2\left(\frac{n\pi}{a}\right)^2\right]^2 \\
&\quad + \alpha a_2\hbar^2k_y^2\left[a_1\hbar^2\left(\frac{l\pi}{b}\right)^2 + a_2\hbar^2k_y^2 + a_3\hbar^2\left(\frac{n\pi}{a}\right)^2\right] \\
&\quad - \alpha a_2'\hbar^2k_y^2\left[a_1\hbar^2\left(\frac{l\pi}{b}\right)^2 + a_2\hbar^2k_y^2 + a_3\hbar^2\left(\frac{n\pi}{a}\right)^2\right] \\
&\quad + \alpha a_2a_2'\hbar^4k_y^4 + a_3e^2B^2\langle x^2\rangle + 2\alpha a_1a_3\hbar^2e^2B^2 \\
&\quad - \alpha\left[2a_1a_3\hbar^2\left(\frac{l\pi}{b}\right)^2 + a_2a_3\hbar^2k_y^2 + 6a_3^2\hbar^2\left(\frac{n\pi}{a}\right)^2\right]e^2B^2\langle x^2\rangle \\
&\quad - \alpha a_3^2e^4B^4\langle x^4\rangle - \alpha a_2'a_3\hbar^2k_y^2e^2B^2\langle x^2\rangle
\end{aligned}
\tag{1.147}
$$

where $\langle x^2\rangle = \left(\frac{b^2}{3} - \frac{b^2}{2l^2\pi^2}\right)$ and $\langle x^4\rangle = \left(\frac{b^4}{5} - \frac{b^4}{l^2\pi^2} + \frac{3b^4}{2l^4\pi^4}\right)$.

The EEM and the DOS function can be calculated from (1.147) and are left for the readers to enjoy the same.

1.2.9 The DR in Quantum Dot of HD IV–VI Semiconductors in the Presence of Magnetic Field

(a) Cohen Model

The Total Hamiltonian be

$$
\begin{aligned}
\hat{H} &= a_1\left(-i\hbar\frac{\partial}{\partial x}\right)^2 + a_2\left(-i\hbar\frac{\partial}{\partial y}\right)^2 + a_3\left(-i\hbar\frac{\partial}{\partial z} - eBx\right)^2 \\
&\quad - \alpha\left[a_1\left(-i\hbar\frac{\partial}{\partial x}\right)^2 + a_2\left(-i\hbar\frac{\partial}{\partial y}\right)^2 + a_3\left(-i\hbar\frac{\partial}{\partial z} - eBx\right)^2\right]^2 \\
&\quad + \alpha a_2\left(-i\hbar\frac{\partial}{\partial y}\right)^2\left[a_1\left(-i\hbar\frac{\partial}{\partial x}\right)^2 + a_2\left(-i\hbar\frac{\partial}{\partial y}\right)^2 + a_3\left(-i\hbar\frac{\partial}{\partial z} - eBx\right)^2\right] \\
&\quad - \alpha a_2'\left(-i\hbar\frac{\partial}{\partial y}\right)^2\left[a_1\left(-i\hbar\frac{\partial}{\partial x}\right)^2 + a_2\left(-i\hbar\frac{\partial}{\partial y}\right)^2 + a_3\left(-i\hbar\frac{\partial}{\partial z} - eBx\right)^2\right] \\
&\quad + \alpha a_2a_2'\left(-i\hbar\frac{\partial}{\partial y}\right)^4
\end{aligned}
\tag{1.148}
$$

Considering non-parabolicity and magnetic field as perturbed, the unperturbed wave equation is

$$\hat{H}_0\psi_0 = E_0\psi_0 \tag{1.149}$$

From (1.149) we can write

$$-a_1\hbar^2\frac{\partial^2\psi_0}{\partial x^2} - a_2\hbar^2\frac{\partial^2\psi_0}{\partial y^2} - a_3\hbar^2\frac{\partial^2\psi_0}{\partial z^2} = E_0\psi_0$$

So that

$$E_0 = E_{nlr}^{(0)} = \frac{\hbar^2}{2m_1}\left(\frac{l\pi}{b}\right)^2 + \frac{\hbar^2}{2m_2}\left(\frac{r\pi}{c}\right)^2 + \frac{\hbar^2}{2m_3}\left(\frac{n\pi}{a}\right)^2 \tag{1.150}$$

$$\psi_0 = \psi_{nlr}^{(0)} = \psi_n = \sqrt{\frac{8}{abc}}\sin\left(\frac{l\pi}{b}x\right)\sin\left(\frac{r\pi}{c}y\right)\sin\left(\frac{n\pi}{a}z\right) \tag{1.151}$$

Considering the perturbing effect, the total Hamiltonian can be written as

$$\begin{aligned}\hat{H} &= \hat{H}_0 + \hat{H}' = -a_1\hbar^2\frac{\partial^2}{\partial x^2} - a_2\hbar^2\frac{\partial^2}{\partial y^2} + a_3\left(-\hbar^2\frac{\partial^2}{\partial z^2} + e^2B^2x^2 + 2i\hbar eBx\frac{\partial}{\partial z}\right)\\ &- \alpha\left[-a_1\hbar^2\frac{\partial^2}{\partial x^2} - a_2\hbar^2\frac{\partial^2}{\partial y^2} + a_3\left(-\hbar^2\frac{\partial^2}{\partial z^2} + e^2B^2x^2 + 2i\hbar eBx\frac{\partial}{\partial z}\right)\right]^2\\ &+ \alpha a_2\left(-\hbar^2\frac{\partial^2}{\partial y^2}\right)\left[-a_1\hbar^2\frac{\partial^2}{\partial x^2} - a_2\hbar^2\frac{\partial^2}{\partial y^2} + a_3\left(-\hbar^2\frac{\partial^2}{\partial z^2} + e^2B^2x^2 + 2i\hbar eBx\frac{\partial}{\partial z}\right)\right]\\ &- \alpha a_2'\left(-\hbar^2\frac{\partial^2}{\partial y^2}\right)\left[-a_1\hbar^2\frac{\partial^2}{\partial x^2} - a_2\hbar^2\frac{\partial^2}{\partial y^2} + a_3\left(-\hbar^2\frac{\partial^2}{\partial z^2} + e^2B^2x^2 + 2i\hbar eBx\frac{\partial}{\partial z}\right)\right]\\ &+ \alpha a_2a_2'\left(\hbar^4\frac{\partial^4}{\partial y^4}\right)\end{aligned} \tag{1.152}$$

Now

$$\begin{aligned}&\left[-a_1\hbar^2\left(\frac{\partial^2}{\partial x^2}\right) - a_2\hbar^2\left(\frac{\partial^2}{\partial y^2}\right) + a_3\left(-\hbar^2\frac{\partial^2}{\partial z^2} + e^2B^2x^2 + 2i\hbar eBx\frac{\partial}{\partial z}\right)\right]\psi_n\\ &= a_1\hbar^2\left(\frac{l\pi}{b}\right)^2\psi_n + a_2\hbar^2\left(\frac{r\pi}{c}\right)^2\psi_n + a_3\hbar^2\left(\frac{n\pi}{a}\right)^2\psi_n\\ &+ a_3e^2B^2x^2\psi_n + 2a_3i\hbar eBx\left(\frac{n\pi}{a}\right)\cot\left(\frac{n\pi}{a}z\right)\psi_n\end{aligned} \tag{1.153}$$

$$\left[-a_1\hbar^2\frac{\partial^2}{\partial x^2} - a_2\hbar^2\frac{\partial^2}{\partial y^2} + a_3\left(-\hbar^2\frac{\partial^2}{\partial z^2} + e^2B^2x^2 + 2i\hbar eBx\frac{\partial}{\partial z}\right)\right]^2\psi_n$$

$$= [a_1^4\hbar^4\left(\frac{l\pi}{b}\right)^4\psi_n + a_1a_2\hbar^4\left(\frac{l\pi}{b}\right)^2\left(\frac{r\pi}{c}\right)^2\psi_n + a_1a_3\hbar^4\left(\frac{l\pi}{b}\right)^2\left(\frac{n\pi}{a}\right)^2\psi_n$$

$$+ a_1a_3\hbar^2\left(\frac{l\pi}{b}\right)^2 e^2B^2x^2\psi_n - 2a_1a_3\hbar^2e^2B^2\psi_n$$

$$+ 2a_1a_3i\hbar^3eBx\left(\frac{l\pi}{b}\right)^2\left(\frac{n\pi}{a}\right)\cot\left(\frac{n\pi}{a}z\right)\psi_n$$

$$+ a_2a_1\hbar^4\left(\frac{l\pi}{b}\right)^2\left(\frac{r\pi}{c}\right)^2\psi_n + a_2^2\hbar^4\left(\frac{r\pi}{c}\right)^4 + a_2a_3\hbar^4\left(\frac{n\pi}{a}\right)^2\left(\frac{r\pi}{c}\right)^2$$

$$+ a_2a_3e^2B^2x^2\hbar^2\left(\frac{r\pi}{c}\right)^2\psi_n + 2a_2a_3i\hbar^3eB\left(\frac{r\pi}{c}\right)^2\left(\frac{n\pi}{a}\right)^2\cot\left(\frac{n\pi}{a}z\right)$$

$$+ a_3a_1\hbar^4\left(\frac{l\pi}{b}\right)^2\left(\frac{n\pi}{a}\right)^2\psi_n + a_3a_2\hbar^4\left(\frac{r\pi}{c}\right)^2\left(\frac{n\pi}{a}\right)^2\psi_n$$

$$+ a_3^2\hbar^4\left(\frac{n\pi}{a}\right)^4\psi_n + a_3^2\hbar^2\left(\frac{n\pi}{a}\right)^2 e^2B^2x^2\psi_n$$

$$+ 2a_3^2i\hbar^3eB\left(\frac{n\pi}{a}\right)^3 x\cot\left(\frac{n\pi}{a}z\right)\psi_n$$

$$+ \left[a_3a_1\hbar^2\left(\frac{l\pi}{b}\right)^2 + a_3a_2\hbar^2\left(\frac{r\pi}{c}\right)^2 + a_3^2\hbar^2\left(\frac{n\pi}{a}\right)^2\right]e^2B^2x^2\psi_n$$

$$+ a_3^2e^4B^4x^4\psi_n + 2a_3^2i\hbar e^3B^3x^3\left(\frac{n\pi}{a}\right)\cot\left(\frac{n\pi}{a}z\right)\psi_n$$

$$+ 2a_3i\hbar eB\left[a_1\hbar^2\left(\frac{l\pi}{b}\right)^2 + a_2\hbar^2\left(\frac{r\pi}{c}\right)^2 + a_3\hbar^2\left(\frac{n\pi}{a}\right)^2\right]\left(\frac{n\pi}{a}\right)x\cot\left(\frac{n\pi}{a}z\right)\psi_n$$

$$+ 2a_3^2i\hbar e^3B^3x^3\left(\frac{n\pi}{a}\right)\cot\left(\frac{n\pi}{a}z\right)\psi_n + 4a_3^2\hbar^2e^2B^2\left(\frac{n\pi}{a}\right)^2\psi_n$$

$$= \left[a_1\hbar^2\left(\frac{l\pi}{b}\right)^2 + a_2\hbar^2\left(\frac{r\pi}{c}\right)^2 + a_3\hbar^2\left(\frac{n\pi}{a}\right)^2\right]^2\psi_n - 2a_1a_3\hbar^2e^2B^2\psi_n$$

$$+ 4a_3i\hbar eB\left[a_1\hbar^2\left(\frac{l\pi}{b}\right)^2 + a_2\hbar^2\left(\frac{r\pi}{c}\right)^2 + a_3\hbar^2\left(\frac{n\pi}{a}\right)^2\right].\left(\frac{n\pi}{a}\right)x\cot\left(\frac{n\pi}{a}z\right)\psi_n$$

$$+ \left[2a_1a_3\hbar^2\left(\frac{l\pi}{b}\right)^2 + 2a_2a_3\hbar^2\left(\frac{r\pi}{c}\right)^2 + 6a_3^2\hbar^2\left(\frac{n\pi}{a}\right)^2\right]e^2B^2x^2\psi_n$$

$$+ 4a_3^2i\hbar e^3B^3\left(\frac{n\pi}{a}\right)x^3\cot\left(\frac{n\pi}{a}z\right)\psi_n + a_3^2e^4B^4x^4\psi_n$$

Again

$$\alpha a_2 a_2' \hbar^4 \frac{\partial^4}{\partial y^4}\psi_n = \alpha a_2 a_2' \hbar^4 \left(\frac{r\pi}{c}\right)^2 \psi_n \tag{1.154}$$

$$\left(-\hbar^2\frac{\partial^2}{\partial y^2}\right)\left[-a_1\hbar^2\frac{\partial^2}{\partial x^2} - a_2\hbar^2\frac{\partial^2}{\partial y^2} + a_3\left(-\hbar^2\frac{\partial^2}{\partial z^2} + e^2B^2x^2 + 2i\hbar eBx\frac{\partial}{\partial z}\right)\right]$$
$$= \hbar^2\left(\frac{r\pi}{c}\right)^2\left[a_1\hbar^2\left(\frac{l\pi}{b}\right)^2 + a_2\hbar^2\left(\frac{r\pi}{c}\right)^2 + a_3\hbar^2\left(\frac{n\pi}{a}\right)^2\right] + a_3\hbar^2\left(\frac{r\pi}{c}\right)^2 e^2B^2x^2\psi_n$$
$$+ 2a_3i\hbar^3\left(\frac{r\pi}{c}\right)^2\left(\frac{n\pi}{a}\right)eBx\cot\left(\frac{n\pi}{a}z\right)\psi_n$$

So

$$\hat{H}\psi_{nlr} = \hat{H}_0\psi_{nlr} + \hat{H}'\psi_{nlr} \tag{1.155}$$

$$\text{or,}\quad \hat{H}\psi_{nlr} = a_1\hbar^2\left(\frac{l\pi}{b}\right)^2\psi_n + a_2\hbar^2\left(\frac{r\pi}{c}\right)^2\psi_n + a_3\hbar^2\left(\frac{n\pi}{a}\right)^2\psi_n + a_3e^2B^2x^2\psi_n$$
$$+ 2a_3i\left(\frac{n\pi}{a}\right)x\cot\left(\frac{n\pi}{a}z\right)\psi_n$$
$$-\alpha\left[a_1\hbar^2\left(\frac{l\pi}{b}\right)^2 + a_2\hbar^2\left(\frac{r\pi}{c}\right)^2 + a_3\hbar^2\left(\frac{n\pi}{a}\right)^2\right]^2\psi_n$$
$$+ 2\alpha a_1a_3\hbar^2e^2B^2\psi_n - 4_3i\hbar eB\left[a_1\hbar^2\left(\frac{l\pi}{b}\right)^2 + a_2\hbar^2\left(\frac{r\pi}{c}\right)^2\right.$$
$$\left.+ a_3\hbar^2\left(\frac{n\pi}{a}\right)^2\right]\left(\frac{n\pi}{a}\right)x\cot\left(\frac{n\pi}{a}z\right)\psi_n$$
$$-\alpha\left[2a_1a_3\hbar^2\left(\frac{l\pi}{b}\right)^2 + 2a_2a_3\hbar^2\left(\frac{r\pi}{c}\right)^2 + 6a_3^2\hbar^2\left(\frac{n\pi}{a}\right)^2\right]e^2B^2x^2\psi_n$$
$$-4_3^2i\hbar e^3B^3\left(\frac{n\pi}{a}\right)x^3\cot\left(\frac{n\pi}{a}z\right)\psi_n - {}_3^2e^4B^4x^4\psi_n + \alpha a_2\hbar^2\left(\frac{r\pi}{c}\right)^2$$
$$\left[a_1\hbar^2\left(\frac{l\pi}{b}\right)^2 + a_2\hbar^2\left(\frac{r\pi}{c}\right)^2 + a_3\hbar^2\left(\frac{n\pi}{a}\right)^2\right]\psi_n$$
$$+ {}_2\,a_3\hbar^2e^2B^2\left(\frac{r\pi}{c}\right)^2x^2\psi_n + 2\alpha a_2a_3i\hbar^3\left(\frac{r\pi}{c}\right)^2\left(\frac{n\pi}{a}\right)eBx\cot\left(\frac{n\pi}{a}z\right)\psi_n$$
$$-\alpha a_2'\hbar^2\left(\frac{r\pi}{c}\right)^2\left[a_1\hbar^2\left(\frac{l\pi}{b}\right)^2 + a_2\hbar^2\left(\frac{r\pi}{c}\right)^2\right.$$
$$\left.+ a_3\hbar^2\left(\frac{n\pi}{a}\right)^2\right]\psi_n - \alpha a_2'a_3\hbar^2\left(\frac{r\pi}{c}\right)^2e^2B^2x^2\psi_n$$
$$-2\alpha a_2'a_3i\hbar^3\left(\frac{r\pi}{c}\right)^2eB\left(\frac{n\pi}{a}\right)x\cot\left(\frac{n\pi}{a}z\right)\psi_n + \alpha a_2a_2'\hbar^4\left(\frac{r\pi}{c}\right)^4\psi_n$$

Applying first order perturbation

$$\left\langle x\cot\left(\frac{n\pi}{a}z\right)\right\rangle = 0,\ \left\langle x^3\cot\left(\frac{n\pi}{a}z\right)\right\rangle = 0,\ \langle x^2\rangle = \frac{2}{b}\int_0^b x^2\sin^2\left(\frac{l\pi}{b}\right)xdx = \left(\frac{b^2}{3} - \frac{b^2}{2l^2\pi^2}\right)$$

$$\langle x^4\rangle = \frac{2}{b}\int_0^b x^4\sin^2\left(\frac{l\pi}{b}x\right)dx = \left(\frac{b^4}{5} - \frac{b^4}{l^2\pi^2} + \frac{3b^4}{2l^4\pi^4}\right)$$

The DR in this case is given by

$$\begin{aligned}
\gamma_3(E,\eta_g) = & \left[a_1\hbar^2\left(\frac{l\pi}{b}\right)^2 + a_2\hbar^2\left(\frac{r\pi}{c}\right)^2 + a_3\hbar^2\left(\frac{n\pi}{a}\right)^2\right] \\
& - \alpha\left[a_1\hbar^2\left(\frac{l\pi}{b}\right)^2 + a_2\hbar^2\left(\frac{r\pi}{c}\right)^2 + a_3\hbar^2\left(\frac{n\pi}{a}\right)^2\right]^2 \\
& + \alpha a_2\hbar^2\left(\frac{r\pi}{c}\right)^2\left[a_1\hbar^2\left(\frac{l\pi}{b}\right)^2 + a_2\hbar^2\left(\frac{r\pi}{c}\right)^2 + a_3\hbar^2\left(\frac{n\pi}{a}\right)^2\right] \\
& - \alpha a_2'\hbar^2\left(\frac{r\pi}{c}\right)^2\left[a_1\hbar^2\left(\frac{l\pi}{b}\right)^2 + a_2\hbar^2\left(\frac{r\pi}{c}\right)^2 + a_3\hbar^2\left(\frac{n\pi}{a}\right)^2\right] \\
& + \alpha a_2 a_2'\hbar^4\left(\frac{r\pi}{c}\right)^2 + a_3 e^2B^2\langle x^2\rangle + 2\alpha a_1 a_3\hbar^2 e^2B^2 \\
& - \alpha\left[2a_1a_3\hbar^2\left(\frac{l\pi}{b}\right)^2 + a_2a_3\hbar^2\left(\frac{r\pi}{c}\right)^2 + 6a_3^2\hbar^2\left(\frac{n\pi}{a}\right)^2\right]e^2B^2\langle x^2\rangle \\
& - \alpha a_2' a_3\hbar^2\left(\frac{r\pi}{c}\right)^2 e^2B^2\langle x^2\rangle - \alpha a_3^2 e^4B^4\langle x^4\rangle
\end{aligned} \tag{1.156a}$$

where $\langle x^2\rangle = \left(\frac{b^2}{3} - \frac{b^2}{2l^2\pi^2}\right)$ and $\langle x^4\rangle = \left(\frac{b^4}{5} - \frac{b^4}{l^2\pi^2} + \frac{3b^4}{2l^4\pi^4}\right)$.

The DOS function in this case is given by

$$N(E) = \frac{2g_v}{d_xd_yd_z}\sum_{n=0}^{n_{\max}}\sum_{l=0}^{l_{\max}}\sum_{r=0}^{r_{\max}}\delta'(E - E_{19,11}) \tag{1.156b}$$

where $E_{19,11}$ is the totally quantized energy in this case.

(b) McClure and Choi Model

In presence of $\vec{B}$ and considering non-parabolicity the total Hamiltonian can be written as

$$\begin{aligned}\hat{H} &= a_1\left(-i\hbar\frac{\partial}{\partial x}\right)^2 + a_2\left(-i\hbar\frac{\partial}{\partial y}\right)^2 + a_3\left(-i\hbar\frac{\partial}{\partial z} - eBx\right)^2 \\ &\quad - \alpha\left[a_1\left(-i\hbar\frac{\partial}{\partial x}\right)^2 + a_2\left(-i\hbar\frac{\partial}{\partial y}\right)^2 + a_3\left(-i\hbar\frac{\partial}{\partial z} - eBx\right)^2\right]^2 \\ &\quad + \alpha a_2 a_2'\left(-i\hbar\frac{\partial}{\partial y}\right)^4 \alpha a_2 a_0\left(-i\hbar\frac{\partial}{\partial y}\right)^2\cdot\left[a_1\left(-i\hbar\frac{\partial}{\partial x}\right)^2 + a_2\left(-i\hbar\frac{\partial}{\partial y}\right)^2 + a_3\left(-i\hbar\frac{\partial}{\partial z} - eBx\right)^2\right] \\ &\quad - \alpha a_1 a_2\left(-i\hbar\frac{\partial}{\partial x}\right)^2\left(-i\hbar\frac{\partial}{\partial y}\right)^2 - \alpha a_2 a_3\left(-i\hbar\frac{\partial}{\partial y}\right)^2\cdot\left[-i\hbar\frac{\partial}{\partial z} - eBx\right]^2\end{aligned} \tag{1.157}$$

Considering non-parabolicity and magnetic field as perturbed quantities, the unperturbed wave equation is

$$\hat{H}_0\psi_0 = E_0\psi_0 \tag{1.158}$$

Or,

$$-a_1\hbar^2\frac{\partial^2\psi_0}{\partial x^2} - a_2\hbar^2\frac{\partial^2\psi_0}{\partial y^2} - a_3\hbar^2\frac{\partial^2\psi_0}{\partial z^2} = E_0\psi_0 \tag{1.159}$$

So that

$$E_0 = E_{nlr}^{(0)} = \frac{\hbar^2}{2m_1}\left(\frac{l\pi}{b}\right)^2 + \frac{\hbar^2}{2m_2}\left(\frac{r\pi}{c}\right)^2 + \frac{\hbar^2}{2m_3}\left(\frac{n\pi}{a}\right)^2 \tag{1.160}$$

$$\psi_0 = \psi_{nlr}^{(0)} = \sqrt{\frac{8}{abc}}\sin\left(\frac{l\pi}{b}\mathrm{x}\right)\sin\left(\frac{r\pi}{c}\mathrm{y}\right)\sin\left(\frac{n\pi}{a}z\right) \tag{1.161}$$

Including perturbing effect, the total Hamiltonian assumes the form

$$\begin{aligned}\hat{H} &= \hat{H}_0 + \hat{H}' = -a_1\hbar^2\frac{\partial^2}{\partial x^2} - a_2\hbar^2\frac{\partial^2}{\partial y^2} + a_3\left(-\hbar^2\frac{\partial^2}{\partial z^2} + e^2B^2x^2 + 2i\hbar eBx\frac{\partial}{\partial z}\right) \\ &\quad - \alpha\left[-a_1\hbar^2\frac{\partial^2}{\partial x^2} - a_2\hbar^2\frac{\partial^2}{\partial y^2} + a_3\left(-\hbar^2\frac{\partial^2}{\partial z^2} + e^2B^2x^2 + 2i\hbar eBx\frac{\partial}{\partial z}\right)\right]^2 + \alpha a_2 a_2'\left(-\hbar^4\frac{\partial^4}{\partial y^4}\right) \\ &\quad + \alpha a_2 a_0\left(-\hbar^2\frac{\partial^2}{\partial y^2}\right)\left[-a_1\hbar^2\frac{\partial^2}{\partial x^2} - a_2\hbar^2\frac{\partial^2}{\partial y^2} + a_3\left(-\hbar^2\frac{\partial^2}{\partial z^2} + e^2B^2x^2 + 2i\hbar eBx\frac{\partial}{\partial z}\right)\right] \\ &\quad - \alpha a_1 a_2\hbar^4\left(\frac{\partial^2}{\partial x^2}\right)\left(\frac{\partial^2}{\partial y^2}\right) - \alpha a_2 a_3\left(\hbar^2\frac{\partial^2}{\partial y^2}\right)\left(-\hbar^2\frac{\partial^2}{\partial z^2} + e^2B^2x^2 + 2i\hbar eBx\frac{\partial}{\partial z}\right)\end{aligned} \tag{1.162}$$

Now
$$\left[-a_1\hbar^2\frac{\partial^2}{\partial x^2} - a_2\hbar^2\frac{\partial^2}{\partial y^2} + a_3\left(-\hbar^2\frac{\partial^2}{\partial z^2} + e^2B^2x^2 + 2i\hbar eBx\frac{\partial}{\partial z}\right)\right]\psi^0$$
$$= a_1\hbar^2\left(\frac{l\pi}{b}\right)^2\psi_0 + a_2\hbar^2\left(\frac{r\pi}{c}\right)^2\psi_0 + a_3\hbar^2\left(\frac{n\pi}{a}\right)^2\psi_0$$
$$+ a_3e^2B^2x^2\psi_0 + 2a_3i\hbar eBx\left(\frac{n\pi}{a}\right)\cot\left(\frac{n\pi}{a}\right)\psi^0 \tag{1.163}$$

$$\left[-a_1\hbar^2\frac{\partial^2}{\partial x^2} - a_2\hbar^2\frac{\partial^2}{\partial y^2} + a_3\left(-\hbar^2\frac{\partial^2}{\partial z^2} + e^2B^2x^2 + 2i\hbar eBx\frac{\partial}{\partial z}\right)\right]\psi^0$$
$$= \left[a_1\hbar^2\left(\frac{l\pi}{b}\right)^2 + a_2\hbar^2\left(\frac{r\pi}{c}\right)^2 + a_3\hbar^2\left(\frac{n\pi}{a}\right)^2\right]^2\psi_0 - 2a_1a_3\hbar^2e^2B^2\psi_0$$
$$+ 4a_3i\left[a_1\hbar^2\left(\frac{l\pi}{b}\right)^2 + a_2\hbar^2\left(\frac{r\pi}{c}\right)^2 + a_3\hbar^2\left(\frac{n\pi}{a}\right)^2\right]$$
$$\left(\frac{n\pi}{a}\right)x\cot\left(\frac{n\pi}{a}z\right)\psi_0 + \left[2a_1a_3\hbar^2\left(\frac{l\pi}{b}\right)^2 + 2a_2a_3\hbar^2\left(\frac{r\pi}{c}\right)^2 + 6a_3^2\hbar^2\left(\frac{n\pi}{a}\right)^2\right]e^2B^2x^2\psi_0$$
$$+ 4a_3^2i\hbar e^3B^3\left(\frac{n\pi}{a}\right)x^3\cot\left(\frac{n\pi}{a}z\right)\psi_0 + a_3^2e^4B^4x^4\psi_0$$
$$+ \alpha a_2a_2'\hbar^4\frac{\partial^4}{\partial y^4}[\psi_0] = \alpha a_2a_2'\hbar^4\left(\frac{r\pi}{c}\right)^4\psi_0 \tag{1.164}$$

So

$$\hat{H}\psi^0_{nlr} = \hat{H}_0\psi^0_{nlr} + \hat{H}'\psi^0_{nlr} \tag{1.165}$$

$$\hat{H}\psi^0_{nlr} = a_1\hbar^2\left(\frac{l\pi}{b}\right)^2\psi^0 + a_2\hbar^2\left(\frac{r\pi}{c}\right)^2\psi^0 + a_3\hbar^2\left(\frac{n\pi}{a}\right)^2\psi^0$$
$$+ a_3e^2B^2x^2\psi^0 + 2a_3i\hbar eB\left(\frac{n\pi}{a}\right)x\cot\left(\frac{n\pi}{a}z\right)\psi^0$$
$$- \alpha\left[a_1\hbar^2\left(\frac{l\pi}{b}\right)^2 + a_2\hbar^2\left(\frac{r\pi}{c}\right)^2 + a_3\hbar^2\left(\frac{n\pi}{a}\right)^2\right]^2\psi^0$$
$$+ 2\alpha a_1a_3\hbar^2e^2B^2\psi^0 - 4\alpha a_3i\hbar eB\left[a_1\hbar^2\left(\frac{l\pi}{b}\right)^2 + a_2\hbar^2\left(\frac{r\pi}{c}\right)^2\right.$$
$$\left. + a_3\hbar^2\left(\frac{n\pi}{a}\right)^2\right]\left(\frac{n\pi}{a}\right)x\cot\left(\frac{n\pi}{a}z\right)\psi^0$$
$$- \alpha\left[2a_1a_3\hbar^2\left(\frac{l\pi}{b}\right)^2 + 2a_2a_3\hbar^2\left(\frac{r\pi}{c}\right)^2 + 6a_3^2\hbar^2\left(\frac{n\pi}{a}\right)^2\right]e^2B^2x^2\psi^0$$

$$
\begin{aligned}
&- 4\alpha a_3^2 i\hbar e^3 B^3 \left(\frac{n\pi}{a}\right) x^3 \cot\left(\frac{n\pi}{a} z\right)\psi^0 \; - \; \alpha a_3^2 e^4 B^4 x^4 \psi^0 \\
&+ \; \alpha a_2 a_2' \hbar^4 \left(\frac{r\pi}{c}\right)^4 \psi^0 \; + \; \alpha a_2 a_0 \hbar^2 \left(\frac{r\pi}{c}\right)^2 \\
&\left[a_1 \hbar^2 \left(\frac{l\pi}{b}\right)^2 + a_2 \hbar^2 \left(\frac{r\pi}{c}\right)^2 + a_3 \hbar^2 \left(\frac{n\pi}{a}\right)^2\right]\psi^0 \\
&+ \; \alpha a_2 a_0 a_3 \hbar^2 \left(\frac{r\pi}{c}\right)^2 e^2 B^2 x^2 \psi^0 \; + \; 2\alpha a_2 a_0 a_3 i\hbar^3 \left(\frac{r\pi}{c}\right)^2 \\
&eBx \cot\left(\frac{n\pi}{a} z\right)\psi^0 \; - \; \alpha a_1 a_2 \hbar^4 \left(\frac{r\pi}{c}\right)^2 \left(\frac{l\pi}{b}\right)^2 \psi^0 \\
&- \; \alpha a_2 a_3 \hbar^4 \left(\frac{r\pi}{c}\right) \left(\frac{n\pi}{a}\right)^2 \psi^0 \; - \; \alpha a_2 a_3 \hbar^2 \left(\frac{r\pi}{c}\right) e^2 B^2 x^2 \psi^0 \\
&- \; 2\alpha a_2 a_3 i\hbar^3 \left(\frac{r\pi}{c}\right)^2 \left(\frac{n\pi}{a}\right) eBx \cot\left(\frac{n\pi}{a} z\right)\psi^0
\end{aligned}
\tag{1.166}
$$

Applying first order perturbation

$$
\left\langle x\cot\left(\frac{n\pi}{a} z\right)\right\rangle = 0, \; \left\langle x^3 \cot\left(\frac{n\pi}{a} z\right)\right\rangle = 0, \; \langle x^2 \rangle = \frac{2}{b}\int_0^b x^2 \sin^2\left(\frac{l\pi}{b} x\right) dx = \left(\frac{b^2}{3} - \frac{b^2}{2l^2\pi^2}\right)
$$

$$
\langle x^4 \rangle = \frac{2}{b}\int_0^b x^4 \sin^2\left(\frac{l\pi}{b} x\right) dx = \left(\frac{b^4}{5} - \frac{b^4}{l^2\pi^2} + \frac{3b^4}{2l^4\pi^4}\right)
$$

Therefore the electron energy spectrum up to the first-order can be written as

$$
\begin{aligned}
E &= E_{nlr}^{(0)} + E_{nlr}^{(1)} \\
E &= \left[a_1 \hbar^2 \left(\frac{l\pi}{b}\right)^2 + a_2 \hbar^2 \left(\frac{r\pi}{c}\right)^2 + a_3 \hbar^2 \left(\frac{n\pi}{a}\right)^2\right] \\
&\quad - \alpha \left[a_1 \hbar^2 \left(\frac{l\pi}{b}\right)^2 + a_2 \hbar^2 \left(\frac{r\pi}{c}\right)^2 + a_3 \hbar^2 \left(\frac{n\pi}{a}\right)^2\right]^2 + \alpha a_2 a_2' \hbar^4 \left(\frac{r\pi}{c}\right)^4 \\
&\quad + \alpha a_2 a_0 \hbar^2 \left(\frac{r\pi}{c}\right)^2 \left[a_1 \hbar^2 \left(\frac{l\pi}{b}\right)^2 + a_2 \hbar^2 \left(\frac{r\pi}{c}\right)^2 + a_3 \hbar^2 \left(\frac{n\pi}{a}\right)^2\right] \\
&\quad - \alpha a_1 a_2 \hbar^4 \left(\frac{l\pi}{b}\right)^2 \left(\frac{r\pi}{c}\right)^2 - \alpha a_2 a_3 \hbar^4 \left(\frac{r\pi}{c}\right)^2 \left(\frac{n\pi}{a}\right)^2 \\
&\quad + a_3 e^2 B^2 \langle x^2 \rangle + 2\alpha a_1 a_3 \hbar^2 e^2 B^2 - \alpha a_3^2 e^4 B^4 \langle x^4 \rangle \\
&\quad - \alpha \left[2a_1 a_3 \hbar^2 \left(\frac{l\pi}{b}\right)^2 + 2a_2 a_3 \hbar^2 \left(\frac{r\pi}{c}\right)^2 + 6a_3^2 \hbar^2 \left(\frac{n\pi}{a}\right)^2\right] e^2 B^2 \langle x^2 \rangle \\
&\quad + \alpha a_2 a_0 a_3 \hbar^2 \left(\frac{r\pi}{c}\right)^2 e^2 B^2 \langle x^2 \rangle - \alpha a_2 a_3 \hbar^2 \left(\frac{r\pi}{c}\right)^2 e^2 B^2 \langle x^2 \rangle
\end{aligned}
\tag{1.167}
$$

Now $a_2 a_0 = (a_2 - a_2')$.
Therefore the DR in this case is given by

$$\begin{aligned}\gamma_3(E,\eta_g) &= a_1\hbar^2\left(\frac{l\pi}{b}\right)^2 + a_2\hbar^2\left(\frac{r\pi}{c}\right)^2 + a_3\hbar^2\left(\frac{n\pi}{a}\right)^2 \\ &- \alpha\left[a_1\hbar^2\left(\frac{l\pi}{b}\right)^2 + a_2\hbar^2\left(\frac{r\pi}{c}\right)^2 + a_3\hbar^2\left(\frac{n\pi}{a}\right)^2\right]^2 \\ &+ \alpha a_2 a_2'\hbar^4\left(\frac{r\pi}{c}\right)^4 + \alpha a_2\hbar^2\left(\frac{r\pi}{c}\right)^2\left[a_1\hbar^2\left(\frac{l\pi}{b}\right)^2 + a_2\hbar^2\left(\frac{r\pi}{c}\right)^2 + a_3\hbar^2\left(\frac{n\pi}{a}\right)^2\right] \\ &- \alpha a_2'\hbar^2\left(\frac{r\pi}{c}\right)^2\left[a_1\hbar^2\left(\frac{l\pi}{b}\right)^2 + a_2\hbar^2\left(\frac{r\pi}{c}\right)^2 + a_3\hbar^2\left(\frac{n\pi}{a}\right)^2\right] \\ &- \alpha a_1 a_2\hbar^4\left(\frac{l\pi}{b}\right)^2\left(\frac{r\pi}{c}\right)^2 - \alpha a_2 a_3\hbar^4\left(\frac{r\pi}{c}\right)^2\left(\frac{n\pi}{a}\right)^2 \\ &+ a_3 e^2 B^2\langle x^2\rangle + 2\alpha a_1 a_2 a_3\hbar^2 e^2 B^2 - \alpha a_3^2 e^4 B^4\langle x^4\rangle \\ &- \alpha\left[2a_1 a_3\hbar^2\left(\frac{l\pi}{b}\right)^2 + 2a_2 a_3\hbar^2\left(\frac{r\pi}{c}\right)^2 + 6a_3^2\hbar^2\left(\frac{n\pi}{a}\right)^2\right] \\ &e^2B^2\langle x^2\rangle - \alpha a_1 a_2'\hbar^2\left(\frac{r\pi}{c}\right)^2 e^2 B^2\langle x^2\rangle\end{aligned} \tag{1.168a}$$

The DOS function in this case is given by

$$N(E) = \frac{2g_v}{d_x d_y d_z}\sum_{n=0}^{n_{\max}}\sum_{l=0}^{l_{\max}}\sum_{r=0}^{r_{\max}}\delta'(E - E_{19,14}) \tag{1.168b}$$

where $E_{19,14}$ is the totally quantized energy in this case.

1.2.10 The DR in Cylindrical Quantum Dot of HD III–V Semiconductors in the Presence of Crossed Electric and Magnetic Fields

In the presence of an external magnetic field, B, and an external electric field (E') along the z-direction, the Schrodinger equation becomes

$$\frac{(p - eA)^2}{2m_c}\psi - \alpha\left(\frac{(p - eA)^2}{2m_c}\right)^2\Psi - eE'z = E\psi \tag{1.169}$$

For a cylindrical quantum dot, the above equation has to be solved for the following boundary conditions:

$$\Psi = 0 \text{ at } \rho = a \text{ and at } z = 0 \text{ and } d$$

Using (ρ, φ, z) as the cylindrical co-ordinates of the system, and a and d are the radius and the width of the cylindrical QD, respectively.

Assuming that the magnetic field is perpendicular to the z-direction, one may choose the vector potential components as $A_\rho = A_\varphi = 0$, and $A_z = B_\rho \sin\varphi$, and consequently the Eq. (1.169) may further be expressed as

$$\begin{aligned} & -\frac{\hbar^2}{2m_c}\left\{\frac{1}{\rho}\frac{\partial}{\partial\rho}\left(\rho\frac{\partial\Psi}{\partial\rho}\right) + \frac{1}{\rho^2}\frac{\partial^2\Psi}{\partial\varphi^2} + \frac{\partial^2\Psi}{\partial z^2}\right\} \\ & + \frac{ie\hbar B}{m_c}\rho\sin\varphi\frac{\partial\Psi}{\partial z} + \frac{e^2B^2}{2m^*}\rho^2\sin^2\varphi\Psi - eE'z\Psi \\ & - \left[-\frac{\hbar^2}{2m_c}\left\{\frac{1}{\rho}\frac{\partial}{\partial\rho}\left(\rho\frac{\partial\Psi}{\partial\rho}\right) + \frac{1}{\rho^2}\frac{\partial^2}{\partial\varphi^2} + \frac{\partial^2}{\partial z^2}\right\} + \frac{ie\hbar B}{m_c}\rho\sin\varphi\frac{\partial}{\partial z} + \frac{e^2B^2}{2m^*}\rho^2\sin^2\varphi\right]^2\Psi \\ & = E\Psi \end{aligned} \tag{1.170}$$

The Eq. (1.170) can be written as

$$(\hat{H}_a - eE'z)\Psi + \left\{\hat{H}_b - \alpha(\hat{H}_a + \hat{H}_b)^2\right\}\Psi = E\Psi \tag{1.171}$$

where

$$\hat{H}_a = -\frac{\hbar^2}{2m_c}\left\{\frac{1}{\rho}\frac{\partial}{\partial\rho}\left(\rho\frac{\partial}{\partial\rho}\right) + \frac{1}{\rho^2}\frac{\partial^2}{\partial\varphi^2} + \frac{\partial^2}{\partial z^2}\right\}$$

and $\hat{H}_b \equiv \frac{ie\hbar B}{m_c}\rho\sin\varphi\frac{\partial}{\partial z} + \frac{e^2B^2}{2m_c}\rho^2\sin^2\varphi$

Assuming that the effect of the magnetic field and of the band non-parabolicity is to perturb the system slightly in the presence of the electric field, we can break up the effective Hamiltonian into two parts:

(i) The unperturbed Hamiltonian $(\hat{H}_a - eE'z)$
(ii) The perturbing Hamiltonian $[\hat{H}_b - \alpha(\hat{H}_a + \hat{H}_b)^2]$

The unperturbed wave function (Ψ_0) can be obtained from the eigen value equation as

$$(\hat{H}_a - eE'z)\Psi_0 = E_0\Psi_0 \tag{1.172}$$

where E_0 is the unperturbed eigen value. Substituting the values of $\hat{H}_a$ into (1.172) and using $\Psi_0 = R(\rho)\phi(\varphi)Z(z)$ for applying the method of separation of variables, we get

$$-\frac{\hbar^2}{2m_c}\left\{\frac{1}{\rho R}\frac{\partial}{\partial\rho}\left(\rho\frac{\partial R}{\partial\rho}\right)+\frac{1}{\rho^2\phi}\frac{\partial^2\phi}{\partial\varphi^2}+\frac{1}{Z}\frac{\partial^2 Z}{\partial z^2}\right\}-eE'z=E_0 \tag{1.173}$$

As the terms depending on z are not functions of ρ and φ, the z-dependent terms can be equated to a constant, say, E_{zn}. So, the z-dependent equation becomes

$$\frac{\partial^2 Z}{\partial z^2}+\frac{2m_c eE'}{\hbar^2}\left(z+\frac{E_{zn}}{eE'}\right)Z=0 \tag{1.174}$$

With the boundary conditions $Z = 0$ at $z = 0$ and d. Making the substitution

$$\xi=\left(\frac{2m_c eE'}{\hbar^2}\right)^{1/3}\left(z+\frac{E_{zn}}{eE'}\right)$$

in the above equation, we can write

$$\frac{\partial^2 Z}{\partial\xi^2}+\xi Z=0 \tag{1.175}$$

with the boundary conditions $Z = 0$ at for $\xi_1 \le \xi \le \xi_2$ where

$$\xi_1=\left(\frac{2m_c eE'}{\hbar^2}\right)^{1/3}\frac{E_{zn}}{eE'}\ \text{and}\ \xi_2=\left(\frac{2m_c eE'}{\hbar^2}\right)^{1/3}\left(d+\frac{E_{zn}}{eE'}\right) \tag{1.176}$$

The general solution of (1.175) is given by

$$Z=CAi(-\xi)+DBi(-\xi) \tag{1.177}$$

where C and D are constants and $Ai(-\xi)$ and $Bi(-\xi)$ are Airy function. The condition to be satisfied for the non-trivial solutions of Z can be shown to be

$$Ai(-\xi_2)Bi(-\xi_1)-Ai(-\xi_1)Bi(-\xi_2)=0 \tag{1.178}$$

For appropriate combinations of ξ_1 and ξ_2, (1.178) can be solved numerically. Then using the relations given by Eq. (1.176), one can get E_{zn}, which is the energy eigen value of an electron inside the sample in the presence of the electric field, considering the motion along the direction of the electric field only. As E_{zn} takes only some discrete values, a suffix n has to be used to denote the quantum number on which the energy E_{zn} depends.

Using the appropriate expressions for $Ai(-\eta)$ and $Bi(-\eta)$ for large arguments, the condition can be reduced to the form

$$\xi_2^{3/2} - \xi_1^{3/2} = \frac{3}{2}n\pi$$

here $n = 1, 2, 3, \ldots\ldots\ldots$ This equation can be solved for ξ_1 and ξ_2 to get E_{zn}.

From (1.173) and (1.174), it can be shown that the $\varphi-$ dependent equation is given by

$$-\frac{1}{\phi}\frac{\partial^2\phi}{\partial\varphi^2} = m^2 \quad (1.179)$$

which gives

$$\phi = C_\phi e \pm im\varphi \quad (1.180)$$

where m is a positive integer including zero, C_φ, being a constant determined by the normalization condition.

Proceeding in the same manner, we can write the radial part of (1.173) as

$$\rho^2\frac{\partial^2 R}{\partial\rho^2} + \rho\frac{\partial R}{\partial\rho} + (\lambda^2\rho^2 - m^2)\mathrm{R} = 0 \quad (1.181)$$

where

$$\lambda^2 = \frac{2m_c(E_0 - E_{zn})}{\hbar^2} \quad (1.182)$$

This leads to the solution for R as

$$R = C_\rho J_m(\lambda\rho)$$

where $J_m(\lambda\rho)$ is the Bessel function of the first kind and order m, C_ρ being a constant obtained by the normalization of the radial wave function R. The boundary condition to be imposed or R is $R = 0$ at $\rho = a$. This gives $\lambda = \alpha_{ml}/a$, where α_{ml} is the l-th zero of the m-th-order Bessel function. Hence, we can write

$$R = C_\rho J_m\left(\frac{\alpha_{ml}}{a}\rho\right) \quad (1.183)$$

Thus, ψ_0 can be obtained using (1.177), (1.180) and (1.183) in the expression for ψ_0 mentioned earlier. Putting $\lambda = \alpha_{ml}/a$ into (1.182), the unperturbed energy eigen value may be shown to be given by

$$E_0 = \frac{\hbar^2}{2m_c}\left(\frac{\alpha_{ml}}{a}\right)^2 + E_{zn} \quad (1.184)$$

We can denote E_0 by E_{lmn}, as E_0 depends on the quantum number l, m and n.

To calculate the influence of the perturbing Hamiltonian, we first split it into two parts:

(i) $\hat{H}_a$ and
(ii) $-\alpha(\hat{H}_a + \hat{H}_b)^2 = \alpha\hat{H}_{ab}^2$, where $\hat{H}_{ab} = (\hat{H}_a + \hat{H}_b)$.

We shall use only corrections arising from the first-order perturbation. The first-order correction due to $\hat{H}_b$ is H_{bss}, where the suffix ss means that both ψ_0^* and ψ_0 correspond to the same s-state of the electron. It can be shown that

$$H_{bss} = \frac{1}{2}\frac{e^2B^2}{2m_c}\langle\rho^2\rangle \tag{1.185}$$

where $\langle\rho^2\rangle$ is the expectation value of ρ^2, given by

$$\langle\rho^2\rangle = \frac{\int_0^a \rho^3 R^2 d\rho}{\int_0^a \rho R^2 d\rho} \tag{1.186}$$

Similarly, the first-order correction due to $-\alpha\hat{H}_{ab}^2$ is H_{abss} where

$$H_{abss} = -\alpha\frac{\int_v \psi_{0s}^* \hat{H}_{ab}^2 \psi_{os} dV}{\int_v \psi_{0s}^* \psi_{os} dV}$$

Using the values of $\hat{H}_{ab}$ and (1.171), where necessary, we can write

$$H_{abss} = I_1 + I_2 + I_3 + I_4 + I_5 + I_6 \tag{1.187}$$

where $I_1, \ldots, I_6$ represent separate integral terms as given below,

$$I_1 = -\alpha E_0\frac{\int_V \psi_{0s}^* \hat{H}_a \psi_{os} dV}{\int_V \psi_{0s}^* \psi_{os} dV}, \quad I_2 = -\alpha\frac{\int_V \psi_{0s}^* \hat{H}_a eE'z\psi_{os} dV}{\int_V \psi_{0s}^* \psi_{os} dV},$$

$$I_3 = -\alpha\frac{\int_V \psi_{0s}^* \hat{H}_a \hat{H}_b \psi_{os} dV}{\int_V \psi_{0s}^* \psi_{os} dV},$$

$$I_4 = -\alpha\frac{\int_V E_0 \psi_{0s}^* \hat{H}_b \psi_{os} dV}{\int_V \psi_{0s}^* \psi_{os} dV}, \quad I_5 = -\alpha\frac{\int_V \psi_{0s}^* \hat{H}_b eE'z\psi_{os} dV}{\int_V \psi_{0s}^* \psi_{os} dV} \quad \text{and}$$

$$I_6 = -\alpha\frac{\int_V \psi_{0s}^* \hat{H}_b \hat{H}_b \psi_{os} dV}{\int_V \psi_{0s}^* \psi_{os} dV}$$

Using the expression for the wave function ψ_{os}, $I_1, \ldots, I_6$ can be evaluated to be given by

$$\begin{aligned}
I_1 &= -\alpha[E_0^2 + E_0 eE\langle z\rangle] \\
I_2 &= -\alpha\left[eE'\langle z\rangle E_0 + e^2E'^2\langle z^2\rangle\right] \\
I_3 &= -\alpha\left[\frac{\hbar^2}{2m_c}\left(\frac{\alpha_{ml}}{a}\right)^2\frac{1}{2}\frac{e^2B^2}{2m_c}\langle\rho^2\rangle + \frac{1}{2}\frac{e^2B^2}{2m_c}\langle\rho^2\rangle(E_{zn} + eE'\langle z\rangle)\right] \\
I_4 &= -\alpha\frac{1}{2}\frac{e^2B^2}{2m^*}\langle\rho^2\rangle E_0 \\
I_5 &= -\alpha\frac{1}{2}\frac{e^2B^2}{2m_c}\langle\rho^2\rangle eE'\langle z\rangle \\
I_6 &= -\alpha\left[4\frac{1}{2}\frac{e^2B^2}{2m_c}\langle\rho^2\rangle(E_{zn} + eE'\langle z\rangle) + \frac{3}{2}\left(\frac{1}{2}\frac{e^2B^2}{2m_c}\right)\langle\rho^4\rangle\right]
\end{aligned}$$

$\langle z\rangle$, $\langle z^2\rangle$ and $\langle\rho^4\rangle$ being the expectation values of z, z^2 and ρ^4 respectively, and may easily be evaluated using the expression for ψ_{os}. For example, $\langle\rho^4\rangle$ may be obtained in a similar manner as shown in (1.186). For the z-dependent function $f(z)$, one may write

$$\langle f(z)\rangle = \frac{\int_0^d z^2 f(z)dz}{\int_0^d z^2 dz} \tag{1.188}$$

Now, inserting the values of $I_1, \ldots, I_6$ we can write, after some tedious algebraic manipulations,

$$\begin{aligned}
H_{abss} = -\alpha\Bigg[&\left(E_0 + \frac{1}{2}\frac{e^2B^2}{2m_c}\langle\rho^2\rangle + eE'\langle z\rangle\right)^2 + \left(\frac{1}{2}\frac{e^2B^2}{2m_c}\right)^2\cdot\left(\frac{3}{2}\langle\rho^4\rangle - \langle\rho^2\rangle^2\right) \\
&+ e^2E'^2\left(\langle z^2\rangle - \langle z\rangle^2\right) + 4\frac{1}{2}\frac{e^2B^2}{2m_c}\langle\rho^2\rangle(E_{zn} + eE'\langle z\rangle)\Bigg]
\end{aligned} \tag{1.189}$$

The energy (E'_{lmn}) after the first-order correction is, therefore, given by

$$\begin{aligned}
E'_{lmn} = E_{lmn} + \frac{1}{2}\frac{e^2B^2}{2m_c}\langle\rho^2\rangle - \Bigg[&\left(E_{lmn} + \frac{1}{2}\frac{e^2B^2}{2m_c}\langle\rho^2\rangle + eE'\langle z\rangle\right)^2 \\
&+ \left(\frac{1}{2}\frac{e^2B^2}{2m_c}\right)^2\cdot\left(\frac{3}{2}\langle\rho^4\rangle - \langle\rho^2\rangle^2\right) + e^2E'^2\left(\langle z^2\rangle - \langle z\rangle^2\right) + 4\frac{1}{2}\frac{e^2B^2}{2m_c}\langle\rho^2\rangle(E_{zn} + eE'\langle z\rangle)\Bigg]
\end{aligned} \tag{1.190}$$

The equation (1.190) was derived for the first time by the group of A.N. Chakravarti et.al. [174]

The DR in cylindrical QD of HD III–V semiconductors in the presence of crossed electric and magnetic fields can be written as

$$\gamma_3(E,\eta_g) = E_{lmn} + \frac{1}{2}\frac{e^2B^2}{2m_c}\langle\rho^2\rangle - \left[\left(E_{lmn} + \frac{1}{2}\frac{e^2B^2}{2m_c}\langle\rho^2\rangle + eE'\langle z\rangle\right)^2 + \left(\frac{1}{2}\frac{e^2B^2}{2m_c}\right)^2\cdot\left(\frac{3}{2}\langle\rho^4\rangle - \langle\rho^2\rangle^2\right)\right.$$
$$\left. + e^2E'\left(\langle z^2\rangle - \langle z\rangle^2\right) + 4\frac{1}{2}\frac{e^2B^2}{2m_c}\langle\rho^2\rangle(E_{zn} + eE'\langle z\rangle)\right] \quad (1.191)$$

The DOS function in this case is given by

$$N(E) = \frac{2g_v}{d_xd_yd_z}\sum_{n=0}^{n_{\max}}\sum_{l=0}^{l_{\max}}\sum_{r=0}^{r_{\max}}\delta'(E - E_{19,15}) \quad (1.191a)$$

where $E_{19,15}$ is the totally quantized energy in this case.

1.2.11 The DR in Quantum Wells of HD III–V Semiconductors in the Presence of Arbitrarily Oriented Magnetic Field

In this case the total Hamiltonian $(\hat{H})$ can be written as

$$\hat{H} = a'(-i\hbar\vec{\nabla} + e\vec{A}) - b'(-i\hbar\vec{\nabla} + e\vec{A})^4 \quad (1.192)$$

In which $a' = \frac{1}{2m_c}, b' = \alpha(a')^2, \alpha = \frac{1}{E_g}, \vec{A} = \vec{i}zB_y + \vec{j}xB_z + \vec{k}_yB_x, \vec{B} = \vec{i}B_x + \vec{j}B_y + \vec{k}B_z, B_x = B\sin\theta\cos\phi, B_y = B\sin\theta\sin\phi, B_z = B\cos\theta, (B,\theta,\phi)$ are the spherical polar coordinates, $\vec{\nabla}\cdot\vec{A} = 0, \vec{\nabla}\times A = \vec{B} = \begin{vmatrix} \vec{i} & \vec{j} & \vec{k} \\ \frac{\partial}{\partial x} & \frac{\partial}{\partial y} & \frac{\partial}{\partial z} \\ zB_y & xB_z & yB_x \end{vmatrix}$ and $(\vec{i},\vec{j},\vec{k})$ are orthogonal triads.

From (1.192) we can write

$$\begin{aligned}\hat{H} &= a'\left[\left(-i\hbar\frac{\partial}{\partial x} + eB_{yz}\right)\vec{i} + \left(-i\hbar\frac{\partial}{\partial y} + exB_z\right)\vec{j} + \vec{k}\left(-i\hbar\frac{\partial}{\partial z} + eyB_x\right)\right]^2 \\ &\quad - b'\left[\left(-i\hbar\frac{\partial}{\partial x} + ezB_y\right)\vec{i} + \left(-i\hbar\frac{\partial}{\partial y} + exB_z\right)\vec{j} + \left(-i\hbar\frac{\partial}{\partial z} + eyB_x\right)\vec{k}\right]^4 \\ &= a'\hat{H}_1^2 - b'\hat{H}_1^4\end{aligned} \quad (1.193)$$

The unperturbed wave function and energy Eigen values in this case assume the forms

$$u_n = \psi_n^0 = \sqrt{\frac{2}{aL_xL_y}}\sin\left(\frac{n\pi}{a}z\right)\exp(ik_xx + ik_yy) \tag{1.194}$$

$$E_n^0 = \frac{\hbar^2}{2m_c}\left(\frac{n\pi}{a}\right)^2 + \frac{\hbar^2k_y^2}{2m_c} + \frac{\hbar^2k_x^2}{2m_c} - \alpha\left[\frac{\hbar^2}{2m_c}\left(\frac{n\pi}{a}\right)^2 + \frac{\hbar^2k_y^2}{2m_c} + \frac{\hbar^2k_x^2}{2m_c}\right]^2 \tag{1.195}$$

Now

$$\begin{aligned}\hat{H}_1\psi_n^0 &= \vec{i}(-i\hbar ik_x + ezB_y)u_n + \vec{j}(-i\hbar ik_y + exB_z)u_n + \vec{k}eyB_xu_n \\ &\quad + \vec{k}\left[-i\hbar\cdot\left(\frac{n\pi}{a}\right)\cot\left(\frac{n\pi}{a}z\right)u_n\right]\end{aligned} \tag{1.196}$$

$$\begin{aligned}\hat{H}_1^2\psi_n^0 &= -i\hbar\frac{\partial}{\partial x}(-i\hbar\cdot ik_x + ezB_y)u_n + eB_yz(-i\hbar\cdot ik_x + ezB_y)u_n + (-i\hbar\cdot ik_y + eB_zx)^2u_n \\ &\quad + eyB_x\left(-i\hbar\frac{\partial}{\partial z} + eyB_x\right)u_n + \left(-i\hbar\frac{\partial}{\partial z} + eyB_x\right)\left(-i\hbar\frac{n\pi}{a}\right)\cot\left(\frac{n\pi}{a}z\right)u_n\end{aligned} \tag{1.197}$$

Therefore (1.197) assumes the form

$$\begin{aligned}\hat{H}_1^2\psi_n^0 &= (\hbar k_x + eB_yz)^2u_n + (\hbar k_y + exB_z)^2u_n + \hbar^2\left(\frac{n\pi}{a}\right)^2u_n \\ &\quad - 2i\hbar\left(\frac{n\pi}{a}\right)eyB_x\cot\left(\frac{n\pi}{a}z\right)u_n + e^2y^2B_x^2u_n\end{aligned} \tag{1.198}$$

$$\begin{aligned}\hat{H}_1^3\psi_n^0 &= \left[\left(-i\hbar\frac{\partial}{\partial x} + eB_yz\right)\vec{i} + \left(-i\hbar\frac{\partial}{\partial y} + exB_z\right)\vec{j} + \left(-i\hbar\frac{\partial}{\partial z} + eyB_x\right)\vec{k}\right]\hat{H}_1^2\psi_n^0 \\ &= \vec{i}\Big[-i\hbar\cdot ik_x\Big\{(\hbar k_x + eB_yz)^2u_n + (\hbar k_y + exB_z)^2u_n + \hbar^2\left(\frac{n\pi}{a}\right)^2u_n \\ &\quad - 2i\hbar\left(\frac{n\pi}{a}\right)eyB_x\cot\left(\frac{n\pi}{a}z\right)u_n + e^2y^2B_x^2u_n\Big\} \\ &\quad + (-i\hbar)(2\hbar k_yeB_z + 2e^2B^2x)u_n\Big] + \vec{i}eB_yz\hat{H}_1^2\psi_n^0 + \vec{j}eB_zx\hat{H}_1^2\psi_n^0 + \vec{k}eyB_x\hat{H}_1^2\psi_n^0 \\ &\quad \& + \vec{j}\Big[-i\hbar ik_y\hat{H}_1^2\psi_n^0 + (-i\hbar)(-2i\hbar) \\ &\quad \cdot\left(\frac{n\pi}{a}\right)eB_x\cot\left(\frac{n\pi}{a}z\right)u_n + (-i\hbar)2e^2B_x^2yu_n\Big]\end{aligned}$$

$$
\begin{aligned}
&+\vec{k}\Big[-i\hbar\Big\{\Big(\frac{n\pi}{a}\Big)\cot\Big(\frac{n\pi}{a}z\Big)\Big\}\hat{H}_1^2\psi_n^0 + (-i\hbar)2eB_y(\hbar k_x + eB_yz)u_n \\
&+\Big(\frac{n\pi}{a}\Big)^2(-i\hbar)(-2i\hbar)eyB_x)(-u_n)\Big] = e\vec{A}\hat{H}_1^2\psi_n^0 + \vec{i}\hbar k_x\hat{H}_1^2\psi_n^0 + \vec{j}\hbar k_y\hat{H}_1^2\psi_n^0 \\
&+\vec{k}\Big(-i\hbar\frac{n\pi}{a}\Big)\cot\Big(\frac{n\pi}{a}z\Big)\hat{H}_1^2\psi_n^0 + i\{-2i\hbar^2k_yeB_z \\
&+\vec{k}\Big[\Big\{-2i\hbar^2k_xeB_y - 2i\hbar e^2B_y^2z + 2\hbar^2\Big(\frac{n\pi}{a}\Big)^2eyB_x\Big\}u_n\Big] \\
=\;&\vec{i}\big[\hbar k_x\hat{H}_1^2\psi_n^0 + eB_yz\hat{H}_1^2\psi_n^0 + (-2i\hbar^2k_yeBz - 2i\hbar e^2B_z^2x)u_n\big] \\
&+\vec{j}\big[\hbar k_y\hat{H}_1^2\psi_n^0 + exB_z\hat{H}_1^2\psi_n^0 \\
&+\Big\{-2\hbar^2\Big(\frac{n\pi}{a}\Big)eB_x\cot\Big(\frac{n\pi}{a}z\Big) - 2i\hbar e^2B_x^2y\Big\}u_n\Big] \\
&+\vec{k}\Big[-i\hbar\Big(\frac{n\pi}{a}\Big)\cot\Big(\frac{n\pi}{a}z\Big)\hat{H}_1^2\psi_n^0 + eyB_x\hat{H}_1^2\psi_n^0 \\
&+\Big\{2\hbar^2\Big(\frac{n\pi}{a}\Big)^2eyB_x - 2i\hbar e^2B_y^2z - 2i\hbar^2k_xeB_y\Big\}u_n\Big]
\end{aligned} \tag{1.199}
$$

Therefore

$$
\hat{H}_1^4\psi_n^0 = R + S + T \tag{1.200}
$$

where

$$
\begin{aligned}
R =\;& \Big(-i\hbar\frac{\partial}{\partial x} + ezB_y\Big)\big[\hbar k_x\hat{H}_1^2\psi_n^0 + ezB_y\hat{H}_1^2\psi_n^0 - 2i\hbar^2k_yeB_zu_n - 2i\hbar e^2B_z^2xu_u\big] \\
=\;& ezB_y\hbar k_x\hat{H}_1^2\psi_n^0 + e^2z^2B_y^2\hat{H}_1^2\psi_n^0 - 2i\hbar^2k_ye^2zB_yB_zu_u - 2i\hbar e^3zxB_yB_z^2u_u \\
&+ (-i\hbar)(-2i\hbar^2)ek_yB_z(ik_x)u_n + (-i\hbar)(-2i\hbar)e^2B_z^2u_u + (-i\hbar)(-2i\hbar)e^2B_z^2x(ik_x)u_u \\
&+ (\hbar k_x + ezB_y)\big[\hbar k_x\hat{H}_1^2\psi_n^0 - 2i\hbar^2k_yeB_zu_u - 2i\hbar e^2B_z^2xu_u\big] \\
=\;& (\hbar k_x + ezB_y)^2\hat{H}_1^2\psi_n^0 - 2i\hbar^3ek_yk_xB_zu_u - 2i\hbar^2e^2B_z^2xk_xu_u - 2\hbar^2e^2B_z^2u_u \\
&- 2i\hbar^3k_xk_yeB_zu_u - 2i\hbar^2k_xe^2B_z^2xu_u - 4i\hbar^2k_ye^2zB_yB_zu_u - 4i\hbar e^3B_z^2B_yzxu_u
\end{aligned} \tag{1.201}
$$

$$
\begin{aligned}
S =\;& \Big(-i\hbar\frac{\partial}{\partial y} + exB_z\Big)\Big[(\hbar k_y + exB_z)\hat{H}_1^2\psi_n^0 - 2\hbar^2\Big(\frac{n\pi}{a}\Big)eB_x\cot\Big(\frac{n\pi}{a}z\Big)u_u - 2i\hbar e^2B_x^2yu_u\Big] \\
=\;& (\hbar k_y + exB_z)\Big[\hbar k_y\hat{H}_1^2\psi_n^0 - 2\hbar^2\Big(\frac{n\pi}{a}\Big)eB_x\cot\Big(\frac{n\pi}{a}z\Big)u_u - 2i\hbar e^2B_x^2yu_u\Big] \\
&+ (-i\hbar ik_y)(-2\hbar^2)\Big(\frac{n\pi}{a}\Big)eB_x\cot\Big(\frac{n\pi}{a}z\Big)u_u \\
&+ (-i\hbar ik_y)(-2i\hbar e^2B_x^2yu_u) + (-i\hbar)(-2i\hbar e^2B_x^2)u_u \\
&+ exB_z(\hbar k_y + exB_z)\hat{H}_1^2\psi_n^0 \\
&+ exB_z\Big[-2\hbar^2\Big(\frac{n\pi}{a}\Big)eB_x\cot\Big(\frac{n\pi}{a}z\Big)u_u - 2i\hbar e^2B_x^2yu_u\Big]
\end{aligned} \tag{1.202}
$$

$$S = (\hbar k_y + exB_z)^2 \hat{H}_1^2 \psi_n^0 + (\hbar k_y + 2xeB_z)\left[-2i\hbar e^2 B_x^2 y u_u - 2\hbar^2 \left(\frac{n\pi}{a}\right) eB_x \cot\left(\frac{n\pi}{a} z\right) u_u\right]$$
$$- 2\hbar^3 k_y \left(\frac{n\pi}{a}\right) eB_x \cot\left(\frac{n\pi}{a} z\right) u_u - 2\hbar^2 e^2 B_x^2 u_u - 2i\hbar^2 k_y e^2 B_x^2 y u_u \tag{1.203}$$

$$T = \left(-i\hbar \frac{\partial}{\partial z} + eyB_x\right)\left[-i\hbar\left(\frac{n\pi}{a}\right)\cot\left(\frac{n\pi}{a} z\right)\hat{H}_1^2 \psi_n^0 + eyB_x \hat{H}_1^2 \psi_n^0\right.$$
$$\left. + 2\hbar^2 \left(\frac{n\pi}{a}\right)^2 eyB_x u_u - 2i\hbar e^2 B_y^2 z u_u - 2i\hbar^2 k_x eB_y u_u\right] \tag{1.204}$$

Thus (1.204) can be written as

$$T = -i\hbar\left(\frac{n\pi}{a}\right)^2 \cot\left(\frac{n\pi}{a} z\right)(-i\hbar)$$
$$\left\{\left(\frac{n\pi}{a}\right)\cot\left(\frac{n\pi}{a} z\right)\hat{H}_1^2 \psi_n^0 + 2(\hbar k_y + eB_y z)eB_y u_n - 2i\hbar\left(\frac{n\pi}{a}\right)^2 eyB_x(-u_n)\right\}$$
$$+ (-i\hbar)\left(\frac{n\pi}{a}\right)\left(-\frac{n\pi}{a}\right)\cos \mathrm{c}\, e^2\left(\frac{n\pi}{a} z\right)(-i\hbar)\hat{H}_1^2 \psi_n^0$$
$$+ eyB_x(-i\hbar)\left\{\left(\frac{n\pi}{a}\right)\cot\left(\frac{n\pi}{a} z\right)\hat{H}_1^2 \psi_n^0 - 2i\hbar\left(\frac{n\pi}{a}\right)^2\right.$$
$$\left. eyB_x(-u_n) + 2(\hbar x + eB_y z)eB_y u_n\right\} + 2\hbar^2\left(\frac{n\pi}{a}\right)^2 eyB_x(-i\hbar)\left(\frac{n\pi}{a}\right)\cot\left(\frac{n\pi}{a} z\right)u_n$$
$$- 2i\hbar e^2 B_y^2 z(-i\hbar)\left(\frac{n\pi}{a}\right)\cot\left(\frac{n\pi}{a} z\right)u_n - 2i\hbar e^2 B_y^2(-i\hbar)u_n$$
$$- 2i\hbar^2 k_x eB_y(-i\hbar)\left(\frac{n\pi}{a}\right)\cot\left(\frac{n\pi}{a} z\right)u_n$$
$$+ eyB_x(-i\hbar)\left(\frac{n\pi}{a}\right)\cot\left(\frac{n\pi}{a} z\right)\hat{H}_1^2 \psi_n^0 + e^2 y^2 B_x^2 \hat{H}_1^2 \psi_n^0$$
$$+ e^2 y^2 B_x^2 2\hbar^2\left(\frac{n\pi}{a}\right)^2 u_n - 2i\hbar e^2 B_y^2 zeyB_x u_n - 2i\hbar^2 ek_x B_y eB_x y u_n \tag{1.205}$$

$$S = (\hbar k_y + exB_z)^2 \hat{H}_1^2 \psi_n^0 - 2\hbar^3 k_y\left(\frac{n\pi}{a}\right)eB_x \cot\left(\frac{n\pi}{a} z\right)u_n$$
$$- 2\hbar^2 e^2 B_x^2 u_n - 2i\hbar^2 k_y e^2 B_x^2 y u_n - 2i\hbar^2 k_y e^2 B_x^2 y u_n$$
$$- 2\hbar^3 k_y\left(\frac{n\pi}{a}\right)eB_x \cot\left(\frac{n\pi}{a} z\right)u_n - 4i\hbar e^3 xyB_x^2 B_z u_n$$
$$- 4\hbar^2\left(\frac{n\pi}{a}\right)e^2 xB_x B_z \cot\left(\frac{n\pi}{a} z\right)u_n \tag{1.206}$$

$$
\begin{aligned}
T = & -\hbar\left(\frac{n\pi}{a}\right)^2\cot^2\left(\frac{n\pi}{a}z\right)\hat{H}_1^2\psi_n^0 - 2\hbar^2\left(\frac{n\pi}{a}\right)\cot\left(\frac{n\pi}{a}z\right)eB_y(\hbar k_x + eB_y z)u_n \\
& - 2i\hbar^3\left(\frac{n\pi}{a}\right)^3\cot\left(\frac{n\pi}{a}z\right)eyB_x u_n + \hbar^2\left(\frac{n\pi}{a}\right)^2\cos ce^2\left(\frac{n\pi}{a}z\right)\hat{H}_1^2\psi_n^0 \\
& - i\hbar\left(\frac{n\pi}{a}\right)eyB_x\cot\left(\frac{n\pi}{a}z\right)\hat{H}_1^2\psi_n^0 \\
& + 2\hbar^2\left(\frac{n\pi}{a}\right)^2 e^2y^2B_x^2u_n - 2i\hbar(\hbar k_x + eB_y z)e^2yB_xB_yu_n \\
& - 2i\hbar^3\left(\frac{n\pi}{a}\right)^3 eyB_x\cot\left(\frac{n\pi}{a}z\right)u_n \\
& - 2\hbar^2\left(\frac{n\pi}{a}\right)e^2B_y^2z\cot\left(\frac{n\pi}{a}z\right)u_n - 2\hbar^2e^2B_y^2u_n \\
& - 2\hbar^3\left(\frac{n\pi}{a}\right)ek_xB_y\cot\left(\frac{n\pi}{a}z\right)u_n \\
& - i\hbar\left(\frac{n\pi}{a}\right)eyB_x\cot\left(\frac{n\pi}{a}z\right)\hat{H}_1^2\psi_n^0 + e^2y^2B_x^2\hat{H}_1^2\psi_n^0 \\
& + 2\hbar^2\left(\frac{n\pi}{a}\right)^2 e^2y^2B_x^2u_n - 2i\hbar e^3zyB_y^2B_xu_n - 2i\hbar^2e^2k_xyB_yB_xu_n
\end{aligned}
\quad (1.207)
$$

$$
\begin{aligned}
T = & \hbar^2\left(\frac{n\pi}{a}\right)^2\hat{H}_1^2\psi_n^0 - 2i\hbar\left(\frac{n\pi}{a}\right)eyB_x\cot\left(\frac{n\pi}{a}z\right)\hat{H}_1^2\psi_n^0 - 2\hbar^3\left(\frac{n\pi}{a}\right)eB_yk_x\cot\left(\frac{n\pi}{a}z\right)u_n \\
& - 2\hbar^2\left(\frac{n\pi}{a}\right)e^2B_y^2z\cot\left(\frac{n\pi}{a}z\right)u_n - 4i\hbar^3\left(\frac{n\pi}{a}\right)^3 eyB_x\cot\left(\frac{n\pi}{a}z\right)u_n + 4\hbar^2\left(\frac{n\pi}{a}\right)^2 e^2y^2B_x^2u_n \\
& - 2i\hbar^2k_xe^2yB_xB_yu_n - 2i\hbar e^3B_y^2B_xzyu_n - 2\hbar^2\left(\frac{n\pi}{a}\right)e^2B_y^2z\cot\left(\frac{n\pi}{a}z\right)u_n \\
& - 2\hbar^3\left(\frac{n\pi}{a}\right)ek_xB_y\cot\left(\frac{n\pi}{a}z\right)u_n - 2\hbar^2e^2B_y^2u_n \\
& + e^2y^2B_x^2\hat{H}_1^2\psi_n^0 - 2i\hbar^3zyB_y^2B_xu_n - 2i\hbar^2e^2k_xyB_yB_xu_n
\end{aligned}
\quad (1.208)
$$

$$
\begin{aligned}
T = & \hbar^2\left(\frac{n\pi}{a}\right)^2\hat{H}_1^2\psi_n^0 + e^2y^2B_x^2\hat{H}_1^2\psi_n^0 - 2i\hbar\left(\frac{n\pi}{a}\right)eyB_x\cot\left(\frac{n\pi}{a}z\right)\hat{H}_1^2\psi_n^0 \\
& - 4\hbar^3\left(\frac{n\pi}{a}\right)ek_xB_y\cot\left(\frac{n\pi}{a}z\right)u_n - 4\hbar^2\left(\frac{n\pi}{a}\right)e^2B_y^2z\cot\left(\frac{n\pi}{a}z\right)u_n \\
& - 4i\hbar^3\left(\frac{n\pi}{a}\right)^3 eyB_x\cot\left(\frac{n\pi}{a}z\right)u_n + 4\hbar^2\left(\frac{n\pi}{a}\right)^2 e^2y^2B_x^2u_n \\
& - 4i\hbar^2e^2k_xyB_yB_xu_n - 4i\hbar e^3zyB_y^2B_xu_n - 2\hbar^2e^2B_y^2u_n
\end{aligned}
\quad (1.209)
$$

$$
\begin{aligned}
R = & (\hbar k_x + ezB_y)^2\hat{H}_1^2\psi_n^0 - 4i\hbar^3k_xk_yeB_zu_n - 4i\hbar^2k_xe^2xB_z^2u_n \\
& - 4i\hbar^2k_ye^2zB_yB_zu_n - 4i\hbar e^3zxB_z^2B_yu_n - 2\hbar^2e^2B_z^2u_n
\end{aligned}
\quad (1.210)
$$

$$
\begin{aligned}
S = {} & (\hbar k_y + exB_z)^2 \hat{H}_1^2 \psi_n^0 - 4\hbar^3 k_y \left(\frac{n\pi}{a}\right) eB_x \cot\left(\frac{n\pi}{a} z\right) u_n - 4i\hbar^2 k_y e^2 B_x^2 y u_n \\
& - 4i\hbar e^3 xy B_x^2 B_z u_n - 4\hbar^2 \left(\frac{n\pi}{a}\right) e^2 x B_x B_z \cot\left(\frac{n\pi}{a} z\right) u_n - 2\hbar^2 e^2 B_x^2 u_n
\end{aligned}
\tag{1.211}
$$

Therefore

$$
\begin{aligned}
\hat{H}_1^4 \psi_n^0 = {} & (\hbar k_x + ezB_y)^2 \hat{H}_1^2 \psi_n^0 + (\hbar k_y + exB_z)^2 \hat{H}_1^2 \psi_n^0 + \hbar^2 \left(\frac{n\pi}{a}\right)^2 \hat{H}_1^2 \psi_n^0 + e^2 y^2 B_x^2 \hat{H}_1^2 \psi_n^0 \\
& - 2i\hbar \left(\frac{n\pi}{a}\right) eyB_x \cot\left(\frac{n\pi}{a} z\right) \hat{H}_1^2 \psi_n^0 - 4\hbar^3 \left(\frac{n\pi}{a}\right) ek_x B_y \cot\left(\frac{n\pi}{a} z\right) u_n \\
& - 4\hbar^3 \left(\frac{n\pi}{a}\right) ek_y B_x \cot\left(\frac{n\pi}{a} z\right) u_n - 4\hbar^2 \left(\frac{n\pi}{a}\right) e^2 B_y^2 z \cot\left(\frac{n\pi}{a} z\right) u_n \\
& - 4\hbar^2 \left(\frac{n\pi}{a}\right) e^2 x B_x B_z \cot\left(\frac{n\pi}{a} z\right) u_n - 4i\hbar^3 \left(\frac{n\pi}{a}\right)^3 eyB_x \cot\left(\frac{n\pi}{a} z\right) u_n \\
& - 4i\hbar^3 k_x k_y e B_y u_n - 4i\hbar^2 k_x e^2 x B_z^2 u_n - 4i\hbar^2 k_y e^2 z B_y B_z u_n - 4i\hbar^2 k_y e^2 y B_x^2 u_n \\
& - 4i\hbar^2 k_x e^2 y B_y B_x u_n - 4i\hbar e^3 zy B_y^2 B_x u_n - 4i\hbar e^3 zx B_z^2 B_y u_n - 4i\hbar e^3 xy B_x^2 B_z u_n \\
& - 2\hbar^2 e^2 B_x^2 u_n - 2\hbar^2 e^2 B_z^2 u_n - 2\hbar^2 e^2 B_y^2 u_n + 4\hbar^2 \left(\frac{n\pi}{a}\right) e^2 y^2 B_x^2 u_n
\end{aligned}
\tag{1.212}
$$

$$
\begin{aligned}
\hat{H}_1^4 \psi_n^0 = {} & \left[\hbar^2 k_x^2 + \hbar^2 k_y^2 + \hbar^2 \left(\frac{n\pi}{a}\right)^2\right]^2 u_n + B_y^4 e^4 z^4 u_n + 4\hbar k_x B_y^3 e^3 z^3 u_n + 4\hbar^2 k_x^2 B_y^2 e^2 z^2 u_n \\
& + B_z^4 e^4 x^4 u_n + 4\hbar k_y B_z^3 e^3 x^3 u_n + 4\hbar^2 k_y^2 B_z^2 e^2 x^2 u_n + B_x^4 e^4 y^4 u_n + 2\left[\hbar^2 k_x^2 + \hbar^2 k_y^2 + \hbar^2 \left(\frac{n\pi}{a}\right)^2\right] \\
& B_y^2 e^2 z^2 u_n + 2\left[\hbar^2 k_x^2 + \hbar^2 k_y^2 + \hbar^2 \left(\frac{n\pi}{a}\right)^2\right] B_z^2 e^2 x^2 u_n + 2\left[\hbar^2 k_x^2 + \hbar^2 k_y^2 + \hbar^2 \left(\frac{n\pi}{a}\right)^2\right] B_x^2 e^2 y^2 u_n \\
& + 4\left[\hbar^2 k_x^2 + \hbar^2 k_y^2 + \hbar^2 \left(\frac{n\pi}{a}\right)^2\right] \hbar k_x B_y ezu_n + 4\left[\hbar^2 k_x^2 + \hbar^2 k_y^2 + \hbar^2 \left(\frac{n\pi}{a}\right)^2\right] \hbar k_y B_z exu_n \\
& + 4i\hbar \left(\frac{n\pi}{a}\right) \left[\hbar^2 k_x^2 + \hbar^2 k_y^2 + \hbar^2 \left(\frac{n\pi}{a}\right)^2\right] B_x ey \cot\left(\frac{n\pi}{a} z\right) u_n + 2B_z^2 B_y^2 e^4 x^2 z^2 u_n \\
& + 4\hbar k_x B_z^2 B_y e^3 x^2 z u_n + 4\hbar k_y B_y^2 B_z e^3 z^2 x u_n + 8\hbar^2 k_y k_x B_y B_z e^2 xz u_n + 2B_y^2 B_x^2 e^4 y^2 z^2 u_n \\
& + 4\hbar k_x B_x^2 B_y e^3 y^2 z u_n + 2B_x^2 B_z^2 e^4 x^2 y^2 u_n + 4\hbar k_y B_x^2 B_z e^3 y^2 x u_n - 4i\hbar \left(\frac{n\pi}{a}\right) B_x^3 e^3 y^3 \cot\left(\frac{n\pi}{a} z\right) u_n \\
& - 4i\hbar \left(\frac{n\pi}{a}\right) B_x B_y^2 e^3 z^2 y \cot\left(\frac{n\pi}{a} z\right) u_n - 4i\hbar \left(\frac{n\pi}{a}\right) B_x B_z^2 e^3 x^2 y \cot\left(\frac{n\pi}{a} z\right) u_n \\
& - 8i\hbar^2 k_x \left(\frac{n\pi}{a}\right) B_x B_y e^2 yz \cot\left(\frac{n\pi}{a} z\right) u_n - 8i\hbar^2 k_y \left(\frac{n\pi}{a}\right) B_x B_z e^2 xy \cot\left(\frac{n\pi}{a} z\right) u_n \\
& - 4\hbar^2 \left(\frac{n\pi}{a}\right)^2 B_x^2 e^2 y^2 \cot^2\left(\frac{n\pi}{a} z\right) u_n - 4\hbar^2 \left(\frac{n\pi}{a}\right) B_y^2 e^2 z \cot\left(\frac{n\pi}{a} z\right) u_n \\
& - 4\hbar^2 \left(\frac{n\pi}{a}\right) B_x B_z e^2 x \cot\left(\frac{n\pi}{a} z\right) u_n - 4i\hbar^3 \left(\frac{n\pi}{a}\right)^3 B_x ey \cot\left(\frac{n\pi}{a} z\right) u_n \\
& - 4\hbar^3 \left(\frac{n\pi}{a}\right) (k_y B_x + k_x B_y) e \cot\left(\frac{n\pi}{a} z\right) u_n - 4i\hbar^2 k_x B_z^2 e^2 x u_n \\
& - 4i\hbar^2 k_y B_x^2 e^2 y u_n - 4i\hbar^2 k_y B_y B_z e^2 z u_n - 4i\hbar^2 k_x B_y B_x e^2 y u_n + 4\hbar^2 \left(\frac{n\pi}{a}\right)^2 B_x^2 e^2 y^2 u_n \\
& - 4i\hbar B_x^2 B_z e^3 xy u_n - 4i\hbar B_y^2 B_x e^3 yz u_n - 4i_z^2 B_y e^3 zx u_n - 2\hbar^2 B_x^2 e^2 u_n \\
& - 2\hbar^2 B_y^2 e^2 u_n - 2\hbar^2 B_z^2 e^2 u_n - 4i\hbar^3 k_x k_y B_z e u_n
\end{aligned}
\tag{1.213}
$$

$$
\begin{aligned}
\hat{H}' = & -b'B_y^4e^4z^4 - 4b'\hbar k_xB_y^3e^3z^3 - 4b'\hbar^2k_x^2B_y^2e^2z^2 - 2b'\left[\hbar^2k_x^2 + \hbar^2k_y^2 + \hbar^2\left(\frac{n\pi}{a}\right)^2\right]B_y^2e^2z^2 \\
& + d'B_y^2e^2z^2 - 4b'\left[\hbar^2k_x^2 + \hbar^2k_y^2 + \hbar^2\left(\frac{n\pi}{a}\right)^2\right]\hbar k_xB_yez + 4bi\hbar^2k_yB_yB_ze^2z + 2a'\hbar k_xB_yez \\
& - b'B_x^4e^4y^4 - 2b'\left[\hbar^2k_x^2 + \hbar^2k_y^2 + \hbar^2\left(\frac{n\pi}{a}\right)^2\right]B_x^2e^2y^2 - 4b'\hbar^2\left(\frac{n\pi}{a}\right)^2B_x^2e^2y^2 + a'B_x^2e^2y^2 \\
& + 4b'i\hbar^2(k_yB_x + k_xB_y)B_xe^2y - b'B_z^4e^4x^4 - 4b'\hbar k_yB_z^3e^3x^3 - 4b'\hbar^2k_y^2B_z^2e^2x^2 \\
& - 2b'\left[\hbar^2k_x^2 + \hbar^2k_y^2 + \hbar^2\left(\frac{n\pi}{a}\right)^2\right]B_z^2e^2x^2 + a'B_z^2e^2x^2 - 4b'\left[\hbar^2k_x^2 + \hbar^2k_y^2 + \hbar^2\left(\frac{n\pi}{a}\right)^2\right]\hbar k_yB_zex \\
& + 4b'i\hbar^2k_xB_z^2e^2x + 2a'\hbar k_yB_zex - 2b'B_z^2B_y^2e^4x^2z^2 - 2b'B_y^2B_x^2e^4y^2z^2 - 4b'\hbar k_yB_y^2B_ze^3xz^2 \\
& - 4b'\hbar k_xB_z^2B_ye^3x^2z + 4b'\hbar(iB_ze - 2\hbar k_yk_x)B_yB_ze^2xz - 4b'\hbar k_xB_x^2B_ye^3y^2z + 4b'i\hbar B_y^2B_xe^3yz \\
& - 2b'B_x^2B_z^2e^4x^2y^2 - 4b'\hbar k_yB_x^2B_ze^3xy^2 + 4b'i\hbar B_x^2B_ze^3xy + 4b'\hbar^2\left(\frac{n\pi}{a}\right)B_y^2e^2z\cot\left(\frac{n\pi}{a}z\right) \\
& + 4b'i\hbar\left(\frac{n\pi}{a}\right)B_xB_y^2e^3yz^2\cot\left(\frac{n\pi}{a}z\right) + 8b'i\hbar^2k_x\left(\frac{n\pi}{a}\right)B_xB_ye^2yz\cot\left(\frac{n\pi}{a}z\right) \\
& + 4b'i\hbar\left(\frac{n\pi}{a}\right)B_x^3e^3y^3\cot\left(\frac{n\pi}{a}z\right) + 4b'i\hbar\left(\frac{n\pi}{a}\right)\left[\hbar^2k_x^2 + \hbar^2k_y^2 + \hbar^2\left(\frac{n\pi}{a}\right)^2\right]B_xey\cot\left(\frac{n\pi}{a}z\right) \\
& - 2a'i\hbar\left(\frac{n\pi}{a}\right)B_xey\cot\left(\frac{n\pi}{a}z\right) + 4b'\hbar^2\left(\frac{n\pi}{a}\right)B_xB_ze^2x\cot\left(\frac{n\pi}{a}z\right) \\
& + 4b'i\hbar\left(\frac{n\pi}{a}\right)B_xB_z^2e^3yx^2\cot\left(\frac{n\pi}{a}z\right) + 8b'i\hbar^2k_y\left(\frac{n\pi}{a}\right)B_xB_ze^2yx\cot\left(\frac{n\pi}{a}z\right) \\
& + 4b'\hbar^2\left(\frac{n\pi}{a}\right)^2B_x^2e^2y^2\cot^2\left(\frac{n\pi}{a}z\right) + 4b'\hbar^3\left(\frac{n\pi}{a}\right)(k_yB_x + k_xB_y)e\cot\left(\frac{n\pi}{a}z\right) \\
& + 2b'\hbar^2e^2(B_x^2 + B_y^2 + B_z^2) + 4b'i\hbar^3k_xk_yB_ze
\end{aligned}
\tag{1.214}
$$

So the first order perturbation to the energy is

$$
E_n^{(1)} = \hat{H}'_{nn} = \frac{\int u_n^*\hat{H}u_n dy}{\int u_n^*u_n dy} = \langle u_n|\hat{H}'|u_n\rangle \tag{1.215}
$$

Now

$$
\begin{aligned}
\int u_nu_n^*dy &= \frac{2}{L_xL_ya}\int_0^a \sin^2\left(\frac{n\pi}{a}z\right)dz\int_0^{l_x}dx\int_0^{l_y}dy = \frac{2}{a}\left[\frac{1}{2}\int_0^a\left\{1 - \cos\left(\frac{2n\pi}{a}z\right)\right\}dz\right] \\
&= \frac{2}{a}\cdot\frac{1}{2}\left[a - \frac{a}{2n\pi}\sin\frac{2n\pi}{a}z\Big|_0^a\right] = \frac{1}{a}\left[a - \frac{a}{2n\pi}\sin 2n\pi\right] = 1
\end{aligned}
\tag{1.216}
$$

Besides various averages are written below:

$$
\begin{aligned}
&\langle z^4\rangle = \frac{a^4}{5} - \frac{a^4}{n^2\pi^2} + \frac{a^4}{2n^4\pi^4},\ \langle z^3\rangle = \frac{a^3}{4} - \frac{3a^3}{4n^2\pi^2},\ \langle z^2\rangle = \frac{a^3}{3} - \frac{a^3}{2n^2\pi^2},\ \langle z\rangle = \frac{a}{2}, \\
&\langle y^4\rangle = \frac{1}{L_y}\int_0^{L_y} y^4dy\int_0^a\left(\frac{2}{a}\right)\sin^2\left(\frac{n\pi}{a}z\right)dz = \frac{L_y^4}{5},\ \langle y^2\rangle = \frac{L_y^2}{3},\ \langle y\rangle = \frac{L_y}{2},
\end{aligned}
$$

$$\langle x^4\rangle = \frac{L_x^4}{5},\ \langle x^3\rangle = \frac{L_x^3}{4},\ \langle x^2\rangle = \frac{L_x^2}{3},\ \langle x\rangle = \frac{L_x}{2},$$

$$\langle x^2z^2\rangle = \frac{1}{L_x}\int_0^{L_x} x^2dx\int_0^a\left(\frac{2}{a}\right)z^2\sin^2\left(\frac{n\pi}{a}z\right)dz = \frac{L_x^2}{3}\langle z^2\rangle,$$

$$\langle y^2z^2\rangle = \frac{L_y^2}{3}\langle z^2\rangle,\ \langle xz^2\rangle = \frac{L_x}{2}\langle z^2\rangle,\ \langle x^2z\rangle = \frac{L_x^2}{3}\langle z\rangle,\ \langle xz\rangle = \frac{L_x}{2}\langle z\rangle$$

$$\langle y^2z\rangle = \frac{L_y^2}{3}\langle z\rangle,\ \langle yz\rangle = \frac{L_y}{2}\langle z\rangle,\ \langle x^2y^2\rangle = \frac{L_x^2L_y^2}{9},\ \langle xy^2\rangle = \frac{L_xL_y^2}{6},\ \langle xy\rangle = \frac{L_xL_y}{4},$$

$$\begin{aligned}\left\langle z\cot\left(\frac{n\pi}{a}z\right)\right\rangle &= \frac{2}{a}\int_0^a z\cos\left(\frac{n\pi}{a}z\right)\sin\left(\frac{n\pi}{a}z\right)dz\\ &=\frac{1}{a}\left[\int_0^a z\sin\frac{2n\pi}{a}zdz\right] = \frac{1}{a}\left[-\frac{za}{2n\pi}\cos\frac{2n\pi}{a}z + \frac{a}{2n\pi}\int\cos\frac{2n\pi}{a}zdz\right]_0^a\\ &=\frac{1}{a}\left[-\frac{za}{2n\pi}\cos\frac{2n\pi}{a}z + \left(\frac{a}{2n\pi}\right)^2\sin\frac{2n\pi}{a}z\right]_0^a\\ &=\frac{1}{a}\left[-\frac{a^2}{2n\pi}\cos 2n\pi + \left(\frac{a}{2n\pi}\right)^2\sin 2n\pi\right] = -\frac{a}{2n\pi},\end{aligned}$$

$$\left\langle yz^2\cot\left(\frac{n\pi}{a}z\right)\right\rangle = \frac{L_y}{2}\left\langle z^2\cot\left(\frac{n\pi}{a}z\right)\right\rangle = \frac{L_ya^2(1 - 2n^2\pi^2)}{8n^3\pi^3}$$

where,

$$\begin{aligned}\left\langle z^2\cot\left(\frac{n\pi}{a}z\right)\right\rangle &= \frac{2}{a}\int_0^a z^2\cos\left(\frac{n\pi}{a}z\right)\sin\left(\frac{n\pi}{a}z\right)dz = \frac{1}{a}\left[\int_0^a z^2\sin\left(\frac{2n\pi}{a}z\right)dz\right]\\ &=\frac{1}{a}\left[-\frac{z^2a}{2n\pi}\cos\frac{2n\pi}{a}z + \frac{2a}{2n\pi}\int_0 z\cos\left(\frac{2n\pi}{a}z\right)dz\right]_0^a = \frac{1}{a}\left[-\frac{z^2a}{2n\pi}\cos\left(\frac{2n\pi}{a}z\right)\right.\\ &\left.+\ \frac{a}{n\pi}\left\{\frac{za}{2n\pi}\sin\frac{2n\pi}{a}z - \frac{a}{2n\pi}\int\sin\frac{2n\pi}{a}zdz\right\}\right]_0^a = \frac{1}{a}\left[-\frac{z^2a}{2n\pi}\cos\left(\frac{2n\pi}{a}z\right)\right.\\ &\left.+\ \left(\frac{a}{n\pi}\right)^2\frac{z}{2}\sin\left(\frac{n\pi}{a}z\right) + \frac{1}{4}\left(\frac{a}{n\pi}\right)^3\cos\frac{2n\pi}{a}z\right]_0^a = -\frac{a^2}{2n\pi}\cos 2n\pi\\ &+\ \frac{a^2}{n^2\pi^2}\frac{1}{2}\sin 2n\pi + \frac{1}{4}\frac{a^2}{n^3\pi^3}\cos 2n\pi = -\frac{a^2}{2n\pi} + \frac{a^2}{4n^3\pi^3} = \frac{a^2(1 - 2n^2\pi^2)}{4n^3\pi^3}\end{aligned}$$

$$\left\langle yz\cot\left(\frac{n\pi}{a}z\right)\right\rangle = \frac{L_y}{2}\left\langle z\cot\left(\frac{n\pi}{a}z\right)\right\rangle = \frac{L_y}{2}\left(-\frac{a}{2n\pi}\right) = -\frac{aL_y}{4n\pi},$$

$$\left\langle y^3\cot\left(\frac{n\pi}{a}z\right)\right\rangle = \frac{L_y^4}{4}\left\langle\cot\frac{n\pi}{a}z\right\rangle = \frac{L_y^4}{4}\left[\frac{2}{a}\int\limits_0^a \sin\left(\frac{n\pi}{a}z\right)\cos\left(\frac{n\pi}{a}z\right)dz\right]$$
$$= \frac{L_y^4}{4}\left[\frac{1}{a}\int\limits_0^a \sin\left(\frac{2n\pi}{a}z\right)dz\right] = \frac{L_y^4}{4a}\left[-\frac{a}{2n\pi}\cos\frac{2n\pi}{a}z\right]_0^a$$
$$= \frac{L_y^4}{4a}\left[-\frac{a}{2n\pi}\{\cos 2n\pi - 1\}\right] = 0,$$

$$\left\langle y\cot\left(\frac{n\pi}{a}z\right)\right\rangle = 0, \left\langle yx^2\cot\left(\frac{n\pi}{a}z\right)\right\rangle = 0, \left\langle x\cot\left(\frac{n\pi}{a}z\right)\right\rangle$$
$$= 0, \left\langle yx\cot\left(\frac{n\pi}{a}z\right)\right\rangle = 0$$

$$\left\langle y^2\cot^2\left(\frac{n\pi}{a}z\right)\right\rangle = \frac{L_y^2}{3}\left[\frac{2}{a}\int\limits_0^a \cos^2\left(\frac{n\pi}{a}z\right)dz\right] = \frac{L_y^2}{3}\frac{1}{a}\left[\int\limits_0^a\left(1 + \cos\frac{2n\pi}{a}\right)dz\right]$$
$$= \frac{L_y^2}{3}\cdot\frac{1}{a}\left[a + \frac{a}{2n\pi}\sin 2n\pi\right] = \frac{L_y^2}{3}$$

So the first order perturbation to the energy is

$$\begin{aligned}
E_n^{(1)} = & -b'B_y^4e^4\langle z^4\rangle - 4b'\hbar k_xB_y^3e^3\langle z^3\rangle - 4b'\hbar^2k_x^2B_y^2e^2\langle z^2\rangle + a'B_y^2e^2\langle z^2\rangle \\
& - 2b'\left[\hbar^2k_x^2 + \hbar^2k_y^2 + \hbar^2\left(\frac{n\pi}{a}\right)^2\right]B_y^2e^2\langle z^2\rangle \\
& - 4b'\left[\hbar^2k_x^2 + \hbar^2k_y^2 + \hbar^2\left(\frac{n\pi}{a}\right)^2\right]\hbar k_xB_ye\langle z\rangle + 2a'\hbar k_xB_ye\langle z\rangle + 4bi\hbar^2k_yB_yB_ze^2\langle z\rangle \\
& - b'B_x^4e^4\frac{L_y^4}{5} - 2b'\left[\hbar^2k_x^2 + \hbar^2k_y^2 + \hbar^2\left(\frac{n\pi}{a}\right)^2\right]B_x^2e^2\frac{L_y^2}{3} - 4b'\hbar^2\left(\frac{n\pi}{a}\right)^2B_x^2e^2\frac{L_y^2}{3} \\
& + a'B_x^2e^2\frac{L_y^2}{3} + 4b'i\hbar^2(k_yB_x + k_xB_y)B_xe^2\frac{L_y}{2} \\
& - b'B_z^4e^4\frac{L_x^4}{5} - 4b'\hbar k_yB_z^3e^3\frac{L_x^3}{4} - 4b'\hbar^2k_y^2B_z^2e^2\frac{L_x^2}{3} \\
& - 2b'\left[\hbar^2k_x^2 + \hbar^2k_y^2 + \hbar^2\left(\frac{n\pi}{a}\right)^2\right]B_z^2e^2\frac{L_x^2}{3} + a'B_z^2e^2\frac{L_x^2}{3} - 4b'\left[\hbar^2k_x^2\right. \\
& \left. + \hbar^2k_y^2 + \hbar^2\left(\frac{n\pi}{a}\right)^2\right]\hbar k_yB_ze\frac{L_x}{2} + 2a'\hbar k_yB_ze\frac{L_x}{2} + 4b'i\hbar^2k_xB_z^2e^2\frac{L_x}{2} \\
& - 2b'B_z^2B_y^2e^4\frac{L_x^2}{3}\langle z^2\rangle - 2b'B_y^2B_x^2e^4\frac{L_y^2}{3}\langle z^2\rangle \\
& - 4b'\hbar k_yB_y^2B_ze^3\frac{L_x}{2}\langle z^2\rangle - 4b'\hbar k_xB_z^2B_ye^3\frac{L_x^2}{3}\langle z\rangle
\end{aligned}$$

$$+ 4b'\hbar(iB_z e - 2\hbar k_y k_x)B_y B_z e^2 \frac{L_x}{2}\langle z\rangle - 4b'\hbar k_x B_x^2 B_y e^3 \frac{L_y^2}{3}\langle z\rangle$$

$$+ 4b' i\hbar B_y^2 B_x e^3 \frac{L_y}{2}\langle z\rangle - 2b' B_x^2 B_z^2 e^4 \left(\frac{L_x^2 L_y^2}{9}\right) - 4b'\hbar k_y B_x^2 B_z e^3 xy^2 \frac{L_x L_y^2}{6}$$

$$+ 4b' i\hbar B_x^2 B_z e^3 \left(\frac{L_x L_y}{4}\right) + 4b'\hbar^2 \left(\frac{n\pi}{a}\right) B_y^2 e^2 \left(-\frac{a}{2n\pi}\right)$$

$$+ 4b' i\hbar \left(\frac{n\pi}{a}\right) B_x B_y^2 e^3 \frac{L_y}{2}\left(-\frac{a^2}{2n\pi} + \frac{a^2}{4n^3\pi^3}\right)$$

$$+ 8b' i\hbar k_x \left(\frac{n\pi}{a}\right) B_x B_y e^2 \left(-\frac{a L_y}{4n\pi}\right) + 4b'\hbar^2 \left(\frac{n\pi}{a}\right)^2 B_x^2 e^2 \frac{L_x^2}{3}$$

$$+ 2b'\hbar^2 e^2 (B_x^2 + B_y^2 + B_z^2) + 4b' i\hbar^3 k_x k_y B_z e$$

$$E_n = E_n^{(0)} + E_n^{(1)}$$

$$a_2 a_0 = (a_2 - a_2')$$

The electron energy spectrum in HD QWs of III–V semiconductors in the presence of an arbitrarily oriented quantizing magnetic field B can be written as

$$\begin{aligned}
\gamma_3(E,\eta_g) &= \frac{1}{2m_c}\left[\hbar^2 k_x^2 + \hbar^2 k_y^2 + \hbar^2\left(\frac{n\pi}{a}\right)^2\right] - \alpha\left[\frac{\hbar^2 k_x^2}{2m_c} + \frac{\hbar^2 k_y^2}{2m_c} + \frac{\hbar^2}{2m_c}\left(\frac{n\pi}{a}\right)^2\right]^2 \\
&+ \frac{e^2 B_y^2}{2m_c}\langle z^2\rangle + \frac{2\hbar k_x e B_y}{2m_c}\langle z\rangle \\
&+ \frac{1}{3}\frac{e^2 B_x^2 L_y^2}{2m_c} + \frac{1}{3}\frac{e^2 B_z^2 L_x^2}{2m_c} + \frac{\hbar k_y e B_z L_x}{2m_c} - \alpha\left(\frac{e^2 B_y^2}{2m_c}\right)^2 \langle z^4\rangle \\
&- 4\alpha\left(\frac{\hbar k_x e^3 B_y^3}{4m_c^2}\right)\langle z^3\rangle - 4\alpha\left(\frac{\hbar k_x e B_y}{2m_c}\right)^2 \langle z^2\rangle \\
&- 2\alpha\left[\hbar^2 k_x^2 + \hbar^2 k_y^2 + \hbar^2\left(\frac{n\pi}{a}\right)^2\right]\left(\frac{eB_y}{2m_c}\right)^2 \langle z^2\rangle \\
&- 4\alpha\left[\hbar^2 k_x^2 + \hbar^2 k_y^2 + \hbar^2\left(\frac{n\pi}{a}\right)^2\right]\frac{\hbar k_x e B_y}{4m_c^2}\langle z\rangle - \frac{\alpha}{5}\left(\frac{e^2 L_y^2 B_x^2}{2m_c}\right)^2 \\
&- \left(\frac{2}{3}\right)\alpha\left[\hbar^2 k_x^2 + \hbar^2 k_y^2 + \hbar^2\left(\frac{n\pi}{a}\right)^2\right]\left(\frac{eL_y B_x}{2m_c}\right)^2 - \frac{4}{3}\alpha\hbar^2\left(\frac{n\pi}{a}\right)^2\left(\frac{eL_y B_x}{2m_c}\right)^2 \\
&- \frac{\alpha}{5}\left(\frac{e^2 L_x^2 B_z^2}{2m_c}\right)^2 - \frac{\alpha\hbar k_y e^3 L_x^3 B_z^3}{4m_c^2} - \frac{4}{3}\alpha\left(\frac{\hbar k_y e L_x B_z}{2m_c}\right)^2
\end{aligned}$$

$$
\begin{aligned}
&-\frac{2}{3}\alpha\left[\hbar^2k_x^2 + \hbar^2k_y^2 + \hbar^2\left(\frac{n\pi}{a}\right)^2\right]\left(\frac{eL_xB_z}{2m_c}\right)^2 - 2\alpha\left[\hbar^2k_x^2 + \hbar^2k_y^2 + \hbar^2\left(\frac{n\pi}{a}\right)^2\right]\\
&\frac{\hbar k_y eL_xB_z}{4m_c^2} - \frac{2}{3}\alpha\left(\frac{e^2L_xB_yB_z}{2m_c}\right)^2\langle z^2\rangle - \frac{2}{3}\alpha\left(\frac{e^2L_yB_yB_x}{2m_c}\right)^2\langle z^2\rangle\\
&-2\alpha\left[\frac{\hbar k_y e^3L_xB_y^2B_z}{4m_c^2}\right]\langle z^2\rangle - \frac{4}{3}\alpha\left[\frac{\hbar k_x e^3L_x^2B_z^2B_y}{4m_c^2}\right]\langle z\rangle\\
&-4\alpha\left[\frac{\hbar^2k_yk_xe^2L_xB_yB_z}{4m_c^2}\right]\langle z\rangle - \frac{4}{3}\alpha\left[\frac{\hbar k_x e^3L_y^2B_x^2B_y}{4m_c^2}\right]\langle z\rangle\\
&-\frac{2}{9}\left(\frac{e^2L_xL_yB_xB_z}{2m_c}\right)^2 - \frac{2}{3}\alpha\hbar k_y e^3L_x\left(\frac{B_xL_y}{2m_c}\right)^2B_z\\
&+\frac{4}{3}\alpha\left[\hbar\left(\frac{n\pi}{a}\right)\frac{eL_yB_x}{2m_c}\right]^2 + 2\alpha\left(\frac{e\hbar}{2m_c}\right)^2[B_x^2 + B_z^2]
\end{aligned}
\tag{1.218}
$$

Under the condition $\alpha \to 0$, (1.218) get simplified as

$$
\frac{1}{2m_c} = \left[\hbar^2k_x^2 + \hbar^2k_y^2 + \hbar^2\left(\frac{n\pi}{a}\right)^2\right] = \gamma_3(E,\eta_g) - a_1k_x - a_2k_y - a_3
\tag{1.219}
$$

where $a_1 = \frac{2\hbar eB_y}{2m_c}\langle z\rangle$, $a_2 = \frac{\hbar eB_zL_x}{2m_c}$, $a_3 = \frac{e^2B_y^2}{2m_c}\langle z^2\rangle + \frac{1}{3}\frac{e^2B_x^2}{2m_c}L_y^2 + \frac{1}{3}\frac{e^2B_z^2}{2m_c}L_x^2$.

Using the method of successive approximation from (1.218) and (1.219) we can write

$$
\begin{aligned}
\gamma_3(E,\eta_g) &= b_1k_x^2 + b_2k_y^2 + b_3 - \alpha[(\gamma_3(E,\eta_g) - a_3) - a_1k_x - a_2k_y]^2\\
&- \alpha(a_7 + a_9 + a_{12})[\gamma_3(E,\eta_g) - a_3 - a_1k_x - a_2k_y]\\
&- \alpha a_8[\gamma_3(E,\eta_g) - a_3 - a_1k_x - a_2k_y]k_x - \alpha a_{13}[\gamma_3(E,\eta_g) - a_3\\
&- a_1k_x - a_2k_y]k_y - \alpha a_6k_x^2 + [a_1 - \alpha(a_5 + a_{15} + a_{17})]k_x\\
&- \alpha a_{16}k_xk_y - \alpha a_{11}k_y^2 + [a_2 - \alpha(a_{10} + a_{14} + a_{18})]k_y + a_3 + \alpha a_4
\end{aligned}
\tag{1.220}
$$

where $b_1 = \dfrac{\hbar^2}{2m_c}$, $b_2 = \dfrac{\hbar^2}{2m_c}$, $b_3 = \dfrac{\hbar^2}{2m_c}\left(\dfrac{n\pi}{a}\right)^2$,

$$a_4 = 2\left(\frac{e\hbar}{2m_c}\right)^2 (B_x^2 + B_z^2) + \frac{4}{3}\left[\frac{\hbar}{2m_c}\left(\frac{n\pi}{a}\right)eL_yB_x\right]^2 - \frac{2}{9}\left(\frac{e^2L_xL_yB_xB_z}{2m_c}\right)^2$$
$$-\frac{2}{3}\left[\left(\frac{e^2L_xB_yB_z}{2m_c}\right)^2 + \left(\frac{e^2L_yB_yB_x}{2m_c}\right)^2\right]\langle z^2\rangle - \frac{4}{3}\hbar^2\left(\frac{n\pi}{a}\right)^2\left(\frac{eL_yB_x}{2m_c}\right)^2$$
$$-\frac{1}{5}\left[\left(\frac{e^2L_x^2B_z^2}{2m_c}\right)^2 + \left(\frac{e^2L_y^2B_x^2}{2m_c}\right)^2\right] - \frac{e^2B_y^2}{2m_c}\langle z^4\rangle$$
$$a_5 = 4\left(\frac{\hbar e^3B_y^3}{4m_c^2}\right)\langle z^3\rangle,\ a_6 = 4\left(\frac{\hbar eB_y}{2m_c}\right)^2\langle z^2\rangle,\ a_7 = 2\left(\frac{eB_y}{2m_c}\right)^2\langle z^2\rangle,\ a_8$$
$$= 4\left(\frac{\hbar eB_y}{2m_c^2}\right)\langle z\rangle,\ a_9 = \frac{2}{3}\left(\frac{eL_yB_x}{2m_c}\right)^2,\ a_{10} = \frac{\hbar e^3L_x^3B_z^3}{4m_c^2},$$
$$a_{11} = \frac{4}{3}\left(\frac{\hbar eL_xB_z}{2m_c}\right)^2,\ a_{12} = \frac{2}{3}\left(\frac{eL_xB_z}{2m_c}\right)^2,\ a_{13} = \frac{2\hbar eL_xB_z}{4m_c^2},$$
$$a_{14} = 2\left[\frac{\hbar e^3L_xB_y^2B_z}{4m_c^2}\right]\langle z^2\rangle,\ a_{15} = \frac{4}{3}\left[\frac{\hbar e^3L_x^2B_z^2B_y}{4m_c^2}\right]\langle z\rangle,$$
$$a_{16} = 4\left[\frac{\hbar^2e^2L_xB_yB_z}{4m_c^2}\right]\langle z\rangle,\ a_{17} = \frac{4}{3}\left[\frac{\hbar e^3L_y^2B_x^2B_y}{4m_c^2}\right]\langle z\rangle,\ a_{18} = \frac{2}{3}\hbar e^3L_x\left(\frac{B_xL_y}{2m_c}\right)^2B_z$$

(1.220) can be written as

$$\begin{aligned}
&[b_1 - \alpha a_1^2 + \alpha a_8a_1 - \alpha a_6]k_x^2 + [b_2 - \alpha a_2^2 + \alpha a_{13}a_2 - \alpha a_{11}]k_y^2 \\
&+ [a_1 - \alpha(a_5 + a_{15} + a_{17}) - \alpha a_8(\gamma_3(E,\eta_g) - a_3) \\
&+ \alpha a_1(a_7 + a_9 + a_{12}) + 2\alpha a_1(\gamma_3(E,\eta_g) - a_3)]k_x \\
&+ [a_2 - \alpha(a_{10} + a_{14} + a_{18}) - \alpha a_{13}(\gamma_3(E,\eta_g) - a_3) \\
&+ \alpha(a_7 + a_9 + a_{12})a_2 + 2\alpha a_2(\gamma_3(E,\eta_g) - a_3)]k_y \\
&+ \alpha(a_{13}a_1 + a_8a_2 - 2a_1a_2 - a_{16})k_xk_y \\
&+ [b_3 - (\gamma_3(E,\eta_g) - \alpha(\gamma_3(E,\eta_g - a_3)^2 \\
&- \alpha(a_7 + a_9 + a_{12})(\gamma_3(E,\eta_g - a_3) + a_3 + \alpha a_4] = 0
\end{aligned} \tag{1.221}$$

(1.221) can be expressed as

$$P_1k_x^2 + Q_1k_y^2 + k_xk_y + P_2k_x + Q_2k_y + c = 0 \tag{1.222}$$

where

$$\begin{aligned}
\mathrm{P}_1 &= [b_1 - \alpha a_1^2 + \alpha a_8 a_1 - \alpha a_6] \qquad Q_1 = [b_2 - \alpha a_2^2 + \alpha a_{13} a_2 - \alpha a_{11}] \\
P_2 &= [a_1 - \alpha(a_5 + a_{15} + a_{17}) - \alpha a_8(\gamma_3(E,\eta_g - a_3) + \alpha a_1(a_7 + a_9 + a_{12}) \\
&\quad + 2\alpha a_1(\gamma_3(E,\eta_g) - a_3))] \\
\mathrm{Q}_2 &= [a_2 - \alpha(a_{10} + a_{14} + a_{18}) - \alpha a_{13}(\gamma_3(E,\eta_g) - a_3) + \alpha(a_7 + a_9 + a_{12})a_2 \\
&\quad + 2\alpha a_2(\gamma_3(E,\eta_g) - a_3))] \\
\mathrm{c} &= [b_3 - (\gamma_3(E,\eta_g) - \alpha(\gamma_3(E,\eta_g) - \mathrm{a}_3)^2 - \alpha(a_7 + a_9 + a_{12}) \\
&(\gamma_3(E,\eta_g) - a_3) + a_3 + {}_4] = 0
\end{aligned}$$

Let us substitute $k_x = k'_x + \alpha_1, \quad k_y = k'_y + \beta_1$ where

$$\alpha_1 = \frac{RQ_2 - 2Q_1P_2}{R^2 - 4Q_1P_1}, \ \beta_1 = \frac{2P_1Q_2 - RP_2}{R^2 - 4Q_1P_1}$$

(1.222) can be expressed as

$$Ak_x'^2 + Bk_y'^2 + 2Hk'_xk'_y = 1 \tag{1.223}$$

where

$$\begin{aligned}
A &= \frac{P_1}{C_1}, \ B = \frac{Q_1}{C_1}, \ 2H = \frac{R}{C_1} \text{ and } C_1 \\
&= C - P_1\alpha_1^2 - Q_1\beta_1^2 + R\alpha_1\beta_1 + P_2\alpha_1 - Q_2\beta_1
\end{aligned}$$

The area of the ellipse is given by

$$A_{19} = \frac{\pi}{\sqrt{AB - H^2}} \tag{1.224}$$

The EEM can be written as

$$m^*(E, n, B_x, B_y, B_z) = \frac{\hbar^2}{2\pi} A'_{19} \tag{1.225}$$

The DOS function in this case is given by

$$N(E) = \frac{2g_v}{d_x d_y d_z} \sum_{n=0}^{n_{\max}} \sum_{l=0}^{l_{\max}} \sum_{r=0}^{r_{\max}} \delta'(E - E_{19,16}) \tag{1.226}$$

where $E_{19,16}$ is the totally quantized energy in this case.

$$c = 0 \tag{1.227}$$

1.3 Results and Discussion

The DR in QWs of HD III–V semiconductors in the presence of magnetic field is given by (1.32). The EEM and DOS function for (1.32) can be expressed through (1.34) and (1.35) respectively. It appears that magnetic field makes the mass quantum number dependent. The DR in NWs of HD III–V Semiconductors in the presence of magnetic field is given by (1.56b) and the EEM and DOS function in this context are expressed through (1.56c) and (1.56d) respectively. The DR and DOS function for QDs of HD III–V semiconductors in the presence of magnetic field are given by (1.70a) and (1.70b) respectively. The DR in QWsof HD III–V semiconductors in the presence of cross fields is given by (1.77). The DR in NWs of HD III–V Semiconductors in the presence of the electric field E_0 is along z direction and the crossed magnetic field B is along y direction has been investigated in (1.78) where as the DR in the present case for the electric field E_0 is along x direction and the crossed magnetic field B is along y direction is given by (1.81). The DR and the DOS function in QDs of HD III–V Semiconductors in the presence of the electric field E_0 is along z direction and the crossed magnetic field B is along y direction are given through (1.82a) and (1.82b) respectively, where as for the electric field E_0 is along x direction and the crossed magnetic field B is along y direction the DR and the DOS function in this case are explored through (1.83a) and (1.83b) respectively. The DR, EEM and the DOS function for QWs of HD IV–VI semiconductors in the presence of magnetic field in accordance with Cohen model can be expressed through (1.99), (1.113) and (1.114) respectively, where as the DR in accordance with the model of McClure and Choi in this case is given by (1.127). The DR in NWs of HD IV–VI semiconductors in the presence of magnetic field in accordance with the model of McClure and Choi is given by (1.138), whereas the same in accordance with Cohen model can be expressed through (1.147). The DR and the DOS function for QDs of HD IV–VI semiconductors in the presence of magnetic field in accordance with Cohen model can be expressed through (1.156a) and (1.156b), where as the said quantities in accordance with the model of McClure and Choi in the present case are given by (1.168a) and (1.168b) respectively. The DR and the DOS function for Cylindrical QDs of HD III–V semiconductors in the presence of crossed electric and magnetic fields are expressed through (1.191) and (1.191a) respectively. The DR, EEM and the DOS function for QWsof HD III–V semiconductors in the presence of arbitrarily oriented magnetic field are given by (1.222), (1.225) and (1.226) respectively.

Using the values of the energy band constants as given in Table 1.1, we have ploted in Fig. 1.1 the zero point energy in QWs of (a) InSb, (b) n-$In_{1-x}As_xGa_yP_{1-y}$ lattice matched InP and (c) $Hg_{1-x}Cd_xTe$ as a function of magnetic field. It appears that the zero point energy increases with increasing magnetic field were the numerical values are totally energy band constant dependent. For InSb the values is the greatest were us $Hg_{1-x}Cd_xTe$ the values of the zero point energy are the least. The values of the zero point energy for QWs of quaternary materials are in between that of InSb and $Hg_{1-x}Cd_xTe$. The Fig. 1.2 exhibits the zero point energy as a

Table 1.1 Numerical values of the energy band constants of few materials

Sl. no	Materials	Numerical values of the energy band constants
1	The conduction electrons of n-Cadmium Germanium Arsenide can be described by three types of band models	1. The values of the energy band constants in accordance with the generalized electron dispersion relation of nonlinear optical materials are as follows $E_{g_0} = 0.57$ eV, $\Delta_{\parallel} = 0.30$ eV, $\Delta_{\perp} = 0.36$ eV, $m_{\parallel}^{*} = 0.034m_0$, $m_{\perp}^{*} = 0.039m_0$, $T = 4$ K, $\delta = -0.21$ eV, $g_v = 1$ [175, 176], $\varepsilon_{sc} = 18.4\varepsilon_0$ [177] (ε_{sc} and ε_0 are the permittivity of the semiconductor material and free space respectively) and $W = 4$ eV [178] 2. In accordance with the three band model of Kane, the spectrum constants are given by $\Delta = (\Delta_{\parallel} + \Delta_{\perp})/2 = 0.33$ eV, $E_{g_0} = 0.57$ eV, $m_c = \left(m_{\parallel}^{*} + m_{\perp}^{*}\right)/2 = 0.0365m_0$ and $\delta = 0$ eV 3. In accordance with two band model of Kane, $E_{g_0} = 0.57$ eV and $m_c = 0.0365m_0$
2	n-Indium Arsenide	The values $E_{g_0} = 1.55$ eV, $\Delta = 0.35$ eV, $m_c = 0.07m_0$, $g_v = 1$, $\varepsilon_{sc} = 12.25\varepsilon_0$ [179] and $W = 5.06$ eV [180] are valid for three band model of Kane
3	n-Gallium Arsenide	The values $E_{g_0} = 1.55$ eV, $\Delta = 0.35$ eV, $m_c = 0.07m_0$, $g_v = 1$, $\varepsilon_{sc} = 12.9\varepsilon_0$ [179] and $W = 4.07$ eV [181] are valid for three band model of Kane. The values $\alpha_{11} = -2132 \times 10^{-40}$ eVm^4, $\alpha_{12} = 9030 \times 10^{-50}$ eVm^5, $\beta_{11} = -2493 \times 10^{-40}$ eVm^4, $\beta_{12} = 12594 \times 10^{-50}$ eVm^5, $\gamma_{11} = 30 \times 10^{-30}$ eVm^3, $\gamma_{12} = -154 \times 10^{-42}$ eVm^4 [183]
4	n-Gallium Aluminium Arsenide	$E_{g_0} = (1.424 + 1.266x + 0.26x^2)$ eV, $\Delta = (0.34 - 0.5x)$ eV, $m_c = [0.066 + 0.088x]m_0$, $g_v = 1$, $\varepsilon_{sc} = [13.18 - 3.12x]\varepsilon_0$ [184] and $W = (3.64 - 0.14x)$ eV [185]
5	n-Mercury Cadmium Telluride	$E_{g_0} = (-0.302 + 1.93x + 5.35 \times 10^{-4}(1 - 2x)T - 0.810x^2 + 0.832x^3)$ eV, $\Delta = (0.63 + 0.24x - 0.27x^2)$ eV, $m_c = 0.1m_0 E_{g_0}(\text{eV})^{-1}$, $g_v = 1$, $\varepsilon_{sc} = [20.262 - 14.812x + 5.22795x^2]\varepsilon_0$ [186] and $W = (4.23 - 0.813(E_{g_0} - 0.083))$ eV [187]
6	n-Indium Gallium Arsenide Phosphide lattice matched to Indium Phosphide	$E_{g_0} = (1.337 - 0.73y + 0.13y^2)$ eV, $\Delta = (0.114 + 0.26y - 0.22y^2)$ eV, $m_c = (0.08 - 0.039y)m_0$, $y = (0.1896 - 0.4052x)/(0.1896 - 0.0123x)$, $g_v = 1$, $\varepsilon_{sc} = [10.65 + 0.1320y]\varepsilon_0$ and $W(x, y) = [5.06(1 - x)y + 4.38(1 - x)(1 - y) + 3.64xy + 3.75\{x(1 - y)\}]$ eV [188]
7	n-Indium Antimonide	

(continued)

Table 1.1 (continued)

Sl. no	Materials	Numerical values of the energy band constants
		$E_{g_0} = 0.2352$ eV, $\Delta = 0.81$ eV, $m_c = 0.01359 m_0$, $gv = 1$, $\varepsilon_{sc} = 15.56\varepsilon_0$ [179] and $W = 4.72$ eV [180]
8	n-Gallium Antimonide	The values of $E_{g_0} = 0.81$ eV, $\Delta = 0.80$ eV, $P = 9.48 \times 10^{-10}$ eVm, $\bar{\zeta}_0 = -2.1$, $\bar{v}_0 = -1.49$, $\bar{\omega}_0 = 0.42$, $g_v = 1$ [189] and $\varepsilon_{sc} = 15.85\varepsilon_0$ [189, 190] are valid for the model of Seiler et al. [189] as given by (3.85a). The values $\bar{E}_1 = 1.024$ eV, $\bar{E}_2 = 0$ eV, $\bar{E}_3 = -1.132$ eV, $\bar{E}_4 = 0.05$ eV, $\bar{E}_5 = 1.107$ eV, $\bar{E}_6 = -0.113$ eV and $\bar{E}_7 = -0.0072$ eV [190] are valid for the model of Zhang [190]
9	n-Cadmium Sulphide	$m^*_{\parallel} = 0.7 m_0$, $m^*_{\perp} = 1.5 m_0$, $\bar{\lambda}_0 = 1.4 \times 10^{-8}$ eVm, $g_v = 1$ [179], $\varepsilon_{sc} = 15.5\varepsilon_0$ [191] and $W = 4.5$ eV [179, 180, 191]
10	n-Lead Telluride	The values $m_t^- = 0.070 m_0$, $m_l^- = 0.54 m_0$, $m_t^+ = 0.010 m_0$, $m_l^+ = 1.4 m_0$, $P_{\parallel} = 141$ meV nm, $P_{\perp} = 486$ meV nm, $E_{g_0} = 190$ meV, $g_v = 4$ [179], $\varepsilon_{sc} = 33\varepsilon_0$ [179, 192] and $W = 4.6$ eV [193] are valid for the Dimmock model [194]. The values $(\overline{R})^2 = 2.3 \times 10^{-19} (\text{eVm})^2$, $E_{g_0} = 0.16$ eV, $(\bar{s})^2 = 4.6(\overline{R})^2$, $\Delta'_c = 3.07$ eV, $(\overline{Q})^2 = 1.3(\overline{R})^2$, $\Delta''_c = 3.28$ eV, $(\overline{A})^2 = 1.66 \times 10^{-19} (\text{eVm})^2$ [195] and $W = 4.21$ eV [180] are valid for the model of Bangert and Kastner [195]. The values $m_{tv} = 0.0965 m_0$, $m_{lv} = 1.33 m_0$, $m_{tc} = 0.088 m_0$, $m_{lc} = 0.83 m_0$ [193] are valid for the model of Foley et al. [193]
11	Stressed n-Indium Antimonide	The values $m^* = 0.048 m_0$, $E_{g_0} = 0.081$ eV, $B_2 = 9 \times 10^{-10}$ eVm, $C_1^c = 3$ eV, $C_2^c = 2$ eV, $a_0 = -10$ eV, $b_0 = -1.7$ eV, $d = -4.4$ eV, $S_{xx} = 0.6 \times 10^{-3} (\text{kbar})^{-1}$, $S_{yy} = 0.42 \times 10^{-3} (\text{kbar})^{-1}$, $S_{zz} = 0.39 \times 10^{-3} (\text{kbar})^{-1}$, $S_{xy} = 0.5 \times 10^{-3} (\text{kbar})^{-1}$, $\varepsilon_{xx} = \sigma S_{xx}$, $\varepsilon_{yy} = \sigma S_{yy}$, $\varepsilon_{zz} = \sigma S_{zz}$, $\varepsilon_{xy} = \sigma S_{xy}$, σ is the stress in kilobar, $g_v = 1$ [198] are valid for the model of Seiler et al. [198]
12	Mercury Telluride	$m_v^* = 0.028 m_0$, $gv = 1$, $\varepsilon_{\infty} = 15.2\varepsilon_0$ [201] and $W = 5.5$ eV [202]
13	Platinum Antimonide	For valence bands, along <111> direction, $\lambda_0 = 0.33$ eV, $l = 1.09$ eV, $\nu = 0.17$ eV, $\bar{\bar{n}} = 0.22$ eV, $\bar{\bar{a}} = 0.643$ nm, $I = 0.30\,(\text{eV})^2$, $\delta_0 = 0.33$ eV, $g_v = 8$ [203], $\varepsilon_{sc} = 30\varepsilon_0$ [204] and $\phi \approx 3.0$ eV [205]

(continued)

Table 1.1 (continued)

Sl. no	Materials	Numerical values of the energy band constants
14	n-Gallium Phosphide	$m_{\parallel}^{*} = 0.92m_0$, $m_{\perp}^{*} = 0.25m_0$, $k_0 = 1.7 \times 10^{19} m^{-1}$, $\lvert V_G \rvert = 0.21$ eV, $g_v = 1$, $g_s = 2$ [206] and $W = 3.75$ eV [180]
15	Germanium	$E_{g_0} = 0.785$ eV, $m_{\parallel}^{*} = 1.57m_0$, $m_{\perp}^{*} = 0.0807m_0$ [181] and $W = 4.14$ eV [180]
16	Tellurium	The values $\psi_1 = 6.7 \times 10^{-16}$ meVm^2, $\psi_2 = 4.2 \times 10^{-16}$ meVm^2, $\psi_3 = 6 \times 10^{-8}$ meVm and $\psi_4 = 3.6 \times 10^{-8}$ meVm [207] are valid for the model of Bouat et al. [207] The values $t_1 = 0.06315$ eV, $t_2 = -10.0\hbar^2/2m_0$, $t_3 = -5.55\hbar^2/2m_0$, $t_4 = 0.3 \times 10^{-36}$ eVm^4, $t_5 = 0.3 \times 10^{-36}$, $t_6 = -5.55\hbar^2/2m_0$, $t_7 = 6.18 \times 10^{-20}$ (eVm)^2 [208] and $W = 1.9708$ eV [209] are valid for the model of Ortenberg and Button [208]
17	Lead Germanium Telluride	The values $\overline{E_{g_0}} = 0.21$ eV, $g_v = 4$ [212] and $\phi \approx 6$ eV [213] are valid for the model of Vassilev [212]
18	Cadmium Antimonide	The values $A_{10} = -4.65 \times 10^{-19}$ eVm^2, $A_{11} = -2.035 \times 10^{-19}$ eVm^2, $A_{12} = -5.12 \times 10^{-19}$ eVm^2, $A_{13} = -0.25 \times 10^{-10}$ eVm, $A_{14} = 1.42 \times 10^{-19}$ eVm^2, $A_{15} = 0.405 \times 10^{-19}$ eVm^2, $A_{16} = -4.07 \times 10^{-19}$ eVm^2, $A_{17} = 3.22 \times 10^{-10}$ eVm, $A_{18} = 1.69 \times 10^{-20}$ eVm^2, $A_{19} = 0.070$ eV [214] and $\phi \approx 2$ eV [215] are valid for the model of Yamada [214]
19	Cadmium Diphosphide	The values $\beta_1 = 8.6 \times 10^{-21}$ eVm^2, $\beta_2 = 1.8 \times 10^{-21}$ (eVm)^2, $\beta_4 = 0.0825$ eV, $\beta_5 = -1.9 \times 10^{-19}$ eVm^2 [216] and $\phi \approx 5$ eV [217] are valid for the model of Chuiko [216]
20	Zinc Diphosphide	The values $\beta_1 = 8.7 \times 10^{-21}$ eVm^2, $\beta_2 = 1.9 \times 10^{-21}$ (eVm)^2, $\beta_4 = 0.0875$ eV, $\beta_5 = -1.9 \times 10^{-19}$ eVm^2 [216] and $W \approx 3.9$ eV [217] are valid for the model of Chuiko [216]
21	Bismuth Telluride	The values $E_{g_0} = 0.145$ eV, $\alpha_{11} = 3.25$, $\alpha_{22} = 4.81$, $\alpha_{33} = 9.02$, $\alpha_{23} = 4.15$, $g_s = 2$, $g_v = 6$ [218] and $\phi = 5.3$ eV [219] are valid for the model of Stordeur et al. [218]
22	Zinc Selenide	$m_2^{*} = 0.16m_0$, $\Delta_2 = 0.42$ eV, $E_{g_{02}} = 2.82$ eV [181] and $W = 3.2$ eV [225]
23	Lead Selenide	$m_t^{-} = 0.23m_0$, $m_l^{-} = 0.32m_0$, $m_t^{+} = 0.115m_0$, $m_l^{+} = 0.303m_0$, $P_{\parallel} \approx 138$ meVnm, $P_{\perp} = 471$ meVnm, $E_{g_0} = 0.28$ eV [226], $\varepsilon_{sc} = 21.0\varepsilon_0$ [206] and $W = 4.2$ eV [227]

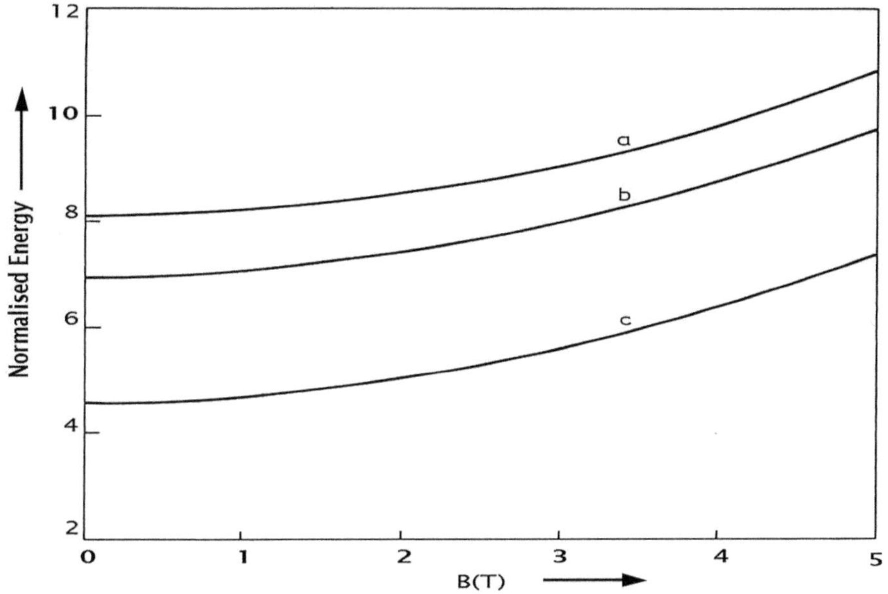

Fig. 1.1 The plot of the zero point energy in QWs of III–V materials (*a*) InSb, (*b*) n-$In_{1-x}As_xGa_yP_{1-y}$ lattice matched InP and (*c*) $Hg_{1-x}Cd_xTe$ as a function of magnetic field

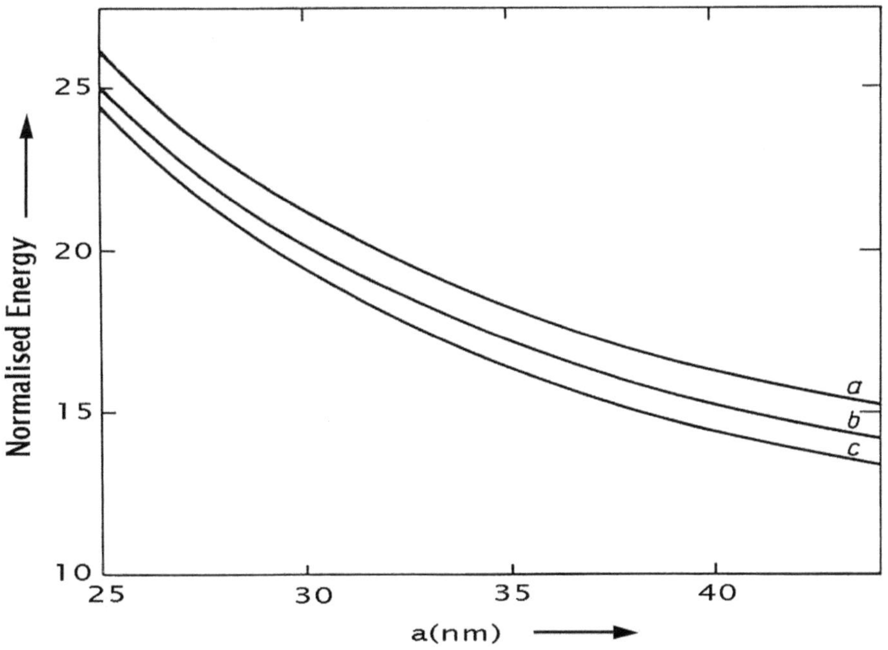

Fig. 1.2 Plot of the zero point energy as a function of film thickness in QWs of IV–VI materials in accordance with (*a*) PbS (using Cohen model), (*b*) PbTe (using MeClure and Choi model) and (*c*) PbSe (using Lax model)

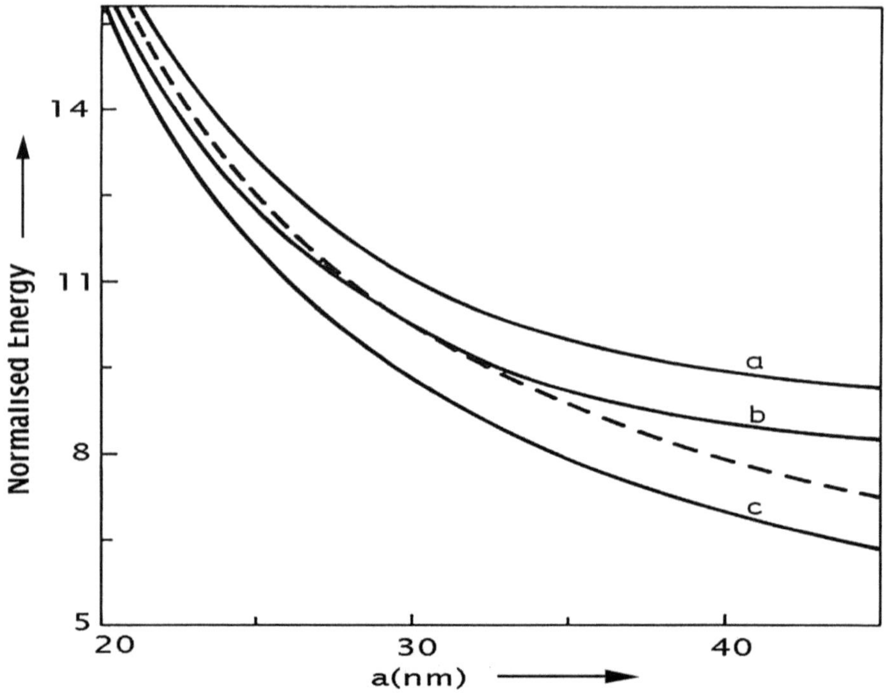

Fig. 1.3 Plot of the zero point energy as a function of film thickness in NWs of IV–VI materials in accordance with (*a*) PbS (using Cohen model), (*b*) PbTe (usingMeClure and Choi model) and (*c*) PbSe (using Lax model). The *dashed plot* shows the same in NWsfor $Pb_{1-x}Sn_xTe$ in accordance with Meclure and Choi model

function of film thickness in QWs of IV–VI materials in accordance with (a) PbS (using Cohen model), (b) PbTe (using MeClure and Choi model) and (c) PbSe (using Lax model). It appears that as film thickness increases, the zero point energy decreases for all the models of QWs of IV–VI materials. The Fig. 1.3 contains the plots of the zero point energy as a function of film thickness in NWs of IV–VI materials in accordance with (a) PbS (using Cohen model), (b) PbTe (using MeClure and Choi model) and (c) PbSe (using Lax model). The dashed plot shows the same in NWs for $Pb_{1-x}Sn_xTe$ in accordance with Meclure and Choi model.It appears that as film thickness increases, the zero point energy decreases for all the models of NWs of IV–VI materials although the rates of change are rather different as compared with the corresponding zero point energy in QWs of IV–VI materials in the presence of magnetic field.

In Fig. 1.4 we have plotted of the zero point energy as a function of magnetic field in NWs of IV–VI materials in accordance with (a) PbS (using Cohen model), (b) PbTe (using MeClure and Choi model) and (c) PbSe (using Lax model) respectively. It appears that the zero point energy increases with increasing magnetic field where the numerical values are totally band structure dependent.In

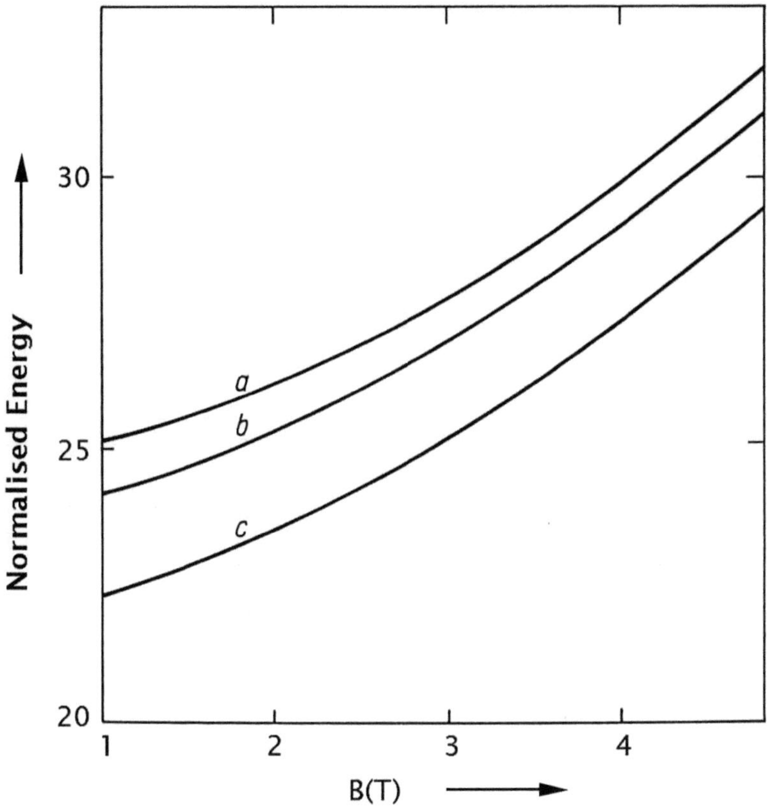

Fig. 1.4 Plot of the zero point energy as a function of magnetic field in NWs of IV–VI materialsin accordance with (*a*) PbS (using Cohen model), (*b*) PbTe (using MeClure and Choi model) and (*c*) PbSe (using Lax model)

Fig. 1.5 we have plotted the zero point energy as a function of magnetic field in QD of IV–VI materials in accordance with (a) Cohen model (taking n-PbS as an example of IV–VI semiconductors), (b) MeClure and Choi model (for PbTe) and (c) Lax model(for PbSe). It appears that the influences of energy band models on the zero point energy are prominent. The zero point energy increases with increasing magnetic field in all the cases but the numerical magnitude is the greatest in accordance with Cohen model although for small values of magnetic field, the zero point energy is invariant with respect to magnetic field for QD of n-PbS in accordance with Cohen model and least for QD of PbSe in accordance with Lax model where the variation is largest. For McClure and Choi model, QD of PbTe exhibits the numerical values of zero point energy which falls in between Cohen model and Lax model.

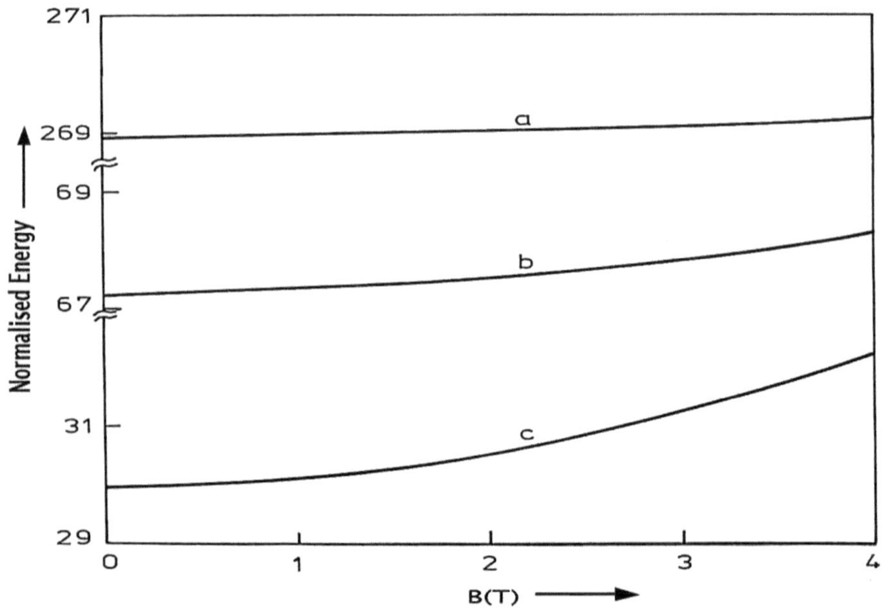

Fig. 1.5 Plot of the zero point energy as a function of magnetic field in Quantum Dot of IV–VI materials in accordance with (*a*) PbS (using Cohen model), (*b*) PbTe (using MeClure and Choi model) and (*c*) PbSe (using Lax model)

In Fig. 1.6 we have plotted the energy eigenvalue corresponding to the lowest quantum state (E'_{101}) in cylindrical quantum dot of n-GaAs (assuming parabolic energy bands) in the presence of cross electric and quantizing magnetic fields as a function of width of the dot for three different combinations of the electric fields and magnetic fields (a) $E' = 0$, B = 3T; (b) 40 kV/m, 3T; and (c) 40 kV/m, B = 0. The radius of the dot is 20 nm. The dashed curve corresponds to the absence of any field. Figure 1.6 shows that the lowest quantized energy level is lowered for larger values of the thickness (d) of the cylindrical quantum dot. If the electric field is increased, the rate of decrease of the energy level at a particular thickness increases whereas larger magnetic fields do not change the rate but change the value of the energy (E'_{101}). The Fig. 1.7 exhibits the plots of the energy eigenvalue corresponding to the lowest quantum state (E'_{101}) in the presence of cross electric and quantizing magnetic fields in cylindrical quantum dot of (a) InSb, (b) GaAs and (c) $Hg_{1-x}Cd_xTe$ as a function of electric field. As the electric field is increased, the energy (E'_{101}) decreases with the rate of decrease being higher at larger values of the electric field. However, the lowest energy level is shifted at a rapid rate at higher values of the magnetic field.

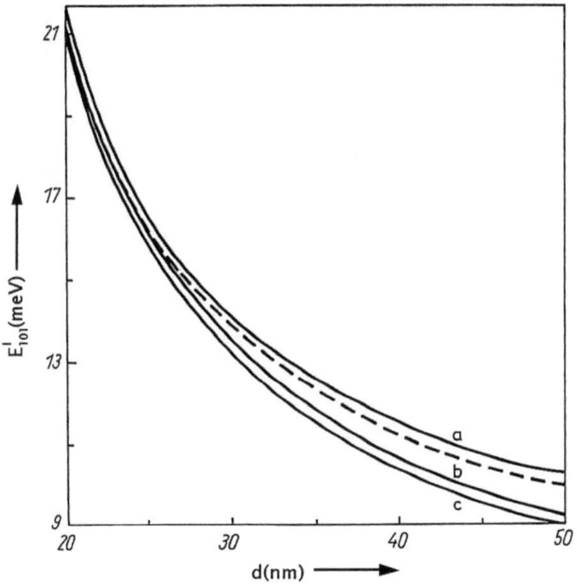

Fig. 1.6 The plot of the energy eigenvalue corresponding to the lowest quantum state (E'_{101}) in cylindrical quantum dot of n-GaAs (assuming parabolic energy bands) in the presence of cross electric and quantizing magnetic fieldsas a function of width of the dot for three different combinations of the electric fields and magnetic fields (*a*) $E' = 0$, B = 3T; (*b*) 40 kV/m, 3T; and (*c*) 40 kV/m, B = 0. The radius of the dot is 20 nm. The *dashed curve* corresponds to the absence of any field

In Fig. 1.8 we have plotted of the energy eigenvalue corresponding to the lowest quantum state (E'_{101}) in the presence of cross electric and quantizing magnetic fields in cylindrical quantum dot of a (a) n-$Hg_{1-x}Cd_xTe$, (b) Gallium Aluminum Arsenide respectively as a function of the radius of the dot. The upper dotted and lower dotted curves are valid for n-InSb and n-GaP respectively. The bulk DR obeys the two band model of Kane. The Fig. 1.8 illustrates the dependence of the energy of the lowest quantized level on the radius (a) of the dot. In general, the level is shifted downwards for larger values of the radius. With the application of higher magnetic fields, the rate of downshifting of the level at a particular radius decreases. The electric field lowers the lowest quantized state but does not change the rate.

In Fig. 1.9 we have plotted the energy eigenvalue corresponding to the lowest quantum state (E'_{101}) in the presence of cross electric and quantizing magnetic fields in cylindrical quantum dot of a (a) n-$Hg_{1-x}Cd_xTe$, (b) Gallium Aluminum Arsenide respectively as a function of the width of the dot. The upper dotted and lower dotted curves are valid for n-InSb and n-GaP respectively. It appears that rate of decrease is also lowered with the increase in the values of d. This figure shows the variation

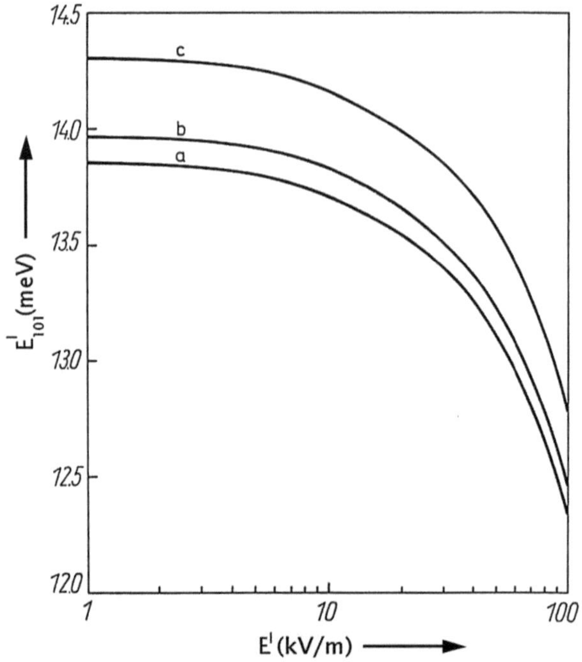

Fig. 1.7 The plot of the energy eigenvalue corresponding to the lowest quantum state (E'_{101}) in the presence of cross electric and quantizing magnetic fields in cylindrical quantum dot of (*a*) InSb, (*b*) GaAs and (*c*) $Hg_{1-x}Cd_xTe$ as a function of electric field

of the lowest quantized state with the width of the dot for various III–V quantum dot materials. The nature of the variation of (E'_{101}) with d for all the materials depends on the values of the energy band constants. The Fig. 1.10 explores the zero point energy as a function of the orientation of the magnetic field in QWs of (a) InSb, (b) GaAs and (c) $Hg_{1-x}Cd_xTe$. The dotted plots correspond to $\alpha = 0$. The rate of change of the eigenvalue with the orientation of the magnetic field is larger when the orientation is near the transverse axis than when it is nearer to the longitudinal axis. It may be noted that similar computations can alsobe extended for all types of quantum wells, wires and dots conveniently for investigating the quantized states beyond the lowest one.

For the purpose of condensation we have plotted very few cases with the hope that the readers will perform all the computer programming for the purpose of creating new physics for effective electron mass, sub band energies and other important transport quantities which are totally DR dependent. The numerical results presented in this chapter would be different for other materials but the nature of variation would be unaltered. The theoretical results as given here would be useful in analyzing various other experimental data related to this phenomenon. We must note that the study of transport phenomena and the formulation of the electronic properties of low dimensional field aidedHD compounds are based on the DRs in such materials. It is worth remarking that this simplified formulation

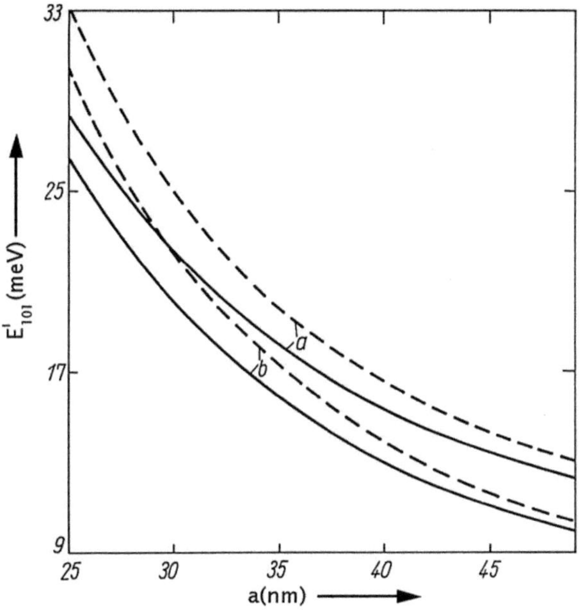

Fig. 1.8 The plot of the energy eigenvalue corresponding to the lowest quantum state (E'_{101}) in the presence of cross electric and quantizing magnetic fields in cylindrical quantum dot of a (*a*) n-$Hg_{1-x}Cd_xTe$, (*b*) Gallium Aluminum Arsenide respectively as a function of the radius of the dot. The *upper dotted* and *lower dotted curves* are valid for n-InSb and n-GaP respectively. The bulk DR obeys the two band model of Kane

exhibits the basic qualitative features of the DRs for field assisted low dimensional materials. The basic objective of this chapter is not solely to demonstrate the influence of quantum confinement on the DRs for quantum confined HD non-parabolic materials in the presence of external fields but also to formulate the appropriate DOS and effective electron mass in the most generalized form, since the transport and other phenomena in such nano structured materials having different band structures and the derivation of the expressions of many important electronic properties are based on the DOS in such compounds. Finally, we can write that the analysis as presented in this chapter can be used to investigate the Burstein Moss shift, the carrier contribution to the elastic constants, the specific heat, activity coefficient, reflection coefficient, Hall coefficient, plasma frequency, various scattering mechanisms and other different transport coefficients of modern HD non-parabolic quantum confined field aided HDdevices operated under different external conditions having varying band structures.

Fig. 1.9 The plot of the energy eigenvalue corresponding to the lowest quantum state (E'_{101}) in thepresence of cross electric and quantizing magnetic fields in cylindrical quantum dot of a (*a*) n-$Hg_{1-x}Cd_xTe$, (*b*) Gallium Aluminum Arsenide respectively as a function of the width of the dot. The *upper dotted* and *lower dotted curves* are valid for n-InSb and n-GaP respectively

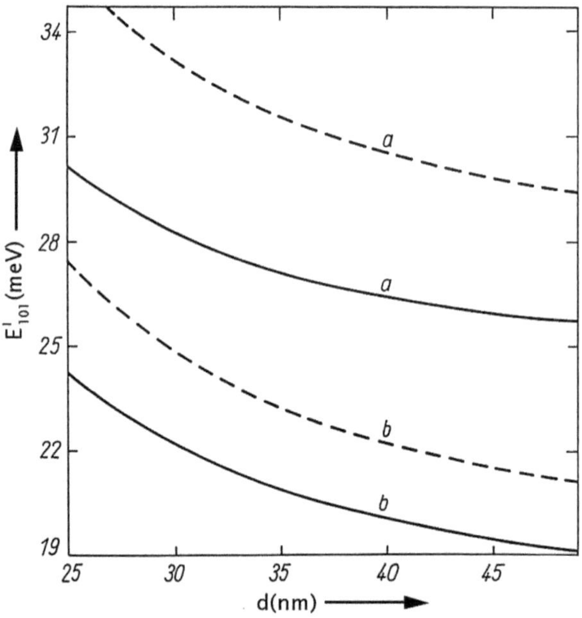

Fig. 1.10 Plot of the zero point energy as a function of the orientation of the magnetic field in QWs of (*a*) InSb, (*b*) GaAs and (*c*) $Hg_{1-x}Cd_xTe$. The *dotted plots* correspond to $\alpha = 0$

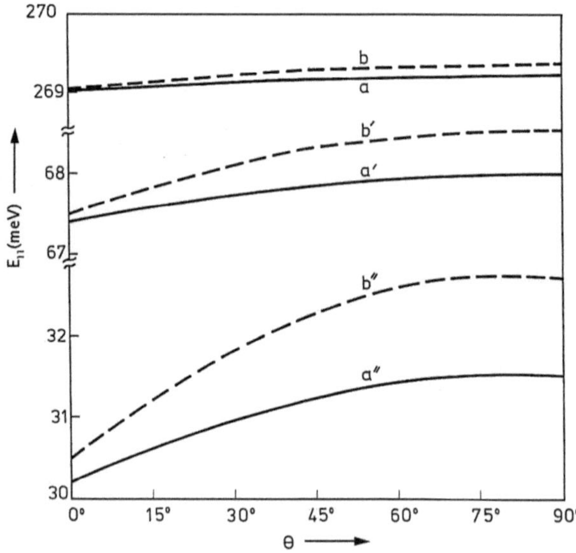

1.4 Open Research Problems

The problems under these sections of this monograph are by far the most important part for the readers and few open research problems are presented from this chapter till end. The numerical values of the energy band constants for various semiconductors are given in Appendix A for the related computer simulations.

(R.1.1) Investigate the DR for the quantum confined HD semiconductors whose respective DRs of the carriers in the absence of band tails are given below:

(a) The electron dispersion law in n-GaP can be written as [228]

$$E = \frac{\hbar^2 k_s^2}{2m_{\parallel}^*} + \frac{\hbar^2 k_s^2}{2m_{\perp}^*} \mp \frac{\overline{\Delta}}{2} \pm \left[\left(\frac{\overline{\Delta}}{2} \right)^2 + P_1 k_z^2 + D_1 k_x^2 k_y^2 \right]^{1/2} \quad \text{(R.1.1)}$$

where, $\overline{\Delta} = 335$ meV, $P_1 = 2 \times 10^{-10}$ eVm, $D_1 = P_1 a_1$ and $a_1 = 5.4 \times 10^{-10}$ m.

(b) The dispersion relation for the conduction electrons for IV–VI semiconductors can also be described by the models of Cohen [229], McClure and Choi [230], Bangert et al. [231], Foley et al. [232] and the models of Takaoka et al. [233] respectively.

(i) In accordance with Cohen [229], the dispersion law of the carriers is given by

$$E(1 + \alpha E) = \frac{p_x^2}{2m_1} + \frac{p_z^2}{2m_3} - \frac{\alpha E p_y^2}{2m_2'} + \left(\frac{\alpha p_y^4}{4m_2 m_2'} \right) + \frac{p_y^2}{2m_2}(1 + \alpha E) \quad \text{(R.1.2)}$$

where m_1, m_2 and m_3 are the effective carrier masses at the band-edge along x, y and z directions respectively and m_2' is the effective- mass tensor component at the top of the valence band (for electrons) or at the bottom of the conduction band (for holes).

(ii) The carrier energy spectra can be written, following McClure and Choi [230] as

$$E(1 + \alpha E) = \frac{p_x^2}{2m_1} + \frac{p_y^2}{2m_2} + \frac{p_z^2}{2m_3} + \frac{p_y^2}{2m_2}\alpha E \left\{ 1 - \left(\frac{m_2}{m_2'} \right) \right\} + \frac{p_y^4 \alpha}{4m_2 m_2'} - \frac{\alpha p_x^2 p_y^2}{4m_1 m_2} - \frac{\alpha p_y^2 p_z^2}{4m_2 m_3} \quad \text{(R.1.3)}$$

(iv) The carrier energy spectrum of IV–VI semiconductors in accordance with Foley et al. [232] can be written as

$$E + \frac{E_g}{2} = E_-(k) + \left[\left[E_+(k) + \frac{E_g}{2}\right]^2 + P_\perp^2 k_s^2 + P_\parallel^2 k_z^2\right]^{1/2} \quad \text{(R.1.4)}$$

where, $E_+ = (\mathrm{k}) = \frac{\hbar^2 k_s^2}{2m_\perp^+} + \frac{\hbar^2 k_z^2}{2m_\parallel^+}$, $E_- = (\mathrm{k}) = \frac{\hbar^2 k_s^2}{2m_\perp^-} + \frac{\hbar^2 k_z^2}{2m_\parallel^-}$ represents the contribution from the interaction of the conduction and the valance band edge states with the more distant bands and the free electron term, $\frac{1}{m_\perp^\pm} = \frac{1}{2}\left[\frac{1}{m_{tc}} \pm \frac{1}{m_{tv}}\right], \frac{1}{m_\parallel^\pm} = \frac{1}{2}\left[\frac{1}{m_{lc}} \pm \frac{1}{m_{lv}}\right],$

For n-PbTe $\mathrm{P}_\perp = 4.61 \times 10^{-10}$ eVm, $\mathrm{P}_\parallel = 4.61 \times 10^{-10}$ eVm, $\frac{m_0}{m_{tv}} = 10.36, \frac{m_0}{m_{tv}} = 0.75, \frac{m_0}{m_{tc}} = 11.36, \frac{m_0}{m_{lc}} = 1.20$, and $g_v = 4$

(v) The carrier DR in accordance with the model of Takaoka et al. [233] can be written as

$$E\left(1 + \frac{E}{E_g}\right) - \frac{\beta\hbar^2 k_y^2}{2M_2} - \frac{\gamma\hbar^4 k_y^4}{4M_2^2 E_g} = \frac{\hbar^2 k_x^2}{2m_1} + \frac{\hbar^2 k_z^2}{2m_3}$$

where $\beta = 1 + \frac{E}{E_g}(1 - \gamma) + \delta, \gamma = \frac{M_2}{M_2'}, \delta = \frac{M_2}{m_2}$ and the notations are defined in [233]

c) The conduction electrons of n-GaSb obey the following two dispersion relations:

(i) In accordance with the model of Seiler et al. [234]

$$E = \left[-\frac{E_g}{2} + \frac{E_g}{2}[1 + \alpha_4 k^2] + \frac{\bar{\zeta}_0 \hbar^2 k^2}{2m_0} + \frac{\bar{v}_0 f_1(k)\hbar^2}{2m_0} \pm \frac{\bar{\omega}_0 f_2(k)\hbar^2}{2m_0}\right] \quad \text{(R.1.5)}$$

where $\alpha_4 \equiv 4P^2\left(E_g + \frac{2}{3}\Delta\right)\left[E_g^2(E_g + \Delta)\right]^{-1}$, P is the isotropic momentum matrix element, $f_1(k) \equiv k^{-2}\left[k_x^2 k_y^2 + k_y^2 k_z^2 + k_z^2 k_x^2\right]$ represents the warping of the Fermi surface, $f_2(k) = \left[\left\{k^2\left(k_x^2 k_y^2 + k_y^2 k_z^2 + k_z^2 k_x^2\right) - 9k_x^2 k_y^2 k_z^2\right\}^{1/2} k^{-1}\right]$ represents the inversion asymmetry splitting of the conduction band and $\bar{\zeta}_0$, $\bar{v}_0$ and $\bar{\omega}_0$ represent the constants of the electron spectrum in this case.

(ii) In accordance with the model of Zhang et al. [235]

$$E = \left[E_2^{(1)} + E_2^{(2)} K_{4,1}\right] k^2 + \left[E_4^{(1)} + E_4^{(2)} K_{4,1}\right] k^4 + k^6\left[E_6^{(1)} + E_6^{(2)} K_{4,1} + E_6^{(3)} K_{6,1}\right] \quad \text{(R.1.6)}$$

where $K_{4,1} \equiv \frac{5}{4}\sqrt{21}\left[\frac{k_x^4 + k_y^4 + k_z^4}{k^4} - \frac{3}{5}\right]$, $K_{6,1} \equiv \sqrt{\frac{639639}{32}}\left[\frac{k_x^2 k_y^2 k_z^2}{k^6} + \frac{1}{22}\left(\frac{k_x^4 + k_y^4 + k_z^4}{k^4} - \frac{3}{5}\right) - \frac{1}{105}\right]$, the coefficients are in eV, the values of k are $10\left(\frac{a}{2\pi}\right)$ times those of k in atomic units (a is the lattice constant), $E_2^{(1)} = 1.0239620$, $E_2^{(2)} = 0$, $E_4^{(1)} = -1.1320772$, $E_4^{(2)} = 0.05658$, $E_6^{(1)} = 1.1072073$, $E_6^{(2)} = -0.1134024$ and $E_6^{(3)} = -0.0072275$.

(d) In addition to the well-known band models of III–V semiconductors as discussed in this monograph, the conduction electrons of such compounds obey the following three dispersion relations:

(i) In accordance with the model of Rossler [236]

$$E = \frac{\hbar^2 k^2}{2m_c} + (\alpha_{11} + \alpha_{12}k)k^4 + (\beta_{11} + \beta_{12}k)\left[k_x^2 k_y^2 + k_y^2 k_z^2 + k_z^2 k_x^2\right] \pm (\gamma_{11} + \gamma_{12}k)\left[k^2\left(k_x^2 k_y^2 + k_y^2 k_z^2 + k_z^2 k_x^2\right) - 9k_x^2 k_y^2 k_z^2\right]^{1/2} \quad \text{(R.1.7)}$$

where, $\alpha_{11} = -2132 \times 10^{-40}\ \text{eVm}^4$, $\alpha_{12} = 9030 \times 10^{-50}\ \text{eVm}^5$, $\beta_{11} = -2493 \times 10^{-40}\ \text{eVm}^4$, $\beta_{12} = 12594 \times 10^{-50}\ \text{eVm}^5$, $\gamma_{11} = 30 \times 10^{-30}\ \text{eVm}^3$ and $\gamma_{12} = -15 \times 10^{-42}\ \text{eVm}^4$.

(ii) In accordance with Johnson and Dickey [237], the electron energy spectrum assumes the form

$$E = \frac{E_g}{2} + \frac{\hbar^2 k^2}{2}\left[\frac{1}{m_0} + \frac{1}{m_{\gamma b}}\right] + \frac{E_g}{2}\left[1 + 4\frac{\hbar^2 k^2 \bar{f}_1(E)}{2m_c'\ E_g}\right]^{1/2}$$

(R1.8)where, $\frac{m_0}{m_2'} \equiv P^2\left[\frac{\left(E_g + \frac{2\Delta}{3}\right)}{E_g(E_g + \Delta)}\right]$, $\bar{f}_1(E) \equiv \frac{(E_g + \Delta)\left(E + E_g + \frac{2\Delta}{3}\right)}{\left(E_g + \frac{2\Delta}{3}\right)(E + E_g + \Delta)}$, $m_c' = 0.139m_0$ and $m_{\gamma b} = \left[\frac{1}{m_c'} - \frac{2}{m_0}\right]^{-1}$.

(iii) In accordance with Agafonov et al. [238], the electron energy spectrum can be written as

$$E = \frac{\bar{\eta} - E_g}{2}\left[1 - \frac{\hbar^2 k^2}{2\bar{\eta} m^*}\left\{\frac{D\sqrt{3} - 3\bar{B}}{2\left(\frac{\hbar^2}{2m^*}\right)}\right\}\left[\frac{k_x^4 + k_y^4 + k_z^4}{k^4}\right]\right] \quad \text{(R.1.9a)}$$

where, $\bar{\eta} \equiv \left(E_g^2 + \frac{8}{3}P^2 k^2\right)^{1/2}$, $\bar{B} \equiv -21\frac{\hbar^2}{2m_0}$ and $D \equiv -40\left(\frac{\hbar^2}{2m_0}\right)$.

(iv) In accordance with the model of Kolodziejczak et al. [239], the electron energy spectrum of III–V compounds can be expressed, taking into account the interaction of the higher bands as

$$E = \frac{\hbar^2 k^2}{2m_0} + \frac{\chi_0 - E_g}{2} + a\left[\frac{\chi_0 + E_g}{2\chi_0}\right]\frac{\hbar^2 k^2}{2m_0} + \frac{\chi_0 - E_g}{2\chi_0}\Bigg]\frac{\hbar^2 k^2}{2m_0}\left[b + \frac{c}{5}\right] \tag{R.1.9b}$$

where $\chi_0 \equiv \left[E_g^2 - 4FE_g\left(\frac{E_g + \frac{2}{3}\Delta}{E_g + \Delta}\right)\left(\frac{\hbar^2 k^2}{2m_0}\right)\right]^{1/2}$
F represents the interaction between the $\Gamma_{25'}$ and Γ_{15} states,
$a \equiv \frac{-p^2[E(\Gamma_{15c}) - E(\Gamma_{15v})] - [E(\Gamma_{15}) - E(\Gamma_{25'})]}{2[E(\Gamma_{15c}) - E(\Gamma_{15v})][E(\Gamma_{15c}) - E(\Gamma_{15v}) - E_g]}$, $b \equiv \frac{1}{3}[M + 4G]$, M represents the interaction between $\Gamma_{25'}$ and Γ_{15} states, G represents the interaction between $\Gamma_{25'}$ and $\Gamma_{12'}$ states,
$c \equiv \frac{1}{2}\left[\frac{(F - G + M)^2 - (F + 2G - M)^2}{F + 2G - M}\right]$ and the other notations are the same as in the above reference.

(e) The dispersion relation of the carriers in n-type $Pb_{1-x}Ga_xTe$ with x = 0.01 can be written following Vassilev [240] as

$$\begin{aligned}&\left[E - 0.606k_s^2 - 0.0722k_z^2\right]\left[E + \bar{E}_g + 0.411k_s^2 + 0.0377k_z^2\right] = 0.23k_s^2 + 0.02k_z^2\\ &\pm\left[0.06\bar{E}_g + 0.061k_s^2 + 0.0066k_z^2\right]k_s\end{aligned} \tag{R.1.10}$$

where, $\bar{E}_g (=0.21\ \text{eV})$ is the energy gap for the transition point, the zero of the energy E is at the edge of the conduction band of the Γ point of the Brillouin zone and is measured positively upwards, k_x, k_y and k_z are in the units of $10^9\ \text{m}^{-1}$.

(f) The energy spectrum of the carriers in the two higher valance bands and the single lower valance band of Te can, respectively, be expressed as [241]

$$\begin{aligned}\bar{E} &= A_{10}k_z^2 + B_{10}k_s^2 \pm \left[\Delta_{10}^2 + (\beta_{10}k_z)^2\right]^{1/2} \text{ and } \bar{E}\\ &= \Delta_{\|} + A_{10}k_z^2 + B_{10}k_s^2 \pm \beta_{10}k_z\end{aligned} \tag{R.1.11a}$$

where, $\bar{E}$ is the energy of the hole as measured from the top of the valance and within it, $A_{10} = 3.77 \times 10^{-19}\ \text{eVm}^2$, $B_{10} = 3.57 \times 10^{-19}\ \text{eVm}^2$, $\Delta_{10} = 0.628\ \text{eV}$, $(\beta_{10})^2 = 6 \times 10^{-20}\ (\text{eVm})^2$ and $\Delta_{\|} = 1004 \times 10^{-5}\ \text{eV}$ are the spectrum constants.

(ii) The dispersion relation of the conduction electrons of Tellurium can be written in accordance with the model of Ortenberg and Button as [242]

$$E = t_1 + t_2k_z^2 + t_3k_s^2 + t_4k_s^4 + t_5k_s^2k_z^2 \pm \left[\left(t_1 + t_6k_s^2\right)^2 + t_7k_z^2\right]^{1/2} \tag{R.1.11b}$$

where t_1, t_2, t_3, t_4, t_5, t_6 and t_7 are the energy band constants.

(g) The dispersion relation of the holes in p-InSb can be written in accordance with Cunningham [243] as

$$\bar{E} = c_4(1 + \gamma_4 f_4)k^2 \pm \frac{1}{3}\left[2\sqrt{2}\sqrt{c_4}\sqrt{16 + 5\gamma_4}\sqrt{E_4}g_4 k\right] \quad \text{(R.1.12)}$$

where, $c_4 \equiv \frac{\hbar^2}{2m_0} + \theta_4$, $\theta_4 \equiv 4.7\frac{\hbar^2}{2m_0}$, $\gamma_4 \equiv \frac{b_4}{c_4}$, $b_4 \equiv \frac{3}{2}b_5 + 2\theta_4$, $b_5 \equiv 2.4\frac{\hbar^2}{2m_0}$, $f_4 \equiv \frac{1}{4}\left[\sin^2 2\theta + \sin^4\theta\sin^2 2\phi\right]$, θ is measured from the positive z-axis, ϕ is measured from positive x-axis, $g_4 \equiv \sin\theta\left[\cos^2\theta + \frac{1}{4}\sin^4\theta\sin^2 2\phi\right]$ and $E_4 = 5 \times 10^{-4}$ eV.

(h) The energy spectrum of the valance bands of CuCl in accordance with Yekimov et al. [244] can be written as

$$E_h = (\gamma_6 - 2\gamma_7)\frac{\hbar^2 k^2}{2m_0} \quad \text{(R.1.13)}$$

and

$$E_{l,s} = (\gamma_6 + \gamma_7)\frac{\hbar^2 k^2}{2m_0} - \frac{\Delta_1}{2} \pm \left[\frac{\Delta_1^2}{4} + \gamma_7\Delta_1\frac{\hbar^2 k^2}{2m_0} + 9\left(\frac{\gamma_7\hbar^2 k^2}{2m_0}\right)^2\right]^{1/2} \quad \text{(R.1.14)}$$

where, $\gamma_6 = 0.53$, $\gamma_7 = 0.07$, $\Delta_1 = 70$ meV.

(i) In the presence of stress, χ_6 along <001> and <111> directions, the energy spectra of the holes in semiconductors having diamond structure valance bands can be respectively expressed following Roman et al. [245] as

$$E = A_6 k^2 \pm \left[\bar{B}_7^2 k^4 + \delta_6^2 + B_7\delta_6\left(2k_z^2 - k_s^2\right)\right]^{1/2} \quad \text{(R.1.15)}$$

and

$$E = A_6 k^2 \pm \left[\bar{B}_7^2 k^4 + \delta_7^2 + \frac{D}{\sqrt{3}}\delta_7\left(2k_z^2 - k_s^2\right)\right]^{1/2} \quad \text{(R.1.16)}$$

where, A_6, B_7, D_6 and C_6 are inverse mass band parameters in which $\delta_6 \equiv l_7(\bar{S}_{11} - \bar{S}_{12})\chi_6$, $\bar{S}_{ij}$ are the usual elastic compliance constants, $\bar{B}_7^2 \equiv \left(B_7^2 + \frac{c_6^2}{5}\right)$ and $\delta_7 \equiv \left(\frac{d_8 S_{44}}{2\sqrt{3}}\right)\chi_6$. For gray tin, $d_8 = -4.1$ eV, $l_7 = -2.3$ eV, $A_6 = 19.2\frac{\hbar^2}{2m_0}$, $B_7 = 26.3\frac{\hbar^2}{2m_0}$, $D_6 = 31\frac{\hbar^2}{2m_0}$ and $c_6^2 = -1112\frac{\hbar^2}{2m_0}$.

(j) The DR of the carriers of cadmium and zinc diphosphides are given by [246]

$$E = \left[\beta_1 + \frac{\beta_2\beta_3(k)}{8\beta_4}\right]k^2 \pm \left\{\left[\beta_4\beta_3(k)x\left(\beta_5 - \frac{\beta_2\beta_3(k)}{8\beta_4}\right)k^2\right] + 8\beta_4^2\left(1 - \frac{\beta_3^2(k)}{4}\right) - \beta_2\left(1 - \frac{\beta_3^2(k)}{4}\right)k^2\right\}^{1/2} \tag{R.1.17}$$

where β_1, β_2, β_4 and β_5 are system constants and $\beta_3(k) = \frac{k_x^2 + k_y^2 - 2k_z^2}{k^2}$

(k) The E-k relation of the conduction electrons in semiconductors in the presence of electron-phonon interaction assumes the form [247]

$$E = \frac{\hbar^2 k^2}{2m_c} - \alpha_c \hbar\omega_0 \frac{p_0}{\hbar k}\tan^{-1}\left[\frac{\hbar^2 k^2}{2m_c(\hbar\omega_0 - E)}\right]^{1/2} \tag{R.1.18}$$

where α_c is the dimensionless coupling constant, $p_0 = (2m_c\hbar\omega_0)^{1/2}$ and ω_0 is the angular frequency of the optical phonon.

(R.1.2) Investigate the DR for bulk specimens of the HD semiconductors in the presences of exponential, Kane, Halperian, Lax and Bonch-Burevich types of band tails [37] for all systems whose unperturbed carrier energy spectra are defined in (R.1.1).

(R.1.3) Investigate the DR for QWs of all the HD semiconductors as considered in (R.1.2).

(R.1.4) Investigate the DR for HD bulk specimens of the negative refractive index, organic, magnetic and other advanced optical materials in the presence of an arbitrarily oriented alternating electric field.

(R.1.5) Investigate the DR for the QWs of HD negative refractive index, organic, magnetic and other advanced optical materials in the presence of an arbitrarily oriented alternating electric field.

(R.1.6) Investigate the DR for the multiple QWs of HD materials whose unperturbed carrier energy spectra are defined in (R.1.1).

(R.1.7) Investigate the DR for all the appropriate HD low dimensional systems of this chapter in the presence of finite potential wells.

(R.1.8) Investigate the DR for all the appropriate HD low dimensional systems of this chapter in the presence of parabolic potential wells.

(R.1.9) Investigate the DR for all the appropriate HD systems of this chapter forming quantum rings

(R.1.10) Investigate the DR for all the above appropriate problems in the presence of elliptical Hill and quantum square rings.

(R.1.11) Investigate the DR for triangular two dimensional systems in the presence of an arbitrarily oriented alternating electric field for all the HD materials whose unperturbed carrier energy spectra are defined in (R.1.1).

(R.1.12) Investigate the DR for HD two dimensional systems of the negative refractive index and other advanced optical materials in the presence of an arbitrarily oriented alternating electric field and non-uniform light waves.

(R.1.13) Investigate the DR for triangular HD two dimensional systems of the negative refractive index, organic, magnetic and other advanced optical materials in the presence of an arbitrarily oriented alternating electric field in the presence of strain.

(R.1.14) (a) Investigate the DR for HD two dimensional systems of the negative refractive index, organic, magnetic and other advanced optical materials in the presence of many body effects
(b) Investigate all the appropriate problems of this chapter for a Dirac electron.

(R.1.15) Investigate all the appropriate problems of this chapter by including the many body, image force, broadening and hot carrier effects respectively.

(R.1.16) Investigate all the appropriate problems of this chapter by removing all the mathematical approximations and establishing the respective appropriate uniqueness conditions.

References

1. P.M. Petroff, A.C. Gossard, W. Wiegmann, Appl. Phys. Lett. **45**, 620 (1984)
2. J.M. Gaines, P.M. Petroff, H. Kroemar, R.J. Simes, R.S. Geels, J.H. English, J. Vac. Sci. Technol., B **6**, 1378 (1988)
3. J. Cilbert, P.M. Petroff, G.J. Dolan, S.J. Pearton, A.C. Gossard, J.H. English, Appl. Phys. Lett. **49**, 1275 (1986)
4. T. Fujui, H. Saito, Appl. Phys. Lett. **50**, 824 (1987)
5. H. Sasaki, Jpn. J. Appl. Phys. **19**, 94 (1980)
6. P.M. Petroff, A.C. Gossard, R.A. Logan, W. Weigmann, Appl. Phys. Lett. **41**, 635 (1982)
7. H. Temkin, G.J. Dolan, M.B. Panish, S.N.G. Chu, Appl. Phys. Lett. **50**, 413 (1988)
8. I. Miller, A. Miller, A. Shahar, U. Koren, P.J. Corvini, Appl. Phys. Lett. **54**, 188 (1989)
9. L.L. Chang, H. Esaki, C.A. Chang, L. Esaki, Phys. Rev. Lett. **38**, 1489 (1977)
10. K. Hess, M.S. Shur, J.J. Drunnond, H. Morkoc, IEEE Trans. Electron. Devices **ED-30**, 07 (1983)
11. G. Bastard, *Wave Mechanics Applied to Semiconductor Hetero-structures, Les Editions de Physique* (Halsted, Les Ulis, New York, 1988)
12. M.J. Kelly, *Low Dimensional Semiconductors: Materials, Physics, Technology, Devices* (Oxford University Press, Oxford, 1995)
13. C. Weisbuch, B. Vinter, *Quantum Semiconductor Structures* (Boston Academic Press, Boston, 1991)
14. N.T. Linch, Festkorperprobleme **23**, 27 (1985)
15. D.R. Sciferes, C. Lindstrom, R.D. Burnham, W. Streifer, T.L. Paoli, Electron. Lett. **19**, 170 (1983)
16. P.M. Solomon, Proc. IEEE **70**, 489 (1982)
17. H. Kroemer, Rev. Mod. Phys. **73**, 783 (2001)
18. D. Kasemet, C.S. Hong, N.B. Patel, P.D. Dapkus, Appl. Phys. Lett. **41**, 912 (1982)
19. K. Woodbridge, P. Blood, E.D. Pletcher, P.J. Hulyer, Appl. Phys. Lett. **45**, 16 (1984)
20. S. Tarucha, H.O. Okamoto, Appl. Phys. Lett. **45**, 16 (1984)
21. V.V. Martin, A.A. Kochelap, M.A. Stroscio, *Quantum Heterostructures* (Cambridge University Press, Cambridge, 1999)

22. B. Mitra, K.P. Ghatak, Phys. Lett. A **357**, 146 (1990)
23. S. Debbarma, A. Bhattacharjee, S. Bhattacharyya, A. Mondal, K.P. Ghatak, J. Adv. Phys. **1**, 84 (2012)
24. U. Kunze, Phys. Rev. B **41**, 1707 (1990)
25. P. Harrison, *Quantum wells, wires and dots* (John Wiley and Sons, Ltd, 2002); B.K. Ridley, *Electrons and Phonons in Semiconductors Multilayers* (Cambridge University Press, Cambridge, 1997)
26. C.S. Lent, D.J. Kirkner, J. Appl. Phys. **67**, 6353 (1990); F. Sols, M. Macucci, U. Ravaioli, K. Hess, Appl. Phys. Lett. **54**, 350 (1980)
27. C.S. Kim, A.M. Satanin, Y.S. Joe, R.M. Cosby, Phys. Rev. B **60**, 10962 (1999)
28. S. Midgley, J.B. Wang, Phys. Rev. B **64**, 153304 (2001)
29. T. Sugaya, J.P. Bird, M. Ogura, Y. Sugiyama, D.K. Ferry, K.Y. Jang, Appl. Phys. Lett. **80**, 434 (2002)
30. B. Kane, G. Facer, A. Dzurak, N. Lumpkin, R. Clark, L. PfeiKer, K. West, Appl. Phys. Lett. **72**, 3506 (1998)
31. C. Dekker, Phys. Today **52**, 22 (1999)
32. A. Yacoby, H.L. Stormer, N.S. Wingreen, L.N. Pfeiffer, K.W. Baldwin, K.W. West, Phys. Rev. Lett. **77**, 4612 (1996)
33. Y. Hayamizu, M. Yoshita, S. Watanabe, H.A.L. PfeiKer, K. West, Appl. Phys. Lett. **81**, 4937 (2002)
34. S. Frank, P. Poncharal, Z.L. Wang, W.A. Heer, *Science* **280**, 1744 (1998)
35. I. Kamiya, I.I. Tanaka, K. Tanaka, F. Yamada, Y. Shinozuka, H. Sakaki, Physica E **13**, 131 (2002)
36. A.K. Geim, P.C. Main, N. LaScala, L. Eaves, T.J. Foster, P.H. Beton, J.W. Sakai, F.W. Sheard, M. Henini, G. Hill et al., Phys. Rev. Lett. **72**, 2061 (1994)
37. A.S. Melinkov, V.M. Vinokur, *Nature* **415**, 60 (2002)
38. K. Schwab, E.A. henriksen, J.M. Worlock, M.L. Roukes, *Nature* **404**, 974 (2000)
39. L. Kouwenhoven, *Nature* **403**, 374 (2000)
40. S. Komiyama, O. Astafiev, V. Antonov, H. Hirai, *Nature* **403**, 405 (2000)
41. E. Paspalakis, Z. Kis, E. Voutsinas, A.F. Terziz, Phys. Rev. B **69**, 155316 (2004)
42. J.H. Jefferson, M. Fearn, D.L.J. Tipton, T.P. Spiller, Phys. Rev. A **66**, 042328 (2002)
43. J. Appenzeller, C. Schroer, T. Schapers, A. Hart, A. Froster, B. Lengeler, H. Luth, Phys. Rev. B **53**, 9959 (1996)
44. J. Appenzeller, C. Schroer, J. Appl. Phys. **87**, 31659 (2002)
45. P. Debray, O.E. Raichev, M. Rahman, R. Akis, W.C. Mitchel, Appl. Phys. Lett. **74**, 768 (1999)
46. T.E. Schlesinger, T. Kuech, Appl. Phys. Lett. **49**, 519 (1986)
47. H. Heiblum, D.C. Thomas, C.M. Knoedler, M.I. Nathan, Appl. Phys. Letts. **47**, 1105 (1985)
48. O. Aina, M. Mattingly, F.Y. Juan, P.K. Bhattacharyya, Appl. Phys. Letts. **50**, 43 (1987)
49. I. Suemune, L.A. Coldren, IEEE. J. Quant. Electronic. **24**, 1178 (1988)
50. D.A.B. Miller, D.S. Chemla, T.C. Damen, J.H. Wood, A.C. Burrus, A.C. Gossard, W. Weigmann, IEEE. J. Quant. Electron. **21**, 1462 (1985)
51. J.S. Blakemore, *Semiconductor Statistics* (Dover, New York, 1987)
52. D. Bimberg, M. Grundmann, N.N. Ledentsov, *Quantum Dot Heterostructures*, John Wiley & Sons, (1999); G.Konstantatos, I. Howard, A. Fischer, S. Howland, J. Clifford, E. Klem, L. Levina, E.H. Sargent, *Nature* **442**, 180 (2006)
53. J.K. Jaiswal, H. Mattoussi, J.M. Mauro, S.M. Simon, Nat. Biotechnol. **21**, 47 (2003)
54. A. Watson, X. Wu, M. Bruchez, Biotechniques **34**, 296 (2003); J. Nakanishi, Y. Kikuchi, T. Takarada, H. Nakayama, K. Yamaguchi, M. Maeda, J. Am. Chem. Soc. **126**, 16314 (2004)
55. X. Michalet, F.F. Pinaud, L.A. Bentolila, J.M. Tsay, S. Doose, J.J. Li, G. Sundaresan, A.M. Wu, S.S. Gambhir, S. Weiss, *Science* **307**, 538 (2005)
56. W.G.J.H.M. van Sark, P.L.T.M. Frederix, D.J. Van den Heuvel, H.C.G.A. Bol, J.N.J. van Lingen, C. de Mello Donegá, A. Meijerink, J. Phys. Chem. **B105**, 8281 (2001)

57. E.J. Sánchez, L. Novotny, X.S. Xie, Phys. Rev. Lett. **82**, 4014 (1999); B. Bailey, D.L. Farkas, D.L. Taylor, F. Lanni, *Nature* **366**, 44 (1993)
58. L.V. Asryan, R.A. Suris in *Selected Topics in Electronics and Systems*, 25, ed. by E. Borovitskaya, M.S. Shur, (World Scientific, Singapore, 2002); L.V. Asryan, R.A. Suris, Int. J. High Speed Electron. Syst., Special Issue on "Quantum Dot Heterostructures—Fabrication, Application, Theory" **12** [1], 111 (2002); L.V. Asryan, S. Luryi, in *Future Trends in Microelectronics: The Nano Millennium*, ed. by S. Luryi, J.M. Xu, A. Zaslavsky, (Wiley Interscience, New York, 2002), p. 219
59. R.A. Freitas, Jr., *J. Comput. Theor. Nanosci.***2**, 1 (2005); A. Ferreira, C. Mavroidis, IEEE Robot. Autom. Mag. **13**, 78 (2006); A. Dubey, G. Sharma, C. Mavroidis, S.M. Tomassone, K. Nikitczuk, M.L. Yarmush, J. Comput. Theor. Nanosci. **1**, 18 (2004); C. Mavroidis, A. Dubey, M.L. Yarmush, Annu. Rev. Biomed. Eng. **6**, 363 (2004)
60. Y. Liu, J.A. Starzyk, Z. Zhu, IEEE Trans. Neural Networks (2008) (In the press); J.A. Starzyk, H. He, IEEE Trans. Neural Networks, **18**(2), 344 (2007); J.A. Starzyk, H. He, IEEE Trans. Circuits Syst. II **54**(2), 176 (2007)
61. E.-S. Hasaneen, E. Heller, R. Bansal, W. Huang, F. Jain, Solid State Electron. **48**, 2055 (2004)
62. T. Kawazoe, S. Tanaka, M. Ohtsu, J. Nanophoton. **2**, 029502 (2008)
63. H.J. Krenner, S. Stufler, M. Sabathil, E.C. Clark, P. Ester, M. Bichler, G. Abstreiter, J. J. Finley, A. Zrenner, New J. Phys. **7**, 184 (2005)
64. A.E. Zhukov, A.R. Kovsh, Quantum Electron. **38**, 409 (2008)
65. M. Sugawara, T. Akiyama, N. Hatori, Y. Nakata, H. Ebe, H. Ishikawa, Meas. Sci. Technol. **13**, 1683 (2002); M. van der Poel, D. Birkedal, J. Hvam, M. Laemmlin, D. Bimberg, Conf. Lasers Electro-Optics (CLEO) **1**, 16 (2004)
66. J.M. Costa-Fernandez, Anal. Bioanal. Chem. **384**, 37 (2006); H.S. Djie, C.E. Dimas, D.-N. Wang, B.-S. Ooi, J.C.M. Hwang, G.T. Dang, W.H. Chang, IEEE Sens. J. **7**, 251 (2007)
67. X.-X. Zhu, Y.-C. Cao, X. Jin, J. Yang, X.-F. Hua, H.-Q. Wang, B. Liu, Z. Wang, J.-H. Wang, L. Yang, Y.-D. Zhao, *Nanotechnology* **19**, 025708 (2008); X. Gao, W.C.W. Chan, S. Nie, J. Biomed. Opt. **7**, 532 (2002)
68. J.K. Jaiswal, E.R. Goldman, H. Mattoussi, S.M. Simon, Nat. Methods **1**, 73 (2004)
69. H. Matsueda, Int. J. Circ. Theo. Appl. **31**, 23 (2003); X. Hu, S. Das Sarma, Phys. Status Solidi (b) 238, 360 (2003)
70. G.-L. Chen, D.M.T. Kuo, W.-T. Lai, P.-W. Li, Nanotechnology **18**, 475402 (2007); A.G. Pogosov, M.V. Budantsev, A.A. Shevyrin, A.E. Plotnikov, A.K. Bakarov, A.I. Toropov, JETP Letts. **87**, 150 (2008)
71. K.W. Johnston, A.G. Pattantyus-Abraham, J.P. Clifford, S.H. Myrskog, D.D. MacNeil, L. Levina, E.H. Sargent, Appl. Phys. Letts. **92**, 151115 (2008); K.S. Leschkies, R. Divakar, J. Basu, E.E. Pommer, J.E. Boercker, C.B. Carter, U.R. Kortshagen, D.J. Norris, E.S. Aydil, Nano Lett. **7**, 1793 (2007)
72. I.-S. Liu, H.-H. Lo, C.-T. Chien, Y.-Y. Lin, C.-W. Chen, Y.-F. Chen, W.-F. Su, S.-C. Liou, J. Mater. Chem. **18**, 675 (2008)
73. N. Hitoshi, Y. Sugimoto, K. Nanamoto, N. Ikeda, Y. Tanaka, Y. Nakamura, S. Ohkouchi, Y. Watanabe, K. Inoue, H. Ishikawa, K. Asakawa, Opt. Express **12**, 6606 (2004); N. Yamamoto, T. Matsuno, H. Takai, N. Ohtani, Jpn. J. Appl. Phys. **44**, 4749 (2005); T. Yamada, Y. Kinoshita, S. Kasai, H. Hasegawa, Y. Amemiya, Jpn. J. Appl. Phys. **40**, 4485 (2001); K. Asakawa, Y. Sugimoto, Y. Watanabe, N. Ozaki, A. Mizutani, Y. Takata, Y. Kitagawa, H. Ishikawa, N. Ikeda, K. Awazu, X. Wang, A. Watanabe, S. Nakamura, S. Ohkouchi, K. Inoue, M. Kristensen, O. Sigmund, P. I Borel, R. Baets, New J. Phys. **8**, 208 (2006)
74. A.R. Clapp, I.L. Medintz, B.R. Fisher, G.P. Anderson, H. Mattoussi, J. Am. Chem. Soc. **127**, 1242 (2005); L. Shi, B. Hernandez, M. Selke, J. Am. Chem. Soc. **128**, 6278 (2006); C. Wu, J. Zheng, C. Huang, J. Lai, S. Li, C. Chen, Y. Zhao, Angewandte Chemie International Edition **46**, 5393 (2007)]
75. A. Rogalski, J. Alloys Comp. **371**, 53 (2004)

76. A. Baumgartner, A. Chaggar, A. Patanè, L. Eaves, M. Henini, Appl. Phys. Lett. **92**, 091121 (2008)
77. J. Devenson, R. Teissier, O. Cathabard, A.N. Baranov, Proc. SPIE **6909**, 69090U (2008)
78. B.S. Passmore, J. Wu, M.O. Manasreh, G.J. Salamo, Appl. Phys. Lett. **91**, 233508 (2007)
79. M. Mikhailova, N. Stoyanov, I. Andreev, B. Zhurtanov, S. Kizhaev, E. Kunitsyna, K. Salikhov, Y. Yakovlev, Proc. SPIE **6585**, 658526 (2007)
80. W. Kruppa, J.B. Boos, B.R. Bennett, N.A. Papanicolaou, D. Park, R. Bass, Electron. Lett. **42**, 688 (2006)
81. L.J. Singh, S. Choudhury, D. Baruah, S.K. Biswas, S. Pahari, K.P. Ghatak, Physica B **368**, 188 (2005)
82. A.N. Chakravarti, K.P. Ghatak, A. Dhar, K.K. Ghosh, S. Ghosh, Acta Phys. Polon A **60**, 151 (1981)
83. G.E. Stillman, C.M. Wolfe, J.O. Dimmock, in *Semiconductors and Semimetals*, 12, ed. by R. K. Willardon, A.C. Beer (Academic Press, New York, 1977), p. 169
84. K.P. Ghatak, J.P. Banerjee, B. Nag, J. Appl. Phys. **83**, 1420 (1998)
85. E.D. Palik, G.S. Picus, S. Teither, R.E. Wallis, Phys. Rev. 475 (1961)
86. P.Y. Lu, C.H. Wung, C.M. Williams, S.N.G. Chu, C.M. Stiles, Appl. Phys. Lett. **49**, 1372 (1986)
87. N.R. Taskar, I.B. Bhat, K.K. Prat, D. Terry, H. Ehasani, S.K. Ghandhi, J. Vac. Sci. Tech. **7A**, 281 (1989)
88. F. Koch, *Springer Series in Solid States Sciences* (Springer, Germany, 1984)
89. L.R. Tomasetta, H.D. Law, R.C. Eden, I. Reyhimy, K. Nakano, IEEE J. Quant. Electron. **14**, 800 (1978)
90. T. Yamato, K. Sakai, S. Akiba, Y. Suematsu, IEEE J. Quantum Electron. **14**, 95 (1978)
91. T.P. Pearsall, B.I. Miller, R.J. Capik, Appl. Phys. Lett. **28**, 499 (1976)
92. M.A. Washington, R.E. Nahory, M.A. Pollack, E.D. Beeke, Appl. Phys. Lett. **33**, 854 (1978)
93. M.I. Timmons, S.M. Bedair, R.J. Markunas, J.A. Hutchby, in *Proceedings of the 16th IEEE Photovoltaic Specialist Conference* (IEEE, San Diego, California 666, 1982)
94. G.P. Agrawal, N.K. Dutta, *Semicond. Lasers* (Van Nostrand Reinhold, New York, 1993)
95. S. Chatterjee, U. Pal, Opt. Eng. (Bellingham) **32**, 2923 (1993)
96. T.K. Chaudhuri, Int. J. Energy Res. **16**, 481 (1992)
97. J.H. Dughaish, Phys. B **322**, 205 (2002)
98. C. Wood, Rep. Prog. Phys. **51**, 459 (1988)
99. K.F. Hsu, S. Loo, F. Guo, W. Chen, J.S. Dyck, C. Uher, T. Hogan, E.K. Polychroniadis, M. G. Kanatzidis, Science **303**, 818–2004
100. J. Androulakis, K.F. Hsu, R. Pcionek, H. Kong, C. Uher, J.J. D'Angelo, A. Downey, T. Hogan, M.G. Kanatzidis, Adv. Mater. **18**, 1170 (2006)
101. P.F.P. Poudeu, J. D'Angelo, A.D. Downey, J.L. Short, T.P. Hogan, M.G. Kanatzidis, Angew. Chem. Int. Ed. **45**, 3835 (2006)
102. P.F. Poudeu, J. D'Angelo, H. Kong, A. Downey, J.L. Short, R. Pcionek, T.P. Hogan, C. Uher, M.G. Kanatzidis, J. Am. Chem. Soc. **128**, 14347 (2006)
103. J.R. Sootsman, R.J. Pcionek, H. Kong, C. Uher, M.G. Kanatzidis, Chem. Mater. **18**, 4993 (2006)
104. A.J. Mountvala, G. Abowitz, J. Am. Ceram. Soc. **48**, 651 (1965)
105. E.I. Rogacheva, I.M. Krivulkin, O.N. Nashchekina, A.Yu. Sipatov, V.A. Volobuev, M.S. Dresselhaus, Appl. Phys. Lett. **78**, 3238 (2001)
106. H.S. Lee, B. Cheong, T.S. Lee, K.S. Lee, W.M. Kim, J.W. Lee, S.H. Cho, J.Y. Huh, Appl. Phys. Lett. **85**, 2782 (2004)
107. K. Kishimoto, M. Tsukamoto, T. Koyanagi, J. Appl. Phys. **92**, 5331 (2002)
108. E.I. Rogacheva, O.N. Nashchekina, S.N. Grigorov, M.A. Us, M.S. Dresselhaus, S.B. Cronin, Nanotechnology **14**, 53 (2003)
109. E.I. Rogacheva, O.N. Nashchekina, A.V. Meriuts, S.G. Lyubchenko, M.S. Dresselhaus, G. Dresselhaus, Appl. Phys. Lett. **86**, 063103 (2005)

110. E.I. Rogacheva, S.N. Grigorov, O.N. Nashchekina, T.V. Tavrina, S.G. Lyubchenko, AYu. Sipatov, V.V. Volobuev, A.G Fedorov, M.S. Dresselhaus, Thin Solid Films **493**, 41 (2005)
111. X. Qiu, Y. Lou, A.C.S. Samia, A. Devadoss, J.D. Burgess, S. Dayal, C. Burda, Angew. Chem. Int. Ed. **44**, 5855 (2005)
112. C. Wang, G. Zhang, S. Fan, Y. Li, J. Phys. Chem. Solids **62**, 1957 (2001)
113. B. Poudel, W.Z. Wang, D.Z. Wang, J.Y. Huang, Z.F. Ren, J. Nanosci. Nanotechnol. **6**, 1050 (2006)
114. B. Zhang, J. He, T.M. Tritt, Appl. Phys. Lett. **88**, 043119 (2006)
115. W. Heiss, H. Groiss, E. KaQWmann, G. Hesser, M. Böberl, G. Springholz, F. Schäffler, K. Koike, H. Harada, M. Yano, Appl. Phys. Lett. **88**, 192109 (2006)
116. B.A. Akimov, V.A. Bogoyavlenskiy, L.I. Ryabova, V.N. Vasil'kov, Phys. Rev. B **61**, 16045 (2000)
117. Ya.A. Ugai, A.M. Samoilov, M.K. Sharov, O.B. Yatsenko, B.A. Akimov, Inorg. Mater. **38**, 12 (2002)
118. Ya.A. Ugai, A.M. Samoilov, S.A. Buchnev, Yu.V. Synorov, M.K. Sharov, Inorg. Mater. **38**, 450 (2002)
119. A.M. Samoilov, S.A. Buchnev, B.L. YuV Synorov, A.M.Khoviv Agapov, Inorg. Mater. **39**, 1132 (2003)
120. A.M. Samoilov, S.A. Buchnev, E.A. Dolgopolova, YuV Synorov, A.M. Khoviv, Inorg. Mater. **40**, 349 (2004)
121. H. Murakami, W. Hattori, R. Aoki, Phys. C **269**, 83 (1996)
122. H. Murakami, W. Hattori, Y. Mizomata, R. Aoki, Phys. C **273**, 41 (1996)
123. H. Murakami, R. Aoki, K. Sakai, Thin Solid Films **27**, 343 (1999)
124. B.A. Volkov, L.I. Ryabova, D.R. Khokhlov, Phys. Usp. **45**, 819 (2002)
125. N. Paitya, K.P. Ghatak, J. Adv. Phys. **1**, 161 (2012); D. De, S. Bhattacharya, S.M. Adhikari, A. Kumar, P.K. Bose, K.P. Ghatak, Beilstein J. Nanotechnol. **2**, 339 (2011)
126. S. Pahari, S. Bhattacharya, S. Roy, A. Saha, D. De, K.P. Ghatak, Superlattices Microstruct. **46**, 760 (2009); D. De, K.P. Ghatak, J. Mater. Sci.: Mater. Electron. **20**, 185 (2009)
127. S.M. Adhikari, D. De, J.K. Baruah, S. Chowdhury, K.P. Ghatak, Adv. Sci. Focus **1**, 54 (2013); S. Bhattacharya, S. Pahari, R. Sarkar, S. Ghosh, K.P. Ghatak, Physica B: Condens. Matter **403**, 3635 (2008)
128. P.K. Chakraborty, S. Bhattacharya, K.P. Ghatak, J. Appl. Phys. **98**, 053517 (2005)
129. P.K. Chakraborty, S. Ghoshal, K.P. Ghatak, Physica B: Condens. Matter **382**, 26 (2006); S. Chowdhary, L.J. Singh, K.P. Ghatak, Physica B: Condens. Matter **365**, 5 (2005); K. P. Ghatak, D.K. Basu, S. Bhattacharya, S. Ghosal, S.K. Biswas, A. Mallik, G. Majumdar, Nonlinear Opt. Quantum Opt. **33**, 221 (2005); L.J. Singh, S. Choudhury, S.S. Roy, K. P. Ghatak, Electr. Eng. **87**, 1 (2005)
130. K.P. Ghatak, S.K. Biswas, D. De, S. Ghosal, S. Chatterjee, Physica B: Condens. Matter **353**, 127 (2004); P.K. Chakraborty, S.K. Biswas, K.P. Ghatak, Physica B: Condens. Matter **352**, 111 (2004); P.K. Chakraborty, L.J. Singh, K.P. Ghatak, J. Appl. Phys. **95**, 5311 (2004); P.K Chakraborty, G.C Datta, K.P Ghatak, Physica B: Condens. Matter **339**, 198 (2003)
131. P.K. Chakraborty, K.P. Ghatak, J. Phys. Chem. Solids **62**, 1061 (2001); K.P. Ghatak, J.Y. Siddiqui, B. Nag, Phys. Lett. A **282**, 428 (2001); P.K. Chakraborty, K.P. Ghatak, J. Appl. Phys. **89**, 1075 (2001)
132. P.K. Chakraborty, K.P. Ghatak, J. Phys. D Appl. Phys. **32**, 2438 (1999); B. Nag, K. P. Ghatak, J. Phys. Chem. Solids **59**, 713 (1998); K.P. Ghatak, B. Nag, J. Phys. Chem. Solids **59**, 411 (**1998)**, K.P. Ghatak, B. Nag, Phys. Status Solidi (b) **205**, 519 (1998)
133. B. Nag, K.P. Ghatak, Nonlinear Opt. **19**, 1 (1998); J.P. Banerjee, S. Ghosh, B. Nag, K. P. Ghatak, Int. J. Electron. **82**, 335 (1997)
134. K.P Ghatak, D.K Basu, B. Nag, J. Phys. Chem. Solids **58**, 133 (1997); B. Mitra, D.K. Basu, B. Nag, K.P. Ghatak, Nonlinear Opt. **17**, 171 (1997); K.P. Ghatak, B. Mitra, B. Nag, Phys. Status Solidi (b) **199**, 95 (1997); B. Nag, K.P. Ghatak, Phys. Scr. **54**, 657 (1996)
135. K.P. Ghatak, D.K. Basu, D. Basu, B. Nag, Il Nuovo Cimento D **18**, 947 (1996); K.P. Ghatak, S. Beta, A. Ali, B. Nag, Fizika A- **5**, 111 (1996); K.P. Ghatak, B. De, B. Nag, P.K.

Chakraborty, Nonlinear Opt. **16**, 221 (1996); K.P. Ghatak, J.P Banerjee, B. Goswami, B. Nag, Nonlinear Opt. **16**, 241 (1996)
136. K.P. Ghatak, J.P. Banerjee, M. Mitra, B. Nag, Nonlinear Opt. **16**, 193 (1996); K.P. Ghatak, M. Mitra, B. Goswami, B. Nag, Nonlinear Opt. **16**, 167 (1996); K.P. Ghatak, B. Nag, Nanostruct. Mater. **5**, 769 (1995); K.P. Ghatak, B. Nag, D. Bhattacharyya, J. Low Temp. Phys. **101**, 983 (1995)
137. K.P Ghatak, D. Bhattacharya, D. Basu, B. Nag, Phys. Status Solidi (b) **191**, 141 (1995); K. P. Ghatak, D. Bhattacharyya, Phys. Scr. **52**, 343 (1995); D. Bhattacharyya, K.P. Ghatak, Phys. Status Solidi (b) **187**, 523 (1995); K.P. Ghatak, S.N. Banik, Fizika A **4**, 33 (1995)
138. K.P Ghatak, M. Mondal, Phys. Status Solidi (b) **185**, K5 (1994); K.P. Ghatak, S.N. Biswas, Solid-State Electron. **37**, 1437 (1994); K.P. Ghatak, Acta Physica Hungarica **74**, 167 (1994); K.P. Ghatak, D. Bhattacharyya, Phys. Lett. A **184**, 366 (1994)
139. S. Banik, K.P. Ghatak, S. Biswas, Fizika A **3**, 77 (1994); K.P. Ghatak, S.N. Banik, Fizika A **3**, 127 (1994); K.P. Ghatak, S.N. Banik, Fizika A **3**, 155 (1994); K.P. Ghatak, D. Bhattacharyya, Phys. Stat. Sol.(b) **181**, 127 (1994); K.P. Ghatak, S. Biswas, Fizika A **3**, 7 (1994); K.P. Ghatak, D. Bhattacharya, Phys. Status Solidi (b) **179**, 383 (1993); P.K. Chakraborty, K.P. Ghatak, J. Appl. Phys. **74**, 3246 (1993)
140. K.P. Ghatak, S.N. Biswas, Nanostruct. Mater. **2**, 91 (1993); K.P. Ghatak, Fizika A. **2**, 133 (1993); K.P. Ghatak, S.N. Biswas, Il Nuovo Cimento D **15**, 87 (1993)
141. R.K. Willardson, ed. by A.C. Beer, **1**, *Semiconductors and Semimetals* (Academic Press, New York, 1966), p. 102
142. E.O. Kane, Phys. Rev. **131**, 79 (1963); Phys. Rev. B. **139**, 343 (1965)
143. V.L. Bonch, Bruevich. Sov. Phys. Sol. Stat. **4**, 53 (1963)
144. R.A. Logan, A.G. Chynoweth, Phys. Rev. **131**, 89 (1963)
145. C.J. Hwang, J. Appl. Phys. **40**, 3731 (1969)
146. J.I. Pankove, Phys. Rev. A **130**, 2059 (1965)
147. B.I. Halperin, M. Lax, Phys. Rev. **148**, 722 (1966)
148. R.A. Abram, G.J. Rees, B.L.H. Wilson, Adv. Phys. **27**, 799 (1978)
149. B.I. Shklovskii, A.L. Efros, in *Electronic Properties of Doped Semiconductors*, Springer series in Solid State Sciences, vol. 45 (Springer, Berlin, 1984)
150. E.O. Kane, Solid State Electron. **28**, 3 (1985)
151. P.K. Chakraborty, J.C. Biswas, J. Appl. Phys. **82**, 3328 (1997)
152. B.R. Nag, in *Electron Transport in Compound Semiconductors*, Springer Series in Solid state Sciences, vol. 11, (Springer, Heidelberg, 1980)
153. P.E. Schmid, Phys. Rev. B **23**, 5531 (1981)
154. G.E. Jellison Jr. , F.A. Modine, C.W. White, R.F. Wood, R.T. Young, Phys. Rev. Lett. **46**, 1414 (1981)
155. V.I. Fistul, *Heavily Doped Semiconductors* (Plenum, New York,1969), Ch 7
156. C.J. Hwang, J. Appl. Phys. **41**, 2668 (1970); W. Sritrakool, H.R. Glyde, V. Sa Yakanit Can. J. Phys. **60**, 373 (1982); H. Ikoma, J. Phys. Soc. Jpn. **27**, 514 (1969)
157. M.H. Chen, C.C. Wu, C.J. Lin, J. Low Temp. Phys. **55**, 127 (1984)
158. S. Adachi, J. Appl. Phys. **58**, R11 (1985)
159. S. Bhattacharya, S. Choudhury, K.P. Ghatak, Superlattices Microstruct. **48**, 257 (2010); K. P. Ghatak, S. Bhattacharya, S. Pahari, D. De, S. Ghosh, M. Mitra, Ann. Phys. **17**, 195 (2008); S. Pahari, S. Bhattacharya, K.P Ghatak, J. Comput. Theor. Nanosci. (Invited Paper), **6**, 2088 (2009)
160. R. Saito, G. Dresselhaus, M.S. Dresselhaus, *Physical Properties of Carbon Nanotubes* (Imperial College Press, London, **1998**); X. Yang, J. Ni, Phys. Rev. B **72**, 195426 (2005)
161. G.L. Bir, G.E. Pikus, *Symmetry and Strain−Induced Effects in Semiconductors* (Nauka, Russia, 1972)
162. A.N. Chakravarti, K.P. Ghatak, K.K. Ghosh, S. Ghosh, H.M. Mukherjee, Phys. Status Solidi b **118**, 843 (1983); S. Bhattacharya, D. De, S.M. Adhikari, K.P. Ghatak, Superlattices Microstruct. **51**, 203 (2012); K.P. Ghatak, M. Mondal, Z.F. Naturforschung **41A**, 821 (1986)
163. P.K. Chakraborty, G.C. Datta, K.P. Ghatak, Phys. Scrip. **68**, 368 (2003)

164. A.N. Chakravarti, K.P. Ghatak, A. Dhar, K.K. Ghosh, S. Ghosh, Appl. Phys. A **26**, 165 (1981)
165. K.P. Ghatak, M. Mondal, Z.F. Physik B. **B69**, 471 (1988)
166. H.A. Lyden, Phys. Rev. **135**, A514 (1964); E.D Palik, G.B. Wright, in *Semiconductors and Semimetals*, ed. by R.K Willardson, A.C Beer **3**, (Academic Press, New York, USA,1967), p. 421
167. B. Mitra, A. Ghoshal, K.P. Ghatak, Nouvo Cimento D **12D**, 891 (1990); K.P.Ghatak, S.N. Biswas, Nonlinear Opt. Quantum Opt. **4**, 347 (1993)
168. B. Mitra, K.P. Ghatak, Solid State Electron. **32**, 177 (1989)
169. G.E. Smith, Phys. Rev. Lett. **9**, 487 (1962); D. Schneider, D. Rurup, A. Plichta, H.-U. Grubert, A. Schlachetzki, K. Hansen, Z. Phys. B, **95**, 281 (1994)
170. V.K. Arora, H. Jeafarian, Phys. Rev. B. **13**, 4457 (1976); S.E. Ostapov, V.V. Zhikharevich, V.G. Deibuk, Semicond. Phys. Quan. Electron. Optoelectron **9**, 29 (2006)
171. M.J. Aubin, L.G. Caron, J.P. Jay, Gerin. Phys. Rev. B **15**, 3872 (1977)
172. K. Tanaka, N. Kotera; in *20th International Conference on Indium Phosphide and Related Materials* (Versailles, France, 2008), pp. 1–4
173. M. Singh, P.R. Wallace, S.D. Jog, J. Erushanov, J. Phys. Chem. Solids **45**, 409 (1984)
174. N.R. Das, A.N. Chakravarti, Phys. Stat. Sol. (b) **170**, 471 (1992); N.R. Das, A.N. Chakravarti, Phys. Stat. Sol. (b) **169**, 97 (1992); N.R. Das, A.N. Chakravarti, Phys. Stat. Sol. (b) **170**, K115 (1992); N.R. Das, A.N. Chakravarti, Phys. Stat. Sol. (b) **176**, 335 (1993); N.R. Das, A.N. Chakravarti, Phys. Stat. Sol. (b) **174**, 45 (1992); M.K. Bose, C. Majumdar, A.B. Maity, A.N. Chakravarti, Phys. Stat. Sol. (b) **143**, 45 (1987); M.K. Bose, C. Majumdar, A.B. Maity, A.N. Chakravarti, Phys. Stat. Sol. (b) **143**, 113 (1987); C. Majumdar, A.B. Maity, A.N. Chakravarti, Phys. Stat. Sol. (b) **144**, K13 (1987); A.B. Maity, C. Majumdar, A.N. Chakravarti, Phys. Stat. Sol. (b) **144**, K93 (1987)
175. J.L. Shay, J.W. Wernik, *Ternary Chalcoprite Semiconductors: Growth Electronic properties and applications* (Pergamon Press, London, 1975)
176. E.A. Arushanov, A.A. Kaynzev, A.N. Natepov, S.I. Radautsan, Sov. Phys. Semicond. **15**, 828 (1981)
177. K.S. Hong, R.F. Speyer, R.A. Condrate, J. Phys. Chem. Solids **51**, 969 (1990)
178. M. Mondal, S. Banik, K.P. Ghatak, J. Low Temp. Phys. **74**, 423 (1989)
179. M. Krieehbaum, P. Kocevar, H. Pascher, G. Bauer, IEEE QE **24**, 1727 (1988); J.J. Hopfield, J.Appl. Phys. **32**, 2277 (1961)
180. S. Adachi, *Properties of Group-IV, III-V and II-VI Semiconductors* (John Wiley and Sons, Chichester, 2005)
181. O. Madelung, *Semiconductors: Data Handbook*, 3rd edn. (Springer-Verlag, Germany, 2003)
182. D.J. Newson, A. Kurobe, Semicond. Sci. Technol. **3**, 786 (1988)
183. K.P. Ghatak, S. Bhattacharya, S.K. Biswas, A. Dey, A.K. Dasgupta, Phys. Scr. **75**, 820 (2007)
184. S. Adachi, J. Appl. Phys. **58**, R1 (1985)
185. S. Adachi, *GaAs and Related Materials: Bulk Semiconductors and Superlattice Properties* (World Scientific, USA, 1994)
186. G.L. Hansen, J.L. Schmit, T.N. Casselman, J. Appl. Phys. **63**, 7079 (1982)
187. J. Wenus, J. Rutkowski, A. Rogalski, I.E.E.E. Trans, Elect. Dev. **48**, 1326 (2001)
188. S. Adachi, J. Appl. Phys. **53**, 8775 (1982)
189. B. Mitra, K.P. Ghatak, Phys. Lett. A **141**, 81 (1989)
190. I.V. Skryabinskii, Yu.I. Ukhanov, Sov. Phys. Solid State **14**, 2838 (1973); P.C. Mathur, S. Jain, Phys. Rev. **19**, 1359 (1979)
191. S. Tiwari, S. Tiwari, Cryst. Res. Technol. **41**, 78 (2006)
192. J.R. Lowney, S.D. Senturia, J. Appl. Phys. **47**, 1771 (1976)
193. W.E. Spicer, G.J. Lapeyre Phys. Rev. **139**, A565 (1965)
194. J.O. Dimmock in *The Physics of Semimetals and Narrowgap Semiconductors*, ed. by D.L. Carter, R.T. Bates (Pergamon Press, Oxford, 1971), p. 319
195. M. Mondal, N. Chattopadhyay, K.P. Ghatak, J. Low Temp. Phys. **66**, 131 (1987)

196. D.R. Lovett, in *Semimetals and Narrow Band Gap Semiconductors* (Pion Limited, London, 1977)
197. B.R. Nag, *Physics of Quantum Well Devices* (Kluwer, Netherlands, 2000)
198. D.G. Seiler, B.D. Bajaj, A.E. Stephens, Phys. Rev. B **16**, 2822 (1977); A.V. Germaneko, G. M. Minkov, Phys. Status Solidi (b) **184**, 9 (1994)
199. C.C. Wu, C.J. Lin, J. Low Temp. Phys. **57**, 469 (1984)
200. S. Takaoka, H. Kawamura, K. Murasa, S. Takano, Phys. Rev. B **13**, 1428 (1976)
201. V.I. Ivanov-Omskii, ASh Mekhtisev, S.A. Rustambekova, E.N. Ukraintsev, Phys. Status Solidi (b) **119**, 159 (1983)
202. H. Kim, K. Cho, H. Song, B. Min, J. Lee, G. Kim, S. Kim, S.H. Kim, T. Noh, Appl. Phys. Lett. **83**, 4619 (2003)
203. P.R. Emtage, Phys. Rev. A **246**, 138 (1965)
204. R.A. Reynolds, M.J. Brau, R.A. Chapman, J. Phys. Chem. Solids **29**, 755 (1968)
205. J. O'Shaughnessy, C. Smith, Solid State Commun. **8**, 481 (1970)
206. G.J. Rees, Physics of compounds, in *Proceedings of the 13th International Conference*, ed. by F.G. Fumi (North Holland Company, 1976), p. 1166
207. P.K. Chakraborty, K.P. Ghatak, Phys. Lett. A **288**, 335 (2001)
208. K.P. Ghatak, B. Mitra, Il Nuovo Cimento D **15**, 97 (1993)
209. G. Haeffler, A.E. Klinkmüller, J. Rangell, U. Berzinsh, D. Hanstorp, Z. Phys. D **38**, 211 (1996)
210. N.B. Brandt, V.N. Davydov, V.A. Kulbachinskii, O.M. Nikitina, Sov. Phys. Sol. Stat. **29**, 1014 (1987)
211. L.M. Viculis, J.J. Mack, O.M. Mayer, H.T. Hahn, R.B. Kaner, J. Mater. Chem. **15**, 974 (2005)
212. S. Takaoka, K. Murase, Phys. Rev. B. Phys. Rev. B **20**, 2823 (1979)
213. D.L. Partin, Superlattices Microstruct. **1**, 131 (1985)
214. Y. Yamada, J. Phys. Jpn. **35**, 1600 (1973)
215. W.J. Turner, A.S. Fischler, W.E. Reese, Phys. Rev. **121**, 759 (1961)
216. K.P. Ghatak, S. Bhattacharya, H. Saikia, D. Baruah, A. Saikia, K.M. Singh, A. Ali, S.N. Mitra, P.K. Bose, A. Sinha, J. Comput. Theor. Nanosci. **3**, 727 (2006)
217. W.E. Swank, P.G. Le Comber, Phys. Rev. **153**, 844 (1967)
218. M. Stordeur, W. Kuhnberger, Phys. Status Solidi (b), **69**, 377 (1975); H. Köhler, Phys. Status Solidi (b), **74**, 591 (1976)
219. D. Haneman, J. Phys. Chem. Solids **11**, 205 (1959)
220. N. Wei, G. Wu, J. Dong, Phys. Lett. A **325**, 403 (2004)
221. S. Reich, J. Maultzsch, C. Thomsen, P. Ordejón, Phys. Rev. B **66**, 035412 (2006)
222. M.S. Lundstrom, J. Guo, *Nanoscale Transistors: Device Physics, Modeling and Simulation* (Springer, USA, 2006); J.W. Mintmire, C.T. White, Phys. Rev. Lett. **81**, 2506 (1998)
223. F. Buonocore, F. Trani, D. Ninno, A. Di Matteo, G. Cantele, G. Iadonisi, Nanotechnology **19**, 025711 (2008)
224. J.B. Ketterson, Phys. Rev. **129**, 18 (1963)
225. R.C. Vilão, J.M. Gil, A. Weidinger, H.V. Alberto, J. Piroto Duarte, N.A. de Campos, R.L. Lichti, K.H. Chow, S.P. Cottrell, S.F.J. Cox, Phys. Rev. B, **77**, 235212 (2008)
226. I. Kang, F.W. Wise, Phys. Rev. B, J. Opt. Soc. Am. B **14**, 1632 (1997)
227. D. Cui, J. Xu, S.-Y. Xu, G. Paradee, B.A. Lewis, M.D. Gerhold, IEEE Trans. Elect. Dev. **5**, 362 (2006)
228. E.L. Ivchenko, G.E. Pikus, Sov. Phys. Semicond. **13**, 579 (1979)
229. M.H. Cohen, Phys. Rev. **121**, 387 (1961)
230. J.W. McClure, K.H. Choi, Solid State Commun. **21**, 1015 (1977)
231. E. Bangert, P. Kastner, Phys. Status Solidi (b) **61**, 503 (1974)
232. G.M.T. Foley, P.N. Langenberg, Phys. Rev. B **15B**, 4850 (1977)
233. S. Takaoka, H. Kawamura, K. Murase, S. Takano, Phys. Rev. **1**, 764 (1970)
234. D.G. Seiler, W.M. Beeker, L.M. Roth, Phys. Rev. **1**, 764 (1970)
235. H.I. Zhang, Phys. Rev. **1B**, 3450 (1970)

236. U. Rossler, Solid State Commun. **49**, 943 (1984)
237. E.J. Johnson, D.H. Dicley, Phys. Rev. **1**, 2676 (1970)
238. V.G. Agafonov, P.M. Valov, B.S. Ryvkin, I.D. Yaroshetskii, Sov. Phys. Semicond. **12**, 1182 (1978)
239. I.V. Kucherenko, YuA Mityagin, L.K. Vodopyanov, A.P. Shotov, Sov. Phys. Semicond. **11**, 282 (1977)
240. L.A. Vassilev, Phys. Status Solidi (b) **121**, 203 (1984)
241. J. Bouat, J.C. Thuillier, Surf. Sci. **73**, 528 (1978)
242. M.V. Ortenberg, K.J. Button, Phys. Rev. **16**, 2618 (1977)
243. R.W. Cunningham, Phys. Rev. **167**, 761 (1968)
244. A.I. Yekimov, A.A. Onushchenko, A.G. Plyukhin, Al.L. Efros, J. Expt. Theor. Phys. **88**, 1490 (1985)
245. B.J. Roman, A.W. Ewald, Phys. Rev. B **5**, 3914 (1972)
246. G.P. Chuiko, Sov. Phys. Semiconduct. **19**, 1381 (1985)
247. S.V. Kryachkov, V.I. Porkhun, Sov. Phys. Semicond. **12**, 1206 (1978)

Part II
Dispersion Relations in HD Quantum Confined Non-parabolic Materials

My greatest glory is not in never falling, but in rising every time I fall.

Chapter 2
The DRs in Ultrathin Films (UFs) of Heavily Doped (HD) Non-parabolic Materials

The secret of a great scholar is to know some new thing that nobody else know.

2.1 Introduction

It is well known that the constant energy 3D wave-vector space of bulk materials becomes 2D wave-vector surface in QWs due to dimensional quantization. Thus, the concept of reduction of symmetry of the wave-vector space and its consequence can unlock the physics of low-dimensional structures. In this chapter, we study the DR in QWs of HD non-parabolic materials having different band structures in the presence of Gaussian band tails. At first we shall investigate the DR in QWs of HD nonlinear optical compounds which are being used in nonlinear optics and light emitting diodes [1]. The quasi-cubic model can be used to investigate the symmetric properties of both the bands at the zone center of wave vector space of the same compound. Including the anisotropic crystal potential in the Hamiltonian, and special features of the nonlinear optical compounds, Kildal [2] formulated the electron dispersion law under the assumptions of isotropic momentum matrix element and the isotropic spin-orbit splitting constant, respectively, although the anisotropies in the two aforementioned band constants are the significant physical features of the said materials [3–5]. In Sect. 2.2.1, the DR in QWs of HD nonlinear optical materials has been investigated on the basis of newly formulated HD DR of the said compound by considering the combined influence of the anisotropies of the said energy band constants together with the inclusion of the crystal field splitting respectively within the framework of $\vec{k} \cdot \vec{p}$ formalism.

In Sect. 2.2.2, the DR in QWs of HD III–V, ternary and quaternary materials has been studied in accordance with the corresponding HD formulation of the band structure and the simplified results for wide gap materials having parabolic energy bands under certain limiting conditions have further been demonstrated as a special case in the absence of band-tails and thus confirming the compatibility test. The II–VI materials are being used in nano-ribbons, blue green diode lasers,

K.P. Ghatak, *Dispersion Relations in Heavily-Doped Nanostructures*,
Springer Tracts in Modern Physics 265, DOI 10.1007/978-3-319-21000-1_2

photosensitive thin films, infrared detectors, ultra-high-speed bipolar transistors, fiber optic communications, microwave devices, solar cells, semiconductor gamma-ray detector arrays, semiconductor detector gamma camera and allow for a greater density of data storage on optically addressed compact discs [6–13]. The carrier energy spectra in II–VI compounds are defined by the Hopfield model [14] where the splitting of the two-spin states by the spin-orbit coupling and the crystalline field has been taken into account. The Sect. 2.2.3 contains the investigation of the DR in QWs of HD II–VI compounds.

Lead Chalcogenides (PbTe, PbSe, and PbS) are IV–VI non-parabolic materials whose studies over several decades have been motivated by their importance in infrared IR detectors, lasers, light-emitting devices, photo-voltaic, and high temperature thermo-electrics [15–25]. PbTe, in particular, is the end compound of several ternary and quaternary high performance high temperature thermoelectric materials [26–30]. It has been used not only as bulk but also as films [31–34], QWs [35] super-lattices [36, 37] nanowires [38] and colloidal and embedded nano-crystals [39–42], and PbTe films doped with various impurities have also been investigated [43–50]. These studies revealed some of the interesting features that had been seen in bulk PbTe, such as Fermi level pinning and, in the case of superconductivity [51]. In Sect. 2.2.4, the 2D DR in QWs of HD IV–VI materials has been studied taking PbTe, PbSe, and PbS as examples. The stressed materials are being investigated for strained silicon transistors, quantum cascade lasers, semiconductor strain gages, thermal detectors, and strained-layer structures [52–55]. The DR in QWs of HD stressed compounds (taking stressed n-InSb as an example) has been investigated in Sect. 2.2.5. The vacuum deposited Tellurium (Te) has been used as the semiconductor layer in thin-film transistors (TFT) [56] which is being used in CO_2 laser detectors [57], electronic imaging, strain sensitive devices [58, 59], and multichannel Bragg cell [60]. Section 2.2.6 contains the investigation of DR in QWs of HD Tellurium. The n-Gallium Phosphide (n-GaP) is being used in quantum dot light emitting diode [61], high efficiency yellow solid state lamps, light sources, high peak current pulse for high gain tubes. The green and yellow light emitting diodes made of nitrogen-doped n-GaP possess a longer device life at high drive currents [62–64]. In Sect. 2.2.7, the DR from QWs of HD n-GaP has been studied. The Platinum Antimonide ($PtSb_2$) finds application in device miniaturization, colloidal nanoparticle synthesis, sensors and detector materials and thermo-photovoltaic devices [65–67]. Section 2.2.8 explores the DR in QWs of HD $PtSb_2$. Bismuth telluride (Bi_2Te_3) was first identified as a material for thermoelectric refrigeration in 1954 [68] and its physical properties were later improved by the addition of bismuth selenide and antimony telluride to form solid solutions. The alloys of Bi_2Te_3 are useful compounds for the thermoelectric industry and have been investigated in the literature [69–73]. In Sect. 2.2.9, the DR in QWs of HD Bi_2Te_3 has been considered. The usefulness of elemental semiconductor Germanium is already well known since the inception of transistor technology and, it is also being used in memory circuits, single photon detectors, single photon avalanche diode, ultrafast optical switch, THz lasers and THz spectrometers [74–77]. In Sect. 2.2.10, the DR has been studied for QWs of HD Ge.

Gallium Antimonide (GaSb) finds applications in the fiber optic transmission window, hetero-junctions, and QWs. A complementary hetero-junction field effect transistor in which the channels for the p-FET device and the n-FET device forming the complementary FET are formed from GaSb. The band gap energy of GaSb makes it suitable for low power operation [78–83]. In Sect. 2.2.11, the DR in QWs of HD GaSb has been studied. The II–V materials have been studied in photovoltaic cells constructed of single crystal semiconductor materials in contact with electrolyte solutions. Cadmium selenide shows an open-circuit voltage of 0.8 V and power conservation coefficients near 6 % for 720-nm light [84]. They are also used in ultrasonic amplification [85]. The development of an evaporated thin film transistor using cadmium selenide as the semiconductor has been reported by Weimer [86, 87]. The DR in HD QWs of II–V materials has been presented in Sect. 2.2.12. In Sect. 2.2.13, the DR in HD QWs of $Pb_{1-x}Ga_xTe$ has been investigated [88]. The diphosphides finds prominent role in biochemistry where the folding and structural stabilization of many important extra-cellular peptide and protein molecules, including hormones, enzymes, growth factors, toxins, and immunoglobulin are concerned [89]. Besides, artificial introduction of extra diphosphides into peptides or proteins can improve biological activity [90] or confer thermal stability [91]. The asymmetric diphosphide bond formation in peptides containing a free thiol group takes place over a wide pH range in aqueous buffers and can be crucially monitored by spectrophotometric titration of the released 3-nitro-2-pyridinethiol [92, 93]. In Sect. 2.2.14, the DR in HD QWs of zinc and cadmium diphosphides has been investigated. Section 2.3 contains the result and discussion pertaining to this chapter. The last Sect. 2.4 contains 16 open research problems.

2.2 Theoretical Background

2.2.1 *The DR in Ultrathin Films (UFs) of HD Nonlinear Optical Materials*

The form of **k·p** matrix for nonlinear optical compounds can be expressed extending Bodnar [3] as

$$H = \begin{bmatrix} H_1 & H_2 \\ H_2^+ & H_1 \end{bmatrix} \tag{2.1}$$

where,

$$H_1 \equiv \begin{bmatrix} E_{g0} & 0 & P_{\parallel}k_z & 0 \\ 0 & (-2\Delta_{\parallel}/3) & (\sqrt{2}\Delta_{\perp}/3) & 0 \\ P_{\parallel}k_z & (\sqrt{2}\Delta_{\perp}/3) & -(\delta + \frac{1}{3}\Delta_{\parallel}) & 0 \\ 0 & 0 & 0 & 0 \end{bmatrix}, \quad H_2 \equiv \begin{bmatrix} 0 & -f_{,+} & 0 & f_{,-} \\ f_{,+} & 0 & 0 & 0 \\ 0 & 0 & 0 & 0 \\ f_{,+} & 0 & 0 & 0 \end{bmatrix}$$

in which E_{g_0} is the band gap in the absence of any field, $P_{\parallel}$ and $P_{\perp}$ are the momentum matrix elements parallel and perpendicular to the direction of crystal axis respectively, δ is the crystal field splitting constant, $\Delta_{\parallel}$ and $\Delta_{\perp}$ are the spin-orbit splitting constants parallel and perpendicular to the C-axis respectively, $f_{,\pm} \equiv \left(P_{\perp}/\sqrt{2}\right)\left(k_x \pm ik_y\right)$ and $i = \sqrt{-1}$. Thus, neglecting the contribution of the higher bands and the free electron term, the diagonalization of the above matrix leads to the DR of the conduction electrons in bulk specimens of nonlinear optical materials as

$$\gamma(E) = f_1(E)k_s^2 + f_2(E)k_z^2 \tag{2.2}$$

where

$$\gamma(E) = E(E + E_{g_0})\left[\left(E + E_{g_0}\right)\left(E + E_{g_0} + \Delta_{\parallel}\right) + \delta\left(E + E_{g_0} + \frac{2}{3}\Delta_{\parallel}\right) + \frac{2}{9}\left(\Delta_{\parallel}^2 - \Delta_{\perp}^2\right)\right],$$

E is the total energy of the electron as measured from the edge of the conduction band in the vertically upward direction in the absence of any quantization, $k_s^2 = k_x^2 + k_y^2$,

$$f_1(E) \equiv \frac{\hbar^2 E_{g_0}\left(E_{g0} + \Delta_{\perp}\right)}{\left[2m_{\perp}^*\left(E_{g_0} + \frac{2}{3}\Delta_{\perp}\right)\right]}\left[\delta\left(E + E_{g0} + \frac{1}{3}\Delta_{\parallel}\right) + \left(E + E_{g_0}\right)\left(E + E_{g_0} + \frac{2}{3}\Delta_{\parallel}\right) + \frac{1}{9}\left(\Delta_{\parallel}^2 - \Delta_{\parallel}^2\right)\right],$$

$$f_2(E) \equiv \frac{\hbar^2 E_{g_0}\left(E_{g0} + \Delta_{\parallel}\right)}{\left[2m_{\parallel}^*\left(E_{g_0} + \frac{2}{3}\Delta_{\parallel}\right)\right]}\left[\left(E + E_{g_0}\right)\left(E + E_{g_0} + \frac{2}{3}\Delta_{\parallel}\right)\right],$$

$\hbar = h/2\pi$, h is Planck's constant and $m_{\parallel}^*$ and $m_{\perp}^*$ are the longitudinal and transverse effective electron masses at the edge of the conduction band respectively.

Thus the generalized unperturbed electron energy spectrum for the bulk specimens of the nonlinear optical materials in the absence of band tails can be expressed following (1.2) as

$$\frac{\hbar^2 k_z^2}{2m_{\parallel}^*} + \left(\frac{b_{\parallel}\, c_{\perp}}{b_{\perp}\, c_{\parallel}}\right)\frac{\hbar^2 k_s^2}{2m_{\perp}^*} = \left\{\frac{E(\alpha E + 1)(b_{\parallel}E + 1)}{(c_{\parallel}E + 1)} + \frac{\alpha b_{\parallel}}{c_{\parallel}}\left[\delta E + \frac{2}{9}\left(\Delta_{\parallel}^2 - \Delta_{\perp}^2\right)\right] - \left(\frac{2}{9}\right)\frac{\alpha b_{\parallel}}{c_{\parallel}}\frac{\left(\Delta_{\parallel}^2 - \Delta_{\perp}^2\right)}{(c_{\parallel}E + 1)}\right\}$$
$$- \left(\frac{\hbar^2 k_s^2}{2m_{\perp}^*}\right)\left\{\left(\frac{b_{\parallel}\, c_{\perp}}{b_{\perp}\, c_{\parallel}}\right)\left[\left(\frac{\delta}{2} + \frac{\Delta_{\parallel}^2 - \Delta_{\perp}^2}{6\Delta_{\parallel}}\right)\frac{\alpha_{\parallel}}{\alpha_{\parallel}E + 1} + \left(\frac{\delta}{2} - \left\{\frac{\Delta_{\parallel}^2 - \Delta_{\perp}^2}{6\Delta_{\parallel}}\right\}\right)\frac{c_{\parallel}}{c_{\parallel}E + 1}\right]\right\} \tag{2.3}$$

where, $b_{\parallel} \equiv 1/\left(E_g + \Delta_{\parallel}\right)$, $c_{\perp} \equiv 1/\left(E_g + \frac{2}{3}\Delta_{\perp}\right)$, $b_{\perp} \equiv 1/\left(E_g + \Delta_{\perp}\right)$, $c_{\parallel} \equiv 1/\left(E_g + \frac{2}{3}\Delta_{\parallel}\right)$ and $\alpha \equiv 1/E_g$.

The Gaussian distribution $F(V)$ of the impurity potential is given by [94]

$$F(V) = \left(\pi\eta_g^2\right)^{-1/2} \exp\left(-V^2/\eta_g^2\right) \quad (2.4)$$

where, η_g is the impurity screening potential. It appears from (2.4) that the variance parameter η_g is not equal to zero, but the mean value is zero. Further, the impurities are assumed to be uncorrelated and the band mixing effect has been neglected in this simplified theoretical formalism.

We have to average the kinetic energy in the order to obtain the DR in nonlinear optical materials in the presence of band tails. Using the (2.3) and (2.4), we get

$$\begin{aligned}
&\left[\frac{\hbar^2 k_z^2}{2m_{\|}^*}\int_{-\infty}^{E} F(V)dV\right] + \left[\left(\frac{b_{\|}}{b_{\perp}}\frac{c_{\perp}}{c_{\|}}\right)\frac{\hbar^2 k_s^2}{2m_{\perp}^*}\int_{-\infty}^{E} F(V)dV\right] \\
&= \left\{\int_{-\infty}^{E} \frac{(E-V)[\alpha(E-V)+1]\left[b_{\|}(E-V)+1\right]}{\left[c_{\|}(E-V)+1\right]} F(V)dV \right. \\
&\quad + \frac{\alpha b_{\|}}{c_{\|}}\left[\delta\int_{-\infty}^{E}(E-V)F(V)dV + \frac{2}{9}\left(\Delta_{\|}^2-\Delta_{\perp}^2\right)\int_{-\infty}^{E} F(V)dV\right] \\
&\quad \left. - \left(\frac{2}{9}\right)\frac{\alpha b_{\|}}{c_{\|}}\left(\Delta_{\|}^2-\Delta_{\perp}^2\right)\int_{-\infty}^{E}\frac{F(V)dV}{\left[c_{\|}(E-V)+1\right]}\right\} \\
&\quad - \left(\frac{\hbar^2 k_s^2}{2m_{\perp}^*}\right)\left\{\left(\frac{\hbar^2 k_s^2}{2m_{\perp}^*}\right)\left[\left(\frac{\delta}{2}+\frac{\Delta_{\|}^2-\Delta_{\perp}^2}{6\Delta_{\|}}\right)\alpha\int_{-\infty}^{E}\frac{F(V)dV}{[\alpha(E-V)+1]}\right.\right. \\
&\quad \left.\left. + \left(\frac{\delta}{2}-\frac{\Delta_{\|}^2-\Delta_{\perp}^2}{6\Delta_{\|}}\right)c_{\|}\int_{-\infty}^{E}\frac{F(V)dV}{\left[c_{\|}(E-V)+1\right]}\right]\right\}
\end{aligned} \quad (2.5)$$

The (2.5) can be rewritten as

$$\begin{aligned}
&\frac{\hbar^2 k_z^2}{2m_{\|}^*} I(1) + \left(\frac{b_{\|}}{b_{\perp}}\frac{c_{\perp}}{c_{\|}}\right)\frac{\hbar^2 k_s^2}{2m_{\perp}^*} I(1) = \left\{I_3(c_{\perp}) + \frac{\alpha b_{\|}}{c_{\|}}\left[\delta I(4) + \frac{2}{9}\left(\Delta_{\|}^2-\Delta_{\perp}^2\right)I(1)\right]\right. \\
&\quad \left. - \left(\frac{2}{9}\right)\frac{\alpha b_{\|}}{c_{\|}}\left(\Delta_{\|}^2-\Delta_{\perp}^2\right)I_6(c_{\|})\right\}\left(\frac{\hbar^2 k_s^2}{2m_{\perp}^*}\right) \\
&\quad \left\{\left(\frac{b_{\|}}{b_{\perp}}\frac{c_{\perp}}{c_{\|}}\right)\left[\left(\frac{\delta}{2}+\frac{\Delta_{\|}^2-\Delta_{\perp}^2}{6\Delta_{\|}}\right)\alpha I(\alpha) + \left(\frac{\delta}{2}-\frac{\Delta_{\|}^2-\Delta_{\perp}^2}{6\Delta_{\|}}\right)c_{\|} I(c_{\|})\right]\right\}
\end{aligned} \quad (2.6)$$

where,

$$I(1) \equiv \int_{-\infty}^{E} F(V)dV \tag{2.7}$$

$$I_3(c_{\|}) \equiv \int_{-\infty}^{E} \frac{(E-V)[\alpha(E-V)+1]\left[b_{\|}(E-V)+1\right]}{\left[c_{\|}(E-V)+1\right]} F(V)dV \tag{2.8}$$

$$I_4(4) \equiv \int_{-\infty}^{E} (E-V)F(V)dV \tag{2.9}$$

$$I(\alpha) \equiv \int_{-\infty}^{E} \frac{F(V)dV}{[\alpha(E-V)+1]} \tag{2.10}$$

Substituting $E - V \equiv x$ and $x/\eta_g \equiv t_0$, we get from (2.7)

$$I(1) = \left(\exp(-E^2/\eta_g^2/\sqrt{\pi})\right) \int_{0}^{\infty} \exp\left[-t_0^2 + (2Et_0/\eta_g)\right] dt_0$$

Thus,

$$I(1) = \left[\frac{1 + Erf(E/\eta_g)}{2}\right] \tag{2.11}$$

where, $Erf(E/\eta_g)$ is the error function of (E/η_g).

From (2.9), one can write

$$I(4) = (1/\eta_g\sqrt{\pi}) \int_{-\infty}^{E} (E-V)\exp\left(-V^2/\eta_g^2\right)dV$$
$$= \frac{E}{2}\left[1 + Erf(E/\eta_g)\right] - \left\{\frac{1}{\sqrt{\pi\eta_g^2}} \int_{-\infty}^{E} V\exp(-V^2/\eta_g^2)dV\right\} \tag{2.12}$$

After computing this simple integration, one obtains

Thus,

$$I(4) = \eta_g \exp\left(-E^2 / \eta_g^2\right)\left(2\sqrt{\pi}\right)^{-1} + \frac{E}{2}\left(1 + Erf(E/\eta_g)\right) = \gamma_0(E, \eta_g) \quad (2.13)$$

From (2.10), we can write

$$I(\alpha) = \frac{1}{\sqrt{\pi\eta_g^2}} \int_{-\infty}^{E} \frac{\exp\left(-V^2 / \eta_g^2\right) dV}{[\alpha(E - V) + 1]} \quad (2.14)$$

When, $V \to \pm\infty$, $\frac{1}{[\alpha(E-V)+1]} \to 0$ and $\exp\left(-V^2 / \eta_g^2\right) \to 0$;
Thus (2.14) can be expressed as

$$I(\alpha) = \left(1/\alpha\eta_g\sqrt{\pi}\right) \int_{-\infty}^{\infty} \exp(-t^2)(u - t)^{-1} dt \quad (2.15)$$

where, $\frac{V}{\eta_g} \equiv t$ and $u \equiv \left(\frac{1 + \alpha E}{\alpha\eta}\right)$.
It is well known that [95, 96]

$$W(Z) = (i/\pi) \int_{-\infty}^{\infty} (Z - t)^{-1} \exp(-t^2) dt \quad (2.16)$$

In which $i = \sqrt{-1}$ and Z, in general, is a complex number.
We also know [95, 96],

$$W(Z) = \exp\left(-Z^2\right) Erfc(-iZ) \quad (2.17)$$

where,

$$Erfc(Z) \equiv 1 - Erf(Z).$$
Thus, $Erfc(-iu) = 1 - Erf(-iu)$
Since, $Erf(-iu) = -Erf(iu)$

Therefore, $Erfc(-iu) = 1 + Erf(iu)$.
Thus,

$$I(\alpha) = \left[-i\sqrt{\pi}/\alpha\eta_g\right] \exp\left(-u^2\right)[1 + Erf(iu)] \quad (2.18)$$

We also know that [95, 96]

$$Erf(x+iy) = Erf(x) + \left(\frac{e^{-x^2}}{2\pi x}\right)\left[(1-\cos(2xy)) + i\sin(2xy) + \frac{2}{\pi}e^{-x^2}\sum_{p=1}^{\infty}\frac{\exp(-p^2/4)}{(p^2+4x^2)}\right]$$
$$\left[f_p(x,y) + ig_p(x,y) + \varepsilon(x,y)\right] \tag{2.19}$$

where, $f_p(x,y) \equiv [2x - 2x\cosh(py)\cos(2xy) + p\sinh(py)\sin(2xy)]$,

$$g_p(x,y) \equiv [2x\cosh(py)\sin(2xy) + p\sinh(py)\cos(2xy)], \quad |\varepsilon(x,y)| \approx 10^{-16}|Erf(x+iy)|$$

Substituting $x = 0$ and $y = u$ in (2.19), one obtains,

$$Erf(iu) = \left(\frac{2i}{\pi}\right)\sum_{p=1}^{\infty}\left\{\frac{\exp(-p^2/4)}{p}\sinh(pu)\right\} \tag{2.20}$$

Therefore, one can write

$$I(\alpha) = C_{21}(\alpha, E, \eta_g) - iD_{21}(\alpha, E, \eta_g) \tag{2.21}$$

where,

$$C_{21}(\alpha, E, \eta_g) \equiv \left[\frac{2}{\alpha\eta_g\sqrt{\pi}}\right]\exp(-u^2)\left[\sum_{p=1}^{\infty}\left\{\frac{\exp(-p^2/4)}{p}\sinh(pu)\right\}\right] \text{ and } D_{21}(\alpha, E, \eta_g) \equiv \left[\frac{\sqrt{\pi}}{\alpha\eta_g}\exp(-u^2)\right].$$

The (2.21) consists of both real and imaginary parts and therefore, $I(\alpha)$ is complex, which can also be proved by using the method of analytic continuation of the subject Complex Analysis.

The integral $I_3(c_{\|})$ in (2.8) can be written as

$$I_3(c_{\|}) = \left(\frac{\alpha b_{\|}}{c_{\|}}\right)I(5) + \left(\frac{\alpha c_{\|} + b_{\|}c_{\|} - \alpha b_{\|}}{c_{\|}^2}\right)I(4) + \frac{1}{c_{\|}}\left(1 - \frac{\alpha}{c_{\|}}\right)\left(1 - \frac{b_{\|}}{c_{\|}}\right)I(1)$$
$$- \left\{\frac{1}{c_{\|}}\left(1 - \frac{\alpha}{c_{\|}}\right)\left(1 - \frac{b_{\|}}{c_{\|}}\right)I(c_{\|})\right\} \tag{2.22}$$

where

$$I(5) \equiv \int_{-\infty}^{E}(E-V)^2F(V)dV \tag{2.23}$$

From (2.23) one can write

$$I(5) = \frac{1}{\sqrt{\pi\eta_g^2}}\left[E^2\int_{-\infty}^{E}\exp\left(\frac{-V^2}{\eta_g^2}\right)dV - 2E\int_{-\infty}^{E}V\exp\left(\frac{-V^2}{\eta_g^2}\right)dV + \int_{-\infty}^{E}V^2\exp\left(\frac{-V^2}{\eta_g^2}\right)dV\right]$$

The evaluations of the component integrals lead us to write

$$I(5) = \frac{\eta_g E}{2\sqrt{\pi}}\exp\left(\frac{-E^2}{\eta_g^2}\right) + \frac{1}{4}\left(\eta_g^2 + 2E^2\right)\left[1 + Erf\left(\frac{E}{\eta_g}\right)\right] = \theta_0(E, \eta_g) \quad (2.24)$$

Thus combining the aforementioned equations, $I_3(c_\parallel)$ can be expressed as

$$I_3(c_\parallel) = A_{21}(E, \eta_g) + iB_{21}(E, \eta_g) \quad (2.25)$$

where,

$$\begin{aligned}
A_{21}(E,\eta) \equiv & \left[\frac{\alpha b_\parallel}{c_\parallel}\left[\frac{\eta_g E}{2\sqrt{\pi}}\exp\left(\frac{-E^2}{\eta_g^2}\right) + \frac{1}{4}\left(\eta_g^2 + 2E^2\right)\left\{1 + Erf\left(\frac{E}{\eta_g}\right)\right\}\right]\right] \\
& + \left[\frac{\alpha c_\parallel + b_\parallel c_\parallel - \alpha b_\parallel}{c_\parallel^2}\right]\left\{\frac{E}{2}[1 + Erf(E/\eta)] + \frac{\eta_g \exp\left(-E^2/\eta_g^2\right)}{2\sqrt{\pi}}\right\} \\
& + \frac{1}{c_\parallel}\left(1 - \frac{\alpha}{c_\parallel}\right)\left(1 - \frac{b_\parallel}{c_\parallel}\right)\frac{1}{2}[1 + Erf(E/\eta)] - \left\{\frac{2}{c_\parallel^2\eta_g\sqrt{\pi}}\left(1 - \frac{\alpha}{c_\parallel}\right)\left(1 - \frac{b_\parallel}{c_\parallel}\right)\exp(-u_1^2)\right\} \\
& \left[\sum_{p=1}^{\infty}\left\{\frac{\exp(-p^2/4)}{p}\sinh(pu_1)\right\}\right],
\end{aligned}$$

$$u_1 \equiv \left[\frac{1 + c_\parallel E}{c_\parallel \eta_g}\right] \quad \text{and} \quad B_{21}(E, \eta_g) \equiv \frac{\sqrt{\pi}}{c_\parallel^2\eta_g}\left(1 - \frac{\alpha}{c_\parallel}\right)\left(1 - \frac{b_\parallel}{c_\parallel}\right)\exp(-u_1^2)$$

Therefore, the combination of all the appropriate integrals together with algebraic manipulations leads to the expression of the DR of the conduction electrons of HD nonlinear optical materials forming Gaussian band tails as

$$\frac{\hbar^2 k_z^2}{2m_\parallel^* T_{21}(E, \eta_g)} + \frac{\hbar^2 k_s^2}{2m_\parallel^* T_{22}(E, \eta_g)} = 1 \quad (2.26)$$

where, $T_{21}(E, \eta_g)$ and $T_{22}(E, \eta_g)$ have both real and complex parts and are given by

$$T_{21}(E,\eta_g) \equiv [T_{27}(E,\eta_g) + iT_{28}(E,\eta_g)],\; T_{27}(E,\eta_g) \equiv \left[\frac{T_{23}(E,\eta_g)}{T_5(E,\eta_g)}\right],$$

$$T_{23}(E,\eta_g) \equiv \left[A_{21}(E,\eta_g) + \frac{\alpha b_\parallel}{c_\parallel}\left[\delta\gamma_0(E,\eta_g) + \frac{1}{9}\left(\Delta_\parallel^2 - \Delta_\perp^2\right)[1 + Erf(E/\eta_g)]\right] - \left\{\frac{2}{9}\left(\frac{\alpha b_\parallel}{c_\parallel}\right)\left(\Delta_\parallel^2 - \Delta_\perp^2\right)G_{21}(c_\parallel, E,\eta_g)\right\}\right],$$

$$G_{21}(E,\eta_g) \equiv \frac{2}{c_\parallel \eta_g\sqrt{\pi}}\exp(-u_1^2)\sum_{p=1}^{\infty}\left\{\frac{\exp(-p^2/4)}{p}\sinh(pu_1)\right\},$$

$$T_5(E,\eta_g) \equiv \frac{1}{2}[1 + Erf(E/\eta_g)],$$

$$T_{28}(E,\eta_g) \equiv \left[\frac{T_{24}(E,\eta_g)}{T_5(E,\eta_g)}\right],$$

$$T_{24}(E,\eta_g) \equiv \left[B_{21}(E,\eta_g) + \frac{2}{9}\frac{\alpha b_\parallel}{c_\parallel}\left(\Delta_\parallel^2 - \Delta_\perp^2\right)H_{21}(E,\eta_g)\right],$$

$$H_{21}(c_\parallel, E,\eta_g) \equiv \left[\frac{\sqrt{\pi}}{\eta_g c_\parallel}\exp(-u_1^2)\right],$$

$$T_{22}(E,\eta_g) \equiv [T_{29}(E,\eta_g) + iT_{30}(E,\eta_g)]$$

$$T_{29}(E,\eta_g) \equiv \frac{T_{23}(E,\eta_g)T_{25}(E,\eta_g) - T_{24}(E,\eta_g)T_{26}(E,\eta_g)}{\left[(T_{25}(E,\eta_g))^2 + (T_{26}(E,\eta_g))^2\right]},$$

$$T_{25}(E,\eta_g) \equiv \left[\left(\frac{b_\parallel\, c_\perp}{b_\perp\, c_\parallel}\right)\frac{1}{2}\left[1 + Erf\left(\frac{E}{\eta_g}\right)\right] + \left(\frac{b_\parallel\, c_\perp}{b_\perp\, c_\parallel}\right)\left(\frac{\delta}{2} + \left[\frac{\Delta_\parallel^2 - \Delta_\perp^2}{6\Delta_\parallel}\right]\right)\alpha_\parallel C_{21}(\alpha_\parallel, E,\eta_g) + \left(\frac{b_\parallel c_\perp}{b_\perp}\right)\left(\frac{\delta}{2} - \left[\frac{\Delta_\parallel^2 - \Delta_\perp^2}{6\Delta_\parallel}\right]\right)G_{21}(\alpha_\parallel, E,\eta_g),\right.$$

$$C_{21}(\alpha, E,\eta_g) \equiv \left[\frac{2}{\alpha\sqrt{\pi}\eta_g}\exp(-u^2)\left[\sum_{p=1}^{\infty}\frac{\exp(-p^2/4)}{p}\sinh(pu)\right]\right],$$

$$T_{26}(E,\eta_g) \equiv \left(\frac{b_\parallel\, c_\perp}{b_\perp\, c_\parallel}\right)\left(\frac{\delta}{2} - \frac{\Delta_\parallel^2 - \Delta_\perp^2}{6\Delta_\parallel}\right)\alpha D_{21}(\alpha, E,\eta_g) + \frac{b_\parallel c_\perp}{b_\perp}\left(\frac{\delta}{2} - \frac{\Delta_\parallel^2 - \Delta_\perp^2}{6\Delta_\parallel}\right)H_{21}(c_\parallel, E,\eta_g) \quad \text{and}$$

$$T_{30}(E,\eta_g) \equiv \frac{T_{24}(E,\eta_g)T_{25}(E,\eta_g) + T_{23}(E,\eta_g)T_{26}(E,\eta_g)}{\left[(T_{25}(E,\eta_g))^2 + (T_{26}(E,\eta_g))^2\right]}$$

From (2.26), it appears that the energy spectrum in HD nonlinear optical materials is complex. **The complex nature of the electron dispersion law in HD materials occurs from the existence of the essential poles in the corresponding electron energy spectrum in the absence of band tails**. It may be noted that the complex band structures have already been studied for bulk materials and super lattices without heavy dopingand bears no relationship with the complex electron dispersion law as indicated by (2.26). The physical picture behind the formulation of the complex energy spectrum in HDS is the interaction of the impurity atoms in the tails with the 0splitting constants of the valance bands. More is the interaction; more is the prominence of the complex part than the other case. In the absence of band tails, $\eta_g \to 0$, and there is no interaction of the impurity atoms in the tails with the spin orbit constants. As a result, there exist no complex energy spectrum and (2.26) gets converted into (2.2) when $\eta_g \to 0$. Besides, the complex spectra are not related to same evanescent modes in the band tails and the conduction bands.

The DOS function is given by

$$N_{HD}(E,\eta_g) = \frac{2g_v m_\perp^* \sqrt{2m_\parallel^*}}{3\pi^2\hbar^3} R_{11}(E,\eta_g)\cos\left[\psi_{11}(E,\eta_g)\right] \tag{2.27}$$

where g_v is the valley degeneracy,

$$R_{11}(E,\eta_g) = \left[\left[\{T_{29}(E,\eta_g)\}'\sqrt{x(E,\eta_g)} + \frac{T_{29}(E,\eta_g)\{x(E,\eta_g)\}'}{2\sqrt{x(E,\eta_g)}} - \{T_{30}(E,\eta_g)\}'\sqrt{y(E,\eta_g)} - \frac{T_{30}(E,\eta_g)\{y(E,\eta_g)\}'}{2\sqrt{y(E,\eta_g)}}\right]^2 + \left[\{T_{29}(E,\eta_g)\}'\sqrt{y(E,\eta_g)} + \frac{T_{29}(E,\eta_g)\{y(E,\eta_g)\}'}{2\sqrt{y(E,\eta_g)}} + \{T_{30}(E,\eta_g)\}'\sqrt{x(E,\eta_g)} - \frac{T_{30}(E,\eta_g)\{x(E,\eta_g)\}'}{2\sqrt{x(E,\eta_g)}}\right]^2\right]^{1/2},$$

$$x(E,\eta_g) \equiv \frac{1}{2}\left[T_{27}(E,\eta_g) + \sqrt{\{T_{27}(E,\eta_g)\}^2 + \{T_{28}(E,\eta_g)\}^2}\right],$$

$$y(E,\eta_g) \equiv \frac{1}{2}\left[\sqrt{\{T_{27}(E,\eta_g)\}^2 + \{T_{28}(E,\eta_g)\}^2} - T_{27}(E,\eta_g)\right] \quad \text{and}$$

$$\psi_{11}(E,\eta_g) \equiv \tan^{-1}\left[\left[\{T_{29}(E,\eta_g)\}'\sqrt{y(E,\eta_g)} + \frac{T_{29}(E,\eta_g)}{2\sqrt{y(E,\eta_g)}} + \{T_{30}(E,\eta_g)\}'\sqrt{x(E,\eta_g)} + \frac{T_{30}\{x(E,\eta_g)\}'}{2\sqrt{x(E,\eta_g)}}\right]\left[\{T_{29}(E,\eta_g)\}'\sqrt{x(E,\eta_g)} + \frac{T_{29}(E,\eta_g)\{x(E,\eta_g)\}'}{2\sqrt{x(E,\eta_g)}} - \{T_{30}(E,\eta_g)\}'\sqrt{y(E,\eta_g)} + \frac{T_{30}\{y(E,\eta_g)\}'}{2\sqrt{y(E,\eta_g)}}\right]^{-1}\right].$$

The oscillatory nature of the DOS for HD nonlinear optical materials is apparent from (2.27). For, $\psi_{11}(E, \eta_g) \geq \pi$, the cosine function becomes negative leading to the negative values of the DOS. The electrons cannot exist for the negative values of the DOS and therefore, this region is forbidden for electrons, which indicates that in the band tail, **there appears a new forbidden zone in addition to the normal band gap of the semiconductor**.

For dimensional quantization along z-direction, the DR of the 2D electrons in this case can be written following (2.26) as

$$\frac{\hbar^2 (\mathrm{n}_z \pi / \mathrm{d}_z)^2}{2m_{\parallel}^* T_{21}(E, \eta_g)} + \frac{\hbar^2 k_s^2}{2m_{\perp}^* T_{22}(E, \eta_g)} = 1 \tag{2.28}$$

where, $n_z (=1, 2, 3, \ldots)$ and d_z are the size quantum number and the nano-thickness along the z-direction respectively.

The general expression of the total 2D DOS $(N_{2DT}(E))$ can, in general, be expressed as

$$N_{2DT}(E) = \frac{2g_v}{(2\pi)^2} \sum_{n_z=1}^{n_{z\max}} \frac{\partial A(E, n_z)}{\partial E} H(E - E_{n_z}) \tag{2.29}$$

where $A(E, n_z)$ is the area of the constant energy 2D wave vector space and in this case it is for QWs, $H(E - E_{n_z})$ is the Heaviside step function and E_{n_z} is the corresponding sub-band energy. Using (2.28) and (2.29), the expression of the $N_{2DT}(E)$ for QWs of HD nonlinear optical materials can be written as

$$N_{2DT}(E) = \frac{m_{\perp}^* g_v}{\pi \hbar^2} \sum_{n_z=1}^{n_{z\max}} T'_{1D}(E, \eta_g, \mathrm{n}_z) H(\mathrm{E} - \mathrm{E}_{n_z D1}) \tag{2.30}$$

where, $T_{1D}(E, \eta_g, \mathrm{n}_z) = \left[1 - \frac{\hbar^2 (\mathrm{n}_z \pi / d_z)^2}{2m_{\parallel}^* T_{21}(E, \eta_g)}\right] T_{22}(E, \eta_g)$ and the sub band energies $\mathrm{E}_{n_z D1}$ in this case is given by the following equation

$$\frac{\hbar^2 (\mathrm{n}_z \pi / \mathrm{d}_z)^2}{2m_{\parallel}^* T_{21}(E_{n_z D1}, \eta_g)} = 1 \tag{2.31}$$

Thus we observe that both the total DOS and sub-band energies of QWs of HD nonlinear opticalmaterials are complex due to the presence of the pole in energy axis of the corresponding materials in the absence of band tails.

The EEM in this case is given by

$$m^*\left(E_{F1HD},\eta_g,\mathrm{n}_z\right)=m_\perp^*\left[\text{Real part of } T_{1D}'(E_{F1HD},\eta_g,\mathrm{n}_z)\right] \tag{2.32}$$

Thus, we observe that the EEM is the function of size quantum number and the Fermi energy due to the combined influence of the crystal filed splitting constant and the anisotropic spin-orbit splitting constants respectively. Besides it is a function of η_g due to which the EEM exists in the band gap, which is otherwise impossible.

In the absence of band-tails, the 2D DR the EEM in the x-y plane at the Fermi level, the total 2D DOS, the sub-band energy $E_{n_{z1}}$, of non-linear optical materials in the absence of band tails can, respectively, be written as

$$\psi_1(E)=\psi_2(E)k_s^2+\psi_3(E)(n_z\pi/d_z)^2 \tag{2.33}$$

$$\begin{aligned}m^*(E_{Fs},n_z)=&\left(\frac{\hbar^2}{2}\right)[\psi_2(E_{Fs})]^{-2}\left[\psi_2(E_{Fs})\left\{\{\psi_1(E_{Fs})\}'-\{\psi_3(E_{Fs})\}'\left(\frac{n_z\pi}{d_z}\right)^2\right\}\right.\\&\left.-\left\{\psi_1(E_{Fs})-\psi_3(E_{Fs})\left(\frac{n_z\pi}{d_z}\right)^2\right\}\{\psi_2(E_{Fs})\}'\right]\end{aligned} \tag{2.34}$$

$$\begin{aligned}N_{2DT}(E)=&\left(\frac{g_v}{2\pi}\right)\sum_{n_z=1}^{n_{z\max}}\left[\psi_2(E)^{-2}\right]\left[\psi_2(E)\left\{\{\psi_1(E)\}'-\{\psi_3(E)\}'\left(\frac{n_z\pi}{d_z}\right)^2\right\}\right.\\&\left.-\left\{\psi_1(E)-\psi_3(E)\left(\frac{n_z\pi}{d_z}\right)^2\right\}\{\psi_2(E)\}'\right]H(E-E_{n_{z1}})\end{aligned} \tag{2.35}$$

$$\psi_1(E_{n_{z1}})=\psi_2(E_{n_{z1}})(n_z\pi/d_z)^2 \tag{2.36}$$

where

$$\psi_1(\mathrm{E})=\gamma(\mathrm{E}),\ \psi_2(\mathrm{E})=\mathrm{f}_1(\mathrm{E}),\ \psi_3(\mathrm{E})=\mathrm{f}_2(\mathrm{E}),$$

In the absence of band-tails, the DOS for bulk specimens of non-linear optical materials is given by

$$D_0(E)=g_v\left(3\pi^2\right)^{-1}\psi_4(E) \tag{2.37}$$

$$\psi_4(E) \equiv \left[\frac{3}{2}\frac{\sqrt{\psi_1(E)}[\psi_1(E)]'}{\psi_2(E)\sqrt{\psi_3(E)}} - \frac{[\psi_2(E)]'[\psi_1(E)]^{3/2}}{[\psi_1(E)]^2\sqrt{\psi_3(E)}} - \frac{1}{2}\frac{[\psi_3(E)]'[\psi_1(E)]^{3/2}}{\psi_2(E)[\psi_3(E)]^{3/2}}\right],$$

$$[\psi_1(E)]' \equiv \left[(2E+E_g)\psi_1(E)\left[E(E+E_g)\right]^{-1} + E(E+E_g)(2E+2E_g+\delta+\Delta_{\|})\right],$$

$$[\psi_1(E)]' \equiv \left[2m_\perp^*\left(E_g+\frac{2}{3}\Delta_\perp\right)\right]^{-1}\left[\hbar^2 E_g(E_g+\Delta_\perp)\right]\left[\delta+2E+2E_g+\frac{2}{3}\Delta_{\|}\right] \text{ and}$$

$$[\psi_3(E)]' \equiv \left[2m_\perp^*\left(E_g+\frac{2}{3}\Delta_{\|}\right)\right]^{-1}\left[\hbar^2 E_g(E_g+\Delta_{\|})\right]\left[2E+2E_g+\frac{2}{3}\Delta_{\|}\right]$$

2.2.2 *The DR in Ultrathin Films (UFs) of HD III–V Materials*

The DR of the conduction electrons of III–V compoundsare described by the models of Kane (both three and two bands) [97, 98], Stillman et al. [99] and Palik et al. [100] respectively. For the purpose of complete and coherent presentation and relative comparison, the DRs in QWs of HD III–V materials have also been investigated.

(a) The Three Band Model of Kane

Under the conditions, $\delta = 0$, $\Delta_{\|} = \Delta_\perp = \Delta$ (isotropic spin orbit splitting constant) and $m_{\|}^* = m_\perp^* = m_c$ (isotropic effective electron mass at the edge of the conduction band), (2.2) gets simplified as

$$\frac{\hbar^2 k^2}{2m_c} = I_{11}(E), \qquad I_{11}(E) \equiv \frac{E(E+E_{g_0})(E+E_{g_0}+\Delta)(E_{g_0}+\frac{2}{3}\Delta)}{E_{g_0}(E_{g_0}+\Delta)(E+E_{g_0}+\frac{2}{3}\Delta)} \tag{2.38}$$

which is known as the three band model of Kane [97, 98] and is often used to investigate the physical properties of III–V materials.

Under the said conditions, the HD electron dispersion law in this case can be written from (2.38) as

$$\frac{\hbar^2 k^2}{2m_c} = T_{31}(E,\eta_g) + iT_{32}(E,\eta_g) \tag{2.39}$$

where,

$$
\begin{aligned}
T_{31}(E,\eta_g) \equiv & \left(\frac{2}{1+Erf(E/\eta_g)}\right)\left[\frac{\alpha b}{c}\theta_0(E,\eta_g)+\left[\frac{\alpha c+bc-\alpha b}{c^2}\right]\gamma_0(E,\eta_g)\right. \\
& +\frac{1}{c}\left(1-\frac{\alpha}{c}\right)\left(1-\frac{b}{c}\right)\frac{1}{2}\left[1+Erf\left(\frac{E}{\eta_g}\right)\right] \\
& \left.-\frac{1}{c}\left(1-\frac{\alpha}{c}\right)\left(1-\frac{b}{c}\right)\frac{2}{c\eta_g\sqrt{\pi}}\exp(-u_2^2)\left[\sum_{p=1}^{\infty}\frac{\exp(-p^2/4)}{p}\sinh(pu_2)\right]\right],
\end{aligned}
$$

$$
b \equiv (E_g+\Delta)^{-1},\ c \equiv \left(E_g+\frac{2}{3}\Delta\right)^{-1},
$$

$$
u_2 \equiv \frac{1+cE}{c\eta_g} \text{ and } T_{32}(E,\eta_g) \equiv \left(\frac{2}{1+Erf(E/\eta_g)}\right)\frac{1}{c}\left(1-\frac{\alpha}{c}\right)\left(1-\frac{b}{c}\right)\frac{\sqrt{\pi}}{c\eta_g}\exp(-u_2^2).
$$

Thus, the complex energy spectrum occurs due to the term $T_{32}(E,\eta_g)$ and this imaginary band is quite different from the forbidden energy band.

The DOS function in this case can be written as

$$
N_{HD}(E,\eta_g) = \frac{g_v}{3\pi^2}\left(\frac{2m_c}{\hbar^2}\right)^{3/2} R_{21}(E,\eta_g)\cos[\vartheta_{21}(E,\eta_g)] \tag{2.40}
$$

where,

$$
R_{21}(E,\eta_g) \equiv \left[\frac{\left[\{\alpha_{11}(E,\eta_g)\}'\right]^2}{4\alpha_{11}(E,\eta_g)}+\frac{\left[\{\beta_{11}(E,\eta_g)\}'\right]^2}{4\beta_{11}(E,\eta_g)}\right]^{1/2},
$$

$$
\alpha_{11}(E,\eta_g) \equiv \frac{1}{2}\left[T_{33}(E,\eta_g)+\sqrt{\{T_{33}(E,\eta_g)\}^2+\{T_{34}(E,\eta_g)\}^2}\right],
$$

$$
T_{33}(E,\eta_g) \equiv \left[\{T_{31}(E,\eta_g)\}^3-3T_{31}(E,\eta_g)\{T_{32}(E,\eta_g)\}^2\right],
$$

$$
T_{34}(E,\eta_g) \equiv \left[3T_{32}(E,\eta_g)\{T_{31}(E,\eta_g)\}^2-\{T_{32}(E,\eta_g)\}^3\right],
$$

$$
\beta_{11}(E,\eta_g) \equiv \frac{1}{2}\left[\sqrt{\{T_{33}(E,\eta_g)\}^2+\{T_{34}(E,\eta_g)\}^2}-T_{33}(E,\eta_g)\right] \text{ and}
$$

$$
\vartheta_{21}(E,\eta_g) \equiv \tan^{-1}\left[\frac{\{\beta_{11}(E,\eta_g)\}'}{\{\alpha_{11}(E,\eta_g)\}'}\sqrt{\frac{\alpha_{11}(E,\eta_g)}{\beta_{11}(E,\eta_g)}}\right].
$$

Thus, the oscillatory DOS function becomes negative for $\vartheta_{21}(E,\eta_g) \geq \pi$ and a new forbidden zonewill appear in addition to the normal band gap.

For dimensional quantization along z-direction, the DR of the 2D electrons in this case can be written following (2.40) as

$$\frac{\hbar^2(\mathrm{n}_z\pi/\mathrm{d}_z)}{2m_c}+\frac{\hbar^2(k_s)^2}{2m_c}=T_{31}(E,\eta_g)+iT_{32}(E,\eta_g) \tag{2.41}$$

The expression of the $N_{2DT}(E)$ in this case assumes the form

$$N_{2DT}(E)=\frac{m_c g_v}{\pi\hbar^2}\sum_{n_z=1}^{n_{z\max}}T_{5D}'(E,\eta_g,\mathrm{n}_z)H(\mathrm{E}-\mathrm{E}_{n_zD5}) \tag{2.42}$$

where, $T_{5D}(E,\eta_g,\mathrm{n}_z)=\left[T_{31}(E,\eta_g)+iT_{32}(E,\eta_g)-\hbar^2(\mathrm{n}_z\pi/d_z)^2(2m_c)^{-1}\right]$ and the sub band energies E_{n_zD5} in this case given by

$$\left\{\hbar^2(n_z/d_z)^2\right\}(2m_c)^{-1}=T_{31}(E_{\mathrm{n}_zD5},\eta_g) \tag{2.43}$$

Thus we observe that both the total DOS in QWs of HD III–V compounds and the sub band energies are complex due to the presence of the pole in energy axis of the corresponding materials in the absence of band tails.

The EEM in this case is given by

$$m^*(E_{F1HD},\eta_g,\mathrm{n}_z)=m_c[T_{31}'(E_{F1HD},\eta_g,\mathrm{n}_z)] \tag{2.44}$$

In the absence of band tails, the 2D DR, EEM in the x-y plane at the Fermi level, the total 2D DOS, the sub-band energy, for QWs of III–V materials assume the following forms

$$\frac{\hbar^2k_s^2}{2m_c}+\frac{\hbar^2}{2m_c}(n_z\pi/d_z)^2=I_{11}(E) \tag{2.45}$$

$$m^*(E_{Fs})=m_c\{I_{11}(E_{Fs})\}' \tag{2.46}$$

It is worth noting that the EEM in this case is a function of Fermi energy alone and is independent of size quantum number.

$$N_{2DT}(E)=\left(\frac{m_c g_v}{\pi\hbar^2}\right)\sum_{n_z=1}^{n_{z\max}}\left\{[I_{11}(E)]'H\left(E-E_{n_{z2}}\right)\right\} \tag{2.47}$$

where, the sub-band energies $E_{n_{z2}}$ can be expressed as

$$I_{11}\left(E_{n_{z_2}}\right) = \frac{\hbar^2}{2m_c}(n_z\pi/d_z)^2 \tag{2.48}$$

In the absence of band tails, the DOS density-of-statesfunction, in bulk III–V, ternary and quaternary materials in accordance with the unperturbed three band model of Kane assume the following forms

$$D_0(E) = 4\pi g_v\left(\frac{2m_c}{\hbar^2}\right)^{3/2}\sqrt{I_{11}(E)}\left[I'_{11}(E)\right] \tag{2.49}$$

Under the inequalities $\Delta \gg E_{g_0}$ or $\Delta \ll E_{g_0}$, (2.38) can be expressed as

$$E(1+\alpha E) = \frac{\hbar^2 k^2}{2m_c} \tag{2.50}$$

where $\alpha \equiv \left(E_{g_0}\right)^{-1}$ and is known as band non-parabolicity.

It may be noted that (2.50) is the well-known two band model of Kane and is used in the literature to study the physical properties of those III–V and opto-electronic materials whose energy band structures obey the aforementioned inequalities.

The DR in HD III–V, ternary and quaternary materials whose energy spectrum in the absence of band tails obeys the two band model of Kaneas defined by (2.50), can be written as

$$\frac{\hbar^2 k^2}{2m_c} = \gamma_2(E,\eta_g) \tag{2.51}$$

where,

$$\gamma_2(E,\eta_g) \equiv \left[\frac{2}{1+Erf(E/\eta_g)}\right]\left[\gamma_0(E,\eta_g) + \alpha\theta_0(E,\eta_g)\right].$$

The EEM in this case can be written as

$$m^*(E_{F_h},\eta_g) = m_c\{\gamma_2(E,\eta_g)\}'\Big|_{E=E_{F_h}} \tag{2.52}$$

Thus, one again observes that the EEM in this case exists in the band gap.

In the absence of band tails, $\eta_g \to 0$ and the EEM assumes the well-known form

$$m^*(E_F) = m_c\{1+2\alpha E\}|_{E=E_F} \tag{2.53}$$

The DOS function in this case can be written as

$$N_{HD}(E,\eta_g) = \frac{g_v}{2\pi^2}\left(\frac{2m_c}{\hbar^2}\right)^{3/2}\sqrt{\gamma_2(E,\eta_g)}\{\gamma_2(E,\eta_g)\}' \tag{2.54}$$

Since, the poles of the original two band Kane model are at infinity and and no finite poles with respect to energy, therefore the HD counterpart will be totally real and the complex band vanishes.

For dimensional quantization along z-direction, the DR of the 2D electrons in this case can be written following (2.51) as

$$\frac{\hbar^2(\mathrm{n}_z\pi/\mathrm{d}_z)^2}{2m_c} + \frac{\hbar^2(k_s)^2}{2m_c} = \gamma_2(E,\eta_g) \tag{2.55}$$

The expression of the $N_{2DT}(E)$ in this case can be written as

$$N_{2DT}(E) = \frac{m_c g_v}{\pi\hbar^2}\sum_{n_z=1}^{n_{z\max}} T'_{7D}(E,\eta_g,\mathrm{n}_z)H(\mathrm{E}-\mathrm{E}_{n_zD7}) \tag{2.56}$$

where,

$$T_{7D}(E,\eta_g,\mathrm{n}_z) = \left[\gamma_2(E,\eta_g) - \hbar^2(\mathrm{n}_z\pi/d_z)^2(2m_c)^{-1}\right],$$

The sub-band energies E_{n_zD7} in this case given by

$$\left\{\hbar^2(\mathrm{n}_z\pi/\mathrm{d}_z)^2\right\}(2m_c)^{-1} = \gamma_2(\mathrm{E}_{n_zD7},\eta_g) \tag{2.57}$$

Thus, we observe that both the total DOS and sub-band energies of QWs of HD III–V compounds in accordance with two band model of Kane are not at all complex since the DR in accordance with the said model has no pole in the finite complex plane.

The EEM in this case is given by

$$m^*(E_{F1HD},\eta_g,\mathrm{n}_z) = m_c\left[\gamma'_2(E_{F1HD},\eta_g,\mathrm{n}_z)\right] \tag{2.58}$$

Under the inequalities $\Delta \gg E_{g_0}$ or $\Delta \ll E_{g_0}$, (2.50) assumes the form

$$E(1+\alpha E) = \frac{\hbar^2 k_s^2}{2m_c} + \frac{\hbar^2}{2m_c}\left(\frac{n_z\pi}{d_z}\right)^2 \tag{2.59}$$

The EEM can be written from (2.59) as

$$m^*(E_{Fs}) = m_c(1 + 2\alpha E_{Fs}) \tag{2.60}$$

The total 2D DOS function assumes the form

$$N_{2DT}(E) = \frac{m_c g_v}{\pi \hbar^2} \sum_{n_z=1}^{n_{z\max}} (1 + 2\alpha E) H(E - E_{n_{z3}}) \tag{2.61}$$

where, the sub-band energy $(E_{n_{z3}})$ can be expressed as

$$\frac{\hbar^2}{2m_c}(n_z \pi/d_z)^2 = E_{n_{z3}}(1 + \alpha E_{n_{z3}}) \tag{2.62}$$

The forms of the DOS, for bulk specimens of III–V materials in the absence of band tails whose energy band structures are defined by the two-band model of Kane can, respectively, be written as

$$D_0(E) = 4\pi g_v \left(\frac{2m_c}{\hbar^2}\right)^{3/2} \sqrt{I_{11e}(E)}\left[I'_{11e}(E)\right] \tag{2.63}$$

where,

$$I_{11e}(E) \equiv E(1 + \alpha E), \quad I'_{11e}(E) \equiv (1 + 2\alpha E),$$

The DR in HDS whose energy spectrum in the absence of band tails obeys the parabolic energy bands is given by

$$\frac{\hbar^2 k^2}{2m_c} = \gamma_3(E, \eta_g) \tag{2.64}$$

where,

$$\gamma_3(E, \eta_g) \equiv \left[\frac{2}{(1 + Erf(E/\eta_g))}\right] \gamma_0(E, \eta_g).$$

Since the DR in accordance with the said model is an all zero function with no pole in the finite complex plane, therefore the HD counterpart will be totally real, which is also apparent form the expression (2.64).

The EEM in this case can be written as

$$m^*(E_{F_h}, \eta_g) = m_c\{\gamma_3(E, \eta_g)\}' \tag{2.65}$$

In the absence of band tails, $\eta_g \to 0$ and the EEM assumes the form

$$m^*(E_F) = m_c \tag{2.66}$$

It is well-known that the EEM in unperturbed parabolic energy bands is a constant quantity in general excluding cross-fields configuration. However, the same mass in the corresponding HD bulk counterpart becomes a complicated function of Fermi energy and the impuritypotential together with the fact that the EEM also exists in the band gap solely due to the presence of finite η_g.

The DOS function in this case can be written as

$$N_{HD}(E,\eta_g) = \frac{g_v}{2\pi^2}\left(\frac{2m_c}{\hbar^2}\right)^{3/2}\sqrt{\gamma_3(E,\eta_g)}\{\gamma_3(E,\eta_g)\}' \tag{2.67}$$

For dimensional quantization along z-direction, the DR of the 2D electrons in this case can be written as

$$\frac{\hbar^2(\mathrm{n_z}\pi/\mathrm{d_z})^2}{2m_c} + \frac{\hbar^2(k_s)^2}{2m_c} = \gamma_3(E,\eta_g) \tag{2.68}$$

the expression of the $N_{2DT}(E)$ in this case can be written as

$$N_{2DT}(E) = \frac{m_c g_v}{\pi\hbar^2}\sum_{n_z=1}^{n_{z\max}} T'_{9D}(E,\eta_g,\mathrm{n_z})H(\mathrm{E}-\mathrm{E}_{n_zD9}) \tag{2.69}$$

where,

$$T_{9D}(E,\eta_g,\mathrm{n_z}) = \left[\gamma_3(E,\eta_g) - \hbar^2(\mathrm{n_z}\pi/\mathrm{d_z})^2(2m_c)^{-1}\right]. \tag{2.70}$$

The sub band energies E_{n_zD9} in this case given by

$$\left\{\hbar^2(n_z\pi/d_z)^2\right\}(2m_c)^{-1} = \gamma_3(E_{n_zD9},\eta_g) \tag{2.71}$$

The EEM in this case can be written as

$$m^*(E_{F1HD},\eta_g,n_z) = m_c\left[\gamma'_3(E_{F1HD},\eta_g)\right] \tag{2.72}$$

Under the condition $\alpha \to 0$, the expressions of total 2D DOS, for materials without forming band tails whose bulk electrons are defined by the isotropic parabolic energy bands can, be written as

$$N_{2DT}(E) = \frac{m_c g_v}{\pi\hbar^2}\sum_{n_z=1}^{n_{z\max}} H\left(E - E_{n_{zp}}\right) \tag{2.73}$$

The sub-band energy $\left(E_{n_{zp}}\right)$, can be expressed as

$$E_{n_{zp}} = \frac{\hbar^2}{2m_c}\left(\frac{n_z \pi}{d_z}\right)^2 \tag{2.74}$$

(b) The Model of Stillman et al.

In accordance with the model of Stillman et al. [99], the electron dispersion law of III–V materials assumes the form

$$E = \bar{t}_{11}k^2 - \bar{t}_{12}k^4 \tag{2.75}$$

where,

$$\bar{t}_{11} \equiv \frac{\hbar^2}{2m_c};$$

$$\bar{t}_{12} \equiv \left(1 - \frac{m_c}{m_0}\right)^2 \left(\frac{\hbar^2}{2m_c}\right)^2 \left[\left(3E_{g_0} + 4\Delta + \frac{2\Delta^2}{E_{g_0}}\right).\left\{\left(E_{g_0} + \Delta\right)\left(2\Delta + 3E_{g_0}\right)\right\}^{-1}\right]$$

and m_0 is the free electron mass.

In the presence of band tails, (2.75) gets transformed as

$$\frac{\hbar^2 k^2}{2m_c} = I_{12}\left(E, \eta_g\right) \tag{2.76}$$

where,

$$I_{12}(E, \eta_g) = a_{11}[1 - (1 - a_{12}\gamma_3(E, \eta_g))^{\frac{1}{2}}],$$

$$a_{11} \equiv \left(\frac{\hbar^2 \bar{t}_{11}}{4m_c \bar{t}_{12}}\right) \text{ and } a_{12} \equiv \frac{4\bar{t}_{12}}{\bar{t}_{11}^2}$$

The EEM can be written as

$$m^*\left(E_{F_h}, \eta_g\right) = m_c\left\{I_{12}\left(E_{F_h}, \eta_g\right)\right\}' \tag{2.77}$$

The DOS function in this case can be written as

$$N_{HD}\left(E, \eta_g\right) = \frac{g_v}{2\pi^2}\left(\frac{2m_c}{\hbar^2}\right)^{3/2} \sqrt{I_{12}\left(E, \eta_g\right)}\left\{I_{12}\left(E, \eta_g\right)\right\}' \tag{2.78}$$

For dimensional quantization along z-direction, the DR of the 2D electrons in this case can be written following (2.76) as

$$\frac{\hbar^2(\mathrm{n}_z\pi/\mathrm{d}_z)^2}{2m_c}+\frac{\hbar^2(k_s)^2}{2m_c}=I_{12}(E,\eta_g) \tag{2.79}$$

the expression of the $N_{\mathrm{2DT}}(E)$ in this case can be written as

$$N_{2DT}(E)=\frac{m_c g_v}{\pi\hbar^2}\sum_{n_z=1}^{n_{z\max}}T'_{11D}(E,\eta_g,\mathrm{n}_z)H(\mathrm{E}-\mathrm{E}_{\mathrm{n}_z D11}) \tag{2.80}$$

where,

$$T_{11D}(E,\eta_g,\mathrm{n}_z)=\left[I_{12}(E,\eta_g)-\hbar^2(\mathrm{n}_z\pi/\mathrm{d}_z)^2(2m_c)^{-1}\right] \tag{2.81}$$

The sub band energies $\mathrm{E}_{\mathrm{n}_z D11}$ in this case given by

$$\left\{\hbar^2(\mathrm{n}_z\pi/\mathrm{d}_z)^2\right\}(2m_c)^{-1}=I_{12}(\mathrm{E}_{n_z D11},\eta_g) \tag{2.82}$$

The EEM in this case assumes the form

$$m^*(E_{F1HD},\eta_g,\mathrm{n}_z)=m_c[I'_{12}(E_{F1HD},\eta_g,\mathrm{n}_z)] \tag{2.83}$$

For unperturbed material, the 2-D EEM can be expressed as

$$m^*(E_{Fs})=m_c\{I_{12}(E_{Fs})\}' \tag{2.84}$$

where

$$I_{12}(E)\equiv a_{11}[1-(1-a_{12}(E))^{\frac{1}{2}}]$$

It appears that the EEM in this case is a function of Fermi energy alone and is independent of size quantum number.

The total 2D DOS function in the absence of band tails in this case can be written as

$$N_{2DT}(E)=\left(\frac{m_c g_v}{\pi\hbar^2}\right)\sum_{n_z=1}^{n_{z\max}}\left\{[I_{12}(E)]'H\left(E-E_{n_{z3}}\right)\right\} \tag{2.85}$$

where, the sub band energies $E_{n_{z3}}$ can be expressed as

$$I_{12}(E_{n_{z3}})=\frac{\hbar^2}{2m_c}(n_z\pi/d_z)^2 \tag{2.86}$$

(c) Model of Palik et al.

The energy spectrum of the conduction electrons in III–V materials up to the fourth order in effective mass theory, taking into account the interactions of heavy hole, light hole and the split-off holescan be expressed in accordance with the model of Palik et al. [100] as

$$E = \frac{\hbar^2 k^2}{2m_c} - \bar{B}_{11} k^4 \tag{2.87}$$

where

$$\bar{B}_{11} = \left[\frac{\hbar^4}{4E_{g_0}(m_c)^2}\right]\left[\frac{1+\frac{x_{11}^2}{2}}{1+\frac{x_{11}}{2}}\right](1-y_{11})^2,\ x_{11} = \left[1+\left(\frac{\Delta}{E_{g_0}}\right)\right]^{-1} \text{ and } y_{11} = \frac{m_c}{m_0}$$

The (2.87) gets simplified as

$$\frac{\hbar^2 k^2}{2m_c} = I_{13}(E) \tag{2.88}$$

where

$$I_{13}(E) = \bar{b}_{12}\left[\bar{a}_{12} - \left((\bar{a}_{12})^2 - 4E\bar{B}_{11}\right)^{1/2}\right],\ \bar{a}_{12} = \left(\frac{\hbar^2}{2m_c}\right) \text{ and } \bar{b}_{12} = \left[\frac{\bar{a}_{12}}{2\bar{B}_{11}}\right]$$

Under the condition of heavy doping forming Gaussian band tails, (2.88) assumes the form

$$\frac{\hbar^2 k^2}{2m_c} = I_{13}(E, \eta_g) \tag{2.89}$$

where,

$$I_{13}(E,\eta_g) = \bar{b}_{12}[\bar{a}_{12} - ((\bar{a}_{12})^2 - 4\bar{B}_{11}\gamma_3(E,\eta_g))^{1/2}]$$

The EEM can be written as

$$m^*(E_{F_h},\eta_g) = m_c\{I_{13}(E_{F_h},\eta_g)\}' \tag{2.90}$$

The DOS function in this case can be expressed as

$$N_{HD}(E,\eta_g) = \frac{gv}{2\pi^2}\left(\frac{2m_c}{\hbar^2}\right)^{3/2}\sqrt{I_{13}(E,\eta_g)}\{I_{13}(E,\eta_g)\}' \tag{2.91}$$

Since, the original band model in this case is a no pole function, in the finite complex plane therefore, the HD counterpart will be totally real and the complex band vanishes.

For dimensional quantization along z-direction, the DR of the 2D electrons in this case can be written following (2.89) as

$$\frac{\hbar^2(\mathrm{n}_z\pi/\mathrm{d}_z)^2}{2m_c}+\frac{\hbar^2(k_s)^2}{2m_c}=I_{13}(E,\eta_g) \tag{2.92}$$

the expression of the $N_{2DT}(E)$ in this case can be written as

$$N_{2DT}(E)=\frac{m_c g_v}{\pi\hbar^2}\sum_{n_z=1}^{n_{z\max}}T'_{13D}(E,\eta_g,n_z)H(E-E_{n_zD13}) \tag{2.93}$$

where,

$$T_{13D}(E,\eta_g,\mathrm{n}_z)=\left[I_{13}(E,\eta_g)-\hbar^2(\mathrm{n}_z\pi/d_z)^2(2m_c)^{-1}\right],$$

The sub band energies E_{n_zD13} in this case given by

$$\left\{\hbar^2(n_z\pi/d_z)^2\right\}(2m_c)^{-1}=I_{13}(E_{n_zD13},\eta_g) \tag{2.94}$$

The EEM in this case can be expressed as

$$m^*(E_{F1HD},\eta_g,n_z)=m_c[I'_{13}(E_{F1HD},\eta_g,n_z)] \tag{2.95}$$

The 2D electron DR in the absence of band tails this case assumes the form

$$\frac{\hbar^2k_s^2}{2m_c}+\frac{\hbar^2}{2m_c}(n_z\pi/d_z)^2=I_{13}(E) \tag{2.96}$$

The EEM in this case can be written from (2.96) as

$$m^*(E_{F_s})=m_c[I_{13}(E_{F_s})]' \tag{2.97}$$

The total 2D DOS function can be written as

$$N_{2DT}(E)=\left(\frac{m_c g_v}{\pi\hbar^2}\right)\sum_{n_z=1}^{n_{z\max}}\left\{[I_{13}(E)]'H\left(E-E_{n_{z4}}\right)\right\} \tag{2.98}$$

where, the sub band energies $E_{n_{z4}}$ can be expressed as

$$I_{13}\left(E_{n_{z4}}\right) = \frac{\hbar^2}{2m_c}(n_z\pi/d_z)^2 \tag{2.99}$$

2.2.3 The DR in Ultrathin Films (UFs) of HD II–VI Materials

The carrier energyspectra in bulk specimens of II–VI compounds in accordance with Hopfield model [14] can be written as

$$E = a_0' k_s^2 + b_0' k_s^2 \pm \bar{\lambda}_0 k_s \tag{2.100}$$

where $a_0' \equiv \hbar^2/2m_\perp^*$, $b_0' \equiv \hbar^2/2m_\parallel^*$, and $\bar{\lambda}_0$ represents the splitting of the two-spin states by the spin orbit coupling and the crystalline field.

Therefore the DR of the carriers in HD II–VI materials in the presence of Gaussian band tails can be expressed as

$$\gamma_3(E,\eta_g) = a_0' k_s^2 + b_0' k_z^2 \pm \bar{\lambda}_0 k_s \tag{2.101}$$

Thus, the energy spectrum in this case is real since the corresponding E-k relation in the absence of band tails as given by (2.101) is a no pole function in the finite complex plane.

The volume in k-space as enclosed (2.101) can be expressed as

$$V(E,\eta_g) = \frac{4\pi}{3a_0'\sqrt{b_0'}}\left[\{\gamma_3(E,\eta_g)\}^{3/2} + \frac{3}{8}\frac{(\bar{\lambda}_0)^2\sqrt{\gamma_3(E,\eta_g)}}{a_0'} \pm \left(\frac{3}{4}\frac{\bar{\lambda}_0}{\sqrt{a_0'}}\right)\left(\gamma_3(E,\eta_g) + \frac{(\bar{\lambda}_0)^2}{4a_0'}\right)\sin^{-1}\left[\frac{\sqrt{\gamma_3(E,\eta_g)}}{\sqrt{\gamma_3(E,\eta_g) + \frac{(\bar{\lambda}_0)^2}{4a_0'}}}\right]\right] \tag{2.102}$$

The DOS in this case is given by

$$N_{HD}(\mathrm{E},\eta_g) = \frac{g_v}{\pi^2 a_0'\sqrt{b_0'}}\frac{\partial}{\partial E}\left[\{\gamma_3(\mathrm{E},\eta_g)\}^{3/2} + \frac{3}{8}\frac{(\bar{\lambda}_0)^2\sqrt{\gamma_3(\mathrm{E},\eta_g)}}{a_0'}\right] \tag{2.103}$$

The DR dispersion relation of the conduction electrons of QWs of HD II–VI materials for dimensional quantization along z-direction can be written following (2.101) as

$$\gamma_3(E,\eta_g) = a_0' k_s^2 + b_0' \left(\frac{\pi n_z}{d_z}\right)^2 \pm \bar{\lambda}_0 k_s \tag{2.104}$$

The EEM can be expressed following (2.104) as

$$m^*(E_{F1HD}, n_z, \eta_g) = m_\perp^* \left[1 \mp \frac{(\bar{\lambda}_0)\gamma_3'(E_{F1HD},\eta_g)}{\left[(\bar{\lambda}_0)^2 - 4a_0'b_0'\left(\frac{n_z\pi}{d_z}\right)^2 + 4a_0'\gamma_3(E_{F1HD},\eta_g)\right]^{1/2}}\right] \tag{2.105}$$

Thus we observe that the doubled valued effective mass in 2-D QWs of HD II–VI materials is a function of Fermi energy, size quantum number and the screening potential respectively together with the fact that the same mass exists in the band gap due to the sole presence of the splitting of the two-spin states by the spin orbit coupling and the crystalline field.

The total DOS function in this case can be written as

$$N_{2D\Gamma}(E) = \frac{g_v}{(2\pi)} \frac{1}{(a_0')} \sum_{n_z=1}^{n_{zmax}} \gamma_3'(E,\eta_g) H(E - E_{n_zD14},\eta_g) \tag{2.106}$$

The sub-band energy in this case is given by

$$\gamma_3(E_{n_zD14},\eta_g) = b_0' \left(\frac{\pi n_z}{d_z}\right)^2 \tag{2.107}$$

The DR of the conduction electrons of QWs of II–VI materials for dimensional quantization along z-direction in the absence of band tails can be written following (2.100) as

$$E = a_0' k_s^2 + b_0' \left(\frac{n_z\pi}{d_z}\right)^2 \pm \bar{\lambda}_0 k_s \tag{2.108}$$

Using (2.108), the EEMin this case can be written as

$$m^*(E_{Fs}, n_z) = m^*_\perp \left[1 \mp \frac{(\bar{\lambda}_0)}{\left[(\bar{\lambda}_0)^2 - 4a'_0 b'_0 \left(\frac{n_z\pi}{d_z}\right)^2 + 4a'_0\gamma_3 E_{Fs}\right]^{1/2}}\right] \quad (2.109)$$

The sub-band energy E_{n_z5} assumes the form

$$E_{n_{z5}} = b'_0 (n_z\pi/d_z)^2 \quad (2.110)$$

The area of constant energy 2D quantized surface in this case is given by where

$$A_\pm(E, n_z) = \left[\frac{\pi}{2(a'_0)^2}\left[(\bar{\lambda}_0)^2 + 2a'_0(E - E_{n_{z5}}) \pm \bar{\lambda}_0\left[(\bar{\lambda}_0)^2 + 4a'_0(E - E_{n_{z5}})\right]^{1/2}\right]\right]$$

The total DOS function in this case can be written as

$$N_{2D\Gamma}(\mathrm{E}) = \frac{g_v}{(2\pi)}\frac{1}{(a'_0)}\sum_{n_z=1}^{n_{z\max}} H(\mathrm{E} - \mathrm{E}_{n_{z5}}) \quad (2.111)$$

2.2.4 *The DR in Ultrathin Films (UFs) of HD IV–VI Materials*

(a) The DR of the conduction electrons in IV–VI materials can be expressed in accordance with Dimmock [99] as

$$\left[\bar{\varepsilon} - \frac{E_{g_0}}{2} - \frac{\hbar^2 k_s^2}{2m_t^-} - \frac{\hbar^2 k_z^2}{2m_l^-}\right]\left[\bar{\varepsilon} + \frac{E_{g_0}}{2} + \frac{\hbar^2 k_s^2}{2m_t^-} + \frac{\hbar^2 k_z^2}{2m_l^-}\right] = P_\perp^2 k_s^2 + P_\parallel^2 k_z^2 \quad (2.112)$$

where, $\bar{\varepsilon}$ is the energy as measured from the center of the band gap E_{g_0}, $m_t^\pm$ and $m_l^\pm$ represent the contributions to the transverse and longitudinal effective masses of the external L_6^+ and L_6^- bands arising from the $\vec{k}\cdot\vec{p}$ perturbations with the other bands taken to the second order.

Substituting, $P_\perp^2 \equiv (\hbar^2 E_g/2m_t^*)$, $P_\parallel^2 \equiv \left(\frac{\hbar^2 E_g}{2m_l^*}\right)$ and $\bar{\varepsilon} \equiv \left[E + \left(\frac{E_g}{2}\right)\right]$ (where, m_t^* and m_l^* are the transverse and the longitudinal effective masses at k = 0), (2.112) gets transformed as

$$\left[E - \frac{\hbar^2 k_s^2}{2m_t^-} - \frac{\hbar^2 k_z^2}{2m_l^-}\right]\left[1 + \alpha E + \alpha\frac{\hbar^2 k_s^2}{2m_t^+} + \alpha\frac{\hbar^2 k_z^2}{2m_l^+}\right] = \frac{\hbar^2 k_s^2}{2m_t^*} + \frac{\hbar^2 k_z^2}{2m_l^*} \tag{2.113}$$

From (2.113), we can write

$$\begin{aligned}&\frac{\alpha\hbar^4 k_s^4}{4m_t^+ m_t^-} + \hbar^2 k_s^2\left[\left(\frac{1}{2m_t^*} - \frac{1}{2m_t^-}\right) + \alpha E\left(\frac{1}{2m_t^-} - \frac{1}{2m_t^+}\right) + \frac{\alpha\hbar^2 k_s^2}{4m_l^- m_t^+}\right]\\ &\quad + \left[\left(\frac{\hbar^2 k_z^2}{2m_l^*} + \frac{\hbar^2 k_z^2}{2m_l^-}\right) + \frac{\alpha E}{2}\hbar^2 k_z^2\left(\frac{1}{m_l^-} - \frac{1}{m_l^+}\right) + \frac{\alpha\hbar^4 k_z^4}{4m_l^+ m_t^-} - E(1+\alpha E)\right] = 0\end{aligned} \tag{2.114}$$

Using (2.114), the DR of the conduction electrons in HD IV–VI materials can be expressed as

$$\begin{aligned}&\frac{\alpha\hbar^4 k_s^4}{4m_t^+ m_l^-} Z_0(E,\eta_g) + \hbar^2 k_s^2\left[\lambda_{71}(E,\eta_g)k_z^2 + \lambda_{72}(E,\eta_g)\right]\\ &\quad + \left[\lambda_{73}(E,\eta_g)k_z^2 + \lambda_{74}(E,\eta_g)k_z^4 - \lambda_{75}(E,\eta_g)\right] = 0\end{aligned} \tag{2.115}$$

where,

$$Z_0(E,\eta_g) \equiv \frac{1}{2}\left[1 + Erf\left(\frac{E}{\eta_g}\right)\right],\quad \lambda_{70}(E,\eta_g) \equiv \frac{\alpha}{4m_t^+ m_t^-} Z_0(E,\eta_g),$$

$$\lambda_{71}(E,\eta_g) \equiv \left[\frac{\alpha\hbar^2}{4m_t^- m_l^+} Z_0(E,\eta_g) + \frac{\alpha\hbar^2}{4m_l^- m_t^+} Z_0(E,\eta_g)\right],$$

$$\lambda_{72}(E,\eta_g) \equiv \left[\left(\frac{1}{2m_t^*} - \frac{1}{2m_t^-}\right) Z_0(E,\eta_g) + \alpha\left(\frac{1}{2m_t^-} - \frac{1}{2m_t^+}\right)\gamma_0(E,\eta_g)\right],$$

$$\lambda_{73}(E,\eta_g) \equiv \left[\left(\frac{\hbar^2}{2m_l^*} + \frac{\hbar^2}{2m_l^-}\right) Z_0(E,\eta_g) + \frac{\alpha\hbar^2}{2}\left(\frac{1}{m_l^-} - \frac{1}{m_l^+}\right)\gamma_0(E,\eta_g)\right],$$

$$\lambda_{74}(E,\eta_g) \equiv \frac{\alpha\hbar^4 Z_0(E,\eta_g)}{4m_l^+ m_l^-} \text{ and } \lambda_{75}(E,\eta_g) \equiv \left[\gamma_0(E,\eta_g) + \alpha\theta_0(E,\eta_g)\right]$$

Thus, the energy spectrum in this case is real since the corresponding DR in the absence of band tails as given by (2.115) is a pole-less function with respect to energy axis in the finite complex plane.

The volume in k-space as enclosed by (2.115) can be written through the integral as

$$V(E,\eta_g) = 2\pi \int_0^{\lambda_{86}(E,\eta_g)} \Big[-\left[\lambda_{79}(E,\eta_g)k_z^2 + \lambda_{80}(E,\eta_g)\right] + \sqrt{\lambda_{81}(E,\eta_g)k_z^4 + \lambda_{82}(E,\eta_g)k_z^2 + \lambda_{83}(E,\eta_g)}\Big] dk_z \tag{2.116}$$

where,

$$\lambda_{86}(E,\eta_g) \equiv \left[\frac{\sqrt{[\lambda_{84}(E,\eta_g)]^2 + 4\lambda_{85}(E,\eta_g)} - \lambda_{84}(E,\eta_g)}{2}\right]^{1/2},$$
$$\lambda_{79}(E,\eta_g) \equiv \frac{\lambda_{71}(E,\eta_g)}{2\hbar^2 Z_0(E,\eta_g)},$$
$$\lambda_{81}(E,\eta_g) \equiv \frac{\lambda_{76}(E,\eta_g)}{4\hbar^4 [Z_0(E,\eta_g)]^2},$$
$$\lambda_{76}(E,\eta_g) \equiv [\lambda_{71}(E,\eta_g)]^2,$$
$$\lambda_{76}(E,\eta_g) \equiv [\lambda_{71}(E,\eta_g)]^2,$$
$$\lambda_{77}(E,\eta_g) \equiv [2\lambda_{71}(E,\eta_g)\lambda_{72}(E,\eta_g) - 4\lambda_{70}(E,\eta_g)\lambda_{73}(E,\eta_g) - 4\lambda_{70}(E,\eta_g)\lambda_{74}(E,\eta_g)],$$
$$\lambda_{83}(E,\eta_g) \equiv \frac{\lambda_{78}(E,\eta_g)}{9\hbar^4 [Z_0(E,\eta_g)]^2} \text{ and}$$
$$\lambda_{78}(E,\eta_g) \equiv [4\lambda_0(E,\eta_g)\lambda_{75}(E,\eta_g)]$$

Thus,

$$V(E,\eta_g) = [\lambda_{87}(E,\eta_g)] \int_0^{\lambda_{86}(E,\eta_g)} \left[\sqrt{k_z^4 + \lambda_{88}(E,\eta_g)k_z^2 + \lambda_{89}(E,\eta_g)} - \lambda_{90}(E,\eta_g)\right] dk_z \tag{2.117}$$

where, $\lambda_{87}(E,\eta_g) \equiv 2\pi\sqrt{\lambda_{81}(E,\eta_g)}$, $\lambda_{88}(E,\eta_g) \equiv \frac{\lambda_{82}(E,\eta_g)}{\lambda_{81}(E,\eta_g)}$, $\lambda_{89}(E,\eta_g) \equiv \frac{\lambda_{83}(E,\eta_g)}{\lambda_{81}(E,\eta_g)}$ and

$$\lambda_{90}(E,\eta_g) \equiv 2\pi\left[\frac{\lambda_{79}(E,\eta_g)\{\lambda_{86}(E,\eta_g)\}^3}{3} + \lambda_{80}(E,\eta_g)\lambda_{89}(E,\eta_g)\right].$$

The (2.117) can be written as

$$V(E,\eta_g) = \left[\lambda_{87}(E,\eta_g)\lambda_{95}(E,\eta_g) - \lambda_{90}(E,\eta_g)\right] \tag{2.118}$$

in which,

$$\lambda_{95}(E,\eta_g) \equiv \Bigg[\frac{\lambda_{91}(E,\eta_g)}{3}\left[-E_i\left[\lambda_{93}(E,\eta_g),\lambda_{94}(E,\eta_g)\right]\right.$$
$$\left[\{\lambda_{91}(E,\eta_g)\}^2 + \{\lambda_{92}(E,\eta_g)\}^2 + 2\{\lambda_{92}(E,\eta_g)\}^2 F_i\left[\lambda_{93}(E,\eta_g),\lambda_{94}(E,\eta_g)\right]\right]$$
$$+\left(\frac{\{\lambda_{86}(E,\eta_g)\}}{3}\right)\left[\{\lambda_{86}(E,\eta_g)\}^2 + \{\lambda_{91}(E,\eta_g)\}^2 + 2\{\lambda_{92}(E,\eta_g)\}^2\right]$$
$$\left[\left[\{\lambda_{91}(E,\eta_g)\}^2 + \{\lambda_{86}(E,\eta_g)\}^2\right]^{1/2}\left[\{\lambda_{92}(E,\eta_g)\}^2 + \{\lambda_{86}(E,\eta_g)\}^2\right]^{-1/2}\right]\Bigg],$$

$$\{\lambda_{91}(E,\eta_g)\}^2 \equiv \frac{1}{2}\left[\sqrt{\{\lambda_{91}(E,\eta_g)\}^2 - 4\lambda_{89}(E,\eta_g)} + \lambda_{88}(E,\eta_g)\right],\ E_i\left[\lambda_{93}(E,\eta_g),\lambda_{94}(E,\eta_g)\right]$$

is the incomplete elliptic integral of the 2nd kind and is given by [95, 96],

$$E_i\left[\lambda_{93}(E,\eta_g),\lambda_{94}(E,\eta_g)\right] \equiv \int_0^{\lambda_{93}(E,\eta_g)} \left[\left\{1 - \{\lambda_{94}(E,\eta_g)\}^2 \sin^2\xi\right\}^{1/2}\right] d\xi,$$

ξ is the variable of integration in this case,

$$\lambda_{93}(E,\eta_g) \equiv \tan^{-1}\left[\frac{\lambda_{86}(E,\eta_g)}{\lambda_{92}(E,\eta_g)}\right],$$

$$\{\lambda_{92}(E,\eta_g)\}^2 \equiv \frac{1}{2}\left[\lambda_{88}(E,\eta_g) - \sqrt{\{\lambda_{88}(E,\eta_g)\}^2 - 4\lambda_{89}(E,\eta_g)}\right],$$

$$\lambda_{94}(E,\eta_g) \equiv \frac{\sqrt{\{\lambda_{91}(E,\eta_g)\}^2 - \{\lambda_{92}(E,\eta_g)\}^2}}{\lambda_{91}(E,\eta_g)},\ F_i\left[\lambda_{93}(E,\eta_g),\lambda_{94}(E,\eta_g)\right]$$

is the incomplete elliptic integral of the 1st kind and is given by [95, 96],

$$F_i\left[\lambda_{93}(E,\eta_g),\lambda_{94}(E,\eta_g)\right] \equiv \int_0^{\lambda_{93}(E,\eta_g)} \left[\left\{1 - \{\lambda_{94}(E,\eta_g)\}^2 \sin^2\xi\right\}^{1/2}\right] d\xi.$$

The DOS function in this case is given by

$$N_{HD}(E,\eta_g) = \frac{g_v}{4\pi^3}\left[\{\lambda_{87}(E,\eta_g)\}'\lambda_{95}(E,\eta_g) + \{\lambda_{95}(E,\eta_g)\}'\lambda_{87}(E,\eta_g) - \{\lambda_{90}(E,\eta_g)\}'\right] \tag{2.119}$$

The 2D DR of the conduction electrons in QWs of IV–VI materials in the absence of band tails for the dimensional quantization along z direction can be expressed as

$$E(1+\alpha E)+\alpha E\left(\frac{\hbar^2k_x^2}{2x_4}+\frac{\hbar^2k_y^2}{2x_5}\right)+\alpha E\frac{\hbar^2}{2x_6}\left(\frac{n_z\pi}{d_z}\right)^2-(1+\alpha E)\left(\frac{\hbar^2k_x^2}{2x_1}+\frac{\hbar^2k_y^2}{2x_2}\right)$$
$$-\alpha\left(\frac{\hbar^2k_x^2}{2x_1}+\frac{\hbar^2k_y^2}{2x_2}\right)\left(\frac{\hbar^2k_x^2}{2x_4}+\frac{\hbar^2k_y^2}{2x_5}\right)-\alpha\left(\frac{\hbar^2k_x^2}{2x_1}+\frac{\hbar^2k_y^2}{2x_2}\right)\frac{\hbar^2}{2x_6}\left(\frac{n_z\pi}{d_z}\right)^2-(1+\alpha E)\frac{\hbar^2}{2x_3}\left(\frac{n_z\pi}{d_z}\right)^2$$
$$-\alpha\frac{\hbar^2}{2x_3}\left(\frac{n_z\pi}{d_z}\right)^2\left(\frac{\hbar^2k_x^2}{2x_4}+\frac{\hbar^2k_y^2}{2x_5}\right)-\alpha\frac{\hbar^2}{2x_3}\left(\frac{n_z\pi}{d_z}\right)^2\frac{\hbar^2}{2x_6}\left(\frac{n_z\pi}{d_z}\right)^2=\frac{\hbar^2k_x^2}{2m_1}+\frac{\hbar^2k_y^2}{2m_2}+\frac{\hbar^2}{2m_3}\left(\frac{n_z\pi}{d_z}\right)^2 \quad (2.120)$$

where

$x_4=m_t^+$, $x_5=\dfrac{m_t^+ + 2m_l^+}{3}$, $x_6=\dfrac{3m_t^+m_l^+}{2m_l^+ + m_t^+}$, $x_1=m_t^-$, $x_2=\dfrac{m_t^- + 2m_l^-}{3}$, $x_3=\dfrac{3m_t^-m_l^-}{2m_l^- + m_t^-}$,
$m_1=m_t^*$, $m_2=\dfrac{m_t^*+2m_l^*}{3}$ and $m_3=\dfrac{3m_t^*m_l^*}{2m_l^*+m_t^*}$.

Therefore, the HD 2-D DR In this case assumes the form

$$\gamma_2(E,\eta_g)+\alpha\gamma_3(E,\eta_g)\left(\frac{\hbar^2k_x^2}{2x_4}+\frac{\hbar^2k_y^2}{2x_5}\right)+\alpha\gamma_3(E,\eta_g)\frac{\hbar^2}{2x_6}\left(\frac{n_z\pi}{d_z}\right)^2$$
$$-(1+\alpha\gamma_3(E,\eta_g))\left(\frac{\hbar^2k_x^2}{2x_1}+\frac{\hbar^2k_y^2}{2x_2}\right)-\alpha\left(\frac{\hbar^2k_x^2}{2x_1}+\frac{\hbar^2k_y^2}{2x_2}\right)\left(\frac{\hbar^2k_x^2}{2x_4}+\frac{\hbar^2k_y^2}{2x_5}\right)$$
$$-\alpha\left(\frac{\hbar^2k_x^2}{2x_1}+\frac{\hbar^2k_y^2}{2x_2}\right)\frac{\hbar^2}{2x_6}\left(\frac{n_z\pi}{d_z}\right)^2-(1+\alpha\gamma_3(E,\eta_g))\frac{\hbar^2}{2x_2}\left(\frac{n_z\pi}{d_z}\right)^2$$
$$-\alpha\frac{\hbar^2}{2x_3}\left(\frac{n_z\pi}{d_z}\right)^2\left(\frac{\hbar^2k_x^2}{2x_4}+\frac{\hbar^2k_y^2}{2x_5}\right)$$
$$-\alpha\frac{\hbar^2}{2x_3}\left(\frac{n_z\pi}{d_z}\right)^2\frac{\hbar^2}{2x_6}\left(\frac{n_z\pi}{d_z}\right)^2=\frac{\hbar^2k_x^2}{2m_1}+\frac{\hbar^2k_y^2}{2m_2}+\frac{\hbar^2}{2m_3}\left(\frac{n_z\pi}{d_z}\right)^2 \quad (2.121)$$

Substituting, $k_x = r\cos\theta$ and $k_y = r\sin\theta$ (where r and θ are 2D polar coordinates in 2D wave vector space) in (2.121), we can write

$$r^4\left[\alpha\frac{1}{4}\left(\frac{\hbar^2\cos^2\theta}{x_1}+\frac{\hbar^2\sin^2\theta}{x_2}\right)\left(\frac{\hbar^2\cos^2\theta}{x_4}+\frac{\hbar^2\sin^2\theta}{x_5}\right)\right]+r^2\frac{1}{2}\left[\left(\frac{\hbar^2\cos^2\theta}{m_1}+\frac{\hbar^2\sin^2\theta}{m_2}\right)\right.$$
$$+\alpha\frac{\hbar^2}{2x_3}\left(\frac{n_z\pi}{d_z}\right)^2\left(\frac{\hbar^2\cos^2\theta}{x_4}+\frac{\hbar^2\sin^2\theta}{x_5}\right)+\alpha\left(\frac{\hbar^2\cos^2\theta}{x_1}+\frac{\hbar^2\sin^2\theta}{x_2}\right)\frac{\hbar^2}{2x_6}\left(\frac{n_z\pi}{d_z}\right)^2$$
$$\left.+\hbar^2(1+\alpha\gamma_3(E,\eta_g))\left(\frac{\cos^2\theta}{x_1}+\frac{\sin^2\theta}{x_2}\right)-\hbar^2\alpha\gamma_3(E,\eta_g)\left(\frac{\cos^2\theta}{x_4}+\frac{\sin^2\theta}{x_5}\right)\right]-\left[\gamma_2(E,\eta_g)\right.$$
$$\left.+\alpha\gamma_3(E,\eta_g)\frac{\hbar^2}{2x_6}\left(\frac{n_z\pi}{d_z}\right)^2-(1+\alpha\gamma_3(E,\eta_g))\frac{\hbar^2}{2x_3}\left(\frac{n_z\pi}{d_z}\right)^2-\alpha\left(\frac{\hbar^4}{4x_3x_6}\left(\frac{n_z\pi}{d_z}\right)^4\right)\right]=0 \quad (2.122)$$

The area $A(E,n_z)$ of the 2D wave vectorspace can be expressed as

$$A(E,n_z)=\bar{J}_1-\bar{J}_2 \quad (2.123)$$

where

$$\bar{J}_1\equiv 2\int_0^{\pi/2}\frac{c_1}{b_1}d\theta \quad (2.124)$$

and

$$\bar{J}_2\equiv 2\int_0^{\pi/2}\frac{ac_1^2}{b_1^3}d\theta \quad (2.125)$$

In which

$$a\equiv\left[\alpha\left(\frac{\hbar^4}{4}\right)\left(\frac{\cos^2\theta}{x_1}+\frac{\sin^2\theta}{x_2}\right)\left(\frac{\cos^2\theta}{x_4}+\frac{\sin^2\theta}{x_5}\right)\right],$$
$$b_1\equiv\left(\frac{\hbar^2}{2}\right)\left[\left(\frac{\cos^2\theta}{m_1}+\frac{\sin^2\theta}{m_2}\right)+\alpha\left(\frac{\hbar^2}{2x_3}\right)\left(\frac{n_z\pi}{d_z}\right)^2\left(\frac{\cos^2\theta}{x_4}+\frac{\sin^2\theta}{x_5}\right)\right.$$
$$+\alpha\left(\frac{\hbar^2}{2x_6}\right)\left(\frac{n_z\pi}{d_z}\right)^2\left(\frac{\cos^2\theta}{m_1}+\frac{\sin^2\theta}{m_2}\right)$$
$$\left.+(1+\alpha\gamma_3(E,\eta_g))\left(\frac{\cos^2\theta}{x_1}+\frac{\sin^2\theta}{x_2}\right)-\alpha\gamma_3(E,\eta_g)\left(\frac{\cos^2\theta}{x_4}+\frac{\sin^2\theta}{x_5}\right)\right]$$

and

$$c_1 \equiv \left[\gamma_2(E,\eta_g) + \alpha\gamma_3(E,\eta_g)\left(\frac{\hbar^2}{2x_6}\right)\left(\frac{n_z\pi}{d_z}\right)^2 - (1+\alpha\gamma_3(E,\eta_g))\left(\frac{\hbar^2}{2x_3}\right)\left(\frac{n_z\pi}{d_z}\right)^2 - \alpha\left(\frac{\hbar^4}{4x_3x_6}\right)\left(\frac{n_z\pi}{d_z}\right)^4\right]$$

The (2.124) can be expressed as

$$\bar{J}_1 = 2\int_0^{\pi/2} \frac{t_{31}(E,n_z)d\theta}{A_{11}(E,n_z)\cos^2\theta + B_{11}(E,n_z)\sin^2\theta} \tag{2.126}$$

where,

$$t_{31}(E,n_z) \equiv c_1,\ A_{11}(E,n_z) \equiv \frac{\hbar^2}{2m_1}t_{11}(E,n_z),$$

$$t_{11}(E,n_z) \equiv \left[1 + m_1\left[\frac{1}{x_4}\frac{\alpha\hbar^2}{2x_3}\left(\frac{n_z\pi}{d_z}\right)^2 + \frac{\alpha\hbar^2}{2x_1x_6}\left(\frac{n_z\pi}{d_z}\right)^2 + \frac{1+\alpha\gamma_2(E,\eta_g)}{x_1} - \frac{\alpha\gamma_3(E,\eta_g)}{x_4}\right]\right]$$

$$B_{11}(E,n_z) \equiv \frac{\hbar^2}{2m_2}t_{21}(E,n_z) \text{ and}$$

$$t_{21}(E,n_z) \equiv \left[1 + m_2\left[\frac{\alpha\hbar^2}{2x_3x_5}\left(\frac{n_z\pi}{d_z}\right)^2 + \frac{\alpha\hbar^2}{2x_2x_6}\left(\frac{n_z\pi}{d_z}\right)^2 + \frac{1+\alpha\gamma_3(E,\eta_g)}{x_2} - \frac{\alpha\gamma_3(E,\eta_g)}{x_5}\right]\right].$$

Performing the integration, we get

$$\bar{J}_1 = \pi t_{31}(E,n_z)[A_{11}(E,n_z)B_{11}(E,n_z)]^{-1/2} \tag{2.127}$$

From (2.125) we can write

$$\bar{J}_2 = \frac{\alpha t_{31}^2(E,n_z)\hbar^4}{2B_{11}^3(E,n_z)}I \tag{2.128}$$

where,

$$I \equiv \int_0^\infty \frac{(a_1+a_2z^2)(a_3+a_4z^2)dz}{\left[(\bar{a})^2+z^2\right]^3},\quad (\bar{a})^2 = \left(\frac{A_{11}(E,n_z)}{B_{11}(E,n_z)}\right), \tag{2.129}$$

in which $a_1 \equiv \frac{1}{x_1}$, $a_2 \equiv \frac{1}{x_2}$, $z = \tan\theta$, θ is a new variable, $a_3 \equiv \frac{1}{x_4}$, $a_4 \equiv \frac{1}{x_5}$ and $(\bar{a})^2 \equiv \left(\frac{A_1(E,n_z)}{B_1(E,n_z)}\right)$.

The use of the Residue theorem leads to the evaluation of the integral in (2.129) as

$$I = \frac{\pi}{4\bar{a}}[a_1 a_4 + 3a_2 a_4], \tag{2.130}$$

Therefore, the 2D area of the 2D wave vector space can be written as

$$A_{HD}(E, n_z) = \frac{\pi t_{31}(E, n_z)}{\sqrt{A_{11}(E, n_z) B_{11}(E, n_z)}} \left[1 - \frac{1}{x_5}\left(\frac{1}{x_1} + \frac{3}{x_3}\right) \frac{\alpha t_{31}(E, n_z)\hbar^4}{8B_{11}^2(E, n_z)}\right] \tag{2.131}$$

The EEM for the HD QWs of IV–VI materials can thus be written as

$$m^*(E, n_z) = \frac{\hbar^2}{2}[\theta_{5HD}(E, n_z)]\Big|_{E=E_{F1HD}} \tag{2.132}$$

where,

$$\begin{aligned}\theta_{5HD}(E, n_z) \equiv & \left[1 - \frac{1}{x_5}\left(\frac{1}{x_1} + \frac{3}{x_2}\right) \frac{\alpha t_{31}(E, n_z)\hbar^4}{8[B_{11}(E, n_z)]^2}\right] [A_{11}(E, n_z) B_{11}(E, n_z)]^{-1} \\ & \left[\sqrt{A_{11}(E, n_z) B_{11}(E, n_z)}\{t_{31}(E, n_z)\}' - t_{31}(E, n_z)\right. \\ & \left.\left\{\frac{1}{2}\{A_{11}(E, n_z)\}'\left[\frac{B_{11}(E, n_z)}{A_{11}(E, n_z)}\right]^{1/2} + \frac{1}{2}\{B_{11}(E, n_z)\}'\left[\frac{A_{11}(E, n_z)}{B_{11}(E, n_z)}\right]^{1/2}\right\}\right] \\ & - \frac{1}{8}\frac{t_{31}(E, n_z)\alpha\hbar^4}{\sqrt{A_{11}(E, n_z) B_{11}(E, n_z)}}\frac{1}{x_5}\left(\frac{1}{x_1} + \frac{1}{x_2}\right)[B_{11}(E, n_z)]^{-4} \\ & \left[\{B_{11}(E, n_z)\}^2\{t_{31}(E, n_z)\}' - 2B_{11}(E, n_z)\{B_{11}(E, n_z)\}' t_{31}(E, n_z)\right].\end{aligned}$$

Thus, the EEM is a function of Fermi energyand the quantum number due to the band non-parabolicity.

The total DOS density-of-states function can be written as

$$N_{2DT}(E) = \left(\frac{g_v}{2\pi}\right) \sum_{n_z=1}^{n_{z\max}} \theta_{5HD}(E, n_z) H\left(E - E_{n_{s7HD}}\right) \tag{2.133}$$

where the sub-band energy $\left(E_{n_{z7}}\right)$ in this case can be written as

$$\gamma_2(E_{n_{z7HD}}, \eta_g) + \alpha\gamma_3(E_{n_{z7HD}}, \eta_g)\frac{\hbar^2}{2x_6}\left(\frac{n_z\pi}{d_z}\right)^2 - (1 + \alpha\gamma_3(E_{n_{z7HD}}, \eta_g))\frac{\hbar^2}{2x_3}\left(\frac{n_z\pi}{d_z}\right)^2$$
$$- \alpha\frac{\hbar^2}{2x_3}\left(\frac{n_z\pi}{d_z}\right)^2\frac{\hbar^2}{2x_6}\left(\frac{n_z\pi}{d_z}\right)^2 - \left[\frac{\hbar^2}{2m_3}\left(\frac{n_z\pi}{d_z}\right)^2\right] = 0 \tag{2.134}$$

In the absence of band-tails the EEM in QWs of IV–VI materials can be written as

$$m^*(E, n_z) = \frac{\hbar^2}{2}[\theta_5(E, n_z)]\Big|_{E=E_{Fs}} \tag{2.135}$$

where,

$$\theta_5(E, n_z) \equiv \left[1 - \frac{1}{x_5}\left(\frac{1}{x_1} + \frac{3}{x_2}\right)\frac{\alpha t_{30}(E, n_z)\hbar^4}{8[B_{10}(E, n_z)]^2}\right][A_{10}(E, n_z)B_{10}(E, n_z)]^{-1}$$
$$\left[\sqrt{A_{10}(E, n_z)B_{10}(E, n_z)}\{t_{30}(E, n_z)\}' - t_{30}(E, n_z)\right.$$
$$\left.\left\{\frac{1}{2}\{A_{10}(E, n_z)\}'\left[\frac{B_{10}(E, n_z)}{A_{10}(E, n_z)}\right]^{1/2} + \frac{1}{2}\{B_{10}(E, n_z)\}'\left[\frac{A_{10}(E, n_z)}{B_{10}(E, n_z)}\right]^{1/2}\right\}\right]$$
$$- \frac{1}{8}\frac{t_{30}(E, n_z)\alpha\hbar^4}{\sqrt{A_{10}(E, n_z)B_{10}(E, n_z)}}\frac{1}{x_5}\left(\frac{1}{x_1} + \frac{3}{x_2}\right)[B_{10}(E, n_z)]^{-4}$$
$$\left[\{B_{10}(E, n_z)\}^2\{t_{30}(E, n_z)\}' - 2B_{10}(E, n_z)\{B_{10}(E, n_z)\}'t_{30}(E, n_z)\right],$$

$$t_{30}(E, n_z) \equiv c_0,$$

$$c_0 \equiv \left[E(1 + \alpha E) + \alpha E\left(\frac{\hbar^2}{2x_6}\right)\left(\frac{n_z\pi}{d_z}\right)^2 - (1 + \alpha E)\left(\frac{\hbar^2}{2x_3}\right)\left(\frac{n_z\pi}{d_z}\right)^2 - \alpha\left(\frac{\hbar^4}{4x_3x_6}\right)\left(\frac{n_z\pi}{d_z}\right)^4\right],$$

$$A_{10}(E, n_z) \equiv \frac{\hbar^2}{2m_1}t_{10}(E, n_z),$$

$$t_{10}(E, n_z) \equiv \left[1 + m_1\left[\frac{1}{x_4}\frac{\alpha\hbar^2}{2x_3}\left(\frac{n_z\pi}{d_z}\right)^2 + \frac{\alpha\hbar^2}{2x_1x_6}\left(\frac{n_z\pi}{d_z}\right)^2 + \frac{1 + \alpha E}{x_1} - \frac{\alpha E}{x_4}\right]\right],$$

$$B_{10}(E, n_z) \equiv \frac{\hbar^2}{2m_1}t_{20}(E, n_z) \text{ and}$$

$$t_{20}(E, n_z) \equiv \left[1 + m_2\left[\frac{\alpha\hbar^2}{2x_3x_5}\left(\frac{n_z\pi}{d_z}\right)^2 + \frac{\alpha\hbar^2}{2x_2x_6}\left(\frac{n_z\pi}{d_z}\right)^2 + \frac{1 + \alpha E}{x_2} - \frac{\alpha E}{x_5}\right]\right],$$

Thus, the EEM is a function of Fermi energyand the quantum number due to the band non-parabolicity.

The total DOS function can be written as

$$N_{2DT}(E) = \left(\frac{g_v}{2\pi}\right) \sum_{n_z=1}^{n_{z\max}} \theta_5(E, n_z) H\left(E - E_{n_{z7}}\right) \tag{2.136}$$

where the sub-band energy $\left(E_{n_{z7}}\right)$ in this case can be written as

$$\begin{aligned} &E_{n_{z7}}\left(1+\alpha E_{n_{z7}}\right)+\alpha E_{n_{z7}}\frac{\hbar^2}{2x_6}\left(\frac{n_z\pi}{d_z}\right)^2-\left(1+\alpha E_{n_{z7}}\right)\frac{\hbar^2}{2x_3}\left(\frac{n_z\pi}{d_z}\right)^2 \\ &-\alpha\frac{\hbar^2}{2x_3}\left(\frac{n_z\pi}{d_z}\right)^2\frac{\hbar^2}{2x_6}\left(\frac{n_z\pi}{d_z}\right)^2-\left[\frac{\hbar^2}{2x_3}\left(\frac{n_z\pi}{d_z}\right)^2\right]=0 \end{aligned} \tag{2.137}$$

For bulk specimens of IV–VI materials, the expression of DOS function at the Fermi level assume the forms

$$N(E) = \frac{g_v}{2\pi^2}\frac{\partial}{\partial E}[\mathrm{M}_{A4}(E)]\bigg|_{E=E_{F_b}} \tag{2.138}$$

where

$$\begin{aligned} M_{A_4}(E_{F_b}) &= \left[\alpha_5 J_{A_1}(E_{F_b}) - \alpha_3(E_{F_b})\bar{\tau}_{A_1}(E_{F_b}) - \frac{\alpha_4}{3}\left[\bar{\tau}_{A_1}(E_{F_b})\right]^3\right], \\ \alpha_5 &= \left[\frac{2m_t^+ m_t^-}{\alpha\hbar^2}\omega_{A_1}\right], \\ \omega_{A_1} &= \left[\frac{\alpha^2}{16}\left[\frac{1}{m_t^- m_l^+} + \frac{1}{m_l^- m_t^+}\right]^2 - \frac{\alpha^2}{4m_l^+ m_t^- m_l^- m_t^+}\right], \\ J_{A_1}(E_{F_b}) &= \frac{A_A(E_{F_b})}{3}\left[-\left(A_A^2(E_{F_b}) + B_A^2(E_{F_b})\right)E(\lambda, q) + 2B_A^2(E_{F_b})F(\lambda, q)\right] \\ &\quad + \frac{\bar{\tau}_{A_1}(E_{F_b})}{3}\left[(\bar{\tau}_{A_1}(E_{F_b}))^2 + A_A^2(E_{F_b}) + 2B_A^2(E_{F_b})\right] \\ &\quad \left[A_A^2(E_{F_b}) + \bar{\tau}_{A_1}^2(E_{F_b})\right]^{1/2}\left[B_A^2(F_{F_b}) + \bar{\tau}_{A_1}^2(E_{F_b})\right]^{-1/2} \end{aligned}$$

$$\lambda = \tan^{-1}\frac{\bar{\tau}_{A_1}(E_{F_b})}{B_A(E_{F_b})},$$

$$q = \left[\frac{\sqrt{A_A^2(E_{F_b}) - B_A^2(E_{F_b})}}{A_A(E_{F_b})}\right],$$

$$A_A(E_{F_b}) = \left[\tau_{A_2}(E_{F_b}) + \sqrt{\tau_{A_2}^2(E_{F_b}) - 4\tau_{A_3}(E_{F_b})}\right]^{1/2} \Big/ \sqrt{2},$$

$$B_A(E_{F_b}) = \left[\tau_{A_2}(E_{F_b}) + \sqrt{\tau_{A_2}^2(E_{F_b}) - 4\tau_{A_3}(E_{F_b})}\right]^{1/2} \Big/ \sqrt{2},$$

$$\tau_{A_2}(E_{F_b}) = \frac{\omega_{A_2}(E_{F_b})}{\omega_{A_1}^2}, \tau_{A_3}(E_{F_b}) = \frac{\omega_{A_3}(E_{F_b})}{\omega_{A_1}^2},$$

$$\omega_{A_2}(E_{F_b}) = \left[\frac{\alpha}{2}\left[\frac{1}{2m_t^*} - \frac{\alpha \cdot E_{F_b}}{2m_t^+} + \frac{1+\alpha \cdot E_{F_b}}{2m_t^-}\right] \cdot \left[\frac{1}{m_t^- m_l^+} + \frac{1}{m_l^- m_t^+}\right]\right.$$
$$\left. - \frac{\alpha}{m_t^+ m_t^-}\left[\frac{1}{2m_l^*} + \frac{\alpha \cdot E_{F_b}}{2m_l^+} + \frac{1+\alpha \cdot E_{F_b}}{2m_l^-}\right]\right]$$

$$\omega_{A_3}(E_{F_b}) = \frac{\alpha \cdot E_{F_b}(1+\alpha \cdot E_{F_b})}{2m_t^+ m_t^-} + \left[\frac{1}{2m_t^*} - \frac{\alpha \cdot E_{F_b}}{2m_t^+} + \frac{1+\alpha \cdot E_{F_b}}{2m_t^-}\right],$$

$$\alpha_2(E_{F_b})\left[\frac{1}{2m_t^*} - \frac{\alpha \cdot E_{F_b}}{2m_t^+} + \frac{1+\alpha \cdot E_{F_b}}{2m_t^-}\right], \alpha_3 = \frac{\alpha\hbar^2}{4}\left[\frac{1}{m_t^- m_l^+} + \frac{1}{m_l^- m_t^+}\right],$$

$$\tau_{A_1}(E_{F_b}) = \left[\frac{2m_l^+ m_l^-}{\alpha\hbar^2}\right]^{1/2}\left[-\left[\frac{1}{2m_l^*} + \frac{1+\alpha \cdot E_{F_b}}{m_l^-} - \frac{\alpha \cdot E_{F_b}}{2m_l^+}\right]\right.$$
$$\left. + \left[\left[\frac{1}{2m_l^*} + \frac{1+\alpha \cdot E_{F_b}}{m_l^-} - \frac{\alpha \cdot E_{F_b}}{2m_l^+}\right]^2 + \frac{\alpha \cdot E_{F_b}(1+\alpha \cdot E_{F_b})}{m_l^- m_l^+}\right]^{1/2}\right]^{1/2},$$

$E(\lambda, q)$ is the in complete Elliptic integral of second kind, $F(\lambda, q)$ is the incomplete Elliptic integral of first kind,

$$I_{17}(E') = \frac{-I_{15}(E') + \sqrt{I_{15}^2(E') + 4I_{16}(E')}}{2},$$

$$I_{15}(E') = \frac{2m_l^+ m_t^-}{\alpha\hbar^2}\left[\frac{1}{m_l^*} + \frac{1}{m_l^-} + \alpha E'\left(\frac{1}{m_l^-} - \frac{1}{m_l^+}\right)\right] \text{ and}$$

$$I_{16}(E') = \frac{4m_l^+ m_t^-}{\alpha\hbar^4} E'(1+\alpha E')$$

(b) The DR of the conduction electrons in bulk specimens of IV–VI materials in accordance with the model of Bangert and Kastner is given by

$$\omega_1(\mathrm{E})k_s^2 + \omega_2(\mathrm{E})k_z^2 = 1 \tag{2.139}$$

where

$$\omega_1(\mathrm{E}) = 2(E)^{-1}\left[\frac{(\bar{\mathrm{R}})^2}{\mathrm{E}_{g_0}(1+\alpha_1\mathrm{E})} + \frac{(\bar{\mathrm{S}})^2}{\Delta'_c(1+\alpha_2\mathrm{E})} + \frac{(\bar{\mathrm{Q}})^2}{\Delta''_c(1+\alpha_3\mathrm{E})}\right] \text{ and}$$

$$\omega_2(\mathrm{E}) = 2(E)^{-1}\left[\frac{(\bar{\mathrm{A}})^2}{\mathrm{E}_{g_0}(1+\alpha_1\mathrm{E})} + \frac{(\bar{\mathrm{S}}+\bar{\mathrm{Q}})^2}{\Delta''_c(1+\alpha_3\mathrm{E})}\right],\ \alpha_1 = \frac{1}{\mathrm{E}_{g0}},\ \alpha_2 = \frac{1}{\Delta'_c},\ \alpha_3 = \frac{1}{\Delta''_c},$$

and the numerical values of the other system constants are given in Table 1.1.

The electron energy spectrum in heavily doped IV–VI materials in accordance with this model can be expressed by using the methods as given in this chapter as

$$\begin{aligned} 2I(4) = k_s^2 & \left[\{C_1(\alpha_1, E, E_g) - iD_1(\alpha_1, E, E_g)\}\frac{(\bar{R})^2}{E_{g_0}}\right. \\ & + \{C_2(\alpha_2, E, E_g) - iD_2(\alpha_2, E, E_g)\}\frac{(\bar{S})^2}{\Delta'_c} \\ & \left. + \{C_3(\alpha_3, E, E_g) - iD_3(\alpha_3, E, E_g)\}\frac{(\bar{Q})^2}{\Delta''_c}\right] \\ & + k_z^2\left[\frac{2(\bar{A})^2}{E_{g_0}}\{C_1(\alpha_1, E, E_g) - iD_1(\alpha_1, E, E_g)\}\right. \\ & \left. + \frac{(\bar{S}+\bar{Q})^2}{\Delta''_c}\{C_3(\alpha_3, E, E_g) - iD_3(\alpha_3, E, E_g)\}\right] \end{aligned} \tag{2.140}$$

where

$$G_i = \frac{1+\alpha_i E}{\eta_g \alpha_i} \tag{2.141}$$

$$c_i(\alpha_i, E, \eta_g) = \left[\frac{2}{\alpha_i \eta_g \sqrt{\pi}}\right]\exp(-u_i^2) \times \left[\sum_{p=1}^{\infty}\{\exp(-p^2/4)(\sinh(pu_i))\}p^{-1}\right], \tag{2.142}$$

$i = 1, 2, 3, 4$ and

$$D_i(\alpha_i, E, \eta_g) = \left[\frac{\sqrt{\pi}}{\alpha_i \eta_g}\right]\exp(-u_i^2) \tag{2.143}$$

Therefore (2.140) can be written as,

$$F_1(E,\eta_g)k_s^2 + F_2(E,\eta_g)k_z^2 = 1 \tag{2.144}$$

where,

$$F_1(E,\eta_g) = [2\gamma_0(E,\eta_g)]^{-1}\left[\frac{(\bar{R})^2}{E_g}\{C_1(\alpha_1,E,E_g) - iD_1(\alpha_1,E,E_g)\}\right.$$
$$+\frac{(\bar{S})^2}{\Delta'_c}\{C_2(\alpha_2,E,E_g) - iD_2(\alpha_2,E,E_g)\}$$
$$\left.+\frac{(\bar{Q})^2}{\Delta''_c}\{C_3(\alpha_3,E,E_g) - iD_3(\alpha_3,E,E_g)\}\right] \text{ and}$$

$$F_2(E,\eta_g) = [2\gamma_0(E,\eta_g)]^{-1}\left[\frac{2(\bar{A})^2}{E_g}\{C_1(\alpha_1,E,E_g) - iD_1(\alpha_1,E,\eta_g)\}\right.$$
$$\left.+\frac{(\bar{S}+\bar{Q})^2}{\Delta''_c}\{C_3(\alpha_3,E,\eta_g) - iD_3(\alpha_3,E,\eta_g)\}\right]$$

Since $F_1(E,\eta_g)$ and $F_2(E,\eta_g)$ are complex, the energy spectrum is also complex in the presence of Gaussian band tails.

The DOS function in this case can be expressed as

$$N_{HD}(\mathrm{E},\eta_g) = \frac{g_v}{3\pi^2}F'_3(\mathrm{E},\eta_g),\quad F_3(\mathrm{E},\eta_g) = \left[F_1(\mathrm{E},\eta_g)\sqrt{F_2(\mathrm{E},\eta_g)}\right]^{-1} \tag{2.145}$$

The 2D DR in this case assumes the form

$$k_s^2 = F_6(E,\eta_g,n_z) \tag{2.146}$$

where,

$$F_6(E,\eta_g,n_z) = \left[\frac{\left[1 - F_2(E,\eta_g)(n_z\pi/d_z)^2\right]}{F_1(E,\eta_g)}\right]$$

The EEM in this case is given by

$$m^*(E_{F1HD},\eta_g,n_z) = \frac{\hbar^2}{2}\text{Real part of }[F'_6(E_{F1HD},\eta_g,n_z)] \tag{2.147}$$

The total DOS function can be written as

$$N_{2DT}(\mathrm{E}) = \frac{g_v}{2\pi}\sum_{n_z=1}^{n_{z\max}} F_6'(E,\eta_g,n_z)H(E - E_{n_{z71HD}}) \tag{2.148}$$

where $E_{n_{z71HD}}$ *is* the quantized energy in this case and is given by

$$1 = F_2(E_{n_{z71HD}},\eta_g)(\pi n_z/d_z)^2 \tag{2.149}$$

In the absence of band-tails the EEMs can be written as

$$m_{\perp}^{*}(E_F) = \left(\frac{\hbar^2}{2}\right)\left(\frac{F_{11}'(E_F)}{F_{12}^2(E_F)}\right) \tag{2.150}$$

$$m_{11}^{*}(E_F) = \left(\frac{\hbar^2}{2}\right)\left(\frac{F_{12}'(E_F)}{F_{12}^2(E_F)}\right) \tag{2.151}$$

where

$$F_{11}(E) = \left[\frac{(\bar{\mathrm{R}})^2}{E_{g0}(1+\alpha_1 E)} + \frac{(\bar{S})^2}{\Delta_c'(1+\alpha_2 E)} + \frac{(\bar{Q})^2}{\Delta_c''(1+\alpha_3 E)}\right][2E]^{-1} \text{ and}$$

$$F_{12}(E) = \left[\frac{(\bar{A})^2}{E_{g0}(1+\alpha_1 E)} + \frac{(\bar{S}+\bar{Q})^2}{\Delta_c''(1+\alpha_2 E)}\right][2E]^{-1}$$

The DOS function in this case can be expressed as

$$N(\mathrm{E}) = \frac{g_v}{3\pi^2}F_{13}'(\mathrm{E}), \quad F_{13}(\mathrm{E}) = [F_{11}(\mathrm{E})\sqrt{F_{12}(\mathrm{E})}]^{-1} \tag{2.152}$$

In the absence of band-tails, the 2D DR in this case assumes the form

$$k_s^2 = F_{16}(E,n_z) \tag{2.153}$$

where,

$$F_{16}(E,n_z) = \left[\frac{\left[1 - F_{12}(E)(n_z\pi/d_z)^2\right]}{F_{11}(E)}\right] \tag{2.154}$$

The EEM in this case is given by

$$m^*(E_{F_s}, n_z) = \frac{\hbar^2}{2}[F'_{16}(E_{F_s}, n_z)] \tag{2.155}$$

The total DOS function can be written as

$$N_{2DT}(E) = \frac{g_v}{2\pi}\sum_{n_z=1}^{n_{z\max}} F'_{16}(E, n_z)H(E - E_{n_{z711}}) \tag{2.156}$$

where $E_{n_{z711}}$ is the quantized energy in this case and is given by

$$1 = F_{12}(E_{n_{z711}}, \eta_g)(\pi n_z/d_z)^2 \tag{2.157}$$

2.2.5 The DR in Ultrathin Films (UFs) of HD Stressed Kane Type Materials

The electron energy spectrum in stressed Kane type materials can be written [101] as

$$\left(\frac{k_x}{\bar{a}_0(E)}\right)^2 + \left(\frac{k_y}{\bar{b}_0(E)}\right)^2 + \left(\frac{k_z}{\bar{c}_0(E)}\right)^2 = 1 \tag{2.158}$$

where, $[\bar{a}_0(E)]^2 \equiv \frac{\bar{K}_0(E)}{\bar{A}_0(E) + \frac{1}{2}\bar{D}_0(E)}$, $\bar{K}_0(E) \equiv \left[E - C_1\varepsilon - \frac{2C_2^2\varepsilon_{xy}^2}{3E'_g}\right]\left(\frac{3E'_g}{2B_2^2}\right)$, C_1 is the conduction band deformation potential, ε is the trace of the strain tensor $\hat{\varepsilon}$ which can be written as $\hat{\varepsilon} = \begin{bmatrix} \varepsilon_{xx} & \varepsilon_{xy} & 0 \\ \varepsilon_{xy} & \varepsilon_{yy} & 0 \\ 0 & 0 & \varepsilon_{zz} \end{bmatrix}$,

C_2 is a constant which describes the strain interaction between the conduction and valance bands, $E'_g \equiv E_g + E - C_1\varepsilon$, B_2 is the momentum matrix element,

$$\bar{A}_0(E) \equiv \left[1 - \frac{(\bar{a}_0 + C_1)}{E'_g} + \frac{3\bar{b}_0\varepsilon_{xx}}{2E'_g} - \frac{\bar{b}_0\varepsilon}{2E'_g}\right],$$

$$\bar{a}_0 \equiv -\frac{1}{3}(\bar{b}_0 + 2\bar{m}),$$

$$\bar{b}_0 \equiv \frac{1}{3}(\bar{l} - \bar{m}),$$

$$\bar{d}_0 \equiv \frac{2\bar{n}}{\sqrt{3}},$$

$\bar{l}, \bar{m}, \bar{n}$ are the matrix elements of the strain perturbation operator, $\bar{D}_0(E) \equiv \left(\bar{d}_0\sqrt{3}\right)\frac{\varepsilon_{xy}}{E'_g}$,

$$[\bar{b}_0(E)]^2 \equiv \frac{\bar{K}_0(E)}{\bar{A}_0(E) - \frac{1}{2}\bar{D}_0(E)},$$
$$[\bar{c}_0(E)]^2 \equiv \frac{\bar{K}_0(E)}{\bar{L}_0(E)} \text{ and}$$
$$\bar{L}_0(E) \equiv \left[1 - \frac{(\bar{a}_0 + C_1)}{E'_g} + \frac{3\bar{b}_0\varepsilon_{zz}}{E'_g} - \frac{\bar{b}_0\varepsilon}{2E'_g}\right]$$

The use of (2.158) can be written as

$$(E - \alpha_1)k_x^2 + (E - \alpha_2)k_y^2 + (E - \alpha_3)k_z^2 = t_1E^3 - t_2E^2 + t_3E + t_4 \tag{2.159}$$

where

$$\alpha_1 \equiv \left[E_g - C_1\varepsilon - (\bar{a}_0 + C_1)\varepsilon + \frac{3}{2}\bar{b}_0\varepsilon_{xx} - \frac{\bar{b}_0}{2}\varepsilon + \left(\sqrt{3}/2\right)\varepsilon_{xy}\bar{d}_0\right],$$
$$\alpha_2 \equiv \left[E_g - C_1\varepsilon - (\bar{a}_0 + C_1)\varepsilon + \frac{3}{2}\bar{b}_0\varepsilon_{xx} - \frac{\bar{b}_0}{2}\varepsilon - \left(\sqrt{3}/2\right)\varepsilon_{xy}\bar{d}_0\right],$$
$$\alpha_3 \equiv \left[E_g - C_1\varepsilon - (\bar{a}_0 + C_1)\varepsilon + \frac{3}{2}\bar{b}_0\varepsilon_{zz} - \frac{\bar{b}_0}{2}\varepsilon\right],$$
$$t_1 \equiv \left(3/2B_2^2\right),$$
$$t_2 \equiv \left(1/2B_2^2\right)[6(E_g - C_1\varepsilon) + 3C_1\varepsilon],$$
$$t_3 \equiv \left(1/2B_2^2\right)\left[3(E_g - C_1\varepsilon)^2 + 6C_1\varepsilon(E_g - C_1\varepsilon) - 2C_2^2\varepsilon_{xy}^2\right] \text{ and}$$
$$t_4 \equiv \left(1/2B_2^2\right)\left[-3C_1\varepsilon(E_g - C_1\varepsilon)^2 + 2C_2^2\varepsilon_{xy}^2\right].$$

The (2.159) can be written as

$$Ek^2 - T_{17}k_x^2 - T_{27}k_y^2 - T_{37}k_z^2 = \left[q_{67}E^3 - R_{67}E^2 + V_{67}E + \rho_{67}\right] \tag{2.160}$$

where

$$T_{17} = \alpha_1, T_{27} = \alpha_2, T_{37} = \alpha_3, t_1 = q_{67}, t_2 = R_{67}, t_3 = V_{67} \text{ and } t_4 = \rho_{67}$$

Under the condition of heavy doping, (2.160) can be written as

$$\begin{aligned} &I(4)k^2 - T_{17}I(1)k_x^2 - T_{27}I(1)k_y^2 - T_{37}k_z^2I(1) \\ &= [q_{67}I(6) - R_{67}I(5) + V_{67}I(4) + \rho_{67}I(1)] \end{aligned} \tag{2.161}$$

where,

$$I(6) = \int_{-\infty}^{E} (E - V)^3 F(V) dV \tag{2.162}$$

The (2.162) can be written as

$$I(6) = E^3 I(1) - 3E^2 I(7) + 3EI(8) - I(9) \tag{2.163}$$

In which,

$$I(7) = \int_{-\infty}^{E} VF(V) dV \tag{2.164}$$

$$I(8) = \int_{-\infty}^{E} V^2 F(V) dV \tag{2.165}$$

$$I(9) = \int_{-\infty}^{E} V^3 F(V) dV \tag{2.166}$$

Using simple algebraic manipulations we can write

$$I(7) = \frac{-\eta_g}{2\sqrt{\pi}} \exp\left(\frac{-E^2}{\eta_g^2}\right) \tag{2.167}$$

$$I(8) = \frac{\eta_g^2}{4}\left[1 + Erf\left(\frac{E}{\eta_g}\right)\right] \tag{2.168}$$

and

$$I(9) = \frac{-\eta_g}{2\sqrt{\pi}} \exp\left(\frac{-E^2}{\eta_g^2}\right)\left[1 + \left(\frac{E^2}{\eta_g^2}\right)\right] \tag{2.169}$$

Thus (1.166) can be written as

$$I(6) = \left[\frac{E}{2}\left[1 + Erf\left(\frac{E}{\eta_g}\right)\right]\left[E^2 + \frac{3}{2}\eta_g^2\right] + \frac{\eta_g}{2\sqrt{\pi}} \exp\left(\frac{-E^2}{\eta_g^2}\right)\left[4E^2 + \eta_g^2\right]\right] \tag{2.170}$$

Thus, combining the appropriate equations, the DR of the conduction electrons in HD stressed materials can be expressed as

$$P_{11}(E,\eta_g)k_x^2 + Q_{11}(E,\eta_g)k_y^2 + S_{11}(E,\eta_g)k_z^2 = 1 \quad (2.171)$$

where,

$$P_{11}(E,\eta_g) \equiv \left[\frac{\gamma_0(E,\eta_g) - (T_{17}/2)[1+Erf(E/\eta_g)]}{\Delta_{14}(E,\eta_g)}\right],$$

$$\Delta_{14}(E,\eta_g) \equiv \left[q_{67}\left\{\frac{E}{2}\left[1+Erf\left(\frac{E}{\eta_g}\right)\right]\left[E^2+\frac{3}{2}\eta_g^2\right] + \frac{\eta_g}{2\sqrt{\pi}}\exp\left(\frac{-E^2}{\eta_g^2}\right)\left[4E^2+\eta_g^2\right]\right\}\right.$$
$$\left. - R_{67}\theta_0(E,\eta_g) + V_{67}\gamma_0(E,\eta_g) + \frac{\rho_{67}}{2}[1+Erf(E/\eta_g)]\right],$$

$$Q_{11}(E,\eta_g) \equiv \left[\frac{\gamma_0(E,\eta_g) - (T_{27}/2)[1+Erf(E/\eta_g)]}{\Delta_{14}(E,\eta_g)}\right] \text{ and}$$

$$S_{11}(E,\eta_g) \equiv \left[\frac{\gamma_0(E,\eta_g) - (T_{37}/2)[1+Erf(E/\eta_g)]}{\Delta_{14}(E,\eta_g)}\right].$$

Thus, the energy spectrum in this case is real since the DR of the corresponding materials in the absence of band tails as given by (2.171) has no poles in the finite complex plane.

The DOS function in this case can be written as

$$N_{HD}(E,\eta_g) = \frac{g_v}{3\pi^2}\Delta_{100}(E,\eta_g)$$

where

$$\Delta_{100}(E,\eta_g) = \{\Delta_{15}(E,\eta_g)\}^{-2}\left[\frac{3}{2}\{\Delta_{15}(E,\eta_g)\}\sqrt{\Delta_{14}(E,\eta_g)}\{\Delta_{14}(E,\eta_g)\}'\right.$$
$$\left. - \{\Delta_{14}(E,\eta_g)\}^{3/2}\{\Delta_{15}(E,\eta_g)\}'\right] \text{ and}$$
$$\Delta_{15}(E,\eta_g) \equiv [[\gamma_0(E,\eta_g) - (T_{17}/2)[1+Erf(E/\eta_g)]]$$
$$[\gamma_0(E,\eta_g) - (T_{27}/2)[1+Erf(E/\eta_g)]]$$
$$[\gamma_0(E,\eta_g) - (T_{37}/2)[1+Erf(E/\eta_g)]]]^{1/2}. \quad (2.172)$$

The DR of the conduction electrons in HD QWs of Kane type materials can be written as

$$P_{11}(E,\eta_g)k_x^2 + Q_{11}(E,\eta_g)k_y^2 + S_{11}(E,\eta_g)(\pi n_z/d_z)^2 = 1 \quad (2.173)$$

The EEM can be expressed as

$$m^*\left(E_{F1HD},\eta_g,n_z\right)=\frac{\hbar^2}{2}A'_{56}\left(E_{F1HD},\eta_g,n_z\right) \tag{2.174}$$

where,

$$A_{56}\left(E,\eta_g,n_z\right)=\frac{\pi\left[1-S_{11}\left(E,\eta_g\right)\left(n_z\pi/d_z\right)^2\right]}{\sqrt{P_{11}\left(E,\eta_g\right)Q_{11}\left(E,\eta_g\right)}} \tag{2.175}$$

From (2.174), it appears that the EEM is a function of Fermi energy, and size quantum number and the same mass exists in the band gap.

Thus, the total 2D DOS function can be expressed as

$$N_{2DT}(E)=\left(\frac{g_v}{2\pi}\right)\sum_{n_z=1}^{n_{z\max}}A'_{56}\left(E_{F1HD},\eta_g,n_z\right) \tag{2.176}$$

The sub band energies $\left(E_{n_{zg}HD}\right)$ are given by

$$S_{11}\left(E_{n_{zg}HD}\right)(\pi n_z/d_z)^2=1 \tag{2.177}$$

In the absence of band tails, the 2D electron energy spectrum in QWs of stressed materials assumes the form

$$\frac{k_x^2}{[\bar{a}_0(E)]^2}+\frac{k_y^2}{[\bar{b}_0(E)]^2}+\frac{1}{[\bar{c}_0(E)]^2}(n_z\pi/d_z)^2=1 \tag{2.178}$$

The area of 2D wave vector space enclosed by (2.178) can be written as

$$A(E,n_z)=\pi P^2(E,n_z)\bar{a}_0(E)\bar{b}_0(E) \tag{2.179}$$

where

$$P^2(E,n_z)=\left[1-[n_z\pi/d_z\bar{c}_0(E)]^2\right].$$

From (2.178), the EEM can be written as

$$m^{*}(E_{F_s}, n_z) = \frac{\hbar^2}{2}\left[P^2(E_{F_s}, n_z)\bar{a}_0(E_{F_s})\bar{b}_0(E_{F_s})\right]' \tag{2.180}$$

Thus, the total 2D DOS function can be expressed as

$$N_{2DT}(E) = \left(\frac{g_v}{2\pi}\right)\sum_{n_z=1}^{n_{z\max}} \theta_6(E, n_z)H(E - E_{n_{z11}}) \tag{2.181}$$

in which,

$$\theta_6(E, n_z) = \left[2P(E, n_z)\{P(E, n_z)\}'\bar{a}_0(E)\bar{b}_0(E) + \{P(E, n_z)\}^2\{\bar{a}_0(E)\}'\bar{b}_0(E) + \{P(E, n_z)\}^2\{\bar{b}_0(E)\}'\bar{a}_0(E)\right] \tag{2.182}$$

The sub band energies $\left(E_{n_{z11}}\right)$ are given by

$$\bar{c}_0(E_{n_{z11}}) = n_z\pi/d_z \tag{2.183}$$

The DOS function for bulk specimens of stressed Kane type materials in the absence of band tail can be written as

$$D_0(E) = g_v(3\pi^2)^{-1}\bar{T}_0(E) \tag{2.184}$$

where

$$\bar{T}_0(E) = \left[\bar{a}_0(E)\bar{b}_0(E)[\bar{c}_0(E)]' + \bar{a}_0(E)[\bar{b}_0(E)]'\bar{c}_0(E) + [\bar{a}_0(E)]'\bar{b}_0(E)\bar{c}_0(E)\right]$$

2.2.6 The DR in Ultrathin Films (UFs) of HD Te

The DR of the conduction electrons in Te can be expressed as [102]

$$E = \psi_1 k_z^2 + \psi_2 k_s^2 \pm [\psi_3^2 k_z^2 + \psi_4^2 k_s^2]^{1/2} \tag{2.185}$$

where, the values of the system constants are given in Table 1.1.

The carrier energy spectrum in HD Te can be written as

$$\gamma_3(E,\eta_g) = \psi_1 k_z^2 + \psi_2 k_s^2 \pm [\psi_3^2 k_z^2 + \psi_4^2 k_s^2]^{1/2} \tag{2.186}$$

The DOS function at the Fermi level can be written as

$$N_{HD}(E_{F_h},\eta_g) = \frac{g_v}{3\pi^2}\frac{\partial}{\partial E_{F_h}}[t_{1HD}(E_{F_h},\eta_g)] \tag{2.187}$$

where,

$$t_{1HD}(E_{F_h},\eta_g) = \left[3\psi_{5HD}(E_{F_h},\eta_g)\Gamma_{3HD}(E_{F_h},\eta_g) - \psi_6\Gamma_{3HD}^3(E_{F_h},\eta_g)\right],$$

$$\psi_{5HD}(E_{F_h},\eta_g) = \left[\frac{\gamma_3(E_{F_h},\eta_g)}{\psi_2} + \frac{\psi_4^2}{2\psi_2^2}\right],$$

$$\Gamma_{3HD}(E_{F_h},\eta_g) = \frac{\sqrt{\psi_3^2 + 4\psi_1\gamma_3(E_{F_h},\eta_g)}}{2\psi_1}, \text{ and}$$

$$\psi_6 = \frac{\psi_1}{\psi_2}$$

The DR in 2D HD Te can be written as

$$k_s^2 = \bar{\psi}_{5HD}(E,\eta_g) \tag{2.188}$$

where

$$\bar{\psi}_{5HD}(E,\eta_g) = \psi_{5HD}(E,\eta_g) - \psi_6\left(\frac{n_z\pi}{d_z}\right)^2 \pm \psi_7\left[\psi_{8HD}^2(E,\eta_g) - \left(\frac{n_z\pi}{d_z}\right)^2\right]^{1/2},$$

$$\psi_7 = \frac{\psi_4\sqrt{\psi_1}}{\psi_2^{3/2}} \text{ and } \psi_{8HD}^2(E,\eta_g) = \left[\frac{\psi_4^4 + 4\gamma_3(E,\eta_g)\psi_2\psi_4^2 + 4\psi_2^2\psi_3^2}{4\psi_1\psi_2\psi_4^2}\right]$$

The EEM in this case is given by

$$m^*(E_{F1HD},\eta_g,n_z) = \frac{\hbar^2}{2}\left[\psi'_{5HD}(E_{F1HD},\eta_g) \pm \frac{\psi_{8HD}(E_{F1HD},\eta_g)\psi'_{8HD}(E_{F1HD},\eta_g)}{\sqrt{\psi_{8HD}^2(E_{F1HD},\eta_g) - (\pi n_z/d_z)^2}}\right] \tag{2.189}$$

The total DOS function in this case can be expressed as

$$N_{2DT}(E) = \left(\frac{g_v}{2\pi}\right) \sum_{n_z=1}^{n_{z\max}} [\bar{\psi}_{5HD}(E, \eta_g)]' H(E - E_{n_{z59HD}}) \tag{2.190}$$

where $E_{n_{z59HD}}$ is the lowest positiveroot of the equation

$$\psi_{5HD}(E_{n_{z59HD}}, \eta_g) - \psi_6\left(\frac{\pi n_z}{d_z}\right)^2 \pm \psi_7\left[\psi_{8HD}^2(E_{n_{z59HD}}, \eta_g) - \left(\frac{\pi n_z}{d_z}\right)^2\right]^{1/2} = 0 \tag{2.191}$$

The 2D electron energy spectrum in QWs of Te in the absence of band tails assumes the form

$$k_s^2 = \psi_5(E) - \psi_6\left(\frac{\pi n_z}{d_z}\right)^2 \pm \psi_7\left[\psi_8^2(E) - \left(\frac{\pi n_z}{d_z}\right)^2\right]^{1/2} \tag{2.192}$$

where,

$$\psi_5(E) = \left[\frac{E}{\psi_2} + \frac{\psi_4^2}{2\psi_2^2}\right] \text{ and } \psi_8^2(E) = \frac{\psi_4^4 + 4E\psi_2\psi_4^2 + 4\psi_2^2\psi_3^2}{4\psi_1\psi_2\psi_4^2}$$

Thus, the total 2D DOS function can be expressed as

$$N_{2DT}(E) = \left(\frac{g_v}{\pi}\right) \sum_{n_z=1}^{n_{z\max}} t_{40}'(E, n_z) H(E - E_{n_{z12}}) \tag{2.193}$$

where,

$$t_{40}(E, n_z) = \left[\psi_5(E) - \psi_6\left(\frac{\pi n_z}{d_z}\right)^2 \pm \psi_7\left[\psi_8^2(E) - \left(\frac{\pi n_z}{d_z}\right)\right]^{1/2}\right]^{1/2}$$

The sub-band energies $(E_{n_{z12}})$ are given by

$$E_{n_{z12}} = \psi_1(n_z\pi/d_z)^2 \pm \psi_3(n_z\pi/d_z) \tag{2.194}$$

Using (2.192) the EEM can be expressed as

$$m^*(E_{F_s}, n_z) = \frac{\hbar^2}{2} t'_{40}(E_{F_s}, n_z) \tag{2.195}$$

The DOS function for bulk specimens of Te in the absence of band tails can be expressed as

$$N(E_F) = \frac{g_v}{3\pi^2} \frac{\partial}{\partial E_F} [M_9(E_F)] \tag{2.196}$$

where,

$$M_9(E_F) = \left[3\psi_5(E_F)\Gamma_3(E_F) - \psi_6 \Gamma_3^3(E_F)\right],$$

$$\psi_5(E_F) = \left[\frac{E_F}{\psi_2} + \frac{\psi_4^2}{2\psi_2^2}\right] \text{ and}$$

$$\Gamma_3(E_F) = [2\psi_1]^{-1}\left[\sqrt{\psi_3^2 + 4\psi_1 E_F} - \psi_3\right]$$

2.2.7 The DR in Ultrathin Films (UFs) of HD Gallium Phosphide

The energy spectrum of the conduction electrons in n-GaP can be written as [103]

$$E = \frac{\hbar^2 k_s^2}{2m_\perp^*} + \frac{\hbar^2}{2m_\parallel^*}[\bar{A}' k_s^2 + k_z^2] - \left[\frac{\hbar^4 k_0^2}{m_\parallel^{*2}}(k_s^2 + k_z^2) + |V_G|^2\right]^{1/2} + |V_G| \tag{2.197}$$

where, k_0 and $|V_G|$ are constants of the energy spectrum and $\bar{A}' = 1$.

The DR of the conduction electrons in HDn-GaP can be expressed as

$$\gamma_3(E, \eta_g) = \frac{\hbar^2 k_s^2}{2m_\perp^*} + \frac{\hbar^2}{2m_\parallel^*}[\bar{A}' k_s^2 + k_z^2] - \left[\frac{\hbar^4 k_0^2}{m_\parallel^{*2}}(k_s^2 + k_z^2) + |V_G|^2\right]^{1/2} - |V_G| \tag{2.198}$$

The EEMs assume the forms as

$$m_z^*(E_{F_h},\eta_g) = \frac{\hbar^2\gamma_3'(E_{F_h},\eta_g)}{b}\left[1 \pm (C+bD)\left[C^2+4bD^2+4bC\gamma_3(E_{F_h},\eta_g) - 4bCD+4b^2\gamma_3(E_{F_h},\eta_g)D\right]^{-1/2}\right] \tag{2.199}$$

and

$$m_s^*(E_{F_h},\eta_g) = \frac{\hbar^2}{2}\left[t_{11}\gamma_3'(E_{F_h},\eta_g) - t_{41}t_5'(E_{F_h},\eta_g)\right] \tag{2.200}$$

where,

$b = \frac{\hbar^2}{2m_{\|}^*}$, $C = (\hbar^2 k_0/m_{\|}^*)^2$, $D = |V_G|$, $t_{11} = \frac{1}{a}$, $a = \frac{\hbar^2}{2m_{\perp}^*} + \bar{A}'b$, $t_{41} = \sqrt{\frac{g_3}{2a^2}}$,
$g_3 = (4abc+4a^2c)$, $t_5^2(E_{F_h},\eta_g) = [g_2 - 4aC\gamma_3(E_{F_h},\eta_g)](g_3)^{-1}$,
$g_2 = (4a^2b^2+C^2+4aCD)$

The DOS function can be expressed as

$$N_{HD}(E_{F_h},\eta_g) = \frac{g_v}{4\pi^2}\frac{\partial}{\partial E_{F_h}}\left[\mathrm{M}_{1HD}(E_{F_h},\eta_g)\right] \tag{2.201}$$

where

$$M_{1HD}(E_{F_h},\eta_g) = \Bigg[2(t_{11}\gamma_3(E_{F_h},\eta_g)+t_{21})\sqrt{t_{81}+t_{91}\gamma_3(E_{F_h},\eta_g)}. + (t_{31}/3)\theta_{,-}^3(E_{F_h},\eta_g) + (t_{41}/2)\left[\theta_{,-}(E_{F_h},\eta_g)\sqrt{\theta_{-}^2(E_{F_h},\eta_g)+t_5(E_{F_h},\eta_g)} - \sqrt{t_5(E_{F_h},\eta_g)}\right] + (t_{41}t_5(E_{F_h},\eta_g)/2)\ln\left|\frac{\theta_{,-}(E_{F_h},\eta_g)\sqrt{\theta_{-}^2(E_{F_h},\eta_g)+t_5(E_{F_h},\eta_g)}}{\sqrt{t_5(E_{F_h},\eta_g)}}\right|\Bigg],$$

$t_{21} = \frac{g_1}{2a^2}, g_1 = -(C + 2aD), t_{81} = [t_{41}^4 + 4t_{41}^2 t_{21} t_{31} + (4t_{31}^2 t_{41}^2 g_2)(g_3)^{-1}], t_{31} = \frac{b}{a},$
$t_{91} = [4t_{11} t_{31} t_{41}^2 + 8t_{11} t_{21} t_{31}^2 - (16t_{31}^2 t_{41}^2 aC)(g_3)^{-1}],$
$\theta_-(E_{F_h}, \eta_g) = (t_{31}\sqrt{2})^{-1}[t_{61} + t_7\gamma_3(E_{F_h}, \eta_g) - \sqrt{t_{81} + t_{91}\gamma_3(E_{F_h}, \eta_g)}], t_{61} = (t_{41}^2 + 2t_{21}t_{31})$ and
$t_{71} = (2t_{11}t_{31})$

The 2D DR in QW of HD GaPcan be expressed following (2.198) as

$$k_s^2 = t_{11}\gamma_3(E, \eta_g) + t_{21} - t_{31}\left(\frac{\pi n_z}{d_z}\right)^2 - t_{41}\left[\left(\frac{\pi n_z}{d_z}\right)^2 + t_5^2(E, \eta_g)\right]^{1/2} \tag{2.202}$$

The EEM in this case can be written following (2.202) as

$$m^*(E_{F1HD}, \eta_g, n_z) = \frac{\hbar^2}{2}\left[t_{11}\gamma_3'(E_{F1HD}, \eta_g) - t_{41}t_5(E_{F1HD}, \eta_g)t_5'(E_{F1HD}, \eta_g)\left[\left(\frac{\pi n_z}{d_z}\right)^2 + t_5^2(E_{F1HD}, \eta_g)\right]^{-1/2}\right] \tag{2.203}$$

The total DOS function assumes the form

$$N_{2DT}(E, \eta_g) = \frac{g_v}{2\pi}\sum_{n_z=1}^{n_{z\max}}\left[t_{11}\gamma_3'(E_{F1HD}, \eta_g) - t_{41}t_5(E_{F1HD}, \eta_g)t_5'(E_{F1HD}, \eta_g)\left[\left(\frac{\pi n_z}{d_z}\right)^2 + t_5^2(E_{F1HD}, \eta_g)\right]^{-1/2}\right]H(E - E_{n_{z8THD}}) \tag{2.204}$$

where, $E_{n_{z8THD}}$ is given by the equation

$$t_{11}\gamma_3(E_{n_{z8THD}}, \eta_g) + t_{21} - t_{31}\left(\frac{\pi n_z}{d_z}\right)^2 - t_{41}\left[\left(\frac{\pi n_z}{d_z}\right)^2 + t_5^2(E_{n_{z8THD}}, \eta_g)\right]^{1/2} = 0 \tag{2.205}$$

The 2D DR in size-quantized n-GaP in the absence of band tails assumes the form

$$E = ak_s^2 + C(n_z\pi/d_z)^2 + |V_G| - \left[Dk_s^2 + |V_G|^2 + D(n_z\pi/d_z)^2\right]^{1/2} \tag{2.206}$$

The sub-band energy $(E_{n_{z13}})$ are given by

$$E_{n_{z13}} = C(n_z\pi/d_z)^2 + |V_G| - \left[|V_G|^2 + D(n_z\pi/d_z)^2\right]^{1/2} \tag{2.207}$$

The (2.215) can be expressed as

$$k_s^2 = t_{42}(E, n_z) \tag{2.208}$$

in which,

$$t_{42} = (E, n_z) \equiv \left[\{2a(E - t_1) + D\} - \left\{[2a(E - t_1) + D]^2 - 4a^2\left[(E - t_1)^2 - t_2\right]\right\}^{1/2}\right],$$
$$t_1 \equiv |V_G| + C(\pi n_z/d_z)^2 \text{ and } t_2 \equiv |V_G|^2 + D(\pi n_z/d_z)^2.$$

The total DOS function is given by

$$N_{2DT}(E) = \frac{g_v}{2\pi}\sum_{n_z=1}^{n_{z\max}} [t_{42}{}'(E, n_z)]H(E - E_{n_{z13}}) \tag{2.209}$$

Using (1.212) the EEM can be expressed as

$$m^*(E_{F_s}, n_z) = \frac{\hbar^2}{2} t'_{42}(E_{F_s}, n_z) \tag{2.210}$$

The DOS function at the Fermi level is given by

$$N(E_F) = \frac{g_v}{4\pi^2}\frac{\partial}{\partial E_F}[M_1(E_F)] \tag{2.211}$$

where

$$M_1(E_F) = \left[2(t_{11}E_F + t_{21})\sqrt{t_{91}E_F + t_{81}} + \frac{t_{31}}{3}\phi^3(E_F) + \frac{t_{41}}{2}\left[\phi(E_F)\sqrt{\phi^2(E_F) + t_5(E_F)}\right.\right.$$
$$\left.\left. + \frac{t_{41}t_5(E_F)}{2}\left[\ln\left|\frac{\phi(E_F)\sqrt{\phi^2(E_F) + t_5(E_F)}}{\sqrt{t_5(E_F)}}\right|\right]\right]\right], \phi(E_F) = (t_{31}\sqrt{2})^{-1}\left[t_{61} + E_F t_{71} - [t_{81} + t_{91}E_F]^{1/2}\right]$$

2.2.8 The DR in Ultrathin Films (UFs) of HD Platinum Antimonide

The DR for the n-type $PtSb_2$ can be written as [104]

$$\left[E + \lambda_0 \frac{a^2}{4} k^2 - l k_s^2 \frac{a^2}{4}\right] \left[E + \delta_0 - v \frac{a^2}{4} k^2 - n' k_s^2 \frac{a^2}{4}\right] = I \left(\frac{a^4}{16}\right) k^4 \tag{2.212}$$

The (2.222) assumes the form

$$[E + \omega_1 k_s^2 + \omega_2 k_z^2][E + \delta_0 + \omega_3 k_s^2 - \omega_4 k_z^2] = I_1 (k_s^2 + k_z^2)^2 \tag{2.213}$$

where

$$\omega_1 = \left[\lambda_0 \frac{a^2}{4} + l \frac{a^2}{4}\right],\ \omega_2 = \lambda_0 \frac{a^2}{4},\ \omega_3 = \left[n' \frac{a^2}{4} - \upsilon \frac{a^2}{4}\right],\ \omega_4 = \upsilon \frac{a^2}{4},\ I_1 = I \left(\frac{a^2}{4}\right)^2,$$

λ_0, l, δ_0, v, n' and a are the band constants.

The carrier dispersion law in HD $PtSb_2$ can be written as

$$T_{11} k_s^4 - k_s^2 [T_{21}(E, \eta_g) - T_{31} k_z^2] + [T_{41} k_z^4 - T_{51}(E, \eta_g) k_z^2 - T_{61}(E, \eta_g)] = 0 \tag{2.214}$$

where,

$$\begin{aligned}
T_{11} &= (I_1 - \omega_2 \omega_3), \\
T_{21}(E, \eta_g) &= [\omega_1 \delta_0 + \omega_1 \gamma_3(E, \eta_g) + \omega_3 \gamma_3(E, \eta_g)], \\
T_{31} &= [2I_1 + \omega_2 \omega_4 - \omega_2 \omega_3], \\
T_{41} &= [2I_1 + \omega_2 \omega_4], \\
T_{51}(E, \eta_g) &= [\omega_2 \gamma_0 - \omega_4 \gamma_3(E, \eta_g) + \omega_2 \gamma_3(E, \eta_g)], \\
T_{61}(E, \eta_g) &= [\gamma_8(E, \eta_g) + \gamma_0 \gamma_3(E, \eta_g)] \text{ and} \\
\gamma_8(E, \eta_g) &= 2\theta_0(E, \eta_g)[1 + Erf(E/\eta_g)]^{-1}
\end{aligned}$$

The DOS function at the Fermi level in this case can be written as

$$N_{HD}(E_{F_h},\eta_g) = \frac{g_v}{3\pi^2}\frac{\partial}{\partial E_{F_h}}[\mathrm{M}_{6HD}(E_{F_h},\eta_g)] \tag{2.215}$$

where

$$M_{6HD}(E_{F_h},\eta_g) = \left[T_{91HD}(E_{F_h},\eta_g)\rho_{2HD}(E_{F_h},\eta_g) - T_{101}\frac{\rho_{2HD}^3(E_{F_h},\eta_g)}{3} - T_{11}J_3(E_{F_h},\eta_g)\right]$$

$$T_{91HD}(E_{F_h},\eta_g) = \frac{T_{21}(E_{F_h},\eta_g)}{2T_{11}},$$

$$\rho_{2HD}(E_{F_h},\eta_g) = \left[(2T_{41})^{-1}\left[T_{51}(E_{F_h},\eta_g) + \sqrt{T_{51}^2(E_{F_h},\eta_g) + 4T_{41}T_{61}(E_{F_h},\eta_g)}\right]\right]^{1/2}$$

$$T_{101} = [T_{31}/2T_{11}],$$

$$J_3(E_{F_h},\eta_g) = \frac{\rho_{2HD}(E_{F_h},\eta_g)}{3}\left[\left[A_{3HD}^2(E_{F_h},\eta_g) + B_{3HD}^2(E_{F_h},\eta_g)\right]E_0(\eta(E_{F_h},\eta_g),t(E_{F_h},\eta_g))\right.$$
$$- \left[A_{3HD}^2(E_{F_h},\eta_g) - B_{3HD}^2(E_{F_h},\eta_g)\right]F_0(\eta(E_{F_h},\eta_g)t(E_{F_h},\eta_g))]$$
$$+ \frac{\rho_{2HD}(E_{F_h},\eta_g)}{3}\left[(A_{3HD}^2(E_{F_h},\eta_g) - \rho_{2HD}^2(E_{F_h},\eta_g))(B_{3HD}^2(E_{F_h},\eta_g) - \rho_{2HD}^2(E_{F_h},\eta_g))\right]^{1/2},$$

$E_0(\eta(E_{F_h},\eta_g),t(E_{F_h},\eta_g))$ and $F_0(\eta(E_{F_h},\eta_g),t(E_{F_h},\eta_g))$ are the incomplete elliptic integrals of second and first respectively,

$$A_{3HD}^2(E_{F_h},\eta_g) = \frac{1}{2}\left[T_{12}(E_{F_h},\eta_g) + \sqrt{T_{12}^2(E_{F_h},\eta_g) - 4T_{13}^2(E_{F_h},\eta_g)}\right], T_{12}(E_{F_h},\eta_g) = \left[T_7(E_{F_h},\eta_g)/\bar{T}_{61}\right]$$
$$\bar{T}_{61} = \left[T_{31}^2 - 4T_{11}T_{41}\right], T_7(E_{F_h},\eta_g) = \left[2T_{31}T_{21}(E_{F_h},\eta_g) - 4T_{11}T_{51}(E_{F_h},\eta_g)\right],$$
$$T_{13}(E_{F_h},\eta_g) = (T_8(E_{F_h},\eta_g)/\bar{T}_8),$$
$$T_8(E_{F_h},\eta_g) = \left[T_{12}^2(E_{F_h},\eta_g) + 4T_{11}T_{61}(E_{F_h},\eta_g)\right],$$
$$B_{3HD}^2(E_{F_h},\eta_g) = \frac{1}{2}\left[T_{12}(E_{F_h},\eta_g) - \sqrt{T_{12}^2(E_{F_h},\eta_g) - 4T_{13}(E_{F_h},\eta_g)}\right], \bar{T}_{11} = \left[\sqrt{\bar{T}_{61}}/2T_{11}\right]$$
$$t(E_{F_h},\eta_g) = [B_3(E_{F_h},\eta_g)/A_3(E_{F_h},\eta_g)], \eta(E_{F_h},\eta_g) = \sin^{-1}\left[\frac{\rho_2(E_{F_h},\eta_g)}{B_3(E_{F_h},\eta_g)}\right]$$

From (2.215) the DR in QWs of HD $PtSb_2$ can be expressed as

$$T_{11}k_s^4 - P_{1HD}(E,\eta_g,n_z)k_s^2 + P_{2HD}(E,\eta_g,n_z) = 0 \tag{2.216}$$

where,

$$P_{1HD}(E,\eta_g,n_z) = [T_{21}(E_{F_h},\eta_g) - T_{31}(\pi n_z/d_z)]$$
$$P_{2HD}(E,\eta_g,n_z) = [T_{41}(\pi n_z/d_z)^4 - T_{51}(E_{F_h},\eta_g)(\pi n_z/d_z)^2 - T_{61}(E_{F_h},\eta_g)]$$

(2.216) can be written as

$$k_s^2 = A_{60}(E, \eta_g, n_z) \tag{2.217}$$

where,

$$A_{60}(E, \eta_g, n_z) = \left[P_{1HD}(E, \eta_g, n_z) - \sqrt{P_{1HD}^2(E, \eta_g, n_z) - 4T_{11}P_{2HD}(E, \eta_g, n_z)}\right]$$

The EEM assumes the form

$$m^*(E_{F1HD}, \eta_g, n_z) = \frac{\hbar^2}{2} A'_{60}(E_{F1HD}, \eta_g, n_z) \tag{2.218}$$

The DOS function in this case is given by

$$N_{2D\Gamma}(E) = \frac{g_v}{2\pi} \sum_{n_z=1}^{n_{z\max}} [A'_{60}(E, \eta_g, n_z)] H(E - E_{n_{z100}}) \tag{2.219}$$

where $E_{n_{z100}}$ is the lowest positive root of the equation

$$P_{2HD}(E_{n_{z100}}, \eta_g, n_z) = 0 \tag{2.220}$$

From (2.217), we can write the expression of the 2D dispersion law in QWs of n-$PtSb_2$ in the absence of band tails as

$$k_s^2 = t_{44}(E, n_z) \tag{2.221}$$

where,

$$t_{44}(E, n_z) = [2A_9]^{-1}\left[-A_{10}(E, n_z) + \sqrt{A_{110}^2(E, n_z) + 4A_9A_{11}(E, n_z)}\right],$$

$$A_9 \equiv [I_1 + \omega_1\omega_3],$$

$$A_{10}(E, n_z) \equiv \left[\omega_3 E + \omega_1\left\{E + \delta_0 - \omega_4\left(\frac{\pi n_z}{d_z}\right)^2\right\} + \omega_2\omega_3\left(\frac{\pi n_z}{d_z}\right)^2 + 2I_1\left(\frac{\pi n_z}{d_z}\right)^2\right]$$

and

$$A_{11}(E, n_z) \equiv \left[E\left[E + \delta_0 - \omega_4\left(\frac{\pi n_z}{d_z}\right)^2\right] + \omega_2\left(\frac{\pi n_z}{d_z}\right)^2\left[E + \delta_0 - \omega_4\left(\frac{\pi n_z}{d_z}\right)^2\right] - I_1\left(\frac{\pi n_z}{d_z}\right)^4\right]$$

The area of k_s space can be expressed as

$$A(E,n_z) = \pi t_{44}(E,n_z) \tag{2.222}$$

The total DOS function assumes the form

$$N_{2DT}(E) = \frac{g_v}{2\pi}\sum_{n_z=1}^{n_{z\max}} [t'_{44}(E,n_z)]H(E - E_{n_{z14}}) \tag{2.223}$$

where the quantized levels $E_{n_{z14}}$ can be expressed through the equation

$$E_{n_{z14}} = (2)^{-1}\left[-\left[\omega_2\left(\frac{\pi n_z}{d_z}\right)^2 + \delta_0 - \omega_4\left(\frac{\pi n_z}{d_z}\right)^2\right] + \left\{\left[\omega_2\left(\frac{\pi n_z}{d_z}\right)^2 + \delta_0 - \omega_4\left(\frac{\pi n_z}{d_z}\right)^2\right]^2 + 4\left[I_1\left(\frac{\pi n_z}{d_z}\right)^4 + \omega_2\omega_4\left(\frac{\pi n_z}{d_z}\right)^4 - \omega_2\delta_0\left(\frac{\pi n_z}{d_z}\right)^2\right]\right\}^{1/2}\right] \tag{2.224}$$

Using (2.221), the EEM in this case can be written as

$$m^*(E_{F_s},n_z) = \frac{\hbar^2}{2} t'_{44}(E_{F_s},n_z) \tag{2.225}$$

2.2.9 *The DR in Ultrathin Films (UFs) of HD Bismuth Telluride*

The DR of the conduction electrons in Bi_2Te_3 can be written as [105]

$$E(1+\alpha E) = \bar{\omega}_1 k_x^2 + \bar{\omega}_2 k_y^2 + \bar{\omega}_3 k_z^2 + 2\bar{\omega}_4 k_z k_y \tag{2.226}$$

where $\bar{\omega}_1 = \frac{\hbar^2}{2m_0}\bar{\alpha}_{11}, \bar{\omega}_2 = \frac{\hbar^2}{2m_0}\bar{\alpha}_{22}, \bar{\omega}_3 = \frac{\hbar^2}{2m_0}\bar{\alpha}_{33}, \bar{\omega}_4 = \frac{\hbar^2}{2m_0}\bar{\alpha}_{23}$ in which $\bar{\alpha}_{11}, \bar{\alpha}_{22}, \bar{\alpha}_{33}$ and $\bar{\alpha}_{23}$ are system constants.

The DR in HD Bi_2Te_3 assumes the form

$$\gamma_2(E,\eta_g) = \bar{\omega}_1 k_x^2 + \bar{\omega}_2 k_y^2 + \bar{\omega}_3 k_z^2 + 2\bar{\omega}_1 k_x k_y \tag{2.227}$$

The DOS function in this case is given by

$$N(E) = 4\pi g_v \left(\frac{2m_0}{\hbar^2}\right)^{3/2} \frac{\sqrt{\gamma_2(E,\eta_g)\gamma_2'(E,\eta_g)}}{\sqrt{\alpha_{11}\alpha_{22}\alpha_{33} - 4\alpha_{11}\alpha_{23}^2}} \tag{2.228}$$

The DR in QWs of HD Bi_2Te_3 can be expressed as

$$\gamma_2(E,\eta_g) = \bar{\omega}_1 \left(\frac{\pi n_x}{d_x}\right)^2 + \bar{\omega}_2 k_y^2 + \bar{\omega}_3 k_z^2 + 2\bar{\omega}_4 k_z k_y \tag{2.229}$$

The EEM can be expressed as

$$m^*(E_{F1HD},\eta_g) = \frac{m_0}{\sqrt{\alpha_{11}\alpha_{33} - 4\alpha_{23}^2}} \gamma_2'(E_{F1HD},\eta_g) \tag{2.230}$$

The DOS function per sub-band can be written as

$$\bar{N}_{2D\Gamma}(E_{F1HD}) = \frac{g_v}{2\pi} R_{60}'(E_{F1HD},\eta_g) \tag{2.231}$$

where

$$R_{60}(E_{F1HD},\eta_g,n_z) = \frac{1}{\sqrt{\alpha_{11}\alpha_{33} - 4\alpha_{23}^2}} \left[\frac{2m_0\gamma_2'(E_{F1HD},\eta_g)}{\hbar^2} - \frac{2m_0}{\hbar^2}\left(\frac{\pi n_x}{d_x}\right)^2 \bar{\alpha}_{11}\right]$$

The 2D DR in QWs of Bi_2Te_3 in the absence of band tails assumes the form

$$E(1+\alpha E) = \bar{\omega}_1 \left(\frac{n_x \pi}{d_x}\right)^2 + \bar{\omega}_2 k_y^2 + \bar{\omega}_3 k_z^2 + 2\bar{\omega}_4 k_z k_y \tag{2.232}$$

The area of the ellipse is given by

$$A_n(E,n_x) = \frac{\pi}{\sqrt{\bar{\alpha}_{22}\bar{\alpha}_{33} - 4\bar{\alpha}_{23}^2}} \left[\frac{2m_0 E(1+\alpha E)}{\hbar^2} - \bar{\omega}_1 \left(\frac{n_x \pi}{d_x}\right)^2\right] \tag{2.233}$$

The total DOS function assumes the form

$$N_{2DT}(E) = \frac{g_v m_0}{\pi\hbar^2 \sqrt{\bar{\alpha}_{22}\bar{\alpha}_{33} - 4\bar{\alpha}_{23}^2}} \sum_{n_x=1}^{n_{x\max}} (1+2\alpha E) H(E - E_{n_{z15}}) \tag{2.234}$$

where, $(E_{n_{z15}})$ can be expressed through the equation

$$E_{n_{z15}}(1+\alpha E_{n_{z15}}) = \bar{\omega}_1 \left(\frac{n_x \pi}{d_x}\right)^2 \tag{2.235}$$

The EEM in this case assumes the form as

$$m^*(E_{F_s}) = \frac{m_0(1+2\alpha(E_{F_s}))}{\sqrt{\bar{\alpha}_{22}\bar{\alpha}_{33} - 4\bar{\alpha}_{23}^2}} \tag{2.236}$$

2.2.10 The DR in Ultrathin Films (UFs) of HD Germanium

It is well known that the conduction electrons of n-Ge obey two different types of dispersion laws since band non-parabolicity has been included in two different ways as given in the literature[106–108].

(a) The DR of the conduction electrons in bulk specimens of n-Ge can be expressed in accordance with Cardona et al. [106] as

$$E = -\frac{E_{g0}}{2} + \frac{\hbar^2 k_z^2}{2m_{\parallel}^*} + \left[\frac{E_{g0}^2}{4} + E_{g0} k_s^2 \left(\frac{\hbar^2}{2m_{\perp}^*}\right)\right]^{1/2} \tag{2.237}$$

where in this case $m_{\parallel}^*$ and $m_{\perp}^*$ are the longitudinal and transverse effective masses along <111> direction at the edge of the conduction band respectively

The (2.237) can be written as

$$\frac{\hbar^2 k_s^2}{2m_{\perp}^*} = E(1+\alpha E) + \alpha\left(\frac{\hbar^2 k_z^2}{2m_{\parallel}^*}\right) - (1+2\alpha E)\left(\frac{\hbar^2 k_z^2}{2m_{\parallel}^*}\right) \tag{2.238}$$

The DR under the condition of heavy doping can be expressed from (2.238) as

$$\frac{\hbar^2 k_s^2}{2m_{\perp}^*} = \gamma_2(E+\eta_g) + \alpha\left(\frac{\hbar^2 k_z^2}{2m_{\parallel}^*}\right)^2 - (1+2\alpha\gamma_3(E+\eta_g))\left(\frac{\hbar^2 k_z^2}{2m_{\parallel}^*}\right) \tag{2.239}$$

The DOS function in this case assumes the form

$$N_{HD}(E_{F_h}, \eta_g) = \frac{8\pi g_v m_{\perp}^* \sqrt{2m_{\parallel}^*}}{\hbar^3} M'_{8HD}(E_{F_h}, \eta_g) \tag{2.240}$$

where

$$M_{8HD}(E_{F_h},\eta_g) = [\gamma_3(E_{F_h},\eta_g)]^{1/2}\left[\gamma_2(E_{F_h},\eta_g) + \frac{\alpha}{5}\gamma_3^2(E_{F_h},\eta_g)\right] - \frac{\gamma_3(E_{F_h},\eta_g)}{3}[1+2\alpha\gamma_3(E_{F_h},\eta_g)]$$

In the presence of size quantization, the dispersion law in QW of HD Ge can be written following (2.239) as

$$\frac{\hbar^2 k_s^2}{2m_\perp^*} = \gamma_2(E+\eta_g) + \alpha\left(\frac{\hbar^2 \left({n_z\pi}/{d_z}\right)^2}{2m_\parallel^*}\right)^2 - (1+2\alpha\gamma_3(E+\eta_g))\frac{\hbar^2\left({n_z\pi}/{d_z}\right)^2}{2m_\parallel^*} \tag{2.241}$$

The EEM assumes the form

$$m_s^*(E_{F1HD},\eta_g,n_z) = m_\perp^*\left[\gamma_2'(E_{F1HD},\eta_g) - \frac{\alpha\hbar^2}{m_\parallel^*}\left(\frac{n_z\pi}{d_z}\right)^2\gamma_3'(E_{F1HD},\eta_g)\right] \tag{2.242}$$

The DOS function per sub-band at the Fermi level in this case can be written as

$$N_{2D\Gamma}(E_{F1HD}) = \frac{g_v m_\perp^*}{\pi\hbar^2}R_1'(E_{F1HD},\eta_g) \tag{2.243}$$

where,

$$R_1(E_{F1HD},\eta_g,n_z) = \left[\gamma_2(E_{F1HD},\eta_g) + \alpha\left(\frac{\hbar^2\left({n_z\pi}/{d_z}\right)^2}{2m_\parallel^*}\right)^2 - (1+2\alpha\gamma_3(E_{F1HD},\eta_g))\frac{\hbar^2\left({n_z\pi}/{d_z}\right)^2}{2m_\parallel^*}\right]$$

The sub-band energy $E_{n_{z100HD}}$ is the lowest positive root of the equation

$$\gamma_2(E_{n_{z100HD}},\eta_g) + \alpha\left[\frac{\hbar^2}{2m_\parallel^*}\left(\frac{n_z\pi}{d_z}\right)^2\right]^2 - [1+2\alpha\gamma_3(E_{n_{z100HD}},\eta_g)]\frac{\hbar^2}{2m_\parallel^*}\left(\frac{n_z\pi}{d_z}\right)^2 = 0 \tag{2.244}$$

In the presence of size quantization along k_z direction, the 2D DR of the conduction relations in QWs of n-Ge in the absence of band tails can be written by extending the method as given in [107] as

$$\frac{\hbar^2 k_x^2}{2m_1^*} + \frac{\hbar^2 k_y^2}{2m_2^*} = \gamma(E, n_z) \tag{2.245}$$

where,

$$m_1^* \equiv m_\perp^*, m_2^* = \frac{m_\perp^* + 2m_\parallel^*}{3},$$

$$\gamma(E, n_z) \equiv \left[E(1+\alpha E) - (1+2\alpha E)\frac{\hbar^2}{2m_3^*}\left(\frac{n_z \pi}{d_z}\right)^2 + \alpha\left[\frac{\hbar^2}{2m_3^*}\left(\frac{n_z \pi}{d_z}\right)^2\right]^2\right] \text{ and}$$

$$m_3^* = \frac{3m_\parallel^* m_\perp^*}{2m_\parallel^* + m_\perp^*}$$

The area of ellipse of the 2D surface as given by (2.245) can be written as

$$A(E, n_z) = \frac{2\pi\sqrt{m_1^* m_2^*}}{\hbar^2}\gamma(E, n_z) \tag{2.246}$$

The EEM in this case can be written as

$$m^*(E_{F_s}, n_z) = \left(\sqrt{m_1^* m_2^*}\right)[\gamma(E_{F_s}, n_z)]' \tag{2.247}$$

The DOS function per sub-band can be expressed as

$$N_{2D}(E) = \frac{4\sqrt{m_1^* m_2^*}}{\pi\hbar^2}\left[1 + 2\alpha E - 2\alpha\left(\frac{\hbar^2}{2m_3^*}\left(\frac{\pi n_z}{d_z}\right)^2\right)\right] \tag{2.248}$$

The total DOS function is given by

$$N_{2DT}(E) = \frac{4}{\pi\hbar^2}\sqrt{m_1^* m_2^*}\sum_{n_z=1}^{n_{z\max}}\left[1 + 2\alpha E - 2\alpha\left(\frac{\hbar^2}{2m_3^*}\left(\frac{\pi n_z}{d_z}\right)^2\right)\right]H(E - E_{n_{z16}}) \tag{2.249}$$

where, $E_{n_{z16}}$ is the positive root of the following equation

$$E_{n_{z16}}\left(1+\alpha E_{n_{z16}}\right)-\left(1+2\alpha E_{n_{z16}}\right)\left(\frac{\hbar^2}{2m_3^*}\left(\frac{\pi n_z}{d_z}\right)^2\right)+\alpha\left(\frac{\hbar^2}{2m_3^*}\left(\frac{\pi n_z}{d_z}\right)^2\right)^2=0 \tag{2.250}$$

The DOS function for bulk specimens of Ge in the absence of band tails can be written following (2.238) as

$$N(E)=4\pi g_v\left(\frac{2m_D^*}{\hbar^2}\right)^{\frac{3}{2}}\left[E^{\frac{1}{2}}-\frac{5}{6}\alpha E^{\frac{3}{2}}+\frac{18\alpha}{5}\left(\frac{m_{11}^*}{\hbar^2}\right)^2E^{\frac{7}{2}}\right];\ m_D=\left(m_\perp^{*2}.m_\parallel^{*2}\right)^{\frac{1}{3}} \tag{2.251}$$

(b) The DR of the conduction electron in bulk specimens of n-Ge can be expressed in accordance with the model of Wang and Ressler [106] can be written as

$$E=\frac{\hbar^2k_z^2}{2m_\parallel^*}+\frac{\hbar^2k_s^2}{2m_\perp^*}-\bar{\alpha}_4\left(\frac{\hbar^2k_s^2}{2m_\perp^*}\right)-\bar{\alpha}_5\left(\frac{\hbar^2k_s^2}{2m_\perp^*}\right)\left(\frac{\hbar^2k_z^2}{2m_\parallel^*}\right)-\bar{\alpha}_6\left(\frac{\hbar^2k_z^2}{2m_\parallel^*}\right)^2 \tag{2.252}$$

where,

$$\bar{\alpha}_4=\beta_4\left(\frac{2m_\perp^*}{\hbar^2}\right),\beta_4=1.4\beta_5,$$

$$\beta_5=\frac{\alpha\hbar^4}{4}\left[\left(m_\perp^*\right)^{-1}-\left(m_0\right)^{-1}\right]^2,\bar{\alpha}_5=\bar{\alpha}_7\left(\frac{4m_\perp^*m_\parallel^*}{\hbar^4}\right),\bar{\alpha}_7=0.8\beta_5\text{ and }\bar{\alpha}_6=\left(0.005\beta_5\right)\left(\frac{2m_\parallel^*}{\hbar^2}\right)^2$$

The energy spectrum under the condition of heavy doping can be written as

$$\gamma_3(E,\eta_g)=\frac{\hbar^2k_z^2}{2m_\parallel^*}+\frac{\hbar^2k_s^2}{2m_\perp^*}-\bar{\alpha}_4\left(\frac{\hbar^2k_s^2}{2m_\perp^*}\right)-\bar{\alpha}_5\left(\frac{\hbar^2k_s^2}{2m_\perp^*}\right)\left(\frac{\hbar^2k_z^2}{2m_\parallel^*}\right)-\bar{\alpha}_6\left(\frac{\hbar^2k_z^2}{2m_\parallel^*}\right)^2 \tag{2.253}$$

The (2.253) can be expressed as

$$\frac{\hbar^2k_s^2}{2m_\perp^*}=\bar{\alpha}_8-\bar{\alpha}_9k_z^2-\bar{\alpha}_{10}\left[k_z^4+\bar{\alpha}_{11}k_z^2+\bar{\alpha}_{12}(E,\eta_g)\right]^{1/2} \tag{2.254}$$

where

$$\bar{\alpha}_8 = \frac{1}{2\bar{\alpha}_4}, \quad \bar{\alpha}_9 = \frac{\bar{\alpha}_5}{2\bar{\alpha}_4}\left(\frac{\hbar^2}{2m_\parallel^*}\right), \quad \bar{\alpha}_{10} = \frac{1}{2\bar{\alpha}_4}\left(\frac{\hbar^2}{2m_\parallel^*}\right)\sqrt{\bar{\alpha}_5^2 - 4\bar{\alpha}_4\bar{\alpha}_6},$$

$$\bar{\alpha}_{11} = \frac{2m_\parallel^*}{\hbar^2}\left[\frac{4\bar{\alpha}_4 - 2\bar{\alpha}_5}{\bar{\alpha}_5 - 4\bar{\alpha}_4\bar{\alpha}_6}\right] \text{ and } \bar{\alpha}_{12}(E,\eta_g) = \left(\frac{2m_\parallel^*}{\hbar^2}\right)^2\left[\frac{(1 - 4\bar{\alpha}_4\gamma_3(E,\eta_g))}{\bar{\alpha}_5^2 - 4\bar{\alpha}_4\bar{\alpha}_6}\right]$$

The DOS function in this case is given by

$$N_{2D}(E_{F_h},\eta_g) = \frac{m_\perp^* g_v}{\pi\hbar^2} I_3'(\mathrm{E}_{F_h},\eta_g) \tag{2.255}$$

where

$$I_3(E_{F_h},\eta_s) = \left[\bar{\alpha}_8\rho_{10}(E_{F_h},\eta_g) - \frac{\bar{\alpha}_9}{3}\rho_{10}^3(E_{F_h},\eta_s) - \bar{\alpha}_{10}J_{10}(E_{F_h},\eta_s)\right],$$

$$\rho_{10}(E_{F_h},\eta_s) = \frac{1}{\hbar}\left[\frac{m_\parallel^*}{\bar{\alpha}_6}\right]^{\frac{1}{2}}\left[1 - \sqrt{1 - 4\bar{\alpha}_6\gamma_3(E_{F_h},\eta_s)}\right]^{\frac{1}{2}}$$

$$\begin{aligned} J_{10}(E_{F_h},\eta_g) = {} & \frac{\bar{A}_1(E_{F_h},\eta_g)}{3}\left[-E_0(\lambda(E_{F_h},\eta_g), q(E_{F_h},\eta_g)).\right. \\ & \left.[\bar{A}_1^2(E_{F_h},\eta_g) + \bar{B}_1^2(E_{F_h},\eta_g)] + 2\bar{B}_1^2(E_{F_h},\eta_g)F_0(\lambda(E_{F_h},\eta_g), q(E_{F_h},\eta_g))\right] \\ & + \frac{\bar{A}_1(E_{F_h},\eta_g)}{3}[\rho_{10}(E_{F_h},\eta_g) + \bar{A}_1{}^2(E_{F_h},\eta_g) + 2\bar{B}_1^2(E_{F_h},\eta_g)] \\ & \left[\frac{\bar{A}_1(E_{F_h},\eta_g) + \rho_{10}^2(E_{F_h},\eta_g)}{\bar{B}_1^2(E_{F_h},\eta_g) + \rho_{10}^2(E_{F_h},\eta_g)}\right]^{\frac{1}{2}} \end{aligned}$$

$$\bar{A}_1{}^2(E_{F_h},\eta_g) = \frac{1}{2}\left[\bar{\alpha}_{11} + \sqrt{\bar{\alpha}_{11}^2 + 4\bar{\alpha}_{12}^2(E_{F_h},\eta_g)}\right],$$

$$\bar{B}_1^2(E_{F_h},\eta_g) = \frac{1}{2}\left[\bar{\alpha}_{11} - \sqrt{\bar{\alpha}_{11}^2 + 4\bar{\alpha}_{12}^2(E_{F_h},\eta_g)}\right],$$

$$\lambda(E_{F_h},\eta_g) = \tan^{-1}\left[\frac{\rho_{10}(E_{F_h},\eta_g)}{\bar{B}_1^2(E_{F_h},\eta_g)}\right],$$

$$q(E_{F_h},\eta_g) = \left[\left[\frac{\bar{A}_1^2(E_{F_h},\eta_g) - \bar{B}_1^2(E_{F_h},\eta_g)}{\bar{A}_1^2(E_{F_h},\eta_g)}\right]\right]$$

The DR in QW of HD Ge can be written as

$$\frac{\hbar^2 k_s^2}{2m_\perp^*} = \bar{a}_8 - \bar{a}_9\left(\frac{n_z\pi}{d_z}\right)^2 - \bar{a}_{10}\left[\left(\frac{n_z\pi}{d_z}\right)^4 + \bar{a}_{11}\left(\frac{n_z\pi}{d_z}\right)^2 + \bar{a}_{12}(E,\eta_g)\right]^{1/2} \quad (2.256)$$

The (2.256) can be expressed as

$$\frac{\hbar^2 k_s^2}{2m_\perp^*} = A_{75}(E,\eta_g,n_z) \quad (2.257)$$

where,

$$A_{75}(E,\eta_g,n_z) = \left[\bar{\alpha}_8 - \bar{\alpha}_9\left(\frac{\pi n_z}{d_z}\right)^2 - \bar{\alpha}_{10}\left[\left(\frac{\pi n_z}{d_z}\right)^4 + \bar{\alpha}_{11}\left(\frac{\pi n_z}{d_z}\right)^2 + \bar{\alpha}_{12}(E,\eta_g)\right]^{1/2}\right]$$

The EEM is given by

$$m^*(E_{F1HD},\eta_g,n_z) = m_\perp^* A'_{75}(E_{F1HD},\eta_g,n_z) \quad (2.258)$$

The DOS function in this case can be written as

$$N_{2D\Gamma}(E) = \frac{m_\perp^* g_v}{\pi\hbar^2}\sum A'_{75}(E,\eta_g,n_z)H(E - E_{n_{z205HD}}) \quad (2.259)$$

where $E_{n_{z205HD}}$ is the lowest positive root of the equation

$$A_{75}(E_{n_{z205HD}},\eta_g,n_z) = 0 \quad (2.260)$$

The 2D dispersion law in the absence of band tails can be expressed as

$$E = A_5(n_z) + A_6(n_z)\beta - \bar{\alpha}_4\beta^2 \quad (2.261)$$

where,

$$A_5(n_z) = \frac{\hbar^2}{2m_3^*}\left(\frac{\pi n_z}{d_z}\right)^2\left[1 - \bar{\alpha}_6\left(\frac{\hbar^2}{2m_3^*}\right)\left(\frac{\pi n_z}{d_z}\right)^2\right],$$

$$A_6(n_z) = \left[1 - \bar{\alpha}_5\left(\frac{\hbar^2}{2m_3^*}\right)\left(\frac{\pi n_z}{d_z}\right)^2\right] \text{ and}$$

$$\beta \equiv \frac{\hbar^2 k_x^2}{2m_1^*} + \frac{\hbar^2 k_y^2}{2m_2^*}.$$

The (2.261) can be written as

$$\frac{\hbar^2 k_x^2}{2m_1^*} + \frac{\hbar^2 k_y^2}{2m_2^*} = I_1(E, n_z) \tag{2.262}$$

where,

$$I_1(E, n_z) = (2\bar{\alpha}_4)^{-1}[A_6(n_z) - [A_6^2(n_z) - 4\bar{\alpha}_4 E + 4\bar{\alpha}_4 A_5(n_z)]^{1/2}]$$

From (2.262), the area of the 2D k_s-space is given by

$$A(E, n_z) = \frac{2\pi\sqrt{m_1^* m_2^*}}{\hbar^2} I_1(E, n_z) \tag{2.263}$$

Using (2.262), the EEM in this case can be expressed as

$$m^*(E_{F_s}, n_z) = \left(\sqrt{m_1^* m_2^*}\right)[I_1(E_{F_s}, n_z)]' \tag{2.264}$$

The DOS function per sub-band can be written as

$$N_{2D}(E) = \frac{4}{\pi}\frac{\sqrt{m_1^* m_2^*}}{\hbar^2}\{I_1(E, n_z)\}' \tag{2.265}$$

where

$$\{I_1(E, n_z)\}' \equiv \frac{\partial}{\partial E}[I_1(E, n_z)]$$

The total DOS function assumes the form

$$N_{2DT}(E) = \frac{4\sqrt{m_1^* m_2^*}}{\pi\hbar^2}\sum_{n_z=1}^{n_{z\max}}\{I_1(E, n_z)\}' H\left(E - E_{n_{z17}}\right) \tag{2.266}$$

where, the sub-band energy $\left(E_{n_{z17}}\right)$ are given by

$$E_{n_{z17}} = \left(\frac{\hbar^2}{2m_3^*}\right)\left(\frac{\pi n_z}{d_z}\right)^2\left[1 - \bar{\alpha}_6\left(\frac{\hbar^2}{2m_3^*}\right)\left(\frac{\pi n_z}{d_z}\right)^2\right] \tag{2.267}$$

2.2.11 The DR in Ultrathin Films (UFs) of HD Gallium Antimonide

The DR of the conduction electrons in n-GaSb can be written as [109]

$$E = \frac{\hbar^2 k^2}{2m_0} - \frac{\bar{E}'_{g_0}}{2} + \frac{\bar{E}'_{g_0}}{2}\left[1 + \frac{2\hbar^2 k^2}{\bar{E}'_{g_0}}\left(\frac{1}{m_c} - \frac{1}{m_0}\right)\right]^{\frac{1}{2}} \tag{2.268}$$

where

$$\bar{E}'_{g_0} = \left[E_{g_0} + \frac{5.10^{-5}T^2}{2(112+T)}\right] \text{eV}$$

The (2.268) can be expressed as

$$\frac{\hbar^2 k^2}{2m_c} = I_{36}(E) \tag{2.269}$$

where

$$I_{36}(E) = [E + \bar{E}'_{g_0} - (m_c/m_0)(\bar{E}'_{g_0}/2) - \Big[(\bar{E}'_{g_0}/2)^2 + [((\bar{E}'_{g_0})^2/2)(1 - (m_c/m_0))] + [(\bar{E}'_{g_0}/2)(1 - (m_c/m_0))]^2 + [E\bar{E}'_{g_0}(1 - (m_c/m_0))]^{1/2}\Big]$$

Under the condition of heavy doping (2.269) assumes the form

$$\frac{\hbar^2 k^2}{2m_c} = I_{36}(E, \eta_g) \tag{2.270}$$

where,

$$I_{36}(E,\eta_g) = \left[\gamma_3(E,\eta_g) + E'_g - \frac{m_c}{m_0}\cdot\frac{E'_g}{2} - \left[\left(\frac{E'_g}{2}\right)^2 + \left[\frac{E'_g}{2}\left(1 - \frac{m_c}{m_0}\right)\right]^2 + \left(\frac{E'_g}{2}\right)^2\left(1 - \frac{m_c}{m_0}\right) + \gamma_3(E,\eta_g)E'_g\left(1 - \frac{m_c}{m_0}\right)\right]^{1/2}\right]$$

The EEM can be written as

$$m^*\left(E_{F_h}, \eta_g\right) = m_c\left\{I_{36}\left(E_{F_h}, \eta_g\right)\right\}' \tag{2.271}$$

The DOS function in this case can be written as

$$N_{HD}(E,\eta_g) = \frac{g_v}{2\pi^2}\left(\frac{2m_c}{\hbar^2}\right)^{3/2}\sqrt{I_{36}(E,\eta_g)}\{I_{36}(E,\eta_g)\}' \quad (2.272)$$

Since, the original band model in this case is a no pole function, therefore, the HD counterpart will be totally real, and the complex band vanishes.

For dimensional quantization along z-direction, the DR of the 2D electrons in QWs of HD GaSb can be written following (2.270) as

$$\frac{\hbar^2(\mathrm{n}_z\pi/\mathrm{d}_z)^2}{2m_c} + \frac{\hbar^2(k_s)^2}{2m_c} = I_{36}(E,\eta_g) \quad (2.273)$$

The expression of the $N_{2DT}(E)$ in this case can be written as

$$N_{2DT}(E) = \frac{m_c g_v}{\pi\hbar^2}\sum_{n_z=1}^{n_{z\max}} T_{119D}'(E,\eta_g,\mathrm{n}_z)H(\mathrm{E}-\mathrm{E}_{n_z D119}) \quad (2.274)$$

where,

$$T_{119D}(E,\eta_g,\mathrm{n}_z) = \left[I_{36}(E_{F_h},\eta_g) - \hbar^2(\mathrm{n}_z\pi/\mathrm{d}_z)^2(2m_c)^{-1}\right], \quad (2.275)$$

The sub band energies $\mathrm{E}_{n_z D119}$ in this case given by

$$\left\{\hbar^2(\mathrm{n}_z\pi/\mathrm{d}_z)^2\right\}(2m_c)^{-1} = I_{36}(\mathrm{E}_{n_z D119},\eta_g) \quad (2.276)$$

The EEM in this case assumes the form

$$m^*(E_{F1HD},\eta_g,\mathrm{n}_z) = m_c[I_{36}'(E_{F1HD},\eta_g,\mathrm{n}_z)] \quad (2.277)$$

The total 2D DOS function in the absence of band tails in this case can be written as

$$N_{2DT}(E) = \left(\frac{m_c g_v}{\pi\hbar^2}\right)\sum_{n_z=1}^{n_{\max}}\{[I_{36}(E)]'H(E-E_{n_{z44}})\} \quad (2.278)$$

where, the sub-band energies $E_{n_{z44}}$ can be expressed as

$$I_{36}(E_{n_{z44}}) = \frac{\hbar^2}{2m_c}(\pi n_z/d_z)^2 \quad (2.279)$$

The EEM in this case can be written as

$$m^{*}(E_{F_s}) = (m_c)[I_{36}(E_{F_s})]' \tag{2.280}$$

2.2.12 The DR in Ultrathin Films (UFs) of HD II–V Materials

The DR of the holes in II–V compounds in accordance with Yamada [110] can be expressed as

$$\begin{aligned} E =& A_{10}k_x^2 + A_{11}k_y^2 + A_{12}k_z^2 + A_{13}k_x \\ &\pm \left[\left(A_{14}k_x^2 + A_{15}k_y^2 + A_{16}k_z^2 + A_{17}k_x \right)^2 + A_{18}k_y^2 + A_{19}^2 \right]^{1/2} \end{aligned} \tag{2.281}$$

where, $A_{10}, A_{11}, A_{12}, A_{13}, A_{14}, A_{15}, A_{16}, A_{17}, A_{18}$ and A_{19} are energy band constants.

The DR under the condition of formation of band tails can be written in this case as

$$\begin{aligned} \gamma_3(E, \eta_g) =& A_{10}k_x^2 + A_{11}k_y^2 + A_{12}k_z^2 + A_{13}k_x \\ &\pm \left[\left(A_{14}k_x^2 + A_{15}k_y^2 + A_{16}k_z^2 + A_{17}k_x \right)^2 + A_{18}k_y^2 + A_{19}^2 \right]^{1/2} \end{aligned} \tag{2.282}$$

The hole energy spectrum in this case assumes the form

$$\begin{aligned} \gamma_3(E, \eta_g) =& A_{10}\left(\frac{n_x\pi}{d_x}\right)^2 + A_{11}k_y^2 + A_{12}k_z^2 + A_{13}\left(\frac{n_x\pi}{d_x}\right) \\ &\pm \left[\left(A_{14}\left(\frac{n_x\pi}{d_x}\right)^2 + A_{15}k_y^2 + A_{16}k_z^2 + A_{17}\left(\frac{n_x\pi}{d_x}\right) \right)^2 + A_{18}k_y^2 + A_{19}^2 \right]^{1/2} \end{aligned} \tag{2.283}$$

The sub band energy $(E_{n_{zHD401}})$ is the lowest positive root of the following equation

$$\gamma_3(E_{n_zHD401}, \eta_g) = A_{10}\left(\frac{n_x\pi}{d_x}\right)^2 + A_{13}\left(\frac{n_x\pi}{d_x}\right) \pm \left[\left(A_{14}\left(\frac{n_x\pi}{d_x}\right)^2 + A_{17}\left(\frac{n_x\pi}{d_x}\right)\right)^2 + A_{19}^2\right]^{1/2} \quad (2.284)$$

The EEM and the DOS function for both the cases should be calculated numerically.

2.2.13 The DR in Ultrathin Films (UFs) of HD Lead Germanium Telluride

The dispersion law of n-type $Pb_{1-x}Ge_xTe$ with x = 0.01 can be expressed following Vassilev [111] as

$$\left[E - 0.606k_s^2 - 0.722k_z^2\right]\left[E + \bar{E}_{g_0} + 0.411k_x^2 + 0.377k_z^2\right] = 0.23k_s^2 + 0.02k_z^2 \pm \left[0.06\bar{E}_{g_0} + 0.061k_s^2 + 0.0066k_z^2\right]k_s \quad (2.285)$$

where $\bar{E}_{g_0} = 0.21$ eV, k_x, k_y and k_z are in the units of 10^9 m^{-1}.

The electron energy spectrum in n-type $Pb_{1-x}Ge_xTe$ under the condition of formation of band tails can be written as

$$\left[\frac{2}{1 + Erf\left(\frac{E}{\eta_g}\right)}\right]\theta_0(E, \eta_g) + \gamma_3(E, \eta_g)[\bar{E}_{g_0} - 0.195k_s^2 - 0.345k_z^2] = [0.23k_s^2 + 0.02k_z^2 \pm [0.06\bar{E}_{g_0} + 0.061k_s^2 + 0.0066k_z^2]k_s + [\bar{E}_{g_0} + 0.411k_s^2 + 0.377k_z^2][0.606k_s^2 + 0.722k_z^2]] \quad (2.286)$$

The $E - k_s$ relation in HD QWs of n-type $Pb_{1-x}Ge_xTe$ assumes the form

$$\left[\frac{2}{1+Erf\left(\frac{E}{\eta_g}\right)}\right]\theta_0(E,\eta_g)+\gamma_3(E,\eta_g)\left[\bar{E}_{g_0}-0.195k_s^2-0.345\left(\frac{n_z\pi}{d_z}\right)^2\right]$$
$$=\left[0.23k_s^2+0.02\left(\frac{n_z\pi}{d_z}\right)^2\pm\left[0.06\bar{E}_{g_0}+0.061k_s^2+0.0066\left(\frac{n_z\pi}{d_z}\right)^2k_s\right]\right.$$
$$\left.+\left[\bar{E}_{g_0}+0.411k_s^2+0.377\left(\frac{n_z\pi}{d_z}\right)^2\right]\left[0.606k_s^2+0.722\left(\frac{n_z\pi}{d_z}\right)^2\right]\right] \tag{2.287}$$

The sub band energy (E_{n_zHD400}) is the lowest positive root of the following equation

$$\left[\frac{2}{1+Erf\left(\frac{E_{n_zHD400}}{\eta_g}\right)}\right]\theta_0(E_{n_zHD400},\eta_g)+\gamma_3(E_{n_zHD400},\eta_g)\left[\bar{E}_{g_0}-0.345\left(\frac{n_z\pi}{d_z}\right)^2\right]$$
$$=\left[0.02\left(\frac{n_z\pi}{d_z}\right)^2\pm\left[\bar{E}_{g_0}+0.377\left(\frac{n_z\pi}{d_z}\right)^2\right]\left[0.722\left(\frac{n_z\pi}{d_z}\right)^2\right]\right] \tag{2.288}$$

The EEM and the DOS function for both the cases should be calculated numerically.

2.2.14 *The DR in Ultrathin Films (UFs) of HD Zinc and Cadmium Diphosphides*

The DR of the holes of Cadmium and Zinc diphosphides can approximately be written following Chuiko [112] as

$$E=\left[\beta_1+\frac{\beta_2\beta_3(k)}{8\beta_4}\right]k^2\pm\left\{\left[\beta_4\beta_3(k)\left(\beta_5-\frac{\beta_2\beta_3(k)}{8\beta_4}\right)k^2\right]\right.$$
$$\left.+8\beta_4^2\left(1-\frac{\beta_3^2(k)}{4}\right)-\beta_2\left(1-\frac{\beta_3^2(k)}{4}\right)k^2\right\}^{1/2} \tag{2.289}$$

where β_1,β_2,β_4 and β_5 are system constants and $\beta_3(k)=\frac{k_x^2+k_y^2-2k_z^2}{k^2}$.

Under the condition of formation of band tail, the above equation assumes the form

$$\gamma_3(E,\eta_g) = \left[\beta_1 + \frac{\beta_2\beta_3(k)}{8\beta_4}\right]k^2 \pm \left\{\left[\beta_4\beta_3(k)\left(\beta_5 - \frac{\beta_2\beta_3(k)}{8\beta_4}\right)k^2\right] + 8\beta_4^2\left(1 - \frac{\beta_3^2(k)}{4}\right) - \beta_2\left(1 - \frac{\beta_3^2(k)}{4}\right)k^2\right\}^{1/2} \quad (2.290)$$

The DR in HD QWs of Zinc and Cadmium diphosphides can be written as

$$\gamma_3(E,\eta_g) = \left[\beta_1 + \frac{\beta_2\beta_{31}(k)}{8\beta_4}\right]\left[k_x^2 + k_y^2 + \left(\frac{n_z\pi}{d_z}\right)^2\right] \pm \left\{\left[\beta_4\beta_{31}(k)\left(\beta_1 - \frac{\beta_2\beta_{31}(k)}{8\beta_4}\right)\right]\left[k_x^2 + k_y^2 + \left(\frac{n_z\pi}{d_z}\right)^2\right] + 8\beta_4^2\left(1 - \frac{\beta_{31}^2(k)}{4}\right) - \beta_2\left(1 - \frac{\beta_{31}^2(k)}{4}\right)\left[k_x^2 + k_y^2 + \left(\frac{n_z\pi}{d_z}\right)^2\right]\right\}^{1/2} \quad (2.291)$$

where

$$\beta_{31}(k) = \left[\frac{k_x^2 + k_y^2 - 2\left(\frac{n_z\pi}{d_z}\right)^2}{k_x^2 + k_y^2 + \left(\frac{n_z\pi}{d_z}\right)^2}\right]$$

The sub band energy $(E_{n_{zHD402}})$ is the lowest positive root of the following equation

$$\gamma_3(E_{n_{zHD402}},\eta_g) = \left[\beta_1 - \frac{\beta_2}{4\beta_4}\right]\left[\left(\frac{n_z\pi}{d_z}\right)^2\right] \pm \left\{\left[-2\beta_4\left[\beta_5 - \frac{\beta_2}{4\beta_4}\right]\left[\left(\frac{n_z\pi}{d_z}\right)^2\right]\right]\right\}^{1/2} \quad (2.292)$$

The EEM and the DOS function for both the cases should be calculated numerically.

Thus, we can summarize the whole mathematical background in the following way.

In this chapter, we have investigated the 3D and 2D DRs in HD bulk and QWs of non-linear optical materials on the basis of a newly formulated electron dispersion law considering the anisotropies of the effective electron masses, the spin orbit splitting constants and the influence of crystal field splitting within the framework of **k·p** formalism. The results for 3D and 2D DRs in HD bulk and QWs of III–V, ternary and quaternary compounds in accordance with the three and two band models of Kane form a special case of our generalized analysis. We have also studied the DR in accordance with the models of Stillman et al. and Palik et al. respectively since these models find use to describe the electron energy spectrum of the aforesaid materials. The 3D and 2DDRs has also been derived for HD bulk and QWs of II–VI, IV–VI, stressed materials, Te, n-GaP, p-$PtSb_2$, Bi_2Te_3, n-Ge and n-GaSb compounds by using the models of Hopfield, Dimmock, Bangert and Kastner, Seiler, Bouat and Thuillier, Rees, Emtage, Kohler, Cardona, Wang et al., Mathuret al., II–V, $Pb_{1-x}Ge_xTe$ and Zinc and Cadmium diphosphides respectively and transforming each on the basis of the appropriate carrier energy spectra. The well-known expressions of the DRs in the absence of band tails for wide gap materials have been obtained as special cases of our generalized analysis under certain limiting conditions. This indirect test not only exhibits the mathematical compatibility of our formulation but also shows the fact that our simple analysis is a more generalized one, since one can obtain the corresponding results for relatively wide gap materials having parabolic energy bands under certain limiting conditions from our present derivation.

2.3 Results and Discussion

Using the appropriate equations together with parameters as given in the Table 1.1, we have plotted the real part of the energy spectrum ($\mathrm{Re}[\theta_1(E,\eta_g)]$) as a function of election energy in Fig. 2.1a, b exhibits the dependence of the imaginary part of the energy spectrum Im $[\theta_1(E,\eta_g)]$ on electron energy for HD n-Cd_3As_2 (an example of tetragonal materials), respectively.

From Fig. 2.1a, it appears that ($\mathrm{Re}[\theta_1(E,\eta_g)]$) has an increasing trend with energy E for positive values of E. Besides for negative values of E, the value of ($\mathrm{Re}[\theta_1(E,\eta_g)]$) is positive indicating its band-tailing nature. Beyond $E = -1.0(\mathrm{eV})$, the value of ($\mathrm{Re}[\theta_1(E,\eta_g)]$) becomes negative and magnitude of the values are insignificant one. It is worth remarking that the band-tailing nature of ($\mathrm{Re}[\theta_1(E,\eta_g)]$) is clearly apparent from the Fig. 2.1a.

From Fig. 2.1b, we observe that Im $[\theta_1(E,\eta_g)]$ has the Gaussian nature of variation with energy E for both positive and negative values of E. The values of Im $[\theta_1(E,\eta_g)]$ are negative for all the values of E as considered in Fig. 2.1b. It may be remarked that the graph of Fig. 2.1b clearly shows the tailing of Im $[\theta_1(E,\eta_g)]$ into conduction band (i.e., for positive values of E) and the tailing within the spin-splitting band (i.e., for negative values of E) respectively. The maximum contribution of Im $[\theta_1(E,\eta_g)]$ appears at $E = -0.25$ (eV) for $\eta_g = 0.8$ (eV) which is beyond the band gap $E_g = 0.095$ (eV). From Fig. 2.2a, we observe that the $\mathrm{Re}[\theta_2(E,\eta_g)]$ has an increasing trend with positive value of E. For negative value of E, the $\mathrm{Re}[\theta_2(E,\eta_g)]$ becomes positive exhibiting clearly the band-tailing nature of it. Besides beyond $E = -1.0$ (eV), the value of $\mathrm{Re}[\theta_2(E,\eta_g)]$ becomes negative. In addition, the band-tailing nature of $\mathrm{Re}[\theta_2(E,\eta_g)]$ is clearly apparent from the Fig. 2.2a.

From Fig. 2.2b, we can write that the nature of variation of the Im $[\theta_2(E,\eta_g)]$ versus E is quite different from Im $[\theta_2(E,\eta_g)]$. Here, Im $[\theta_2(E,\eta_g)]$ has a positive alues with positive variations of E. Also, the positive values of Im $[\theta_2(E,\eta_g)]$ for negative E indicate the band-tailing nature. Beyond $E = -0.5$ (eV), Im $[\theta_2(E, \eta_g)]$ becomes negative and the magnitude of the negative values are significant. The nature of variation of Im $[\theta_2(E, \eta_g)]$ is not Gaussian type but can roughly be approximated to it. This occurs for $\delta = 0.085$ (eV) (i.e., positive value of δ). Because for $\delta = -0.21$ (eV) for $CdGeAs_2$, the nature of variation of Im $[\theta_2(E, \eta_g)]$ is Gaussian. The contribution of (Re$[\theta_2(E, \eta_g)]$) is of significant value as compared to Im $[\theta_2(E, \eta_g)]$. The band-tailing is clearly shown in the variation of Im $[\theta_2(E, \eta_g)]$.

From Fig. 2.3a, it appears that the nature of variation of the real part of the electron energy spectrum for $CdGeAs_2$ (an example of non-linear optical materials) is more or les same as that for the tetragonal materials as given by Fig. 2.1a. In this case, the band tailing nature of variation of Re$[\theta_1(E, \eta_g)]$ for $CdGeAs_2$ is clearly shown in the graph. From Fig. 2.3b, it appears that the nature of variation of the imaginary part of the electron energy spectrum for non-linear optical materials is more or less the same as that for the tetragonal material as given by Fig. 2.1b. The Gaussian distribution of Im $[\theta_1(E, \eta_g)]$ with respect of E is apparent from the graph and the band tailing is clearly shown.

From Fig. 2.4a, we can write that the band tailing effect is clearly shown in the graph. From Fig.2.4b, it appears that the variation of Im $[\theta_2(E, \eta_g)]$ with respect to E is a Gaussian type with negative From Fig. 2.5, we can write that the curve (a) is valid for $\delta \neq 0$ and $\Delta_{\|} \neq \Delta_{\perp}$ for the case of Cd_3As_2 and the curve (b) is valid for $\delta = 0$, $\Delta = (\Delta_{\|} + \Delta_{\perp})/2$ and $\Delta_{\|} = \Delta_{\perp}$ to obtain the corresponding three band model of Cd_3As_2. The curve (a) shows that the DOS increases with the increase in the positive values of E. The band tailing is clearly being observed from the graph. The variation of Im $[\theta_2(E, \eta_g)]$ with respect to $E$are unlike that with respect of Im $[\theta_2(E, \eta_g)]$ for Cd_3As_2 because of the negative value of $\delta(-0.21(\text{eV}))$ in $CdGeAs_2$. The curve (b) also shows the same nature and finally the curve (b) merges with the curve (a). For negative value of E, the curves (a) and (b) exhibit tailing in the DOS together with the oscillations. It is worth remarking in this context that at the value of E corresponding to point MN, curve (a) shows that the DOS becomes negative indicating the formation of a **new forbidden zone in the material**. The value of E at those points lie between -1.45 (eV) at M and -1.65 (eV) at N which corresponds to the region away or near to the spin-orbit splitting band of the material. Besides, beyond -1.65 (eV), the DOS becomes positive.

The oscillatory nature of the DOS for negative values of E has been predicted by the present theory. **Also for $\psi_{11}(E, \eta_g) \geq \pi$, the cosine function becomes negative indicating the negative value of the DOS and hence the creation of a new forbidden zone**. The curve (b) is widely separated from the curve (a) for negative values of E. Also, it shows oscillation in this region with two points, P and Q where the DOS for curve (b) shows negative value. For this value of the DOS, the new forbidden zone appears for the curve (b). The value of E at P is -1.45 (eV) and at Q is -1.58 (eV). Thereafter, the DOS again becomes positive. **The oscillatory nature of the DOS for the Curve (b), is also predicted by the present formulation**. For

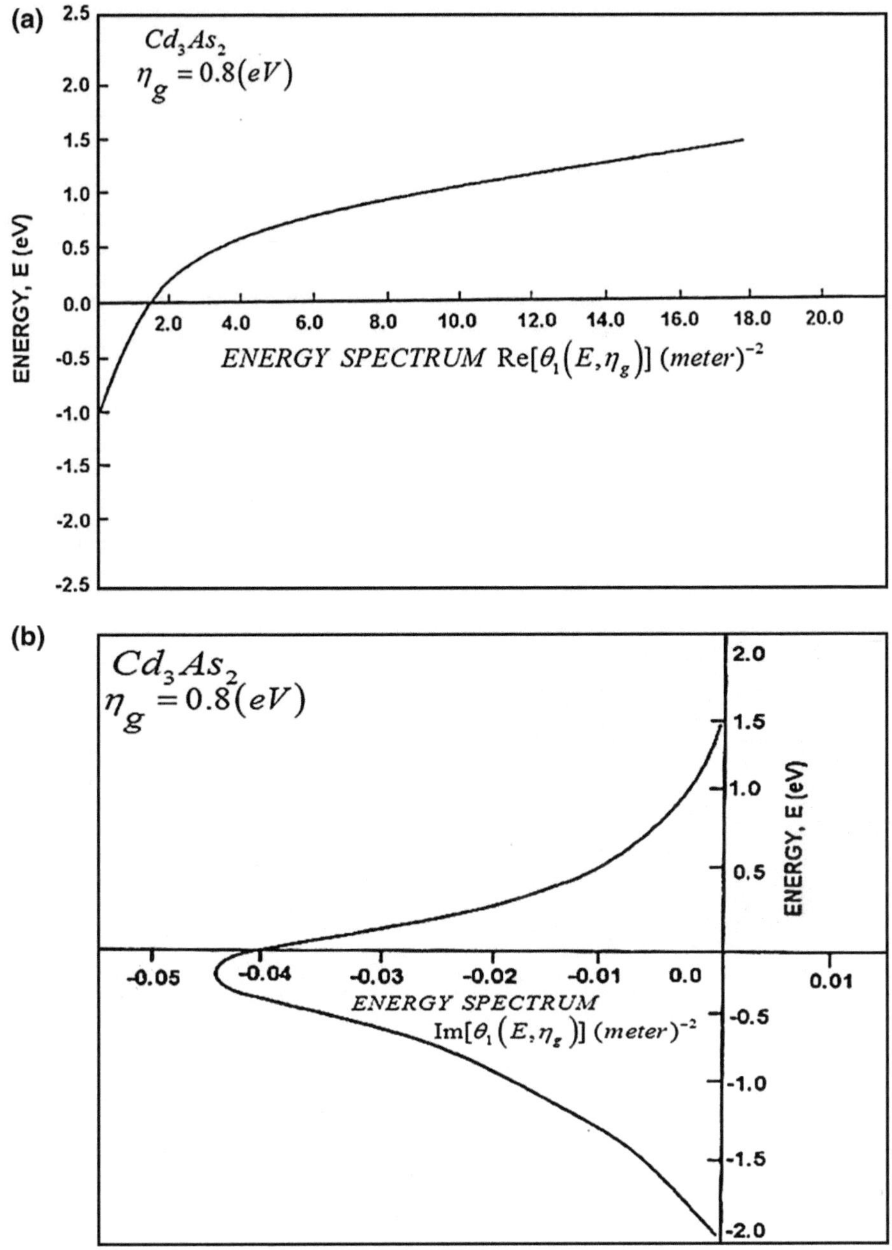

Fig. 2.1 **a** Plot of the electron energy spectrum of $\mathrm{Re}[\theta_1(E, \eta_g)]$ (in m^{-2}) versus energy, E (eV) for Cd_3As_2. **b** Plot of the electron energy spectrum of $\mathrm{Im}[\theta_1(E, \eta_g)]$ (in m^{-2}) versus energy, E (eV) for Cd_3As_2

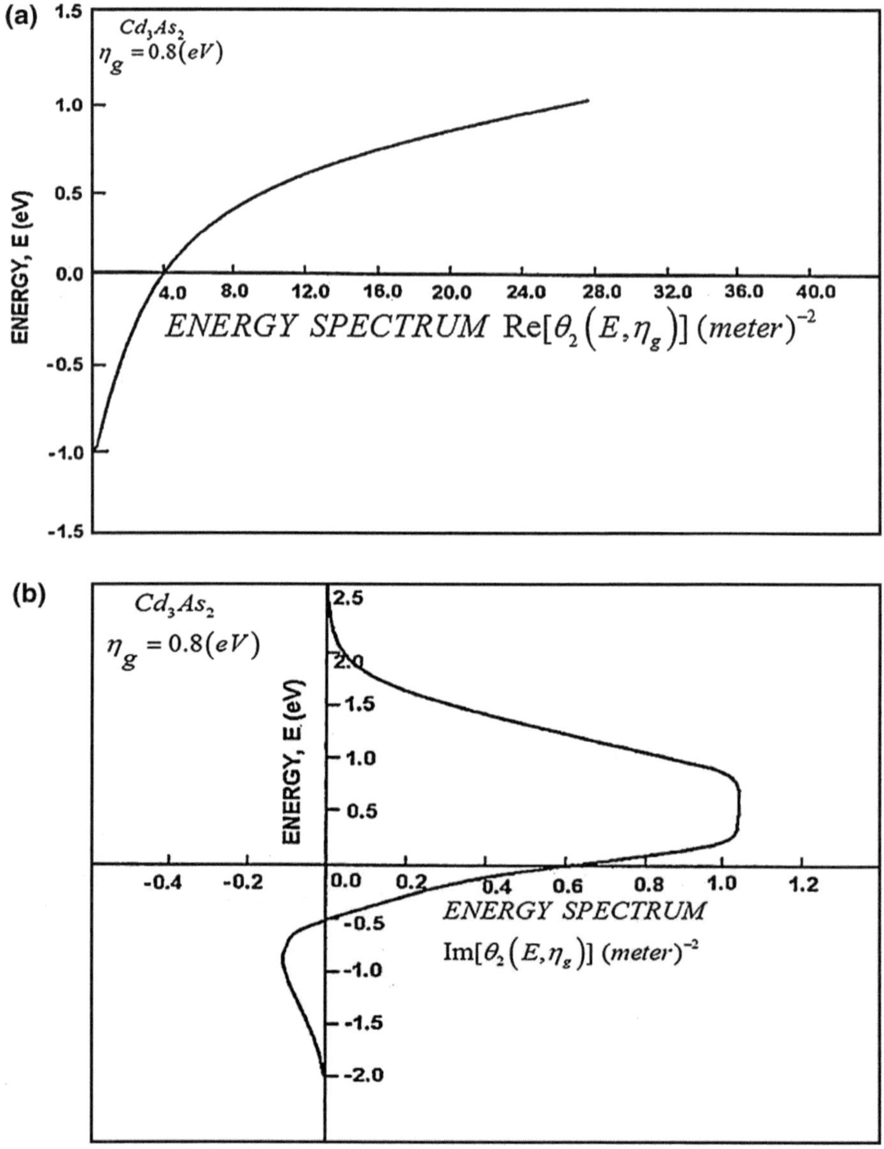

Fig. 2.2 **a** Plot of the electron energy spectrum of $\text{Re}[\theta_2(E,\eta_g)]$ (in m^{-2}) versus energy, E (eV) for Cd_3As_2. **b** Plot of the electron energy spectrum of $\text{Im}[\theta_2(E,\eta_g)]$ (in m^{-2}) versus energy, E (eV) for Cd_3As_2

the curve (b), the new forbidden zone (i.e., PQ) appears earlier than the same for the curve (a). The band tailing in the DOS for the curves (a) and (b) are clearly indicated in the graph for negative values of E. In the plot, the value of the DOS has been normalized by (1.0×10^{20}) factor.

From Fig. 2.6, it appears that both the curves (a) and (b) of the DOS increase with increase in E and finally the curve (b) converges with the curve (a). For negative value of E, the curve (a) exhibits oscillations with positive values of the DOS. Two oscillation peaks have been shown to appear over the region of study of energy, E. For the curve (a), the value of the DOS becomes negative as indicated by the region (X, Y), where the new forbidden zone has appeared. Besides, beyond the point Y, the DOS becomes positive with an oscillatory nature. The value of E at X is ≈ -1.68 (eV) and at Y is ≈ -1.8 (eV). The curve (a) is being widely separated from the curve (b) for the negative values of E. In addition, the appearance of oscillations with the new forbidden zone in the DOS has also been predicted by the theory. The curve (b) also shows oscillation in the DOS as indicated in the figure. Between the point (R, S), the curve (b) shows negative values of the DOS indicating the formation of a new forbidden zone for the three band Kane model representation of $CdGeAs_2$. The value of E at point R is E = −1.7(eV) and at point Sis E $\approx$ −1.8(eV). Besides beyond point S, the DOS becomes positive with oscillatory nature. The curve (b) is widely separated from the curve (a) for negative values of E. The band tailing in the DOS for the curves (a) and (b) is clearly indicated in the graph.

In Fig. 2.7, we have plotted the energy spectra of n-InSb where the graph 3_R indicates the real part $T_{31}(E, \eta_g)$ for the perturbed three-band model of Kane in which the curve 3_{Im} exhibits the imaginary part $T_{32}(E, \eta_g)$. The curve 3_{up} indicates the unperturbed three-band model of Kane. The curve (2) has been drawn for the perturbed two-band model of Kane in which 2up indicates the corresponding unperturbed DR). In Fig. 2.7, I indicate the perturbed parabolic band model and I_{up} exhibits the energy spectrum for unperturbed parabolic energy bands (i.e. $E = \hbar^2 k^2 / 2m_c$).

Using the other material constants form the Table of the Appendix we have plotted the DRs for n-InAs, $In_{1-x}Ga_xAs_yP_{1-y}$ lattice matched to InP and $Hg_{1-x}Cd_xTe$ for all the cases of Fig. 2.7 in Figs. 2.8, 2.9, 2.10, respectively. In Figs. 2.11, 2.12, 2.13 and 2.14 we have plotted the DOS function for all cases of Fig. 2.7 for n-InSb, n-InAs, $In_{1-x}Ga_xAs_yP_{1-y}$ lattice matched to InP and $Hg_{1-x}Cd_xTe$, respectively. In this context, it may be noted that we have taken the first fifteen terms of the finite series in $T_{31}(E, \eta_g)$ for the purpose of numerical evaluations since we have observed that the contributions of the higher terms in the infinite series of $T_{31}(E, \eta_g)$ become negligible after it.

This statement is valid for all the cases in general. From Figs. 2.7, 2.8, 2.9, 2.10, respectively. In Figs. 2.11, 2.12, 2.13 and 2.14, the following points can be noted:

1. When $E_g < \Delta$ (e.g. InSb and $Hg_{1-x}Cd_xTe$), the imaginary part of the energy spectrum (3_{Im}) is most prominent as compared to $In_{1-x}Ga_xAs_yP_{1-y}$ lattice matched to InP where $E_g < \Delta$. The imaginary part of energy spectrum enters in the conduction band where $E > 0$ (i.e. tails in the conduction band) as compared to the cases of n-InAs and $In_{1-x}Ga_xAs_yP_{1-y}$ lattice matched to InP where the tails are very small.

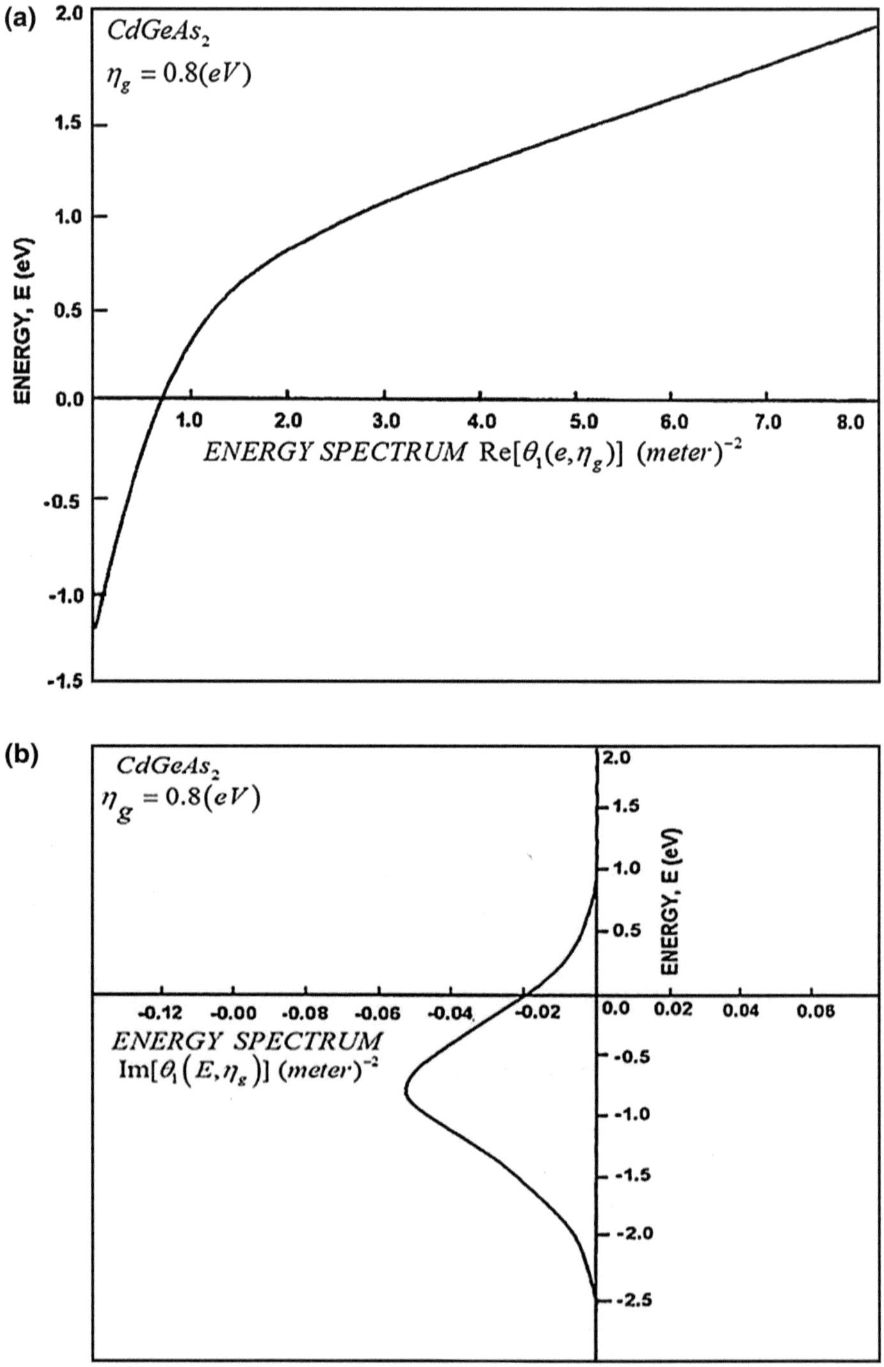
(a)
CdGeAs2
ηg = 0.8(eV)
ENERGY, E (eV)
2.0
1.5
1.0
0.5
0.0
-0.5
-1.0
-1.5
1.0
2.0
3.0
4.0
5.0
6.0
7.0
8.0
ENERGY SPECTRUM Re[θ1(e,ηg)] (meter)-2
(b)
CdGeAs2
ηg = 0.8(eV)
2.0
1.5
1.0
0.5
-0.5
-1.0
-1.5
-2.0
-2.5
ENERGY, E (eV)
-0.12
-0.00
-0.08
-0.06
-0.04
-0.02
0.0
0.02
0.04
0.06
ENERGY SPECTRUM
Im[θ1(E,ηg)] (meter)-2

◀ **Fig. 2.3** **a** Plot of the electron energy spectrum of $\mathrm{Re}[\theta_1(E,\eta_g)]$ (in m^{-2}) versus energy, E (eV) for $CdGeAs_2$. **b** Plot of the electron energy spectrum of $\mathrm{Im}[\theta_1(E,\eta_g)]$ (in m^{-2}) versus energy, E (eV) for $CdGeAs_2$

2. When the imaginary part (3_{Im}) is prominent (Figs. 2.1 and 2.3), the tails of the real part is shortened. The curves 3_{Im} exhibits tail of the conduction band and with this tail, the imaginary band enters into the region of conduction band ($E > 0$). For $Hg_{1-x}Cd_xTe$ and n-InSb, the 3_{Im} tails into the split-off band.
3. When the DRs of the conduction electrons of the materials are defined by the perturbed two-band model of Kane, the imaginary part of the energy spectrum vanishes. The same is also true for perturbed parabolic band model. The unperturbed bands never exhibit tails in the energy spectrum.
4. The $N_{HD}(E,\eta_g)$ as given by (1.43) shows oscillations with E for $E < 0$. The oscillatory part is not seen in Fig. 2.11 for $E > 0$. This is because, for $E<0, |T_{32}(E,\eta_g) \ll T_{31}(E,\eta_g)|$. So $\vartheta_{21}(E,\eta_g)$ is equal to zero leading to the non-oscillatory result $\cos\vartheta_{21}(E,\eta_g)$ for $E > 0$. For $E<0$, the value of $\vartheta_{21}(E,\eta_g)$ is significant and oscillations are found for $N_{HD}(E,\eta_g)$ for $E > 0$, as evident from Fig. 2.11. For $\vartheta_{21}(E,\eta_g) \geq \pi$, $\cos\vartheta_{21}(E,\eta_g)$ becomes negative leading to the negative values of the DOS. The boundary points M and N in the graph mark the points where the DOS become negative. The electrons cannot exist for negative values of the DOS so this region is forbidden for electrons, which implies that, in the band-tails, there appears a new forbidden zone in addition to the normal band-gap of the semiconductor. It appears beyond the spin-orbit splitting band. No oscillations are found for the perturbed two band model of Kane and perturbed parabolic energy bands respectively although there is tailing in the DOS. For $\eta_g \to 0$, the oscillations and the appearance of new band gap in the tailed zone of the DOS are absent.

Form the above discussion, we observed that $T_{31}(E,\eta_g)$ has a tail in the forbidden band and extended further. As this tail crosses the forbidden band (when $T_{31}(E,\eta_g)T_{31}(E,\eta_g)$ does not vanish in the forbidden band for small E_g) and enters into the split-off band where Δ exists and in the split-off band the tail vanishes, leaving behind some part of it to cross; then the remaining part of the split-off band interacts with the impurity atoms of the doped materials to produce a complex band. This interaction produces complex energy spectrum in the heavily doped materials whose unperturbed conduction electron are defined by the three-band model of Kane. Obviously for $E_g \ll \Delta$, the imaginary part of the complex energy spectrum will be most prominent and depth of the tail will be small. In the case when $E_g \gg \Delta$, the tails due to $T_{31}(E,\eta_g)$ (i.e. the real part) lies almost within the forbidden region; only a very thin tail enters into the spin-split off band. This causes a prominent tail of $T_{31}(E,\eta_g)$ and imaginary part is slightly present. For $E_g \approx \Delta$ (e.g., n-InAs shown in Fig. 2.8), the tail of $T_{31}(E,\eta_g)$ is present considerably in the forbidden band, but $T_{31}(E,\eta_g)$ does not vanishes to zero at the edge of the

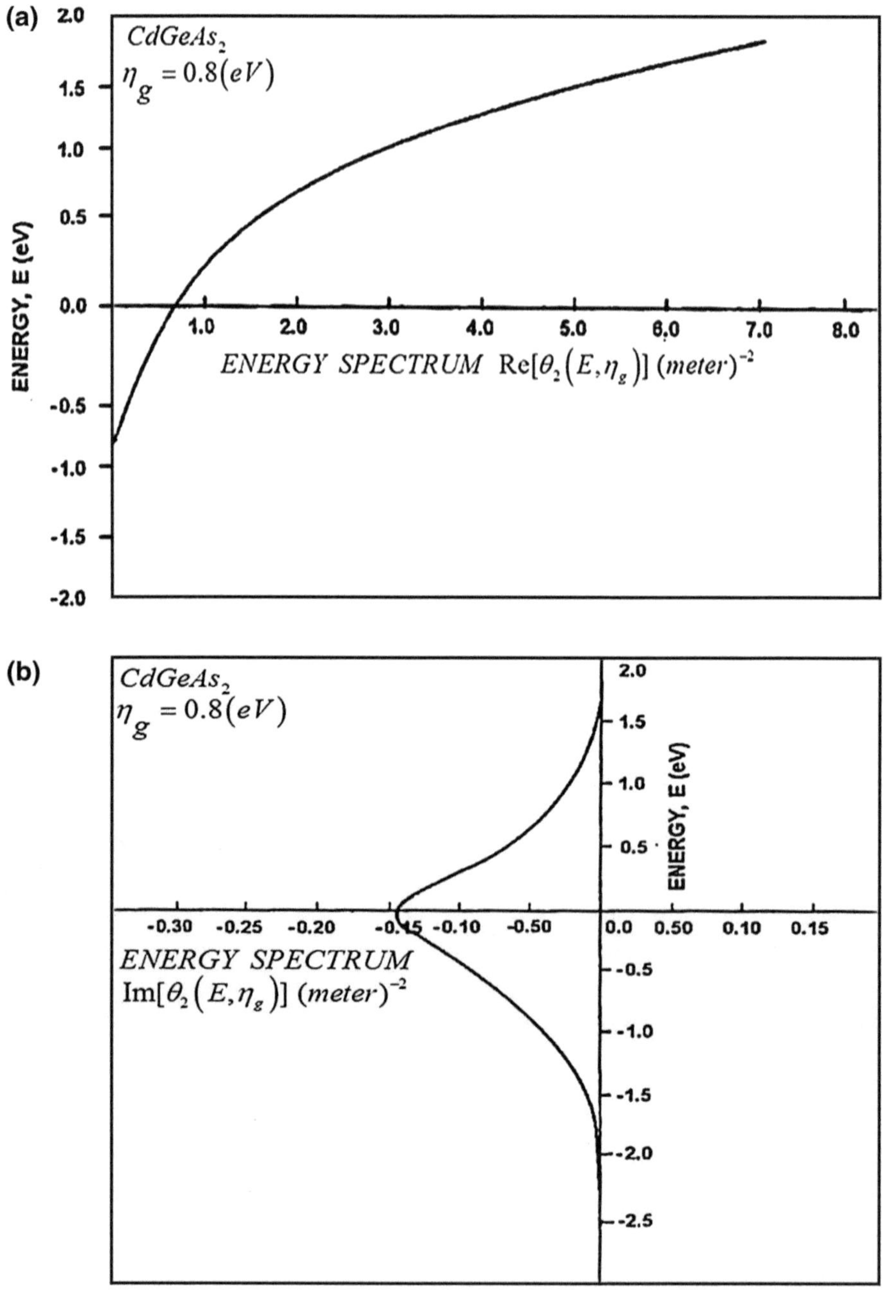

Fig. 2.4 **a** Plot of the electron energy spectrum of $\mathrm{Re}[\theta_2(E,\eta_g)]$ (in m^{-2}) versus energy, E (eV) for $CdGeAs_2$. **b** Plot of the electron energy spectrum of $\mathrm{Im}[\theta_2(E,\eta_g)]$ (in m^{-2}) versus energy, E (eV) for $CdGeAs_2$

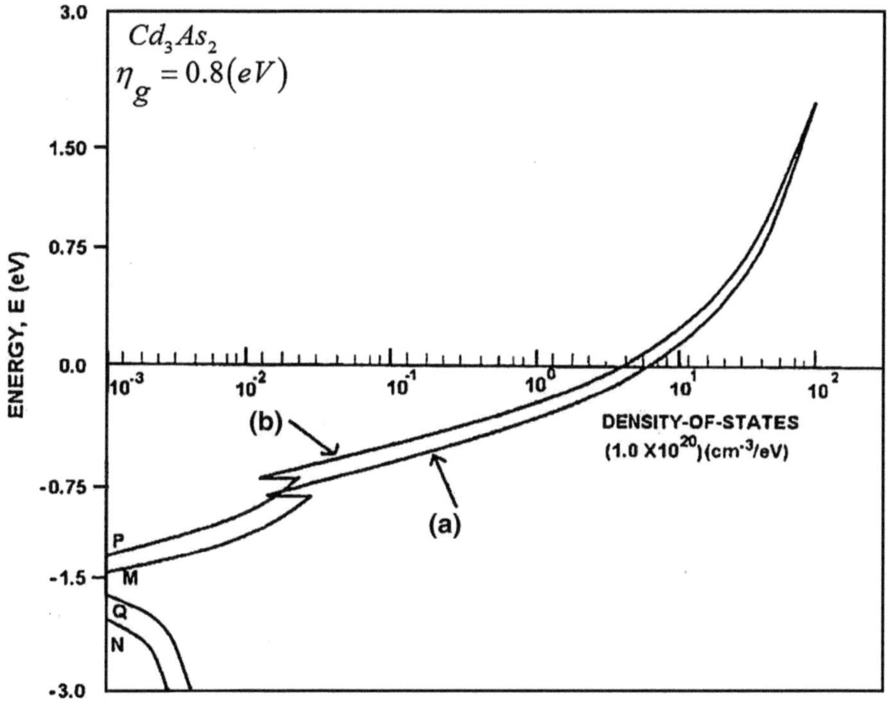

Fig. 2.5 The plots of the DOS, (cm^{-3}/eV) (normalized by 1.0×10^{20}) versus Energy, *E* (eV) for Cd_3As_2 (the *curve* (**a**)), and also for the HD three band Kane model of Cd_3As_2 (*curve* (**b**)) respectively

light-hole valance band where forbidden band ends; rather enters into the split-off band. In the split-off band (when Δ is comparable to E_g), the tail covers the band considerably. So the tail due to $T_{31}(E, \eta_g)$ and the imaginary energy spectrum due for to $T_{32}(E, \eta_g)$ are present prominently.

So from Figs. 2.7, 2.8, 2.9 and 2.10, we observe that the complex energy spectrum is due to the interaction of the impurity atoms with the spin orbit splitting constant of the valance band for the three-band model of Kane where no real energy band as well as impurity band exist. More is the interaction (depends on the cross over region of Δ by the tail) causes more prominence of the imaginary part than the other case. Under undoped condition the band-tailing vanishes and there is no interaction with the splitting band. As a result, there exist no complex energy spectrum i.e. $T_{32}(E, \eta_g)$ approaches to zero as $\eta_g \to 0$.

From the 2D DR in QWs of HD nonlinear optical and tetragonal materials (2.28), we observe that constant energy 2D wave vector surfaces are the series of concentric quantized circles in the complex energy plane which is the consequence of non-removable poles in the corresponding DR in the absence of band tails. The (2.140) represents the 2D DR of HD IV–VI materials in accordance with the model

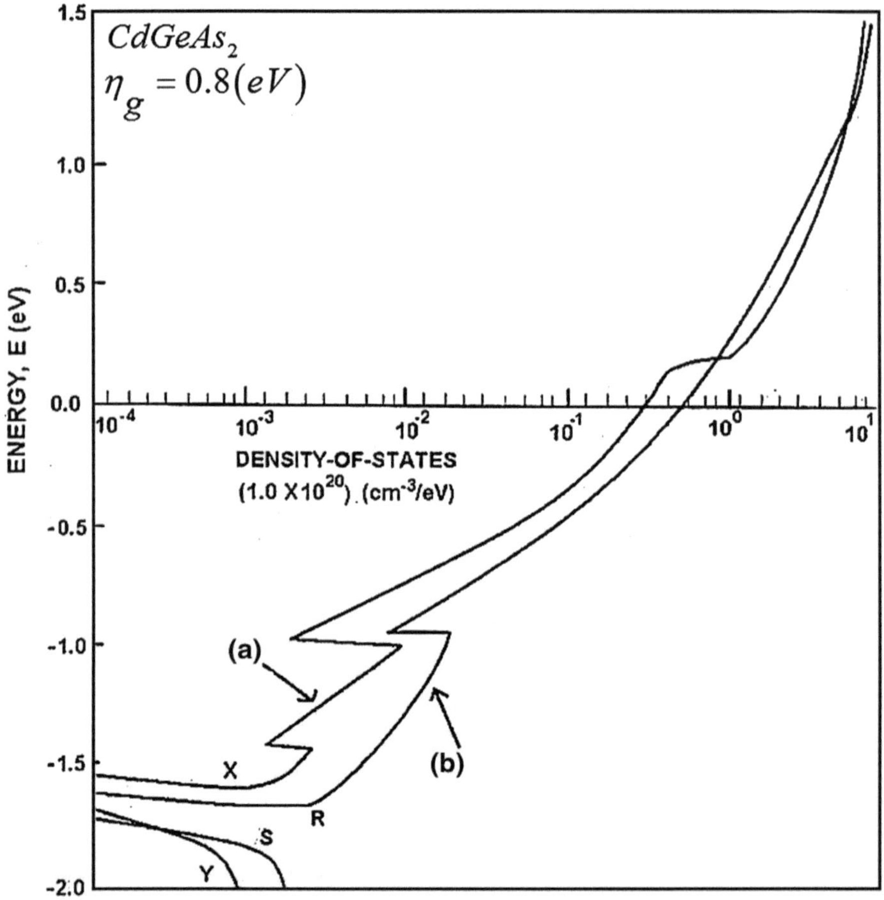

Fig. 2.6 The plots of the DOS, (cm^{-3}/eV) (normalized by 1.0×10^{20}) versus energy, E (eV) for CdGeAs2 (the *curve* (**a**)) and also for the HD three band Kane model of $CdGeAs_2$ (*curve* (**b**)) respectively

of Bangert and Kastner and same conclusion is also valid. From (2.41) we have the same inference for QWs of HD III–V materials whose unperturbed conduction electrons obey the three band model of Kane, which contains one non-removal pole in energy axis. The 3D electrons in HD III–V materials are also described by two band model of Kane, parabolic energy bands, model of Stillman et al. and the model of Palik et al. with the 2D DRs as given by (2.55), (2.68), (2.79) and (2.92) respectively. Besides the 2D DRs of Te, GaP, $PtSb_2$, Bi_2Te_3, Ge and GaSb are given by (2.288), (2.202), (2.217), (2.229), (2.241, in accordance with the model of Cardona et al. of Ge), (2.256, in accordance with the model of Wang and Ressler of Ge) and (2.223) respectively. Since all the said DRs possess no poles in the finite energy planes, the constant energy of 2D wave vector surfaces are the series of concentric quantized circles in the real plane instead of the complex one. The 2D

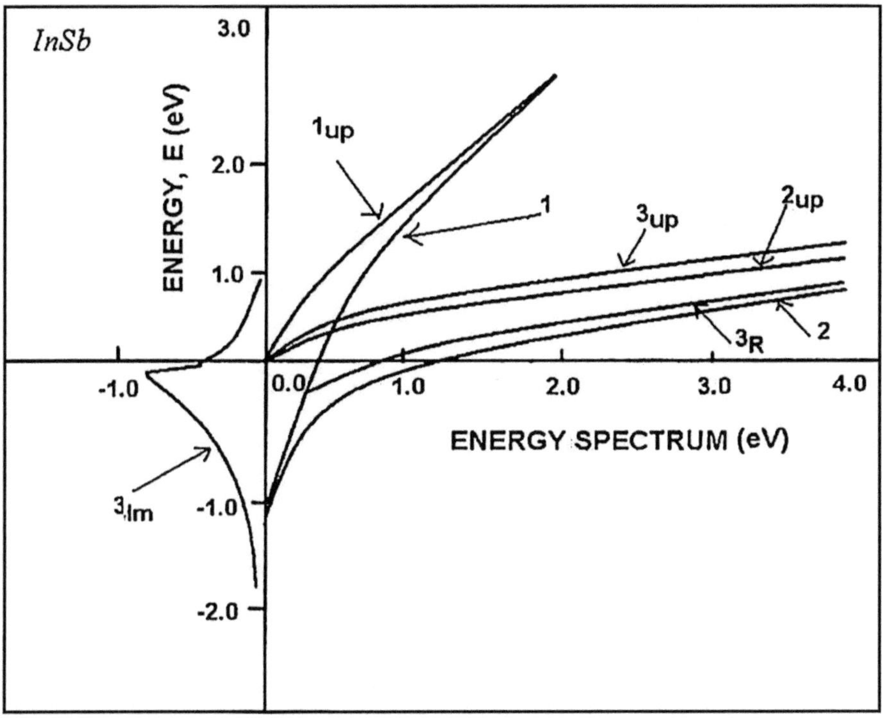

Fig. 2.7 The energy spectrum of InSb has been plotted with the following notations: **a** 3_R indicates the real part $T_{31}(E, \eta_g)$ for the perturbed three-band model of Kane; **b** 3_{Im} exhibits the imaginary part $T_{32}(E, \eta_g)$; **c** 3_{up} indicates the unperturbed three-band model of Kane; **d** (2) has been drawn for the perturbed two-band model of Kane; **e** 2_{up} indicates the corresponding unperturbed DR; **f** I indicates the perturbed parabolic band model; **g** 1_{up} exhibits the energy spectrum for unperturbed parabolic energy bands (i.e. $E = \hbar k^2/2m_c$)

DR (2.104) in HD II–VI materials reflects the fact that the constant energy 2D surface is series of concentric displaced quantized circles in the real plane. The 2D DR (2.121) of HD IV–VI materials represents constant energy 2D wave vector surface as the series of concentric quantized closed surfaces in accordance with Dimmock model. The 2D DRs for II–V, $Pb_{1-x}Ge_xTe$, ZnP_2 and CdP_2 are given by (2.283), (2.287) and (2.291) respectively and the aforementioned conclusion is also true in this case. The 2D DR (2.273) in QWs of HD stressed Kane type materials reflects the fact that the constant energy 2D wave vector surfaces are the series of concentric ellipses in the real plane.

The influence of quantum confinement is immediately apparent from the said 2D equations since the DR depends strongly on the thickness of the quantum-confined materials in contrast with the corresponding bulk specimens. The energy decreases with increasing film thickness in an quantized way with different numerical magnitudes for QWs of HD materials. It appears that any electronic property exhibits spikes for particular values of film thickness which, in turn, depends on the

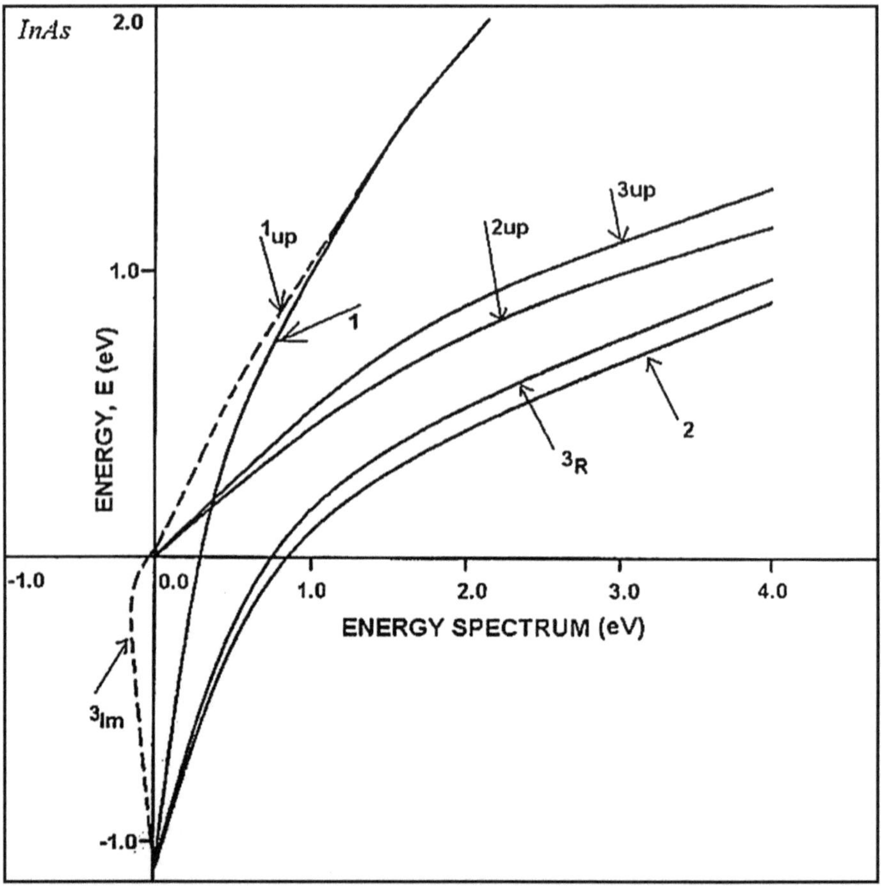

Fig. 2.8 The energy spectrum of InAs has been plotted for all the cases of Fig. 2.7

particular band structure of the specific material. Moreover, the electron energy in QWs of HD compounds can become several orders of magnitude larger than of bulk specimens of the same HD materials, which is also a direct signature of quantum confinement.

It may be noted that with the advent of MBE and other experimental techniques, it is possible to fabricate quantum-confined structures with an almost defect-free surface. If the direction normal to the film was taken differently from that as assumed in this work, the expressions for the DR for QWs of HD materials would be different analytically, since the basic DRs for many materials are anisotropic. In formulating the generalized electron energy spectrum for non-linear optical materials, we have considered the crystal-field splitting parameter, the anisotropies in the momentum-matrix elements, and the spin-orbit splitting parameters, respectively. In the absence of the crystal field splitting parameter together with the assumptions of isotropic effective electron mass and isotropic spin orbit splitting, our basic relation

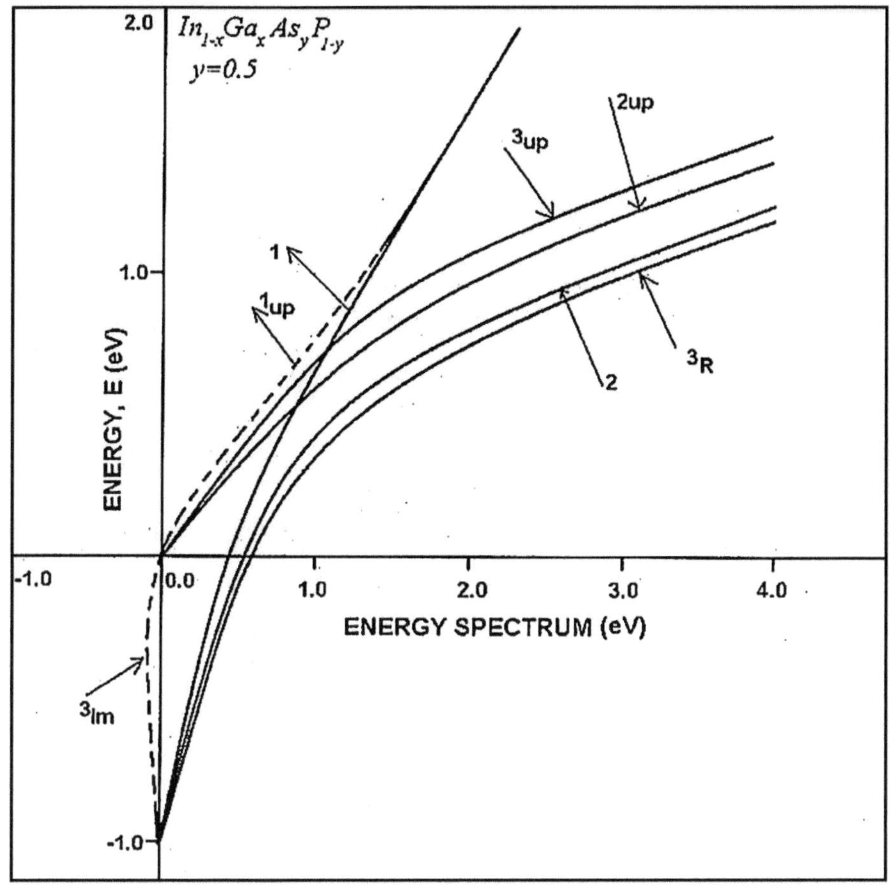

Fig. 2.9 The energy spectrum of $In_{1-x}Ga_xAs_yP_{1-y}$ lattice matched to InPhas been plotted for all the cases of Fig. 2.7

as given by (1.2) converts into the well-known three-band Kane model and is valid for III–V compounds, in general. It should be used as such for studying the electronic properties of n-InAs where the spin-orbit splitting parameter (Δ) is of the order of band gap (E_g). For many important materials $\Delta \gg E_g$ and under this inequality, the three band model of Kane assumes the form $E(1 + EE_g^{-1}) = \hbar^2k^2/2m_c$ which is the well-known two-band Kane model. Also under the condition, $E_g \to \infty$, the above equation gets simplified to the well-known form of parabolic energy bands as $E = \hbar^2k^2/2m_c$. It is important to note that under certain limiting conditions, all the results for all the models as derived here have transformed into the well-known expression of the DR for size quantized materials having parabolic bands. We have not considered other types of compounds or

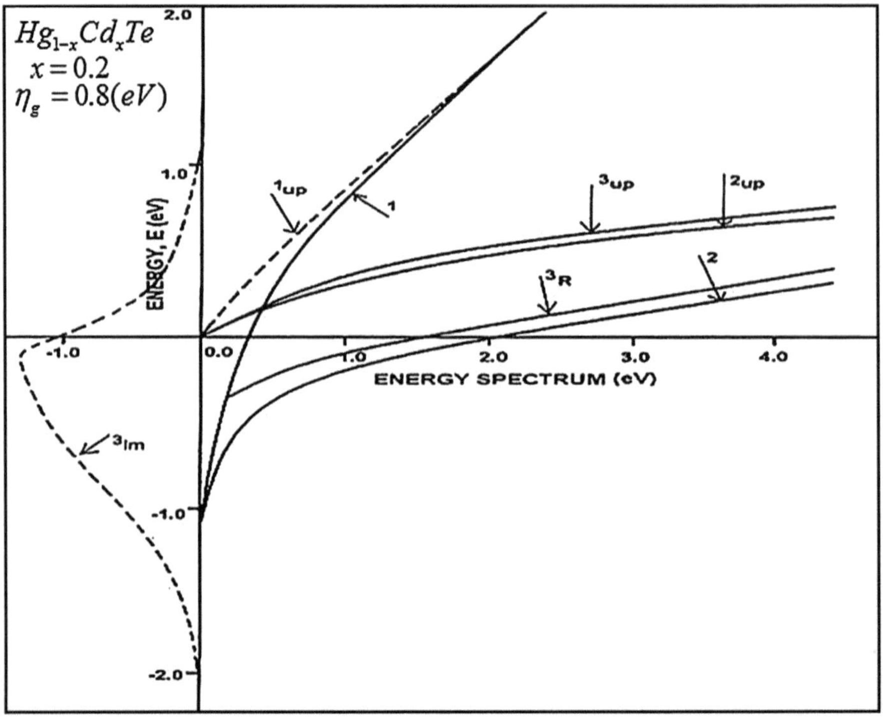

Fig. 2.10 The energy spectrum of $Hg_{1-x}Cd_xTe$has been plotted for all the cases of Fig. 2.7

external physical variables for numerical computations in order to keep the presentation brief.

It may be noted that the complex band structures have already been studied for bulk materials and super lattices without heavy doping [113, 114] and bear no relationship with the complex energy spectrum as formulated in this chapter. The physical picture behind the formulation of complex energy spectrum in doped small gap materials, whose unperturbed conduction band is defined by the three band model of Kane, is the interaction of the impurity atoms in the tails with the spin-orbit splitting constant of the valence band as already noted. Besides, the complex spectra are not related to same evanescent modes (extinction modes) in the band tails and the conduction band. In this context, we wish to further note that many band tails models are proposed using the Gaussian distribution of the impurity potential variation. In this chapter we have used the Gaussian distribution function of the impurity potential and obtained an exact DR for heavily doped nonlinear optical and tetragonal compounds and other materials forming band tails. Our method is not at all related with the DOS technique as used in the aforementioned works. From the DR we can obtain the DOS but the DOS technique as used in the literature [113, 114] cannot provide the DR. Therefore our study is more fundamental than those in the existing literature because the Boltzmann transport

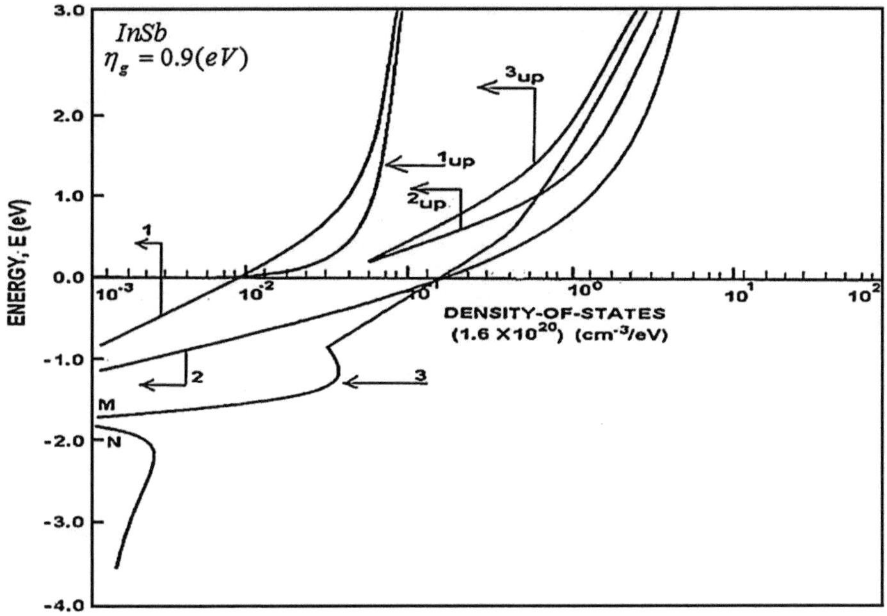

Fig. 2.11 The DOS functions have been plotted for all cases of Fig. 2.7 for n-InSb

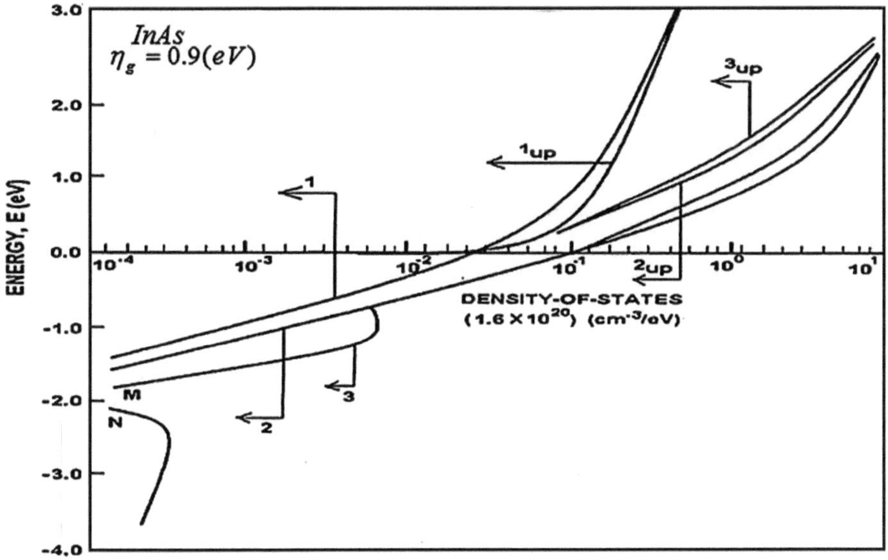

Fig. 2.12 The DOS functions have been plotted for all cases of Fig. 2.8 for n-InAs

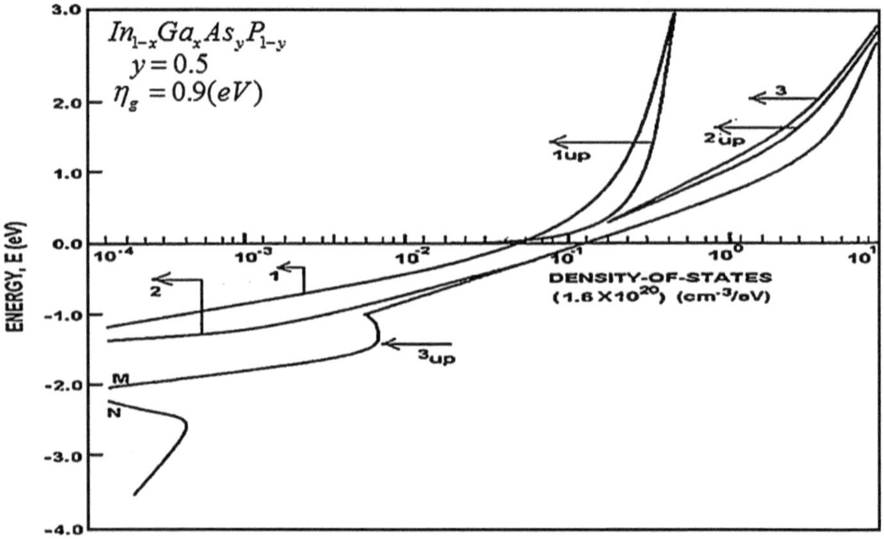

Fig. 2.13 The DOS functions have been plotted for all cases of Fig. 2.9 for $In_{1-x}Ga_xAs_yP_{1-y}$ lattice matched to InP

equation, which controls the study of the charge transport properties of semiconductor devices, can be solved if and only if the DR is known. In this context it may be noted that although we have used the Gaussian model to explain the band tailing in heavily doped materials but it is not the only well-established one.

It may be noted that the presence of non-removable poles in the DR of the undoped material creates the complex energy spectrum of the corresponding heavily doped sample. All investigations of the transport properties of modern electronic devices made of heavily doped materials should be reformulated since the Boltzmann transport equation which controls all the transport properties should be solved at first for complex energy spectrum which is altogether a new field of research. Consequently, all the band structure dependent properties of all the electronic devices made of heavily doped materials will change leading to new physical ideas and new experimental findings under different physical conditions. We have not considered the many body effects in this simplified theoretical formalism due to the lack of availability in the literature of proper analytical techniques for including them in the generalized system as considered in this chapter. Our simplified approach will be useful for the purpose of comparison when methods of tackling the formidable problem after inclusion of the many body effects for the generalized systems appear. It is worth remarking in this context that the results of our simple theory, in the limit the band gap tends to infinity, get transformed to the well-known formulation for wide gap materials having parabolic energy bands. This indirect test not only exhibits the mathematical compatibility of our formulation but also shows the fact that our simple analysis is a more generalized one, since one can obtain the corresponding results for the relatively wide

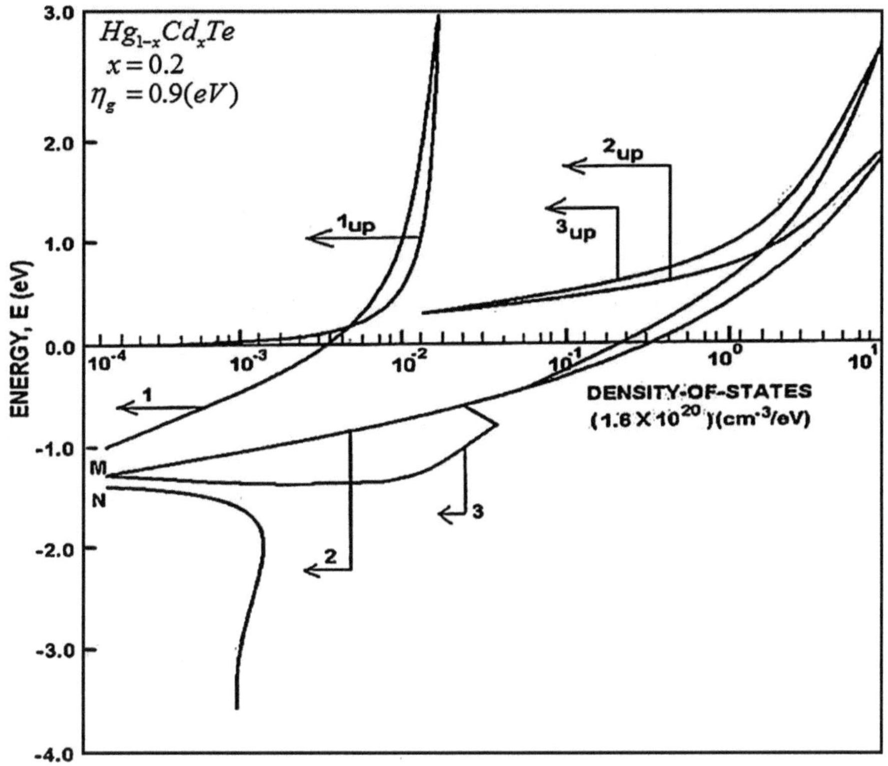

Fig. 2.14 The DOS functions have been plotted for all cases of Fig. 2.7 for $Hg_{1-x}Cd_xTe$

gap materials having parabolic energy bands under certain limiting conditions from our present derivation. The experimental results for the verification of the theoretical analyses of this chapter are still not available in the literature. It may be noted in this context that our theoretical formulation will be useful to analyze the experimental data when they appear. The inclusion of the said effect would certainly increase the accuracy of the results, although the qualitative features of the DOS discussed in this chapter would not change in the presence of the aforementioned effect. An important feature of the present work is that the influence of the energy band parameters on the DR and the DOS can be determined for various types of HD materials as considered in this chapter.

2.4 Open Research Problems

(R.2.1) Investigate the DR for bulk specimens of the HD materials in the presences of exponential, Kane, Halperian, Lax and Bonch-Burevich types of band tails [115] for all systems whose unperturbed carrier energy spectra are defined in this chapter.

(R.2.2) Investigate the DR for QWs of all the HD materials as considered in R.2.1.

(R.2.3) Investigate the DR for HD bulk specimens of the negative refractive index, organic, magnetic and other advanced optical materials in the presence of an arbitrarily oriented alternating electric field.

(R.2.5) Investigate the DR for the QWs of HD negative refractive index, organic, magnetic and other advanced optical materials in the presence of an arbitrarily oriented alternating electric field.

(R.2.6) Investigate the DR for the multiple QWs of HD materials whose unperturbed carrier energy spectra are defined in R.2.1.

(R.2.7) Investigate the DR for all the appropriate HD low dimensional systems of this chapter in the presence of finite potential wells.

(R.2.8) Investigate the DR for all the appropriate HD low dimensional systems of this chapter in the presence of parabolic potential wells.

(R.2.9) Investigate the DR for all the appropriate HD systems of this chapter forming quantum rings.

(R.2.10) Investigate the DR for all the above appropriate problems in the presence of elliptical Hill and quantum square rings.

(R.2.11) Investigate the DRfor triangular two dimensional systems in the presence of an arbitrarily oriented alternating electric field for all the HD materials whose unperturbed carrier energy spectra are defined in R.2.1.

(R.2.12) Investigate the DR for HD two dimensional systems of the negative refractive index and other advanced optical materials in the presence of an arbitrarily oriented alternating electric field and non-uniform light waves.

(R.2.13) Investigate the DR for triangular HD two dimensional systems of the negative refractive index, organic, magnetic and other advanced optical materials in the presence of an arbitrarily oriented alternating electric field in the presence of strain.

(R.2.14) (a) Investigate the DR for HD two dimensional systems of the negative refractive index, organic, magnetic and other advanced optical materials in the presence of many body effects

(b) Investigate all the appropriate problems of this chapter for a Dirac electron.

(R.2.15) Investigate all the appropriate problems of this chapter by including the many body, image force, broadening and hot carrier effects respectively.

(R.2.16) Investigate all the appropriate problems of this chapter by removing all the mathematical approximations and establishing the respective appropriate uniqueness conditions.

References

1. J.W. Rowe, J.L. Shay, Phys. Rev. B **3**, 451 (1973)
2. H. Kildal, Phys. Rev. B **10**, 5082 (1974)
3. J. Bodnar, in *Proceedings of the International Conference on Physics of Narrow-gap Semiconductors* (Polish Science Publishers, Warsaw, 1978)
4. G.P. Chuiko, N.N. Chuiko, Sov. Phys. Semicond. **15**, 739 (1981)
5. K.P. Ghatak, S.N. Biswas, Proc. SPIE **1484**, 149 (1991)
6. J.A. Zapien, Y.K. Liu, Y.Y. Shan, H. Tang, C.S. Lee, S.T. Lee, Appl. Phys. Lett. **90**, 213114 (2007)
7. M. Park, Proc. SPIE **2524**, 142 (1995)
8. S.-G. Hur, E.T. Kim, J.H. Lee, G.H. Kim, S.G. Yoon, Electrochem. Solid-State Lett. **11**, H176 (2008)
9. H. Kroemer, Rev. Mod. Phys. **73**, 783 (2001)
10. T. Nguyen Duy, J. Meslage, G. Pichard, J. Crys. Growth **72**, 490 (1985)
11. T. Aramoto, F. Adurodija, Y. Nishiyama, T. Arita, A. Hanafusa, K. Omura, A. Morita, Solar Energy Mater. Solar Cells **75**, 211 (2003)
12. H.B. Barber, J. Electron. Mater. **25**, 1232 (1996)
13. S. Taniguchi, T. Hino, S. Itoh, K. Nakano, N. Nakayama, A. Ishibashi, M. Ikeda, Electron. Lett. **32**, 552 (1996)
14. J.J. Hopfield, J. Appl. Phys. **32**, 2277 (1961)
15. K.P. Ghatak, J.Y. Siddiqui, B. Nag, Phys. Lett. A **282**, 428 (2001)
16. K.P. Ghatak, S. Karmakar, D. De, S. Pahari, S.K. Charaborty, S.K. Biswas, S. Chowdhury, J. Comput. Theor. Nanosci. **3**, 153 (2006)
17. K.P. Ghatak, M. Mondal, J. Appl. Phys. **66**, 3056 (1989)
18. K.P. Ghatak, M. Mondal, Thin Solid Films **148**, 219 (1987)
19. K.P. Ghatak, P.K. Bose, S. Bhattacharya, A. Bhattacharjee, D. De, S. Ghosh, S. Debbarma, N. Paitya, Quantum Matter **2**, 83 (2013)
20. A. Sinha, A.K. Sharma, R. Barui, A.R. Ghatak, S. Bhattacharya, K.P. Ghatak, Physica B **391**, 141 (2007)
21. G.P. Agrawal, N.K. Dutta, *Semicond. Lasers* (Van Nostrand Reinhold, New York, 1993)
22. S. Chatterjee, U. Pal, Opt. Eng. (Bellingham) **32**, 2923 (1993)
23. T.K. Chaudhuri, Int. J. Energy Res. **16**, 481 (1992)
24. J.H. Dughaish, Phys. B **322**, 205 (2002)
25. C. Wood, Rep. Prog. Phys. **51**, 459 (1988)
26. K.-F. Hsu, S. Loo, F. Guo, W. Chen, J.S. Dyck, C. Uher, T. Hogan, E.K. Polychroniadis, M. G. Kanatzidis, Science **303**, 818 (2004)
27. J. Androulakis, K.F. Hsu, R. Pcionek, H. Kong, C. Uher, J.J. D'Angelo, A. Downey, T. Hogan, M.G. Kanatzidis, Adv. Mater. **18**, 1170 (2006)
28. P.F.P. Poudeu, J. D'Angelo, A.D. Downey, J.L. Short, T.P. Hogan, M.G. Kanatzidis, Angew. Chem. Int. Ed. **45**, 3835 (2006)
29. P.F. Poudeu, J. D'Angelo, H. Kong, A. Downey, J.L. Short, R. Pcionek, T.P. Hogan, C. Uher, M.G. Kanatzidis, J. Am. Chem. Soc. **128**, 14347 (2006)
30. J.R. Sootsman, R.J. Pcionek, H. Kong, C. Uher, M.G. Kanatzidis, Chem. Mater. **18**, 4993 (2006)
31. A.J. Mountvala, G. Abowitz, J. Am. Ceram. Soc. **48**, 651 (1965)
32. E.I. Rogacheva, I.M. Krivulkin, O.N. Nashchekina, AYu. Sipatov, V.A. Volobuev, M.S. Dresselhaus. Appl. Phys. Lett. **78**, 3238 (2001)
33. H.S. Lee, B. Cheong, T.S. Lee, K.S. Lee, W.M. Kim, J.W. Lee, S.H. Cho, J.Y. Huh, Appl. Phys. Lett. **85**, 2782 (2004)
34. K. Kishimoto, M. Tsukamoto, T. Koyanagi, J. Appl. Phys. **92**, 5331 (2002)
35. E.I. Rogacheva, O.N. Nashchekina, S.N. Grigorov, M.A. Us, M.S. Dresselhaus, S.B. Cronin, Nanotechnology **14**, 53 (2003)

36. E.I. Rogacheva, O.N. Nashchekina, A.V. Meriuts, S.G. Lyubchenko, M.S. Dresselhaus, G. Dresselhaus, Appl. Phys. Lett. **86**, 063103 (2005)
37. E.I. Rogacheva, S.N. Grigorov, O.N. Nashchekina, T.V. Tavrina, S.G. Lyubchenko, AYu. Sipatov, V.V. Volobuev, A.G. Fedorov, M.S. Dresselhaus. Thin Solid Films **493**, 41 (2005)
38. X. Qiu, Y. Lou, A.C.S. Samia, A. Devadoss, J.D. Burgess, S. Dayal, C. Burda, Angew. Chem. Int. Ed. **44**, 5855 (2005)
39. C. Wang, G. Zhang, S. Fan, Y. Li, J. Phys. Chem. Solids **62**, 1957 (2001)
40. B. Poudel, W.Z. Wang, D.Z. Wang, J.Y. Huang, Z.F. Ren, J. Nanosci. Nanotechnol. **6**, 1050 (2006)
41. B. Zhang, J. He, T.M. Tritt, Appl. Phys. Lett. **88**, 043119 (2006)
42. W. Heiss, H. Groiss, E. KaQWmann, G. Hesser, M. Böberl, G. Springholz, F. Schäffler, K. Koike, H. Harada, M. Yano, Appl. Phys. Lett. **88**, 192109 (2006)
43. B.A. Akimov, V.A. Bogoyavlenskiy, L.I. Ryabova, V.N. Vasil'kov, Phys. Rev. B **61**, 16045 (2000)
44. Ya. A. Ugai, A.M. Samoilov, M.K. Sharov, O.B. Yatsenko, B.A. Akimov, Inorg. Mater. **38**, 12 (2002)
45. Ya. A. Ugai, A.M. Samoilov, S.A. Buchnev, Yu. V. Synorov, M.K. Sharov, Inorg. Mater. **38**, 450 (2002)
46. A.M. Samoilov, S.A. Buchnev, YuVSynorov, B.L. Agapov, A.M. Khoviv. Inorg. Mater. **39**, 1132 (2003)
47. A.M. Samoilov, S.A. Buchnev, E.A. Dolgopolova, YuVSynorov, A.M. Khoviv. Inorg. Mater. **40**, 349 (2004)
48. H. Murakami, W. Hattori, R. Aoki, Phys. C **269**, 83 (1996)
49. H. Murakami, W. Hattori, Y. Mizomata, R. Aoki, Phys. C **273**, 41 (1996)
50. H. Murakami, R. Aoki, K. Sakai, Thin Solid Films **27**, 343 (1999)
51. B.A. Volkov, L.I. Ryabova, D.R. Khokhlov, Phys. Usp. **45**, 819 (2002)
52. F. Hüe, M. Hÿtch, H. Bender, F. Houdellier, A. Claverie, Phys. Rev. Lett. **100**, 156602 (2008)
53. S. Banerjee, K.A. Shore, C.J. Mitchell, J.L. Sly, M. Missous, I.E.E. Proc, Circuits Devices Syst. **152**, 497 (2005)
54. M. Razeghi, A. Evans, S. Slivken, J.S. Yu, J.G. Zheng, V.P. Dravid, Proc. SPIE **5840**, 54 (2005)
55. R.A. Stradling, Semicond. Sci. Technol. **6**, C52 (1991)
56. P.K. Weimer, Proc. IEEE **52**, 608 (1964)
57. G. Ribakovs, A.A. Gundjian, IEEE J. Quant. Electron. **QE-14**, 42 (1978)
58. S.K. Dey, J. Vac. Sci. Technol. **10**, 227 (1973)
59. S.J. Lynch, Thin Solid Films **102**, 47 (1983)
60. V.V. Kudzin, V.S. Kulakov, D.R. Pape', S.V. Kulakov, V.V. Molotok, IEEE. Ultrason. Symp. **1**, 749 (1997)
61. F. Hatami, V. Lordi, J.S. Harris, H. Kostial, W.T. Masselink, J. Appl. Phys. **97**, 096106 (2005)
62. B.W. Wessels, J. Electrochem. Soc. **722**, 402 (1975)
63. D.W.L. Tolfree, J. Sci. Instrum. **41**, 788 (1964)
64. P.B. Hart, Proc. IEEE **61**, 880 (1973)
65. M.A. Hines, G.D. Scholes, Adv. Mater. **15**, 1844 (2003)
66. C.A. Wang, R.K. Huang, D.A. Shiau, M.K. Connors, P.G. Murphy, P.W. O'Brien, A.C. Anderson, D.M. DePoy, G. Nichols, M.N. Palmisiano, Appl. Phys. Lett. **83**, 1286 (2003)
67. C.W. Hitchcock, R.J. Gutmann, J.M. Borrego, I.B. Bhat, G.W. Charache, I.E.E.E. Trans, Electron. Devices **46**, 2154 (1999)
68. H.J. Goldsmid, R.W. Douglas, Br. J. Appl. Phys. **5**, 386 (1954)
69. F.D. Rosi, B. Abeles, R.V. Jensen, J. Phys. Chem. Sol. **10**, 191 (1959)
70. T.M. Tritt (ed.), *Semiconductors and Semimetals*, vols. 69, 70 and 71: *Recent Trends in Thermoelectric Materials Research I, II and III* (Academic Press, New York, 2000)
71. D.M. Rowe (ed.), *CRC Handbook of Thermoelectrics* (CRC Press, Boca Raton, 1995)

72. D.M. Rowe, C.M. Bhandari, *Modern Thermoelectrics* (Reston Publishing Company, Virginia, 1983)
73. D.M. Rowe (ed.), *Thermoelectrics Handbook: Macro to Nano* (CRC Press, Boca Raton, 2006)
74. H. Choi, M. Chang, M. Jo, S.J. Jung, H. Hwang, Electrochem. Solid-State Lett. **11**, H154 (2008)
75. S. Cova, M. Ghioni, A. Lacaita, C. Samori, F. Zappa, Appl. Opt. **35**, 1956 (1996)
76. H.W.H. Lee, B.R. Taylor, S.M. Kauzlarich, *Nonlinear Optics: Materials, Fundamentals, and Applications* (Technical Digest, 2000), p. 12
77. E. Brundermann, U. Heugen, A. Bergner, R. Schiwon, G.W. Schwaab, S. Ebbinghaus, D.R. Chamberlin, E.E. Haller, M. Havenith, in *29th International Conference on Infrared and MillimeterWaves and 12th International Conference on Terahertz, Electronics*, vol. 283 (2004)
78. A.N. Baranov, T.I. Voronina, N.S. Zimogorova, L.M. Kauskaya, Y.P. Yakoviev, Sov. Phys. Semicond. **19**, 1676 (1985)
79. M. Yano, Y. Suzuki, T. Ishii, Y. Matsushima, M. Kimata, Jpn. J. Appl. Phys. **17**, 2091 (1978)
80. F.S. Yuang, Y.K. Su, N.Y. Li, Jpn. J. Appl. Phys. **30**, 207 (1991)
81. F.S. Yuang, Y.K. Su, N.Y. Li, K.J. Gan, J. Appl. Phys. **68**, 6383 (1990)
82. Y.K. Su, S.M. Chen, J. Appl. Phys. **73**, 8349 (1993)
83. S.K. Haywood, A.B. Henriques, N.J. Mason, R.J. Nicholas, P.J. Walker, Semicond. Sci. Technol. **3**, 315 (1988)
84. G.C. Young, W.W. Anderson, L.B. Anderson, Trans. Electron Dev. IEEE **24**, 492 (1977)
85. R.L. Gordon, V.I. Neeley, H.R. Curtin, Proc. IEEE **54**, 2014 (1966)
86. P.K. Weimer, Proc. IRE **50**, 1462 (1962)
87. M.J. Lee, S. W. Wright, C. P. Judge, P. Y. Cheung, in *Display Research Conference, International Conference Record*, p. 211 (1991)
88. L.A. Vassilev, Phys. Stat. Sol. (b) **121**, 203 (1984)
89. J.M. Thornton, J. Mol. Biol. **151**, 261 (1981); T.E. Creighton, Methods Enzymol. **131**, 83 (1986); T.E. Creighton, Bio Essays **8**, 57 (1988)
90. V.J. Hruby, Life Sci. **31**, 189 (1982); V.J. Hruby, F. Al-Obeidi, W. Kazmierski, Biochem. J. **268**, 249 (1990)
91. R. Wetzel, Trends Biochem. Sci. **12**, 478 (1987)
92. T. Kimura, R. Matsueda, Y. Nakagawa, E.T. Kaiser, Anal. Biochem. **122**, 274 (1982)
93. D. Andreu, F. Albericio, N.A. Sole, M.C. Munson, M. Ferrer, G. Barnay, in *Methods in Molecular Biology, 35 Peptide Synthesis Protocols*, ed. by M.W. Pennington, B.M. Dunna (Humana Press Inc., USA, 1994)
94. E.O. Kane, Phys. Rev. **131**, 79 (1963). Phys. Rev. B. **139**, 343 (1965)
95. M. Abramowitz, I.A. Stegun, *Handbook of Mathematical Functions with Formulas, Graphs and Mathematical Tables* (Wiley, New York, 1964)
96. I.S. Gradshteyn, I.M. Ryzhik, *Tables of Integrals, Series and Products* (Academic, New York, 1965)
97. E.O. Kane, in *Semiconductors and Semimetals*, vol. 1, ed. by R.K. Willirdson, A.C. Beer (Academic Press, New York, 1966), p. 75
98. E.D Palik, G.B. Wright, in *Semiconductors and Semimetals* vol. 3, ed. by R.K Willardson, A. C Beer, 3 (Academic Press, New York, USA, 1967), p. 421; H.I. Zhang, Phys. Rev, **1B**, 3450 (1970)
99. G.E. Stillman, C.M. Wolfe, J.O. Dimmock, in *Semiconductors and Semimetals*, vol. 12, ed. by R.K. Willardon, A.C. Beer (Academic Press, New York, 1977), p. 169
100. E.D. Palik, G.S. Picus, S. Teither, R.E. Wallis, Phys. Rev. 475 (1961)
101. D.G. Seiler, B.D. Bajaj, A.E. Stephens, Phys. Rev. **B 16,** 2822 (1977); A.V. Germaneko, G. M. Minkov, Phys. Stat. Sol. (b) **184**, 9 (1994); G.L. Bir, G.E. Pikus, *Symmetry and Strain–Induced effects in Semiconductors Nauka*, Russia (1972). (in Russian); M. Mondal, K. P. Ghatak, Phys. Stat. Sol. (b) **135**, K21 (1986)

102. J. Bouat, J.C. Thuillier, Surf. Sci. **73**, 528 (1978)
103. G.J. Rees, *Phys. of Compounds, Proc. of the 13th Inter. Nat. Conf.*, ed. by F.G. Fumi (North Holland Company, 1976), p. 1166
104. P.R. Emtage, Phys. Rev. **138**, A246 (1965)
105. M. Stordeur, W. Kuhnberger, Phys. Stat. sol. (b), **69**, 377 (1975); D.R. Lovett, *Semimetals and Narrow-Bandgap Semiconductor* (Pion Limited, UK, 1977); H. Kohler, Phys. Stat. Sol. (b), **74**, 591 (1976)
106. M. Cardona, W. Paul, H. Brooks Helv, Acta Phys. **33**, 329 (1960); A.F. Gibson in *Proceeding of International School of Physics "ENRICO FERMI"* course XIII, ed. by R.A Smith (Academic Press, 1963), p. 171
107. C.C. Wang, N.W. Ressler, Phys. Rev. **2**, 1827 (1970)
108. M. Zalazny, Phys. B **124**, 352 (1984)
109. P.C. Mathur, S. Jain, Phys. Rev. **19**, 1359 (1979)
110. Y. Yamada, J. Phys. Japan **35**, 1600 (1973); M. Singh, P.R. Wallace, S.D. Jog, E. Arushanov, J. Phys. Chem. Solids **45**, 409 (1984)
111. S. Takaoka, K. Murase, Phys. Rev. B, Phys. Rev. B **20**, 2823 (1979)
112. G.P. Chuiko, Sov. Phys. Semi. **19**(12), 1381 (1985)
113. V. Heine, Proc. Phys. Soc. **81**, 300 (1963)
114. J.N. Schulman, Y.C. Chang, Phys. Rev. B **24**, 4445 (1981)
115. B.R. Nag, *Electron Transport in Compound Semiconductors* (Springer, Heidelberg, 1980)

Chapter 3
The DRs in Quantum Wires (QWs) of Heavily Doped (HD) Non-parabolic Materials

The greatest thing in life is to keep the MIND YOUNG.

3.1 Introduction

It is well-known that in quantum wires (QWs), the restriction of the motion of the carriers along two directions may be viewed as carrier confinement by two infinitely deep 1D rectangular potential wells, along any two orthogonal directions leading to quantization of the wave vectors along the said directions, allowing 1D carrier transport [1]. With the help of modern experimental techniques, such one dimensional quantized structures have been experimentally realized and enjoy an enormous range of important applications in the realm of nano-science in the quantum regime. They have generated much interest in the analysis of nano-structured devices for investigating their electronic, optical and allied properties [2–4]. Examples of such new applications are based on the different transport properties of ballistic charge carriers which include quantum resistors [5–10], resonant tunneling diodes and band filters [11, 12], quantum switches [13], quantum sensors [14–16], quantum logic gates [17, 18], quantum transistors and sub tuners [19–21], heterojunction FETs [22], high-speed digital networks [23], high-frequency microwave circuits [24], optical modulators [25], optical switching systems [26, 27], and other devices.

In this chapter in Sects. 3.2.1–3.2.14, we have investigated the DRs in NWs of HD non-linear optical, III–V, II–VI, stressed Kane type, Te, GaP, $PtSb_2$, Bi_2Te_3, Ge, GaAs, II–V, lead germanium telluride and zinc and cadmium phosphides respectively. The Sect. 3.3 contains the summary and conclusion pertaining to this chapter. The Sect. 3.4 presents 19 open research problems.

K.P. Ghatak, *Dispersion Relations in Heavily-Doped Nanostructures*,
Springer Tracts in Modern Physics 265, DOI 10.1007/978-3-319-21000-1_3

3.2 Theoretical Background

3.2.1 *The DR in Quantum Wires (QWs) of HD Nonlinear Optical Materials*

The DR of the 1D electrons in this case can be written following (2.32) as

$$\frac{\hbar^2(\mathrm{n}_z\pi/d_z)^2}{2m_{\parallel}^* T_{21}(\mathrm{E},\eta_g)}+\frac{\hbar^2(\mathrm{n}_y\pi/\mathrm{d}_y)^2}{2m_{\parallel}^* T_{22}(\mathrm{E},\eta_g)}+\frac{\hbar^2k_x^2}{2m_{\parallel}^* T_{21}(\mathrm{E},\eta_g)}=1 \tag{3.1}$$

where, $n_z(=1,2,3,\ldots)$, d_z are the size quantum number and the nano-thickness along the z-direction respectively, $n_y(=1,2,3,\ldots)$ and d_y are the size quantum number and the nano-thickness along the y-direction respectively.

The 1D DOS function per sub-band is given by

$$N_{1D}(E)=\frac{2g_v}{\pi}\frac{\partial k_x}{\partial E} \tag{3.2}$$

Thus by using (3.1) and (3.2) the total DOS function in this case can be written as

$$N_{1DHD\Gamma}(\mathrm{E},\eta_g)=\frac{2g_v}{\pi}\text{Real part of}\sum_{n_y=1}^{n_{y\max}}\sum_{n_z=1}^{n_{z\max}}T_1'(\mathrm{E},\mathrm{n}_y,\mathrm{n}_z,\eta_g)\mathrm{H}(\mathrm{E}-\mathrm{E}'_{1HDNW}) \tag{3.3}$$

where E'_{1HDNW} is the complex sub-band energy which can be expressed in this case as

$$\frac{\hbar^2(\mathrm{n}_z\pi/\mathrm{d}_z)^2}{2m_{\parallel}^* T_{21}(\mathrm{E}'_{1HDNW},\eta_g)}+\frac{\hbar^2(\mathrm{n}_y\pi/\mathrm{d}_y)^2}{2m_{\parallel}^* T_{22}(\mathrm{E}'_{1HDNW},\eta_g)}=1 \tag{3.4}$$

and $T'_{1HDNW}(\mathrm{E},\mathrm{n}_y,\mathrm{n}_z,\eta_g)=\left[\left[1-\frac{\hbar^2(\mathrm{n}_z\pi/\mathrm{d}_z)^2}{2m_{\parallel}^* T_{21}(\mathrm{E},\eta_g)}+\frac{\hbar^2(\mathrm{n}_y\pi/\mathrm{d}_y)^2}{2m_{\parallel}^* T_{22}(\mathrm{E},\eta_g)}\right]\frac{2m_{\parallel}^* T_{21}(\mathrm{E},\eta_g)}{\hbar^2}\right]^{1/2}$.

The EEM in this case in given by

$$m^*(\mathrm{E}_{F1HDNW},\mathrm{n}_y,\mathrm{n}_z,\eta_g)=\frac{\hbar^2}{2}\left[\text{Real part of}\frac{\partial}{\partial(\mathrm{E}_{F1HDNW})}\left[T_{1HDNW}(\mathrm{E},\mathrm{n}_y,\mathrm{n}_z,\eta_g)\right]^2\right] \tag{3.5}$$

where E_{F1HDNW} in the Fermi energy in this case.

Thus, we observe that the EEM is the function of size quantum numbers in both the directions and the Fermi energy due to the combined influence of the crystal filed splitting constant and the anisotropic spin-orbit splitting constants respectively. Besides it is a function of η_g due to which the EEM exists in the band gap, which is otherwise impossible.

In the absence of band-tails, for electron motion along x-direction only, the 1D electron dispersion law in this case can be written following (2.2) as

$$\gamma(E) = f_1(E)k_x^2 + f_1(E)(\pi n_y/d_y)^2 + f_2(E)(\pi n_z/d_z)^2 \tag{3.6}$$

Using (3.2) and (3.6) the total DOS function in this case.

$$N_{1D\Gamma}(\mathrm{E}) = \frac{2g_v}{\pi}\sum_{n_y=1}^{n_{y\max}}\sum_{n_z=1}^{n_{z\max}} f'_{10}(\mathrm{E}, \mathrm{n_y}, \mathrm{n_z})\mathrm{H}(\mathrm{E} - \mathrm{E}'_1) \tag{3.7}$$

where $f_{10}(\mathrm{E}, \mathrm{n_y}, \mathrm{n_z}) = [\mathrm{f_1(E)}]^{\frac{-1}{2}}\left[\gamma(\mathrm{E}) - \mathrm{f_1(E)}\left(\frac{n_y\pi}{d_y}\right)^2 - \mathrm{f_2(E)}\left(\frac{n_z\pi}{d_z}\right)^2\right]^{\frac{1}{2}}$

In (3.7), the sub-band energy (E'_1) are given by the equation

$$\gamma(E'_1) = f_1(E'_1)(\pi n_y/d_y)^2 + f_2(E'_1)(\pi n_z/d_z)^2 \tag{3.8a}$$

The EEM in this case assumes the form

$$m^*(E_{F1D}, n_y, n_z) = \frac{\hbar^2}{2} f'_{10}(E_{F1D}, n_y, n_z) \tag{3.8b}$$

where E_{F1D} is the Fermi energy in this case.

3.2.2 The DR In Quantum Wires (QWs) of HD III–V Materials

(i) Three Band Model of Kane

The dispersion relation of the 1D electrons in this case can be written following (2.45) as

$$\frac{\hbar^2(\mathrm{n_z}\pi/\mathrm{d_z})^2}{2m_c} + \frac{\hbar^2(\mathrm{n_y}\pi/\mathrm{d_y})^2}{2m_c} + \frac{\hbar^2 k_x^2}{2m_c} = T_{31}(\mathrm{E}, \eta_g) + iT_{32}(\mathrm{E}, \eta_g) \tag{3.9}$$

The DOS function in this case assumes the form

$$N_{1DHD\Gamma}(\mathrm{E}, \eta_g) = \frac{2g_v}{\pi}\text{Real part of} \sum_{n_y=1}^{n_{y\max}}\sum_{n_z=1}^{n_{z\max}} T'_2(\mathrm{E}, \mathrm{n_y}, \mathrm{n_z}, \eta_g)\mathrm{H}(\mathrm{E} - \mathrm{E}'_{2HDNW}) \tag{3.10}$$

where

$T_2(\mathrm{E},\, \mathrm{n}_y, \mathrm{n}_z, \mathrm{n}_g) = \left[\left[T_{31}(\mathrm{E}, \eta_g) + iT_{32}(\mathrm{E}, \eta_g) - \left[\frac{\hbar^2(\mathrm{n}_z\pi/\mathrm{d}_z)^2}{2m_c} = \frac{\hbar^2(\mathrm{n}_y\pi/\mathrm{d}_y)^2}{2m_c}\right]\right]\frac{2m_c}{\hbar^2}\right]^{\frac{1}{2}}$

In (3.10), E'_{2HDNW} is the sub-band energy in this case which can be expressed as

$$\frac{\hbar^2(\mathrm{n}_z\pi/\mathrm{d}_z)^2}{2m_c} + \frac{\hbar^2(\mathrm{n}_y\pi/\mathrm{d}_y)^2}{2m_c} + T_{31}(E'_{2HDNW}, \eta_g) + iT_{32}(E'_{2HDNW}, \eta_g) \tag{3.11}$$

The EEM in this case is given by

$$m^*(\mathrm{E}_{F1HDNW}, \eta_g) = \mathrm{m}_c[\mathrm{T}'_{31}(\mathrm{E}_{F1HDNW}, \eta_g)] \tag{3.12}$$

In the absence of band tails the DR in this case assumes the form

$$I_{11}(\mathrm{E}) = \frac{\hbar^2 k_x^2}{2m_c} + G_2(\mathrm{n}_y, \mathrm{n}_z) \tag{3.13}$$

where $\mathrm{G}_2(\mathrm{n}_y, \mathrm{n}_z) = \frac{\hbar^2\pi^2}{2m_c}\left[\left(\frac{n_y\pi}{d_y}\right)^2 + \left(\frac{n_z\pi}{d_z}\right)^2\right]$

The DOS function can be expressed as

$$N_{1D\Gamma}(\mathrm{E}) = \frac{2g_v}{\pi}\sum_{n_y=1}^{n_{y\max}}\sum_{n_z=1}^{n_{z\max}} f'_{11}(\mathrm{E}, \mathrm{n}_y, \mathrm{n}_z)\mathrm{H}(\mathrm{E} - \mathrm{E}'_2) \tag{3.14}$$

where $f_{11}(\mathrm{E}, \mathrm{n}_y, \mathrm{n}_z) = \left[\frac{2m_c}{\hbar^2}\left[I_{11}(\mathrm{E}) - G_2(\mathrm{n}_y, \mathrm{n}_z)\right]\right]^{\frac{1}{2}}$

In (3.14), the sub band energy E'_2 can be written as

$$G_2(n_y, n_z) = I_{11}(E'_2) \tag{3.15}$$

The EEM in this case assumes the form

$$m^*(E_{F1D}) = m_c I'_{11}(E_{F1D}) \tag{3.16}$$

(ii) Two Band Model of Kane

The DR of the 1D electrons in this case can be written as

$$\frac{\hbar^2(\mathrm{n}_z\pi/\mathrm{d}_z)^2}{2m_c} + \frac{\hbar^2(\mathrm{n}_y\pi/\mathrm{d}_y)^2}{2m_c} + \frac{\hbar^2 k_x^2}{2m_c} = \gamma_2(\mathrm{E}, \eta_g) \tag{3.17}$$

The DOS function in this case assumes the form

$$N_{1DHD\Gamma}(\mathrm{E},\eta_g)=\frac{2g_v}{\pi}\sum_{n_y=1}^{n_{y\max}}\sum_{n_z=1}^{n_{z\max}}T_3'(\mathrm{E},\mathrm{n}_y,\mathrm{n}_z,\eta_g)\mathrm{H}(\mathrm{E}-E'_{3HDNW}) \quad (3.18)$$

where $T_3(\mathrm{E},\mathrm{n}_y,\mathrm{n}_z,\eta_g)=\left[\left[\gamma_2(\mathrm{E},\eta_g)-\left[\frac{\hbar^2(\mathrm{n}_z\pi/\mathrm{d}_z)^2}{2m_c}+\frac{\hbar^2(\mathrm{n}_y\pi/\mathrm{d}_y)^2}{2m_c}\right]\right]\frac{2m_c}{\hbar^2}\right]^{\frac{1}{2}}$

In (3.18), E'_{3HDNW} is the sub-band energy in this case which can be expressed as

$$\frac{\hbar^2(\mathrm{n}_z\pi/\mathrm{d}_z)^2}{2m_c}+\frac{\hbar^2(\mathrm{n}_y\pi/\mathrm{d}_y)^2}{2m_c}=\gamma_2(E'_{3HDNW},\eta_g) \quad (3.19)$$

The EEM in this case is given by

$$m^*(\mathrm{E}_{F1HDNW},\eta_g)=\mathrm{m}_c\left[\gamma_2'(\mathrm{E}_{F1HDNW},\eta_g)\right] \quad (3.20)$$

The expression of 1D DR, for NWs of III–V materials whose energy band structures are defined by the two-band model of Kane in the absence of band tailing assumes the form

$$E(1+\alpha E)=\frac{\hbar^2k_x^2}{2m_c}+G_2(n_y,n_z) \quad (3.21)$$

The DOS function can be expressed as

$$N_{1D\Gamma}(\mathrm{E})=\frac{2g_v}{\pi}\sum_{n_y=1}^{n_{y\max}}\sum_{n_z=1}^{n_{z\max}}f_{12}'(\mathrm{E},\mathrm{n}_y,\mathrm{n}_z)\mathrm{H}(\mathrm{E}-\mathrm{E}_3') \quad (3.22)$$

where $f_{12}(\mathrm{E},\mathrm{n}_y,\mathrm{n}_z)=\left[\frac{2m_c}{\hbar^2}\left[\mathrm{E}(1+\alpha\mathrm{E})-G_2(\mathrm{n}_y,\mathrm{n}_z)\right]\right]^{\frac{1}{2}}$

In (3.22), the sub band energy E_3' can be written as

$$E_3'=(2\alpha)^{-1}\left[-1+\sqrt{1+4\alpha G_2(n_y,n_z)}\right] \quad (3.23)$$

The EEM in this case assumes the form

$$m^*(E_{F1D})=m_c(1+2\alpha\mathrm{E}_{F1D}) \quad (3.24)$$

(iii) Parabolic Energy Bands

The DR of the 1D electrons in this case can be written as

$$\frac{\hbar^2(\mathrm{n}_z\pi/\mathrm{d}_z)^2}{2m_c}+\frac{\hbar^2(\mathrm{n}_y\pi/\mathrm{d}_y)^2}{2m_c}+\frac{\hbar^2k_x)^2}{2m_c}=\gamma_3(\mathrm{E},\eta_g) \quad (3.25)$$

The DOS function in this case assumes the form

$$N_{1DHD\Gamma}(\mathrm{E},\eta_g) = \frac{2g_v}{\pi}\sum_{n_y=1}^{n_{y\max}}\sum_{n_z=1}^{n_{z\max}} T_4'(\mathrm{E},\mathrm{n_y},\mathrm{n_z},\eta_g)\mathrm{H}(\mathrm{E}-\mathrm{E}'_{4HDNW}) \tag{3.26}$$

where $T_4(\mathrm{E},\mathrm{n_y},\mathrm{n_z},\eta_g) = \left[\left[\gamma_3(\mathrm{E},\eta_g) - \left[\frac{\hbar^2(\mathrm{n_z}\pi/\mathrm{d_z})^2}{2m_c} + \frac{\hbar^2(\mathrm{n_y}\pi/\mathrm{d_y})^2}{2m_c}\right]\right]\frac{2m_c}{\hbar^2}\right]^{\frac{1}{2}}$

In (3.26), E'_{3HDNW} is the sub-band energy in this case which can be expressed as

$$\frac{\hbar^2(\mathrm{n_z}\pi/\mathrm{d_z})^2}{2m_c} + \frac{\hbar^2(\mathrm{n_y}\pi/\mathrm{d_y})^2}{2m_c} = \gamma_3(E'_{4HDNW},\eta_g) \tag{3.27}$$

The EEM in this case is given by

$$m^*(\mathrm{E}_{F1HDNW},\eta_g) = \mathrm{m}_c[\gamma_3'(\mathrm{E}_{F1HDNW},\eta_g)] \tag{3.28}$$

The expression of 1D DR, for NWs of III–V materials whose energy band structures are defined by the parabolic energy bandsin the absence of band tailing assumes the form

$$E = \frac{\hbar^2 k_x^2}{2m_c} + G_2(n_y,n_z) \tag{3.29}$$

The DOS function can be expressed as

$$N_{1D\Gamma}(\mathrm{E}) = \frac{2g_v}{\pi}\sum_{n_y=1}^{n_{y\max}}\sum_{n_z=1}^{n_{z\max}} f_{13}'(\mathrm{E},\mathrm{n_y},\mathrm{n_z})\mathrm{H}(\mathrm{E}-\mathrm{E}_4') \tag{3.30}$$

where $f_{13}(\mathrm{E},\mathrm{n_y},\mathrm{n_z}) = \left[\frac{2m_c}{\hbar^2}[\mathrm{E} - G_2(\mathrm{n_y},\mathrm{n_z})]\right]^{\frac{1}{2}}$

In (3.30), the sub band energy E_4' can be written as

$$E_4' = G_2(\mathrm{n_y},\mathrm{n_z}) \tag{3.31}$$

The EEM in this case assumes the form

$$\mathrm{m}^*(E_{F1D}) = m_c \tag{3.32}$$

(iv) The Model of Stillman et al.

The DR of the 1D electrons in this case can be written as

$$\frac{\hbar^2(\mathrm{n_z}\pi/\mathrm{d_z})^2}{2m_c} + \frac{\hbar^2(\mathrm{n_y}\pi/\mathrm{d_y})^2}{2m_c} + \frac{\hbar^2 k_x^2}{2m_c} = \theta_4(\mathrm{E},\eta_g) \tag{3.33}$$

where $\theta_4(\mathrm{E}, \eta_g) = \mathrm{I}_{12}(\mathrm{E}, \eta_g)$

The DOS function in this case assumes the form

$$N_{1DHD\Gamma}(\mathrm{E}, \eta_g) = \frac{2g_v}{\pi} \sum_{n_y=1}^{n_{ymax}} \sum_{n_z=1}^{n_{zmax}} T_5'(\mathrm{E}, \mathrm{n}_y, \mathrm{n}_z, \eta_g) \mathrm{H}(\mathrm{E} - \mathrm{E}_{5HDNW}') \tag{3.34}$$

where $T_5(\mathrm{E}, \mathrm{n}_y, \mathrm{n}_z, \eta_g) = \left[[\theta_4(\mathrm{E}, \eta_g) - [G_2(\mathrm{n}_y, \mathrm{n}_z)]] \frac{2m_c}{\hbar^2} \right]^{\frac{1}{2}}$

In (3.34), E_{5HDNW}' is the sub-band energy in this case which can be expressed as

$$G_2(\mathrm{n}_y, \mathrm{n}_z) = \theta_4(E_{5HDNW}', \eta_g) \tag{3.35}$$

The EEM in this case is given by

$$m^*(\mathrm{E}_{F1HDNW}, \eta_g) = \mathrm{m}_c[\theta_4'(\mathrm{E}_{F1HDNW}, \eta_g)] \tag{3.36}$$

The expression of 1D DR, for NWs of III–V materials whose energy band structures are defined by the model of Stillman et al. in the absence of band tailing assumes the form

$$I_{12}(E) = \frac{\hbar^2 k_x^2}{2m_c} + G_2(n_y, n_z) \tag{3.37}$$

In this case, the quantized energy E_9' is given by

$$I_{12}(E_9') = G_2(n_y, n_z) \tag{3.38}$$

The DOS function can be expressed as

$$N_{1D\Gamma}(\mathrm{E}) = \frac{2g_v}{\pi} \sum_{n_y=1}^{n_{ymax}} \sum_{n_z=1}^{n_{zmax}} f_{14}'(\mathrm{E}, \mathrm{n}_y, \mathrm{n}_z) \mathrm{H}(\mathrm{E} - \mathrm{E}_9') \tag{3.39}$$

where $f_{14}(\mathrm{E}, \mathrm{n}_y, \mathrm{n}_z) = \left[\frac{2m_c}{\hbar^2} [\mathrm{I}_{12}(\mathrm{E}) - G_2(\mathrm{n}_y, \mathrm{n}_z)] \right]^{\frac{1}{2}}$

The EEM in this case assumes the form

$$\mathrm{m}^*(E_{F1D}) = m_c I_{12}'(E_{F1D}) \tag{3.40}$$

(v) The Model of Palik et al.

The DR of the 1D electrons in this case can be written as

$$\frac{\hbar^2(\mathrm{n}_z\pi/\mathrm{d}_z)^2}{2m_c} + \frac{\hbar^2(\mathrm{n}_y\pi/\mathrm{d}_y)^2}{2m_c} + \frac{\hbar^2 k_x^2}{2m_c} = \theta_5(\mathrm{E}, \eta_g) \tag{3.41}$$

where $\theta_5(\mathrm{E},\eta_g) = \mathrm{I}_{13}(\mathrm{E},\eta_g)$

The DOS function in this case assumes the form

$$N_{1DHD\Gamma}(\mathrm{E},\eta_g) = \frac{2g_v}{\pi}\sum_{n_y=1}^{n_{ymax}}\sum_{n_z=1}^{n_{zmax}} T_6'(\mathrm{E},\mathrm{n}_y,\mathrm{n}_z,\eta_g)\mathrm{H}(\mathrm{E}-\mathrm{E}'_{6HDNW}) \quad (3.42)$$

where $T_6(\mathrm{E},\mathrm{n}_y,\mathrm{n}_z,\eta_g) = \left[\left[\theta_5(\mathrm{E},\eta_g) - \left[G_2(\mathrm{n}_y,\mathrm{n}_z)\right]\right]\frac{2m_c}{\hbar^2}\right]^{\frac{1}{2}}$

In (3.42), E'_{6HDNW} is the sub-band energy in this case which can be expressed as

$$G_2(\mathrm{n}_y,\mathrm{n}_z) = \theta_5(E'_{6HDNW},\eta_g) \quad (3.43)$$

The EEM in this case is given by

$$m^*(\mathrm{E}_{F1HDNW},\eta_g) = \mathrm{m}_c[\theta_5'(\mathrm{E}_{F1HDNW},\eta_g)] \quad (3.44)$$

The expression of 1D DR, for NWs of III–V materials whose energy band structures are defined by the model of Palik et al. in the absence of band tailing assumes the form

$$I_{13}(E) = \frac{\hbar^2 k_x^2}{2m_c} + G_2(n_y,n_z) \quad (3.45)$$

In this case, the quantized energy E'_{10} is given by

$$I_{13}(E'_{10}) = G_2(n_y,n_z) \quad (3.46)$$

The DOS function can be expressed as

$$N_{1D\Gamma}(\mathrm{E}) = \frac{2g_v}{\pi}\sum_{n_y=1}^{n_{ymax}}\sum_{n_z=1}^{n_{zmax}} f_{15}'(\mathrm{E},\mathrm{n}_y,\mathrm{n}_z)\mathrm{H}(\mathrm{E}-\mathrm{E}'_{10}) \quad (3.47)$$

where $f_{15}(\mathrm{E},\mathrm{n}_y,\mathrm{n}_z) = \left[\frac{2m_c}{\hbar^2}[\mathrm{I}_{13}(\mathrm{E}) - G_2(\mathrm{n}_y,\mathrm{n}_z)]\right]^{\frac{1}{2}}$

The EEM in this case assumes the form

$$\mathrm{m}^*(E_{F1D}) = m_c I_{13}'(E_{F1D}) \quad (3.48)$$

3.2.3 The DR in Quantum Wires (QWs) of HD II–VI Materials

The 1D DR in NW of HD II–VI materials can be written as

$$\gamma_3(\mathrm{E},\eta_g) = a_0'\left[\left(\frac{n_x\pi}{d_x}\right)^2 + \left(\frac{n_y\pi}{d_y}\right)^2\right] \pm \bar{\lambda}_0\left[\left(\frac{n_x\pi}{d_x}\right)^2 + \left(\frac{n_y\pi}{d_y}\right)^2\right]^{1/2} + \frac{\hbar^2 k_z^2}{2m_\parallel^*} \quad (3.49)$$

The DOS function in this case assumes the form

$$N_{1DHD\Gamma}(\mathrm{E},\eta_g) = \frac{g_v}{\pi}\sum_{n_x=1}^{n_{x\max}}\sum_{n_y=1}^{n_{y\max}} T_7'(\mathrm{E},\mathrm{n}_x,\mathrm{n}_y,\eta_g)\mathrm{H}(\mathrm{E}-\mathrm{E}_{13HDNW}') \quad (3.50)$$

where $T_7(\mathrm{E},\mathrm{n}_y,\mathrm{n}_z,\eta_g) = \left[\left[\gamma_3(\mathrm{E},\eta_g) - \left[G_{3,\pm}(\mathrm{n}_x,\mathrm{n}_y)\right]\right]\frac{2m_\parallel^*}{\hbar^2}\right]^{\frac{1}{2}}$ and

$$G_{3,\pm}(n_x,n_y) = a_0'\left[\left(\frac{n_x\pi}{d_x}\right)^2 + \left(\frac{n_y\pi}{d_y}\right)^2\right] \pm \bar{\lambda}_0\left[\left(\frac{n_x\pi}{d_x}\right)^2 + \left(\frac{n_y\pi}{d_y}\right)^2\right]^{1/2}$$

and E_{13HDNW}' is the sub-band energy in this case which can be expressed as

$$\gamma_3(E_{13HDNW}',\eta_g) = a_0'\left[\left(\frac{n_x\pi}{d_x}\right)^2 + \left(\frac{n_y\pi}{d_y}\right)^2\right] \pm \bar{\lambda}_0\left[\left(\frac{n_x\pi}{d_x}\right)^2 + \left(\frac{n_y\pi}{d_y}\right)^2\right]^{1/2} \quad (3.51)$$

The EEM in this case is given by

$$m^*(E_{F1HDNW},\eta_g) = m_\parallel^*\gamma_3'(E_{F1HDNW},\eta_g) \quad (3.52)$$

The 1D DR for NWs of II–VI materials in the absence of band-tailscan be written as

$$E = b_0' k_z^2 + G_{3,\pm}(n_x,n_y) \quad (3.53)$$

The DOS function can be expressed as

$$N_{1D\Gamma}(\mathrm{E}) = \frac{g_v}{\pi}\sum_{n_x=1}^{n_{x\max}}\sum_{n_z=1}^{n_{y\max}} f_{20}'(\mathrm{E},\mathrm{n}_x,\mathrm{n}_y)\mathrm{H}(\mathrm{E}-\mathrm{E}_{20}') \quad (3.54)$$

where $f_{20}(\mathrm{E},\mathrm{n}_x,\mathrm{n}_y) = \left[\frac{2m_\parallel^*}{\hbar^2}\left[\mathrm{E}-G_{3,\pm}(\mathrm{n}_x,\mathrm{n}_y)\right]\right]^{\frac{1}{2}}$

In Eq. (3.54), the sub band energy E'_{20} can be written as

$$E'_{20} = G_{3,\pm}(\mathrm{n}_x, \mathrm{n}_y) \tag{3.55}$$

The EEM in this case assumes the form

$$\mathrm{m}^*(E_{F1D}) = m^*_{\|} \tag{3.56}$$

3.2.4 *The DR in Quantum Wires (QWs) of HD IV–VI Materials*

(i) Dimmock Model

The 1D electron dispersion law in NW of HD IV–VI materials can be expressed as

$$\begin{aligned}
&\gamma_2(E,\eta_g) + \alpha\gamma_3(E,\eta_g)\left(\frac{\hbar^2}{2x_4}\left(\frac{n_x\pi}{d_x}\right)^2 + \frac{\hbar^2}{2x_5}\left(\frac{n_y\pi}{d_y}\right)^2\right) \\
&+ \alpha\gamma_3(E,\eta_g)\frac{\hbar^2}{2x_6}k_z^2 - (1+\alpha\gamma_3(E,\eta_g))\left(\frac{\hbar^2}{2x_1}\left(\frac{n_x\pi}{d_x}\right)^2 + \frac{\hbar^2}{2x_2}\left(\frac{n_y\pi}{d_y}\right)^2\right) \\
&- \alpha\left(\frac{\hbar^2}{2x_1}\left(\frac{n_x\pi}{d_x}\right)^2 + \frac{\hbar^2}{2x_2}\left(\frac{n_y\pi}{d_y}\right)^2\right)\left(\frac{\hbar^2}{2x_4}\left(\frac{n_x\pi}{d_x}\right)^2 + \frac{\hbar^2}{2x_5}\left(\frac{n_y\pi}{d_y}\right)^2\right) \\
&- \alpha\left(\frac{\hbar^2}{2x_1}\left(\frac{n_x\pi}{d_x}\right)^2 + \frac{\hbar^2}{2x_2}\left(\frac{n_y\pi}{d_y}\right)^2\right)\frac{\hbar^2}{2x_6}k_z^2 - (1+\alpha\gamma_3(E,\eta_g))\frac{\hbar^2}{2x_3}k_z^2 \\
&- \alpha\frac{\hbar^2}{2x_3}k_z^2\left(\frac{\hbar^2}{2x_4}\left(\frac{n_x\pi}{d_x}\right)^2 + \frac{\hbar^2}{2x_5}\left(\frac{n_y\pi}{d_y}\right)^2\right) - \alpha\frac{\hbar^4k_z^4}{4x_3x_6} \\
&= \frac{\hbar^2}{2m_1}\left(\frac{n_x\pi}{d_x}\right)^2 + \frac{\hbar^2}{2m_2}\left(\frac{n_y\pi}{d_y}\right)^2 + \frac{\hbar^2}{2m_3}k_z^2
\end{aligned} \tag{3.57}$$

Equation (3.57) can be written as

$$k_z = T_{36}(E, n_x, n_y) \tag{3.58}$$

where

$$\begin{aligned}
T_{36}(E,n_g,n_x,n_y) = &\left[(2C_{22})^{-1}\left[-B_{HD}(E,n_g,n_x,n_y)..\right.\right. \\
&\left.\left.+\sqrt{B_{HD}^2(E,n_g,n_x,n_y)+4C_{22}A_{HD}(E,n_g,n_x,n_y)}\right]\right]^{1/2}
\end{aligned}$$

$$C_{22} = \left(\alpha \frac{\hbar^2}{4x_3x_6}\right) B_{HD}(E, \eta_g, n_x, n_y)$$
$$= \left[\alpha\left(\frac{\hbar^2}{2x_1}\left(\frac{n_x\pi}{d_x}\right)^2 + \frac{\hbar^2}{2x_2}\left(\frac{n_y\pi}{d_y}\right)^2\right)\frac{\hbar^2}{2x_6} + (1+\alpha\gamma_3(E,\eta_g))\frac{\hbar^2}{2x_3} - \alpha\gamma_3(E,\eta_g)\frac{\hbar^2}{2x_6} + \frac{\hbar^2}{2m_3} + \alpha\frac{\hbar^2}{2x_3}\left(\frac{\hbar^2}{2x_4}\left(\frac{n_x\pi}{d_x}\right)^2 + \frac{\hbar^2}{2x_5}\left(\frac{n_y\pi}{d_y}\right)^2\right)\right] \text{ and}$$

$$A_{HD}(E, \eta_g, n_x, n_y) = \left[-\left[\frac{\hbar^2}{2m_1}\left(\frac{n_x\pi}{d_x}\right)^2 + \frac{\hbar^2}{2m_2}\left(\frac{n_y\pi}{d_y}\right)^2\right]\gamma_2(E,\eta_g) + \alpha\gamma_3(E,\eta_g)\left(\frac{\hbar^2}{2x_4}\left(\frac{n_x\pi}{d_x}\right)^2 + \frac{\hbar^2}{2x_5}\left(\frac{n_y\pi}{d_y}\right)^2\right) - \alpha\left(\frac{\hbar^2}{2x_1}\left(\frac{n_x\pi}{d_x}\right)^2 + \frac{\hbar^2}{2x_2}\left(\frac{n_y\pi}{d_y}\right)^2\right)\left(\frac{\hbar^2}{2x_4}\left(\frac{n_x\pi}{d_x}\right)^2 + \frac{\hbar^2}{2x_5}\left(\frac{n_y\pi}{d_y}\right)^2\right) - (1+\alpha\gamma_3(E,\eta_g))\left(\frac{\hbar^2}{2x_1}\left(\frac{n_x\pi}{d_x}\right)^2 + \frac{\hbar^2}{2x_2}\left(\frac{n_y\pi}{d_y}\right)^2\right)\right]$$

The DOS function in this case can be written as

$$N_{1DHD\Gamma}(\mathrm{E}, \eta_g) = \frac{2g_v}{\pi}\sum_{n_x=1}^{n_{x\max}}\sum_{n_y=1}^{n_{y\max}} T'_{36}(\mathrm{E}, \mathrm{n}_x, \mathrm{n}_y, \eta_g)\mathrm{H}(\mathrm{E} - \mathrm{E}'_{14HDNW}) \tag{3.59}$$

In (3.59), E'_{14HDNW} is the sub-band energy in this case which can be expressed as

$$0 = T_{36}(E'_{14HDNW}, \eta_g, n_x, n_y) \tag{3.60}$$

The EEM in this case is given by

$$m^*(E_{F1HDNW}, \eta_g, n_x, n_y) = \frac{\hbar^2}{2}\frac{\partial}{\partial E}[T^2_{36}(E_{F1HDNW}, \eta_g, n_x, n_y)] \tag{3.61}$$

The 1D DR in NW of IV–VI materials in the absence of band tails can be expressed as

$$
\begin{aligned}
&E(1+\alpha E)+\alpha E\left(\frac{\hbar^2}{2x_4}\left(\frac{n_x\pi}{d_x}\right)^2+\frac{\hbar^2}{2x_5}\left(\frac{n_y\pi}{d_y}\right)^2\right)+\alpha E\frac{\hbar^2}{2x_6}k_z^2-(1+\alpha E)\\
&\left(\frac{\hbar^2}{2x_1}\left(\frac{n_x\pi}{d_x}\right)^2+\frac{\hbar^2}{2x_2}\left(\frac{n_y\pi}{d_y}\right)^2\right)-\alpha\left(\frac{\hbar^2}{2x_1}\left(\frac{n_x\pi}{d_x}\right)^2+\frac{\hbar^2}{2x_2}\left(\frac{n_y\pi}{d_y}\right)^2\right)\\
&\left(\frac{\hbar^2}{2x_4}\left(\frac{n_x\pi}{d_x}\right)^2+\frac{\hbar^2}{2x_5}\left(\frac{n_y\pi}{d_y}\right)^2\right)-\alpha\left(\frac{\hbar^2}{2x_1}\left(\frac{n_x\pi}{d_x}\right)^2+\frac{\hbar^2}{2x_2}\left(\frac{n_y\pi}{d_y}\right)^2\right)\frac{\hbar^2}{2x_6}k_z^2\\
&-(1+\alpha E)\frac{\hbar^2}{2x_3}k_z^2-\alpha\frac{\hbar^2}{2x_3}k_z^2\left(\frac{\hbar^2}{2x_4}\left(\frac{n_x\pi}{d_x}\right)^2+\frac{\hbar^2}{2x_5}\left(\frac{n_y\pi}{d_y}\right)^2\right)-\alpha\frac{\hbar^4k_z^4}{4x_3x_6}\\
&=\frac{\hbar^2}{2m_1}\left(\frac{n_x\pi}{d_x}\right)^2+\frac{\hbar^2}{2m_2}\left(\frac{n_y\pi}{d_y}\right)^2+\frac{\hbar^2}{2m_3}k_z^2
\end{aligned}
\tag{3.62}
$$

Equation (3.62) can be written as

$$
k_z = T_{40}(E, n_x, n_y) \tag{3.63}
$$

where $T_{40}(E,n_x,n_y)=[(2C_{22})^{-1}[-B_0(E,n_x,n_y)+\sqrt{B_0^2(E,n_x,n_y)+4C_{22}A_0(E,n_x,n_y)}]]^{1/2}$

where
$$
\begin{aligned}
B_0(E,n_x,n_y)=&\left[\alpha\left(\frac{\hbar^2}{2x_1}\left(\frac{n_x\pi}{d_x}\right)^2+\frac{\hbar^2}{2x_2}\left(\frac{n_y\pi}{d_y}\right)^2\right)\frac{\hbar^2}{2x_6}+(1+\alpha E)\frac{\hbar^2}{2x_3}-\alpha E\frac{\hbar^2}{2x_6}\right.\\
&\left.+\frac{\hbar^2}{2m_3}+\alpha\frac{\hbar^2}{2x_3}\left(\frac{\hbar^2}{2x_4}\left(\frac{n_x\pi}{d_x}\right)^2+\frac{\hbar^2}{2x_5}\left(\frac{n_y\pi}{d_y}\right)^2\right)\right] \text{ and}
\end{aligned}
$$

$$
\begin{aligned}
A_0(E,n_x,n_y)=&\left[-\left[\frac{\hbar^2}{2m_1}\left(\frac{n_x\pi}{d_x}\right)^2+\frac{\hbar^2}{2m_2}\left(\frac{n_y\pi}{d_y}\right)^2\right]E(1+\alpha E)\right.\\
&+\alpha E\left(\frac{\hbar^2}{2x_4}\left(\frac{n_x\pi}{d_x}\right)^2+\frac{\hbar^2}{2x_5}\left(\frac{n_y\pi}{d_y}\right)^2\right)\\
&-\alpha\left(\frac{\hbar^2}{2x_1}\left(\frac{n_x\pi}{d_x}\right)^2+\frac{\hbar^2}{2x_2}\left(\frac{n_y\pi}{d_y}\right)^2\right)\left(\frac{\hbar^2}{2x_4}\left(\frac{n_x\pi}{d_x}\right)^2+\frac{\hbar^2}{2x_5}\left(\frac{n_y\pi}{d_y}\right)^2\right)\\
&\left.-(1+\alpha E)\left(\frac{\hbar^2}{2x_1}\left(\frac{n_x\pi}{d_x}\right)^2+\frac{\hbar^2}{2x_2}\left(\frac{n_y\pi}{d_y}\right)^2\right)\right]
\end{aligned}
$$

The DOS function can be expressed as

$$
N_{1D\Gamma}(\mathrm{E})=\frac{g_v}{\pi}\sum_{n_x=1}^{n_{x\max}}\sum_{n_y=1}^{n_{y\max}}T_{40}'(\mathrm{E},\mathrm{n}_x,\mathrm{n}_y)\mathrm{H}(\mathrm{E}-\mathrm{E}_{22}') \tag{3.64}
$$

In (3.64), E'_{22} is the sub-band energy in this case which can be expressed as

$$0 = T_{40}(E'_{22}, n_x, n_y) \tag{3.65}$$

The EEM in this case is given by

$$m^*(E_{F1D}, n_x, n_y) = \frac{\hbar^2}{2}\frac{\partial}{\partial E}[T_{40}^2(E_{F1D}, n_x, n_y)] \tag{3.66}$$

(ii) Bangert and Kastner Model

The 1D DR in NW of IV–VI materials in accordance with the present model can be written as

$$F_1(E,\eta_g)\left[\left(\frac{n_x\pi}{d_x}\right)^2 + \left(\frac{n_y\pi}{d_y}\right)^2\right] + F_2(E,\eta_g)\mathrm{k}_z^2 = 1 \tag{3.67}$$

The (3.67) can be written as

$$k_z = T_{60}(E, \eta_g, n_x, n_y) \tag{3.68}$$

where

$$T_{60}(E,\eta_g,n_x,n_y) = \left[\left[1 - F_1(E,\eta_g)\left[\left(\frac{n_x\pi}{d_x}\right)^2 + \left(\frac{n_y\pi}{d_y}\right)^2\right]\right]\right.$$
$$\left.[F_2(E,\eta_g)]^{-1}\right]^{1/2}$$

The DOS function in this case can be written as

$$N_{1DHD\Gamma}(\mathrm{E},\eta_g) = \frac{2g_v}{\pi}\sum_{n_x=1}^{n_{x\max}}\sum_{n_y=1}^{n_{y\max}} T'_{40}(\mathrm{E},\mathrm{n}_x,\mathrm{n}_y,\eta_g)\mathrm{H}(\mathrm{E} - \mathrm{E}'_{15HDNW}) \tag{3.69}$$

In (3.69), E'_{15HDNW} is the sub-band energy in this case which can be expressed as

$$0 = T_{60}(E'_{15HDNW}, \eta_g, n_x, n_y) \tag{3.70}$$

The EEM in this case is given by

$$m^*(E_{F1HDNW}, \eta_g, n_x, n_y) = \frac{\hbar^2}{2}\frac{\partial}{\partial E}[T_{40}^2(E_{F1HDNW}, \eta_g, n_x, n_y)] \tag{3.71}$$

The 1D DR in the absence of band tailing can be written in this case as

$$\omega_1(\mathrm{E})\left[\left(\frac{\pi n_x}{d_x}\right)^2+\left(\frac{\pi n_y}{d_y}\right)^2\right]+\omega_2(\mathrm{E})k_z^2 = 1 \tag{3.72}$$

The (3.72) can be written as

$$k_z = T_{61}(E, n_x, n_y) \tag{3.73}$$

where $T_{61}(E,n_x,n_y)=\left[\left[1-\omega_1(E)\left[\left(\frac{n_x\pi}{d_x}\right)^2+\left(\frac{n_y\pi}{d_y}\right)^2\right]\right][\omega_2(E)]^{-1}\right]^{1/2}$

The DOS function can be expressed as

$$N_{1D\Gamma}(\mathrm{E})=\frac{2g_v}{\pi}\sum_{n_x=1}^{n_{x\max}}\sum_{n_y=1}^{n_{y\max}}T'_{61}(\mathrm{E},\mathrm{n}_x,\mathrm{n}_y)\mathrm{H}(\mathrm{E}-\mathrm{E}'_{24}) \tag{3.74}$$

In (3.74), E'_{24} is the sub-band energy in this case which can be expressed as

$$0 = T_{40}(E'_{24}, n_x, n_y) \tag{3.75}$$

The EEM in this case is given by

$$m^*(E_{F1D},n_x,n_y)=\frac{\hbar^2}{2}\frac{\partial}{\partial E}[T_{61}^2(E_{F1D},n_x,n_y)] \tag{3.76}$$

3.2.5 The DR in QWs of HD Stressed Kane Type Materials

The 1D DR in this case can be written as

$$P_{11}(E,\eta_g)\left(\frac{\pi n_x}{d_x}\right)^2+Q_{11}(E,\eta_g)\left(\frac{\pi n_y}{d_y}\right)^2+S_{11}(E,\eta_g)k_z^2 = 1 \tag{3.77}$$

The (3.77) can be written as

$$k_z = T_{70}(E,\eta_g,n_x,n_y) \tag{3.78}$$

where $t_{72}(\mathrm{E},\mathrm{n}_y,\mathrm{n}_z,\eta_g)=\left[-\left(\frac{n_y\pi}{d_y}\right)^2+\psi_{5HD}(\mathrm{E},\eta_g)-\psi_6\left(\frac{\pi n_z}{d_z}\right)^2\pm\psi_7\left[\psi_{8HD}^2(\mathrm{E},\eta_g)-\left(\frac{\pi n_z}{d_z}\right)^2\right]^{1/2}\right]^{1/2}$

The DOS function in this case can be written as

$$N_{1DHD\Gamma}(\mathrm{E},\eta_g) = \frac{2g_v}{\pi}\sum_{n_x=1}^{n_{x\max}}\sum_{n_y=1}^{n_{y\max}} T'_{70}(\mathrm{E},\mathrm{n}_x,\mathrm{n}_y,\eta_g)\mathrm{H}(\mathrm{E} - \mathrm{E}'_{30HDNW}) \tag{3.79}$$

In (3.79), E'_{30HDNW} is the sub-band energy in this case which can be expressed as

$$0 = T_{70}(E'_{30HDNW},\eta_g,n_x,n_y) \tag{3.80}$$

The EEM in this case is given by

$$m^*(E_{F1HDNW},\eta_g,n_x,n_y) = \frac{\hbar^2}{2}\frac{\partial}{\partial E}[T^2_{70}(E_{F1HDNW},\eta_g,n_x,n_y)] \tag{3.81}$$

In the absence of band tailing the 1D DR in this case assumes the form

$$k_z = t_{70}(E,n_x,n_y) \tag{3.82}$$

where $t_{70}(E,n_x,n_y) = \left[\left[\bar{\mathrm{c}}_0(\mathrm{E})[1 - \left(\frac{\pi n_x}{d_x\bar{a}_0(\mathrm{E})}\right)^2 - \left(\frac{\pi n_y}{d_y\bar{b}_0(\mathrm{E})}\right)^2\right]\right]^{1/2}$

$$N_{1D\Gamma}(\mathrm{E}) = \frac{2g_v}{\pi}\sum_{n_x=1}^{n_{x\max}}\sum_{n_y=1}^{n_{y\max}} t'_{70}(\mathrm{E},\mathrm{n}_x,\mathrm{n}_y)\mathrm{H}(\mathrm{E} - \mathrm{E}'_{26}) \tag{3.83}$$

In (3.83), E'_{26} is the sub-band energy in this case which can be expressed as

$$0 = t_{70}(E'_{26},n_x,n_y) \tag{3.84}$$

The EEM in this case is given by

$$m^*(E_{F1d},n_x,n_y) = \frac{\hbar^2}{2}\frac{\partial}{\partial E}[\mathrm{t}^2_{60}(E_{F1d},n_x,n_y)] \tag{3.85}$$

3.2.6 The DR in Quantum Wires (QWs) of HD Te

The 1D DR may be written in this case as

$$k_x = t_{72}(\mathrm{E},\mathrm{n}_y,\mathrm{n}_z,\eta_g) \tag{3.86}$$

where $t_{72}(\mathrm{E}, \mathrm{n_y}, \mathrm{n_z}, \eta_g) = \left[-\left(\frac{n_y\pi}{d_y}\right)^2 + \psi_{5HD}(\mathrm{E}, \eta_g) - \psi_6\left(\frac{\pi n_z}{d_z}\right)^2 \pm \psi_7\left[\psi_{8HD}^2(\mathrm{E}, \eta_g) - \left(\frac{\pi n_z}{d_z}\right)^2\right]^{1/2}\right]^{1/2}$

The DOS function in this case can be written as

$$N_{1DHD\Gamma}(\mathrm{E}, \eta_g) = \frac{2g_v}{\pi}\sum_{n_y=1}^{n_{y\max}}\sum_{n_z=1}^{n_{z\max}} t_{72}'(\mathrm{E}, \mathrm{n_y}, \mathrm{n_z}, \eta_g)\mathrm{H}(\mathrm{E} - \mathrm{E}_{31HDNW}') \tag{3.87}$$

In (3.87), E_{31HDNW}' is the sub-band energy in this case which can be expressed as

$$0 = t_{72}(E_{31HDNW}', \eta_g, n_y, n_z) \tag{3.88}$$

The EEM in this case is given by

$$m^*(E_{F1HDNW}, \eta_g, n_y, n_z) = \frac{\hbar^2}{2}\frac{\partial}{\partial E}\left[t_{72}^2(E_{F1HDNW}, \eta_g, n_y, n_z)\right] \tag{3.89}$$

In the absence of band tailing the 1D DR in this case assumes the form

$$k_x = H_{70}(\mathrm{E}, \mathrm{n_y}, \mathrm{n_z}) \tag{3.90}$$

where

$$H_{70}(\mathrm{E}, \mathrm{n_y}, \mathrm{n_z}) = \left[-\left(\frac{n_y\pi}{d_y}\right)^2 + \psi_5(\mathrm{E}) - \psi_6\left(\frac{\pi n_z}{d_z}\right)^2 \pm \psi_7\left[\psi_8^2(\mathrm{E}) - \left(\frac{\pi n_z}{d_z}\right)^2\right]^{1/2}\right]^{1/2}$$

$$N_{1D\Gamma}(\mathrm{E}) = \frac{2g_v}{\pi}\sum_{n_x=1}^{n_{x\max}}\sum_{n_y=1}^{n_{y\max}} H_{70}'(\mathrm{E}, \mathrm{n_y}, \mathrm{n_z})\mathrm{H}(\mathrm{E} - \mathrm{E}_{44}') \tag{3.91}$$

In (3.91), E_{44}' is the sub-band energy in this case which can be expressed as

$$0 = H_{70}(E_{44}', n_y, n_z) \tag{3.92}$$

The EEM in this case is given by

$$m^*(E_{F1D}, n_y, n_z) = \frac{\hbar^2}{2}\frac{\partial}{\partial E}[\mathrm{H}_{70}^2(E_{F1D}, n_y, n_z)] \quad (3.93)$$

3.2.7 The DR in Quantum Wires (QWs) of HD Gallium Phosphide

The 1D DR may be written in this case as

$$k_x = u_{70}(\mathrm{E}, \mathrm{n}_y, \mathrm{n}_z, \eta_g) \quad (3.94)$$

where $u_{70}(\mathrm{E}, \mathrm{n}_y, \mathrm{n}_z, \eta_g) = \left[-\left(\frac{n_y\pi}{d_y}\right)^2 + t_{11}\gamma_3(\mathrm{E}, \eta_g) + t_{21} - t_{31}\left(\frac{n_z\pi}{d_z}\right)^2 - t_{41}\left[\left(\frac{n_z\pi}{d_z}\right)^2 + t_5^2(\mathrm{E}, \eta_g)\right]^{1/2}\right]^{1/2}$

The DOS function in this case can be written as

$$N_{1DHD\Gamma}(\mathrm{E}, \eta_g) = \frac{2g_v}{\pi}\sum_{n_y=1}^{n_{y\max}}\sum_{n_z=1}^{n_{z\max}} u'_{70}(\mathrm{E}, \mathrm{n}_y, \mathrm{n}_z, \eta_g)\mathrm{H}(\mathrm{E} - \mathrm{E}'_{32HDNW}) \quad (3.95)$$

In (3.95), E'_{32HDNW} is the sub-band energy in this case which can be expressed as

$$0 = u_{70}(E'_{32HDNW}, \eta_g, n_y, n_z) \quad (3.96)$$

The EEM in this case is given by

$$m^*(E_{F1HDNW}, \eta_g, n_y, n_z) = \frac{\hbar^2}{2}\frac{\partial}{\partial E}\left[\mathrm{u}_{70}^2(E_{F1HDNW}, \eta_g, n_y, n_z)\right] \quad (3.97)$$

In the absence of band tailing the 1D DR in this case can be written as

$$k_x = X_{71}(\mathrm{E}, \mathrm{n}_y, \mathrm{n}_z) \quad (3.98)$$

where $X_{71}(\mathrm{E}, \mathrm{n}_y, \mathrm{n}_z) = \left[-\left(\frac{n_y\pi}{d_y}\right)^2 + t_{42}(\mathrm{E}, \mathrm{n}_z)\right]^{1/2}$

The DOS function is given by

$$N_{1D\Gamma}(\mathrm{E}) = \frac{2g_v}{\pi}\sum_{n_y=1}^{n_{y\max}}\sum_{n_z=1}^{n_{z\max}} X'_{71}(\mathrm{E}, \mathrm{n_y}, \mathrm{n_z})\mathrm{H}(\mathrm{E} - \mathrm{E}'_{46}) \tag{3.99}$$

In (3.99), E'_{46} is the sub-band energy in this case which can be expressed as

$$0 = X_{71}(E'_{46}, n_y, n_z) \tag{3.100}$$

The EEM in this case is given by

$$m^*(E_{F1D}, n_y, n_z) = \frac{\hbar^2}{2}\frac{\partial}{\partial E}\left[\mathrm{X}^2_{71}(E_{F1D}, n_y, n_z)\right] \tag{3.101}$$

3.2.8 *The DR in Quantum Wires (QWs) of HD Platinum Antimonide*

The 1D DR may be written in this case as

$$k_x = V_{70}(\mathrm{E}, \mathrm{n_y}, \mathrm{n_z}, \eta_g) \tag{3.102}$$

where $V_{70}(\mathrm{E}, \mathrm{n_y}, \mathrm{n_z}, \eta_g) = \left[-\left(\frac{n_y\pi}{d_y}\right)^2 + A_{60}(\mathrm{E}, \eta_g, \mathrm{n_y})\right]^{1/2}$

The DOS function in this case can be written as

$$N_{1DHD\Gamma}(\mathrm{E}, \eta_g) = \frac{2g_v}{\pi}\sum_{n_y=1}^{n_{y\max}}\sum_{n_z=1}^{n_{z\max}} V'_{70}(\mathrm{E}, \mathrm{n_y}, \mathrm{n_z}, \eta_g)\mathrm{H}(\mathrm{E} - \mathrm{E}'_{34HDNW}) \tag{3.103}$$

In (3.103), E'_{34HDNW} is the sub-band energy in this case which can be expressed as

$$0 = V_{70}(E'_{34HDNW}, \eta_g, n_y, n_z) \tag{3.104}$$

The EEM in this case is given by

$$m^*(E_{F1HDNW}, \eta_g, n_y, n_z) = \frac{\hbar^2}{2}\frac{\partial}{\partial E}\left[\mathrm{V}^2_{70}(E_{F1HDNW}, \eta_g, n_y, n_z)\right] \tag{3.105}$$

In the absence of band tailing the 1D DR in this case can be written as

$$k_x = D_{71}(\mathrm{E}, \mathrm{n_y}, \mathrm{n_z}) \tag{3.106}$$

where $D_{71}(\mathrm{E}, \mathrm{n_y}, \mathrm{n_z}) = \left[-\left(\frac{n_y \pi}{d_y}\right)^2 + t_{44}(\mathrm{E}, \mathrm{n_z})\right]^{1/2}$

The DOS function is given by

$$N_{1D\Gamma}(\mathrm{E}) = \frac{2g_v}{\pi} \sum_{n_y=1}^{n_{y\max}} \sum_{n_z=1}^{n_{z\max}} D'_{71}(\mathrm{E}, \mathrm{n_y}, \mathrm{n_z}) \mathrm{H}(\mathrm{E} - \mathrm{E}'_{48}) \tag{3.107}$$

In (3.107), E'_{48} is the sub-band energy in this case which can be expressed as

$$0 = D_{71}(E'_{48}, n_y, n_z) \tag{3.108}$$

The EEM in this case is given by

$$m^*(E_{F1D}, n_y, n_z) = \frac{\hbar^2}{2} \frac{\partial}{\partial E} \left[\mathrm{D}^2_{71}(E_{F1D}, n_y, n_z)\right] \tag{3.109}$$

3.2.9 The DR in Quantum Wires (QWs) of HD Bismuth Telluride

The DR in this case can be written as

$$k_x = J_{70}(\mathrm{E}, \mathrm{n_y}, \mathrm{n_z}, \eta_g) \tag{3.110}$$

where

$J_{70}(\mathrm{E}, \mathrm{n_y}, \mathrm{n_z}, \eta_g) = \left[\left[\gamma_2(\mathrm{E}, \eta_g) - \bar{\omega}_2 \left(\frac{n_y \pi}{d_y}\right)^2 - \bar{\omega}_3 \left(\frac{n_z \pi}{d_z}\right)^2 - 2\bar{\omega}_4 \frac{n_y n_z \pi^2}{d_y d_z}\right] (\bar{\omega}_1)^{-1}\right]^{1/2}$

The DOS function in this case can be written as

$$N_{1DHD\Gamma}(\mathrm{E}, \eta_g) = \frac{2g_v}{\pi} \sum_{n_y=1}^{n_{y\max}} \sum_{n_z=1}^{n_{z\max}} J'_{70}(\mathrm{E}, \mathrm{n_y}, \mathrm{n_z}, \eta_g) \mathrm{H}(\mathrm{E} - \mathrm{E}'_{50HDNW}) \tag{3.111}$$

and E'_{50HDNW} is the sub-band energy in this case which can be expressed as

$$0 = J_{70}(E'_{50HDNW}, \eta_g, n_y, n_z) \tag{3.112}$$

The EEM in this case is given by

$$m^{*}(E_{F1HDNW},\eta_g,n_y,n_z)=\frac{\hbar^2}{2}\frac{\partial}{\partial E}[\mathrm{J}_{70}^{2}(E_{F1HDNW},\eta_g,n_y,n_z)] \tag{3.113}$$

In the absence of band tailing the 1D DR in this case can be written as

$$k_x = B_{71}(\mathrm{E},\mathrm{n_y},\mathrm{n_z}) \tag{3.114}$$

where

$$B_{71}(\mathrm{E},\mathrm{n_y},\mathrm{n_z})=\left[\left[\mathrm{E}(1+\alpha\mathrm{E})-\bar{\omega}_2\left(\frac{n_y\pi}{d_y}\right)^2-\bar{\omega}_3\left(\frac{n_z\pi}{d_z}\right)^2-2\bar{\omega}_4\frac{n_yn_z\pi^2}{d_yd_z}\right](\bar{\omega}_1)^{-1}\right]^{1/2}$$

The DOS function is given by

$$N_{1D\Gamma}(\mathrm{E})=\frac{2g_v}{\pi}\sum_{n_y=1}^{n_{y\max}}\sum_{n_z=1}^{n_{z\max}}B'_{71}(\mathrm{E},\mathrm{n_y},\mathrm{n_z})\mathrm{H}(\mathrm{E}-\mathrm{E}'_{50}) \tag{3.115}$$

In (3.115) E'_{50} is the sub-band energy in this case which can be expressed as

$$0=B_{71}(E'_{50},n_y,n_z) \tag{3.116}$$

The EEM in this case is given by

$$m^{*}(E_{F1D},n_y,n_z)=\frac{\hbar^2}{2}\frac{\partial}{\partial E}[\mathrm{B}_{71}^{2}(E_{F1D},\eta_g,n_y,n_z)] \tag{3.117}$$

3.2.10 The DR in Quantum Wires (QWs) of HD Germanium

(a) Model of Cardona et al.

The DR in accordance with this model in the present case can be written as

$$k_x = L_{70}(\mathrm{E},\mathrm{n_y},\mathrm{n_z},\eta_g) \tag{3.118}$$

where $$L_{70}(\mathrm{E},\mathrm{n_y},\mathrm{n_z},\eta_g)=\left[\left[\gamma_2(\mathrm{E},\eta_g)+\alpha\left[\frac{\hbar^2}{2m^{*}_{\parallel}}\left(\frac{n_z\pi}{d_z}\right)^2\right]^2-(1+2\alpha\gamma_3(\mathrm{E},\eta_g))\frac{\hbar^2}{2m^{*}_{\parallel}}\left(\frac{n_z\pi}{d_z}\right)^2\right]\left(\frac{2m^{*}_{\parallel}}{\hbar^2}\right)\right]^{1/2}$$

The DOS function in this case can be written as

$$N_{1DHD\Gamma}(\mathrm{E},\eta_g)=\frac{2g_v}{\pi}\sum_{n_y=1}^{n_{y\max}}\sum_{n_z=1}^{n_{z\max}}L'_{70}(\mathrm{E},\mathrm{n}_y,\mathrm{n}_z,\eta_g)\mathrm{H}(\mathrm{E}-\mathrm{E}'_{52HDNW}) \qquad (3.119)$$

In (3.119) E'_{52HDNW} is the sub-band energy in this case which can be expressed as

$$0=L_{70}(E'_{52HDNW},\eta_g,n_y,n_z) \qquad (3.120)$$

The EEM in this case is given by

$$m^*(E_{F1HDNW},\eta_g,n_y,n_z)=\frac{\hbar^2}{2}\frac{\partial}{\partial E}[\mathrm{L}^2_{70}(E_{F1HDNW},\eta_g,n_y,n_z)] \qquad (3.121)$$

In the absence of band tailing the 1D DR in this case can be written as

$$k_x=B_{77}(\mathrm{E},\mathrm{n}_y,\mathrm{n}_z) \qquad (3.122)$$

where

$$B_{77}(\mathrm{E},\mathrm{n}_y,\mathrm{n}_z)=\left[\left[\left[E(1+\alpha\mathrm{E})+\alpha\left[\frac{\hbar^2}{2m^*_\parallel}\left(\frac{n_z\pi}{d_z}\right)^2\right]\right]^2-(1+2\alpha E)\frac{\hbar^2}{2m^*_\parallel}\left(\frac{n_z\pi}{d_z}\right)^2\right]\frac{2m^*_\parallel}{\hbar^2}\right]^{1/2}$$

The DOS function is given by

$$N_{1D\Gamma}(\mathrm{E})=\frac{2g_v}{\pi}\sum_{n_y=1}^{n_{y\max}}\sum_{n_z=1}^{n_{z\max}}B'_{77}(\mathrm{E},\mathrm{n}_y,\mathrm{n}_z)\mathrm{H}(\mathrm{E}-\mathrm{E}'_{60}) \qquad (3.123)$$

In (3.123) E'_{60} is the sub-band energy in this case which can be expressed as

$$0=B_{77}(E'_{60},n_y,n_z) \qquad (3.124)$$

The EEM in this case is given by

$$m^*(E_{F1D},n_y,n_z)=\frac{\hbar^2}{2}\frac{\partial}{\partial E}[\mathrm{B}^2_{77}(E_{F1D},n_y,n_z)] \qquad (3.125)$$

(b) Model of Wang et al.

The DR in accordance with this model in the present case can be written as

$$k_x=\beta_{70}(\mathrm{E},\mathrm{n}_y,\mathrm{n}_z,\eta_g) \qquad (3.126)$$

where $\beta_{70}(\mathrm{E}, \mathrm{n}_y, \mathrm{n}_z, \eta_g) = \left[-\left(\frac{n_y\pi}{d_y}\right)^2 + \frac{2m_\perp^*}{\hbar^2}\left[\bar{\alpha}_8 - \bar{\alpha}_9\left(\frac{\pi n_z}{d_z}\right)^2 - \bar{\alpha}_{10}\left[\left(\frac{\pi n_z}{d_z}\right)^4 + \bar{\alpha}_{11}\left(\frac{\pi n_z}{d_z}\right)^2 + \bar{\alpha}_{12}(\mathrm{E}, \eta_g)\right]^{1/2}\right]\right]^{1/2}$

The DOS function in this case can be written as

$$N_{1DHD\Gamma}(\mathrm{E}, \eta_g) = \frac{2g_v}{\pi}\sum_{n_y=1}^{n_{y\max}}\sum_{n_z=1}^{n_{z\max}} \beta'_{70}(\mathrm{E}, \mathrm{n}_y, \mathrm{n}_z, \eta_g)\mathrm{H}(\mathrm{E} - \mathrm{E}'_{54HDNW}) \tag{3.127}$$

In (3.127), E'_{54HDNW} is the sub-band energy in this case which can be expressed as

$$0 = \beta_{70}(E'_{54HDNW}, \eta_g, n_y, n_z) \tag{3.128}$$

The EEM in this case is given by

$$m^*(E_{F1HDNW}, \eta_g, n_y, n_z) = \frac{\hbar^2}{2}\frac{\partial}{\partial E}[\beta_{70}^2(E_{F1HDNW}, \eta_g, n_y, n_z)] \tag{3.129}$$

In the absence of band tailing the 1D DR in this case can be written as

$$k_x = P_{77}(\mathrm{E}, \mathrm{n}_y, \mathrm{n}_z) \tag{3.130}$$

where $P_{77}(\mathrm{E}, \mathrm{n}_y, \mathrm{n}_z) = \left[\left[\mathrm{I}_1(\mathrm{E}, \mathrm{n}_z) - \frac{\hbar^2}{2m_2^*}\left(\frac{n_y\pi}{d_y}\right)^2\right]\left(\frac{2m_1^*}{\hbar^2}\right)\right]^{1/2}$

The DOS function is given by

$$N_{1D\Gamma}(\mathrm{E}) = \frac{2g_v}{\pi}\sum_{n_y=1}^{n_{y\max}}\sum_{n_z=1}^{n_{z\max}} P'_{77}(\mathrm{E}, \mathrm{n}_y, \mathrm{n}_z)\mathrm{H}(\mathrm{E} - \mathrm{E}'_{80}) \tag{3.131}$$

and E'_{80} is the sub-band energy in this case which can be expressed as

$$0 = P_{77}(E'_{80}, n_y, n_z) \tag{3.132}$$

The EEM in this case is given by

$$m^*(E_{F1D}, \eta_y, n_z) = \frac{\hbar^2}{2}\frac{\partial}{\partial E}[P_{77}^2(E_{F1D}, n_y, n_z)] \tag{3.133}$$

3.2.11 *The DR in Quantum Wires (QWs) of HD Gallium Antimonide*

The DR of the 1D electrons in this case can be written as

$$\frac{\hbar^2 (n_z \pi/d_z)^2}{2m_c} + \frac{\hbar^2 (n_y \pi/d_y)^2}{2m_c} + \frac{\hbar^2 k_x^2}{2m_c} = I_{36}(E, \eta_g) \tag{3.134}$$

The DOS function in this case can be written as

$$N_{1DHD\Gamma}(E, \eta_g) = \frac{2g_v}{\pi} \sum_{n_y=1}^{n_{y\max}} \sum_{n_z=1}^{n_{z\max}} N'_{100}(E, n_y, n_z, \eta_g) H(E - E'_{100HDNW}) \tag{3.135}$$

where $N_{100}(E, n_y, n_z, \eta_g) = \left[[I_{36}(E, \eta_g) - G_2(n_y, n_z)] \left(\frac{2m_c}{\hbar^2} \right) \right]^{\frac{1}{2}}$
and $E'_{100HDNW}$ is the sub-band energy in this case which can be expressed as

$$\frac{\hbar^2 (n_z \pi/d_z)^2}{2m_c} + \frac{\hbar^2 (n_y \pi/d_y)^2}{2m_c} = I_{36}(E'_{100HDNW}, \eta_g) \tag{3.136}$$

The EEM in this case is given by

$$m^*(E_{F1HDNW}, \eta_g) = m_c [I'_{36}(E_{F1HDNW}, \eta_g)] \tag{3.137}$$

The expression of 1D DR, for NWs of GaSb whose energy band structures in the absence of band tailing assumes the form

$$I_{36}(E) = \frac{\hbar^2 k_x^2}{2m_c} + G_2(n_y, n_z) \tag{3.138}$$

The DOS function can be expressed as

$$N_{1D\Gamma}(E) = \frac{2g_v}{\pi} \sum_{n_y=1}^{n_{y\max}} \sum_{n_z=1}^{n_{z\max}} f'_{151}(E, n_y, n_z) H(E - E'_{101}) \tag{3.139}$$

where $f_{151}(E, n_y, n_z) = \left[\frac{2m_c}{\hbar^2} [I_{36}(E) - G_2(n_y, n_z)] \right]^{\frac{1}{2}}$
In this case, the quantized energy E'_{101} is given by

$$I_{36}(E'_{101}) = G_2(n_y, n_z) \tag{3.140}$$

3.2.12 The DR in Quantum Wells (QWs) of HD II–V Materials

The DR of the 1D holes in II–V compounds can be expressed as

$$\gamma_3(E,\eta_g) = A_{10}\left(\frac{n_x\pi}{d_x}\right)^2 + A_{11}\left(\frac{n_y\pi}{d_y}\right)^2 + A_{12}k_z^2 + A_{13}\left(\frac{n_x\pi}{d_x}\right)$$
$$\pm\left[\left(A_{14}\left(\frac{n_x\pi}{d_x}\right)^2 + A_{15}\left(\frac{n_y\pi}{d_y}\right)^2 + A_{16}k_z^2 + A_{17}\left(\frac{n_x\pi}{d_x}\right)^2\right) + A_{18}\left(\frac{n_y\pi}{d_y}\right)^2 + A_{19}^2\right]^{1/2} \quad (3.141)$$

where the numerical values of the energy band constants are given in Appendix A.

The sub-band energy $(E_{n_{zHD401}})$ is the lowest positive root of the following equation

$$\gamma_3(E_{n_{zHD401}},\eta_g) = A_{10}\left(\frac{n_x\pi}{d_x}\right)^2 + A_{13}\left(\frac{n_x\pi}{d_x}\right)$$
$$\pm\left[\left(A_{14}\left(\frac{n_x\pi}{d_x}\right)^2 + A_{17}\left(\frac{n_x\pi}{d_x}\right)\right)^2 + A_{19}^2\right]^{1/2} \quad (3.142)$$

(3.142) can be expressed as

$$k_z = \Delta_{27}(E,\eta_g,n_x,n_y) \quad (3.143)$$

where

$$\Delta_{27}(E,\eta_g,n_x,n_y) = \Big[(A_{12}^2 - A_{16}^2)^{-1}[[\Delta_{21}(E,\eta_g,n_x,n_y)A_{12} + A_{16}\Delta_{22}(n_x,n_y)] + [[\Delta_{21}(E,\eta_g,n_x,n_y)A_{12}$$
$$+ A_{16}\Delta_{22}(n_x,n_y)]^2 - (A_{12}^2 - A_{16}^2)\Delta_{25}(n_x,n_y,E,\eta_g)\Big]^{1/2}\Big]\Big],$$

$$\Delta_{21}(E,\eta_g,n_x,n_y) = \left[\gamma_3(E,\eta_g) - A_{10}\left(\frac{n_x\pi}{d_x}\right)^2 - A_{11}\left(\frac{n_y\pi}{d_y}\right)^2 - A_{13}\left(\frac{n_x\pi}{d_x}\right)\right],$$

$$\Delta_{22}(n_x,n_y) = \left[A_{14}\left(\frac{n_x\pi}{d_x}\right)^2 + A_{15}\left(\frac{n_y\pi}{d_y}\right)^2 + A_{17}\left(\frac{n_x\pi}{d_x}\right)\right],$$

$$\Delta_{25}(n_x,n_y,E,\eta_g) = [\Delta_{21}^2(E,\eta_g,n_x,n_y) - \Delta_{24}(n_x,n_y)],$$

$$\Delta_{24}(n_x,n_y) = [\Delta_{22}^2(n_x,n_y) + \Delta_{23}(n_y)] \text{ and } \Delta_{23}(n_y) = \left[A_{18}\left(\frac{n_y\pi}{d_y}\right)^2 + A_{19}^2\right]$$

The DOS function in this case can be written as

$$N_{1DHD\Gamma}(\mathrm{E},\eta_g)=\frac{g_v}{\pi}\sum_{n_x=1}^{n_{x\max}}\sum_{n_y=1}^{n_{y\max}}\Delta'_{27}(\mathrm{E},\mathrm{n}_x,\mathrm{n}_y,\eta_g)\mathrm{H}(\mathrm{E}-\mathrm{E}'_{200HDNW}) \tag{3.144}$$

In (3.144), $E'_{200HDNW}$ is the sub-band energy in this case which can be expressed as

$$0=\Delta_{27}(E'_{200HDNW},\mathrm{n}_x,\mathrm{n}_y,\eta_g) \tag{3.145}$$

The EEM in this case is given by

$$m^*(E_{F1HDNW},\eta_g,n_x,n_y)=\frac{\hbar^2}{2}\frac{\partial}{\partial E}[\Delta_{27}^2(E_{F1HDNW},\eta_g,n_x,n_y)] \tag{3.146}$$

In the absence of band-tailing, the 1D hole energy spectrum in this case assumes the form

$$E=A_{10}\left(\frac{n_x\pi}{d_x}\right)^2+A_{11}\left(\frac{n_y\pi}{d_y}\right)^2+A_{12}k_z^2+A_{13}\left(\frac{n_x\pi}{d_x}\right)$$
$$\pm\left[\left(A_{14}\left(\frac{n_x\pi}{d_x}\right)^2+A_{15}\left(\frac{n_y\pi}{d_y}\right)^2+A_{16}k_z^2+A_{17}\left(\frac{n_x\pi}{d_x}\right)\right)^2+A_{18}\left(\frac{n_y\pi}{d_y}\right)^2+A_{19}^2\right]^{1/2} \tag{3.147}$$

The subband energy (E'_{300}) is the lowest positive root of the following equation

$$E'_{300}=A_{10}\left(\frac{n_x\pi}{d_x}\right)^2+A_{13}\left(\frac{n_x\pi}{d_x}\right)^2\pm\left[\left(A_{14}\left(\frac{n_x\pi}{d_x}\right)^2+A_{17}\left(\frac{n_x\pi}{d_x}\right)\right)^2+A_{19}^2\right]^{1/2} \tag{3.148}$$

(3.147) can be expressed as

$$k_z=\Delta_{271}(E,n_x,n_y) \tag{3.149}$$

where

$$\Delta_{271}(E,n_x,n_y) = \Big[(A_{12}^2 - A_{16}^2)^{-1}\big[[\Delta_{211}(E,n_x,n_y)A_{12} + A_{16}\Delta_{22}(n_x,n_y)] + \Big[[\Delta_{211}(E,n_x,n_y)A_{12} + A_{16}\Delta_{22}(n_x,n_y)]^2 - (A_{12}^2 - A_{16}^2)\Delta_{251}(n_x,n_y,E)\Big]^{1/2}\big]\Big],$$

$$\Delta_{211}(E,n_x,n_y) = \left[E - A_{10}\left(\frac{n_x\pi}{d_x}\right)^2 - A_{11}\left(\frac{n_y\pi}{d_y}\right)^2 - A_{13}\left(\frac{n_x\pi}{d_x}\right)\right],$$

$$\Delta_{22}(n_x,n_y) = \left[A_{14}\left(\frac{n_x\pi}{d_x}\right)^2 + A_{15}\left(\frac{n_y\pi}{d_y}\right)^2 + A_{17}\left(\frac{n_x\pi}{d_x}\right)\right],$$

$$\Delta_{251}(n_x,n_y,E) = [\Delta_{211}^2(E,n_x,n_y) - \Delta_{24}(n_x,n_y)],$$

$$\Delta_{24}(n_x,n_y) = [\Delta_{22}^2(n_x,n_y) + \Delta_{23}(n_y)] \text{ and } \Delta_{23}(n_y) = \left[A_{18}\left(\frac{n_y\pi}{d_y}\right)^2 + A_{19}^2\right]$$

The DOS function in this case can be written as

$$N_{1D\Gamma}(E) = \frac{g_v}{\pi}\sum_{n_x=1}^{n_{x\max}}\sum_{n_y=1}^{n_{y\max}} \Delta'_{271}(E,n_x,n_y)H(E - E'_{300}) \tag{3.150}$$

In (3.150), E'_{300} is the sub-band energy in this case which can be expressed as

$$0 = \Delta_{271}(E'_{300},n_x,n_y) \tag{3.151}$$

The EEM in this case is given by

$$m^*(E_{F1D},n_x,n_y) = \frac{\hbar^2}{2}\frac{\partial}{\partial E}[\Delta_{271}^2(E_{F1D},n_x,n_y)] \tag{3.152}$$

3.2.13 The DR in Quantum Wells (QWs) of HD Lead Germanium Telluride

The 1D electron energy spectrum in n-type $Pb_{1-x}Ge_xTe$ under the condition of formation of band tails can be written as

$$\left[\frac{2}{1+Erf\left(\frac{E}{\eta_g}\right)}\right]\theta_0(E,\eta_g)+\gamma_3(E,\eta_g)\left[\bar{E}_{g_0}-0.195\left[\left(\frac{n_x\pi}{d_x}\right)^2+\left(\frac{n_y\pi}{d_y}\right)^2\right]\right.$$
$$\left.-0.345k_z^2\right]=\left[0.23\left[\left(\frac{n_x\pi}{d_x}\right)^2+\left(\frac{n_y\pi}{d_y}\right)^2\right]+0.02k_z^2\right.$$
$$\pm\left[0.06\bar{E}_{g_0}+0.061\left[\left(\frac{n_x\pi}{d_x}\right)^2+\left(\frac{n_y\pi}{d_y}\right)^2\right]+0.0066k_z^2\right]\left[\left(\frac{n_x\pi}{d_x}\right)^2+\left(\frac{n_y\pi}{d_y}\right)^2\right]^{1/2}$$
$$+\left[\bar{E}_{g_0}+0.411\left[\left(\frac{n_x\pi}{d_x}\right)^2+\left(\frac{n_y\pi}{d_y}\right)^2\right]+0.377k_z^2\right]\left[0.606\left[\left(\frac{n_x\pi}{d_x}\right)^2\right.\right.$$
$$\left.\left.+\left(\frac{n_y\pi}{d_y}\right)^2\right]+0.722k_z^2\right] \tag{3.153}$$

The subband energy $(E_{n_{zHD500}})$ is the lowest positive root of the following equation

$$\left[\frac{2}{1+Erf\left(\frac{E_{n_{zHD500}}}{\eta_g}\right)}\right]\theta_0(E_{n_{zHD500}},\eta_g)+\gamma_3(E_{n_{zHD500}},\eta_g)\left[\bar{E}_{g_0}-0.195\left[\left(\frac{n_x\pi}{d_x}\right)^2+\left(\frac{n_y\pi}{d_y}\right)^2\right]\right]$$
$$=\left[0.23\left[\left(\frac{n_x\pi}{d_x}\right)^2+\left(\frac{n_y\pi}{d_y}\right)^2\right]\pm\left[0.06\bar{E}_{g_0}+0.061\left[\left(\frac{n_x\pi}{d_x}\right)^2+\left(\frac{n_y\pi}{d_y}\right)^2\right]\right]\left[\left(\frac{n_x\pi}{d_x}\right)^2+\left(\frac{n_y\pi}{d_y}\right)^2\right]^{1/2}\right.$$
$$\left.+\left[\bar{E}_{g_0}+0.411\left[\left(\frac{n_x\pi}{d_x}\right)^2+\left(\frac{n_y\pi}{d_y}\right)^2\right]\right]\left[0.606\left[\left(\frac{n_x\pi}{d_x}\right)^2+\left(\frac{n_y\pi}{d_y}\right)^2\right]\right]\right] \tag{3.154}$$

The EEM and the DOS function for both the cases should be calculated numerically.

The 1D dispersion law of n-type $Pb_{1-x}Ge_xTe$ with $x = 0.01$ in the absence of band-tails can be expressed as

$$\left[E-0.606\left[\left(\frac{n_x\pi}{d_x}\right)^2+\left(\frac{n_y\pi}{d_y}\right)^2\right]-0.722k_z^2\right]$$
$$\left[E+\overline{E}_{g_0}+0.411\left[\left(\frac{n_x\pi}{d_x}\right)^2+\left(\frac{n_y\pi}{d_y}\right)^2\right]+0.377k_z^2\right]$$
$$=0.23\left[\left(\frac{n_x\pi}{d_x}\right)^2+\left(\frac{n_y\pi}{d_y}\right)^2\right]+0.02k_z^2\pm\left[0.06\overline{E}_{g_0}+0.061\left[\left(\frac{n_x\pi}{d_x}\right)^2+\left(\frac{n_y\pi}{d_y}\right)^2\right]\right.$$
$$\left.+0.0066k_z^2\right]\left[\left(\frac{n_x\pi}{d_x}\right)^2+\left(\frac{n_y\pi}{d_y}\right)^2\right]^{1/2} \tag{3.155}$$

The sub-band energy $\bar{E}_{500}$ in this can be written as

$$\left[\bar{E}_{500}-.606\left[\left(\frac{n_x\pi}{d_x}\right)^2+\left(\frac{n_y\pi}{d_y}\right)^2\right]\right]\left[\bar{E}_{500}+\overline{E}_{g_0}+0.411\left[\left(\frac{n_x\pi}{d_x}\right)^2+\left(\frac{n_y\pi}{d_y}\right)^2\right]\right]$$
$$=0.23\left[\left(\frac{n_x\pi}{d_x}\right)^2+\left(\frac{n_y\pi}{d_y}\right)^2\right]\pm\left[0.06\overline{E}_{g_0}+0.061\left[\left(\frac{n_x\pi}{d_x}\right)^2+\left(\frac{n_y\pi}{d_y}\right)^2\right]\right]$$
$$\left[\left(\frac{n_x\pi}{d_x}\right)^2+\left(\frac{n_y\pi}{d_y}\right)^2\right]^{1/2} \tag{3.156}$$

3.2.14 The DR in Quantum Wires (QWs) of HD Zinc and Cadmium Diphosphides

The DR in HD NWs of Zinc and Cadmium diphosphides can be written as

$$\gamma_3(E,\eta_g) = \left[\beta_1 + \frac{\beta_2\beta_{31}(k_x,n_y,n_z)}{8\beta_4}\right]\left[k_x^2 + \left(\frac{n_y\pi}{d_y}\right)^2 + \left(\frac{n_z\pi}{d_z}\right)^2\right]$$
$$\pm\left\{\left[\beta_4\beta_{31}(k_x,n_y,n_z)\left(\beta_5 - \frac{\beta_2\beta_{31}(k_x,n_y,n_z)}{8\beta_4}\right)\left[k_x^2 + \left(\frac{n_y\pi}{d_y}\right)^2 + \left(\frac{n_z\pi}{d_z}\right)^2\right]\right]\right.$$
$$\left.+\,8\beta_4^2\left(1 - \frac{\beta_{31}^2(k_x,n_y,n_z)}{4}\right) - \beta_2\left(1 - \frac{\beta_{31}^2(k_x,n_y,n_z)}{4}\right)\left[k_x^2 + (\frac{n_y\pi}{d_y})^2 + (\frac{n_z\pi}{d_z})^2\right]\right\}^{1/2} \tag{3.157}$$

where $\beta_{31}(k_x,n_y,n_z) = \left[\dfrac{k_x^2 + \left(\frac{n_y\pi}{d_y}\right)^2 - 2\left(\frac{n_z\pi}{d_z}\right)^2}{\left[k_x^2 + \left(\frac{n_y\pi}{d_y}\right)^2 + \left(\frac{n_z\pi}{d_z}\right)^2\right]}\right]$

The sub-band energy E_{n_zHD600} in this case assumes the form

$$\gamma_3(E_{n_zHD600},\eta_g) = \left[\beta_1 + \frac{\beta_2\beta_{31}(0,n_y,n_z)}{8\beta_4}\right]\left[\left(\frac{n_y\pi}{d_y}\right)^2 + \left(\frac{n_z\pi}{d_z}\right)^2\right]$$
$$\pm\left\{\left[\beta_4\beta_{31}(0,n_y,n_z)\left(\beta_5 - \frac{\beta_2\beta_{31}(0,n_y,n_z)}{8\beta_4}\right)\left[\left(\frac{n_y\pi}{d_y}\right)^2 + \left(\frac{n_z\pi}{d_z}\right)^2\right]\right]\right.$$
$$\left.+\,8\beta_4^2\left(1 - \frac{\beta_{31}^2(0,n_y,n_z)}{4}\right) - \beta_2\left(1 - \frac{\beta_{31}^2(0,n_y,n_z)}{4}\right)\left[\left(\frac{n_y\pi}{d_y}\right)^2 + \left(\frac{n_z\pi}{d_z}\right)^2\right]\right\}^{1/2} \tag{3.158}$$

where

$$\beta_{31}(0,n_y,n_z) = \left[\frac{\left(\frac{n_y\pi}{d_y}\right)^2 - 2\left(\frac{n_z\pi}{d_z}\right)^2}{\left[\left(\frac{n_y\pi}{d_y}\right)^2 + \left(\frac{n_z\pi}{d_z}\right)^2\right]}\right]$$

The EEM and the DOS function should be obtained numerically.

The 1D DR in NWs of Zinc and Cadmium diphosphides in the absence of band-tails can be written as

$$E = \left[\beta_1 + \frac{\beta_2\beta_{311}(k_x, n_y, n_z)}{8\beta_4}\right]\left[k_x^2 + \left(\frac{n_y\pi}{d_y}\right)^2 + \left(\frac{n_z\pi}{d_z}\right)^2\right] \pm \left\{\left[\beta_4\beta_{311}(k_x, n_y, n_z)\left(\beta_5 - \frac{\beta_2\beta_{311}(k_x, n_y, n_z)}{8\beta_4}\right)\left[k_x^2 + \left(\frac{n_y\pi}{d_y}\right)^2 + \left(\frac{n_z\pi}{d_z}\right)^2\right]\right] + 8\beta_4^2\left(1 - \frac{\beta_{311}^2(k_x, n_y, n_z)}{4}\right) - \beta_2\left(1 - \frac{\beta_{311}^2(k_x, n_y, n_z)}{4}\right)\left[k_x^2 + \left(\frac{n_y\pi}{d_y}\right)^2 + \left(\frac{n_z\pi}{d_z}\right)^2\right]\right\}^{1/2} \quad (3.159)$$

The subband energy((E_{700}) is the lowest positive root of the following equation

$$E_{700} = \left[\beta_1 + \frac{\beta_2\beta_{311}(0, n_y, n_z)}{8\beta_4}\right]\left[\left(\frac{n_y\pi}{d_y}\right)^2 + \left(\frac{n_z\pi}{d_z}\right)^2\right] \pm \left\{\left[\beta_4\beta_{311}(0, n_y, n_z)\left(\beta_5 - \frac{\beta_2\beta_{311}(0, n_y, n_z)}{8\beta_4}\right)\left[\left(\frac{n_y\pi}{d_y}\right)^2 + \left(\frac{n_z\pi}{d_z}\right)^2\right]\right] + 8\beta_4^2\left(1 - \frac{\beta_{311}^2(0, n_y, n_z)}{4}\right) - \beta_2\left(1 - \frac{\beta_{311}^2(0, n_y, n_z)}{4}\right)\left[\left(\frac{n_y\pi}{d_y}\right)^2 + \left(\frac{n_z\pi}{d_z}\right)^2\right]\right\}^{1/2} \quad (3.160)$$

The EEM and the DOS function for both the cases should be calculated numerically.

3.3 Summary and Conclusion

From the 1D DR in NWs of HD nonlinear optical and tetragonal materials (3.1), we observe that the real part of k_x versus E plot are the series of quantized non-parabolae. The (3.68) represents the 1D DR in NWs of HD IV–VI materials in accordance with the model of Bangert and Kastner and same conclusion is also valid. From (3.9) we have the same inference for NWs of HD III–V materials whose unperturbed conduction electrons obey the three band model of Kane, which contains one non removal pole in energy axis. The 1D electrons in HD III–V materials are also described by two band model of Kane, parabolic energy bands, model of Stillman et al. and the model of Palik et al. with the 1D DRs as given by (3.17), (3.25), (3.33) and (3.41) respectively. Besides the 1D DRs of Te, GaP, $PtSb_2$, Bi_2Te_3, Ge and GaSb are given by (3.86), (3.94), (3.102), (3.110), (3.118 in accordance with the model of Cardona et al. of Ge), (3.126 in accordance with the model of Wang and Ressler of Ge) and (3.134) respectively. All the said DRs possess no poles in the finite energy planes, the k_x versus E plot are the series of quantized non-parabolae. The 1D DR (3.49) in HD II–VI materials reflects the fact that the k_x versus E plot are the series of quantized non-parabolae. The 1D DR (3.58) of HD IV–VI materials represents the fact that the k_x versus E plot are the series of quantized non-parabolae in accordance with Dimmock model. The 1D DRs for II–V, $Pb_{1-x}Ge_xTe$, ZnP_2 and CdP_2 are given by (3.143), (3.153) and (3.157) respectively and the aforementioned conclusion is also true in this case. The 1D DR (3.78) in NWs of HD stressed Kane type materials reflects the fact identical conclusion in this context.

The influence of quantum confinement is immediately apparent from the 1D DRs since the DR depends strongly on the thickness of the quantum-confined materials in contrast with the corresponding bulk specimens. Finally, it may be noted that the basic aim of this chapter is not solely to demonstrate the influence of quantum confinement on the DR from NWs of HD non-linear optical, III–V, II–VI, IV–VI, n-GaP, n-Ge, $PtSb_2$ and stressed compounds respectively but also to formulate the appropriate DOS in the most generalized form, since the transport and other phenomena in quantized structures having different band structures and the derivation of the expressions of many important electronic properties are based on the DOS in such materials.

3.4 Open Research Problems

(R.3.1) Investigate the DR for NWs of all of the HD materials in the presences of Gaussian, exponential, Kane, Halperian, Lax and Bonch-Burevich types of band tails for all systems whose unperturbed carrier energy spectra are defined in (R.1.1).

(R.3.2) Investigate the DR in the presence of strain for NWs of all of the HDmaterials of the negative refractive index, organic, magnetic and other advanced optical materials.

(R.3.3) Investigate the DR for the NWs of HD negative refractive index, organic, magnetic and other advanced optical materials in the presence of an arbitrarily oriented alternating electric field.

(R.3.4) Investigate the DR for the multiple NWs of HD materials whose unperturbed carrier energy spectra are defined in (R.1.1).

(R.3.5) Investigate the DR for all the appropriate HD one dimensional systems of this chapter in the presence of finite potential wells.

(R.3.6) Investigate the DR for all the appropriate HD one dimensional systems of this chapter in the presence of parabolic potential wells.

(R.3.7) Investigate the DR for HD one dimensional systems of the negative refractive index and other advanced optical materials in the presence of an arbitrarily oriented alternating electric field and non-uniform light waves and in the presence of strain.

(R.3.8) Investigate the DR for triangular HD one dimensional systems of the negative refractive index, organic, magnetic and other advanced optical materials in the presence of an arbitrarily oriented alternating electric field in the presence of strain.

(R.3.9) Investigate the DR for all the problems of (R.1.1) in the presence of arbitrarily oriented magnetic field.

(R.3.10) Investigate the DR for all the problems of (R.1.1) in the presence of alternating electric field.

(R.3.11) Investigate the DR for all the problems of (R.1.1) in the presence of alternating magnetic field.

(R.3.12) Investigate the DR for all the problems of (R.1.1) in the presence of crossed electric field and quantizing magnetic fields.

(R.3.13) Investigate the DR for all the problems of (R.1.1) in the presence of crossed alternating electric field and alternating quantizing magnetic fields.

(R.3.14) Investigate the DR for HD NWs of the negative refractive index, organic and magnetic materials.

(R.3.15) Investigate the DR for HD NWs of the negative refractive index, organic and magnetic materials in the presence of alternating time dependent magnetic field.

(R.3.16) Investigate the DR for HD NWs of the negative refractive index, organic and magnetic materials in the presence of in the presence of crossed alternating electric field and alternating quantizing magnetic fields.

(R.3.17) (a) Investigate the DR for HD NWs of the negative refractive index, organic, magnetic and other advanced optical materials in the presence of an arbitrarily oriented alternating electric field considering many body effects.

(b) Investigate all the appropriate problems of this chapter for a Dirac electron.

(R.3.18) Investigate all the appropriate problems of this chapter by including the many body, image force, broadening and hot carrier effects respectively.

(R.3.19) Investigate all the appropriate problems of this chapter by removing all the mathematical approximations and establishing the respective appropriate uniqueness conditions.

References

1. P. Harrison, *Quantum Wells, Wires and Dots*. (Wiley, New York, 2002); B.K. Ridley, *Electrons and Phonons in Semiconductors Multilayers*. (Cambridge University Press, Cambridge, 1997); G. Bastard, Wave mechanics applied to semiconductor heterostructures, Halsted; Les Ulis, Les Editions de Physique, New York (1988); V.V. Martin, A.A. Kochelap, M.A. Stroscio, *Quantum Heterostructures*. (Cambridge University Press, Cambridge, 1999)

2. C. S. Lent and D. J. Kirkner, *J. Appl. Phys.*67, 6353 (1990); F. Sols, M. Macucci, U. Ravaioli and K. Hess, *Appl. Phys. Lett.* **54**, 350 (1980)
3. C.S. Kim, A.M. Satanin, Y.S. Joe, R.M. Cosby, Phys. Rev. B **60**, 10962 (1999)
4. S. Midgley, J.B. Wang, Phys. Rev. B **64**, 153304 (2001)
5. T. Sugaya, J.P. Bird, M. Ogura, Y. Sugiyama, D.K. Ferry, K.Y. Jang, Appl. Phys. Lett. **80**, 434 (2002)
6. B. Kane, G. Facer, A. Dzurak, N. Lumpkin, R. Clark, L. PfeiKer, K. West, Appl. Phys. Lett. **72**, 3506 (1998)
7. C. Dekker, Phys. Today **52**, 22 (1999)
8. A. Yacoby, H.L. Stormer, N.S. Wingreen, L.N. Pfeiffer, K.W. Baldwin, K.W. West, Phys. Rev. Lett. **77**, 4612 (1996)
9. Y. Hayamizu, M. Yoshita, S. Watanabe, H.A.L. PfeiKer, K. West, Appl. Phys. Lett. **81**, 4937 (2002)
10. S. Frank, P. Poncharal, Z.L. Wang, W.A. Heer, Science **280**, 1744 (1998)
11. I. Kamiya, I.I. Tanaka, K. Tanaka, F. Yamada, Y. Shinozuka, H. Sakaki, Physica E **13**, 131 (2002)
12. A.K. Geim, P.C. Main, N. LaScala, L. Eaves, T.J. Foster, P.H. Beton, J.W. Sakai, F.W. Sheard, M. Henini, G. Hill et al., Phys. Rev. Lett. **72**, 2061 (1994)
13. A.S. Melinkov, V.M. Vinokur, Nature **415**, 60 (2002)
14. K. Schwab, E.A. henriksen, J. M. Worlock, M.L. Roukes, Nature **404**, 974 (2000)
15. L. Kouwenhoven, Nature **403**, 374 (2000)
16. S. Komiyama, O. Astafiev, V. Antonov, H. Hirai, Nature **403**, 405 (2000)
17. E. Paspalakis, Z. Kis, E. Voutsinas, A.F. Terziz, Phys. Rev. B **69**, 155316 (2004)
18. J.H. Jefferson, M. Fearn, D.L.J. Tipton, T.P. Spiller, Phys. Rev. A **66**, 042328 (2002)
19. J. Appenzeller, C. Schroer, T. Schapers, A. Hart, A. Froster, B. Lengeler, H. Luth, Phys. Rev. B **53**, 9959 (1996)
20. J. Appenzeller, C. Schroer, J. Appl. Phys. **87**, 31659 (2002)
21. P. Debray, O.E. Raichev, M. Rahman, R. Akis, W.C. Mitchel, Appl. Phys. Lett. **74**, 768 (1999)
22. P.M. Solomon, Proc. IEEE **70**, 489 (1982); T.E. Schlesinger, T. Kuech, Appl. Phys. Lett. **49**, 519 (1986)
23. D. Kasemet, C.S. Hong, N.B. Patel, P.D. Dapkus, Appl. Phys. Lett. **41**, 912 (1982); K. Woodbridge, P. Blood, E.D. Pletcher, P.J. Hulyer, Appl. Phys. Lett. **45**, 16 (1984); S. Tarucha, H.O. Okamoto, Appl. Phys. Lett. **45**, 16 (1984); H. Heiblum, D.C. Thomas, C.M. Knoedler, M.I Nathan, Appl. Phys. Letts. **47**, 1105 (1985)
24. O. Aina, M. Mattingly, F.Y. Juan, P.K. Bhattacharyya, Appl. Phys. Lett. **50**, 43 (1987)
25. I. Suemune, L.A. Coldren, IEEE. J. Quant. Electron. **24**, 1178 (1988)
26. D.A.B. Miller, D.S. Chemla, T.C. Damen, J.H. Wood, A.C. Burrus, A.C. Gossard, W. Weigmann, IEEE. J. Quant. Electron. **21**, 1462 (1985)
27. J.S. Blakemore, Semiconductor Statistics, Dover, New York (1987); K.P. Ghatak, S. Bhattacharya, S. K. Biswas, A. Deyand, A.K. Dasgupta. Phys. Scr. **75**, 820 (2007)

Chapter 4
The DRs in Quantum Dots (QDs) of Heavily Doped (HD) Non-parabolic Materials

I must dream as I live forever and I must live as I die to-day.

4.1 Introduction

It is well known that as the dimension of the UFs increases from 1D to 3D, the degree of freedom of the free carriers decreases drastically and the density-of-states function changes from the Heaviside step function in QWs to the Dirac's delta function in Quantum Dots (QDs) [1].

The QDs can be used for visualizing and tracking molecular processes in cells using standard fluorescence microscopy [2–4]. They display minimal photo-bleaching [5], thus allowing molecular tracking over prolonged periods and consequently, single molecule can be tracked by using optical fluorescence microscopy [6]. The salient features of quantum dot (QD) lasers [7] include low threshold currents, higher power, and great stability as compared with the conventional one and the QDs find extensive applications in nano-robotics [8], neural networks [9] and high density memory or storage media [10]. The QDs are also used in nano-photonics [11] because of their theoretically high quantum yield and have been suggested as implementations of Q-bits for quantum information processing [12]. The QDs also find applications in diode lasers [13], amplifiers [14], and optical sensors [15]. High-quality QDs are well suited for optical encoding [16] because of their broad excitation profiles and narrow emission spectra. The new generations of QDs have far-reaching potential for the accurate investigations of intracellular processes at the single-molecule level, high-resolution cellular imaging, long-term in vivo observation of cell trafficking, tumor targeting, and diagnostics [17]. The QD nanotechnology is one of the most promising candidates for use in solid-state quantum computation [18]. It may also be noted that the QBs are being used in single electron transistors [19], photovoltaic devices [20], photoelectrics [21], ultrafast all-optical switches and logic gates [22], organic dyes [23] and in other types of nano devices.

K.P. Ghatak, *Dispersion Relations in Heavily-Doped Nanostructures*,
Springer Tracts in Modern Physics 265, DOI 10.1007/978-3-319-21000-1_4

In this chapter in Sects. 4.2.1–4.2.14, we have investigated the DR in QDs of HD non-linear optical, III–V, II–VI, stressed Kane type, Te, GaP, $PtSb_2$, Bi_2Te_3, Ge, GaSb, II–V, lead Germanium Telluride, Zinc and Cadmium Diphosphides respectively and corresponding DOS functions. The Sect. 4.3 contains the summary and conclusion pertaining to this chapter. The Sect. 4.4 presents 22 open research problems.

4.2 Theoretical Background

4.2.1 *The DR in Quantum Dot (QD) of HD Nonlinear Optical Materials*

The DR in this case can be written following (2.32) as

$$\frac{\hbar^2 (n_z \pi/d_z)^2}{2m_{\parallel}^* T_{21}(E_{1QBHD}, \eta_g)} + \frac{\hbar^2 (n_y \pi/d_y)^2}{2m_{\parallel}^* T_{22}(E_{1QBHD}, \eta_g)} + \frac{\hbar^2 (n_x \pi/d_x)^2}{2m_{\parallel}^* T_{21}(E_{1QBHD}, \eta_g)} = 1 \tag{4.1}$$

where E_{1QBHD} is the totally quantized energy in this case.

The total DOS function in this case is given by

$$N_{ODT}(E) = \frac{2g_v}{d_x d_y d_z} \sum_{n_x=1}^{n_{x\max}} \sum_{n_y=1}^{n_{y\max}} \sum_{n_z=1}^{n_{z\max}} \delta'(E - E_{1QBHD}) \tag{4.2}$$

where $\delta'(E - E_{1QBHD})$ is the Dirac's Delta function.

For the purpose of comparison we shall also formulate the DR in the absence of band tails is this case.

Let E_{n_i} $(i = x, y, \text{ and } z)$ be the quantized energy levels due to infinitely deep potential well along ith-axis with n_i $(=1, 2, 3, \ldots)$ as the size quantum numbers. Therefore, from (2.2), one can write

$$\gamma(E_{n_x}) = f_1(E_{n_x}) \left(\frac{\pi n_x}{d_x}\right)^2 \tag{4.3}$$

$$\gamma(E_{n_y}) = f_1(E_{n_y}) \left(\frac{\pi n_y}{d_y}\right)^2 \tag{4.4}$$

$$\gamma(E_{n_z}) = f_2(E_{n_z}) \left(\frac{\pi n_z}{d_z}\right)^2 \tag{4.5}$$

From (2.2), the totally quantized energy (E_{QD1}) can be expressed as

$$\gamma(E_{QD1}) = f(E_{QD1})\left[\left(\frac{\pi n_x}{d_x}\right)^2 + \left(\frac{\pi n_y}{d_y}\right)^2\right] + f_2(E_{QD1})\left[\left(\frac{\pi n_z}{d_z}\right)^2\right] \quad (4.6)$$

The total DOS in this case is given by

$$N_{ODT}(E) = \frac{2g_v}{d_x d_y d_z}\sum_{n_x=1}^{n_{x\max}}\sum_{n_y=1}^{n_{y\max}}\sum_{n_z=1}^{n_{z\max}}\delta'(E - E_{QD1}) \quad (4.7)$$

4.2.2 The DR in Quantum Dot (QD) of HD III–V Materials

The DR of the conduction electrons of III–V materials are described by the models of Kane (both three and two bands) Stillman et al. Palik et al. respectively. For the purpose of complete and coherent presentation, the DRs in QDs of HD III–V compounds have also been investigated in accordance with the aforementioned different DRs for relative comparison as follows:

(a) The HD three band model of Kane

The DR in this case can be written following (2.45) as

$$\frac{\hbar^2(n_z\pi/d_z)^2}{2m_c} + \frac{\hbar^2(n_y\pi/d_y)^2}{2m_c} + \frac{\hbar^2(n_x\pi/d_x)^2}{2m_c} = T_{44}(E_{2QBHD}, \eta_g) \quad (4.8)$$

where $T_{44}(E_{2QBHD}, \eta_g) = T_{31}(E_{2QBHD}, \eta_g) + iT_{31}(E_{2QBHD}, \eta_g)$ and E_{2QBHD} is the totally quantized energy in this case

The DOS function is given by

$$N_{ODT}(E) = \frac{2g_v}{d_x d_y d_z}\sum_{n_x=1}^{n_{x\max}}\sum_{n_y=1}^{n_{y\max}}\sum_{n_z=1}^{n_{z\max}}\delta'(E - E_{2QBHD}) \quad (4.9)$$

The quantized energy levels (E_{n_x}, E_{n_y} and E_{n_z} along x, y, and z directions respectively) in the absence of band tails in QDs of III–V materials in accordance with the three band model of Kane can be expressed as

$$I_{11}(E_{n_x}) = \frac{\hbar^2}{2m_c}\left(\frac{\pi n_x}{d_x}\right)^2 \quad (4.10)$$

$$I_{11}(E_{n_y}) = \frac{\hbar^2}{2m_c}\left(\frac{\pi n_y}{d_y}\right)^2 \tag{4.11}$$

$$\text{and } I_{11}(E_{n_z}) = \frac{\hbar^2}{2m_c}\left(\frac{\pi n_z}{d_z}\right)^2 \tag{4.12}$$

The totally quantized energy (E_{QD2}) (changed) in this case assumes the form

$$I_{11}(E_{QD2}) = \frac{\hbar^2\pi^2}{2m_c}\left[\left(\frac{n_x}{d_x}\right)^2 + \left(\frac{n_y}{d_y}\right)^2 + \left(\frac{n_z}{d_z}\right)^2\right] \tag{4.13}$$

The DOS function is given by

$$N_{ODT}(E) = \frac{2g_v}{d_x d_y d_z}\sum_{n_x=1}^{n_{x\max}}\sum_{n_y=1}^{n_{y\max}}\sum_{n_z=1}^{n_{z\max}}\delta'(E - E_{QD2})$$

(b) The HD two band model of Kane

The DR in this case can be written following (2.59) as

$$\frac{\hbar^2(n_z\pi/d_z)^2}{2m_c} + \frac{\hbar^2(n_y\pi/d_y)^2}{2m_c} + \frac{\hbar^2(n_x\pi/d_x)^2}{2m_c} = \gamma_2(E_{3QBHD}, \eta_g) \tag{4.14}$$

and E_{3QBHD} is the totally quantized energy in this case

The DOS function is given by

$$N_{ODT}(E) = \frac{2g_v}{d_x d_y d_z}\sum_{n_x=1}^{n_{x\max}}\sum_{n_y=1}^{n_{y\max}}\sum_{n_z=1}^{n_{z\max}}\delta'(E - \mathrm{E}_{3QBHD}) \tag{4.15}$$

In the absence of band-tails, the totally quantized energy (E_{QD3}) in this case is given by

$$E_{QD3}(1 + \alpha E_{QD3}) = \frac{\hbar^2\pi^2}{2m_c}\left[\left(\frac{n_x}{d_x}\right)^2 + \left(\frac{n_y}{d_y}\right)^2 + \left(\frac{n_z}{d_z}\right)^2\right] \tag{4.16}$$

The DOS function is given by

$$N_{ODT}(E) = \frac{2g_v}{d_x d_y d_z}\sum_{n_x=1}^{n_{x\max}}\sum_{n_y=1}^{n_{y\max}}\sum_{n_z=1}^{n_{z\max}}\delta'(E - E_{QD2}) \tag{4.17}$$

(c) The HD parabolic energy bands

The DR in this case can be written following (2.72) as

$$\frac{\hbar^2(\mathrm{n}_z\pi/\mathrm{d}_z)^2}{2m_c}+\frac{\hbar^2(\mathrm{n}_y\pi/\mathrm{d}_y)^2}{2m_c}+\frac{\hbar^2(\mathrm{n}_x\pi/\mathrm{d}_x)^2}{2m_c}=\gamma_3(E_{4QBHD},\eta_g) \tag{4.18}$$

and E_{4QBHD} is the totally quantized energy in this case

The DOS function is given by

$$N_{ODT}(E)=\frac{2g_v}{d_xd_yd_z}\sum_{n_x=1}^{n_{x\max}}\sum_{n_y=1}^{n_{y\max}}\sum_{n_z=1}^{n_{z\max}}\delta'(E-\mathrm{E}_{4QBHD}) \tag{4.19}$$

In the absence of band-tails, the totally quantized energy $(\bar{E}_{QD3})$ in this case is given by

$$\bar{E}_{QD3}=\frac{\hbar^2\pi^2}{2m_c}\left[\left(\frac{n_x}{d_x}\right)^2+\left(\frac{n_y}{d_y}\right)^2+\left(\frac{n_z}{d_z}\right)^2\right] \tag{4.20}$$

The DOS function is given by

$$N_{ODT}(E)=\frac{2g_v}{d_xd_yd_z}\sum_{n_x=1}^{n_{x\max}}\sum_{n_y=1}^{n_{y\max}}\sum_{n_z=1}^{n_{z\max}}\delta'(E-\bar{E}_{QD3}) \tag{4.21a}$$

(d) The HD Model of Stillman et al.

The DR of the electrons in this case can be written following (2.83) as

$$\frac{\hbar^2(\mathrm{n}_z\pi/\mathrm{d}_z)^2}{2m_c}+\frac{\hbar^2(\mathrm{n}_y\pi/\mathrm{d}_y)^2}{2m_c}+\frac{\hbar^2(\mathrm{n}_x\pi/\mathrm{d}_x)^2}{2m_c}=\theta_4(\mathrm{E}_{5QBHD},\eta_g) \tag{4.21b}$$

and E_{5QBHD} is the totally quantized energy in this case

The DOS function is given by

$$N_{ODT}(E)=\frac{2g_v}{d_xd_yd_z}\sum_{n_x=1}^{n_{x\max}}\sum_{n_y=1}^{n_{y\max}}\sum_{n_z=1}^{n_{z\max}}\delta'(E-\mathrm{E}_{5QBHD}) \tag{4.22}$$

In the absence of band-tails, the E_{QD5} in this case can be defined as

$$I_{12}\left(E_{QD5}\right) = \frac{\hbar^2\pi^2}{2m_c}\left[\left(\frac{n_x}{d_x}\right)^2 + \left(\frac{n_y}{d_y}\right)^2 + \left(\frac{n_z}{d_z}\right)^2\right] \tag{4.23}$$

The DOS function is given by

$$N_{ODT}(E) = \frac{2g_v}{d_x d_y d_z}\sum_{n_x=1}^{n_{x\max}}\sum_{n_y=1}^{n_{y\max}}\sum_{n_z=1}^{n_{z\max}}\delta'(E - E_{QD5}) \tag{4.24}$$

(e) The HD model of Palik et al.

The DR of the electrons in this case can be written following (2.96) as

$$\frac{\hbar^2(n_z\pi/d_z)^2}{2m_c} + \frac{\hbar^2(n_y\pi/d_y)^2}{2m_c} + \frac{\hbar^2(n_x\pi/d_x)^2}{2m_c} = \theta_5(E_{6QBHD}, \eta_g) \tag{4.25}$$

and E_{6QBHD} is the totally quantized energy in this case
The DOS function is given by

$$N_{ODT}(E) = \frac{2g_v}{d_x d_y d_z}\sum_{n_x=1}^{n_{x\max}}\sum_{n_y=1}^{n_{y\max}}\sum_{n_z=1}^{n_{z\max}}\delta'(E - E_{6QBHD}) \tag{4.26}$$

In the absence of band-tails, the E_{QD5} in this case can be defined as

$$I_{13}\left(E_{QD7}\right) = \frac{\hbar^2}{2m_c}\left[\left(\frac{\pi n_x}{d_x}\right)^2 + \left(\frac{\pi n_y}{d_y}\right)^2 + \left(\frac{\pi n_z}{d_z}\right)^2\right] \tag{4.27}$$

The DOS function is given by

$$N_{ODT}(E) = \frac{2g_v}{d_x d_y d_z}\sum_{n_x=1}^{n_{x\max}}\sum_{n_y=1}^{n_{y\max}}\sum_{n_z=1}^{n_{z\max}}\delta'(E - E_{QD7}) \tag{4.28}$$

(f) The HD model of Rossler

The DR of the conduction electrons in this case in accordance with the model of Rossler can be written as [24]

$$E = \frac{\hbar^2 k^2}{2m_c} + [\alpha_{11} + \alpha_{12}k]k^4 + (\beta_{11} + \beta_{12}k)\left[k_x^2 k_y^2 + k_y^2 k_z^2 + k_z^2 k_x^2\right] \\ \pm [\gamma_{11} + \gamma_{12}k]\left[k^2\left(k_x^2 k_y^2 + k_y^2 k_z^2 + k_z^2 k_x^2\right) - 9k_x^2 k_y^2 k_z^2\right]^{1/2} \tag{4.29}$$

where α_{11}, α_{12}, β_{11}, β_{12}, γ_{11} and γ_{12} are energy-band constants.

The HD DR assumes the form

$$\begin{aligned}\gamma_3(\bar{E}_{8QBHD}, \eta_g) \equiv & \frac{\hbar^2 \pi^2}{2m^*}\left[\left(\frac{n_x}{d_x}\right)^2 + \left(\frac{n_y}{d_y}\right)^2 + \left(\frac{n_z}{d_z}\right)^2\right] \\ & + \left[\alpha_{11} + \alpha_{12}\left[\left(\frac{\pi n_x}{d_x}\right)^2 + \left(\frac{\pi n_y}{d_y}\right)^2 + \left(\frac{\pi n_z}{d_z}\right)^2\right]^{1/2}\right] \\ & \times \left[\left(\frac{\pi n_x}{d_x}\right)^2 + \left(\frac{\pi n_y}{d_y}\right)^2 + \left(\frac{\pi n_z}{d_z}\right)^2\right] \\ & + \left[\beta_{11} + \beta_{12}\left[\left(\frac{\pi n_x}{d_x}\right)^2 + \left(\frac{\pi n_y}{d_y}\right)^2 + \left(\frac{\pi n_z}{d_z}\right)^2\right]^{1/2}\right] \\ & \times \left[\pi^4\left(\frac{n_x n_y}{d_x d_y}\right)^2 + \pi^4\left(\frac{n_y n_z}{d_y d_z}\right)^2 + \pi^4\left(\frac{n_z n_x}{d_z d_x}\right)^2\right] \\ & \pm \left[\gamma_{11} + \gamma_{12}\left[\left(\frac{\pi n_x}{d_x}\right)^2 + \left(\frac{\pi n_y}{d_y}\right)^2 + \left(\frac{\pi n_z}{d_z}\right)^2\right]^{1/2}\right] \\ & \times \left[\left[\left(\frac{\pi n_x}{d_x}\right)^2 + \left(\frac{\pi n_y}{d_y}\right)^2 + \left(\frac{\pi n_z}{d_z}\right)^2\right]\left[\pi^4\left(\frac{n_x n_y}{d_x d_y}\right)^2 + \pi^4\left(\frac{n_y n_z}{d_y d_z}\right)^2 + \pi^4\left(\frac{n_z n_x}{d_z d_x}\right)^2\right]\right. \\ & \left. - 9\pi^6\left(n_x n_y n_z / d_x d_y d_z\right)^6\right]\end{aligned} \tag{4.30}$$

The DOS function can be expressed as

$$N_{ODT}(E) = \frac{g_v}{d_x d_y d_z} \sum_{n_{x=1}}^{n_{x\max}} \sum_{n_{y=1}}^{n_{y\max}} \sum_{n_{z=1}}^{n_{z\max}} \delta'(E - \bar{E}_{8QBHD}) \tag{4.31}$$

In the absence of band-tails, the electron energy spectrum assumes the form

$$
\begin{aligned}
E_{QD6\pm} \equiv {} & \frac{\hbar^2\pi^2}{2m_c}\left[\left(\frac{n_x}{d_x}\right)^2+\left(\frac{n_y}{d_y}\right)^2+\left(\frac{n_z}{d_z}\right)^2\right] \\
& + \left[\alpha_{11}+\alpha_{12}\left[\left(\frac{\pi n_x}{d_x}\right)^2+\left(\frac{\pi n_y}{d_y}\right)^2+\left(\frac{\pi n_z}{d_z}\right)^2\right]^{1/2}\right] \\
& \times \left[\left(\frac{\pi n_x}{d_x}\right)^2+\left(\frac{\pi n_y}{d_y}\right)^2+\left(\frac{\pi n_z}{d_z}\right)^2\right] \\
& + \left[\beta_{11}+\beta_{12}\left[\left(\frac{\pi n_x}{d_x}\right)^2+\left(\frac{\pi n_y}{d_y}\right)^2+\left(\frac{\pi n_z}{d_z}\right)^2\right]^{1/2}\right] \\
& \times \left[\pi^4\left(\frac{n_x n_y}{d_x d_y}\right)^2+\pi^4\left(\frac{n_y n_z}{d_y d_z}\right)^2+\pi^4\left(\frac{n_z n_x}{d_z d_x}\right)^2\right] \\
& \pm \left[\gamma_{11}+\gamma_{12}\left[\left(\frac{\pi n_x}{d_x}\right)^2+\left(\frac{\pi n_y}{d_y}\right)^2+\left(\frac{\pi n_z}{d_z}\right)^2\right]^{1/2}\right] \\
& \times \left[\left[\left(\frac{\pi n_x}{d_x}\right)^2+\left(\frac{\pi n_y}{d_y}\right)^2+\left(\frac{\pi n_z}{d_z}\right)^2\right]\left[\pi^4\left(\frac{n_x n_y}{d_x d_y}\right)^2+\pi^4\left(\frac{n_y n_z}{d_y d_z}\right)^2+\pi^4\left(\frac{n_z n_x}{d_z d_x}\right)^2\right]\right. \\
& \left. - 9\pi^6\left(n_x n_y n_z / d_x d_y d_z\right)^6\right]
\end{aligned}
\tag{4.32}
$$

The DOS function can be expressed as

$$
N_{ODT}(E) = \frac{2g_v}{d_x d_y d_z}\sum_{n_x=1}^{n_{x\max}}\sum_{n_y=1}^{n_{y\max}}\sum_{n_z=1}^{n_{z\max}} \delta'(E - E_{QD6\pm}) \tag{4.33}
$$

(g) The HD model of Agafonov et al.

In accordance with the model of Agafonov et al. [25], the electron dispersion law can be expressed as

$$
E = \frac{\bar{y} - E_{g0}}{2}\left[1 - T_5\left(\frac{k_x^4 + k_y^4 + k_z^4}{\bar{y}k^2}\right)\right] \tag{4.34}
$$

where, $(\bar{y})^2 = \left[E_{g_0}^2 + \frac{8}{3}P_0^2k^2\right]$, $T_5 = \frac{\bar{D}\sqrt{3} - 3\bar{B}}{2}$, $\bar{B} = -21\left(\frac{\hbar^2}{2m_0}\right)$, $\bar{D} = -40\left(\frac{\hbar^2}{2m_0}\right)$ and P_0 is the momentum matrix element.

The HD electron dispersion law can be written as

$$\gamma_3(E,\eta_g) = \frac{\bar{y} - E_{g_0}}{2}\left[1 - T_5\left(\frac{k_x^4 + k_y^4 + k_z^4}{\bar{y}k^2}\right)\right] \quad (4.35)$$

The totally quantized energy can be written as

$$\gamma_3(\bar{E}_{8QBHD1},\eta_g) = \left(\frac{\psi_{30} - E_{g_0}}{2}\right)\left[1 - \left[T_5\left[\left(\frac{\pi n_x}{d_x}\right)^4 + \left(\frac{\pi n_y}{d_y}\right)^4 + \left(\frac{\pi n_z}{d_z}\right)^4\right]\right].\right.$$
$$\left.\left[\psi_{30}\left[\left(\frac{\pi n_x}{d_x}\right)^2 + \left(\frac{\pi n_y}{d_y}\right)^2 + \left(\frac{\pi n_z}{d_z}\right)^2\right]\right]^{-1}\right] \quad (4.36)$$

where, $\psi_{30} = \left[E_{g_0}^2 + \frac{8}{3}P^2\left[\left(\frac{\pi n_x}{d_x}\right)^2 + \left(\frac{\pi n_y}{d_y}\right)^2 + \left(\frac{\pi n_z}{d_z}\right)^2\right]\right]^{1/2}$

The DOS function can be expressed as

$$N_{ODT}(E) = \frac{2g_v}{d_x d_y d_z}\sum_{n_x=1}^{n_{x\max}}\sum_{n_y=1}^{n_{y\max}}\sum_{n_z=1}^{n_{z\max}}\delta'(E - \bar{E}_{8QBHD1}) \quad (4.37)$$

In the absence of band-tails the totally quantized energy E_{QD9} assumes the form

$$E_{QD9} = \left(\frac{\psi_{30} - E_{g_0}}{2}\right)\left[1 - \left[T_5\left[\left(\frac{\pi n_x}{d_x}\right)^4 + \left(\frac{\pi n_y}{d_y}\right)^4 + \left(\frac{\pi n_z}{d_z}\right)^4\right]\right].\right.$$
$$\left.\left[\psi_{30}\left[\left(\frac{\pi n_x}{d_x}\right)^2 + \left(\frac{\pi n_y}{d_y}\right)^2 + \left(\frac{\pi n_z}{d_z}\right)^2\right]\right]^{-1}\right] \quad (4.38)$$

The DOS function can be expressed as

$$N_{ODT}(E) = \frac{g_v}{d_x d_y d_z}\sum_{n_x=1}^{n_{x\max}}\sum_{n_y=1}^{n_{y\max}}\sum_{n_z=1}^{n_{z\max}}\delta'(E - E_{QD9}) \quad (4.39)$$

4.2.3 The DR in Quantum Dot (QD) of HD II–VI Materials

The 0D electron dispersion law in QD of HD II–VI materials can be written following (2.108) as

$$\gamma_3(E_{7QBHD},\eta_g) = d_0'\left[\left(\frac{n_x\pi}{d_x}\right)^2+\left(\frac{n_y\pi}{d_y}\right)^2\right] \pm \bar{\lambda}_0\left[\left(\frac{n_x\pi}{d_x}\right)^2+\left(\frac{n_y\pi}{d_y}\right)^2\right]^{1/2} + \frac{\hbar^2(n_z\pi/\mathrm{d}_z)^2}{2m_{\parallel}^*} \quad (4.40)$$

where E_{7QBHD} is the totally quantized energy in this case.

The DOS function is given by

$$N_{0DT}(E) = \frac{g_v}{d_xd_yd_z}\sum_{n_x=1}^{n_{x\max}}\sum_{n_y=1}^{n_{y\max}}\sum_{n_z=1}^{n_{z\max}}\delta'(E-\mathrm{E}_{7QBHD}) \quad (4.41)$$

In the absence of band tails the totally quantized energy $E_{QD10,\pm}$ in this case can be expressed as

$$E_{QD10,\pm} = d_0'\left[\left(\frac{\pi n_x}{d_x}\right)^2+\left(\frac{\pi n_y}{d_y}\right)^2\right] + \frac{1}{2m_{\parallel}^*}\left(\frac{\hbar\pi n_z}{d_z}\right)^2 \pm \bar{\lambda}_0\left[\left(\frac{\pi n_x}{d_x}\right)^2+\left(\frac{\pi n_y}{d_y}\right)^2\right]^{1/2} \quad (4.42)$$

The DOS function is given by

$$N_{0DT}(E) = \frac{g_v}{d_xd_yd_z}\sum_{n_x=1}^{n_{x\max}}\sum_{n_y=1}^{n_{y\max}}\sum_{n_z=1}^{n_{z\max}}\delta'(E-E_{QD10,\pm}) \quad (4.43)$$

4.2.4 The DR in Quantum Dot (QD) of HD IV–VI Materials

(a) Dimmock Model

In this case the DR of the electrons can be written as the following (2.124) as

$$\frac{n_z\pi}{d_z} = T_{36}(\mathrm{E}_{81QBHD},\eta_g,n_x,n_y) \quad (4.44)$$

where E_{81QBHD} is the totally quantized energy in this case.

The DOS function is given by

$$N_{0DT}(E) = \frac{2g_v}{d_x d_y d_z} \sum_{n_x=1}^{n_{x\max}} \sum_{n_y=1}^{n_{y\max}} \sum_{n_z=1}^{n_{z\max}} \delta'(E - \mathrm{E}_{81QBHD}) \tag{4.45}$$

In the absence of band tailing, the electron DR in this case can be written as

$$\frac{n_z \pi}{d_z} = T_{40}(E_{QD11}, n_x, n_y) \tag{4.46}$$

where E_{QD11} is the totally quantized energy in this case.
The DOS function is given by

$$N_{0DT}(E) = \frac{2g_v}{d_x d_y d_z} \sum_{n_x=1}^{n_{x\max}} \sum_{n_y=1}^{n_{y\max}} \sum_{n_z=1}^{n_{z\max}} \delta'(E - \mathrm{E}_{QD11}) \tag{4.47}$$

(b) Bangert and Kastner Model

The electron DR in this case is given by following (2.150) as

$$\frac{n_z \pi}{d_z} = T_{60}(\mathrm{E}_{9QBHD}, \eta_g, n_x, n_y) \tag{4.48}$$

where E_{9QBHD} is the totally quantized energy in this case.
The DOS function is given by

$$N_{0DT}(E) = \frac{2g_v}{d_x d_y d_z} \sum_{n_x=1}^{n_{x\max}} \sum_{n_y=1}^{n_{y\max}} \sum_{n_z=1}^{n_{z\max}} \delta'(E - \mathrm{E}_{9QBHD}) \tag{4.49}$$

In the absence of band-tails the DR is given by

$$\frac{n_z \pi}{d_z} = T_{61}(\mathrm{E}_{12QD}, \eta_g, n_x, n_y) \tag{4.50}$$

where E_{12QD} is the totally quantized energy in this case.
The DOS function is given by

$$N_{0DT}(E) = \frac{2g_v}{d_x d_y d_z} \sum_{n_x=1}^{n_{x\max}} \sum_{n_y=1}^{n_{y\max}} \sum_{n_z=1}^{n_{z\max}} \delta'(E - \mathrm{E}_{12QD}) \tag{4.51}$$

(iii) In accordance with Foley et al. [26], the electron DR assumes the form

$$E+\frac{E_{g0}}{2}=\frac{\hbar^2k_s^2}{2m_\perp^-}+\frac{\hbar^2k_z^2}{2m_\parallel^-}+\left[\left[\frac{\hbar^2k_s^2}{2m_\perp^+}+\frac{\hbar^2k_z^2}{2m_\parallel^+}+\frac{E_{g0}}{2}\right]^2+P_\parallel^2k_z^2+P_\perp^2k_s^2\right]^{1/2} \tag{4.52}$$

where $\frac{1}{m_\perp^\pm}=\frac{1}{2}\left[\frac{1}{m_{tc}}\pm\frac{1}{m_{tv}}\right]$, $\frac{1}{m_\parallel^\pm}=\frac{1}{2}\left[\frac{1}{m_{lc}}\pm\frac{1}{m_{lv}}\right]$, m_{tc} and m_{lc} are the transverse and longitudinal effective electron masses of the conduction electrons at the edge of the conduction band, m_{tv} and m_{lv} are the transverse and longitudinal effective hole masses of the holes at the edge of the valence band.

The totally quantized energy $\bar{E}_{10QBHD}$ for this model is given by

$$\begin{aligned}\gamma_3\left(\bar{E}_{10QBHD},\eta_g\right)=&-\frac{E_{g0}}{2}+\frac{\hbar^2}{2m_\perp^-}\left[\left(\frac{n_x\pi}{d_x}\right)^2+\left(\frac{n_y\pi}{d_y}\right)^2\right]+\frac{\hbar^2}{2m_\parallel^-}\left(\frac{n_z\pi}{d_z}\right)^2\\&+\left[\left[\frac{E_{g0}}{2}+\frac{\hbar}{2m_\parallel^+}\left(\frac{n_z\pi}{d_z}\right)^2+\frac{\hbar}{2m_\perp^+}\left\{\left(\frac{n_x\pi}{d_x}\right)^2+\left(\frac{n_y\pi}{d_y}\right)^2\right\}\right]^2\right.\\&\left.+P_\parallel^2\left(\frac{n_z\pi}{d_z}\right)^2+P_\perp^2\left\{\left(\frac{n_x\pi}{d_x}\right)^2+\left(\frac{n_y\pi}{d_y}\right)^2\right\}\right]^{1/2}\end{aligned} \tag{4.53}$$

The DOS function is given by

$$N_{ODT}(E)=\frac{2g_v}{d_xd_yd_z}\sum_{n_x=1}^{n_{x\max}}\sum_{n_y=1}^{n_{y\max}}\sum_{n_z=1}^{n_{z\max}}\delta'(E-\bar{E}_{10QBHD}) \tag{4.54}$$

The totally quantized energy E_{QD30} in the absence of band-tails for this model is given by

$$\begin{aligned}E_{QD30}=&-\frac{E_{g0}}{2}+\frac{\hbar^2}{2m_\perp^-}\left[\left(\frac{n_x\pi}{d_x}\right)^2+\left(\frac{n_y\pi}{d_y}\right)^2\right]+\frac{\hbar^2}{2m_\parallel^-}\left(\frac{n_z\pi}{d_z}\right)^2\\&+\left[\left[\frac{E_{g0}}{2}+\frac{\hbar}{2m_\parallel^+}\left(\frac{n_z\pi}{d_z}\right)^2+\frac{\hbar}{2m_\perp^+}\left\{\left(\frac{n_x\pi}{d_x}\right)^2+\left(\frac{n_y\pi}{d_y}\right)^2\right\}\right]^2\right.\\&\left.+P_\parallel^2\left(\frac{n_z\pi}{d_z}\right)^2+P_\perp^2\left\{\left(\frac{n_x\pi}{d_x}\right)^2+\left(\frac{n_y\pi}{d_y}\right)^2\right\}\right]^{1/2}\end{aligned} \tag{4.55}$$

The DOS function is given by

$$N_{0DT}(E) = \frac{2g_v}{d_x d_y d_z} \sum_{n_x=1}^{n_{x\max}} \sum_{n_y=1}^{n_{y\max}} \sum_{n_z=1}^{n_{z\max}} \delta'(E - \mathrm{E}_{QD30}) \tag{4.56}$$

4.2.5 The DR in Quantum Dot (QD) of HD Stressed Kane Type Materials

The electron DR in this case is given by following (2.177) as

$$\frac{n_z \pi}{d_z} = T_{70}(\mathrm{E}_{10QBHD}, \eta_g, n_x, n_y) \tag{4.57}$$

where E_{10QBHD} is the totally quantized energy in this case.

The DOS function is given by

$$N_{0DT}(E) = \frac{2g_v}{d_x d_y d_z} \sum_{n_x=1}^{n_{x\max}} \sum_{n_y=1}^{n_{y\max}} \sum_{n_z=1}^{n_{z\max}} \delta'(E - \mathrm{E}_{10QBHD}) \tag{4.58}$$

In the absence of band-tails, the totally quantized energy $\bar{E}_{QD23}$ in this case assumes the form

$$\left(\frac{\pi n_x}{d_x}\right)^2 [\bar{a}_0(\bar{E}_{QD23})]^{-2} + \left(\frac{\pi n_y}{d_y}\right)^2 [\bar{b}_0(\bar{E}_{QD23})]^{-2} + \left(\frac{\pi n_z}{d_z}\right)^2 [\bar{c}_0(\bar{E}_{QD23})]^{-2} = 1 \tag{4.59}$$

The DOS function is given by

$$N_{0DT}(E) = \frac{2g_v}{d_x d_y d_z} \sum_{n_x=1}^{n_{x\max}} \sum_{n_y=1}^{n_{y\max}} \sum_{n_z=1}^{n_{z\max}} \delta'(E - \bar{\mathrm{E}}_{QD23}) \tag{4.60}$$

4.2.6 The DR in Quantum Dot (QD) of HD Te

The 0D DR may be written in this case following (2.196) as

$$\frac{n_x \pi}{d_x} = t_{72}(\mathrm{E}_{11QBHD}, n_y, n_z, \eta_g) \tag{4.61}$$

where E_{11QBHD} is the totally quantized energy in this case

The DOS function is given by

$$N_{0DT}(E) = \frac{g_v}{d_x d_y d_z} \sum_{n_x=1}^{n_{x\max}} \sum_{n_y=1}^{n_{y\max}} \sum_{n_z=1}^{n_{z\max}} \delta'(E - \mathrm{E}_{11QBHD}) \tag{4.62}$$

The totally quantized energy in the absence of band-tails can be written as

$$E_{QD14,\pm} = \psi_1 \left(\frac{\pi n_z}{d_z}\right)^2 + \psi_2 \left[\left(\frac{\pi n_x}{d_x}\right)^2 + \left(\frac{\pi n_y}{d_y}\right)^2\right] \pm \left[\psi_3^2 \left(\frac{\pi n_z}{d_z}\right)^2 + \psi_4^2 \left[\left(\frac{\pi n_x}{d_x}\right)^2 + \left(\frac{\pi n_y}{d_y}\right)^2\right]\right]^{1/2} \tag{4.63}$$

The DOS function is given by

$$N_{0DT}(E) = \frac{g_v}{d_x d_y d_z} \sum_{n_x=1}^{n_{x\max}} \sum_{n_y=1}^{n_{y\max}} \sum_{n_z=1}^{n_{z\max}} \delta'(E - E_{QD14,\pm}) \tag{4.64}$$

4.2.7 The DR in Quantum Dot (QD) of HD Gallium Phosphide

The 0D DR may be written in this case following (2.206) as

$$\frac{n_x \pi}{d_x} = u_{70}(\mathrm{E}_{14QBHD}, \mathrm{n}_y, \mathrm{n}_z, \eta_g) \tag{4.65}$$

where E_{14QBHD} is the totally quantized energy in this case
The DOS function is given by

$$N_{0DT}(E) = \frac{2g_v}{d_x d_y d_z} \sum_{n_x=1}^{n_{x\max}} \sum_{n_y=1}^{n_{y\max}} \sum_{n_z=1}^{n_{z\max}} \delta'(E - \mathrm{E}_{14QBHD}) \tag{4.66}$$

In the absence of doping, the totally quantized energy (E_{QD16}) in this case can be written as

$$E_{QD16} = \frac{\hbar^2}{2m_\perp^*}\left[\left(\frac{\pi n_x}{d_x}\right)^2 + \left(\frac{\pi n_y}{d_y}\right)^2\right] + \frac{\hbar^2}{2m_\parallel^*}\left[\left(\frac{\pi n_x}{d_x}\right)^2 + \left(\frac{\pi n_y}{d_y}\right)^2 + \left(\frac{\pi n_z}{d_z}\right)^2\right] - \left[\frac{\hbar^2 k_0^2}{m_\parallel^{*2}}\left[\left(\frac{\pi n_x}{d_x}\right)^2 + \left(\frac{\pi n_y}{d_y}\right)^2 + \left(\frac{\pi n_z}{d_z}\right)^2\right] + |V_G|^2\right]^{1/2} + |V_G| \tag{4.67}$$

The DOS function is given by

$$N_{ODT}(E) = \frac{2g_v}{d_x d_y d_z} \sum_{n_x=1}^{n_{x\max}} \sum_{n_y=1}^{n_{y\max}} \sum_{n_z=1}^{n_{z\max}} \delta'(E - E_{QD16}) \tag{4.68}$$

4.2.8 The DR in Quantum Dot (QD) of HD Platinum Antimonide

The 0D DR may be written in this case following (2.221) as

$$\frac{n_x \pi}{d_x} = V_{70}(\mathrm{E}_{15QBHD}, \mathrm{n}_y, \mathrm{n}_z, \eta_g) \tag{4.69}$$

where E_{15QBHD} is the totally quantized energy in this case

The DOS function is given by

$$N_{ODT}(E) = \frac{2g_v}{d_x d_y d_z} \sum_{n_x=1}^{n_{x\max}} \sum_{n_y=1}^{n_{y\max}} \sum_{n_z=1}^{n_{z\max}} \delta'(E - \mathrm{E}_{15QBHD}) \tag{4.70}$$

In the absence of band tailing the 0D DR in this case can be written as

$$\frac{n_x \pi}{d_x} = D_{71}(E_{QD17}, n_y, n_z) \tag{4.71}$$

where E_{QD17} is the totally quantized energy in this case

The DOS function is given by

$$N_{ODT}(E) = \frac{2g_v}{d_x d_y d_z} \sum_{n_x=1}^{n_{x\max}} \sum_{n_y=1}^{n_{y\max}} \sum_{n_z=1}^{n_{z\max}} \delta'(E - E_{QD17}) \tag{4.72}$$

4.2.9 The DR in Quantum Dot (QD) of HD Bismuth Telluride

The DR in this case can be written following (2.233) as

$$\frac{n_x \pi}{d_x} = J_{70}(E_{18QBHD}, n_y, n_z, \eta_g) \tag{4.73}$$

where E_{18QBHD} is the totally quantized energy in this case

The DOS function is given by

$$N_{ODT}(E) = \frac{2g_v}{d_x d_y d_z} \sum_{n_x=1}^{n_{x\max}} \sum_{n_y=1}^{n_{y\max}} \sum_{n_z=1}^{n_{z\max}} \delta'(E - \mathrm{E}_{18QBHD}) \tag{4.74}$$

The totally quantizedenergy E_{QD33}, in the absence of band-tailscan be written as

$$E_{QD33}(1 + \alpha E_{QD33}) = \left[\bar{\omega}_1 \left(\frac{n_x \pi}{d_x}\right)^2 + \bar{\omega}_2 \left(\frac{n_y \pi}{d_y}\right)^2 + \bar{\omega}_3 \left(\frac{n_z \pi}{d_z}\right)^2 + 2\bar{\omega}_4 \left(\frac{\pi^2 n_y n_z}{d_y d_z}\right)\right] \tag{4.75}$$

The DOS function is given by

$$N_{ODT}(E) = \frac{2g_v}{d_x d_y d_z} \sum_{n_x=1}^{n_{x\max}} \sum_{n_y=1}^{n_{y\max}} \sum_{n_z=1}^{n_{z\max}} \delta'(E - E_{QD33}) \tag{4.76}$$

4.2.10 *The DR in Quantum Dot (QD) of HD Germanium*

(a) Model of Cardona et al.

The DR in accordance with this model in the present case can be written following (2.245) as

$$\frac{n_x \pi}{d_x} = L_{70}(E_{20QBHD}, n_y, n_z, \eta_g) \tag{4.77}$$

where E_{20QBHD} is the totally quantized energy in this case
The DOS function is given by

$$N_{ODT}(E) = \frac{2g_v}{d_x d_y d_z} \sum_{n_x=1}^{n_{x\max}} \sum_{n_y=1}^{n_{y\max}} \sum_{n_z=1}^{n_{z\max}} \delta'(E - \mathrm{E}_{20QBHD}) \tag{4.78}$$

In the absence of doping the totally quantized energy E_{QD30} in this case can be written as

$$E_{QD30} = -\frac{E_{g_0}}{2} + \frac{\hbar^2}{2m_{\parallel}^*}\left(\frac{\pi n_z}{d_z}\right)^2 + \left[\frac{E_{g_0}^2}{4} + E_{g_0}\left(\frac{\hbar^2}{2m_{\perp}^*}\right)\left\{\left(\frac{\pi n_x}{d_x}\right)^2 + \left(\frac{\pi n_y}{d_y}\right)^2\right\}\right]^{1/2} \tag{4.79}$$

The DOS function is given by

$$N_{ODT}(E) = \frac{2g_v}{d_x d_y d_z} \sum_{n_x=1}^{n_{x\max}} \sum_{n_y=1}^{n_{y\max}} \sum_{n_z=1}^{n_{z\max}} \delta'(E - E_{QD30}) \quad (4.80)$$

(b) Model of Wang and Ressler

The DR in accordance with this model in the present case can be written following (2.261) as

$$\frac{n_x \pi}{d_x} = \beta_{70}(E_{24QBHD}, n_y, n_z, \eta_g) \quad (4.81)$$

where E_{24QBHD} is the totally quantized energy in this case

The DOS function is given by

$$N_{ODT}(E) = \frac{2g_v}{d_x d_y d_z} \sum_{n_x=1}^{n_{x\max}} \sum_{n_y=1}^{n_{y\max}} \sum_{n_z=1}^{n_{z\max}} \delta'(E - E_{24QBHD}) \quad (4.82)$$

In the absence of doping, the totally quantized energy E_{QD40} in this case is given by

$$\begin{aligned} E_{QD40} = & \frac{\hbar^2}{2m_\parallel^*}\left(\frac{\pi n_z}{d_z}\right)^2 + \frac{\hbar^2}{2m_\perp^*}\left\{\left(\frac{\pi n_x}{d_x}\right)^2 + \left(\frac{\pi n_y}{d_y}\right)^2\right\} \\ & - \bar{c}_1\left(\frac{\hbar^2}{2m_\perp^*}\right)\left\{\left(\frac{\pi n_x}{d_x}\right)^2 + \left(\frac{\pi n_y}{d_y}\right)^2\right\}^2 \\ & - \bar{d}_1\left(\frac{\hbar^2}{2m_\perp^*}\right)\left\{\left(\frac{\pi n_x}{d_x}\right)^2 + \left(\frac{\pi n_y}{d_y}\right)^2\right\}\left(\frac{\hbar^2}{2m_\parallel^*}\right)\left(\frac{\pi n_z}{d_z}\right)^2 \\ & - \bar{e}_1\left(\frac{\hbar^2}{2m_\parallel^*}\right)^2\left(\frac{\pi n_z}{d_z}\right)^4 \end{aligned} \quad (4.83)$$

The DOS function is given by

$$N_{ODT}(E) = \frac{2g_v}{d_x d_y d_z} \sum_{n_x=1}^{n_{x\max}} \sum_{n_y=1}^{n_{y\max}} \sum_{n_z=1}^{n_{z\max}} \delta'(E - E_{QD40}) \quad (4.84)$$

4.2.11 The DR in Quantum Dot (QD) of HD Gallium Antimonide

The conduction electrons of n-GaSb obey the three dispersion relations as provided by Seiler et al. [27], Mathur et al. [28] and Zhang [29] respectively. The DRof QBs of GaSb is being presented in accordance with the aforementioned models for the joint purpose of coherent presentation and relative assessment.

(i) In accordance with the model of Seiler et al. [27] the DR of the conduction electrons in n-GaSb assume the form

$$E = \left[-\frac{E_{g0}}{2} + \frac{E_{g0}}{2}\left[1+\alpha_4 k^2\right]^{1/2} + \frac{\overline{\zeta_0}\hbar^2 k^2}{2m_0} + \frac{\overline{v_0}\theta_1(k)\hbar^2}{2m_0} \pm \frac{\overline{w_0}\theta_2(k)\hbar^2}{2m_0}\right] \quad (4.85a)$$

where $\alpha_4 = 4P_0^2\left(E_{g_0} + \frac{2}{3}\Delta\right)\left[E_{g_0}^2\left(E_{g_0} + \Delta\right)\right]^{-1}$, $\theta_1(k) = k^{-2}\left(k_x^2k_y^2 + k_y^2k_z^2 + k_z^2k_x^2\right)$ represents the warping of the Fermi surface, $\theta_2(k) = \left[\left\{k^2\left(k_x^2k_y^2 + k_y^2k_z^2 + k_z^2k_x^2\right)\right.\right.$ $\left.\left.-9k_x^2k_y^2k_z^2\right\}^{1/2}\cdot k^{-1}\right]$ represents the inversion asymmetry splitting of the conduction band, $\overline{\zeta_0}(=-2.1)$, $\overline{v_0}(=-1.49)$ and $\overline{w_0}(=0.42)$ represent the constants of the spectrum.

The DR in HD GaSb in accordance with the model of Seiler et al. can be written as

$$\gamma_3(\bar{E}_{30QBHD1}, \eta_g) = \left[-\frac{E_{g0}}{2} + \frac{E_{g0}}{2}\left[1+\alpha_4 k_1^2\right]^{1/2} + \frac{\overline{\zeta_0}\hbar^2 k_1^2}{2m_0} + \frac{\overline{v_0}\theta_1(k_1)\hbar^2}{2m_0} \pm \frac{\overline{w_0}\theta_2(k_1)\hbar^2}{2m_0}\right] \quad (4.85b)$$

where $\bar{E}_{30QBHD1}$ is the totally quantized energy, $k_1^2 = \left[\left(\frac{n_x\pi}{d_x}\right)^2 + \left(\frac{n_y\pi}{d_y}\right)^2 + \left(\frac{n_z\pi}{d_z}\right)^2\right]$, $\theta_1(k_1) = k_1^{-2}\left[\left(\frac{n_x\pi}{d_x}\right)^2\left(\frac{n_y\pi}{d_y}\right)^2 + \left(\frac{n_y\pi}{d_y}\right)^2\left(\frac{n_z\pi}{d_z}\right)^2 + \left(\frac{n_z\pi}{d_z}\right)^2\left(\frac{n_x\pi}{d_x}\right)^2\right]$ and $\theta_2(k_1) = \left[\left\{k_1^2\right.\right.$ $\left[\left(\frac{n_x\pi}{d_x}\right)^2\left(\frac{n_y\pi}{d_y}\right)^2 + \left(\frac{n_y\pi}{d_y}\right)^2\left(\frac{n_z\pi}{d_z}\right)^2 + \left(\frac{n_z\pi}{d_z}\right)^2\left(\frac{n_x\pi}{d_x}\right)^2\right] - 9\left(\frac{n_x\pi}{d_x}\right)^2\left(\frac{n_y\pi}{d_y}\right)^2\left(\frac{n_z\pi}{d_z}\right)^2\left.\left.\right\}^{1/2}\cdot k_1^{-1}\right]$

The DOS function is given by

$$N_{ODT}(E) = \frac{2g_v}{d_x d_y d_z}\sum_{n_x=1}^{n_{xmax}}\sum_{n_y=1}^{n_{ymax}}\sum_{n_z=1}^{n_{zmax}}\delta'(E - \bar{E}_{30QBHD1}) \quad (4.85c)$$

(ii) in accordance with the model of Mathur et al., the DR of the 0D electrons in this case can be written following (2.277) as

$$\frac{\hbar^2(\mathrm{n}_z\pi/\mathrm{d}_z)^2}{2m_c} + \frac{\hbar^2(\mathrm{n}_y\pi/\mathrm{d}_y)^2}{2m_c} + \frac{\hbar^2\left(\frac{n_x\pi}{d_x}\right)^2}{2m_c} = I_{36}(\bar{\mathrm{E}}_{30QBHD}, \eta_g) \quad (4.85d)$$

where $\bar{E}_{30QBHD}$ is the totally quantized energy in this case

The DOS function is given by

$$N_{ODT}(E) = \frac{2g_v}{d_x d_y d_z} \sum_{n_x=1}^{n_{x\max}} \sum_{n_y=1}^{n_{y\max}} \sum_{n_z=1}^{n_{z\max}} \delta'(E - \bar{E}_{30QBHD}) \tag{4.86}$$

The totally quantized energy E_{QD60} in the absence of band-tails assumes the form

$$I_{36}(E_{QD60}) = \frac{\hbar^2}{2m_c}\left[\left(\frac{\pi n_x}{d_x}\right)^2 + \left(\frac{\pi n_y}{d_y}\right)^2 + \left(\frac{\pi n_z}{d_z}\right)^2\right] \tag{4.87}$$

The DOS function is given by

$$N_{ODT}(E) = \frac{2g_v}{d_x d_y d_z} \sum_{n_x=1}^{n_{x\max}} \sum_{n_y=1}^{n_{y\max}} \sum_{n_z=1}^{n_{z\max}} \delta'(E - E_{QD60}) \tag{4.88a}$$

(iii) The dispersion relation of the conduction electrons in n-GaSb can be expressed in accordance with Zhang [29] as

$$E = \left[\overline{E_1} + \overline{E_2} Z_1(k)\right]k^2 + \left[\overline{E_3} + \overline{E_4} Z_1(k)\right]k^4 + k^6\left[\overline{E_5} + \overline{E_6} Z_1(k) + \overline{E_7} Z_2(k)\right] \tag{4.88b}$$

where $Z_1(k) = \frac{5\sqrt{21}}{4}\left[\frac{k_x^4 + k_y^4 + k_z^4}{k^4} - \frac{3}{5}\right]$, $Z_2(k) = \left[\frac{639639}{32}\right]^{1/2}\left[\frac{k_x^2 k_y^2 k_z^2}{k^6} + \frac{1}{12}\left(\frac{k_x^4 + k_y^4 + k_z^4}{k^4} - \frac{3}{5}\right) - \frac{1}{105}\right]$, the coefficients are in eV, the values of k are in $10\left(\frac{a}{2\pi}\right)$ times of k in atomic units (a is lattice constant), $\overline{E_1}$, $\overline{E_2}$, $\overline{E_3}$, $\overline{E_4}$, $\overline{E_5}$, $\overline{E_6}$ and $\overline{E_7}$ are energy-band constants.

The DR in HD GaSb in accordance with the model of Zhang [29], can be written as

$$\begin{aligned}\gamma_3(\bar{E}_{30QBHD2}, \eta_g) = {} & \left[\overline{E_1} + \overline{E_2} Z_1(k_1)\right]k_1^2 + \left[\overline{E_3} + \overline{E_4} Z_1(k_1)\right]k_1^4 \\ & + k_1^6\left[\overline{E_5} + \overline{E_6} Z_1(k_1) + \overline{E_7} Z_2(k_1)\right]\end{aligned} \tag{4.88c}$$

where $\bar{E}_{30QBHD2}$ is the totally quantized energy, $Z_1(k_1) = \frac{5\sqrt{21}}{4}\left[\frac{\left(\frac{n_x\pi}{d_x}\right)^4 + \left(\frac{n_y\pi}{d_y}\right)^4 + \left(\frac{n_z\pi}{d_z}\right)^4}{k_1^4} - \frac{3}{5}\right]$ and $Z_2(k_1) = \left[\frac{639639}{32}\right]^{1/2}\left[\frac{\left(\frac{n_x\pi}{d_x}\right)^2\left(\frac{n_y\pi}{d_y}\right)^2\left(\frac{n_z\pi}{d_z}\right)^2}{k_1^6} + \frac{1}{12}\left(\frac{\left(\frac{n_x\pi}{d_x}\right)^4 + \left(\frac{n_y\pi}{d_y}\right)^4 + \left(\frac{n_z\pi}{d_z}\right)^4}{k_1^4} - \frac{3}{5}\right) - \frac{1}{105}\right]$

The DOS function is given by

$$N_{ODT}(E) = \frac{2g_v}{d_x d_y d_z} \sum_{n_x=1}^{n_{x\max}} \sum_{n_y=1}^{n_{y\max}} \sum_{n_z=1}^{n_{z\max}} \delta'(E - \bar{E}_{30QBHD2}) \tag{4.88d}$$

4.2.12 The DR in Quantum Dot (QD) of HD II–V Semiconductors

The DR of the holes in QDs of HD II–V compounds can be expressed as

$$\begin{aligned} \gamma_3(E_{100QBHD,\pm}, \eta_g) &= A_{10}\left(\frac{n_x\pi}{d_x}\right)^2 + A_{11}\left(\frac{n_y\pi}{d_y}\right)^2 + A_{12}\left(\frac{n_z\pi}{d_z}\right)^2 + A_{13}\left(\frac{n_x\pi}{d_x}\right) \\ &\pm \left[\left(A_{14}\left(\frac{n_x\pi}{d_x}\right)^2 + A_{15}\left(\frac{n_y\pi}{d_y}\right)^2 + A_{16}\left(\frac{n_z\pi}{d_z}\right)^2 + A_{17}\left(\frac{n_x\pi}{d_x}\right)^2\right) + A_{18}\left(\frac{n_y\pi}{d_y}\right)^2 + A_{19}^2\right]^{1/2} \end{aligned} \tag{4.89}$$

where $E_{100QBHD,\pm}$ is the totally quantized energy in this case.

The DOS function is given by

$$N_{ODT}(E) = \frac{g_v}{d_x d_y d_z} \sum_{n_x=1}^{n_{x\max}} \sum_{n_y=1}^{n_{y\max}} \sum_{n_z=1}^{n_{z\max}} \delta'(E - E_{100QBHD,\pm}) \tag{4.90}$$

In the absence of band-tailing, the 0D hole energy spectrum in this case assumes the form

$$\begin{aligned} E_{QD70,\pm} &= A_{10}\left(\frac{n_x\pi}{d_x}\right)^2 + A_{11}\left(\frac{n_y\pi}{d_y}\right)^2 + A_{12}\left(\frac{n_z\pi}{d_z}\right)^2 + A_{13}\left(\frac{n_x\pi}{d_x}\right)^2 \\ &\pm \left[\left(A_{14}\left(\frac{n_x\pi}{d_x}\right)^2 + A_{15}\left(\frac{n_y\pi}{d_x}\right)^2 + A_{16}\left(\frac{n_z\pi}{d_z}\right)^2 + A_{17}\left(\frac{n_x\pi}{d_x}\right)^2\right) + A_{18}\left(\frac{n_y\pi}{d_y}\right)^2 + A_{19}^2\right]^{1/2} \end{aligned} \tag{4.91}$$

where $E_{QD70,\pm}$ is the totally quantized energy in this case.

The DOS function is given by

$$N_{ODT}(E) = \frac{g_v}{d_x d_y d_z} \sum_{n_x=1}^{n_{x\max}} \sum_{n_y=1}^{n_{y\max}} \sum_{n_z=1}^{n_{z\max}} \delta'(E - E_{QD70,\pm}) \tag{4.92}$$

4.2.13 The DR in Quantum Dot (QD) of HD Lead Germanium Telluride

The 0D electron energy spectrum in n-type $Pb_{1-x}Ge_xTe$ under the condition of formation of band tails can be written as

$$\left[\frac{2}{1+Erf\left(\frac{E_{101QBHD,\pm}}{\eta_g}\right)}\right]\theta_0(E_{101QBHD,\pm},\eta_g)+\gamma_3(E_{101QBHD,\pm},\eta_g)$$
$$\left[\bar{E}_{g_0}-0.195\left[\left(\frac{n_x\pi}{d_x}\right)^2+\left(\frac{n_y\pi}{d_y}\right)^2\right]-0.345\left(\frac{n_z\pi}{d_z}\right)^2\right]$$
$$=\left[0.23\left[\left(\frac{n_x\pi}{d_x}\right)^2+\left(\frac{n_y\pi}{d_y}\right)^2\right]+0.02\left(\frac{n_z\pi}{d_z}\right)^2\right]$$
$$\pm\left[0.06\bar{E}_{g_0}+0.061\left[\left(\frac{n_x\pi}{d_x}\right)^2+\left(\frac{n_y\pi}{d_y}\right)^2\right]+0.0066\left(\frac{n_z\pi}{d_z}\right)^2\right]\left[\left(\frac{n_x\pi}{d_x}\right)^2+\left(\frac{n_y\pi}{d_y}\right)^2\right]^{1/2}$$
$$+\left[\bar{E}_{g_0}+0.411\left[\left(\frac{n_x\pi}{d_x}\right)^2+\left(\frac{n_y\pi}{d_y}\right)^2\right]+0.377\left(\frac{n_z\pi}{d_z}\right)^2\right]\left[0.606\left[\left(\frac{n_x\pi}{d_x}\right)^2+\left(\frac{n_y\pi}{d_y}\right)^2+0.722\left(\frac{n_z\pi}{d_z}\right)^2\right]\right] \tag{4.93}$$

where $E_{101QBHD,\pm}$ is the quantized energy in this case.

The DOS function is given by

$$N_{ODT}(E)=\frac{g_v}{d_xd_yd_z}\sum_{n_x=1}^{n_{xmax}}\sum_{n_y=1}^{n_{ymax}}\sum_{n_z=1}^{n_{zmax}}\delta'(E-E_{101QBHD,\pm}) \tag{4.94}$$

The 0D dispersion law of n-type $Pb_{1-x}Ge_xTe$ with x = 0.01 in the absence of band-tails can be expressed as

$$\left[E_{QD71,\pm}-0.606\left[\left(\frac{n_x\pi}{d_x}\right)^2+\left(\frac{n_y\pi}{d_y}\right)^2\right]-0.722\left(\frac{n_z\pi}{d_z}\right)^2\right]$$
$$\left[E_{QD71,\pm}+\bar{E}_{g_0}+0.411\left[\left(\frac{n_x\pi}{d_x}\right)^2+\left(\frac{n_y\pi}{d_y}\right)^2\right]+0.377\left(\frac{n_z\pi}{d_z}\right)^2\right]$$
$$=0.23\left[\left(\frac{n_x\pi}{d_x}\right)^2+\left(\frac{n_y\pi}{d_y}\right)^2\right]+0.02\left(\frac{n_z\pi}{d_z}\right)^2\pm\left[0.06\bar{E}_{g_0}+0.061\left[\left(\frac{n_x\pi}{d_x}\right)^2+\left(\frac{n_y\pi}{d_y}\right)^2\right]+0.0066\left(\frac{n_z\pi}{d_z}\right)^2\right]$$
$$\left[\left(\frac{n_x\pi}{d_x}\right)^2+\left(\frac{n_y\pi}{d_y}\right)^2\right]^{1/2} \tag{4.95}$$

where $E_{QD71,\pm}$ is the quantized energy in this case.

The DOS function is given by

$$N_{ODT}(E) = \frac{g_v}{d_x d_y d_z} \sum_{n_x=1}^{n_{x_{\max}}} \sum_{n_y=1}^{n_{y_{\max}}} \sum_{n_z=1}^{n_{z_{\max}}} \delta'(E - E_{QD71,\pm}) \tag{4.96}$$

4.2.14 *The DR in Quantum Dot (QD) of HD Zinc and Cadmium Diphosphides*

The DR in HD QDs of Zinc and Cadmium diphosphides can be written as

$$\begin{aligned}\gamma_3(E_{102QBHD,\pm}, \eta_g) = {} & \left[\beta_1 + \frac{\beta_2 \beta_{31}(n_x, n_y, n_z)}{8\beta_4}\right]\left[\left(\frac{n_x\pi}{d_x}\right)^2 + \left(\frac{n_y\pi}{d_y}\right)^2 + \left(\frac{n_z\pi}{d_z}\right)^2\right] \\ & \pm \left\{\left[\beta_4 \beta_{31}(n_x, n_y, n_z)\left(\beta_5 - \frac{\beta_2 \beta_{31}(n_x, n_y, n_z)}{8\beta_4}\right)\right.\right. \\ & \left.\left[\left(\frac{n_x\pi}{d_x}\right)^2 + \left(\frac{n_y\pi}{d_y}\right)^2 + \left(\frac{n_z\pi}{d_z}\right)^2\right]\right] \\ & + 8\beta_4^2\left(1 - \frac{\beta_{31}^2(n_x, n_y, n_z)}{4}\right) - \beta_2\left(1 - \frac{\beta_{31}^2(n_x, n_y, n_z)}{4}\right) \\ & \left.\left[\left(\frac{n_x\pi}{d_x}\right)^2 + \left(\frac{n_y\pi}{d_y}\right)^2 + \left(\frac{n_z\pi}{d_z}\right)^2\right]\right\}^{1/2}\end{aligned} \tag{4.97}$$

where $\beta_{31}(n_x, n_y, n_z) = \left[\frac{\left(\frac{n_x\pi}{dx}\right)^2 + \left(\frac{n_y\pi}{dy}\right)^2 - 2\left(\frac{n_z\pi}{dz}\right)^2}{\left[\left(\frac{n_x\pi}{dx}\right)^2 + \left(\frac{n_y\pi}{dy}\right)^2 + \left(\frac{n_z\pi}{dz}\right)^2\right]}\right]$ and $E_{102QBHD,\pm}$ is the quantized energy in this case

The DOS function is given by

$$N_{ODT}(E) = \frac{g_v}{d_x d_y d_z} \sum_{n_x=1}^{n_{x_{\max}}} \sum_{n_y=1}^{n_{y_{\max}}} \sum_{n_z=1}^{n_{z_{\max}}} \delta'(E - E_{102QBHD,\pm}) \tag{4.98}$$

The 0D DR in QDs of Zinc and Cadmium diphosphides in the absence of band-tails can be written as

$$E_{72QB,\pm} = \left[\beta_1 + \frac{\beta_2\beta_{31}(n_x,n_y,n_z)}{8\beta_4}\right]\left[\left(\frac{n_x\pi}{d_x}\right)^2 + \left(\frac{n_y\pi}{d_y}\right)^2 + \left(\frac{n_z\pi}{d_z}\right)^2\right]$$
$$\pm\left\{\left[\beta_4\beta_{31}(n_x,n_y,n_z)\left(\beta_5 - \frac{\beta_2\beta_{31}(n_x,n_y,n_z)}{8\beta_4}\right)\left[\left(\frac{n_x\pi}{d_x}\right)^2 + \left(\frac{n_y\pi}{d_y}\right)^2 + \left(\frac{n_z\pi}{d_z}\right)^2\right]\right]\right.$$
$$\left. + 8\beta_4^2\left(1 - \frac{\beta_{31}^2(n_x,n_y,n_z)}{4}\right) - \beta_2\left(1 - \frac{\beta_{31}^2(n_x,n_y,n_z)}{4}\right)\left[\left(\frac{n_x\pi}{d_x}\right)^2 + \left(\frac{n_y\pi}{d_y}\right)^2 + \left(\frac{n_z\pi}{d_z}\right)^2\right]\right\}^{1/2} \quad (4.99)$$

where $E_{QD72,\pm}$ is the quantized energy in this case.

The DOS function is given by

$$N_{ODT}(E) = \frac{g_v}{d_x d_y d_z}\sum_{n_x=1}^{n_{x_{\max}}}\sum_{n_y=1}^{n_{y_{\max}}}\sum_{n_z=1}^{n_{z_{\max}}} \delta'(E - E_{QD72,\pm}) \quad (4.100)$$

4.3 Summary and Conclusion

- The DRs for QDs of HD materials exhibit the fact that the total energy is quantized since the corresponding wave vector space is totally quantized.
- The DOS functions for all the materials in this case are series of non-uniformly distributed Dirac's Delta functions at specified quantized points in the respective energy axis. The spacing between the consecutive Delta functions are functions of energy band constants and quantization of the wave vector space of a particular material.
- It may be noted that the HD QDs lead to the discrete energy levels, somewhat like atomic energy levels, which produce very large changes. This follows from the inherent nature of the quantum confinement of the carrier gas dealt with here. In QDs, there remain no free carrier states in between any two allowed sets of size-quantized levels unlike that found for UFs and NWs where the quantum confinements are 1D and 2D, respectively. Consequently, the crossing of the Fermi level by the size-quantized levels in HD QDs would have much greater impact on the redistribution of the carriers among the allowed levels, as compared to that found for UFs and NWs respectively. The quantum signature of HD QDs for the DR is rather prominent as compared to the same from UFs and NWs.
- It is the band structure which changes in a fundamental way and consequently all the physical properties of all the electronic materials changes radically leading to new physical concepts.

4.4 Open Research Problems

(R.4.1) Investigate the DR for QDs ofthe HDS in the presences of Gaussian, exponential, Kane, Halperian, Lax and Bonch-Burevich types of band tails for all systems whose unperturbed carrier energy spectra are defined in (R.1.1).

(R.4.2) Investigate the DR for QDs of all the HD materials as considered in (R.4.1) under non uniform strain.

(R.4.3) Investigate the DR in the presence of non uniform strain for QDs of HD negative refractive index, organic, magnetic and other advanced optical materials in the presence of an alternating electric field.

(R.4.4) Investigate the DR for the QDs of HD negative refractive index, organic, magnetic and other advanced optical materials in the presence of an arbitrarily oriented alternating electric field.

(R.4.5) Investigate the DR for the multiple QDs of HD materials whose unperturbed carrier energy spectra are defined in (R.1.1).

(R.4.6) Investigate the DR for all the appropriate HD zero dimensional systems of this chapter in the presence of finite potential wells.

(R.4.7) Investigate the DR for all the appropriate HD zero dimensional systems of this chapter in the presence of parabolic potential wells.

(R.4.8) Investigate the DR for all the above appropriate problems in the presence of elliptical Hill and quantum square rings in the presence of strain.

(R.4.9) Investigate the DR for parabolic cylindrical HD zero dimensional systems in the presence of an arbitrarily oriented alternating electric field for all the HD materials whose unperturbed carrier energy spectra are defined in (R.1.1) in the presence of strain.

(R.4.10) Investigate the DR for HD zero dimensional systems of the negative refractive index and other advanced optical materials in the presence of an arbitrarily oriented alternating electric field and non-uniform light waves and in the presence of strain.

(R.4.11) Investigate the DR for triangular HD zero dimensional systems of the negative refractive index, organic, magnetic and other advanced optical materials in the presence of an arbitrarily oriented alternating electric field in the presence of strain.

(R.4.12) Investigate the DR for all the problems of (R.4.1) in the presence of arbitrarily oriented magnetic field.

(R.4.13) Investigate the DR for all the problems of (R.4.1) in the presence of alternating electric field.

(R.4.14) Investigate the DR for all the problems of (R.4.1) in the presence of alternating magnetic field.

(R.4.15) Investigate the DR for all the problems of (R.4.1) in the presence of crossed electric field and quantizing magnetic fields.

(R.4.16) Investigate the DR for all the problems of (R.4.1) in the presence of crossed alternating electric field and alternating quantizing magnetic fields.

(R.4.17) Investigate the DR for HD QDs of the negative refractive index, organic and magnetic materials.

(R.4.18) Investigate the DR for HD QDs of the negative refractive index, organic and magnetic materials in the presence of alternating time dependent magnetic field.

(R.4.19) Investigate the Dr for HD QDs of the negative refractive index, organic and magnetic materials in the presence of in the presence of crossed alternating electric field and alternating quantizing magnetic fields.

(R.4.20) (a) Investigate the DR for HD QDs of the negative refractive index, organic, magnetic and other advanced optical materials in the presence of an arbitrarily oriented alternating electric field considering many body effects.

(b) Investigate all the appropriate problems of this chapter for a Dirac electron.

(R.4.21) Investigate all the appropriate problems of this chapter by including the many body, image force, broadening and hot carrier effects respectively.

(R.4.22) Investigate all the appropriate problems of this chapter by removing all the mathematical approximations and establishing the respective appropriate uniqueness conditions.

References

1. D. Bimberg, M. Grundmann, N.N. Ledentsov, *Quantum Dot Heterostructures* (Wiley, New York, 1999); G. Konstantatos, I. Howard, A. Fischer, S. Howland, J. Clifford, E. Klem, L. Levina, E.H. Sargent, Nature **442**, 180 (2006)
2. J.K. Jaiswal, H. Mattoussi, J.M. Mauro, S.M. Simon, Nat Biotechnol **21**, 47 (2003)
3. A. Watson, X. Wu, M. Bruchez, Biotechniques **34**, 296 (2003); J. Nakanishi, Y. Kikuchi, T. Takarada, H. Nakayama, K.Yamaguchi, M. Maeda, J. Am. Chem. Soc. **126**, 16314 (2004)
4. X. Michalet, F.F. Pinaud, L.A. Bentolila, J.M. Tsay, S. Doose, J.J. Li, G. Sundaresan, A.M. Wu, S.S. Gambhir, S. Weiss, Science **307**, 538 (2005)
5. W.G.J.H.M. van Sark, P.L.T.M. Frederix, D.J. Van den Heuvel, H.C.G.A. Bol, J.N.J. van Lingen, C. de Mello Donegá, A. Meijerink, J. Phys. Chem. B **105**, 8281 (2001)
6. E.J. Sánchez, L. Novotny, X.S. Xie, Phys. Rev. Lett. **82**, 4014 (1999); B. Bailey, D.L. Farkas, D.L. Taylor, F. Lanni, Nature **366**, 44 (1993)
7. L.V. Asryan, R.A. Suris, in *Selected Topics in Electronics and Systems*, ed. by E. Borovitskaya, M.S. Shur, vol. 25 (World Scientific, Singapore, 2002); L.V. Asryan, R.A. Suris, *Int. J. High* Speed Electron. Syst. Special Issue on "Quantum Dot Heterostructures—Fabrication, Application, Theory" **12** [1], 111 (2002); L.V. Asryan, S. Luryi, *Future Trends in Microelectronics: The Nano Millennium*, ed. by S. Luryi, J.M. Xu, A. Zaslavsky, vol. 219 (Wiley Interscience, New York, 2002)
8. R.A. Freitas, Jr., J. Comput. Theor. Nanosci. **2**, 1 (2005); A. Ferreira, C. Mavroidis, IEEE Robot. Autom. Mag. **13**, 78 (2006); A. Dubey, G. Sharma, C. Mavroidis, S.M. Tomassone, K.

Nikitczuk, M.L. Yarmush, J. Comput. Theor. Nanosci. **1**, 18 (2004); C. Mavroidis, A. Dubey, M.L. Yarmush, Ann. Rev. Biomed. Eng. **6**, 363 (2004)
9. Y. Liu, J.A. Starzyk, Z. Zhu, IEEE Trans. Neural Networks (2008) [In the press]; J.A. Starzyk, H. He, IEEE Trans. Neural Networks, **18**(2), 344 (2007); J.A. Starzyk, H. He, IEEE Trans. Circuits Syst. II **54**(2), 176 (2007)
10. E.–S. Hasaneen, E. Heller, R. Bansal, W. Huang, F. Jain, Solid State Electron. **48**, 2055 (2004)
11. T. Kawazoe, S. Tanaka, M. Ohtsu, J. Nanophoton. **2**, 029502 (2008)
12. H.J. Krenner, S. Stufler, M. Sabathil, E.C. Clark, P. Ester, M. Bichler, G. Abstreiter, J. J. Finley, A. Zrenner, New J. Phys. **7**, 184 (2005)
13. A.E. Zhukov, A.R. Kovsh, Quantum Electron. **38**, 409 (2008)
14. M. Sugawara, T. Akiyama, N. Hatori, Y. Nakata, H. Ebe, H. Ishikawa, Meas. Sci. Technol. **13**, 1683 (2002); M. van der Poel, D. Birkedal, J. Hvam, M. Laemmlin, D. Bimberg, Conf Lasers Electro-Optics (CLEO) **1**, 16 (2004)
15. J.M. Costa-Fernandez, Anal. Bioanal. Chem. **384**, 37 (2006); H.S. Djie, C.E. Dimas, D.–N. Wang, B.–S. Ooi, J.C.M. Hwang, G.T. Dang, W.H. Chang, IEEE Sens. J. **7**, 251 (2007)
16. X.-X. Zhu, Y.-C. Cao, X. Jin, J. Yang, X.-F. Hua, H.-Q. Wang, B. Liu, Z. Wang, J.-H. Wang, L. Yang, Y.-D. Zhao, Nanotechnology **19**, 025708 (2008); X. Gao, W.C.W. Chan, S. Nie, J. Biomed. Opt. **7**, 532 (2002)
17. J.K. Jaiswal, E.R. Goldman, H. Mattoussi, S.M. Simon, Nat. Methods **1**, 73 (2004)
18. H. Matsueda, Int. J. Circuit Theory Appl. **31**, 23 (2003); X. Hu, S. Das Sarma, Phys. Stat. Sol. (b) **238**, 360 (2003)
19. G.-L. Chen, D.M.T. Kuo, W.-T. Lai, P.-W. Li, Nanotechnology **18**, 475402 (2007); A.G. Pogosov, M.V. Budantsev, A.A. Shevyrin, A.E. Plotnikov, A.K. Bakarov, A.I. Toropov, JETP Letts. **87**, 150 (2008)
20. K.W. Johnston, A.G. Pattantyus-Abraham, J.P. Clifford, S.H. Myrskog, D.D. MacNeil, L. Levina, E.H. Sargent, Appl. Phys. Letts. **92**, 151115 (2008); K.S. Leschkies, R. Divakar, J. Basu, E.E. Pommer, J.E. Boercker, C.B. Carter, U.R. Kortshagen, D.J. Norris, E.S. Aydil, Nano Lett. **7**, 1793 (2007)
21. I.-S. Liu, H.-H. Lo, C.-T. Chien, Y.-Y. Lin, C.-W. Chen, Y.-F. Chen, W.-F. Su, S.-C. Liou, J. Mater. Chem. **18**, 675 (2008)
22. N. Hitoshi, Y. Sugimoto, K. Nanamoto, N. Ikeda, Y. Tanaka, Y. Nakamura, S. Ohkouchi, Y. Watanabe, K. Inoue, H. Ishikawa, K. Asakawa, Opt. Express **12**, 6606 (2004); N. Yamamoto, T. Matsuno, H. Takai, N. Ohtani, Jpn. J. Appl. Phys. **44**, 4749 (2005); T. Yamada, Y. Kinoshita, S. Kasai, H. Hasegawa, Y. Amemiya, Jpn. J. Appl. Phys. **40**, 4485 (2001); K. Asakawa, Y. Sugimoto, Y. Watanabe, N. Ozaki, A. Mizutani, Y. Takata, Y. Kitagawa, H. Ishikawa, N. Ikeda, K. Awazu, X. Wang, A. Watanabe, S. Nakamura, S. Ohkouchi, K. Inoue, M. Kristensen, O. Sigmund, P. I Borel, R. Baets, New J. Phys. **8**, 208 (2006)
23. A.R. Clapp, I.L. Medintz, B.R. Fisher, G.P. Anderson, H. Mattoussi, J. Am. Chem. Soc. **127**, 1242 (2005); L. Shi, B. Hernandez, M. Selke, J. Am. Chem. Soc. **128**, 6278 (2006); C. Wu, J. Zheng, C. Huang, J. Lai, S. Li, C. Chen, Y. Zhao, Angew. Chem. Int. Ed. **46**, 5393 (2007)]
24. U. Rossler, Solid State Commun. **49**, 943 (1984)
25. V.G. Agafonov, P.M. Valov, B.S. Ryvkin, I.D. Yaroshetskii, Sov. Phys. Semi-cond. **12**, 1182 (1978)
26. G.M.T. Foley, P.N. Langenberg, Phys. Rev. B **15B**, 4850 (1977)
27. D.G. Seiler, W.M. Beeker, K.M. Roth, Phys. Rev. **1**, 764 (1970)
28. P.C. Mathur, S. Jain, Phys. Rev. **19**, 1359 (1979)
29. H.I. Zhang, Phys. Rev. B **1**, 3450 (1970)

Chapter 5
The DR in Doping Superlattices of HD Non-parabolic Semiconductors

I must lower my mind to the dust of all persons' feet so that I can enjoy the GOD's Laugh from every heart.

5.1 Introduction

The technological importance of super-lattices in general, and specifically doping super-lattices [1–20] has already been stated in the preface and also in the references of this chapter. In Sect. 5.2.1, of the theoretical background, the DR in doping superlattices of HD non-linear optical semiconductors has been investigated. The Sect. 5.2.2 contains the results for doping superlattices of HD III–V, ternary and quaternary semiconductors in accordance with the three and the two band models of Kane together with parabolic energy bands and they form the special cases of Sect. 5.2.1. The Sects. 5.2.3–5.2.5 contain the study of the DR for doping superlattices of HD II–VI, IV–VI and stressed Kane type semiconductors respectively. The Sects. 5.3 and 5.4 contain the summary and conclusion and 5 open research problems for this chapter.

5.2 Theoretical Background

5.2.1 The DR in Doping Superlattices of HD Nonlinear Optical Semiconductors

The DR of the conduction electrons in doping superlattices of HD nonlinear optical materials can be expressed by using (2.2) and following the method as given in [19, 20] as

K.P. Ghatak, *Dispersion Relations in Heavily-Doped Nanostructures*,
Springer Tracts in Modern Physics 265, DOI 10.1007/978-3-319-21000-1_5

$$\frac{\left(n_i+\frac{1}{2}\right)}{\hbar T_{21}\left(E,\eta_g\right)}\omega_{8HD}\left(E,\eta_g\right)+\frac{\hbar^2 k_s^2}{2m_\perp^* T_{22}\left(E,\eta_g\right)}=1 \tag{5.1}$$

where

$$\omega_{8HD}\left(E,\eta_g\right)\equiv \text{Real part of}\left(\frac{n_0|e|^2}{d_0\varepsilon_{sc}\left[m_\parallel^* T'_{21}\left(E,\eta_g\right)\right]}\right)^{1/2},\ n_i\,(=0,1,2\ldots)$$

is the mini-band index for nipi structures and d_0 is the superlattice period.

The EEM in this case assumes the form

$$m^*\left(\mathrm{E}_{FnHD},\mathrm{n}_i,\eta_g\right)\ =\ \text{Real part of}\left(\frac{\hbar^2}{2}\right)G'_{21HD}(\mathrm{E}_{FnHD},\mathrm{n}_i,\eta_g) \tag{5.2}$$

where,

$$G_{21HD}(E,\eta_g,\mathrm{n}_i)=\frac{2m_\perp^* T_{22}\left(E,\eta_g\right)}{\hbar^2}\left[1-\frac{\left(n_i+\frac{1}{2}\right)}{\hbar T_{21}\left(E,\eta_g\right)}\omega_{8HD}\left(E,\eta_g\right)\right]$$

and $\bar{E}_{FnHD}$ is the Fermi energy in the present case as measured from the edge of the conduction band in vertically upward direction in the absence of any quantization.

From (5.2), we observe that the EEM is a function of the Fermi energy, nipi subband index, scattering potential and the other material constants which is the characteristic feature of doping superlattices of HD non-linear optical materials.

The subband energy (E_{1n_iHD}) can be written as

$$\frac{\left(n_i+\frac{1}{2}\right)}{\hbar T_{21}\left(E_{1n_iHD},\eta_g\right)}\omega_{8HD}\left(E_{1n_iHD},\eta_g\right)=1 \tag{5.3}$$

The DOS function for doping superlattices of HD non-linear optical materials can be expressed as

$$N_{nipiHD}(E,\eta_g)=\frac{g_v}{2\pi}\sum_{n_i=0}^{n_{i\max}}G'_{21HD}(E,\eta_g,n_i)H(E-E_{1n_iHD}) \tag{5.4}$$

The DR of the conduction electrons in doping superlattices of nonlinear optical materials in the absence of band tails assumes the form

$$\psi_1(E)=\psi_2(E)k_s^2+\psi_3(E)\left(n_i+\frac{1}{2}\right)\frac{2m_\parallel^*}{\hbar}\omega_8(E) \tag{5.5}$$

where

$$\omega_8(E) \equiv \left(\frac{n_0|e|^2}{d_0\varepsilon_{sc}[\theta_1(E)]}\right)^{1/2} \text{ and } \theta_1(E) = \frac{\hbar^2}{2}\left\{\frac{\psi_3(E)[\psi_1(E)]' - \psi_1(E)[\psi_3(E)]'}{[\psi_3(E)]^2}\right\}$$

The EEM in this case can be written as

$$m^*(E_{Fn}, n_i) = \left(\frac{\hbar^2}{2}\right) R_{81}(E, n_i)\Big|_{E=\bar{E}_{Fn}} \tag{5.6}$$

where,

$$\begin{aligned} R_{81}(E, n_i) = [\psi_2(E)]^{-2}\Bigg[\psi_2(E)\Bigg\{[\psi_1(E)]' - \left(\frac{2m_\parallel^*}{\hbar}\right)[\psi_3(E)]'\left(n_i + \frac{1}{2}\right)[\omega_8(E)] \\ - \left(\frac{2m_\parallel^*}{\hbar}\right)[\psi_3(E)]\left(n_i + \frac{1}{2}\right)[\omega_8(E)]'\Bigg\} \\ - \left\{[\psi_2(E)] - \left(\frac{2m_\parallel^*}{\hbar}\right)[\psi_3(E)]\left(n_i + \frac{1}{2}\right)[\omega_8(E)]\right\}[\psi_3(E)]'\Bigg] \end{aligned}$$

and $\bar{E}_{Fn}$ is the Fermi energy in the present case as measured from the edge of the conduction band in vertically upward direction in the absence of any quantization. The sub-band energy (E_{1n_i}) can be written as

$$\psi_1(E_{1n_i}) = \psi_3(E_{1n_i})\left(n_i + \frac{1}{2}\right)\left(\frac{2m_\parallel^*}{\hbar}\right)\omega_8(E_{1n_i}) \tag{5.7}$$

The DOS function for doping superlattices of nonlinear optical materials can be expressed as

$$N_{nipi}(E) = \frac{g_v}{2\pi}\sum_{n_i=0}^{n_{imax}} R_{81}(E, n_i)H(E - E_{1ni}) \tag{5.8}$$

5.2.2 *The DR in Doping Superlattices of HD III–V, Ternary and Quaternary Semiconductors*

(a) The electron energy spectrum in doping superlattices of HD III–V, ternary and quaternary materials can be expressed from (2.1) under the conditions $\Delta_\parallel = \Delta_\perp = \Delta$, $\delta = 0$ and $m_\parallel^* = m_\perp^* = m_c$, as

$$\frac{\hbar^2 k_s^2}{2m_c} = \left[T_{31}(\mathrm{E}, \eta_g) + \mathrm{i}T_{32}(\mathrm{E}, \eta_g) - \left(n_i + \frac{1}{2}\right)\hbar\omega_{9HD}(\mathrm{E}, \eta_g)\right] \tag{5.9}$$

where

$$\omega_{9HD}(E, \eta_g) \equiv \left(\frac{n_0|e|^2}{d_0 \varepsilon_{sc} T'_{31}(E, \eta_g) m_c}\right)^{1/2}$$

The EEM in this case assumes the form

$$m^*(E_{FnHD}, n_i, \eta_g) = \text{Real part of}\left(\frac{\hbar^2}{2}\right) G'_{23HD}(E_{FnHD}, \eta_g, n_i) \tag{5.10}$$

where

$$G_{23HD}(E_{FnHD}, \eta_g, n_i) = \frac{2m_c}{\hbar^2}\left[T_{31}(E_{FnHD}, \eta_g) + \mathrm{i}T_{32}(E_{FnHD}, \eta_g) - \left(n_i + \frac{1}{2}\right)\hbar\omega_{9HD}(E_{FnHD}, \eta_g)\right]$$

The sub-band energy E_{2n_iHD} can be written as

$$\left[T_{31}(E_{2n_iHD}, \eta_g) + \mathrm{i}T_{32}(E_{2n_iHD}, \eta_g) - \left(n_i + \frac{1}{2}\right)\hbar\omega_{9HD}(E_{2n_iHD}, \eta_g)\right] = 0 \tag{5.11}$$

The DOS function for doping superlattics of HD III–V, ternary and quaternary materials can be expressed as

$$N_{nipiHD}(E, \eta_g) = \frac{g_v m_c}{\pi\hbar^2} \sum_{n_i=0}^{n_{imax}} G'_{23HD}(E, \eta_g, n_i) H(E - E_{2niHD}) \tag{5.12}$$

In the absence of band tails, the DR in this case assumes the form

$$I_{11}(E) = \left(n_i + \frac{1}{2}\right)\hbar\omega_{19}(E) + \frac{\hbar^2 k_s^2}{2m_c} \tag{5.13}$$

where

$$\omega_{19}(E) \equiv \left(\frac{n_0|e|^2}{d_0 \varepsilon_{sc} I'_{11}(E) m_c}\right)^{1/2}.$$

The EEM in this case can be written as

$$m^*(E_{Fn}, n_i) = m_c R_{82}(E, n_i)|_{E=E_{Fn}} \tag{5.14}$$

in which,

$$R_{82}(E, n_i) \equiv \left\{[I_{11}(E)]' - \left(n_i + \frac{1}{2}\right)\hbar[\omega_{19}(E)]'\right\}.$$

From (5.14), we observe that the EEM in this case is a function of the Fermi energy, nipi subband index and the other material constants which is the characteristic feature of doping superlattices of III–V, ternary and quaternary compounds whose bulk DRs is defined by the three band model of Kane.

The subband energies (E_{2ni}) can be written as

$$I_{11}(E_{2ni}) = \left(n_i + \frac{1}{2}\right)\hbar\omega_{19}(E_{2ni}) \tag{5.15}$$

The DOS function in this case can be expressed as

$$N_{nipi}(E) = \frac{m_c g_v}{\pi\hbar^2} \sum_{n_i=0}^{n_{imax}} R_{82}(E, n_i) H(E - E_{2ni}) \tag{5.16}$$

(b) The electron energy spectrum in doping superlattices of HD III–V, ternary and quaternary materials whose energy band structures in the absence of band tails are described by the two band model of Kane can be expressed from (5.13) under the conditions $\Delta \gg E_g$ or $\Delta \ll E_g$, as

$$\frac{\hbar^2 k_s^2}{2m_c} = \left[\gamma_2(E, \eta_g) - \left(n_i + \frac{1}{2}\right)\hbar^2\omega_{10HD}(E, \eta_g)\right] \tag{5.17}$$

where

$$\omega_{10HD}(E) \equiv \left(\frac{n_0|e|^2}{d_0 \varepsilon_{sc} \gamma_2'(E, \eta_g) m_c}\right)^{1/2}$$

The EEM in this case assumes the form

$$m^*(E_{FnHD}, n_i, \eta_g) = \left(\frac{\hbar^2}{2}\right) G'_{25HD}(E_{FnHD}, \eta_g, n_i) \tag{5.18}$$

where

$$G_{25HD}(E_{FnHD},\eta_g,n_i)=\frac{2m_c}{\hbar^2}\left[\gamma_2(E_{FnHD},\eta_g)-\left(n_i+\frac{1}{2}\right)\hbar\omega_{10HD}(E_{FnHD},\eta_g)\right]$$

The subband energy E_{3n_iHD} can be written as

$$\left[\gamma_2(E_{3n_iHD},\eta_g)-\left(n_i+\frac{1}{2}\right)\hbar\omega_{9HD}(E_{3n_iHD},\eta_g)\right]=0 \quad (5.19)$$

The DOS function in this case is given by

$$N_{nipiHD}(E,\eta_g)=\frac{g_v m_c}{\pi\hbar^2}\sum_{n_{i=0}}^{n_{imax}}G'_{25HD}(E,\eta_g,n_i)H(E-E_{3n_iHD}) \quad (5.20)$$

In the absence of band tails, the DR in this case assumes the form

$$E(1+\alpha E)=\left(n_i+\frac{1}{2}\right)\hbar\omega_{20}(E)+\frac{\hbar^2k_s^2}{2m_c} \quad (5.21)$$

where

$$\omega_{20}(E)\equiv\left(\frac{n_0|e|^2}{d_0\varepsilon_{sc}(1+2\alpha E)m_c}\right)^{1/2}.$$

The EEM in this case can be written as

$$m^*(E_{Fn},n_i)=m_cR_{182}(E,n_i)|_{E=E_{Fn}} \quad (5.22)$$

in which,

$$R_{182}(E,n_i)\equiv\left\{[1+2\alpha E]-\left(n_i+\frac{1}{2}\right)\hbar[\omega_{19}(E)]'\right\}.$$

From (5.22), we observe that the EEM in this case is a function of the Fermi energy, nipi subband index and the other material constants which is the characteristic feature of doping superlattices of III–V, ternary and quaternary compounds whose bulk DRs is defined by the three band model of Kane.

The subband energies (E_{3ni}) can be written as

$$E_{3ni}(1+\alpha E_{3ni})=\left(n_i+\frac{1}{2}\right)\hbar\omega_{20}(E_{3ni}) \quad (5.23)$$

The DOS function in this case can be expressed as

$$N_{nipi}(E) = \frac{m_c g_v}{\pi \hbar^2} \sum_{n_i=0}^{n_{imax}} R_{182}(E, n_i) H(E - E_{3ni}) \tag{5.24}$$

(c) The electron energy spectrum in nipi structures of HD III–V, ternary and quaternary materials whose energy band structures in the absence of band tails are described by the parabolic energy bands can be expressed as

$$\frac{\hbar^2 k_s^2}{2m_c} = \left[\gamma_3(\mathrm{E}, \eta_g) - \left(n_i + \frac{1}{2}\right)\hbar\omega_{11HD}(\mathrm{E}, \eta_g)\right] \tag{5.25}$$

where

$$\omega_{11HD}(E) \equiv \left(\frac{n_0 |e|^2}{d_0 \varepsilon_{sc} \gamma'(E, \eta_g) m_c}\right)^{1/2}$$

The EEM in this case assumes the form

$$m^*(\mathrm{E}_{FnHD}, n_i, \eta_g) = \left(\frac{\hbar^2}{2}\right) G'_{27HD}(E_{FnHD}, \eta_g, n_i) \tag{5.26}$$

where

$$G_{27HD}(E_{FnHD}, \eta_g, n_i) = \frac{2m_c}{\hbar^2}\left[\gamma_3(\mathrm{E}_{FnHD}, \eta_g) - \left(n_i + \frac{1}{2}\right)\hbar\omega_{11HD}(\mathrm{E}_{FnHD}, \eta_g)\right]$$

The subband energy E_{4n_iHD} can be expressed as

$$\left[\gamma_3(E_{4n_iHD}, \eta_g) - \left(n_i + \frac{1}{2}\right)\hbar\omega_{11HD}(E_{4n_iHD}, \eta_g)\right] = 0 \tag{5.27}$$

The DOS function in this case is given by

$$N_{nipiHD}(E, \eta_g) = \frac{g_v m_c}{\pi \hbar^2} \sum_{n_i=0}^{n_{imax}} G'_{27HD}(E, \eta_g, n_i) H(E - E_{4n_iHD}) \tag{5.28}$$

In the absence of band tails, the DR in this case assumes the form

$$E = \left(n_i + \frac{1}{2}\right)\hbar\omega_{21} + \frac{\hbar^2 k_s^2}{2m_c} \tag{5.29}$$

where

$$\omega_{21} \equiv \left(\frac{n_0|e|^2}{d_0\varepsilon_{sc}m_c}\right)^{1/2}.$$

The EEM in this case can be written as

$$m^*(E_{Fn}, n_i) = m_c \tag{5.30}$$

Thus the EEM in this case is a constant quantity.
The subband energies (E_{4ni}) can be written as

$$E_{4n_i} = \left(n_i + \frac{1}{2}\right)\hbar\omega_{21} \tag{5.31}$$

The DOS function in this case can be expressed as

$$N_{nipi}(E) = \frac{m_c g_v}{\pi\hbar^2}\sum_{n_i=0}^{n_{imax}} H(E - E_{4n_i}) \tag{5.32}$$

5.2.3 The DR in Doping Superlattices of HD II–VI Semiconductors

The 2D electron dispersion law in doping superlattices of HD II–VI semiconductors can be expressed as

$$\gamma_3(\mathrm{E}, \eta_g) = a_0' k_s^2 + \left(n_i + \frac{1}{2}\right)\hbar\omega_{30}(\mathrm{E}, \eta_g) \pm \bar{\lambda}_0 k_s,$$
$$\omega_{30}(\mathrm{E}, \eta_g) \equiv \left(\frac{n_0|e|^2}{d_0\gamma_3'(\mathrm{E}, \eta_g)\varepsilon_{sc}m_\parallel^*}\right)^{1/2} \tag{5.33}$$

The EEM in this case assumes the form as

$$m^*(\mathrm{E}_{FnHD}, n_i, \eta_g) = m_\perp^*\left\{1 - \bar{\lambda}_0\left[(\bar{\lambda}_0)^2 + 4a_0'\gamma_3(\mathrm{E}_{FnHD}, \eta_g) - 4a_0'\left(n_i + \frac{1}{2}\right)\hbar\omega_{30}(\mathrm{E}_{FnHD}, \eta_g)\right]^{-1/2}\right\}\gamma_3'(\mathrm{E}_{FnHD}, \eta_g) \tag{5.34}$$

The subband energy can be written as

$$\gamma_3(E_{6n_iHD}, \eta_g) = \left(n_i + \frac{1}{2}\right)\hbar\omega_{30}\left(E_{6n_iHD}, \eta_g\right) \tag{5.35}$$

The DOS function in this case is given by

$$N_{nipiHD}(E) = \frac{g_v}{4\pi(a_0')^2}\sum_{n_i=0}^{n_{imax}}[G_{30HD}(E, \eta_g, n_i)]'\mathrm{H}(E - E_{6n_iHD}) \tag{5.36}$$

where

$$G_{30HD}(E, \eta_g, n_i) = \left[(\bar{\lambda}_0) - 2d_0\left\{\left(n_i + \frac{1}{2}\right)\hbar\omega_{30}(E, \eta_g, n_i) - \gamma_3(E, \eta_g, n_i)\right\}\right]$$

In the absence of band-tails, the carrier dispersion law in doping superlattices of II–VI compounds can be expressed as

$$E = a_0'k_s^2 + \left(n_i + \frac{1}{2}\right)\hbar\bar{\omega}_{10} \pm \bar{\lambda}_0 k_s, \qquad \bar{\omega}_{10} = \left(\frac{n_0|e|^2}{d_0\varepsilon_{sc}m_{\parallel}^*}\right)^{\frac{1}{2}} \tag{5.37}$$

Using (5.37), the EEM in this case can be written as

$$m^*(E_{Fn}, n_i) = m_{\perp}^*\left\{1 - \bar{\lambda}_0\left[(\bar{\lambda}_0)^2 + 4a_0'E_{Fn} - 4a_0'\left(n_i + \frac{1}{2}\right)\hbar\bar{\omega}_{10}\right]^{-1/2}\right\} \tag{5.38}$$

Thus, the EEM in this case is a function of the Fermi energy, the nipi subband index number and the energy spectrum constants due to the only presence of $\bar{\lambda}_0$.

The subband energies (E_{8ni}) assume the form as

$$E_{8ni} = \left(n_i + \frac{1}{2}\right)\hbar\bar{\omega}_{10} \tag{5.39}$$

The DOS function in this case can be expressed as

$$N_{nipi}(E) = \frac{m_{\perp}^* g_v}{\pi\hbar^2}\sum_{n_i=0}^{n_{imax}}\left[1 - \frac{a_{81}}{\sqrt{E + b_{81}(n_i)}}\right]H(E - E_{8ni}) \tag{5.40}$$

in which,

$$a_{81} \equiv \frac{\bar{\lambda}_0}{2\sqrt{a_0'}}$$

and

$$b_{81}(n_i) \equiv \left[\frac{1}{4a_0'}\left[(\bar{\lambda}_0)^2 - 4a_0'\left(n_i + \frac{1}{2}\right)\hbar\bar{\omega}_{10}\right]\right].$$

5.2.4 *The DR in Doping Superlattices of HD IV–VI Semiconductors*

The 2D electron dispersion law in this case is given by

$$k_s^2 = \delta_{15}(E, \eta_g, n_i) \tag{5.41}$$

where

$$\begin{aligned}
\delta_{15}(E, \eta_g, n_i) &= [2\delta_{12}(E, \eta_g)]^{-1}[-\delta_{13}(E, \eta_g, n_i) \\
&\quad + \sqrt{\delta_{13}^2(E, \eta_g, n_i) - 4\delta_{12}(E, \eta_g)\delta_{14}(E, \eta_g, n_i)}], \\
\delta_{12}(E, \eta_g) &= \frac{\alpha\hbar^4 Z_0(E, \eta_g)}{4m_t^+ m_l^-}, \\
\delta_{13}(E, \eta_g, n_i) &= \hbar^2[\lambda_{71}(E, \eta_g)\delta_{11}(E, \eta_g, n_i) + \lambda_{12}(E, \eta_g)], \\
\delta_{14}(E, \eta_g, n_i) &= [\lambda_{73}(E, \eta_g)\delta_{11}^2(E, \eta_g, n_i) \\
&\quad + \lambda_{74}(E, \eta_g)\delta_{11}^4(E, \eta_g, n_i) - \lambda_{74}(E, \eta_g)], \\
\delta_{11}(E, \eta_g, n_i) &= \frac{2}{\hbar} m_{HD}^*(0, \eta_g)\left(n_i + \frac{1}{2}\right)\left[\frac{e^2 n_o}{d_0 \varepsilon_{sc} m_{HD}^*(E, \eta_g)}\right]^{1/2}
\end{aligned}$$

and

$$\begin{aligned}
m_{HD}^*(E, \eta_g) = \frac{\hbar^2}{4\lambda_{76}^2(E, \eta_g)}\Bigg[&2\lambda_{74}(E, \eta_g)\Bigg\{-\lambda_{73}'(E, \eta_g) \\
&+ \frac{\lambda_{73}(E, \eta_g)\lambda_{73}'(E, \eta_g) + 2\lambda_{74}'(E, \eta_g)\lambda_{75}(E, \eta_g) + 2\lambda_{74}(E, \eta_g)\lambda_{75}'(E, \eta_g)}{\sqrt{\lambda_{73}^2(E, \eta_g) + 4\lambda_{74}(E, \eta_g)\lambda_{75}(E, \eta_g)}}\Bigg\} \\
&- 2\lambda_{74}'(E, \eta_g)\left\{-\lambda_{73}(E, \eta_g) + \sqrt{\lambda_{73}^2(E, \eta_g) + 4\lambda_{74}(E, \eta_g)\lambda_{75}(E, \eta_g)}\right\}\Bigg]
\end{aligned}$$

The EEM in this case assumes the form

$$m^*(E_{FnHD}, n_i, \eta_g) = \left(\frac{\hbar^2}{2}\right)\delta'_{15}(E_{FnHD}, \eta_g, n_i) \tag{5.42}$$

The sub-band energy E_{9n_iHD} can be expressed as in this case as

$$0 = \delta_{15}(E_{9n_iHD}, \eta_g, n_i) \tag{5.43}$$

The DOS function in this case is given by

$$N_{nipiHD}(E) = \frac{gv}{2\pi}\sum_{n_{i=0}}^{n_{imax}} [\delta_{15}(E, \eta_g, n_i)]'\mathrm{H}(E - E_{9n_iHD}) \tag{5.44}$$

The carrier energy spectrum in doping superlattices of IV–VI compounds in the absence of band tails can be written as

$$k_s^2 = \left(\hbar^2 S_{19}\right)^{-1}\left[-S_{20}(E, n_i) + \sqrt{S_{20}^2(E, n_i) + 4S_{19}S_{21}(E, n_i)}\right] \tag{5.45}$$

In which,

$$S_{19} \equiv \left(\frac{\alpha}{m_l^+ m_t^-}\right), S_{20}(E, n_i) \equiv \left\{\frac{1}{m_t^*} - \left(\frac{\alpha E}{m_t^+}\right) + \frac{1+\alpha E}{m_t^-} + \frac{\alpha\hbar^2}{2m_l^+ m_t^-}\left(n_i + \frac{1}{2}\right)T(E) + \frac{\alpha\hbar^2}{2m_l^- m_t^+}\left(n_i + \frac{1}{2}\right)T(E)\right\}$$

$$T(E) = \frac{2m^*(0)}{\hbar}\omega_{11}(E),\ m^*(0) \equiv \left(\frac{m_l^* m_l^-}{m_l^* + m_l^-}\right), \quad \omega_{11}(E) \equiv \left(\frac{n_0|e|^2}{d_0\varepsilon_{sc}m^*(E)}\right)^{\frac{1}{2}},$$

$$m^*(E) \equiv \frac{1}{4t_1}\left[-(t_2(E))' + \frac{t_2(E)(t_2(E))' + 2t_1(1+2\alpha E)}{\sqrt{t_2^2(E) + 4Et_1(1+\alpha E)}}\right],$$

$$t_1 \equiv \left(\frac{\alpha}{4m_l^+ m_l^-}\right),\ t_2(E) \equiv \frac{1}{2}\left[\left(\frac{1}{m_l^*}\right) - \left(\frac{\alpha E}{m_l^+}\right) + \left(\frac{1+\alpha E}{m_l^-}\right)\right],$$

$$(t_2(E))' \equiv \frac{\alpha}{2}\left(\frac{1}{m_l^-} - \left(\frac{1}{m_l^+}\right)\right)$$

and

$$S_{21}(E, n_i) = \left[E(1+\alpha E) + \frac{\alpha E\hbar^2}{2m_l^+}\left(n_i + \frac{1}{2}\right)T(E) + \frac{\hbar^2}{2m_l^-}\left(n_i + \frac{1}{2}\right)T(E)(1+\alpha E) + \frac{\hbar^4}{4m_l^- m_l^+}\left(n_i + \frac{1}{2}\right)T(E) - \left(\frac{\hbar^2}{2m_l^*}\right)T(E)\left(n_i + \frac{1}{2}\right)\right].$$

Using (5.45) the EEM in this case can be written as

$$m^*(E_{Fn}, n_i) = R_{84}(E, n_i)|_{E=E_{Fn}} \tag{5.46}$$

where,

$$R_{84}(E, n_i) \equiv (2S_{19})^{-1}\left[-(S_{20}(E, n_i))' + \frac{S_{20}(E, n_i)[S_{20}(E, n_i)]' + 2S_{19}[S_{21}(E, n_i)]'}{\left[\{[S_{20}(E, n_i)]'\}^2 + 4S_{19}S_{21}(E, n_i)\right]}\right].$$

Thus, one can observe that the EEM in this case is a function of both the Fermi energy and the nipi sub-band index number together with the spectrum constants of the system due to the presence of band non-parabolicity.

The sub-band energies (E_{10ni}) can be written as

$$\left[E_{10ni} - \frac{\hbar^2}{2m_l^-}T(E_{10ni})\left(n_i + \frac{1}{2}\right)\right]\left[1 + \alpha E_{10ni} + \alpha\frac{\hbar^2}{2m_l^+}T(E_{10ni})\left(n_i + \frac{1}{2}\right)\right] = \left[\frac{\hbar^2}{2m_l^*}T(E_{10ni})\left(n_i + \frac{1}{2}\right)\right] \tag{5.47}$$

The DOS function in this case assumes the form as

$$N_{nipi}(\mathrm{E}) = \frac{g_v}{\pi\hbar^2}\sum_{n_i=0}^{n_{imax}} R_{84}(E, n_i)H(E - E_{10ni}) \tag{5.48}$$

5.2.5 *The DR in Doping Superlattices of HD Stressed Kane Type Semiconductors*

The 2D DR in this case is given by

$$P_{11}(E, \eta_g)\mathrm{k}_x^2 + Q_{11}(E, \eta_g)\mathrm{k}_y^2 + S_{11}(E, \eta_g)\delta_{19}(E, \eta_g, n_i) = 1 \tag{5.49}$$

where

$$\delta_{19}(E, \eta_g, n_i) = \frac{2}{\hbar}m_{zz}^*(0, \eta_g)\left(n_i + \frac{1}{2}\right)\left[\frac{n_0 e^2}{d_0 \varepsilon_{sc} m_{zz}(E, \eta_g)}\right]^{1/2}$$

The EEM in this case assumes the form

$$m^*(\mathrm{E}_{FnHD}, n_i, \eta_g) = \left(\frac{\hbar^2}{2}\right)\delta'_{20}(E_{FnHD}, \eta_g, n_i) \tag{5.50}$$

where

$$\delta_{20}(E_{FnHD}, \eta_g, n_i) = \frac{[1 - S_{11}(E_{FnHD}, \eta_g)\delta_{19}(E_{FnHD}, \eta_g, n_i)]}{\sqrt{P_{11}(E_{FnHD}, \eta_g)Q_{11}(E_{FnHD}, \eta_g)}}$$

The sub-band energy E_{15n_iHD} can be expressed as in this case as

$$S_{11}(E_{15n_iHD}, \eta_g)\delta_{19}(E_{15n_iHD}, \eta_g, n_i) = 1 \tag{5.51}$$

The DOS function in this case is given by

$$N_{nipiHD}(E) = \frac{g_v}{2\pi}\sum_{n_{i=0}}^{n_{imax}}[\delta_{20}(E, \eta_g, n_i)]'\mathrm{H}(E - E_{15n_{iHD}}) \tag{5.52}$$

The electron dispersion law in the doping superlattices of stressed Kane type semiconductors can be written as

$$\frac{k_x^2}{[\bar{a}_0(E)]^2} + \frac{k_y^2}{[\bar{b}_0(E)]^2} + \frac{k_z^2}{[\bar{c}_0(E)]^2}\frac{2m_z^*(0)}{\hbar}\left(n_i + \frac{1}{2}\right)\omega_{12}(E) = 1 \tag{5.53}$$

where

$$\omega_{12}(E) \equiv \left(\frac{n_0|e|^2}{d_0\varepsilon_{sc}m_z^*(E)}\right)^{1/2}$$

and

$$m_z^*(E) \equiv \hbar^2\bar{c}_0(E)\frac{\partial}{\partial E}[\bar{c}_0(E)].$$

The use of (5.53) leads to the expression of the EEM as

$$m^*(E_{Fn}, n_i) = \left(\frac{\hbar^2}{2}\right)R_{85}(E, n_i)\bigg|_{E=E_{Fn}} \tag{5.54}$$

where,

$$R_{85}(E,n_i) \equiv \left[\left[(\bar{a}_0(E))'b_0(E) + (\bar{b}_0(E))'\bar{a}_0(E)\right]\right.$$
$$\left[1 - \frac{1}{[\bar{c}_0(E)]^2}\frac{2m_z^*(0)}{\hbar}\left(n_i + \frac{1}{2}\right)\omega_{12}(E)\right]$$
$$- \left[\frac{\bar{a}_0(E)\bar{b}_0(E)}{[\bar{c}_0(E)]^2}\frac{2m_z^*(0)}{\hbar}\left(n_i + \frac{1}{2}\right)[\omega_{12}(E)]'\right]$$
$$\left. + \left[\frac{\bar{a}_0(E)\bar{b}_0(E)[\bar{c}_0(E)]'}{[\bar{c}_0(E)]^3}\frac{4m_z^*(0)}{\hbar}\left(n_i + \frac{1}{2}\right)[\omega_{12}(E)]\right]\right] \quad (5.55)$$

Thus, the EEM is a function of the Fermi energy and the nipi subband index due to the presence of stress and band non-parabolicity only.

The subband energies (E_{25ni}) can be written as

$$\frac{1}{[\bar{c}_0(E_{25ni})]^2}\frac{2m_z^*(0)}{\hbar}\left(n_i + \frac{1}{2}\right)\omega_{12}(E_{25n_i}) = 1 \quad (5.56)$$

The DOS function can be written as

$$N_{nipi}(E) = \frac{g_v}{\pi\hbar^2}\sum_{n_i=0}^{n_{imax}} R_{85}(E,n_i)H(E - E_{25ni}) \quad (5.57)$$

5.3 Summary and Conclusion

From the 2D DR in doping superlattices of HD nonlinear optical and tetragonal materials (5.1), we observe that constant energy 2D wave vector surfaces are the series of concentric quantized circles in the complex energy plane which is the consequence of non removable poles in the corresponding DR in the absence of band tails. From (5.9) we have the same inference for doping superlattices of HD III–V materials whose unperturbed conduction electrons obey the three band model of Kane, which contains one non removal pole in energy axis. The 2D electrons in HD doping superlattices of III–V materials are also described by two band model of Kane and parabolic energy bands with the 2D DRs as given by (5.17) and (5.25) respectively. Besides the 2D DRs in this case for IV–VI materials is given by (5.41). Since all the said DRs possess no poles in the finite energy planes, the constant energy of 2D wave vector surfaces are the series of concentric quantized circles in the real plane instead of the complex one. The 2D DR (5.33) in HD doping superlattices of II–VI materials reflects the fact that the constant energy 2D surface is series of concentric displaced quantized circles in the real plane. The 2D DR (5.53) in doping superlattices of HD stressed Kane type semiconductors reflects the fact that the constant energy 2D wave vector surfaces are the series of concentric ellipses in the real plane.

5.4 Open Research Problems

(R.5.1) Investigate the DR in the presence of an arbitrarily oriented non-quantizing magnetic field for nipi structures of HD nonlinear optical semiconductors by including the electron spin. Study all the special cases for HD III–V, ternary and quaternary materials in this context.

(R.5.2) Investigate the DRs in nipi structures of HD IV–VI, II–VI and stressed Kane type compounds in the presence of an arbitrarily oriented non-quantizing magnetic field by including the electron spin.

(R.5.3) Investigate the DR for HD nipi structures of all the materials as stated in this chapter in the presence of non-uniform strain.

(R.5.4) Investigate the DR for all the problems from (R.5.1) to (R.5.3) in the presence of an additional arbitrarily oriented electric field.

(R.5.5) Investigate the DR for all the problems from (R.5.1) to (R.5.5) in the presence of arbitrarily oriented crossed electric and magnetic fields.

References

1. N.G. Anderson, W.D. Laidig, R.M. Kolbas, Y.C. Lo, J. Appl. Phys. **60**, 2361 (1986)
2. F. Capasso, Semiconductors Semimetals **22**, 2 (1985)
3. F. Capasso, K. Mohammed, A.Y. Cho, R. Hull, A.L. Hutchinson, Appl. Phys. Lett. **47**, 420 (1985)
4. F. Capasso, R.A. Kiehl, J. Appl. Phys. **58**, 1366 (1985)
5. K. Ploog, G.H. Doheler, Adv. Phys. **32**, 285 (1983)
6. F. Capasso, K. Mohammed, A.Y. Cho, Appl. Phys. Lett. **48**, 478 (1986)
7. R. Grill, C. Metzner, G.H. Döhler, Phys. Rev. B **63**, 235316 (2001)
8. A.R. Kost, M.H. Jupina, T.C. Hasenberg, E.M. Garmire, J. Appl. Phys. **99**, 023501 (2006)
9. A.G. Smirnov, D.V. Ushakov, V.K. Kononenko, Proc. SPIE **4706**, 70 (2002)
10. D.V. Ushakov, V.K. Kononenko, I.S. Manak, Proc. SPIE **4358**, 171 (2001)
11. J.Z. Wang, Z.G. Wang, Z.M. Wang, S.L. Feng, Z. Yang, Phys. Rev. B **62**, 6956 (2000)
12. A.R. Kost, L. West, T.C. Hasenberg, J.O. White, M. Matloubian, G.C. Valley, Appl. Phys. Lett. **63**, 3494 (1993)
13. S. Bastola, S.J. Chua, S.J. Xu, J. Appl. Phys. **83**, 1476 (1998)
14. Z.J. Yang, E.M. Garmire, D. Doctor, J. Appl. Phys. **82**, 3874 (1997)
15. G.H. Avetisyan, V.B. Kulikov, I.D. Zalevsky, P.V. Bulaev, Proc. SPIE **2694**, 216 (1996)
16. U. Pfeiffer, M. Kneissl, B. Knüpfer, N. Müller, P. Kiesel, G.H. Döhler, J.S. Smith, Appl. Phys. Lett. **68**, 1838 (1996)
17. H.L. Vaghjiani, E.A. Johnson, M.J. Kane, R. Grey, C.C. Phillips, J. Appl. Phys. **76**, 4407 (1994)

18. P. Kiesel, K.H. Gulden, A. Hoefler, M. Kneissl, B. Knuepfer, S.U. Dankowski, P. Riel, X.X. Wu, J.S. Smith, G.H. Doehler, Proc. SPIE **1985**, 278 (1993)
19. G.H. Doheler, Phys. Script. **24**, 430 (1981)
20. S. Mukherjee, S.N. Mitra, P.K. Bose, A.R. Ghatak, A. Neoigi, J.P. Banerjee, A. Sinha, M. Pal, S. Bhattacharya, K.P. Ghatak, J. Compu. Theor. Nanosc. **4**, 550 (2007); N. Paitya, K. P. Ghatak, J. Adv. Phys. **1**, 161 (2012); N. Paitya, S. Bhattacharya, D. De, K.P. Ghatak Adv. Sci. Engg. Medi. **4**, 96 (2012); S. Bhattacharya, D. De, S.M. Adhikari, K.P. Ghatak Superlatt. Microst. **51**, 203 (2012); D. De, S. Bhattacharya, S.M. Adhikari, A. Kumar, P.K. Bose, K. P. Ghatak, Beilstein Jour. Nanotech. **2**, 339 (2012); D. De, A. Kumar, S.M. Adhikari, S. Pahari, N. Islam, P. Banerjee, S.K. Biswas, S. Bhattacharya, K.P. Ghatak, Superlatt. Microstruct. **47**, 377 (2010); S. Pahari, S. Bhattacharya, S. Roy, A. Saha, D. De, K.P. Ghatak, Superlatt. Microstruct. **46**, 760 (2009); S. Pahari, S. Bhattacharya, K.P. Ghatak J. Comput. Theor. Nanosci. **6**, 2088 (2009); S.K. Biswas, A. R. Ghatak, A. Neogi, A. Sharma, S. Bhattacharya, K.P. Ghatak, Phys. E: Low-dimen. Syst. Nanostruct. **36**, 163 (2007); L.J. Singh, S. Choudhury, D. Baruah, S.K. Biswas, S. Pahari, K.P. Ghatak, Phys. B: Conden. Matter **368**, 188 (2005); S. Chowdhary, L.J. Singh, K.P. Ghatak, Phys. B: Conden. Matter **365**, 5 (2005); L.J. Singh, S. Choudhary, A. Mallik, K.P. Ghatak, J. Comput. Theor. Nanosci. **2**, 287 (2005); K.P. Ghatak, J. Mukhopadhyay, J.P. Banerjee, SPIE Proc. Ser. **4746**, 1292 (2002); K. P. Ghatak, S. Dutta, D.K. Basu, B. Nag, Il Nuovo Cimento. **D 20**, 227 (1998); K.P. Ghatak, D. K. Basu, B. Nag, J. Phys. Chem. Solids **58**, 133 (1997); K.P. Ghatak, B. De, Mat. Resc. Soc. Proc. **300**, 513 (1993); K.P. Ghatak, B. Mitra, Il Nuovo Cimento. **D 15**, 97 (1993); K. P. Ghatak, Int. Soci. Opt. Photon. Proc. Soc. Photo Opt. Instru. Eng. **1626**, 115 (1992); K. P. Ghatak, A. Ghoshal, Phys. Stat. Sol. (b) **170**, K27 (1992); K.P. Ghatak, S. Bhattacharya, S. N. Biswas, Proc. Soc. Photo Opt. instru. Eng. **836**, 72 (1988); K.P. Ghatak, A Ghoshal, S.N. Biswas, M. Mondal, Proc. Soc. Photo Opt. Instru. Eng. **1308**, 356 (1990); K.P. Ghatak, G. Mazumder, Proc. Matt. Res. Soc. **484**, 679 (1998); K.P. Ghatak, B. De, Defect Eng. Semi. Growth, Process. Device Tech. Symp. Mat. Res. Soc. **262**, 911 (1992); S.N. Biswas, K. P. Ghatak, Int. J. Electron. Theor. Exp. **70**, 125 (1991); B Mitra, K.P. Ghatak, Phys. Lett. A. **146**, 357 (1990); B. Mitra, K.P. Ghatak, Phys. Lett. A. **142**, 401 (1989); K.P. Ghatak, B. Mitra, A. Ghoshal, Phys. Stat. Sol. (b) **154**, K121 (1989); B. Mitra, K.P. Ghatak, Phys. Stat. Sol. (b). **149**, K117 (1988); K.P. Ghatak, S.N. Biswas, Proc. Soc. Photo Opt. Instru. Eng. **792**, 239 (1987); S. Bhattacharyya, K.P. Ghatak, S. Biswas, OE/Fibers' 87, Int. Soc. Opt. Photon **836**, 73 (1988); M. Mondal, K.P. Ghatak, Czech. J. Phys. B. **36**, 1389 (1986); K.P. Ghatak, A. N. Chakravarti, Phys. Stat. Sol. (b). **117**, 707 (1983)

Chapter 6
The DR in Accumulation and Inversion Layers of Non-parabolic Semiconductors

Having A smile on my face is a good compliment to life, but putting a smile on others face by my own tremendous effort is the best compliment to life.

6.1 Introduction

It is well known that the electrons in bulk semiconductors in general, have three dimensional freedom of motion. When, these electrons are confined in a one dimensional potential well whose width is of the order of the carrier wavelength, the motion in that particular direction gets quantized while that along the other two directions remains as free. Thus, the energy spectrum appears in the shape of discrete levels for the one dimensional quantization, each of which has a continuum for the two dimensional free motion. The transport phenomena of such one dimensional confined carriers have recently studied [1–20] with great interest. For the metal-oxide-semiconductor (MOS) structures, the work functions of the metal and the semiconductor substrate are different and the application of an external voltage at the metal-gate causes the change in the charge density at the oxide semiconductor interface leading to a bending of the energy bands of the semiconductor near the surface. As a result, a one dimensional potential well is formed at the semiconductor interface. The spatial variation of the potential profile is so sharp that for considerable large values of the electric field, the width of the potential well becomes of the order of the de Broglie wavelength of the carriers. The Fermi energy, which is near the edge of the conduction band in the bulk, becomes nearer to the edge of the valance band at the surface creating accumulation layers. The energy levels of the carriers bound within the potential well get quantized and form electric subbands. Each of the subband corresponds to a quantized level in a plane perpendicular to the surface leading to a quasi two dimensional electron gas. Thus, the extreme band bending at low temperature allows us to observe the quantum effects at the surface. Though considerable work has already been done, nevertheless it appears from the literature that the DR in

K.P. Ghatak, *Dispersion Relations in Heavily-Doped Nanostructures*,
Springer Tracts in Modern Physics 265, DOI 10.1007/978-3-319-21000-1_6

accumulation layers of non-parabolic semiconductors has yet to be investigated in details. For the purpose of comparison we shall also study the DR for inversion layers of non-parabolic compounds.

In what follows in Sect. 6.2.1, of the theoretical background, the DR in accumulation and Inversion layers of nonlinear optical semiconductors has been studied under weak electric field limit. The Sect. 6.2.2 contains the results for accumulation and Inversion layers of III–V, ternary and quaternary semiconductors for the weak electric field limit whose bulk electrons obey the three and the two band models of Kane together with parabolic energy bands and they form the special cases of Sect. 6.2.1. The Sect. 6.2.3 contains the study of the DR for accumulation and Inversion layers of II–VI semiconductors, which is valid for all values of electric field. The Sects. 6.2.4 and 6.2.5 contain the study of the DR in accumulation and Inversion layers of IV–VI and stressed semiconductors respectively. The Sect. 6.2.6 contains the study of the DR in accumulation and Inversion layers of Ge. The Sect. 6.3 contains the summary and conclusion of this chapter. The last Sect. 6.4 contains 12 open research problems of this chapter.

6.2 Theoretical Background

6.2.1 The DR in Accumulation and Inversion Layers of Non-linear Optical Semiconductors

In the presence of a surface electric field F_s along z direction and perpendicular to the surface, (2.2) assumes the form

$$\frac{\hbar^2 k_z^2}{2m_{\parallel}^*} + \frac{\hbar^2 k_s^2}{2m_{\perp}^*} \frac{T_{21}\left(\mathrm{E} - |e|\mathrm{F_s z}, \eta_g\right)}{T_{22}\left(\mathrm{E} - |e|\mathrm{F_s z}, \eta_g\right)} = T_{21}\left(\mathrm{E} - |e|\mathrm{F_s z}, \eta_g\right) \tag{6.1}$$

where, for this chapter, E represents the electron energy as measured from the edge of the conduction band at the surface in the vertically upward direction.

The quantization rule for 2D carriers in this case, is given by [5]

$$\int_0^{z_t} k_z dz = \frac{2}{3}(S_i)^{3/2} \tag{6.2}$$

where, z_t is the classical turning point and S_i is the zeros of the Airy function$(Ai(-S_i) = 0)$.

Using (6.1) and (6.2) leads to the DR of the 2D electrons in accumulation layers of HD non-linear optical materials under the condition of weak electric field limit as

$$\frac{\hbar^2 k_s^2}{2m_{\parallel}^*} = L_6(E, i, \eta_g) \tag{6.3}$$

where $L_6(E,i,\eta_g) = \frac{T_{21}(E,\eta_g) - L_3(E,i\eta_g)}{L_4(E,i,\eta_g)}$, $L_3(E,i,\eta_g) = S_i\left[T'_{21}(E,\eta_g)\right]^{2/3}\left[\frac{\hbar|e|F_s}{\sqrt{2m_{\parallel}^*}}\right]^{2/3}$

and $L_4(E,i,\eta_g) = \left[\frac{T_{21}(E,\eta_g)}{T_{22}(E,\eta_g)} + L_3(E,i,\eta_g)\frac{T_{21}(E,\eta_g)}{T'_{21}(E,\eta_g)T_{22}(E,\eta_g)} \cdot \frac{2}{3}\left\{\frac{T'_{21}(E,\eta_g)}{T_{21}(E,\eta_g)} - \frac{T'_{22}(E,\eta_g)}{T_{22}(E,\eta_g)}\right\}\right]$

The EEM in this case can be written as

$$m^*(E'_f, i, \eta_g) = m_{\parallel}^* \text{ Real part of } \left[L'_6\left(E'_f, i, \eta_g\right)\right] \tag{6.4}$$

where $E'_f = eV_g - \frac{e^2 n_s d_{ox}}{\varepsilon_{ox}} + E_{FB}$, V_g is the gate voltage, n_s is the surface electron concentration, d_{ox} is the thickness of the oxide layer, ε_{ox} is the permittivity of the oxide layer, $F_s = \frac{en_s}{\varepsilon_{sc}}$, ε_{sc} is the semiconductor permittivity and E_{FB} should be determined from the equation

$$n_B = \frac{2g_v}{(2\pi)^3}\frac{2m_{\perp}^*\sqrt{2m_{\parallel}^*}}{\hbar^3} \text{ Real part of } \left[T_{22}(E_{FB},\eta_g)\sqrt{T_{21}(E_{FB},\eta_g)}\right] \tag{6.5}$$

and n_B is the bulk electron concentration.

The sub-band energy E_i can be determined from the equation

$$0 = \text{Real part of } L_6(E_i, i, \eta_g) \tag{6.6}$$

The surface electron concentration in the regime of very low temperatures where the quantum effects become prominent can be written as

$$n_s = 2g_v \text{ Real part of the } \sum_{i=0}^{i_{\max}}\left[\left[\frac{m_{\perp}^*}{2\pi\hbar^2}L_6\left(E'_f, i, \eta_g\right)\right] + \frac{1}{(2\pi)^3}\frac{2m_{\perp}^*\sqrt{2m_{\parallel}^*}}{\hbar^3}t_i \left[T_{22}(E_{FB},\eta_g)\sqrt{T_{21}(E_{FB},\eta_g)}\right]\right] \tag{6.7}$$

where $t_i = \frac{E_{i\max}}{eF_s(1+i_{\max})}$, $E_{i\max}$ is the root of the Real part of the equation

$$T_{21}(E_{i\max},\eta_g) - L_3(E_{i\max}, i_{\max}, \eta_g) = 0 \tag{6.8a}$$

In what follows, we shall discuss the DR in inversion layers of non-linear optical materials for the purpose of relative comparison. In the presence of a surface electric field F_s along z direction and perpendicular to the surface, the (2.2) assumes the form

$$\psi_1(E - |e|F_s z) = \psi_2(E - |e|F_s z)k_s^2 + \psi_3(E - |e|F_s z)k_z^2 \tag{6.8b}$$

where $\psi_1(E) = \gamma(E),\ \psi_2(E) = f_1(E)$ and $\psi_3(E) = f_2(E)$

Using (6.2) and (6.8b), under the weak electric field limit, one can write,

$$\int_0^{z_t} \sqrt{A_7(E) - |e|F_s z D_7(E) dz} = \frac{2}{3}(S_i)^{3/2} \tag{6.9}$$

in which, $A_7(E) \equiv \left[\frac{\psi_1(E) - \psi_2(E)k_s^2}{\psi_3(E)}\right]$ $D_7(E) \equiv [B_7(E) - A_7(E)C_7(E)]$, $B_7(E) \equiv \left[\frac{(\psi_1(E))' - (\psi_2(E))' k_s^2}{\psi_3(E)}\right]$ and $C_7(E) \equiv \left[\frac{(\psi_3(E))'}{\psi_3(E)}\right]$.

Thus, the 2D electron dispersion law in inversion layers of nonlinear optical materials under the weak electric field limit can approximately be written as

$$\psi_1(E) = P_7(E,i)k_s^2 + Q_7(E,i) \tag{6.10}$$

where, $P_7(E,i) \equiv \left[\psi_2(E) - \left(\frac{2t_2(E)}{3[t_1(E)]^{1/3}}\right)\psi_3(E)S_i(|e|F_s)^{2/3}\right]$, $t_2(E) \equiv \left[\frac{[\psi_2(E)]'}{\psi_3(E)} - \left(\frac{\psi_2(E)[\psi_3(E)]'}{[\psi_3(E)]^2}\right)\right]$, $t_1(E) \equiv \left[\frac{[\psi_1(E)]'}{\psi_3(E)} - \left(\frac{\psi_1(E)[\psi_3(E)]'}{[\psi_3(E)]^2}\right)\right]$ and $Q_7(E,i) \equiv S_i\psi_3(E)[|e|F_s t_1(E)]^{2/3}$.

The EEM in the x-y plane can be expressed as

$$m^*(E_{Fiw}, i) = \left(\frac{\hbar^2}{2}\right) G_7(E,i)\Big|_{E=E_{Fiw}} \tag{6.11}$$

where $G_7(E,i) \equiv [P_7(E,i)]^{-2}[P_7(E,i)\{(\psi_1(E))' - (Q_7(E,i))'\} - \{\psi_1(E) - (Q_7(E_i))\}(P_7(E,i))']$ and E_{Fiw} is the Fermi energy under the weak electric field limit as measured from the edge of the conduction band at the surface in the vertically upward direction. Thus, we observe that the EEM is the function of subband index, the Fermi energy and other band constants due to the combined influence of the crystal filed splitting constant and the anisotropic spin-orbit splitting constants respectively.

The subband energy $(E_{n_{iwl}})$ in this case can be obtained from (6.10) as

$$\psi_1(E_{n_{iwl}}) = Q_7(E_{n_{iwl}}, i) \tag{6.12}$$

The general expression of the 2D total DOS function in this case can be written as

$$N_{2D_i}(E) = \frac{2g_v}{(2\pi)^2}\sum_{i=0}^{i_{\max}} \frac{\partial}{\partial E}[A(E,i)H(E - E_{n_i})] \tag{6.13}$$

where, $A(E,i)$ is the area of the constant energy 2D wave vector space for inversion layers and E_{n_i} is the corresponding subband energy.

Using (6.10) and (6.13), the total 2D DOS function under the weak electric field limit can be expressed as

$$N_{2D_i}(E) = \frac{g_v}{(2\pi)} \sum_{i=0}^{i_{\max}} [G_7(E,i)H(E - E_{n_{iw1}})] \tag{6.14}$$

6.2.2 *The DR in Accumulation and Inversion Layers of III–V, Ternary and Quaternary Semiconductors*

(a) Using the substitutions $\delta = 0$, $\Delta_{\|} = \Delta_{\perp} = \Delta$ and $m_{\|}^* = m_{\perp}^* = m_c$, (6.3) under the condition of weak electric field limit, assumes the form

$$T_{90}(E, \eta_g) = \frac{\hbar^2 k_s^2}{2m_c} + S_i \left[\frac{\hbar |e| F_s [T_{90}(E, \eta_g)]'}{\sqrt{2m_c}} \right]^{2/3} \tag{6.15}$$

where,

$$T_{90}(E, \eta_g) = T_{31}(E, \eta_g) + iT_{32}(E, \eta_g)$$

(6.15) represents the DR of the 2D electrons in accumulation layers of HD III–V, ternary and quaternary materials under the weak electric field limit whose bulk electrons obey the HD three band model of Kane. Since the electron energy spectrum in accordance with the HD three-band model of Kane is complex in nature, the (6.15) will also be complex. The both complexities occur due to the presence of poles in the finite complex plane of the dispersion relation of the materials in the absence of band tails.

The EEM can be expressed as

$$m^*(E_f', i, \eta_g) = m_c \text{ Real part of } P_{3HD}'(E_f', i, \eta_g) \tag{6.16}$$

where,

$$P_{3HD}\left(E_f', i, \eta_g\right) = \left[T_{90}\left(E_f', \eta_g\right) - S_i \left[\frac{\hbar |e| F_s \left[T_{90}\left(E_f', \eta_g\right)\right]'}{\sqrt{2m_c}} \right]^{2/3} \right]$$

Thus, one can observe that the EEM is a function of the sub-band index, surface electric field, the Fermi energy and the other spectrum constants due to the combined influence of E_g and Δ.

The sub-band energy E_{i1} is given by

$$0 = \text{Real part of} \left[T_{90}(E_{i1}, \eta_g) - S_i \left[\hbar |e| F_s [T_{90}(E_{i1}, \eta_g)]' \cdot (2m_c)^{-1/2} \right]^{2/3} \right] \quad (6.17)$$

The DOS function can be written as

$$N_{2D_i}(E) = \frac{m_c g_v}{\pi \hbar^2} \sum_{i=0}^{i_{\max}} \left[P_{3HD}(E, i, \eta_g) H(E - E_{i1}) \right] \quad (6.18)$$

Thus the DOS function is complex in nature.

The surface electron concentration is given by

$$n_s = g_v \,\text{Real part of the} \sum_{i=0}^{i_{\max}} \left[\left[\frac{m_c}{\pi \hbar^2} P_{3HD}\left(E'_f, i, \eta_g\right) \right] + \frac{1}{3\pi^2} \left(\frac{2m_c}{\hbar^2} \right)^{3/2} t_i \left[T_{90}(\mathrm{E_{FB}}, \eta_g) \right]^{3/2} \right] \quad (6.19)$$

where $\mathrm{E_{FB}}$ should be determined from the following equation

$$n_B = \frac{g_v}{3\pi^2} \left(\frac{2m_c}{\hbar^2} \right)^{3/2} \text{Real part of} \left[\mathrm{T_{90}}(\mathrm{E_{FB}}, \eta_g) \right]^{3/2} \quad (6.20)$$

Using the substitutions $\delta = 0$, $\Delta_{\|} = \Delta_{\perp} = \Delta$ and $m^*_{\|} = m^*_{\perp} = m_c$, (6.10) under the condition of weak electric field limit, assumes the form

$$I_{11}(E) = \frac{\hbar^2 k_s^2}{2m_c} + S_i \left[\frac{\hbar |e| F_s [I_{11}(E)]'}{\sqrt{2m_c}} \right]^{2/3} \quad (6.21)$$

(6.21) represents the dispersion relation of the 2D electrons in inversion layers of III–V, ternary and quaternary materials under the weak electric field limit whose bulk electrons obey the three band model of Kane.

The EEM can be expressed as

$$m^*(E_{Fiw}, i) = m_c [P_3(E, i)]|_{E = E_{Fiw}} \quad (6.22)$$

where, $P_3(E, i) \equiv \left\{ [I_{11}(E)]' - \left\{ \frac{2}{3} S_i \left[\frac{\hbar |e| F_s}{\sqrt{2m_c}} \right]^{2/3} \{ [I_{11}(E)]' \}^{-1/3} [I_{11}(E)]'' \right\} \right\}$.

Thus, one can observe that the EEM is a function of the subband index, surface electric field, the Fermi energy and the other spectrum constants due to the combined influence of E_g and Δ.

The subband energy $(E_{n_{iw2}})$ in this case can be obtained from the (6.21) as

$$I_{11}(E_{n_{iw2}}) = S_i \left[\frac{\hbar |e| F_s [I_{11}(E_{n_{iw2}})]'}{\sqrt{2m_c}} \right]^{2/3} \tag{6.23}$$

Thus the 2D total DOS function in weak electric field limit can be expressed as

$$N_{2D_i}(E) = \frac{m_c g_v}{\pi \hbar^2} \sum_{i=0}^{i_{max}} [P_3(E, i) H(E - E_{n_{iw2}})] \tag{6.24}$$

(b) Using the constraints $\Delta \gg E_g$ or $\Delta \ll E_g$, the (6.21) under the low electric field limit assumes the form

$$\gamma_2(E, \eta_g) = \frac{\hbar^2 k_s^2}{2m_c} + S_i \left[\frac{\hbar |e| F_s [\gamma_2(E, \eta_g)]'}{\sqrt{2m_c}} \right]^{2/3} \tag{6.25}$$

The (6.25) represents the dispersion relation of the 2D electrons in accumulation layers of HD III–V, ternary and quaternary materials under the weak electric field limit whose bulk electrons obey the HD two band model of Kane.

The EEM can be expressed as

$$m^*\left(E'_f, i, \eta_g\right) = m_c P'_{3HD1}\left(E'_f, i, \eta_g\right) \tag{6.26}$$

where,

$$P_{3HD1}\left(E'_f, i, \eta_g\right) = \left[\gamma_2\left(E'_f, \eta_g\right) - S_i \left[\frac{\hbar |e| F_s \left[\gamma_2\left(E'_f, \eta_g\right)\right]'}{\sqrt{2m_c}} \right]^{2/3} \right]$$

Thus, one can observe that the EEM is a function of the sub-band index, surface electric field, the Fermi energy and the other spectrum constants due to the combined influence of E_g and Δ.

The sub-band energy E_{i1} is given by

$$0 = \left[\gamma_2(E_{i2}, \eta_g) - S_i \left[\hbar |e| F_s [\gamma_2(E_{i2}, \eta_g)]' \cdot (2m_c)^{-1/2} \right]^{2/3} \right] \tag{6.27}$$

The DOS function can be written as

$$N_{2D_i}(E) = \frac{m_c g_v}{\pi \hbar^2} \sum_{i=0}^{i_{max}} \left[P_{3HD1}(E, i, \eta_g) H(E - E_{i2}) \right] \tag{6.28}$$

Thus the DOS function is complex in nature.

The surface electron concentration is given by

$$n_s = g_v \sum_{i=0}^{i_{\max}} \left[\left[\frac{m_c}{\pi \hbar^2} P_{3HD}\left(E'_f, i, \eta_g\right) \right] + \frac{1}{3\pi^2} \left(\frac{2m_c}{\hbar^2} \right)^{3/2} t_i \left[\gamma_2(\mathrm{E_{FB}}, \eta_g) \right]^{3/2} \right] \quad (6.29)$$

where $\mathrm{E_{FB}}$ should be determined from the following equation

$$n_B = \frac{g_v}{3\pi^2} \left(\frac{2m_c}{\hbar^2} \right)^{3/2} \left[\gamma_2(\mathrm{E_{FB}}, \eta_g) \right]^{3/2} \quad (6.30)$$

Using the constraints $\Delta \gg E_g$ or $\Delta \ll E_g$, the (6.21) under the low electric field limit assumes the form

$$E(1 + \alpha E) = \frac{\hbar^2 k_s^2}{2m_c} + S_i \left[\frac{\hbar |e| F_s (1 + 2\alpha E)}{\sqrt{2m_c}} \right]^{2/3} \quad (6.31)$$

For large values of i, $S_i \to \left[\frac{3\pi}{2} \left(i + \frac{3}{4} \right) \right]^{2/3}$ [5], and the (6.31) gets simplified as

$$E(1 + \alpha E) = \frac{\hbar^2 k_s^2}{2m_c} + \left[\frac{3\pi \hbar |e| F_s}{2} \left(i + \frac{3}{4} \right) \frac{(1 + 2\alpha E)}{\sqrt{2m_c}} \right]^{2/3} \quad (6.32)$$

The (6.32) was derived *for the first time by Antcliffe et al.* [3].
The EEM in this case is given by

$$m^*(E_{Fiw}, i) = m_c [P_6(E, i)]|_{E = E_{Fiw}} \quad (6.33)$$

where,

$$P_6(E, i) \equiv \left\{ 1 + 2\alpha E - \frac{4\alpha}{3} S_i \left[\frac{\hbar |e| F_s}{\sqrt{2m_c}} \right]^{2/3} \{1 + 2\alpha E\}^{-1/3} \right\}.$$

Thus, one can observe that the EEM is a function of the subband index, surface electric field and the Fermi energy due to the presence of band non-parabolicity only.

The subband energies $(E_{n_{iw3}})$ are given by

$$E_{n_{iw3}} (1 + \alpha E_{n_{iw3}}) = S_i \left[\frac{\hbar |e| F_s (1 + 2\alpha E_{n_{iw3}})}{\sqrt{2m_c}} \right]^{2/3} \quad (6.34)$$

The total 2D DOS function can be written as

$$N_{2D}(E) = \frac{m_c g_v}{\pi\hbar^2}\sum_{i=0}^{i_{\max}}\left\{\left[1 + 2\alpha E - \frac{4\alpha}{3}S_i\left[\frac{\hbar|e|F_s}{\sqrt{2m_c}}\right]^{2/3}(1 + 2\alpha E)^{-1/3}\right]H(E - E_{n_{iw3}})\right\} \tag{6.35}$$

(c) Using the constraints $\alpha \to 0$, the (6.25) under the low electric field limit assumes the form

$$\gamma_3(E,\eta_g) = \frac{\hbar^2 k_s^2}{2m_c} + S_i\left[\frac{\hbar|e|F_s[\gamma_3(E,\eta_g)]'}{\sqrt{2m_c}}\right]^{2/3} \tag{6.36}$$

The (6.36) represents the dispersion relation of the 2D electrons in accumulation layers of HD III–V, ternary and quaternary materials under the weak electric field limit whose bulk electrons obey the HD parabolic band model.

The EEM can be expressed as

$$m^*\left(E_f', i, \eta_g\right) = m_c P_{3HD2}'\left(E_f', i, \eta_g\right) \tag{6.37}$$

where,

$$P_{3HD2}\left(E_f', i, \eta_g\right) \equiv \left[\gamma_3\left(E_f', \eta_g\right) - S_i\left[\frac{\hbar|e|F_s\left[\gamma_3\left(E_f', \eta_g\right)\right]'}{\sqrt{2m_c}}\right]^{2/3}\right]$$

Thus, one can observe that the EEM is a function of the sub-band index, surface electric field, the Fermi energy and the other spectrum constants due to the combined influence of E_g and Δ.

The sub-band energy E_{i1} is given by

$$0 = \left[\gamma_3(E_{i2},\eta_g) - S_i\left[\hbar|e|F_s[\gamma_3(E_{i2},\eta_g)]' \cdot (2m_c)^{-1/2}\right]^{2/3}\right] \tag{6.38}$$

The DOS function can be written as

$$N_{2D_i}(E) = \frac{m_c g_v}{\pi\hbar^2}\sum_{i=0}^{i_{\max}}\left[P_{3HD2}(E,i,\eta_g)H(E - E_{i3})\right] \tag{6.39}$$

The surface electron concentration is given by

$$n_s = g_v \sum_{i=0}^{i_{\max}} \left[\left[\frac{m_c}{\pi \hbar^2} P_{3HD2}\left(E'_f, i, \eta_g\right) \right] + \frac{1}{3\pi^2} \left(\frac{2m_c}{\hbar^2} \right)^{3/2} t_i \left[\gamma_3 (\mathrm{E_{FB}}, \eta_g) \right]^{3/2} \right] \quad (6.40)$$

where $\mathrm{E_{FB}}$ should be determined from the following equation

$$n_B = \frac{g_v}{3\pi^2} \left(\frac{2m_c}{\hbar^2} \right)^{3/2} \left[\gamma_3 (\mathrm{E_{FB}}, \eta_g) \right]^{3/2} \quad (6.41)$$

For $\alpha \to 0$, as for inversion layers, whose bulk electrons are defined by the parabolic energy bands, from (6.32), we can write,

$$E = \frac{\hbar^2 k_s^2}{2m_c} + S_i \left[\frac{\hbar |e| F_s}{\sqrt{2m_c}} \right]^{2/3} \quad (6.42)$$

The (5.42) is valid for all values of the surface electric field [1].
The electric subband energy $(E_{n_{i4}})$ assumes the form, from (6.42) as

$$E_{n_{i4}} = S_i \left[\frac{\hbar |e| F_s}{\sqrt{2m_c}} \right]^{2/3} \quad (6.43)$$

The total DOS function can be written using (6.42) as

$$N_{2D}(E) = \frac{m_c g_v}{\pi \hbar^2} \sum_{i=0}^{i_{\max}} H(E - E_{n_{i4}}) \quad (6.44)$$

6.2.3 The DR in Accumulation and Inversion Layers of II–VI Semiconductors

The use of (2.105) and (6.2) leads to the expression of the quantization integral as

$$\frac{\sqrt{2m_{\parallel}^*}}{\hbar} \int_0^{z_t} \left[\gamma_3(E, \eta_g) - |e| F_s z \gamma'_3(E, \eta_g) - a'_0 k_s^2 \mp (\bar{\lambda}_0) k_s \right]^{1/2} dz = \frac{2}{3} (s_i)^{3/2} \quad (6.45)$$

where,

$$z_t \equiv \left(|e|F_s\gamma_3'(E,\eta_g)\right)^{-1}\left[\gamma_3(E,\eta_g) - a_0'k_s^2 \mp (\bar{\lambda}_0)k_s\right].$$

Therefore, the 2D electron dispersion law for accumulation layers of HD II–VI semiconductors can be expressed as

$$\gamma_3(\mathrm{E},\eta_g) = a_0'k_s^2 \pm (\bar{\lambda}_0)k_s + S_i\left(\frac{\hbar|e|F_s\gamma_3'(\mathrm{E},\eta_g)}{\sqrt{2m_\parallel^*}}\right)^{2/3} \tag{6.46}$$

The area of the 2D surface as enclosed by the (6.46) can be expressed as

$$A(E,\eta_g,i) = \frac{\pi}{a_0'^2}\Delta_{10}(E,\eta_g,i) \tag{6.47}$$

where

$$\Delta_{10}(E,\eta_g,i) = \left[(\bar{\lambda}_0)^2 - 2a_0'\left\{-\gamma_3(E,\eta_g) + S_i\left(\frac{\hbar e F_s\gamma_3'(E,\eta_g)}{\sqrt{2m_\parallel^*}}\right)^{2/3}\right\}\right]$$

The EEM in this case assumed the form

$$m^*\left(E_f',\eta_g,i\right) = m_\perp^*\Delta_{10}'\left(E_f',\eta_g,i\right) \tag{6.48}$$

The sub-band energy E_{i2} can be written as

$$\gamma_3(E_{i2},\eta_g) = S_i\left(\frac{\hbar|e|F_s\gamma_3'(E_{i2},\eta_g)}{\sqrt{2m_\parallel^*}}\right)^{2/3} \tag{6.49}$$

The surface electron concentration can be written as

$$\begin{aligned} n_s = g_v \sum_{i=0}^{i_{\max}} &\left[\left\{\left(\frac{m_\perp^*}{\pi\hbar^2}\right)\left[\Delta_{10}\left(E_f',i,\eta_g\right) + \Delta_{11}\left(E_f',i,\eta_g\right)\right]\right\}\right. \\ &\left. + \frac{t_i}{2}\left(\frac{k_0T}{\pi b_0'}\right)^{\frac{3}{2}}\left(\frac{b_0'}{a_0'}\right)\left[F_{\frac{1}{2}}\left(\frac{E_{FB}}{k_BT}\right) + \frac{(\bar{\lambda}_0)^2}{2a_0'k_BT}F_{\frac{-1}{2}}\left(\frac{E_{FB}}{k_BT}\right)\right]\right] \end{aligned} \tag{6.50}$$

The E_{FB} can be determined from the following equation

$$n_B = \frac{g_v}{2}\left(\frac{k_0 T}{\pi b'_0}\right)^{\frac{3}{2}}\left(\frac{b'_0}{a'_0}\right)\left[F_{\frac{1}{2}}\left(\frac{E_{FB}}{k_B T}\right) + \frac{(\bar{\lambda}_0)^2}{2a'_0 k_B T} F_{\frac{-1}{2}}\left(\frac{E_{FB}}{k_B T}\right)\right] \tag{6.51}$$

The use of (2.104) and (6.2) leads to the expression of the quantization integral in this case as

$$\frac{\sqrt{2m_{\parallel}^*}}{\hbar}\int_0^{z_t}\left[E - |e|F_s z - a'_0 k_s^2 \mp (\bar{\lambda}_0)k_s\right]^{1/2} dz = \frac{2}{3}(s_i)^{3/2} \tag{6.52}$$

where,

$$z_t \equiv (|e|F_s)^{-1}\left[E - a'_0 k_s^2 \mp (\bar{\lambda}_0)k_s\right].$$

Therefore, the 2D electron dispersion law for inversion layers of II–VI semiconductors can be expressed for all values of F_s as

$$E = a'_0 k_s^2 \pm (\bar{\lambda}_0)k_s + S_i\left(\frac{\hbar|e|F_s}{\sqrt{2m_{\parallel}^*}}\right)^{2/3} \tag{6.53}$$

The area of the 2D surface as enclosed by the (6.53) can be expressed as

$$A(E,i) = \frac{\pi(m_\perp^*)^2}{\hbar^4}\left[\left\{2(\bar{\lambda}_0)^2 - \frac{2\hbar^2}{m_\perp^*}S_i\left(\frac{\hbar|e|F_s}{\sqrt{2m_{\parallel}^*}}\right)^{2/3} + \frac{2\hbar^2 E}{m_\perp^*}\right\} - 2(\bar{\lambda}_0)\left[(\bar{\lambda}_0)^2 - \frac{2\hbar^2}{m_\perp^*}S_i\left(\frac{\hbar|e|F_s}{\sqrt{2m_{\parallel}^*}}\right)^{2/3} + \frac{2\hbar^2 E}{m_\perp^*}\right]^{1/2}\right] \tag{6.54}$$

The EEM is given by

$$m^*(E_{Fi}, i) = m_\perp^*\left[1 - \frac{\rho_{71}}{\sqrt{E_{Fi} + \rho_{72}}}\right] \tag{6.55}$$

where, E_{Fi} is the Fermi energy in this case, $\rho_{71} \equiv \frac{\bar{\lambda}_0}{2\sqrt{a'_0}}$ and $\rho_{72} \equiv \left[(\rho_{71})^2 - \left(\frac{\hbar|e|F_s}{\sqrt{2m_{\parallel}^*}}\right)^{2/3}\right]$.

Thus, the EEM depends on both the Fermi energy and the sub-band index due to the presence of the term $\bar{\lambda}_0$.

The subband energy (E_{ni6}) can be written as

$$E_{ni6} = S_i \left(\frac{\hbar |e| F_s}{\sqrt{2m_{\parallel}^*}} \right)^{2/3} \tag{6.56}$$

6.2.4 The DR in Accumulation and Inversion Layers of IV–VI Semiconductors

The 2D electron dispersion relation in accumulation layers of IV–VI semiconductors can be written as

$$\theta_1(E, i, \eta_g) k_x^2 + \theta_2(E, i, \eta_g) k_y^2 = \theta_3(E, i, \eta_g) \tag{6.57}$$

where

$$\begin{aligned}
\theta_1(E, i, \eta_g) &= \left[\mathrm{F}_1(E, \eta_g) + \mathrm{S}_i (\mathrm{eF}_s \mathrm{a}_1(E, \eta_g))^{2/3} F_2(E, \eta_g) \right] \\
\mathrm{a}_1(E, \eta_g) &= \frac{1}{\mathrm{F}_2(E, \eta_g)} \left[\frac{F_2'(E, \eta_g)}{F_2(E, \eta_g)} F_1(E, \eta_g) - F_1'(E, \eta_g) \right] \\
\theta_2(E, i, \eta_g) &= \left[\left[\mathrm{F}_1(E, \eta_g) + \frac{2}{3} \frac{\mathrm{a}_2(E, \eta_g)}{\mathrm{a}_1(E, \eta_g)} (\mathrm{eF}_s a_1(E, \eta_g))^{2/3} \mathrm{S}_i F_1(E, \eta_g) \right] \right] \\
\mathrm{a}_2(E, \eta_g) &= \frac{1}{F_2(E, \eta_g)} \left[\frac{F_2'(E, \eta_g)}{F_2(E, \eta_g)} F_1(E, \eta_g) - F_1'(E, \eta_g) \right] \\
\theta_3(E, i, \eta_g) &= \left[1 + \frac{2}{3} \frac{\mathrm{C}(E, \eta_g)}{\mathrm{a}_1(E, \eta_g)} \mathrm{S}_i (\mathrm{eF}_s \mathrm{a}_1(E, \eta_g))^{2/3} F_2(E, \eta_g) \right] \text{and}\, \mathrm{C}(E, \eta_g) = \left[\frac{\mathrm{F}_2'(E, \eta_g)}{F_2^2(E, \eta_g)} \right]
\end{aligned}$$

The EEM can be expressed as

$$m^* \left(E_f', i, \eta_g \right) = \frac{\hbar^2}{2} \theta_4' \left(E_f', i, \eta_g \right) \tag{6.58}$$

where

$$\theta_4\left(E'_f, i, \eta_g\right) = \frac{\theta_3\left(E'_f, i, \eta_g\right)}{\sqrt{\theta_1\left(E'_f, i, \eta_g\right)\theta_2\left(E'_f, i, \eta_g\right)}}$$

The sub band energy E_{i3} is given by

$$\theta_3\left(E_{i3}, i, \eta_g\right) = 0 \tag{6.59}$$

The 2D electron concentration in accumulation layer of IV–VI materials under the condition of extreme degeneracy and low electric field limit can be written as

$$n_s = g_v \text{ Real part of } \sum_{i=0}^{i_{\max}} \left[\theta_4\left(E'_f, i, \eta_g\right) + \frac{t_i}{3\pi^2}\left[F_1\left(E_{FB}, \eta_g\right)\sqrt{F_2\left(E_{FB}, \eta_g\right)}\right]^{-1}\right] \tag{6.60}$$

where E_{FB} can be determined from the equation

$$n_B = g_v \text{ Real part of } \left[\frac{1}{3\pi^2}\left[F_1\left(E_{FB}, \eta_g\right)\sqrt{F_2\left(E_{FB}, \eta_g\right)}\right]^{-1}\right] \tag{6.61}$$

The 2D electron dispersion relation of the inversion layers of IV–VI semiconductors in the low electric field limit can be written as

$$k_s^2 = \beta_3(E, i) \tag{6.62}$$

where

$$\beta_3(E, i) = \frac{\beta_1(E, i)}{\beta_2(E, i)},$$

$$\beta_1(E, i) = 1 - \left[\frac{eF_s V'_2(E)}{V_2^2(E)}\right]^{\frac{2}{3}} S_i V_2(E),\; V_2(E) = \left[\frac{2(\bar{A})^2}{E_{g_0}(1+\alpha_1 E)} + \frac{(\bar{S}+\bar{Q})^2}{\Delta''_c(1+\alpha_3 E)}\right](2E)^{-1},$$

$$\beta_2(E, i) = \left[V_1(E) + \left[\frac{eF_s V'_2(E)}{V_2^2(E)}\right]^{\frac{2}{3}} S_i V_2(E) \frac{2}{3}\frac{V_2^2(E)}{V'_2(E)}\left[\frac{V_1(E)V'_2(E)}{V_2^2(E)} - \frac{V'_1(E)}{V_2(E)}\right]\right] \text{ and}$$

$$V_1(E) = \left[\frac{(\bar{R})^2}{E_{g_0}(1+\alpha_1 E)} + \frac{(\bar{S})^2}{\Delta'_c(1+\alpha_2 E)} + \frac{(\bar{Q})^2}{\Delta''_c(1+\alpha_3 E)}\right](2E)^{-1}$$

The EEM can be expressed as

$$m^{*}(E_{Fi}, i) = \frac{\hbar^{2}}{2}\beta_{3}'(E_{Fi}, i) \tag{6.63}$$

The sub-band energy (E_{i4}) can be written as

$$0 = \beta_{3}(E_{i4}, i) \tag{6.64}$$

6.2.5 The DR in Accumulation and Inversion Layers of Stressed Kane Type Semiconductors

The 2D electron DR in accumulation layers of stressed III–V semiconductors can be written as

$$\theta_{13}(E, i, \eta_{g})k_{x}^{2} + \theta_{23}(E, i, \eta_{g})k_{y}^{2} = \theta_{33}(E, i, \eta_{g}) \tag{6.65}$$

where

$$\theta_{13}(E, i, \eta_{g}) = \left[f_{1}(E, \eta_{g}) + S_{i}(\mathrm{eF_{s}a_{13}}(E, \eta_{g}))^{2/3}f_{3}(E, \eta_{g})\right]$$

$$\mathrm{a}_{13}(E, \eta_{g}) = \frac{1}{f_{3}(E, \eta_{g})}\left[\frac{f_{3}'(E, \eta_{g})}{f_{3}(E, \eta_{g})}f_{1}(E, \eta_{g}) - f_{1}'(E, \eta_{g})\right]$$

$$\theta_{23}(E, i, \eta_{g}) = \left[\left[f_{2}(E, \eta_{g}) + \frac{2}{3}\frac{\mathrm{a}_{13}(E, \eta_{g})}{\mathrm{a}_{13}(E, \eta_{g})}(\mathrm{eF_{s}}a_{13}(E, \eta_{g}))^{2/3}S_{i}f_{2}(E, \eta_{g})\right]\right]$$

$$\mathrm{a}_{23}(E, \eta_{g}) = \frac{1}{f_{3}(E, \eta_{g})}\left[\frac{f_{3}'(E, \eta_{g})}{f_{3}(E, \eta_{g})}f_{2}(E, \eta_{g}) - f_{2}'(E, \eta_{g})\right]$$

$$\theta_{33}(E, i, \eta_{g}) = \left[1 + \frac{2}{3}\frac{\mathrm{C}_{3}(E, \eta_{g})}{\mathrm{a}_{13}(E, \eta_{g})}S_{i}(\mathrm{eF_{s}a_{13}}(E, \eta_{g}))^{2/3}f_{3}(E, \eta_{g})\right]$$

$$\mathrm{C}_{3}(E, \eta_{g}) = \left[\frac{f_{3}'(E, \eta_{g})}{f_{3}^{2}(E, \eta_{g})}\right], \text{ and}$$

$f_{1}(E, \eta_{g}), f_{2}(E, \eta_{g}), f_{3}(E, \eta_{g}), P_{11}(E, \eta_{g}), Q_{11}(E, \eta_{g})$ and $S_{11}(E, \eta_{g})$ are defined in Chap. 2 respectively.

The EEM can be expressed as

$$m^{*}\left(E_{f}', i, \eta_{g}\right) = \frac{\hbar^{2}}{2}\theta_{43}'\left(E_{f}', i, \eta_{g}\right) \tag{6.66}$$

where

$$\theta_{43}\left(E'_f, i, \eta_g\right) = \frac{\theta_{33}\left(E'_f, i, \eta_g\right)}{\sqrt{\theta_{13}\left(E'_f, i, \eta_g\right)\theta_{23}\left(E'_f, i, \eta_g\right)}}$$

The sub-band energy E_{i33} is given by

$$\theta_{33}\left(E_{i33}, i, \eta_g\right) = 0 \tag{6.67}$$

The 2D electron concentration in accumulation layers of stressed III–V materials under the condition of extreme degeneracy and low electric field limit can be written as

$$n_s = g_v \,\text{Real part of} \sum_{i=0}^{i_{\max}} \left[\theta_{43}\left(E'_f, i, \eta_g\right) + \frac{t_i}{3\pi^2}\left[f_1(E_{FB}, \eta_g) f_2(E_{FB}, \eta_g) f_3(E_{FB}, \eta_g)\right]^{-1/2}\right] \tag{6.68}$$

The E_{FB} can be determined from the following equation

$$n_B = \frac{g_v}{3\pi^2}\left[f_1(E_{FB}, \eta_g) f_2(E_{FB}, \eta_g) f_3(E_{FB}, \eta_g)\right]^{-1/2} \tag{6.69}$$

The expression of the DR of the 2D electrons in inversion layers of stressed III–V materials under the low electric field limit as

$$[T_{57}(E,i)]k_x^2 + [T_{67}(E,i)]k_y^2 = T_{77}(E,i) \tag{6.70}$$

where,

$$T_{57}(E,i) = \left[E - \alpha_1 + \frac{2}{3} S_i \left(\frac{|e|^2}{\varepsilon_{sc}}\right)^{2/3} (n_{2Dw})^{2/3} L_{17}(E)\right],$$

$$L_{17}(E) = \left[\frac{(E-\alpha_1)}{(E-\alpha_3)^{2/3}[\bar{T}_{47}(E)]^{1/3}} - (E-\alpha_3)^{1/3}[\bar{T}_{47}(E)]^{-1/3}\right],$$

$$[\bar{T}_{47}(E)] = \left[\{\rho_5(E)\}' - \left(\frac{\rho_5(E)}{E-\alpha_3}\right)\right],$$

$$T_{67}(E,i) = \left[E - T_2 + \frac{2}{3} S_i \left(\frac{|e|^2}{\varepsilon_{sc}}\right)^{2/3} (n_{2Dw})^{2/3} L_{27}(E)\right],$$

$$L_{27}(E) = \left[\frac{(E-\alpha_2)}{(E-\alpha_3)^{2/3}[\bar{T}_{47}(E)]^{1/3}} - \left(\frac{(E-\alpha_3)^{1/3}}{[\bar{T}_{47}(E)]^{-1/3}}\right)\right],$$

$$T_{77}(E,i) = \left[\rho_5(E) - S_i \left(\frac{|e|^2}{\varepsilon_{sc}}\right)^{2/3} (n_{2Dw})^{2/3} L_{37}(E)\right],$$

$$L_{37}(E) \equiv (E-\alpha_3)^{1/3}[\bar{T}_{47}(E)]^{2/3}$$

and $\rho_5(E) \equiv [t_1E^3 - t_2E^2 + t_3E + t_4]$.

The area of the 2D surface under the weak electric field limit can be written as

$$A(E,i) = \frac{\pi T_{77}(E,i)}{\sqrt{T_{57}(E,i)T_{67}(E,i)}} \tag{6.71}$$

The sub-band energies $(E_{n_{iw8}})$ in this case are defined by

$$T_{47}(E_{n_{iw8}}) = S_i\left(\frac{|e|^2}{\varepsilon_{sc}}\right)^{2/3}(n_{2Dw})^{2/3}L_{37}(E_{n_{iw8}}) \tag{6.72}$$

The expression of the EEM in this case can be written as

$$m^*(E_{Fiw},i) = \frac{\hbar^2}{2}L_{47}(E,i)\bigg|_{E=E_{Fiw}} \tag{6.73}$$

where,

$$L_{47}(E,i) \equiv \left[\frac{1}{T_{57}(E,i)T_{67}(E,i)}\right]\left[\{T_{77}(E,i)\}'[T_{57}(E,i)T_{67}(E,i)]^{1/2} - \left(\frac{T_{77}(E,i)}{2}\right)\right.$$
$$\left.\left\{\{T_{57}(E,i)\}'\left[\frac{T_{67}(E,i)}{T_{57}(E,i)}\right]^{1/2} + \{T_{67}(E,i)\}'\left[\frac{T_{57}(E,i)}{T_{67}(E,i)}\right]^{1/2}\right\}\right].$$

The total 2D DOS function can be expressed as

$$N_{2D}(E) = \frac{g_v}{2\pi}\sum_{i=0}^{i_{\max}}\{L_{47}(E,i)H(E-E_{n_{iw8}})\} \tag{6.74}$$

6.2.6 The DR in Accumulation and Inversion Layers of Germanium

The 2D DR in accumulation layers of Ge can be written as

$$\frac{\hbar^2k_x^2}{2m_1^*} + \frac{\hbar^2k_y^2}{2m_2^*} = \gamma_{10}(E,i,\eta_g) \tag{6.75}$$

where

$$\gamma_{10}(E,i,\eta_g) = \left[\gamma_3(E,\eta_g)[1+\alpha\gamma_3(E,\eta_g)] - S_i\left[\frac{\hbar eF_s\gamma_3'(E,\eta_g)}{\sqrt{2m_3^*}}\right]^{\frac{2}{3}}[1+2\alpha\gamma_3(E,\eta_g)] + \alpha\left[S_i\left[\frac{\hbar eF_s\gamma_3'(E,\eta_g)}{\sqrt{2m_3^*}}\right]^{\frac{2}{3}}\right]^2\right]$$

The EEM can be expressed as

$$m^*\left(E_f',\mathrm{i},\eta_g\right) = \sqrt{m_1^*m_2^*}[\gamma_{10}'(E,\mathrm{i},\eta_g)] \tag{6.76}$$

The band non-parabolicity and heavy doping makes the mass quantum number dependent.

The sub band energy E_{i14} can be written as

$$\gamma_{10}(E_{i14},i,\eta_g) = 0 \tag{6.77}$$

The surface electron concentration in accumulation layers can be written as

$$n_s = g_v\sum_{i=0}^{i_{\max}}\left[\frac{\sqrt{m_1^*m_2^*}}{\pi\hbar^2}\left[\gamma_{10}\left(E_f',i,\eta_g\right)\right] + t_i\frac{8\pi m_\perp^*\sqrt{m_\parallel^*}}{h^3}[2\gamma_3(E_{FB},\eta_g)]^{\frac{3}{2}}\left[1+\frac{4\alpha}{5}\gamma_3(E_{FB},\eta_g)\right]\right] \tag{6.78}$$

where E_{FB} can be determined from the following equation

$$n_B = g_v\left[\frac{8\pi m_\perp^*\sqrt{m_\parallel^*}}{h^3}[2\gamma_3(E_{FB},\eta_g)]^{\frac{3}{2}}\left[1+\frac{4\alpha}{5}\gamma_3(E_{FB},\eta_g)\right]\right] \tag{6.79}$$

The 2D electron dispersion law in inversion layers of Ge at low electric field limit can be expressed as

$$\frac{\hbar^2k_x^2}{2m_1} + \frac{\hbar^2k_y^2}{2m_2} = \left[E(1+\alpha E) + \alpha E_{i20}^2 - E_{i20}(1+2\alpha E)\right] \tag{6.80}$$

where,

$$E_{i20} = S_i \left(\frac{\hbar e F_s}{\sqrt{2m_3}} \right)^{2/3}$$

The area of 2D space is

$$A = \frac{2\pi\sqrt{m_1 m_2}}{\hbar^2} \left[E(1 + \alpha E) + \alpha E_{i20}^2 - E_{i20}(1 + 2\alpha E) \right] \quad (6.81)$$

The EEM assumes the form

$$m^*(E_{Fiw}, i) = \sqrt{m_1 m_2}[1 + 2\alpha E_{Fiw} - E_{i20} 2\alpha] \quad (6.82)$$

Thus the EEM is the function of both Fermi energy and quantum number due to band non-parabolicity.

The DOS function is given by

$$N_{2D}(E) = \frac{2g_v}{(2\pi)^2} \cdot \frac{2\pi\sqrt{m_1 m_2}}{\hbar^2} \sum_{i=0}^{i_{max}} [1 + 2\alpha E - 2\alpha E_{i20}] H(E - E_{i20}) \quad (6.83)$$

6.3 Summary and Conclusion

From the 2D DR of accumulation (6.3) and inversion layers (6.10) of HD nonlinear optical and tetragonal materials, we observe that constant energy 2D wave vector surfaces are the series of concentric quantized circles in the complex energy plane which is the consequence of non removable poles in the corresponding DR in the absence of band tails. Equations (6.15), (6.25) and (6.36) represent 2D DR in accumulations layers and (6.21), (6.31) and (6.42) inversion layers of III–V materials and the represent concentric quantized circles in complex energy plane in accordance with three band model of Kane and the same in real plane for two band model of Kane and that of parabolic energy bands. Equations (6.46) and (6.53) represent 2D DR in accumulation and inversion layers of II–VI materials and they represent series of concentric quantized circles in the real energy plane. Equations (6.57) and (6.62) represent 2D DR in accumulation and inversion layers of IV–VI materials and they represent series of concentric quantized ellipses in the real energy plane. Equations (6.65) and (6.70) represent 2D DR in accumulation and inversion layers of stressed Kane type materials and they represent series of concentric quantized ellipses in the real energy plane. Equations (6.75) and (6.80) represent 2D DR in accumulation and inversion layers of Ge and they represent series of concentric quantized ellipses in the real energy plane.

It may be noted that if the direction of application of the surface electric field applied perpendicular to the surface be taken as either k_x or k_y and not as k_z as assumed in the present work, the DR would be different analytically for both the limits. Nevertheless, the arbitrary choice of the direction normal to the surface would not result in a change of the basic qualitative feature of the DR in accumulation layers of semiconductors. The approximation of the potential well at the surface by a triangular well introduces some errors, as for instance the omission of the free charge contribution to the potential. This kind of approach is reasonable if there are only few charge carriers in the accumulation layer, but is responsible for an overestimation of the splitting when the accumulation carrier density exceeds that of the depletion layer. It has been observed that the maximum error due to the triangular potential well is tolerable in the practical sense because for actual calculations, one need a self consistent solution which is a formidable problem, for the present generalized systems due to the non availability of the proper analytical techniques, without exhibiting a widely different qualitative behavior. The second assumption of electric quantum limit in the numerical calculation is valid in the range of low temperatures, where the quantum effects become prominent. The errors which are being introduced for these assumptions are found not to be serious enough at low temperatures. Thus, whenever the condition of the electric quantum limit has been applied, the temperature has been assumed to be low enough so that the assumption becomes well grounded because at low temperature, one can assume that nearly all electrons are at the lowest electric subband. We wish to note that the many body effects, the hot electron effects, the formation of band tails, arbitrary orientation of the direction of the electric quantization and the effects of surface of states have been neglected in our simplified theoretical formalism due to the lack of availability in the literature of the proper analytical techniques for including them for the generalized systems as considered in this paper. Our simplified approach will be useful for the purpose of comparison, when, the methods of tackling of the aforementioned formidable problems for the present generalized system appear. The inclusion of the said effects would certainly increase the accuracy of our results, and the qualitative features of the DR as discussed in this chapter would not change in the presence of the aforementioned influences.

6.4 Open Research Problems

(R.6.1) Investigate the DR in the presence of an arbitrarily oriented electric quantization for accumulation layers of tetragonal semiconductors. Study all the special cases for III–V, ternary and quaternary materials in this context.

(R.6.2) Investigate the DR in accumulation layers of IV–VI, II–VI and stressed Kane type compounds in the presence of an arbitrarily oriented quantizing electric field.

(R.6.3) Investigate the DR in accumulation layers of all the materials as stated in (R.1.1) of Chap. 1 in the presence of an arbitrarily oriented quantizing electric field.

(R.6.4) Investigate the DR in the presence of an arbitrarily oriented non-quantizing magnetic field in accumulation layers of tetragonal semiconductors by including the electron spin. Study all the special cases for III–V, ternary and quaternary materials in this context.

(R.6.5) Investigate the DR in accumulation layers of IV–VI, II–VI and stressed Kane type compounds in the presence of an arbitrarily oriented non-quantizing magnetic field by including the electron spin.

(R.6.6) Investigate the DR in accumulation layers of all the materials as stated in (R.1.1) of Chap. 1 in the presence of an arbitrarily oriented non-quantizing magnetic field by including electron spin.

(R.6.7) Investigate the DR in accumulation layers for all the problems from (R.6.1) to (R.6.6) in the presence of an additional arbitrarily oriented electric field.

(R.6.8) Investigate the DR in accumulation layers for all the problems from (R.6.1) to (R.6.3) in the presence of arbitrarily oriented crossed electric and magnetic fields.

(R.6.9) Investigate the DR in accumulation layers for all the problems from (R.6.1) to (R.6.8) in the presence of surface states.

(R.6.10) Investigate the DR in accumulation layers for all the problems from (R.6.1) to (R.6.8) in the presence of hot electron effects.

(R.6.11) Investigate the DR in accumulation layers for all the problems from (R.6.1) to (R.6.6) by including the occupancy of the electrons in various electric subbands.

(R.6.12) Investigate the problems from (R.6.1) to (R.6.11) for the appropriate p-channel accumulation layers.

References

1. T. Ando, H. Fowler, F. Stern, Rev. Mod. Phys. **54**, 437 (1982)
2. J.J. Quinn, P.J. Styles (eds.), *Electronic Properties of Quasi Two Dimensional Systems* (North Holland, Amsterdam, 1976)
3. G.A. Antcliffe, R.T. Bate, R.A. Reynolds, in *Proceedings of the International Conference*, ed. by D.L. Carter, R.T. Bate. Physics of Semi-metals and Narrow-Gap semiconductors (Pergamon Press, Oxford, 1971), p. 499
4. Z.A. Weinberg, Solid State Electron. **20**, 11 (1977)
5. G. Paasch, T. Fiedler, M. Kolar, I. Bartos, Phys. Status Solidi B **118**, 641 (1983)
6. S. Lamari, Phys. Rev. B **64**, 245340 (2001)
7. T. Matsuyama, R. Kürsten, C. Meißner, U. Merkt, Phys. Rev. B **61**, 15588 (2000)
8. P.V. Santos, M. Cardona, Phys. Rev. Lett. **72**, 432 (1994)
9. L. Bu, Y. Zhang, B.A. Mason, R.E. Doezema, J.A. Slinkman, Phys. Rev. B **45**, 11336 (1992)

10. P.D. Dresselhaus, C.M. Papavassiliou, R.G. Wheeler, R.N. Sacks, Phys. Rev. Lett. **68**, 106 (1992)
11. U. Kunze, Phys. Rev. B **41**, 1707 (1990)
12. E. Yamaguchi, Phys. Rev. B **32**, 5280 (1985)
13. Th. Lindner, G. Paasch, J. Appl. Phys. **102**, 054514 (2007)
14. S. Lamari, J. Appl. Phys. **91**, 1698 (2002)
15. K.P. Ghatak, M. Mondal, J. Appl. Phys. **70**, 299 (1991)
16. K.P. Ghatak, S.N. Biswas, J. Vac. Sc. and Technol. **7B**, 104 (1989)
17. B. Mitra, K.P. Ghatak, Solid State Electron. **32**, 177 (1989)
18. K.P. Ghatak, M. Mondal, J. Appl. Phys. **62**, 922 (1987)
19. M. Mondal, K.P. Ghatak, J. Magnet, Magnetic Mat. **62**, 115 (1986); M. Mondal, K.P. Ghatak, Phys. Script. **31**, 613 (1985)
20. K.P. Ghatak, M. Mondal, Z. fur, Physik B, **64**, 223 (1986); K.P. Ghatak, S.N. Biswas, Sol. State Electron. **37**, 1437 (1994); D.R. Choudhury, A.K. Chowdhury, K.P. Ghatak, A.N. Chakravarti, Phys. Stat. Sol. B **98**, (K141) (1980); A.N. Chakravarti, A.K. Chowdhury, K. P. Ghatak, Phys. Stat. Sol. A **63**, K97 (1981); M. Mondal, K.P. Ghatak, Acta Phys. Polon. A **67**, 983 (1985); M. Mondal, K.P. Ghatak, Phys. Stat. Sol. B **128**, K21 (1985); M. Mondal, K.P. Ghatak, Phys. Stat. Sol. A **93**, 377 (1986); K.P. Ghatak, M. Mondal, Phys. Stat. Sol. B **135**, 819 (1986); M. Mondal, K.P. Ghatak, Phys. Stat. Sol. B **139**, 185 (1987); K.P. Ghatak, N. Chattopadhyay, S.N. Biswas, OE/Fibers' **87**, 203 (1987); K.P. Ghatak, N. Chatterjee, M. Mondal, Phys. Stat. Sol. B **139**, K25 (1987); K.P. Ghatak, M. Mondal, Phys. Stat. Sol. B **148**, 645 (1988); K.P. Ghatak, A. Ghosal Phys. Stat. Sol. B **151**, K135 (1989); K.P. Ghatak, N. Chattopadhyay, M. Mondal, Appl. Phys. A **48**, 365 (1989)

Chapter 7
The DR in Heavily Doped (HD) Non-parabolic Semiconductors Under Magnetic Quantization

Team work is more "WE" and less "ME".

7.1 Introduction

It is well known that the band structure of semiconductors can be dramatically changed by applying the external fields. The effects of the quantizing magnetic field on the band structure of compound semiconductors are more striking and can be observed easily in experiments [1–3]. Under magnetic quantization, the motion of the electron parallel to the magnetic field remains unaltered while the area of the wave vector space perpendicular to the direction of the magnetic field gets quantized in accordance with the Landau's rule of area quantization in the wave-vector space [3]. The energy levels of the carriers in a magnetic field (with the component of the wave-vector parallel to the direction of magnetic field be equated with zero) are termed as the Landau levels and the quantized energies are known as the Landau sub-bands. It is important to note that the same conclusion may be arrived either by solving the single-particle time independent Schrödinger differential equation in the presence of a quantizing magnetic field or by using the operator method. The quantizing magnetic field tends to remove the degeneracy and increases the band gap. A semiconductor, placed in a magnetic field B, can absorb radiative energy with the frequency $(\omega_0 = (|e|B/m_c))$. This phenomenon is known as cyclotron or diamagnetic resonance. The effect of energy quantization is experimentally noticeable when the separation between any two consecutive Landau levels is greater than $k_B T$. A number of interesting transport phenomena originate from the change in the basic band structure of the semiconductor in the presence of quantizing magnetic field. These have been widely been investigated and also served as diagnostic tools for characterizing the different materials having various band structures [4–7]. The discreteness in the Landau levels leads to a whole crop of magneto-oscillatory phenomena, important among which are (i) Shubnikov-de Haas oscillations in magneto-resistance; (ii) De Haas-van Alphen

K.P. Ghatak, *Dispersion Relations in Heavily-Doped Nanostructures*,
Springer Tracts in Modern Physics 265, DOI 10.1007/978-3-319-21000-1_7

oscillations in magnetic susceptibility; (iii) magneto-phonon oscillations in thermoelectric power, etc.

In this chapter in Sect. 7.2.1, of the theoretical background, the DR has been investigated in HD non linear optical semiconductors in the presence of a quantizing magnetic field. The Sect. 7.2.2 contains the results for HD III–V, ternary and quaternary compounds in accordance with the three and the two band models of Kane. In the same section the DR in accordance with the models of Stillman et al. and Palik et al. have also been studied for the purpose of relative comparison. The Sect. 7.2.3 contains the study of the DR for HD II–VI semiconductors under magnetic quantization. In Sect. 7.2.4, the DR in HD IV–VI materials has been discussed in accordance with the models of Cohen, Lax, Dimmock, Bangert and Kastner and Foley and Landenberg respectively. In Sect. 7.2.5, the magneto-DR for the stressed HD Kane type semiconductors has been investigated. In Sect. 7.2.6, the DR in HD Te has been studied under magnetic quantization. In Sect. 7.2.7, the magneto-DR in n-GaP has been studied. In Sect. 7.2.8, the DR in HD $PtSb_2$ has been explored under magnetic quantization. In Sect. 7.2.9, the magneto-DR in HD Bi_2Te_3 has been studied. In Sect. 7.2.10, the DR in HD Ge has been studied under magnetic quantization in accordance with the models of Cardona et al. and Wang and Ressler respectively. In Sects. 7.2.11 and 7.2.12, the magneto-DR in HD n-GaSb and II–V compounds has respectively been studied. In Sects. 7.2.13 the magneto DR in HD $Pb_{1-x}Ge_xTe$ *has been discussed.* The Sect. 7.3 explores the discussion and the last Sect. 7.4 contains 15 open research problems.

7.2 Theoretical Background

7.2.1 The DR in HD Nonlinear Optical Semiconductors Under Magnetic Quantization

The DR under magnetic quantization in non-linear optical materials can be written as

$$\gamma(E) = \frac{\hbar^2 k_s^2}{2m_\perp^*} f_3(E) + \frac{\hbar^2 k_z^2}{2m_\parallel^*} f_4(E) \pm \frac{eB\hbar E_g}{6}\left[\frac{(E_{g_0} + \Delta_\perp)}{(E_{g_0} + \frac{2}{3}\Delta_\perp)}\right]\left[E + E_{g_0} + \delta + \frac{\Delta_\parallel^2 - \Delta_\perp^2}{3\Delta_\parallel}\right] \tag{7.1}$$

where $f_3(E) = \frac{E_g(E_g + \Delta_\perp)}{(E_g + \frac{2}{3}\Delta_\perp)}\left[(E + E_g)(E + E_g + \frac{2}{3}\Delta_\parallel) + \delta(E + E_g + \frac{1}{3}\Delta_\parallel) + \frac{1}{9}(\Delta_\parallel^2 - \Delta_\perp^2)\right]$ and $f_4(E) = \frac{E_g(E_g + \Delta_\parallel)}{(E_g + \frac{2}{3}\Delta_\parallel)}\left[(E + E_g)(E + E_g + \frac{2}{3}\Delta_\parallel)\right]$

The (7.1) can be expressed as

$$\frac{\hbar^2 k_z^2}{2m_\parallel^*} + \left(\frac{b_\parallel c_\perp}{b_\perp c_\parallel}\right)\left(\frac{\hbar^2 k_s^2}{2m_\perp^*}\right) = \left\{\left[\frac{ab_\parallel}{c_\parallel}E^2 + \frac{(ac_\parallel + b_\parallel c_\parallel - ab_\parallel)}{c_\parallel^2}E\right.\right.$$
$$+\frac{1}{c_\parallel}\left(1-\frac{a}{c_\parallel}\right)\left(1-\frac{b_\parallel}{c_\parallel}\right)-\frac{1}{c_\parallel}\left(1-\frac{a}{c_\parallel}\right)\left(1-\frac{b_\parallel}{c_\parallel}\right)\frac{1}{(c_\parallel E+1)}\Bigg]$$
$$+\frac{ab_\parallel}{c_\parallel}\left[\delta E + \frac{2}{9}(\Delta_\parallel^2 - \Delta_\perp^2)\right] - \frac{2}{9}\frac{ab_\parallel}{c_\parallel}\frac{(\Delta_\parallel^2 - \Delta_\perp^2)}{(c_\parallel E+1)}\Bigg\} - \left(\frac{\hbar^2 k_s^2}{2m_\perp^*}\right)$$
$$\left\{\left(\frac{b_\parallel c_\perp}{b_\perp c_\parallel}\right)\left[\left(\frac{\delta}{2}+\frac{\Delta_\parallel^2 - \Delta_\perp^2}{6\Delta_\parallel}\right)\frac{a}{(aE+1)} + \left(\frac{\delta}{2}-\frac{\Delta_\parallel^2 - \Delta_\perp^2}{6\Delta_\parallel}\right)\frac{c_\parallel}{(c_\parallel E+1)}\right]\right\}$$
$$\pm e_1\left[\frac{\rho_1}{E+E_g} + \frac{\rho_2}{E+E_g+\frac{2}{3}\Delta_\parallel}\right] \tag{7.2}$$

where $e_1 = \left[\frac{eB\hbar\left(E_g + \frac{2}{3}\Delta_\parallel\right)}{6(E_g + \Delta_\parallel)}\frac{(E_g + \Delta_\perp)}{\left(E_g + \frac{2}{3}\Delta_\perp\right)}\right]$, $\rho_1 = \frac{(-E_g + G_1)}{\left(\frac{2}{3}\Delta_\parallel\right)}$, $G_1 = \left[E_{g0} + \delta + \frac{\Delta_\parallel^2 - \Delta_\perp^2}{3\Delta_\parallel}\right]$ and $\rho_2 = \frac{3}{2\Delta_\parallel}\left[\frac{\Delta_\parallel}{3} + \frac{\Delta_\perp^2}{3\Delta_\parallel} - \delta\right]$.

Therefore, the DR of the conduction electrons in heavily doped non-linear optical semiconductors in the presence of a quantizing magnetic field B can be written following the methods as developed in Chap. 1 as

$$\frac{\hbar^2 k_z^2}{2m_\parallel^*} = U_{1,\pm}(E,n,\eta_g) + iU_{2,\pm}(E,n,\eta_g) \tag{7.3}$$

where

$$U_{1,\pm}(E,n,\eta_g) = \left[\frac{-eB}{m_\perp^*}\left(n+\tfrac{1}{2}\right)\left(\frac{b_\parallel c_\perp}{c_\parallel b_\perp}\right) + \left[\frac{1+Erf(E/\eta_g)}{2}\right]^{-1}\left[\left\{\left[\frac{ab_\parallel}{c_\parallel}\theta_0(E,\eta_g) + \frac{ac_\parallel + b_\parallel c_\parallel - ab_\parallel}{c_\parallel^2}\gamma_0(E,\eta_g)\right.\right.\right.$$
$$+\frac{1}{c_\parallel}\left(1-\frac{a}{c_\parallel}\right)\left(1-\frac{b_\parallel}{c_\parallel}\right)\left[\frac{1+Erf(E/\eta_g)}{2}\right] - \frac{1}{c_\parallel}\left(1-\frac{a}{c_\parallel}\right)\left(1-\frac{b_\parallel}{c_\parallel}\right)c(\beta_1,E,\eta_g) + \frac{ab_\parallel}{c_\parallel}\left[\delta\gamma_0(E,\eta_g) + \frac{2}{9}(\Delta_\parallel^2 - \Delta_\perp^2)\right.$$
$$\left.\left.\left[\frac{1+Erf(E/\eta_g)}{2}\right]\right] - \frac{2}{9}\frac{ab_\parallel}{c_\parallel}(\Delta_\parallel^2 - \Delta_\perp^2)c(\beta_1,E,\eta_g)\right\} - \frac{\hbar eB}{m_\perp^*}\left(n+\tfrac{1}{2}\right)\left\{\left(\frac{b_\parallel c_\perp}{c_\parallel b_\perp}\right)\left[\left(\frac{\delta}{2}+\frac{\Delta_\parallel^2-\Delta_\perp^2}{6\Delta_\parallel}\right)ac(\beta_2,E,\eta_g)\right.\right.$$
$$\left.\left.\left.+\left(\frac{\delta}{2}-\frac{\Delta_\parallel^2-\Delta_\perp^2}{6\Delta_\parallel}\right)c_\parallel c(\beta_1,E,\eta_g)\right]\right\} \pm \frac{\rho_1 e_1}{E_g}c(\beta_2,E,\eta_g) \pm \frac{\rho_2 e_1}{\left(E_g + \frac{2}{3}\Delta_\parallel\right)}c(\beta_3,E,\eta_g)\right],$$

$$U_{2,\pm}(E,n,\eta_g)=\left[\frac{1+Erf(E/\eta_g)}{2}\right]^{-1}$$
$$\left[\frac{1}{c_\parallel}\left(1-\frac{a}{c_\parallel}\right)\left(1-\frac{b_\parallel}{c_\parallel}\right)D(\beta_1,E,\eta_g)+\frac{2}{9}\frac{ab_\parallel}{c_\parallel}(\Delta_\parallel^2-\Delta_\perp^2)D(\beta_1,E,\eta_g)\right.$$
$$+\frac{\hbar eB}{m_\perp^*}\left(n+\frac{1}{2}\right)\left\{\left(\frac{b_\parallel c_\perp}{c_\parallel b_\perp}\right)\left[\left(\frac{\delta}{2}+\frac{\Delta_\parallel^2-\Delta_\perp^2}{6\Delta_\parallel}\right)aD(\beta_2,E,\eta_g)+\left(\frac{\delta}{2}-\frac{\Delta_\parallel^2-\Delta_\perp^2}{6\Delta_\parallel}\right)c_\parallel D(\beta_1,E,\eta_g)\right]\right\}$$
$$\left.\mp\left[\frac{\rho_1 e_1}{E_g}D(\beta_2,E,\eta_g)\pm\frac{\rho_2 e_1}{(E_g+\frac{2}{3}\Delta_\parallel)}D(\beta_3,E,\eta_g)\right]\right]$$

$C(\beta_i,E,\eta_g)=\left[\frac{2}{\beta_i\eta_g\sqrt{\pi}}\right]\exp(-u_i^2)\times\left[\sum_{p=1}^{\infty}\left\{\frac{\exp\left(\frac{-p^2}{4}\right)}{p}\right\}\sinh(pu_i)\right]$, $u_i=\frac{1+\beta_i E}{\beta_i\eta_g}$ and

$D(\beta_i,E,\eta_g)=\left[\frac{\sqrt{\pi}}{\beta_i\eta_g}\exp(-u_i^2)\right]$

The EEM at the Fermi Level can be written from (6.3) as

$$m_\pm^*(E_{FBHD},n,\eta_g)=m_\parallel^* U'_{1,\pm}(E_{FBHD},n,\eta_g)\tag{7.4}$$

where E_{FBHD} is the Fermi energy in this case.

Therefore the double valued EEM in this case is a function of Fermi energy, magnetic field, quantum number and the scattering potential together with the fact that the EEM exists in the band gap which is the general characteristics of HD materials.

The complex density of states function under magnetic quantization is given by

$$\begin{aligned}N_B(E)&=N_{BR}(E)+iN_{BI}(E)\\&=\frac{eB}{2\pi^2\hbar^2}\sqrt{2m_\parallel^*}\sum_{n=0}^{n_{\max}}\left[\frac{x'}{2\sqrt{x}}+\frac{iy'}{2\sqrt{y}}\right]\end{aligned}\tag{7.5}$$

where

$$x=\frac{\sqrt{(U_{1,\pm}(E,n,\eta_g))^2+(U_{2,\pm}(E,n,\eta_g))^2}+(U_{1,\pm}(E,n,\eta_g))}{2},$$
$$y=\frac{\sqrt{(U_{1,\pm}(E,n,\eta_g))^2+(U_{2,\pm}(E,n,\eta_g))^2}+(U_{1,\pm}(E,n,\eta_g))}{2}$$

and x' and y' are the differentiations of x and y with respect to energy E.

Therefore, from (7.5) we can write

$$N_{BR}(E)=\frac{eB\sqrt{2m_\parallel^*}}{4\pi^2\hbar^2}\sum_{n=0}^{n_{\max}}\frac{x'}{\sqrt{x}}\tag{7.6}$$

$$\text{and } N_{BI}(E) = \frac{eB\sqrt{2m_{\|}^{*}}}{4\pi^2\hbar^2}\sum_{n=0}^{n_{\max}}\frac{y'}{\sqrt{y}} \tag{7.7}$$

where E_{nHD} is the complex Landau sub-band energy which can be obtained from (6. 3) by substituting $k_z = 0$ and $E = E_{nHD}$.

7.2.2 The DR in HD III–V Semiconductors Under Magnetic Quantization

(a) The electron energy spectrum in III–V semiconductors under magnetic quantization is given by

$$\frac{E(E+E_g)(E+E_g+\Delta)(E_g+\frac{2}{3}\Delta)}{E_g(E_g+\Delta)(E+E_g+\frac{2}{3}\Delta)} = \left(n+\frac{1}{2}\right)\hbar\omega_0 + \frac{\hbar^2k_z^2}{2m_c} \pm \frac{eB\hbar\Delta}{6m_c(E+E_g+\frac{2}{3}\Delta)} \tag{7.8}$$

The (7.8) can be written as

$$\left[\frac{ab}{c}E^2 + \left(\frac{ac+bc-ab}{c^2}\right)E + \frac{1}{c}\left(1-\frac{a}{c}\right)\left(1-\frac{b}{c}\right) - \frac{1}{c}\left(1-\frac{a}{c}\right)\left(1-\frac{b}{c}\right)\frac{1}{(1+cE)}\right.$$
$$\left. = \left(n+\frac{1}{2}\right)\hbar\omega_0 + \frac{\hbar^2k_z^2}{2m_c} \pm \frac{eB\hbar\Delta}{6m_c(1+cE)(E_g+\frac{2}{3}\Delta)}\right]$$

where $a = \frac{1}{E_g}$, $b = \frac{1}{E_g+\Delta}$ and $c = \frac{1}{E_g+\frac{2}{3}\Delta}$

Therefore

$$\frac{ab}{c}I(5) + \left(\frac{ab+bc-ab}{c^2}\right)I(4) + \left[\frac{1}{c}\left(1-\frac{a}{c}\right)\left(1-\frac{b}{c}\right) - \left(n+\frac{1}{2}\right)\hbar\omega_0\right]I(1)$$
$$- g_{\pm}\left[G(C,E,\eta_g) - iH(C,E,\eta_g)\right] = \frac{\hbar^2k_z^2}{2m_c}I(1) \tag{7.9}$$

where

$$g_{\pm} = \left[\frac{1}{c}\left(1-\frac{a}{c}\right)\left(1-\frac{b}{c}\right) \pm \frac{eB\hbar\Delta}{6m_c(E_g+\frac{2}{3}\Delta)}\right],$$

$$G(C,E,\eta_g) = \left[\frac{2}{C\eta_g\sqrt{\pi}}\right]\exp(-u^2)\times\left[\sum_{p=1}^{\infty}\left\{\frac{\exp\left(\frac{-p^2}{4}\right)}{p}\right\}\sinh(pu)\right],\ u = \frac{1+CE}{C\eta_g},$$

$$H(C,E,\eta_g) = \left[\frac{\sqrt{\pi}}{C\eta_g}\exp(-u^2)\right]$$

Therefore

$$\begin{aligned}\frac{\hbar^2 k_z^2}{2m_c} = &\left[\frac{ab}{c}\right]\theta_0(E,\eta_g)\left[\frac{1+Erf(E/\eta_g)}{2}\right]^{-1} \\ &+\left(\frac{ac+bc-ab}{c^2}\right)\gamma_0(E,\eta_g)\left[\frac{1+Erf(E/\eta_g)}{2}\right]^{-1} \\ &+\left[\frac{1}{c}\left(1-\frac{a}{c}\right)\left(1-\frac{b}{c}\right)-\left(n+\frac{1}{2}\right)\hbar\omega_0\right] \\ &-g_{\pm}\left[\frac{1+Erf(E/\eta_g)}{2}\right]^{-1}[G(C,E,\eta_g)-iH(C,E,\eta_g)]\end{aligned} \tag{7.10}$$

Therefore the DR is given by

$$\frac{\hbar^2 k_z^2}{2m_c} = U_{3,\pm}(E,n,\eta_g)+iU_{4,\pm}(E,\eta_g) \tag{7.11}$$

where

$$\begin{aligned}U_{3,\pm}(E,n,\eta_g) = &\left[\frac{ab}{c}\theta_0(E,\eta_g)\times\left[\frac{1+Erf(E/\eta_g)}{2}\right]^{-1}\right. \\ &+\left(\frac{ac+bc-ab}{c^2}\right)\gamma_0(E,\eta_g)\times\left[\frac{1+Erf(E/\eta_g)}{2}\right]^{-1} \\ &+\frac{1}{c}\left(1-\frac{a}{c}\right)\left(1-\frac{b}{c}\right)-\left(n+\frac{1}{2}\right)\hbar\omega_0 \\ &\left.+g_{\pm}\left[\frac{1+Erf(E/\eta_g)}{2}\right]^{-1}G(C,E,\eta_g)\right]\end{aligned}$$

and

$$U_{4,\pm}(E,\eta_g) = g_\pm \left[\frac{1+Erf(E/\eta_g)}{2}\right]^{-1} H(C,E,\eta_g)$$

The complex Landau energy E_{nHD1} in this case can be obtained by substituting $k_z = 0$ and $E = E_{nHD}$ in (6.11). The EEM at the Fermi Level can be written from (7. 11) as

$$m^*_\pm(E_{FBHD},n,\eta_g) = m^*_\parallel U'_{3,\pm}(E_{FBHD},n,\eta_g) \tag{7.12}$$

Thus the EEM is a function of Fermi energy, Landau quantum number and scattering potential together with the fact it is double valued due to spin.

The complex density of states function under magnetic quantization is given by

$$\begin{aligned} N_B(E) &= N_{BR1}(E) + iN_{BI1}(E) \\ &= \frac{eB}{2\pi^2\hbar^2}\sqrt{2m_c}\sum_{n=0}^{n_{\max}}\left[\frac{x'_1}{2\sqrt{x_1}} + \frac{iy'_1}{2\sqrt{y_1}}\right] \end{aligned} \tag{7.13}$$

where

$$x = \frac{\sqrt{(U_{3,\pm}(E,n,\eta_g))^2 + (U_{4,\pm}(E,n,\eta_g))^2} + (U_{3,\pm}(E,n,\eta_g))}{2},$$

$$y = \frac{\sqrt{(U_{3,\pm}(E,n,\eta_g))^2 + (U_{4,\pm}(E,n,\eta_g))^2} - (U_{3,\pm}(E,n,\eta_g))}{2}$$

and x'_1 and y'_1 are the differentiations of x and y with respect to energy E.

From (7.13) we can write

$$N_{BR1}(E) = \frac{eB\sqrt{2m_c}}{4\pi^2\hbar^2}\sum_{n=0}^{n_{\max}}\frac{x'_1}{\sqrt{x_1}} \tag{7.14}$$

$$\text{and } N_{BI}(E) = \frac{eB\sqrt{2m_c}}{4\pi^2\hbar^2}\sum_{n=0}^{n_{\max}}\frac{y'_1}{\sqrt{y_1}} \tag{7.15}$$

(b) Two band model of Kane

The magneto-dispersion law in this case is given by

$$\frac{\hbar^2 k^2}{2m_c} = \gamma_2(E, \eta_g) - \left(n + \frac{1}{2}\right)\hbar\omega_0 \mp \frac{1}{2} g^* \mu_0 B \tag{7.16}$$

where g^* is the magnitude of the effective g factor at the edge of the conduction band and μ_0 is the Bohr magnetron.

The EEM at the Fermi Level can be written from (7.16) as

$$m^*(E_{FBHD}, \eta_g) = m_c \gamma_2'(E_{FBHD}, \eta_g) \tag{7.17}$$

Thus EEM is independent of quantum number.

(c) Parabolic Energy Bands

The magneto-dispersion law in this case is given by

$$\frac{\hbar^2 k^2}{2m_c} = \gamma_3(E, \eta_g) - \left(n + \frac{1}{2}\right)\hbar\omega_0 \mp \frac{1}{2} g^* \mu_0 B \tag{7.18}$$

The EEM at the Fermi Level can be written from (7.18) as

$$m^*(E_{FBHD}, \eta_g) = m_c \gamma_3'(E_{FBHD}, \eta_g) \tag{7.19}$$

Thus the EEM in heavily doped parabolic energy bands is a function of Fermi energy and scattering potential whereas in the absence of heavy doping the same mass is a constant quantity invariant of any variables.

(d) The model of Stillman et al.

The (2.79) under the condition of band tailing assumes the form

$$k^2 = \frac{\left[\bar{t}_{11} - \sqrt{(\bar{t}_{11})^2 - 4\bar{t}_{12}\gamma_3(E, \eta_g)}\right]}{2\bar{t}_{12}} \tag{7.20}$$

Therefore the magneto DR is given by

$$k_z^2 = U_7(E, n, \eta_g) \tag{7.21}$$

$$\text{where}\ \ U_7(E,n,\eta_g) = \left[\frac{\left[\bar{t}_{11} - \sqrt{(\bar{t}_{11})^2 - 4\bar{t}_{12}\gamma_3(E,\eta_g)}\right]}{2\bar{t}_{12}} - \frac{2eB}{\hbar}\left(n+\frac{1}{2}\right)\right]$$

The EEM at the Fermi Level can be written from (7.21) as

$$m^*(E_{FBHD},\eta_g) = \frac{\hbar^2}{2}U_7'(E_{FBHD},\eta_g) \tag{7.22}$$

(e) The model of Palik et al.

To the fourth order in effective mass theory and taking into account the interactions of the conduction, light hole, heavy-hole and split-off hole bands, the electron energy spectrum in III–V semiconductors in the presence of a quantizing magnetic field $\vec{B}$ can be written as

$$\begin{aligned} E = {} & J_{31} + \left(n+\frac{1}{2}\right)\hbar\omega_0 + \frac{\hbar^2k_z^2}{2m_c} \pm \frac{1}{4}\left(\frac{m_c}{m_0}\right)\hbar\omega_0 g_0^* \\ & \pm k_{30}\alpha\left(n+\frac{1}{2}\right)(\hbar\omega_0)^2 \pm k_{31}\alpha\hbar\omega_0\left(\frac{\hbar^2k_z^2}{2m_c}\right) \\ & + k_{32}\alpha\left[\hbar\omega_0\left(n+\frac{1}{2}\right) + \left(\frac{\hbar^2k_z^2}{2m_c}\right)\right]^2 \end{aligned} \tag{7.23}$$

where

$$J_{31} = -\frac{1}{2}\alpha\hbar\omega_0\left[(1-y_{11})/(2+x_{11})^2\right]\cdot J_{32},$$

$$J_{32} = \left\{\left[\frac{1}{3}(1-x_{11})^2 - (2+x_{11})^2\right](2+x_{11})\cdot y_{11} + \frac{1}{2}(1-x_{11}^2)(1+x_{11})(1+y_{11})\right\},$$

$$g_0^* = 2\left\{1 - \left[\frac{(1-x_{11})}{(2+x_{11})}\right]\left[\frac{(1-y_{11})}{y_{11}}\right]\right\},$$

$$k_{30} = (1-y_{11})(1-x_{11})\left\{\left[\left(2+\frac{3}{2}x_{11}+x_{11}^2\right)\cdot\frac{(1-y_{11})}{(2+x_{11})^2}\right] - \frac{2}{3}y_{11}\right\},$$

$$k_{31} = (1-y_{11})\left[\frac{(1-x_{11})}{(2+x_{11})}\right]$$
$$\cdot\left\{\left[\left(2+\frac{3}{2}x_{11}+x_{11}^2\right)\cdot\frac{(1-y_{11})}{(2+x_{11})}\right]-\frac{2}{3}(1-x_{11})y_{11}\right\},$$

$k_{32} = -\left[\left(1+\frac{1}{2}x_{11}^2\right)/\left(1+\frac{1}{2}x_{11}\right)\right](1-y_{11})^2,\ x_{11} = \left[1+\left(\frac{\Delta}{E_g}\right)\right]^{-1}$ and $y_{11} = \frac{m_c}{m_0}$

Under the condition of heavy doping, the (7.23) assumes the form

$$J_{34}k_z^4 + J_{35,\pm}(n)k_z^2 + J_{36,\pm}(n) - \gamma_3(E,\eta_g) = 0 \tag{7.24}$$

where

$$J_{34} = \alpha k_{32}\left(\hbar^2/2m_c\right)^2,$$
$$J_{35,\pm}(n) = \left[\frac{\hbar^2}{2m_c} \pm \alpha k_{31}\hbar\omega_0\cdot\frac{\hbar^2}{2m_c} + \alpha k_{32}\hbar\omega_0\cdot\frac{\hbar^2}{2m_c}\left(n+\frac{1}{2}\right)\right],$$

$$J_{36,\pm}(n) = \left[J_{31} \pm \frac{1}{4}\left(\frac{m_c}{m_0}\right)\hbar\omega_0 g_0^* \pm k_{30}\alpha(\hbar\omega_0)^2\left(n+\frac{1}{2}\right) + k_{32}\alpha\left[(\hbar\omega_0)\left(n+\frac{1}{2}\right)\right]^2\right]$$

The (7.24) can be written as

$$k_z^2 = A_{35,\pm}(E,n,\eta_g) \tag{7.25}$$

where $A_{35,\pm}(E,n,\eta_g) = (2J_{34})^{-1}\left[-J_{35,\pm}(n) + \sqrt{(J_{35,\pm}(n))^2 - 4J_{34}\left[J_{36,\pm}(n) - \gamma_3(E,\eta_g)\right]}\right]$

The EEM at the Fermi Level can be written from (7.25) as

$$m_\pm^*(E_{FBHD},n,\eta_g) = \frac{\hbar^2}{2}A'_{35,\pm}(E_{FBHD},n,\eta_g) \tag{7.26}$$

Thus, the EEM is a function of Fermi energy, Landau quantum number and the scattering potential.

7.2.3 The DR in HD II–VI Semiconductors Under Magnetic Quantization

The magneto dispersion relation of the carriers in heavily doped II–VI semiconductors are given by

$$\gamma_3(E,n,\eta_g)=a_0'\frac{2eB}{\hbar}\left(n+\frac{1}{2}\right)+b_0'k_z^2\pm\lambda_0'\left[\frac{2eB}{\hbar}\left(n+\frac{1}{2}\right)\right]^{\frac{1}{2}} \tag{7.27}$$

The (7.27) can be written as

$$k_z^2=U_{8\pm}(E,n,\eta_g) \tag{7.28}$$

where $U_{8\pm}(E,n,\eta_g)=(\mathrm{b}_0')^2\left[\gamma_3(E,\eta_g)-\frac{2eBa_0'}{\hbar}\left(n+\frac{1}{2}\right)\mp\bar{\lambda}_0\left[\frac{2eB}{\hbar}\left(n+\frac{1}{2}\right)\right]^{\frac{1}{2}}\right]$

The EEM at the Fermi Level can be written from (7.28) as

$$m^*(E_{FBHD},\eta_g)=\frac{\hbar^2}{2}U_{8\pm}'(E_{FBHD},n,\eta_g) \tag{7.29}$$

7.2.4 The DR in HD IV–VI Semiconductors Under Magnetic Quantization

The electron energy spectrum in IV–VI semiconductors are defined by the models of Cohen, Lax, Dimmock and Bangert and Kastner respectively. The magneto DR in HD IV–VI semiconductors is discussed in accordance with the said model for the purpose of relative comparison.

(a) Cohen Model

In accordance with the Cohen model, the dispersion law of the carriers in IV–VI semiconductors is given by

$$E(1+\alpha E)=\frac{p_x^2}{2m_1}+\frac{p_z^2}{2m_3}-\frac{\alpha E p_y^2}{2m_2'}+\frac{p_y^2(1+\alpha E)}{2m_2}+\frac{\alpha p_y^4}{4m_2m_2'} \tag{7.30}$$

where, $p_i \equiv \hbar k_i$, $i = x, y, z$, m_1, m_2, and m_3 are the effective carrier masses at the band-edge along x, y and z directions respectively and m_2' is the effective-mass tensor component at the top of the valence band (for electrons) or at the bottom of the conduction band (for holes).

The magneto electron energy spectrum in IV–VI semiconductors in the presence of quantizing magnetic field B along z-direction can be written as

$$E(1+\alpha E) = \left(n+\frac{1}{2}\right)\hbar\omega(E) \pm \frac{1}{2}g_0^*\mu_0 B + \frac{3}{8}\alpha\left(n^2+n+\frac{1}{2}\right)\hbar^2\omega^2(E) + \frac{\hbar^2 k_z^2}{2m_3} \tag{7.31}$$

where,

$$\omega(E) \equiv \frac{|e|B}{\sqrt{m_1 m_2}}\left[1+\alpha E\left(1-\frac{m_2}{m_2'}\right)\right]^{1/2}$$

Therefore the magneto dispersion law in heavily doped IV–VI materials can be expressed as

$$\frac{\hbar^2 k_z^2}{2m_3} = U_{16,\pm}(E, n, \eta_g) \tag{7.32}$$

where

$$\begin{aligned} U_{16,\pm}(E,n,\eta_g) = \Bigg[& \gamma_2(E,\eta_g) - \left(n+\frac{1}{2}\right)\frac{\hbar eB}{\sqrt{m_1 m_2}} \mp \frac{1}{2}g^*\mu_0 B \\ & - \frac{3\alpha}{8}\left(n^2+n+\frac{1}{2}\right)\left(\frac{\hbar eB}{\sqrt{m_1 m_2}}\right)^2 \\ & - \gamma_3(E,\eta_g)\left[\frac{\alpha}{2}\left(n+\frac{1}{2}\right)\frac{\hbar eB}{\sqrt{m_1 m_2}}\left(1-\frac{m_2}{m'_2}\right) \right. \\ & \left. + \frac{3\alpha^2}{8}\left(n^2+n+\frac{1}{2}\right)\left(\frac{\hbar eB}{\sqrt{m_1 m_2}}\right)^2\left(1-\frac{m_2}{m'_2}\right)\right]\Bigg] \end{aligned}$$

The EEM at the Fermi Level can be written from (7.32) as

$$m_\pm^*(E_{FBHD}, n, \eta_g) = m_3 U'_{16,\pm}(E_{FBHD}, n, \eta_g) \tag{7.33}$$

Thus, the EEM is a function of Fermi energy, Landau quantum number and the scattering potential.

(b) Lax Model

In accordance with this model, the magneto dispersion relation assumes the form

$$E(1+\alpha E) = \left(n+\frac{1}{2}\right)\hbar\omega_{03}(E) + \frac{\hbar^2 k_z^2}{2m_3} \pm \frac{1}{2}\mu_0 g^* B \tag{7.34}$$

where,

$$\omega_{03}(E) = \frac{eB}{\sqrt{m_1 m_2}}$$

The magneto dispersion relation in heavily doped IV–VI materials, can be written following (7.34) as

$$\gamma_2(E,\eta_g) = \left(n+\frac{1}{2}\right)\hbar\omega_{03}(E) + \frac{\hbar^2 k_z^2}{2m_3} \pm \frac{1}{2}g^*\mu_0 B \tag{7.35}$$

$$\text{(7.35) can be written as } \frac{\hbar^2 k_z^2}{2m_3} = U_{17,\pm}(E,n,\eta_g) \tag{7.36a}$$

where

$$U_{17,\pm}(E,n,\eta_g) = \gamma_2(E,\eta_g) - \left(n+\frac{1}{2}\right)\hbar\omega_{03} \pm \frac{1}{2}g^*\mu_0 B$$

The EEM at the Fermi Level can be written from (7.36a) as

$$m^*(E_{FBHD},\eta_g) = m_3 U'_{17,\pm}(E_{FBHD},n,\eta_g) \tag{7.36b}$$

(c) Dimmock Model

The dispersion relation under magnetic quantization in HD IV–VI semiconductors can be expressed in accordance with Dimmock model as

$$\begin{aligned}
&\gamma_2(E,\eta_g) + \alpha\gamma_3(E,\eta_g)\frac{2eB}{\hbar}\left(n+\frac{1}{2}\right)\frac{\hbar^2}{2}\left(\frac{1}{m_t^+}-\frac{1}{m_t^-}\right) \\
&+\alpha\gamma_3(E,\eta_g)x\frac{\hbar^2}{2}\left(\frac{1}{m_l^+}-\frac{1}{m_l^-}\right) = \frac{\hbar^2 k_s^2}{2m_t^*} + \frac{\hbar^2 k_z^2}{2m_l^*} \\
&+\frac{\hbar^2 k_s^2}{2m_t^-} + \frac{\hbar^2 k_z^2}{2m_l^-} + \alpha\left[\frac{\hbar^4 k_s^4}{4m_t^- m_t^+} + \frac{\hbar^4 k_s^2 k_z^2}{4m_l^- m_l^+} + \frac{\hbar^4 k_z^2 k_s^2}{4m_t^+ m_l^-} + \frac{\hbar^4 k_z^4}{4m_l^- m_l^+}\right] \\
&= \frac{2eB}{\hbar}\left(n+\frac{1}{2}\right)\frac{\hbar^2}{2}\left(\frac{1}{m_t^*}-\frac{1}{m_t^-}\right) + x\frac{\hbar^2}{2}\left(\frac{1}{m_l^*}-\frac{1}{m_l^-}\right) \\
&+\alpha\left[\frac{\hbar^4}{4m_t^- m_t^+}\left(\frac{2eB}{\hbar}\left(n+\frac{1}{2}\right)\right)^2 + x\left[\frac{\hbar^4 eB}{2m_l^+ m_t^- \hbar} + \frac{\hbar^4 eB}{2m_t^+ m_t^- \hbar}\right]\left(n+\frac{1}{2}\right) + \frac{\hbar^4}{4m_l^- m_t^+}x^2\right]
\end{aligned} \tag{7.37}$$

where

$$x = k_z^2$$

Therefore the magneto dispersion relation in heavily doped IV–VI materials, whose unperturbed carriers obey the Dimmock Model can be expressed as

$$k_z^2 = U_{170}(E, n, \eta_g) \tag{7.38}$$

where $U_{170}(E, n, \eta_g) = [2p_9]^{-1}\left[-q_9(E, n, \eta_g) + \left[q_9^2(E, n, \eta_g) + 4p_9R_9(E, n, \eta_g)\right]^{\frac{1}{2}}\right]$,

$$p_9 = \frac{\alpha\hbar^4}{4m_l^- m_l^+},\quad q_9(E, n, \eta_g) = \left[\frac{\hbar^2}{2}\left(\frac{1}{m_l^*} + \frac{1}{m_l^-}\right) + \frac{\alpha\hbar^3 eB}{2}\left(n + \frac{1}{2}\right)\left(\frac{1}{m_t^- m_l^+} + \frac{1}{m_l^- m_t^+}\right) - \alpha\gamma_3(E, \eta_g)\left(\frac{1}{m_l^+} + \frac{1}{m_l^-}\right)\right]$$

and

$$R_9(E, n, \eta_g) = \left[\gamma_2(E, \eta_g) + \alpha eB\gamma_3(E, \eta_g)\left(n + \frac{1}{2}\right)\hbar\left(\frac{1}{m_t^+} - \frac{1}{m_t^-}\right) - \hbar eB\left(n + \frac{1}{2}\right)\left(\frac{1}{m_t^*} + \frac{1}{m_t^-}\right) - \frac{\alpha\hbar^2}{m_t^- m_t^+}\left[eB\left(n + \frac{1}{2}\right)\right]^2\right]$$

The EEM at the Fermi Level can be written from (7.38) as

$$m^*(E_{FBHD}, n, \eta_g) = \frac{\hbar^2}{2} U'_{170}(E_{FBHD}, n, \eta_g) \tag{7.39}$$

Thus, the EEM is a function of Fermi energy, Landau quantum number and the scattering potential.

(d) Model of Bangert and Kastner

In accordance with this model [8], the carrier energy spectrum in HD IV–VI semiconductors can be written following (2.143) as

$$\frac{k_s^2}{\rho_{11}^2(E, \eta_g)} + \frac{k_y^2}{\rho_{12}^2(E, \eta_g)} = 1 \tag{7.40}$$

where

$$\rho_{11}(E,\eta_g) = \frac{1}{\sqrt{S_1(E,\eta_g)}}, \quad \rho_{12}(E,\eta_g) = \frac{1}{\sqrt{S_2(E,\eta_g)}},$$

$$\begin{aligned} S_1(E,\eta_g) = & [2\gamma_0(E,\eta_g)]^{-1}\left[\frac{(\bar{R})^2}{E_g}\{c_1(\alpha_1,E,E_g) - iD_1(\alpha_1,E,E_g)\}\right. \\ & + \frac{(\bar{S})^2}{\Delta'_c}\{c_2(\alpha_2,E,E_g) - iD_2(\alpha_2,E,E_g)\} \\ & \left. + \frac{(\bar{Q})^2}{\Delta''_c}\{c_3(\alpha_3,E,E_g) - iD_3(\alpha_3,E,E_g)\}\right] \end{aligned}$$

and

$$\begin{aligned} S_2(E,\eta_g) = & [2\gamma_0(E,\eta_g)]^{-1}\left[\frac{2(\bar{A})^2}{E_g}\{c_1(\alpha_1,E,\eta_g) - iD_1(\alpha_1,E,\eta_g)\}\right. \\ & \left. + \frac{(\bar{S}+\bar{Q})^2}{\Delta''_c}\{c_3(\alpha_3,E,\eta_g) - iD_3(\alpha_3,E,\eta_g)\}\right] \end{aligned}$$

Since $S_1(E,\eta_g)$ and $S_2(E,\eta_g)$ are complex, the energy spectrum is also complex in the presence of Gaussian band tails.

Therefore the magneto dispersion law in the presence of a quantizing magnetic field B which makes an angle θ with k_z axis can be written as

$$k_z^2 = U_{18}(E,n,\eta_g) \tag{7.41}$$

where

$$\begin{aligned} U_{18}(E,n,\eta_g) = & [\rho_{11}{}^2(E,\eta_g)\sin^2\theta + \rho_{12}{}^2(E,\eta_g)\cos^2\theta] \\ & \left[\frac{2eB}{\hbar}\left(n+\frac{1}{2}\right)\left[(\rho_{11}{}^2(E,\eta_g)\rho_{12}(E,\eta_g))^{-1}\right.\right. \\ & \left.\left.\{\rho_{11}{}^2(E,\eta_g)\sin^2\theta + \rho_{12}{}^2(E,\eta_g)\cos^2\theta\}^{\frac{3}{2}}\right]\right] \end{aligned}$$

The EEM at the Fermi Level can be written from (7.41) as

$$m^*(E_{FBHD}, n, \eta_g) = \frac{\hbar^2}{2}\text{Real part of } \left[U_{18}(E_{FBHD}, n, \eta_g)\right]' \tag{7.42}$$

Thus,the EEM is a function of Fermi energy, Landau quantum number and the scattering potential and the orientation of the applied quantizing magnetic field.

(e) Model of Foley and Langenberg

The dispersion relation of the conduction electrons of IV–VI semiconductors in accordance with Foley et al. can be written as [9]

$$E + \frac{Eg}{2} = E_-(k) + \left[\left[E_+(k) + \frac{E_g}{2}\right]^2 + P_\perp^2 k_s^2 + P_\parallel^2 k_z^2\right]^{\frac{1}{2}} \tag{7.43}$$

where $E_+(k) = \frac{\hbar^2 k_s^2}{2m_\perp^+} + \frac{\hbar^2 k_z^2}{2m_\parallel^+}$, $E_-(k) = \frac{\hbar^2 k_s^2}{2m_\perp^-} + \frac{\hbar^2 k_z^2}{2m_\parallel^-}$ represents the contribution from the interaction of the conduction and the valence band edge states with the more distant bands and the free electrons term, $\frac{1}{m_\perp^\pm} = \frac{1}{2}\left[\frac{1}{m_{tc}} \pm \frac{1}{m_{tv}}\right]$, $\frac{1}{m_\parallel^\pm} = \frac{1}{2}\left[\frac{1}{m_{lc}} \pm \frac{1}{m_{lv}}\right]$

Following the methods as given in Chap. 1, the dispersion relation in heavily doped IV–VI materials in the present case is given by

$$\begin{aligned}&\left[\left[\gamma_3(E,\eta_g) + \frac{E_{g0}}{2}\right] - \left[\frac{\hbar^2 k_s^2}{2m_\perp^-} + \frac{\hbar^2 k_z^2}{2m_\parallel^-}\right]\right]^2 = \left[\frac{\hbar^2 k_s^2}{2m_\perp^+} + \frac{\hbar^2 k_z^2}{2m_\parallel^+}\right]^2 + \frac{E_{g0}^2}{4}\\&+E_{g0}\left[\frac{\hbar^2 k_s^2}{2m_\perp^+} + \frac{\hbar^2 k_z^2}{2m_\parallel^+}\right] + P_\parallel^2 k_z^2 + P_\perp^2 k_s^2\end{aligned} \tag{7.44}$$

Therefore the magneto-dispersion relation in heavily doped IV–VI materials can be written as

$$\begin{aligned}&\gamma_3^2(E,\eta_g) + \frac{E_g^2}{4} + E_g\gamma_3(E,\eta_g) + \left[\frac{\hbar eB}{m_\perp^-}\left(n+\frac{1}{2}\right) + \frac{\hbar^2 x}{2m_\parallel^-}\right]^2\\&-2\left[\gamma_3(E,\eta_g) + \frac{E_g}{2}\right]\left[\frac{\hbar eB\left(n+\frac{1}{2}\right)}{m_\perp^-} + \frac{\hbar^2 x}{2m_\parallel^-}\right]\\&=\left[\frac{\hbar eB\left(n+\frac{1}{2}\right)}{m_\perp^+} + \frac{\hbar^2 x}{2m_\parallel^+}\right]^2 + E_g\left[\frac{\hbar eB}{m_\perp^+}\left(n+\frac{1}{2}\right) + \frac{\hbar^2 x}{2m_\parallel^+}\right]\\&+P_\parallel^2 x + P_\perp^2\frac{2eB}{\hbar}\left(n+\frac{1}{2}\right)\end{aligned} \tag{7.45}$$

where

$$k_z^2 = x$$

Therefore the magneto dispersion relation in IV–VI heavily doped materials, where unperturbed carriers follow the model of Foley et al. can be expressed as

$$k_z^2 = U_{19}(E, n, \eta_g) \tag{7.46}$$

where $U_{19}(E, n, \eta_g) = [2p_{91}]^{-1}\left[-q_{91}(E, n, \eta_g) + \{q_{91}^2(E, n, \eta_g) + 4p_{91}R_{91}(E, n, \eta_g)\}^{\frac{1}{2}}\right]$

$$p_{91} = \frac{\hbar^4}{4}\left[\frac{1}{(m_\parallel^+)^2} - \frac{1}{(m_\parallel^-)^2}\right],$$

$$q_{91}(E, n, \eta_g) = \left[\frac{\hbar^3 eB}{m_\perp^- m_\parallel^+}\left(n + \frac{1}{2}\right) + P_\parallel^2 + \frac{\hbar^3 E_g}{2m_\parallel^+} - \frac{\hbar^3 eB\left(n + \frac{1}{2}\right)}{m_\perp^- m_\parallel^-} + \frac{\hbar^2}{m_\parallel^-}\left(\gamma_3(E, \eta_g) + \frac{E_g}{2}\right)\right]$$

$$R_{91}(E, \eta_g, n) = \left[\gamma_3^2(E, \eta_g) + E_g\gamma_3(E, \eta_g) - \frac{2\hbar eB}{m_\perp^-}\left(\gamma_3(E, \eta_g) + \frac{E_g}{2}\right)\left(n + \frac{1}{2}\right)\right.$$
$$\left. - E_g\frac{\hbar eB}{m_\perp^+}\left(n + \frac{1}{2}\right) - P_\perp^2 \cdot \frac{2eB}{\hbar}\left(n + \frac{1}{2}\right)\right]$$

The EEM at the Fermi Level can be written from (7.46) as

$$m^*(E_{FBHD}, n, \eta_g) = \frac{\hbar^2}{2}U_{19}'(E_{FBHD}, n, \eta_g) \tag{7.47}$$

Thus, as noted already in this case also the EEM is a function of Fermi energy, Landau quantum number and the scattering potential.

7.2.5 The DR in HD Stressed Kane Type Semiconductors Under Magnetic Quantization

The DR of the conduction electrons in heavily doped Kane type semiconductors can be written following (2.175) of Chap. 2 as

$$\frac{k_x^2}{a_{\|}^2(E,\eta_g)}+\frac{k_y^2}{b_{\|}^2(E,\eta_g)}+\frac{k_z^2}{c_{\|}^2(E,\eta_g)}=1$$

where $a_{\|}(E,\eta_g)=\dfrac{1}{\sqrt{P_{\|}(E,\eta_g)}}$, $\mathrm{b}_{\|}(E,\eta_g)=\dfrac{1}{\sqrt{Q_{\|}(E,\eta_g)}}$ and $c_{\|}(E,\eta_g)$

$$=\frac{1}{\sqrt{S_{\|}(E,\eta_g)}} \tag{7.48a}$$

The electron energy spectrum in heavily doped Kane type semiconductors in the presence of an arbitrarily oriented quantizing magnetic field B which makes an angle $\bar{\alpha}_1$, $\bar{\beta}_1$ and $\bar{\gamma}_1$ with k_x, k_y and k_z axes respectively, can be written as

$$(k_z')^2 = U_{41}(\mathrm{E},n,\eta_g) \tag{7.48b}$$

$$\text{where}\quad U_{41}(\mathrm{E},n,\eta_g)=I_2(\mathrm{E},\eta_g)[1-I_3(\mathrm{E},n,\eta_g)]$$

$$I_2(\mathrm{E},\eta_g)=\left[[a_{11}(\mathrm{E},\eta_g)]^2\cos^2\bar{\alpha}_1+[b_{11}(\mathrm{E},\eta_g)]^2\cos^2\bar{\beta}_1+[c_{11}(\mathrm{E},\eta_g)]^2\cos^2\bar{\gamma}_1\right]$$

and

$$I_3(\mathrm{E},n,\eta_g)=\frac{2eB}{\hbar}\left(n+\frac{1}{2}\right)[a_{11}(\mathrm{E},\eta_g)b_{11}(\mathrm{E},\eta_g)c_{11}(\mathrm{E},\eta_g)]^{-1}[I_2(\mathrm{E},\eta_g)]^{1/2}$$

The EEM at the Fermi Level can be written from (7.48b) as

$$m^*(E_{FBHD},n,\eta_g)=\frac{\hbar^2}{2}U_{41}'(E_{FBHD},n,\eta_g) \tag{7.48c}$$

From (7.48c) we observe that the EEM is a function of Fermi energy, Landau quantum number, the scattering potential and the orientation of the applied quantizing magnetic field.

7.2.6 The DR in HD Te Under Magnetic Quantization

The magneto dispersion relation of the conduction electrons in HD Te can be expressed following (2.190) of Chap. 2 as

$$k_z^2 = U_{42\pm}(E, n, \eta_g) \tag{7.49a}$$

where

$$U_{42,\pm}(E, n, \eta_g) = (2\psi_1^2)^{-1}\left[\left\{2\gamma_3(E, \eta_g)\psi_1 + \psi_3^2 - 4\psi_1\psi_2\frac{eB}{\hbar}\left(n + \frac{1}{2}\right)\right\}\right.$$
$$\left. - \left\{\psi_3^4 + 4\psi_1\psi_3^2\gamma_3(E, \eta_g) + \frac{8eB}{\hbar}\left(n + \frac{1}{2}\right)(\psi_1^2\psi_4^2 - \psi_1\psi_2\psi_3^2)\right\}^{-1/2}\right]$$

The EEM at the Fermi Level can be written from (7.49a) as

$$m_\pm^*(E_{FBHD}, n, \eta_g) = \frac{\hbar^2}{2}U'_{42\pm}(E_{FBHD}, n, \eta_g) \tag{7.49b}$$

Thus from (7.49b) we note that the EEM is a function of three variables namely Fermi energy, Landau quantum number and the scattering potential.

7.2.7 The DR in HD Gallium Phosphide Under Magnetic Quantization

The magneto dispersion relation in HD GaP can be written following (2.202) of Chap. 2 as

$$k_z^2 = U_{43}(E, n, \eta_g) \tag{7.50a}$$

where

$$U_{43}(\mathrm{E},n,\eta_g) = (2b^2)^{-1}\left[\left\{2\gamma_3(\mathrm{E},\eta_g)b + c - 2Db - 4ab\frac{eB}{\hbar}\left(n+\frac{1}{2}\right)\right\}\right.$$
$$+\left\{\left[c^2 + 4bc\gamma_3(\mathrm{E},\eta_g) + 4D^2b^2 - 4cDb\right] - \frac{8eB}{\hbar}\left(n+\frac{1}{2}\right)(2ab^2D + 4\gamma_3(\mathrm{E},\eta_g)b^2a)\right.$$
$$\left.\left. + abc - 2b^2a\gamma_3(\mathrm{E},\eta_g) - b^2c\right\}^{-1/2}\right]$$

The EEM at the Fermi Level can be expressed from (7.50a) as

$$m^*(E_{FBHD},n,\eta_g) = \frac{\hbar^2}{2}U'_{43}(E_{FBHD},n,\eta_g) \tag{7.50b}$$

Thus, from (7.50b) it appears that the EEM is the function of Fermi energy, Landau quantum number and the scattering potential.

7.2.8 The DR in HD Platinum Antimonide Under Magnetic Quantization

The magneto dispersion relation in HD $PtSb_2$ can be written following (2.218) as

$$k_z^2 = U_{44}(E,n,\eta_g) \tag{7.51a}$$

where $U_{44}(\mathrm{E},n,\eta_g) = \frac{1}{2T_{41}}\left[T_{71}(\mathrm{E},n,\eta_g) + \sqrt{T_{71}^2(\mathrm{E},n,\eta_g) + 4T_{41}T_{71}(\mathrm{E},n,\eta_g)}\right]$,

$$T_{71}(\mathrm{E},n,\eta_g) = \left[T_{51}(\mathrm{E},\eta_g) - T_{31}\frac{2eB}{\hbar}\left(n+\frac{1}{2}\right)\right] \text{ and}$$
$$T_{72}(\mathrm{E},n,\eta_g) = \left[T_{61}(\mathrm{E},\eta_g) + \frac{2eB}{\hbar}\left(n+\frac{1}{2}\right)T_{21}(\mathrm{E},\eta_g)\right]$$

The EEM at the Fermi Level can be written from (7.51a) as

$$m^*(E_{FBHD},n,\eta_g) = \frac{\hbar^2}{2}U'_{44}(E_{FBHD},n,\eta_g) \tag{7.51b}$$

Thus, from the above equation we infer that the EEM is a function of Landau quantum number, the Fermi energy and the scattering potential.

7.2.9 The DR in HD Bismuth Telluride Under Magnetic Quantization

The magneto dispersion relation in HD Bi_2Te_3 can be written following (2.231) as

$$k_x^2 = U_{45}(E, \eta_g, n) \tag{7.52a}$$

where

$$U_{45}(E, \eta_g, n) = \frac{\gamma_2(E, \eta_g) - \left(n + \frac{1}{2}\right)\frac{e\hbar B}{M_{31}}}{\bar{\omega}_1} \text{ and } M_{31} = \frac{m_0}{\left(\bar{\alpha}_{22}\bar{\alpha}_{33} - \frac{(\bar{\alpha}_{23})^2}{4}\right)^{\frac{1}{2}}}$$

The EEM at the Fermi Level can be written from (7.52a) as

$$m^*(E_{FBHD}, \eta_g) = \frac{\hbar^2}{2} U'_{45}(E_{FBHD}, n, \eta_g) \tag{7.52b}$$

7.2.10 The DR in HD Germanium Under Magnetic Quantization

a. Model of Cardona et al.

The magneto dispersion relation in HD Ge can be written following (2.243) as

$$k_x^2 = U_{46}(E, \eta_g, n) \tag{7.53a}$$

where

$$U_{46}(\mathrm{E}, n, \eta_g) = \frac{2m_\parallel^*}{\hbar^2}\left[\gamma_3(\mathrm{E}, \eta_g) + \frac{\mathrm{E}_{g0}}{2} - \left[\frac{\mathrm{E}_{g0}^2}{4} + \frac{\mathrm{E}_{g0}\hbar^2}{m_\perp^*}\frac{2eB}{\hbar}\left(n + \frac{1}{2}\right)\right]\right]^{1/2}$$

The EEM at the Fermi Level can be written from (7.53a) as

$$m^*(E_{FBHD}, n, \eta_g) = \frac{\hbar^2}{2} U'_{46}(E_{FBHD}, n, \eta_g) \tag{7.53b}$$

From (7.53b) it appears that the EEM is a function of Fermi energy and Landau quantum number due to band non-parabolicity.

b. Model of Wang and Ressler

The magneto dispersion relation in HD Ge can be written following (2.258) as

$$k_z^2 = U_{47}(\mathrm{E}, n, \eta_g) \tag{7.54a}$$

where

$$U_{47}(\mathrm{E}, n, \eta_g) = \left(\frac{m_{\|}^*}{\hbar^2 \bar{\alpha}_6}\right)\left[1 - \bar{\alpha}_5\left(n + \frac{1}{2}\right)\hbar\omega_{\perp} - \left\{\theta_7(n) - 4\bar{\alpha}_6\gamma_3(\mathrm{E}, \eta_g)\right\}^{1/2}\right],$$

$$\omega_{\perp} = \frac{eB}{m_{\perp}^*} \text{ and}$$

$$\theta_7(n) = \left[1 + (\bar{\alpha}_5)^2\left\{\left(n + \frac{1}{2}\right)\hbar\omega_{\perp}\right\}^2 - 2\bar{\alpha}_5\left(n + \frac{1}{2}\right)\hbar\omega_{\perp} + 4\bar{\alpha}_6\left(n + \frac{1}{2}\right)\hbar\omega_{\perp} - 4\bar{\alpha}_6\bar{\alpha}_4\left\{\left(n + \frac{1}{2}\right)\hbar\omega_{\perp}\right\}^2\right]$$

The EEM at the Fermi Level can be written from (7.54a) as

$$m^*(E_{FBHD}, n, \eta_g) = \frac{\hbar^2}{2} U_{47}'(E_{FBHD}, n, \eta_g) \tag{7.54b}$$

From (7.54b) we note that the mass is a function of Fermi energy and quantum number due to band non-parabolicity.

7.2.11 *The DR in HD Gallium Antimonide Under Magnetic Quantization*

The magneto dispersion relation in HD GaSb can be written following (2.274) as

$$k_z^2 = U_{48}(\mathrm{E}, n, \eta_g) \tag{7.55a}$$

where

$$U_{48}(E,n,\eta_g) = \left[\frac{-2eB}{\hbar}\left(n+\frac{1}{2}\right) + (2\alpha_9^2)^{-1}\left[\left\{2\alpha_9\gamma_3^2(E,\eta_g) + \alpha_9\bar{\mathrm{E}}'_{g0} + \frac{\alpha_{10}(\bar{\mathrm{E}}'_{g0})^2}{4}\right\}\right.\right.$$
$$\left.\left. - \left\{\alpha_9^2(\bar{\mathrm{E}}'_{g0})^2 + \frac{\alpha_{10}^2(\bar{\mathrm{E}}'_{g0})^4}{16} + \alpha_9\alpha_{10}\gamma_3(E,\eta_g)(\bar{\mathrm{E}}'_{g0})^2 + \frac{\alpha_9\alpha_{10}(\bar{\mathrm{E}}'_{g0})^3}{2}\right\}^{1/2}\right]\right],$$

$$\alpha_9 = \frac{\hbar^2}{2m_0} \text{ and } \alpha_{10} = \frac{2\hbar^2}{\bar{\mathrm{E}}'_{g0}}\left(\frac{1}{m_c} - \frac{1}{m_0}\right)$$

The EEM at the Fermi Level can be written from (7.55a) as

$$m^*(E_{FBHD}, \eta_g) = \frac{\hbar^2}{2} U'_{48}(E_{FBHD}, n, \eta_g) \tag{7.55b}$$

7.2.12 The DR in HD II–V Materials Under Magnetic Quantization

The dispersion relation of the holes are given by [10]

$$E = \theta_1 k_x^2 + \theta_2 k_y^2 + \theta_3 k_z^2 + \delta_4 k_x$$
$$\mp \left[\left\{\theta_5 k_x^2 + \theta_6 k_y^2 + \theta_7 k_z^2 + \delta_5 k_x\right\}^2 + G_3^2 k_y^2 + \Delta_3^2\right]^{\frac{1}{2}} \pm \Delta_3 \tag{7.56}$$

where, k_x, k_y and k_z are expressed in the units of $10^{10}\ \mathrm{m}^{-1}$,

$$\theta_1 = \frac{1}{2}(a_1 + b_1), \theta_2 = \frac{1}{2}(a_2 + b_2), \theta_3 = \frac{1}{2}(a_3 + b_3), \delta_4 = \frac{1}{2}(A + B),$$
$$\theta_5 = \frac{1}{2}(a_1 - b_1), \theta_6 = \frac{1}{2}(a_2 - b_2), \theta_7 = \frac{1}{2}(a_3 - b_3), \delta_5 = \frac{1}{2}(A + B),$$

$a_i (i = 1, 2, 3, 4), b_i, A, B, G_3$ and Δ_3 are systems constants

The hole energy spectrum in HD II–V semiconductors can be expressed following the method of Chap. 1 as

$$\gamma_3(E,\eta_g) = \theta_1 k_x^2 + \theta_2 k_y^2 + \theta_3 k_z^2 + \delta_4 k_x \mp \left[\left\{\theta_5 k_x^2 + \theta_6 k_y^2 + \theta_7 k_z^2 + \delta_5 k_x\right\}^2 + G_3^2 k_y^2 + \Delta_3^2\right]^{\frac{1}{2}} \pm \Delta_3 \quad (7.57)$$

the magneto dispersion law in HD II–V semiconductors assumes the form

$$k_y^2 = U_{49,\pm}(\mathrm{E}, n, \eta_g) \quad (7.58)$$

where, $U_{49,\pm}(\mathrm{E},n,\eta_g) = \left[I_{35}\gamma_3(\mathrm{E},\eta_g) + I_{36,\pm}(n) \pm \left[\gamma_3^2(\mathrm{E},\eta_g) + \gamma_3(\mathrm{E},\eta_g) I_{38,\pm}(n) + I_{39,\pm}(n)\right]^{\frac{1}{2}}\right.$,

$$I_{35} = \frac{\theta_2}{(\theta_2^2 - \theta_5^2)}, I_{36,\pm}(n) = \frac{I_{33,\pm}(n)}{2(\theta_2^2 - \theta_5^2)}, I_{38,\pm}(n) = (4\theta_5^2)^{-1}[4\theta_2 I_{33,\pm}(n) + 8\theta_2^2 I_{31,\pm}(n) - \theta_5^2 I_{31,\pm}(n)],$$

$$I_{39,\pm}(n) = (4\theta_5^2)^{-1}[I_{33,\pm}^2(n) + 4\theta_2^3 I_{34,\pm}(n) - 4\theta_5^3 I_{34,\pm}(n)],$$
$$I_{33,\pm}(n) = [G_3^2 + 2\theta_5 I_{32}(n) - 2\theta_2 I_{31,\pm}(n)],$$

$$I_{34,\pm}(n) = [I_{32,\pm}^2(n) + \Delta_3^2 - I_{31,\pm}(n)],$$
$$I_{31,\pm}(n) = \left[\left(n + \frac{1}{2}\right)\hbar\omega_{31} - \frac{\delta_4^2}{4\theta_1} \pm \Delta_3\right],$$
$$I_{32}(n) = \left[\left(n + \frac{1}{2}\right)\hbar\omega_{32} - \frac{\delta_5^2}{4\theta_5}\right],$$

$$\omega_{31} = \frac{eB}{\sqrt{M_{31}M_{32}}}, \ \omega_{32} = \frac{eB}{\sqrt{M_{33}M_{34}}}, \ M_{31} = \frac{\hbar^2}{2\theta_1},$$
$$M_{32} = \frac{\hbar^2}{2\theta_3}, \ M_{33} = \frac{\hbar^2}{2\theta_5} \text{ and } M_{34} = \frac{\hbar^2}{2\theta_7}.$$

The EEM at the Fermi Level can be written from (7.58) as

$$m_\pm^*(E_{FBHD}, n, \eta_g) = \frac{\hbar^2}{2} U'_{49,\pm}(E_{FBHD}, n, \eta_g) \quad (7.59)$$

From (7.59) we note that the EEM is a function of Fermi energy, Landau quantum number and the scattering potential.

7.2.13 The DR in HD Lead Germanium Telluride Under Magnetic Quantization

The DR of the carriers in n-type $Pb_{1-x}Ga_xTee$ with x = 0.01 can be written following Vassilev [11] as

$$\begin{aligned}&\left[E - 0.606k_s^2 - 0.0722k_z^2\right]\left[E + \bar{E}_g + 0.411k_s^2 + 0.0377k_z^2\right]\\&= 0.23k_s^2 + 0.02k_z^2 \pm \left[0.06\bar{E}_g + 0.061k_s^2 + 0.0066k_z^2\right]k_s\end{aligned} \quad (7.60)$$

where, $\bar{E}_g(= 0.21\ \text{eV})$ is the energy gap for the transition point, the zero of the energy E is at the edge of the conduction band of the Γ point of the Brillouin zone and is measured positively upwards, k_x, k_y and k_z are in the units of $10^9\ \text{m}^{-1}$.

The magneto dispersion law in HD $Pb_{1-x}Ge_xTe$ can be expressed following the methods as given in Chap. 1 as

$$\begin{aligned}&\left[\frac{2\theta_0(E,\eta_g)}{1+Erf(E/\eta_g)}\right] + \gamma_3(E,\eta_g)\\&\left[\overline{E_{g0}} - 0.345x - 0.390\frac{eB}{\hbar}\left(n+\frac{1}{2}\right)\right]\\&= \frac{0.46eB}{\hbar}\left(n+\frac{1}{2}\right) + 0.02x \pm\\&\left[0.06\overline{E_{g0}} + 0.122\frac{eB}{\hbar}\left(n+\frac{1}{2}\right) + 0.0066x\right]\left(\frac{2eB}{\hbar}\left(n+\frac{1}{2}\right)\right)^{\frac{1}{2}}\\&+\left[\overline{E_{g0}} + \frac{0.822eB}{\hbar}\left(n+\frac{1}{2}\right) + 0.377x\right]\\&\left[\frac{1.212eB}{\hbar}\left(n+\frac{1}{2}\right) + 0.722x\right]\end{aligned} \quad (7.61)$$

The (7.61) assumes the form

$$k_z^2 = U_{50,\mp}(E,n,\eta_g) \quad (7.62)$$

where $U_{50,\mp}(E,n,\eta_g) = (2p_{10})^{-1}\left[q_{10}(E,n,\eta_g) - \left[q_{10}^2(E,n,\eta_g) + 4p_{10}R_{10,\mp}(E,n,\eta_g)\right]^{\frac{1}{2}}\right]$

$$p_{10} = (0.377 \times 0.722), q_{10}(E, n, \eta_g) = \left[0.02 + 0.345\gamma_3(E, \eta_g) \pm 0.0066\left(\frac{2eB}{\hbar}\left(n+\frac{1}{2}\right)\right)^{\frac{1}{2}} + 0.377 \times \frac{1.212eB}{\hbar}\left(n+\frac{1}{2}\right) + 0.722\left[\bar{E}_{g_0} + 0.822\frac{eB}{\hbar}\left(n+\frac{1}{2}\right)\right]\right]$$

and

$$R_{10,\mp}(E, n, \eta_g) = \left[\frac{2\theta_0(E, \eta_g)}{1 + Eof(E/\eta_g)} + \gamma_3(E, \eta_g) \left[\bar{E}_{g_0} - 0.390\frac{eB}{\hbar}\left(n+\frac{1}{2}\right)\right] \mp \left(0.06\bar{E}_{g_0} + 0.122\frac{eB}{\hbar}\left(n+\frac{1}{2}\right)\right) \left(\frac{2eB}{\hbar}\left(n+\frac{1}{2}\right)\right)^{\frac{1}{2}} - \left(\bar{E}_{g_0} + 0.822\frac{eB}{\hbar}\left(n+\frac{1}{2}\right)\right)\frac{1.212eB}{\hbar}\left(n+\frac{1}{2}\right) - \frac{0.46eB}{\hbar}\left(n+\frac{1}{2}\right)\right].$$

The EEM at the Fermi Level can be written from (7.62) as

$$m^*_{\mp}(E_{FBHD}, \eta_g) = \frac{\hbar^2}{2} U'_{50,\mp}(E_{FBHD}, n, \eta_g) \tag{7.63}$$

Thus from (7.63) we note that the EEM is a function of the Fermi energy, Landau quantum number and the scattering potential.

7.3 Discussion

- It appears that the magneto DRs are in general quantized non-parabolas and the consecutive different between two curves are not constant but varies with other physical constants together with scattering potentials.
- The magneto DR is concentration dependent, a fact only possible for HD materials forming band-tails
- The Landau sub-bands are concentration dependent

It may be noted that initial works based on Kane's density-of-states technique [12] in deriving the DOS have been made by Tsitsishvili [13] and Dyakonov et al. [14]. Also, Dyakonov et al. found the distribution of impurity potential as the Gaussian distribution [14]. The basic limitations of Kane's model of density-of-states technique, as mentioned above, have also strongly influenced the models proposed by Tsitsishivili [13] and Dyakonov et al. [14].

Dyakonov et al. [14] considered the basic equation for the present case as

$$E = V(\bar{r}) + \frac{\hbar^2 k_z^2}{2m_c} + \left(n + \frac{1}{2}\right)\hbar\omega_0 B \pm \frac{1}{2} g\mu_B B \quad (7.64)$$

where $V(\bar{r})$ is a slowly varying impurity potential in the crystal.

From (7.64), we find the density-of-states, $\rho(E_{Is}; \bar{r})$, at the local point $(\bar{r})$ in space as

$$\rho(E_{Is}, \bar{r}) = \frac{1}{4\pi^2}\left(\frac{2m_c}{\hbar^2}\right)^{3/2} \hbar\omega_0 \sum_{Is} [E_{Is} - V(\bar{r})]^{-1/2} \quad (7.65)$$

where $E_{Is} = E - \left(n + \frac{1}{2}\right)\hbar\omega_0 \mp \frac{1}{2} g\mu_B B$ and s represents the spin + or −.

In (7.65), each elemental volume is considered independently so that the density-of-states, $\rho_D(E_{I,s}, \eta_g)$, for the whole system is obtained by averaging $\rho(E_{I,s}, \bar{r})$ over the Gaussian potential distribution of the local potential fluctuations, as did Dyakonov et al. [14], using the relation:

$$\rho_D(E_{I,s}, \eta_g) = \int_{-\infty}^{E_{Is}} \rho_D(E_{Is}, \bar{r}) F(V) dV \quad (7.66)$$

We name this method as density-of-states technique of calculating the DOS. Thus we get

$$\rho_D(E_{I,s}, \eta_g) = \frac{1}{4\pi^2}\left(\frac{2m_c}{\hbar^2}\right)^{3/2} \frac{\hbar\omega_0}{(\pi\eta_g^2)^{1/2}} \sum_{I,s} I(I, s) \quad (7.67)$$

where

$$I(I, s) = \pi^{1/2} 2^{-1/4} \eta_g^{1/2} \exp\left(-\frac{E_{I,s}^2}{2\eta_g^2}\right) D_{-1/2}\left(\frac{\sqrt{2}E_{I,s}}{\eta_g}\right) \quad (7.68)$$

It may be noted from (7.68) that the parabolic cylindrical function, $D_{-1/2}(-\sqrt{2}E_{I,s}/\eta_g)$, exists for $(E_{I,s}/\eta_g) > 0$ (representing the quantized Landau sub-bands) and for $(E_{I,s}/\eta_g) < 0$ (representing the forbidden band). This indicates that the DOS function has been tailed into the forbidden region. We now write the asymptotic expansion of $D_{-1/2}(-\sqrt{2}E_{I,s}/\eta_g)$ under the following cases:

Case A: when $(E_{I,s}/\eta_g) < 0$ and $|E_{I,s}/\eta_g| \to \infty$

These conditions indicate the region deep into the forbidden band. Under these limiting condition, we find [15]

$$D_{-1/2}\left(-\sqrt{2}E_{I,s}\big/\eta_g\right) = \pi^{-1/2}\left(\frac{E_{I,s}}{\sqrt{2}\eta_g}\right)^{1/2} K_{1/4}\left(\frac{1}{2}\frac{E_{I,s}^2}{\eta_g^2}\right) \tag{7.69}$$

where $K_{1/4}(x)$ is the real valued MacDonald function.

Therefore, combining (7.67)–(7.69), the expression for the DOS function can be written as

$$\rho_D(E_{I,s},\eta_g) = \frac{1}{4\pi^2}\left(\frac{2m_c}{\hbar^2}\right)^{3/2}\frac{\hbar\omega_0}{\eta_g}\sum_{Is}\left(\frac{E_{I,s}}{2\pi}\right)^{1/2}\cdot\exp\left(-\frac{E_{I,s}^2}{2\eta_g^2}\right)K_{1/4}\left(\frac{E_{I,s}^2}{2\eta_g^2}\right) \tag{7.70}$$

The (7.70) is exactly the same as that of Dyakonov et al. [14].

The asymptotic expansion [15] of $D_{1/2}\left(-\sqrt{2}E_{I,s}/\eta_g\right)$ provides us

$$\rho_D(E_{I,s},\eta_g) = \frac{1}{4\pi^2}\left(\frac{2m_c}{\hbar^2}\right)^{3/2}\hbar\omega_0\sum_{I,s}\frac{1}{\sqrt{2E_{I,s}}}\exp\left(-\frac{E_{I,s}^2}{\eta_g^2}\right)\cdot\left\{1-\frac{3}{16}\frac{\eta_g^2}{E_{I,s}^2}+\frac{105}{512}\frac{\eta_g^4}{E_{I,s}^4}-\cdots\infty\right\} \tag{7.71}$$

Equations (7.70) and (7.71) represent the DOS function deep into the forbidden band. Also, this shows that the DOS function has been tailed into the forbidden region. In the limiting case. $\eta_g \to 0$, the density-of-states, $\rho_D(E_{I,s},\eta_g) \to 0$. This implies that at a very low doping condition, the tail of the density-of-states disappears.

Case B: when $(E_{I,s}/\eta_g) > 0$ and $|E_{I,s}/\eta_g| \to \infty$

This region lies far above the CB edge. In this limiting case, we find [15]

$$D_{-1/2}\left(-\sqrt{2}E_{I,s}\big/\eta_g\right) = \frac{\pi^{1/2}}{2}\left(\frac{E_{I,s}}{2\eta_g}\right)^{1/2} I_{1/4}\left(\frac{1}{2}\frac{E_{I,s}^2}{\eta_g^2}\right) \tag{7.72}$$

where $I_{1/4}(x)$ is the imaginary part of the MacDonald function.

Combining the appropriate equations we get

$$\rho_D(E_{I,s},\eta_g) = \frac{1}{4\pi^2}\left(\frac{2m_c}{\hbar^2}\right)^{3/2}\frac{\hbar\omega_0}{2\eta_g}\sum_{I,s}\left(\frac{\pi E_{I,s}}{2}\right)^{1/2}\cdot\exp\left(-\frac{E_{I,s}^2}{2\eta_g^2}\right)I_{1/4}\left(\frac{E_{I,s}^2}{2\eta_g^2}\right) \tag{7.73}$$

The asymptotic expansion [15] of $D_{-1/2}\left(-\sqrt{2}E_{I,s}/\eta_g\right)$ for this case, yields

$$\rho_D(E_{I,s},\eta_g) = \frac{1}{4\pi^2}\left(\frac{2m_c}{\hbar^2}\right)^{3/2}\hbar\omega_0\sum_{I,s}\left(\frac{1}{\sqrt{2E_{I,s}}}\right)\left\{1+\frac{3}{16}\frac{\eta_g^2}{E_{I,s}^2}+\frac{105}{512}\frac{\eta_g^4}{E_{I,s}^4}+\cdots\infty\right\} \tag{7.74}$$

The (7.74) represents the density-of-states function in the region far away from the CB edge. It has a singularity at the point $E_{I,s} = 0$. In the Limiting case, when $\eta_g \to 0$, for low doping condition, we find from (7.74)

$$\rho_D(E_{I,s},0) = \frac{1}{4\pi^2}\left(\frac{2m_c}{\hbar^2}\right)^{3/2}\hbar\omega_0\sum_{I,s}\frac{1}{\sqrt{E_{I,s}}} \tag{7.75}$$

The (7.75) shows the magneto DOS function in the normal case of a parabolic band, in the absence of band-tails.

We now find the expression for DOS,$\rho_D(E_{I,s},\eta_g)$, when energy, $E_{I,s}$ is small. This means that the region is very near to the CB edge. This case is possible, when $I = 0$ and $s = -1$. It may be noted here that for a small value of E, neither Tsitsishvili [13] nor Dyakonov et al. [14] provided any expression for DOS function. For this case, we have to find the asymptotic expansion of $D_{-1/2}(-\sqrt{2}E_{I,s}/\eta_g)$ appearing in (7.68), when $I = 0$ and $s = -1$ and E_0

Case C: when $(E_{0-1}/\eta_g)<0$ and $|E_{0-1}/\eta_g \to 0|$

These limits correspond to the region very near to the CB edge and in the forbidden band. In this limit, we find [15]

$$D_{-1/2}\left(-\frac{\sqrt{2}E_{0,-1}}{\eta_g}\right) = \pi^{1/2}\exp\left(-\frac{E_{0,-1}^2}{2\eta_g^2}\right)\times\left(\frac{2^{1/4}}{\eta_{g3/4}}-\frac{2^{3/4}}{\eta_{g1/4}}\frac{E_{0,-1}}{\eta_g}\right) \tag{7.76}$$

Thus, we get

$$\rho_D(E_{0,-1},\eta_g) = \frac{1}{2\pi^2}\left(\frac{2m_c}{\hbar^2}\right)^{3/2}\frac{\hbar\omega_0}{\sqrt{\eta_g}}\exp\left(-\frac{E_{0,-1}^2}{\eta_g^2}\right) \cdot\left[0.51138 - 0.34568\left(\frac{E_{0,-1}}{\eta_g}\right)\right] \tag{7.77}$$

From (7.77) exhibits the exponential variation of the DOS, without any singularity point, very near to the CB edge in the forbidden region.

Case D: when $(E_{0-1}/\eta_g) > 0$ and $|E_{0-1}/\eta_g \to 0|$

This limiting case indicates the region very near to and above the CB edge, i.e. the lowest Landau sub-band. In this limiting condition, we find [15]

$$D_{-1/2}\left(-\frac{\sqrt{2}E_{0,-1}}{\eta_g}\right)=\pi^{1/2}\exp\left(-\frac{E_{0,-1}^2}{2\eta_g^2}\right)\cdot\left(\frac{2^{-1/4}}{\eta_{g3/4}}+\frac{2^{3/4}}{\eta_{g1/4}}\frac{E_{0,-1}}{\eta_g}\right) \tag{7.78}$$

Thus, we find

$$\rho_D(E_{0,-1},\eta_g)=\frac{1}{2\pi^2}\left(\frac{2m_c}{\hbar^2}\right)^{3/2}\frac{\hbar\omega_0}{\sqrt{\eta_g}}\exp\left(-\frac{E_{0,-1}^2}{\eta_g^2}\right)\times\left[0.51139-0.34568\left(\frac{E_{0,-1}}{\eta_g}\right)\right] \tag{7.79}$$

From (7.79) shows the exponential variation of the DOS function without any singularity in the region very near to the lowest Landau sub-band. As noted above, neither Tsitsishvili [13] nor Dyakonov et al. [14] provided the expression of the DOS function in this region.

From (7.77) and (7.79), we find the limiting value of the DOS function at the lowest sub-band, when $E_{0,-1}=0$, as

$$\rho_D(0,\eta_g)=0.0732755\left(\frac{m_c}{\hbar^2}\right)^{3/2}\frac{\hbar\omega_0}{\sqrt{\eta_g}} \tag{7.80}$$

The (7.80) shows that in the heavy doping condition, the DOS at the lowest sub-band edge is not infinitely large, but has a definite value. This implies that the band gets perturbed because of the heavy doping. In the case, when there is no tail (7.75), the DOS function is infinitely large at $E_{0,-1}=0$. In such a case, we call it an unperturbed conduction band.

It may be noted further that (7.77) and (7.79) represent the DOS function near to the CB edge, where $E_{0,-1}\to\pm\varepsilon$, and $\varepsilon\to 0$. We are to find now the limiting values of ε, so that (7.77) and (7.79) are valid. The limiting values are obtained from the derivatives of (7.76) and (7.78), when they are to be equated to the values $\left(-\frac{\pi^{1/2}2^{1/4}}{\eta_{g1/4}}\right)$ and $\left(\frac{\pi^{1/2}2^{1/4}}{\eta_{g1/4}}\right)$ respectively, where $U'(0,0)=-\frac{\pi^{1/2}2^{1/4}}{\eta_{g1/4}}$ and $V'(0,0)=\frac{\pi^{1/2}2^{1/4}}{\eta_{g1/4}}$.

$U'(0,0)$ and $V'(0,0)$ are the derivatives of parabolic cylinder functions at $E_{0,-1}\to 0$. Therefore, the limiting values of $E_{0,-1}$ for the DOS function.

Table 7.1 Limiting values of energies ($E_{0,-1}$) for various cases of DOS function

DOS $\rho_D(E_{0,-1},\eta_g)$	Workable range of $E_{0,-1}$
(6.71)	$-\infty$ to $-0.98623\eta_g$
(6.74)	0 to ∞
(6.77)	$-0.98623\eta_g$ to ∞
(6.79)	0

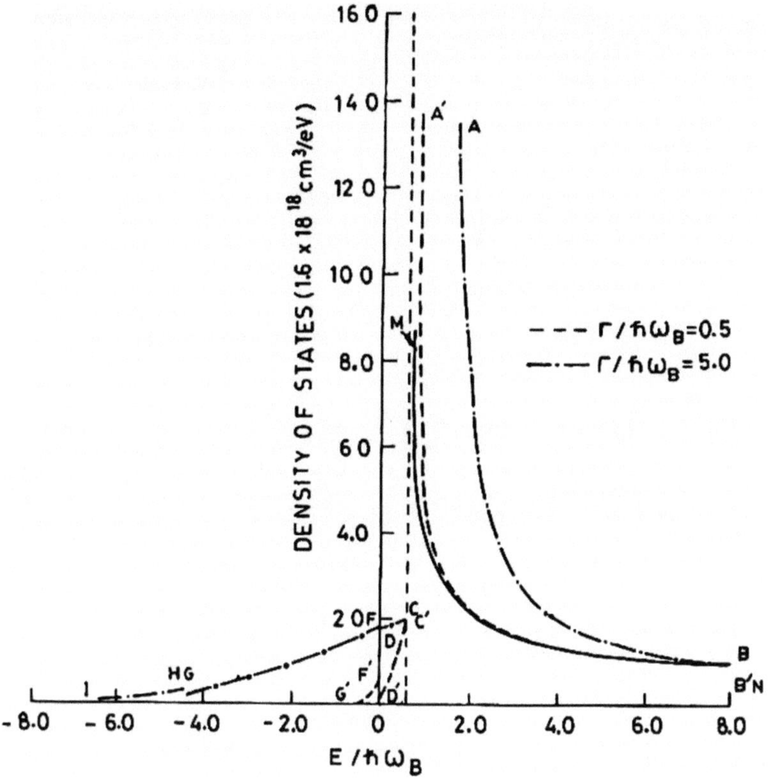

Fig. 7.1 Graphs $AB(A'B')$, $CD(C'D')$, $FG(F'G')$, HI are the representations of density-of-states $\rho_D(E_{0,-1};\eta_g)$ against $(E/\hbar\omega_0)$ for $E/\hbar\omega_0 = 5.0\ (0.5)$ in the perturbed band at $B = 3.0$ T for HD GaAs material and extended for $(E/\hbar\omega_0) \geq 0$ and <0. Graph MN is drawn for the density-of-states $\rho_c(E, 0)$ against $(E/\hbar\omega_0)$ at $B = 3.0$ T for GaAs material. The expressions used in the plots are given in Table 7.2. For this plot $\eta_g = \Gamma$ *and* $\omega_0 = \omega_B$

$\rho_D(E_{0,-1}, \eta_g)$, can be obtained for values of $E_{0,-1}$ from $-0.98623\eta_g$ to 0, and that of (6.79) as $E_{0,-1} = 0$. For the purpose of condensed presentation in Table 7.1 we have written the limiting values for energies which are valid for four cases of the DOS functions (7.71), (7.74), (7.77) and (7.79) in the perturbed band. It may be noted that the theories of Dyakonov et al. [14] and Tsitsishvili [13] were unable to show the singularity in the DOS function in the forbidden region.

In Figs. 7.1 and 7.3 we have plotted the DOS function for GaAs, taking $|g| = 0.44, B = 3.0$ T, $l = 0$ and $s = -1$ for the lowest Landau sub-band, $\eta_g = 5.0\hbar\omega_0$ in order to cover a wider range of η_g.

In Fig. 7.1, graphs AB(A′B′), CD(C′D′), FG(F′G′) and HI are drawn for the DOS, $\rho_D(E_{0,-1}, \eta_g)$ versus $(E/\hbar\omega_0)$ in the perturbed band with the DOS technique proposed by Dyakonov et al. [14] The graph MN is for the DOS, $\rho_D(E_{0,-1}, \eta_g)$(7.75) with the unperturbed band. The approximate expressions that are used in the plots

Table 7.2 The approximate expressions for various cases of DOS function that are used in the plots of Fig. 7.1 for GaAs

Density-of-states (approximate)	Curve in Fig. 7.1	Workable range of $E_{0,-1}$	Equation number
$\rho_D(E_{I,s},\eta_g)=\frac{1}{4\pi^2}\left(\frac{2m_c}{\hbar^2}\right)^{3/2}\hbar\omega_0\frac{1}{\sqrt{E_{0,-1}}}\cdot\left\{1+\frac{3}{16}\frac{\eta_g^2}{(E_{0,-1})^2}\right\}$	$AB(A'B')$	0 *to* ∞	(7.74)
$\rho_D(E_{0,-1},\eta_g)=\frac{1}{2\pi^2}\left(\frac{2m_c}{\hbar^2}\right)^{3/2}\frac{\hbar\omega_0}{\sqrt{\eta_g}}\exp\left(-\frac{E_{0,-1}^2}{\eta_g^2}\right)\cdot\left[0.51139+0.34568\left(\frac{E_{0,-1}}{\eta_g}\right)\right]$	$CD(C'D')$	0	(7.79)
$\rho_D(E_{0,-1},\eta_g)=\frac{1}{2\pi^2}\left(\frac{2m_c}{\hbar^2}\right)^{3/2}\frac{\hbar\omega_0}{\sqrt{\eta_g}}\exp\left(-\frac{E_{0,-1}^2}{\eta_g^2}\right)\cdot\left[0.51139-0.34568\left(\frac{E_{0,-1}}{\eta_g}\right)\right]$	$FG(F'G')$	$0.98623\eta_g$ *to* 0	(7.77)
$\rho_D(E_{0,-1},\eta_g)=\frac{1}{4\pi^2}\left(\frac{2m_c}{\hbar^2}\right)^{3/2}\hbar\omega_0\frac{1}{\sqrt{2E_{0,-1}}}\exp\left(-\frac{E_{0,-1}^2}{\eta_g^2}\right)\cdot\left\{1-\frac{3}{16}\frac{\eta_g^2}{(E_{0,-1})^2}\right\}$	HI	$-\infty$ *to* $0.98623\eta_g$	(7.71)
$\rho_D(E_{0,-1},\eta_g)=\frac{1}{4\pi^2}\left(\frac{2m_c}{\hbar^2}\right)^{3/2}\frac{\hbar\omega_0}{\sqrt{E_{0,-1}}}$	MN		(7.75)

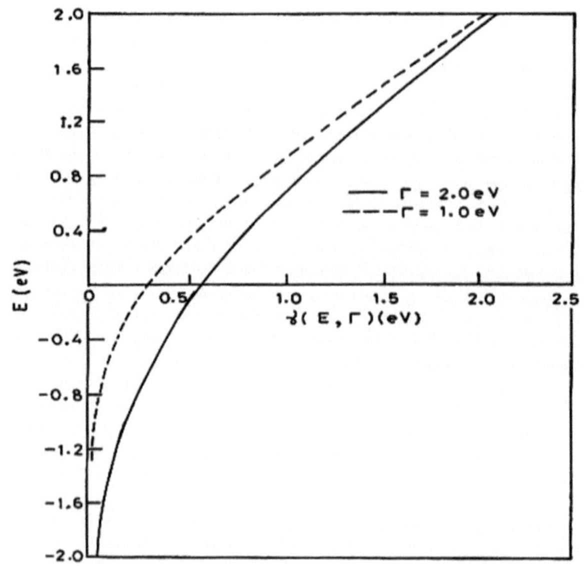

Fig. 7.2 Graphs are plotted for $\gamma(E,\eta_g)$ (in eV) versus electron energy E (ineV) in the case of impurity screening energy $\eta_g=2.0$ eV (*solid line*) and $\eta_g=1.0$ eV (*dotted line*), for $E\geq 0$ and $E<0$. For this plot $\eta_g=\Gamma$ and $\omega_0=\omega_B$

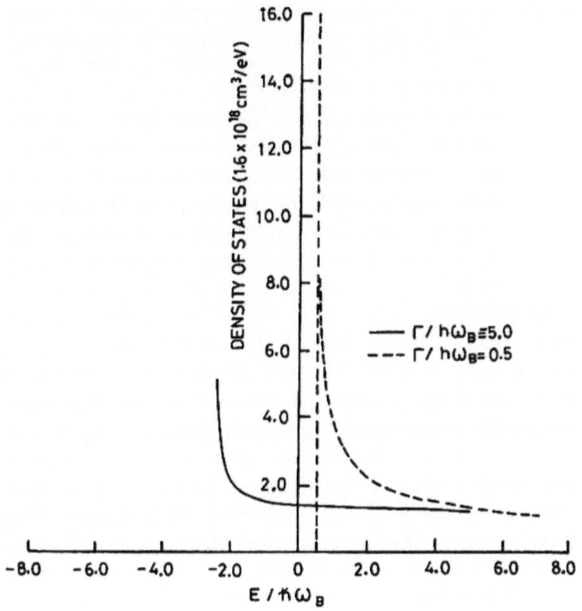

Fig. 7.3 Graph are plotted for density-of-states $\rho(E, \eta_g)$ against $(E/\hbar\omega_0)$ for $\eta_g/\hbar_{\omega_B} = 5.0$ (*solid line*) and 0.5 (*dotted line*) at $B = 3.0$ T, for GaAs material. Graphs are extended for $(E/\hbar\omega_0) \geq 0$ and $(E/\hbar\omega_0) \leq 0$. For this plot $\eta_g = \Gamma$ *and* $\omega_0 = \omega_B$

are given in Table 7.2. It appears from Fig. 7.1 that CD(C′D′), FG(F′G′) and HI exhibit the tail region in the DOS for $\eta_g = 5.0\hbar\omega_0$. The unperturbed DOS (graph MN) never shows such a tail rather singularity at $(E/\hbar\omega_0) = 0.50737$. The break points in the curves AB(A′B′), CD(C′D′), FG(F′G′) and HI are due to the limiting ranges of $(E_{0,-1}, \eta_g)$ in the different models for the DOS, mentioned in Table 7.2. It may be noted that our results presented above are more simple to handle as compared to others [13, 14].

In Fig. 7.2 we have plotted $\gamma(E, \eta_g)$ against E for $\eta_g = 1.0$ eV (dotted curve) and $\eta_g = 2.0$ (solid curve) in order to show the effect of η_g on the tailing of band. These curves are for the values of E extending $-\infty$ to ∞. For $E < 0$, the curve show a positive value of $\gamma(E, \eta_g)$. This implies that the energy band is tailing into the forbidden region. For $E > 0$, the graph is no more a straight line and for every large value of E, the graphs becomes linear. This implies that under band-tailing states, the $E - \bar{k}$ relation remains no more related to its parabolic feature. The curve for $\gamma(E, \eta_g)$ cuts the $E = 0$ axis at different points, depending on the value of η_g, as shown in Fig. 7.2.

In Fig. 7.3, we have presented the plots of DOS, $\rho(E, \eta_g)$(solid line for $(\eta_g/\hbar\omega_0) = 5.0$) and dotted line for $((\eta_g/\hbar\omega_0) = 5.0)$, against $(E/\hbar\omega_0) > 0$ and $(E/\hbar\omega_0) < 0$ in order to compare the effect of $(\eta_g/\hbar\omega_0)$ on it. Figure 7.3 shows that the tailing of the band in the $E - \bar{k}$ relation curves (Fig. 7.2) does not exist in the DOS, rather a singularity point, that has been shifted to the forbidden region for $(\eta_g/\hbar\omega_0) = 0.5$. For $(\eta_g/\hbar\omega_0) = 0.5$, the singularity point remains in the CB without any tailing in the curve. Figure 7.3, does not show any break point in it and the tailing in the forbidden band region, as in Fig. 7.1 for $\rho_D(E_{0,-1}, \eta_g)$. We wish to remark that the singularity for all energies must exist in the DOS function in HDS under quantizing magnetic field in order to explain the experimentally observed oscillations in various semiconductor phenomena. We therefore conclude from the above analysis that there is not enough basis for consideration of the DOS technique (e.g. the approaches due to Tsitsishvili [13] and Dyakonov et al. [14]) in the form of the basic equation for the development of band-tailing theory in the presence of an external magnetic field. Our study on band-tailing using the $E - \bar{k}$ relationship in the presence of a magnetic field is more appropriate and can be used in a more general way. Also, by band-tailing, we mean that energy band has a tail in the forbidden band but not the case when the DOS function has a tail in this region.

We have not plotted any other cases and leave the computer plots on the shoulders of our readers with the hope that they will enjoy the new physics and the related numerical computations in the own way

7.4 Open Research Problems

(R.7.1) Investigate the DR in the presence of an arbitrarily oriented quantizing magnetic field for all the materials as given in problems in (R.1.1) of Chap. 1 in the presence of the Gaussian type band tails

(R.7.2) Investigate the DR in the presence of an arbitrarily oriented quantizing magnetic field in HD nonlinear optical semiconductors by including broadening and the electron spin. Study all the special cases for HD III–V, ternary and quaternary materials in this context

(R.7.3) Investigate the DRs for HD IV–VI, II–VI and stressed Kane type compounds in the presence of an arbitrarily oriented quantizing magnetic field by including broadening and electron spin

(R.7.4) Investigate the DR for all the materials as stated in (R.1.1) of Chap. 1 in the presence of an arbitrarily oriented quantizing magnetic field by including broadening and electron spin under the condition of heavily doping

(R.7.5) Investigate the DR in the presence of an arbitrarily oriented quantizing magnetic field and crossed electric fields in HD nonlinear optical semiconductors by including broadening and the electron spin. Study all the special cases for HD III–V, ternary and quaternary materials in this context

(R.7.6) Investigate the DRs for HD IV–VI, II–VI and stressed Kane type compounds in the presence of an arbitrarily oriented quantizing magnetic field and crossed electric field by including broadening and electron spin

(R.7.7) Investigate the DR for all the materials as stated in (R.1.1) of Chap. 1 in the presence of an arbitrarily oriented quantizing magnetic field and crossed electric field by including broadening and electron spin under the condition of heavy doping

(R.7.8) Investigate the 2D DR in QWs of HD nonlinear optical, III–V, II–VI, IV–VI and stressed Kane type semiconductors in the presence of magnetic quantization

(R.7.9) Investigate the 2D DR for HD QWs of all the materials as considered in problems (R.1.1) in the presence of magnetic quantization

(R.7.10) Investigate the 2D DR in the presence of an arbitrarily oriented non-quantizing magnetic field for the QWs of HD nonlinear optical semiconductors by including the electron spin. Study all the special cases for III–V, ternary and quaternary materials in this context

(R.7.11) Investigate the DRs in QWs of HD IV–VI, II–VI and stressed Kane type compounds in the presence of an arbitrarily oriented non-quantizing magnetic field by including the electron spin

(R.7.12) Investigate the 2D DR for HD QWs of all the materials as stated in (R.1.1) of Chap. 1 in the presence of an arbitrarily oriented magnetic field by including electron spin and broadening

(R.7.13) Investigate the DR for all the problems of (R.1.1) under an additional arbitrarily oriented non uniform magnetic field in the presence of heavy doping

(R.7.14) Investigate the DR for all the problems of (R.1.1) under the arbitrarily oriented crossed electric and magnetic fields in the presence of heavy doping

(R.7.15) Investigate all the problems of this section by removing all the mathematical approximations and establishing the respective appropriate uniqueness conditionsUniqueness conditions.

References

1. N. Miura, *Physics of Semiconductors in High Magnetic Fields*. Series on Semiconductor Science and Technology (Oxford University Press, New York, 2007); K.H.J. Buschow, F.R. de Boer, *Physics of Magnetism and Magnetic Materials* (Springer, New York, 2003); D. Sellmyer, R. Skomski (eds.), *Advanced Magnetic Nanostructures* (Springer, New York, 2005); J.A.C. Bland, B. Heinrich (eds.), *Ultrathin Magnetic Structures III: Fundamentals of Nanomagnetism (Pt. 3)* (Springer, Germany, 2005); B.K. Ridley, *Quantum Processes in semiconductors*, 4th edn (Oxford Publications, Oxford, 1999); J.H. Davies, *Physics of Low Dimensional Semiconductors* (Cambridge University Press, London, 1998); S. Blundell, *Magnetism in Condensed Matter*. Oxford Master Series in Condensed Matter Physics (Oxford University Press, New York, 2001); C. Weisbuch, B. Vinter, *Quantum Semiconductor Structures: Fundamentals and Applications* (Academic Publishers, New York, 1991); D. Ferry, *Semiconductor Transport* (CRC, New York, 2000); M. Reed (ed.), *Semiconductors and Semimetals: Nanostructured Systems* (Academic Press, New York, 1992); T. Dittrich, *Quantum Transport and Dissipation* (Wiley-VCH Verlag GmbH, Germany, 1998); A.Y. Shik,

Quantum Wells: Physics & Electronics of Two Dimensional Systems (World Scientific, New York, 1997)
2. K.P. Ghatak, M. Mondal, *Zietschrift fur Naturforschung A***41a**, 881 (1986); K.P. Ghatak, M. Mondal, J. Appl. Phys. **62**, 922 (1987); K.P. Ghatak, S.N. Biswas, Phys. Stat. Sol. (b) **140**, K107 (1987); K.P. Ghatak, M. Mondal, J. Mag. Mag. Mat. **74**, 203 (1988); K.P. Ghatak, M. Mondal, Phys. Stat. Sol. (b)**139**, 195 (1987); K.P. Ghatak, M. Mondal, Phys. Stat. Sol. (b) **148**, 645 (1988); K.P. Ghatak, B. Mitra, A. Ghoshal, Phys. Stat. Sol. (b) **154**, K121 (1989); K. P. Ghatak, S.N. Biswas, J. Low Temp. Phys. **78**, 219 (1990); K.P. Ghatak, M. Mondal, Phys. Stat. Sol. (b) **160**, 673 (1990); K.P. Ghatak, B. Mitra, Phys. Letts. A **156**, 233 (1991); K. P. Ghatak, A. Ghoshal, B. Mitra, Nouvo Cimento D **13D**, 867 (1991); K.P. Ghatak, M. Mondal, Phys. Stat. Sol. (b) **148**, 645 (1989); K.P. Ghatak, B. Mitra, Int. J. Electr. **70**, 345 (1991); K.P. Ghatak, S.N. Biswas, J. Appl. Phys. **70**, 299 (1991); K.P. Ghatak, A. Ghoshal, Phys. Stat. Sol. (b) **170**, K27 (1992); K.P. Ghatak, Nouvo Cimento D **13D**, 1321 (1992); K. P. Ghatak, B. Mitra, Int. J. Electr. **72**, 541 (1992); K.P. Ghatak, S.N. Biswas, Nonlinear Opt. **4**, 347 (1993); K.P. Ghatak, M. Mondal, Phys. Stat. Sol. (b) **175**, 113 (1993); K.P. Ghatak, S.N. Biswas, Nonlinear Opt. **4**, 39 (1993); K.P. Ghatak, B. Mitra, Nouvo Cimento **15D**, 97 (1993); K.P. Ghatak, S.N. Biswas, Nanostructured Mater. **2**, 91 (1993); K.P. Ghatak, M. Mondal, Phys. Stat. Sol. (b) **185**, K5 (1994); K.P. Ghatak, B. Goswami, M. Mitra, B. Nag, Nonlinear Opt. **16**, 9 (1996); K.P. Ghatak, M. Mitra, B. Goswami, B. Nag, Nonlinear Opt. **16**, 167 (1996); K.P. Ghatak, B. Nag, Nanostructured Mater. **10**, 923 (1998)
3. D. Roy Choudhury, A.K. Choudhury, K.P. Ghatak, A.N. Chakravarti, Phys. Stat. Sol. (b) **98**, K141 (1980); A.N. Chakravarti, K.P. Ghatak, A. Dhar, S. Ghosh, Phys. Stat. Sol. (b) **105**, K55 (1981); A.N. Chakravarti, A.K. Choudhury, K.P. Ghatak, Phys. Stat. Sol. (a) **63**, K97 (1981); A.N. Chakravarti, A.K. Choudhury, K.P. Ghatak, S. Ghosh, A. Dhar, Appl. Phys. **25**, 105 (1981); A.N. Chakravarti, K.P. Ghatak, G.B. Rao, K.K. Ghosh, Phys. Stat. Sol. (b) **112**, 75 (1982); A.N. Chakravarti, K.P. Ghatak, K.K. Ghosh, H.M. Mukherjee, Phys. Stat. Sol. (b) **116**, 17 (1983); M. Mondal, K.P. Ghatak, Phys. Stat. Sol. (b) **133**, K143 (1984); M. Mondal, K.P. Ghatak, Phys. Stat. Sol. (b) **126**, K47 (1984); M. Mondal, K.P. Ghatak, Phys. Stat. Sol. (b) **126**, K41 (1984); M. Mondal, K.P. Ghatak, Phys. Stat. Sol. (b) **129**, K745 (1985); M. Mondal, K.P. Ghatak, Phys. Scr. **31**, 615 (1985); M. Mondal, K.P. Ghatak, Phys. Stat. Sol. (b) **135**, 239 (1986); M. Mondal, K. P. Ghatak, Phys. Stat. Sol. (b) **93**, 377 (1986); M. Mondal, K.P. Ghatak, Phys. Stat. Sol. (b) **135**, K21 (1986); M. Mondal, S. Bhattacharyya, K.P. Ghatak, Appl. Phys. A **42A**, 331 (1987); S.N. Biswas, N. Chattopadhyay,K.P. Ghatak, Phys. Stat. Sol. (b) **141**, K47 (1987); B. Mitra, K.P. Ghatak, Phys. Stat. Sol. (b) **149**, K117 (1988); B. Mitra, A. Ghoshal, K.P. Ghatak, Phys. Stat. Sol. (b) **150**, K67 (1988); M. Mondal, K.P. Ghatak, Phys. Stat. Sol. (b) **147**, K179 (1988); M. Mondal, K.P. Ghatak, Phys. Stat. Sol. (b) **146**, K97 (1988); B. Mitra, A. Ghoshal, K.P. Ghatak, Phys. Stat. Sol. (b) **153**, K209 (1989); B. Mitra, K.P. Ghatak, Phys. Letts. **142A**, 401 (1989); B. Mitra, A. Ghoshal,K. P. Ghatak, Phys. Stat. Sol. (b) **154**, K147 (1989); B. Mitra, K.P. Ghatak, Sol. State Elect. **32**, 515 (1989); B. Mitra, A. Ghoshal, K.P. Ghatak, Phys. Stat. Sol. (b) **155**, K23 (1989); B. Mitra, K.P. Ghatak, Phys. Letts. **135A**, 397 (1989); B. Mitra, K.P. Ghatak, Phys. Letts. A **146A**, 357 (1990); B. Mitra, K. P. Ghatak, Phys. Stat. Sol. (b) **164**, K13 (1991); S.N. Biswas, K. P. Ghatak, Int. J Electr. **70**, 125 (1991)
4. P.R. Wallace, Phys. Stat. Sol. (b) **92**, 49 (1979)
5. B.R. Nag, *Electron Transport in Compound Semiconductors*. Springer Series in Solid-State Sciences, vol. 11 (Springer, Germany, 1980
6. K.P. Ghatak, S. Bhattacharya, D. De, *Einstein Relation in Compound Semiconductors and Their Nanostructures*. Springer Series in Materials Science, vol. 116 (Springer, Germany, 2009)
7. C.C. Wu, C.J. Lin, J. Low Temp. Phys. **57**, 469 (1984); M.H. Chen, C.C. Wu, C.J. Lin, J. Low Temp. Phys. **55**, 127 (1984)
8. E. Bangert, P. Kastner, Phys. Stat. Sol (b) **61**, 503 (1974)
9. G.M.T. Foley, P.N. Langenberg, Phys. Rev. B **15B**, 4850 (1977)

10. M. Singh, P.R. Wallace, S.D. Jog, E. Arushanov, J. Phys. Chem. Solids, **45**, 409 (1984); Y. Yamada, Phys. Soc. Japan **35**, 1600 (1973); **37**, 606 (1974)
11. L.A. Vassilev, Phys. Stat. Sol. (b) **121**, 203 (1984)
12. B.I. Shklovshii, A.L. Efros, *Electronic Properties of Doped Semiconductors*, vol. 45 (Springer, Berlin, 1980)
13. E.G. Tsitsishvili, Sov. Phys. Solid State **8**, 950 (1966)
14. M.I. Dyakonov, A.L. Efros, D.L. Mitchell, Phys. Rev. **180**, 813 (1969)
15. M. Abramovitz, I.A. Stegun, *Handook of Mathematical Functions with Formulas, Graphs and Mathematical Tables*, (John Wiley, New Yourk, 1964)

Chapter 8
The DR in HDs Under Cross-Fields Configuration

Nothing is interesting until I am interested.

8.1 Introduction

The influence of crossed electric and quantizing magnetic fields on the transport properties of semiconductors having various band structures are relatively less investigated as compared with the corresponding magnetic quantization, although, the cross-fields are fundamental with respect to the addition of new physics and the related experimental findings. In 1966, Zawadzki and Lax [1] formulated the electron dispersion law for III–V semiconductors in accordance with the two band model of Kane under cross fields configuration which generates the interest to study this particular topic of semiconductor science in general [2–14].

In Sect. 8.2.1 of theoretical background, the DR in HD nonlinear optical materials in the presence of crossed electric and quantizing magnetic fields has been investigated by formulating the electron dispersion relation. The Sect. 8.2.2 reflects the study of the DR in HD III–V, ternary and quaternary compounds as a special case of Sect. 8.2.1. The Sect. 8.2.3 contains the study of the DR for the HD II–VI semiconductors in the present case. In Sect. 8.2.4, the DR under cross field configuration in HD IV–VI semiconductors has been investigated in accordance with the models of the Cohen, the Lax nonparabolic ellipsoidal and the parabolic ellipsoidal respectively. In the Sect. 8.2.5, the DR for the HD stressed Kane type semiconductors has been investigated. The Sect. 8.3 contains the summary and conclusion of this chapter and the last Sect. 8.4 presents three open research problems.

K.P. Ghatak, *Dispersion Relations in Heavily-Doped Nanostructures*,
Springer Tracts in Modern Physics 265, DOI 10.1007/978-3-319-21000-1_8

8.2 Theoretical Background

8.2.1 *The DR in HD Nonlinear Optical Semiconductors Under Cross-Fields Configuration*

The (2.26) of Chap. 2 can be expressed as

$$T_{22}(\mathrm{E}, \eta_g) = \frac{p_s^2}{2m_\perp^*} + \frac{p_z^2}{2M_\parallel} T_{22}(\mathrm{E}, \eta_g)\left[T_{21}(\mathrm{E}, \eta_g)\right]^{-1} \tag{8.1}$$

where, $p_s = \hbar k_s$ and $p_z = \hbar k_z$

We know that from electromagnetic theory that,

$$\vec{B} = \nabla \times \vec{A} \tag{8.2}$$

where, $\vec{A}$ is the vector potential. In the presence of quantizing magnetic field B along z direction, the (8.2) assumes the form

$$0\hat{i} + 0\hat{j} + B\hat{k} = \begin{vmatrix} \hat{i} & \hat{j} & \hat{k} \\ \frac{\partial}{\partial x} & \frac{\partial}{\partial y} & \frac{\partial}{\partial z} \\ A_x & A_y & A_z \end{vmatrix} \tag{8.3}$$

where $\hat{i}$, $\hat{j}$ and $\hat{k}$ are orthogonal triads. Thus, we can write

$$\begin{aligned} \frac{\partial A_z}{\partial y} - \frac{\partial A_y}{\partial z} &= 0 \\ \frac{\partial A_x}{\partial z} - \frac{\partial A_z}{\partial x} &= 0 \\ \frac{\partial A_y}{\partial x} - \frac{\partial A_x}{\partial y} &= B \end{aligned} \tag{8.4}$$

This particular set of equations is being satisfied for $A_x = 0$, $A_y = Bx$ and $A_z = 0$.

Therefore in the presence of the electric field E_o along x axis and the quantizing magnetic field B along z axis for the present case following (8.1) one can approximately write,

$$T_{22}(E, \eta_g) + |e|E_0\hat{x}\rho(E, \eta_g) = \frac{\hat{p}_x^2}{2m_\perp^*} + \frac{(\hat{p}_y - |e|B\hat{x})^2}{2m_\perp^*} + \frac{\hat{p}_z^2}{2a(E, \eta_g)} \tag{8.5}$$

where

$$\rho(E) \equiv \frac{\partial}{\partial E}\left[\mathrm{T}_{22}(E,\eta_g)\right] \text{ and } a(E,\eta_g) \equiv m_{\parallel}^{*}\left[\mathrm{T}_{22}(E,\eta_g)\right]^{-1}\left[\mathrm{T}_{21}(E,\eta_g)\right]$$

Let us define the operator $\hat{\theta}$ as

$$\hat{\theta} = -\hat{p}_y + |e|B\hat{x} - \frac{m_{\perp}^{*}E_o\rho(E,\eta_g)}{B} \tag{8.6}$$

Eliminating the operator $\hat{x}$, between (8.5) and (8.6) the dispersion relation of the conduction electron in tetragonal semiconductors in the presence of cross fields configuration is given by

$$\mathrm{T}_{22}(E,\eta_g) = \left[\left(n+\frac{1}{2}\hbar\omega_{01}\right) + \left(\frac{[\hbar k_z(E)]^2}{2a(E,\eta_g)}\right) - \left(\frac{E_o\hbar k_y\rho(E,\eta_g)}{B}\right) - \left(\frac{M_{\perp}\rho^2(E,\eta_g)E_o^2}{2B^2}\right)\right] \tag{8.7}$$

where,

$$\omega_{01} = \frac{|e|B}{m_{\perp}^{*}}$$

The EEMs along Z and Y directions can, respectively be expressed from (8.7) as

$$\begin{aligned} m_z^{*}(\bar{E}_{FBHD},\eta_g,n,E_0) = \text{Real part of } &\left[a'(\bar{E}_{FBHD},\eta_g)\left[T_{22}(\bar{E}_{FBHD},\eta_g) - \left(n+\frac{1}{2}\right)\hbar\omega_{01}\right.\right. \\ &\left.+\frac{M_{\perp}\rho^2(\bar{E}_{FBHD},\eta_g)E_0^2}{2B^2}\right] + a(\bar{E}_{FBHD},\eta_g) \\ &\left.\left[T'_{22}(\bar{E}_{FBHD},\eta_g) + \frac{M_{\perp}\rho(\bar{E}_{FBHD},\eta_g)\rho'(\bar{E}_{FBHD},\eta_g)E_0^2}{B^2}\right]\right] \end{aligned} \tag{8.8}$$

and

$$\begin{aligned} m_y^{*}(\bar{E}_{FBHD},\eta_g,n,E_0) = \left(\frac{B}{E_0}\right)^2 &\text{Real part of } \left[\rho(\bar{E}_{FBHD},\eta_g)\right]^{-3} \\ &\left[T_{22}(\bar{E}_{FBHD},\eta_g) - \left(n+\frac{1}{2}\right)\hbar\omega_{01} + \frac{M_{\perp}\rho^2(\bar{E}_{FBHD},\eta_g)E_0^2}{2B^2}\right] \\ &\left[\left[\rho(\bar{E}_{FBHD},\eta_g)\left[T'_{22}(\bar{E}_{FBHD},\eta_g) + \frac{M_{\perp}\rho(\bar{E}_{FBHD},\eta_g)\rho'(\bar{E}_{FBHD},\eta_g)E_0^2}{B^2}\right]\right.\right. \\ &-\left[T_{22}(\bar{E}_{FBHD},\eta_g) - \left(n+\frac{1}{2}\right)\hbar\omega_{01} + \frac{M_{\perp}\rho^2(\bar{E}_{FBHD},\eta_g)E_0^2}{2B^2}\right] \\ &\left.\left.\rho'(\bar{E}_{FBHD},\eta_g)\right]\right] \end{aligned} \tag{8.9}$$

where $\bar{E}_{FBHD}$ is the Fermi energy in the presence of cross-fields configuration and heavy doping as measured from the edge of the conduction band in the vertically upward direction in the absence of any quantization.

When $E_0 \to 0$, $m_z^*(\bar{E}_{FBHD}, \eta_g, n, E_0) \to \infty$, which is a physically justified result. The dependence of the EEM along y direction on the Fermi energy, electric field, magnetic field and the magnetic quantum number is an intrinsic property of cross fields together with the fact in the present case of heavy doping, the EEM exists in the band gap. Another characteristic feature of cross field is that various transport coefficients will be sampled dimension dependent. These conclusions are valid for even isotropic parabolic energy bands and cross fields introduce the index dependent anisotropy in the effective mass.

The formulation of DR requires the expression of the electron concentration which can, in general, be written excluding the electron spin as

$$n_0 = \frac{-g_v}{L_x \pi^2} \sum_{n=0}^{n_{\max}} \int_{\bar{E}_0}^{\infty} I(E, \eta_g) \frac{\partial f_0}{\partial E} dE \quad (8.10)$$

where L_x is the sample length along x direction, $\bar{E}_0$ is determined by the equation

$$I(\bar{E}_0, \eta_g) = 0$$

where

$$I(E, \eta_g) = \int_{x_l(E,\eta_g)}^{x_h(E,\eta_g)} k_z(E) dk_y \quad (8.11)$$

in which, $x_l(E, \eta_g) \equiv \frac{-E_0 M_\perp \rho(E,\eta_g)}{\hbar B}$ and $x_h(E, \eta_g) \equiv \frac{|e| B L_x}{\hbar} + x_l(E, \eta_g)$.

Thus we get

$$I(E, \eta_g) = \frac{2}{3} \left[\frac{B\sqrt{2a(E, \eta_g)}}{\hbar^2 E_0 \rho(E, \eta_g)} \right] \left[\left[T_{22}(E, \eta_g) - \left(n + \frac{1}{2}\right) \frac{\hbar |e| B}{m_\perp^*} + |e| E_0 L_x \rho(E, \eta_g) - \frac{m_\perp^* E_0^2 [\rho(E, \eta_g)]^2}{2B^2} \right]^{\frac{3}{2}} - \left[T_{22}(E, \eta_g) - \left(n + \frac{1}{2}\right) \frac{\hbar |e| B}{m_\perp^*} - \frac{m_\perp^* E_0^2 [\rho(E, \eta_g)]^2}{2B^2} \right]^{\frac{3}{2}} \right] \quad (8.12)$$

Therefore the electron concentration is given by

$$n_0 = \left(\frac{2g_v B\sqrt{2}}{3L_x \pi^2 \hbar^2 E_0}\right) \text{Real part of} \sum_{n=0}^{n_{\max}} \left[T_{41HD}\left(n, \bar{E}_{FBHD}, \eta_g\right) + T_{42HD}\left(n, \bar{E}_{FBHD}, \eta_g\right)\right] \tag{8.13}$$

where

$$T_{41}\left(n, \bar{E}_{FBHD}, \eta_g\right) \equiv \frac{\sqrt{a\left(\bar{E}_{FBHD}, \eta_g\right)}}{\rho\left(\bar{E}_{FBHD}, \eta_g\right)} \left[\left[T_{22}\left(\bar{E}_{FBHD}, \eta_g\right) - \left(n + \frac{1}{2}\right)\frac{\hbar|e|B}{M_\perp} + |e|E_0 L_x \rho\left(\bar{E}_{FBHD}, \eta_g\right) \cdot \right.\right.$$
$$\left. - \frac{m_\perp^* E_0^2 \left[\rho\left(\bar{E}_{FBHD}, \eta_g\right)\right]^2}{2B^2}\right]^{\frac{3}{2}}$$
$$\left. - \left[T_{22}\left(\bar{E}_{FBHD}, \eta_g\right) - \left(n + \frac{1}{2}\right)\frac{\hbar|e|B}{m_\perp^*} - \frac{m_\perp^* E_0^2 \left[\rho\left(\bar{E}_{FBHD}, \eta_g\right)\right]^2}{2B^2}\right]^{\frac{3}{2}}\right]$$

where $\bar{E}_{FBHD}$ is the Fermi energy in this case and

$$T_{42HD}\left(n, \bar{E}_{FBHD}, \eta_g\right) \equiv \sum_{r=1}^{s} \left[L(r) T_{41HD}\left(n, \bar{E}_{FBHD}, \eta_g\right)\right]$$

Therefore the DOS function can be written as

$$N(E_0, B) = \left(\frac{2g_v B\sqrt{2}}{3L_x \pi^2 \hbar^2 E_0}\right) \sum_{n=0}^{n_{\max}} \left[T_{41HD}\left(n, E, \eta_g\right)\right]' H\left(E - \bar{E}_{71}\right) \tag{8.14}$$

where $\bar{E}_{71}$ is the complex Landau level energy in this case.

8.2.2 The DR in HD Kane Type III–V Semiconductors Under Cross-Fields Configuration

(a) Under the conditions $\delta = 0$, $\Delta_\| = \Delta_\perp = \Delta$ and $m_\|^* = m_\perp^* = m_c$, (8.7) assumes the form

$$T_{33}(E,\eta_g) = \left(n+\frac{1}{2}\right)\hbar\omega_0 + \frac{[\hbar k_z(E)]^2}{2m_c} - \frac{E_0}{B}\hbar k_y\{T_{33}(E,\eta_g)\}' - \frac{m_c E_0^2\left[\{T_{33}(E,\eta_g)\}'\right]^2}{2B^2} \tag{8.15}$$

where

$$\mathrm{T}_{33}(\mathrm{E},\eta_g) = \mathrm{T}_{31}(\mathrm{E},\eta_g) + i\mathrm{T}_{32}(\mathrm{E},\eta_g)$$

The use of (8.15) leads to the expressions of the EEMs' along z and y directions as

$$m_z^*(\bar{E}_{FBHD},\eta_g,n,E_0) = m_c \text{ Real part of } \left[\{T_{33}(\bar{E}_{FBHD},\eta_g)\}'' + \frac{m_c E_0^2\{T_{33}(\bar{E}_{FBHD},\eta_g)\}'\{T_{33}(\bar{E}_{FBHD},\eta_g)\}''}{B^2}\right] \tag{8.16}$$

$$\begin{aligned} m_y^*(\bar{E}_{FBHD},\eta_g,n,E_0) = {} & \left(\frac{B}{E_0}\right)^2 \text{Real part of } \left[\{T_{33}(\bar{E}_{FBHD},\eta_g)\}'\right]^{-1} \\ & \left[T_{33}(\bar{E}_{FBHD},\eta_g) - \left(n+\frac{1}{2}\right)\hbar\omega_0 + \frac{m_c E_0^2\left[\{T_{33}(\bar{E}_{FBHD},\eta_g)\}'\right]^2}{2B^2}\right] \\ & \left[\frac{-\{T_{33}(\bar{E}_{FBHD},\eta_g)\}''}{\left[\{T_{33}(\bar{E}_{FBHD},\eta_g)\}'\right]^2}\left[T_{33}(\bar{E}_{FBHD},\eta_g) - \left(n+\frac{1}{2}\right)\hbar\omega_0 + \frac{m_c E_0^2\left[\{T_{33}(\bar{E}_{FBHD},\eta_g)\}'\right]^2}{2B^2}\right]\right. \\ & \left. +1+\frac{m_c E_0^2\{T_{33}(\bar{E}_{FBHD},\eta_g)\}''}{B^2}\right] \end{aligned} \tag{8.17}$$

The Landau energy $(\bar{E}_{n_1})$ can be written as

$$T_{33}(\bar{E}_{n_1},\eta_g) = \left(n+\frac{1}{2}\right)\hbar\omega_0 - \frac{m_c E_0^2\left[\{T_{33}(\bar{E}_{n_1},\eta_g)\}'\right]^2}{2B^2} \tag{8.18}$$

The electron concentration in this case assumes the form

$$n_0 = \left(\frac{2g_v B\sqrt{2m_c}}{3L_x\pi^2\hbar^2 E_0}\right) \text{Real part of} \sum_{n=0}^{n_{\max}} \left[T_{43HD}\left(n, \bar{E}_{FB}, \eta_g\right) + T_{44HD}\left(n, \bar{E}_{FB}, \eta_g\right)\right] \tag{8.19}$$

where,

$$T_{43HD}\left(n, \bar{E}_{FBHD}, \eta_g\right) = \left[T_{33}\left(\bar{E}_{FB}, \eta_g\right) - \left(n+\frac{1}{2}\right)\hbar\omega_0 - \frac{m_c E_0^2}{2B^2}\left[\left\{T_{33}\left(\bar{E}_{FB}, \eta_g\right)\right\}'\right]^2 .\right.$$
$$\left. + |e|E_0 L_x\left[\left\{T_{33}\left(\bar{E}_{FB}, \eta_g\right)\right\}'\right]\right]^{\frac{3}{2}}$$
$$- \left[T_{33}\left(\bar{E}_{FB}, \eta_g\right) - \left(n+\frac{1}{2}\right)\hbar\omega_0 - \frac{m_c E_0^2}{2B^2}\left[\left\{T_{33}\left(\bar{E}_{FB}, \eta_g\right)\right\}'\right]^2\right]^{\frac{3}{2}}$$

$$\frac{1}{\left[\left\{T_{33}\left(\bar{E}_{FB}, \eta_g\right)\right\}'\right]} \quad \text{and}$$

$$T_{44HD}\left(n, \bar{E}_{FBHD}, \eta_g\right) \equiv \sum_{r=1}^{s}\left[L(r) T_{44HD}\left(n, \bar{E}_{FBHD}, \eta_g\right)\right].$$

Therefore the DOS function can be written as

$$N(E_0, B) = \left(\frac{2g_v B\sqrt{2m_c}}{3L_x\pi^2\hbar^2 E_0}\right) \sum_{n=0}^{n_{\max}} \left[T_{43HD}\left(n, E, \eta_g\right)\right]' H\left(E - \bar{E}_{n_1}\right) \tag{8.20}$$

(b) Under the condition$\Delta \gg E_g$, (8.16) assumes the form

$$\gamma_2\left(E, \eta_g\right) = \left(n+\frac{1}{2}\right)\hbar\omega_0 - \frac{E_0}{B}\hbar k_y \gamma_2'\left(E, \eta_g\right) - \frac{m_c E_0^2}{2B^2}\left(\gamma_2'\left(E, \eta_g\right)\right)^2 + \frac{\left[\hbar k_z(E)\right]^2}{2m_c} \tag{8.21}$$

The use of (8.21) leads to the expressions of the EEMs' along z and y directions as

$$m_z^*\left(\bar{E}_{FBHD}, \eta_g, n, E_0\right) = m_c\left[\left\{\gamma_2\left(\bar{E}_{FBHD}, \eta_g\right)\right\}'' + \frac{m_c E_0^2\left\{\gamma_2\left(\bar{E}_{FBHD}, \eta_g\right)\right\}'\left\{\gamma_2\left(\bar{E}_{FBHD}, \eta_g\right)\right\}''}{B^2}\right] \tag{8.22}$$

$$m_y^*(\bar{E}_{FBHD},\eta_g,n,E_0)=\left(\frac{B}{E_0}\right)^2\frac{1}{\left[\{\gamma_2(\bar{E}_{FBHD},\eta_g)\}'\right]}$$
$$\left[\gamma_2(\bar{E}_{FBHD},\eta_g)-\left(n+\frac{1}{2}\right)\hbar\omega_0+\frac{m_cE_0^2\left[\{\gamma_2(\bar{E}_{FBHD},\eta_g)\}'\right]^2}{2B^2}\right]$$
$$\left[\frac{-\{\gamma_2(\bar{E}_{FBHD},\eta_g)\}''}{\left[\{\gamma_2(\bar{E}_{FBHD},\eta_g)\}'\right]^2}\right.$$
$$\left[\gamma_2(\bar{E}_{FBHD},\eta_g)-\left(n+\frac{1}{2}\right)\hbar\omega_0+\frac{m_cE_0^2\left[\{\gamma_2(\bar{E}_{FBHD},\eta_g)\}'\right]^2}{2B^2}\right]$$
$$\left.+1+\frac{m_cE_0^2\{\gamma_2(\bar{E}_{FBHD},\eta_g)\}''}{B^2}\right] \tag{8.23}$$

The Landau energy $(\bar{E}_{n_2})$ can be written as

$$\gamma_2(\bar{E}_{n_2},\eta_g)=\left(n+\frac{1}{2}\right)\hbar\omega_0-\frac{m_cE_0^2}{2B^2}\left(\gamma_2'(\bar{E}_{n_2},\eta_g)\right)^2 \tag{8.24}$$

The expressions for n_0 in this case assume the forms

$$n_0=\frac{2g_vB\sqrt{2m_c}}{3L_x\pi^2\hbar^2E_0}\sum_{n=0}^{n_{\max}}\left[T_{47HD}(n,\bar{E}_{FBHD},\eta_g)+T_{48HD}(n,\bar{E}_{FBHD},\eta_g)\right] \tag{8.25}$$

where

$$T_{47HD}(n,\bar{E}_{FBHD},\eta_g)\equiv\left[\left[\gamma_2(\bar{E}_{FBHD},\eta_g)-\left(n+\frac{1}{2}\right)\hbar\omega_0\right.\right.$$
$$\left.+|e|E_0L_x(\gamma_2'(\bar{E}_{FBHD},\eta_g))-\frac{m_cE_0^2}{2B^2}\left(\gamma_2'(\bar{E}_{FBHD},\eta_g)\right)^2\right]^{\frac{3}{2}}$$
$$\left.-\left[(\gamma_2(\bar{E}_{FBHD},\eta_g))-\left(n+\frac{1}{2}\right)\hbar\omega_0-\frac{m_cE_0^2}{2B^2}\left(\gamma_2'(\bar{E}_{FBHD},\eta_g)\right)^2\right]^{\frac{3}{2}}\right]$$
$$\left[\gamma_2'(\bar{E}_{FBHD},\eta_g)\right]^{-1}$$

and $T_{48HD}\left(n, \bar{E}_{FBHD}, \eta_g\right) \equiv \sum_{r=0}^{s} L(r)\left[T_{47HD}\left(n, \bar{E}_{FBHD}, \eta_g\right)\right]$.

Therefore the DOS function can be written as

$$N(\mathrm{E}_0, B) = \frac{2g_v B\sqrt{2m_c}}{3L_x\pi^2\hbar^2 E_0}\sum_{n=0}^{n_{\max}}\left[T_{47HD}\left(n, E, \eta_g\right)\right]' H(\mathrm{E} - \bar{\mathrm{E}}_{n_2}) \tag{8.26}$$

(c) For $\alpha \to 0$ and we can write,

$$\gamma_3\left(E, \eta_g\right) = \left(n + \frac{1}{2}\right)\hbar\omega_0 - \frac{E_0}{B}\hbar k_y \gamma_3'\left(E, \eta_g\right) - \frac{m_c E_0^2}{2B^2}\left(\gamma_3'\left(E, \eta_g\right)\right)^2 + \frac{\left[\hbar k_z(E)\right]^2}{2m_c} \tag{8.27}$$

The use of (8.26) leads to the expressions of the EEMs' along z and y directions as

$$m_z^*\left(\bar{E}_{FBHD}, \eta_g, n, E_0\right) = m_c\left[\left\{\gamma_3\left(\bar{E}_{FBHD}, \eta_g\right)\right\}'' + \frac{m_c E_0^2\left\{\gamma_3\left(\bar{E}_{FBHD}, \eta_g\right)\right\}'\left\{\gamma_3\left(\bar{E}_{FBHD}, \eta_g\right)\right\}''}{B^2}\right] \tag{8.28}$$

$$\begin{aligned} m_y^*\left(\bar{E}_{FBHD}, \eta_g, n, E_0\right) = & \left(\frac{B}{E_0}\right)^2 \frac{1}{\left[\left\{\gamma_3\left(\bar{E}_{FBHD}, \eta_g\right)\right\}'\right]} \\ & \left[\gamma_3\left(\bar{E}_{FBHD}, \eta_g\right) - \left(n + \frac{1}{2}\right)\hbar\omega_0 + \frac{m_c E_0^2\left[\left\{\gamma_3\left(\bar{E}_{FBHD}, \eta_g\right)\right\}'\right]^2}{2B^2}\right] \\ & \left[\frac{-\left\{\gamma_3\left(\bar{E}_{FBHD}, \eta_g\right)\right\}''}{\left[\left\{\gamma_3\left(\bar{E}_{FBHD}, \eta_g\right)\right\}'\right]^2}.\right. \\ & \left[\gamma_3\left(\bar{E}_{FBHD}, \eta_g\right) - \left(n + \frac{1}{2}\right)\hbar\omega_0 + \frac{m_c E_0^2\left[\left\{\gamma_3\left(\bar{E}_{FBHD}, \eta_g\right)\right\}'\right]^2}{2B^2}\right] \\ & \left. + 1 + \frac{m_c E_0^2\left\{\gamma_3\left(\bar{E}_{FBHD}, \eta_g\right)\right\}''}{B^2}\right] \end{aligned} \tag{8.29}$$

The Landau energy $(\bar{E}_{n_3})$ can be written as

$$\gamma_3\left(\bar{E}_{n_3}, \eta_g\right) = \left(n + \frac{1}{2}\right)\hbar\omega_0 - \frac{m_c E_0^2}{2B^2}\left(\gamma_3'\left(E_{n_3}, \eta_g\right)\right)^2 \tag{8.30}$$

The expressions for n_0 in this case assume the forms

$$n_0 = \frac{2g_v B\sqrt{2m_c}}{3L_x\pi^2\hbar^2 E_0}\sum_{n=0}^{n_{\max}}\left[T_{49HD}\left(n, \bar{E}_{FBHD}, \eta_g\right) + T_{50HD}\left(n, \bar{E}_{FBHD}, \eta_g\right)\right] \tag{8.31}$$

where

$$\begin{aligned} T_{49HD}\left(n, \bar{E}_{FBHD}, \eta_g\right) \equiv \Bigg[\Bigg[&\gamma_3\left(\bar{E}_{FBHD}, \eta_g\right) - \left(n+\frac{1}{2}\right)\hbar\omega_0 \\ &+ |e|E_0L_x\left(\gamma_3'\left(\bar{E}_{FBHD}, \eta_g\right)\right) - \frac{m_c E_0^2}{2B^2}\left(\gamma_3'\left(\bar{E}_{FBHD}, \eta_g\right)\right)^2\Bigg]^{\frac{3}{2}} \\ &- \left[\left(\gamma_3\left(\bar{E}_{FBHD}, \eta_g\right)\right) - \left(n+\frac{1}{2}\right)\hbar\omega_0 - \frac{m_c E_0^2}{2B^2}\left(\gamma_3'\left(\bar{E}_{FBHD}, \eta_g\right)\right)^2\right]^{\frac{3}{2}}\Bigg] \\ &\left[\gamma_3'\left(\bar{E}_{FBHD}, \eta_g\right)\right]^{-1} \end{aligned}$$

and $T_{50HD}\left(n, \bar{E}_{FBHD}, \eta_g\right) \equiv \sum_{r=0}^{s} L(r)\left[T_{49HD}\left(n, \bar{E}_{FBHD}, \eta_g\right)\right]$.

Therefore the DOS function can be written as

$$N(\mathrm{E}_0, \mathrm{B}) = \frac{2g_v B\sqrt{2m_c}}{3L_x\pi^2\hbar^2 E_0}\sum_{n=0}^{n_{\max}}\left[T_{49HD}\left(n, E, \eta_g\right)\right]' H\left(E - \bar{E}_{n_3}\right) \tag{8.32}$$

8.2.3 The DR in HD II–VI Semiconductors Under Cross-Fields Configuration

The electron energy spectrum in HD II–VI semiconductors in the presence of electric field E_0 along x direction and quantizing magnetic field B along z direction can approximately be written as

$$\gamma_3\left(E, \eta_g\right) = \beta_1(n, E_0) - \frac{E_0}{B}\hbar k_y \gamma_3'\left(E, \eta_g\right) - \frac{m_\parallel^* E_0^2}{2B^2}\left(\gamma_3'\left(E, \eta_g\right)\right)^2 + \frac{\left[\hbar k_z(E)\right]^2}{2m_\parallel^*} \tag{8.33}$$

where $\beta_1(n, E_0) \equiv \left[\left(n+\frac{1}{2}\right)\hbar\omega_{02} - \left(\frac{E_0^2 m_\perp^*}{2B^2}\right) + D\left\{\left(n+\frac{1}{2}\right)\hbar\omega_{02} + \left(\frac{E_0^2 m_\perp^*}{2B^2}\right)\right\}^{\frac{1}{2}}\right]$, $\omega_{02} \equiv \frac{|e|B}{m_\perp^*}$.

and

$$D \equiv \pm\frac{\bar{\lambda}_0\sqrt{2m_\perp^*}}{\hbar}$$

The use of (8.31) leads to the expressions of the EEMs' along z and y directions as

$$m_z^*\left(\bar{E}_{FBHD},\eta_g,n,E_0\right) = m_{\|}^*\left[\left\{\gamma_3\left(\bar{E}_{FBHD},\eta_g\right)\right\}''. + \frac{m_{\|}^* E_0^2 \left\{\gamma_3\left(\bar{E}_{FBHD},\eta_g\right)\right\}'\left\{\gamma_3\left(\bar{E}_{FBHD},\eta_g\right)\right\}''}{B^2}\right] \quad (8.34)$$

$$\begin{aligned}m_y^*\left(\bar{E}_{FBHD},\eta_g,n,E_0\right) = & \left(\frac{B}{E_0}\right)^2 \frac{1}{\left[\left\{\gamma_3\left(\bar{E}_{FBHD},\eta_g\right)\right\}'\right]} \\ & \left[\gamma_3\left(\bar{E}_{FBHD},\eta_g\right) - \beta_1(n,E_0) + \frac{m_{\|}^* E_0^2\left[\left\{\gamma_3\left(\bar{E}_{FBHD},\eta_g\right)\right\}'\right]^2}{2B^2}\right] \\ & \left[\frac{-\left\{\gamma_3\left(\bar{E}_{FBHD},\eta_g\right)\right\}''}{\left[\left\{\gamma_3\left(\bar{E}_{FBHD},\eta_g\right)\right\}'\right]^2} \right. \\ & \left[\gamma_3\left(\bar{E}_{FBHD},\eta_g\right) - \beta_1(n,E_0) + \frac{m_{\|}^* E_0^2\left[\left\{\gamma_3\left(\bar{E}_{FBHD},\eta_g\right)\right\}'\right]^2}{2B^2}\right] \\ & \left. + 1 + \frac{m_{\|}^* E_0^2\left\{\gamma_3\left(\bar{E}_{FBHD},\eta_g\right)\right\}''}{B^2}\right]\end{aligned} \quad (8.35)$$

The Landau energy $\left(\bar{E}_{n_4}\right)$ can be written as

$$\gamma_3\left(\bar{E}_{n_4},\eta_g\right) = \beta_1(n,E_0) - \frac{m_{\|}^* E_0^2}{2B^2}\left(\gamma_3'\left(\bar{E}_{n_4},\eta_g\right)\right)^2 \quad (8.36)$$

The expression for n_0 in this case assumes the form

$$n_0 = \frac{2g_v B\sqrt{2m_{\|}^*}}{3L_x\pi^2\hbar^2 E_0}\sum_{n=0}^{n_{\max}}\left[T_{53HD}\left(n,\bar{E}_{FBHD},\eta_g\right) + T_{54HD}\left(n,\bar{E}_{FBHD},\eta_g\right)\right] \quad (8.37)$$

where

$$T_{53HD}(n, \bar{E}_{FBHD}, \eta_g) \equiv \Bigg[\big[\gamma_3(\bar{E}_{FBHD}, \eta_g) - \beta_1(n, E_0) + |e|E_0 L_x(\gamma_3'(\bar{E}_{FBHD}, \eta_g)). - \frac{m_{\parallel}^* E_0^2}{2B^2} (\gamma_3'(\bar{E}_{FBHD}, \eta_g))^2 \Bigg]^{\frac{3}{2}} - \Bigg[(\gamma_3(\bar{E}_{FBHD}, \eta_g)) - \beta_1(n, E_0) - \frac{m_{\parallel}^* E_0^2}{2B^2} (\gamma_3'(\bar{E}_{FBHD}, \eta_g))^2 \Bigg]^{\frac{3}{2}} \big[\gamma_3'(\bar{E}_{FBHD}, \eta_g)\big]^{-1}$$

and $T_{54HD}(n, \bar{E}_{FBHD}, \eta_g) \equiv \sum_{r=0}^{s} L(r)\big[T_{53HD}(n, \bar{E}_{FBHD}, \eta_g)\big]$.

Therefore the DOS function can be written as

$$N(\mathrm{E}_0, \mathrm{B}) = \frac{2g_v B\sqrt{2m_c}}{3L_x \pi^2 \hbar^2 E_0} \sum_{n=0}^{n_{\max}} \big[T_{53HD}(n, E, \eta_g)\big]' H(E - \bar{E}_{n_4}) \tag{8.38}$$

8.2.4 The DR in HD IV–VI Semiconductors Under Cross-Fields Configuration

The (2.143) can be written as

$$\frac{p_s^2}{2M_1^*(E, \eta_g)} + \frac{p_z^2}{2M_3^*(E, \eta_g)} = g(E, \eta_g) \tag{8.39}$$

where

$$M_1^*(E, \eta_g) = \Big[\frac{(\bar{R})^2}{E_g}\{c_1(\alpha_1, E, E_g) - iD_1(\alpha_1, E, E_g)\} + \frac{(\bar{S})^2}{\Delta_c'}\{c_2(\alpha_2, E, E_g) - iD_2(\alpha_2, E, E_g)\} + \frac{(\bar{Q})^2}{\Delta_c'}\{c_3(\alpha_3, E, E_g) - iD_3(\alpha_3, E, E_g)\}\Big]^{-1}$$

$M_3^*(E, \eta_g) = \Big[\frac{2(\bar{A})^2}{E_g}\{c_1(\alpha_1, E, E_g) - iD_1(\alpha_1, E, E_g)\} + \frac{(\bar{S}+\bar{Q})^2}{\Delta_c''}\{c_3(\alpha_3, E, E_g) - iD_3(\alpha_3, E, E_g)\}\Big]^{-1}$ and $g^*(E, \eta_g) = 2\hbar^2\gamma_0(E, \eta_g)$

In the presence of quantizing magnetic field B along z direction and the electric field along x-axis, from above equation one obtains

$$\frac{\hat{p}_x^2}{2M_1^*(E,\eta_g)}+\frac{(\hat{p}_y-|e|B\hat{x})^2}{2M_1^*(E,\eta_g)}+\frac{\hat{p}_z^2}{2M_3^*(E,\eta_g)}=g^*(E,\eta_g)+|e|E_0\hat{x}\rho_1^*(E,\eta_g) \quad (8.40)$$

$$\text{where } \rho_1^*(E,\eta_g)=\frac{\partial}{\partial E}[g^*(E,\eta_g)]$$

Let us define the operator $\hat{\theta}$ as

$$\hat{\theta}=-\hat{p}_y+|e|B\hat{x}-\frac{\rho_1^*(E,\eta_g)E_0[M_1^*(E,\eta_g)]}{B} \quad (8.41)$$

Eliminating $\hat{x}$, between the above two equations, the dispersion relation of the conduction electrons in HD stressed Kane type semiconductors in the presence of cross fields configuraration can be expressed as

$$g^*(E,\eta_g)=(n+\tfrac{1}{2})\hbar\overline{\omega_{i1}}(E,\eta_g)+\frac{\hbar^2k_z^2}{2M_3^*(E,\eta_g)}-\frac{E_0}{B}\rho_1^*(E,\eta_g)\hbar k_y-\frac{E_0^2}{2B^2}[\rho_1^*(E,\eta_g)]^2M_1^*(E,\eta_g)$$
$$\text{where } \overline{\omega_{i1}}(E,\eta_g)=eB[M_1^*(E,\eta_g)]^{-1} \quad (8.42)$$

The use of (8.42) leads to the expressions of the EEM s' along z and y directions as

$$\begin{aligned}m_z^*(\bar{E}_{FBHD},\eta_g,n,E_0)=\text{Real part of }&\Big[[M_3^*(\bar{E}_{FBHD},\eta_g)]'\\&\Big[g^*(\bar{E}_{FBHD},\eta_g)-\left(n+\frac{1}{2}\right)\hbar\overline{\omega_{i1}}(\bar{E}_{FBHD},\eta_g)\\&+\frac{E_0^2}{2B^2}[\rho_1^*(\bar{E}_{FBHD},\eta_g)]^2\\&M_1^*(\bar{E}_{FBHD},\eta_g)]+[M_3^*(\bar{E}_{FBHD},\eta_g)]\\&\Big[[g^*(\bar{E}_{FBHD},\eta_g)]'-\left(n+\frac{1}{2}\right)\hbar[\overline{\omega_{i1}}(\bar{E}_{FBHD},\eta_g)]'\\&+\frac{E_0^2}{2B^2}\Big[[\rho_1^*(\bar{E}_{FBHD},\eta_g)]^2[M_1^*(\bar{E}_{FBHD},\eta_g)]'\Big]\\&+2[M_3^*(\bar{E}_{FBHD},\eta_g)][\rho_1^*(\bar{E}_{FBHD},\eta_g)][\rho_1^*(\bar{E}_{FBHD},\eta_g)]'\Big]\Big]\end{aligned} \quad (8.43)$$

$$m_y^*(\bar{E}_{FBHD},\eta_g,n,E_0) = (B/E_0)^2 \text{Real part of } \left[\rho_1^*(\bar{E}_{FBHD},\eta_g)\right]^{-3}$$
$$\left[g^*(\bar{E}_{FBHD},\eta_g) - \left(n+\frac{1}{2}\right)\hbar\overline{\omega_{i1}}(\bar{E}_{FBHD},\eta_g) + \frac{E_0^2}{2B^2}\left[\rho_1^*(\bar{E}_{FBHD},\eta_g)\right]^2 M_1^*(\bar{E}_{FBHD},\eta_g)\right]\left[\rho_1^*(\bar{E}_{FBHD},\eta_g)\right]$$
$$\left[\left[g^*(\bar{E}_{FBHD},\eta_g)\right]' - \left(n+\frac{1}{2}\right)\hbar\left[\overline{\omega_{i1}}(\bar{E}_{FBHD},\eta_g)\right]' + \frac{E_0^2}{2B^2}\left[\left[\rho_1^*(\bar{E}_{FBHD},\eta_g)\right]^2\left[M_1^*(\bar{E}_{FBHD},\eta_g)\right]'\right]\right.$$
$$\left. - \left[\rho_1^*(\bar{E}_{FBHD},\eta_g)\right]'\left[g^*(\bar{E}_{FBHD},\eta_g) - \left(n+\frac{1}{2}\right)\hbar\overline{\omega_{i1}}(\bar{E}_{FBHD},\eta_g) \frac{E_0^2}{2B^2}\left[\rho_1^*(\bar{E}_{FBHD},\eta_g)\right]^2 M_1^*(\bar{E}_{FBHD},\eta_g)\right]\right] \tag{8.44}$$

The Landau level energy (E_{n9}) in this case can be expressed through the equation

$$g^*(E,\eta_g) = \left(n+\frac{1}{2}\right)\hbar\overline{\omega_{i1}}(\bar{E}_{n_9},\eta_g) - \frac{E_0^2}{2B^2}\left[\rho_1^*(\bar{E}_{n_9},\eta_g)\right]^2 M_1^*(\bar{E}_{n_9},\eta_g) \tag{8.45}$$

The electron concentration can be written as

$$n_0 = \frac{2B}{3L_x\pi^2\hbar^2 E_0}\text{Real part of}\sum_{n=0}^{n_{\max}}\left[T_{4131HD}(n,\bar{E}_{FBHD},\eta_g) + T_{4141HD}(n,\bar{E}_{FBHD},\eta_g)\right] \tag{8.46a}$$

where

$$T_{4131HD}(n,\bar{E}_{FBHD},\eta_g) \equiv \left[\frac{\sqrt{2M_3^*(\bar{E}_{FBHD},\eta_g)}}{\rho_1^*(\bar{E}_{FBHD},\eta_g)}\right]$$
$$\left[\left[T_{51}(n,\bar{E}_{FBHD},\eta_g) + \frac{E_0}{B}\rho_1^*(\bar{E}_{FBHD},\eta_g)\hbar x_{hHD1}(\bar{E}_{FBHD},\eta_g)\rho_1^*(\bar{E}_{FBHD},\eta_g)\right]^{\frac{3}{2}}\right.$$
$$\left. - \left[T_{51}(n,\bar{E}_{FBHD},\eta_g) + \frac{E_0}{B}\rho_1^*(\bar{E}_{FBHD},\eta_g)\hbar x_{hHD1}(\bar{E}_{FBHD},\eta_g)\rho_1^*(\bar{E}_{FBHD},\eta_g)\right]^{\frac{3}{2}}\right],$$

$$T_{51}(n,\bar{E}_{FBHD},\eta_g) = \left[g^*(\bar{E}_{FBHD},\eta_g) - \left(n+\frac{1}{2}\right)\hbar\overline{\omega_{i1}}(\bar{E}_{FBHD},\eta_g).\right.$$
$$\left. + \frac{M_1^*(\bar{E}_{FBHD},\eta_g)E_0^2}{2B^2}\left[\rho_1^*(\bar{E}_{FBHD},\eta_g)\right]^2\right]$$

$$x_{lHD1}(\bar{E}_{FBHD}, \eta_g) = \frac{-M_1^*(\bar{E}_{FBHD}, \eta_g)E_0\left[\rho_1^*(\bar{E}_{FBHD}, \eta_g)\right]}{B},$$

$$x_{hHD1}(\bar{E}_{FBHD}, \eta_g) = \frac{|e|BL_x}{\hbar} + x_{lHD1}(\bar{E}_{FBHD}, \eta_g) \quad \text{and}$$

$$T_{4141HD}(n, \bar{E}_{FBHD}, \eta_g) \equiv \sum_{r=1}^{s} L(r)T_{4131HD}(n, \bar{E}_{FBHD}, \eta_g)$$

The DOS in this case is given by

$$N(E_0, B) = \frac{2B}{3L_x\pi^2\hbar^2E_0}\sum_{n=0}^{n_{\max}}\left[T_{4131HD}(n, E, \eta_g)\right]' H(E - E_{n9}) \tag{8.46b}$$

8.2.5 The DR in HD Stressed Kane Type Semiconductors Under Cross-Fields Configuration

The use of (2.175) can be written as

$$\frac{p_x^2}{2m_1^*(E, \eta_g)} + \frac{p_y^2}{2m_2^*(E, \eta_g)} + \frac{p_z^2}{2m_3^*(E, \eta_g)} = G^*(E, \eta_g) \tag{8.47}$$

where

$$m_1^*(E, \eta_g) = \left[2\hbar^2\left[\gamma_0(E, \eta_g) - I(1)T_{17}\right]\right]^{-1},$$

$$T_{17} = \left[E_g - C_1\varepsilon - (\bar{a}_0 + C_1)\varepsilon + \frac{3}{2}\bar{b}_0\varepsilon_{xx} - \frac{\bar{b}_0}{2}\varepsilon + \left(\frac{\sqrt{3}}{2}\right)\varepsilon_{xy}\bar{d}_0\right],$$

$$m_2^*(E, \eta_g) = \left[2\hbar^2\left[\gamma_0(E, \eta_g) - I(1)T_{27}\right]\right]^{-1},$$

$$T_{27} = \left[E_g - C_1\varepsilon - (\bar{a}_0 + C_1)\varepsilon + \frac{3}{2}\bar{b}_0\varepsilon_{xx} - \frac{\bar{b}_0}{2}\varepsilon - \left(\frac{\sqrt{3}}{2}\right)\varepsilon_{xy}\bar{d}_0\right],$$

$$m_3^*(E, \eta_g) = \left[2\hbar^2\left[\gamma_0(E, \eta_g) - I(1)T_{37}\right]\right]^{-1},$$

$$T_{37} = \left[E_g - C_1\varepsilon - (\bar{a}_0 + C_1)\varepsilon + \frac{3}{2}\bar{b}_0\varepsilon_{xx} - \frac{\bar{b}_0}{2}\varepsilon\right],$$

and the other symbols are written in Chap. 2.

In the presence of quantizing magnetic field B along z direction and the electric field along x-axis, from (8.47) one obtains

$$\frac{\hat{p}_x^2}{2m_1^*(E,\eta_g)} + \frac{(\hat{p} - |e|B\hat{x})^2}{2m_2^*(E,\eta_g)} + \frac{\hat{p}_z^2}{2m_3^*(E,\eta_g)} = G^*(E,\eta_g) + |e|E_0\hat{x}\left[\frac{m_1^*(E,\eta_g)}{m_2^*(E,\eta_g)}\right]^{\frac{1}{2}} \rho^*(E,\eta_g) \tag{8.48}$$

where

$$\rho^*(E,\eta_g) = \frac{\partial}{\partial E}\left[G^*(E,\eta_g)\right]$$

Let us define the operator $\hat{\theta}$ as

$$\hat{\theta} = -\hat{p}_y + |e|B\hat{x} - \frac{\rho^*(E,\eta_g)E_0\left[m_1^*(E,\eta_g)m_2^*(E,\eta_g)\right]^{\frac{1}{2}}}{B} \tag{8.49a}$$

Eliminating $\hat{x}$, between the above two equations, the dispersion relation of the conduction electrons in HD stressed Kane type semiconductors in the presence of cross fields configuration can be expressed as

$$G^*(E,\eta_g) = \left(n+\frac{1}{2}\right)\hbar\bar{\omega}_i(E,\eta_g) + \frac{\hbar^2k_z^2}{2m_3^*(E,\eta_g)} - \frac{E_0}{B}\rho^*(E,\eta_g)\left[\frac{m_1^*(E,\eta_g)}{m_2^*(E,\eta_g)}\right]^{\frac{1}{2}}\hbar k_y - \frac{E_0^2}{2B^2}\left[\rho^*(E,\eta_g)\right]^2 m_1^*(E,\eta_g) \tag{8.49b}$$

$$\text{where } \bar{\omega}_i(E,\eta_g) = eB\left[m_1^*(E,\eta_g)m_2^*(E,\eta_g)\right]^{-\frac{1}{2}}$$

The use of (8.49b) leads to the expressions of the EEMs' along z and y directions as

$$\begin{aligned} m_z^*(\bar{E}_{FBHD},\eta_g,n,E_0) = {} & \Big[\left[m_3^*(\bar{E}_{FBHD},\eta_g)\right]'\Big[G^*(\bar{E}_{FBHD},\eta_g) - \left(n+\frac{1}{2}\right)\hbar\overline{\omega_i}(\bar{E}_{FBHD},\eta_g) \\ & + \frac{E_0^2}{2B^2}\left[\rho^*(\bar{E}_{FBHD},\eta_g)\right]^2 m_1^*(\bar{E}_{FBHD},\eta_g)\Big] \\ & + \left[m_3^*(\bar{E}_{FBHD},\eta_g)\right]\Big[\left[G^*(\bar{E}_{FBHD},\eta_g)\right]' - \left(n+\frac{1}{2}\right)\hbar\left[\overline{\omega_i}(\bar{E}_{FBHD},\eta_g)\right]' \\ & + \frac{E_0^2}{2B^2}\Big[2\left[\rho^*(\bar{E}_{FBHD},\eta_g)\right]\left[\rho^*(\bar{E}_{FBHD},\eta_g)\right]' \\ & \left[m_1^*(\bar{E}_{FBHD},\eta_g)\right]\left[m_1^*(\bar{E}_{FBHD},\eta_g)\right]'\left[\rho^*(\bar{E}_{FBHD},\eta_g)\right]^2\Big]\Big]\Big] \end{aligned} \tag{8.50}$$

$$m_y^*\left(\bar{E}_{FBHD},\eta_g,n,E_0\right)=(B/E_0)^2\left[m_4^*(\bar{E}_{FBHD},\eta_g)\right]^{-3}\left[G^*\left(\bar{E}_{FBHD},\eta_g\right)-\left(n+\frac{1}{2}\right)\hbar\overline{\omega_i}\left(\bar{E}_{FBHD},\eta_g\right)\right.$$
$$\left.+\frac{E_0^2}{2B^2}\left[\rho^*\left(\bar{E}_{FBHD},\eta_g\right)\right]^2 m_1^*\left(\bar{E}_{FBHD},\eta_g\right)\right]$$
$$\left[m_4^*(\bar{E}_{FBHD},\eta_g)\right]\left[\left[G^*\left(\bar{E}_{FBHD},\eta_g\right)\right]'-\left(n+\frac{1}{2}\right)\hbar\left[\overline{\omega_{i1}}\left(\bar{E}_{FBHD},\eta_g\right)\right]'\right.$$
$$+\frac{E_0^2}{2B^2}\left[\left[\rho^*\left(\bar{E}_{FBHD},\eta_g\right)\right]^2\left[m_1^*\left(\bar{E}_{FBHD},\eta_g\right)\right]'\right]$$
$$-\left[m_4^*\left(\bar{E}_{FBHD},\eta_g\right)\right]'\left[G^*\left(\bar{E}_{FBHD},\eta_g\right)\right.$$
$$-\left(n+\frac{1}{2}\right)\hbar\overline{\omega_{i1}}\left(\bar{E}_{FBHD},\eta_g\right)+\frac{E_0^2}{2B^2}\left[\rho^*\left(\bar{E}_{FBHD},\eta_g\right)\right]^2$$
$$\left.\left. m_1^*\left(\bar{E}_{FBHD},\eta_g\right)\right]\right] \tag{8.51}$$

where

$$m_4^*\left(\bar{E}_{FBHD},\eta_g\right)=\left[\left[\rho^*\left(\bar{E}_{FBHD},\eta_g\right)\right]\left[\frac{m_1^*\left(\bar{E}_{FBHD},\eta_g\right)}{m_2^*\left(\bar{E}_{FBHD},\eta_g\right)}\right]^{\frac{1}{2}}\right]$$

The Landau level energy (E_{n_g}) in this case can be expressed through the equation

$$G^*\left(E,\eta_g\right)=\left(n+\frac{1}{2}\right)\hbar\bar{\omega}_i\left(E_{n_g},\eta_g\right)-\frac{E_0^2}{2B^2}\left[\rho^*\left(E,\eta_g\right)\right]^2 m_1^*\left(E_{n_g},\eta_g\right) \tag{8.52}$$

The electron concentration can be written as

$$n_0=\frac{2B}{3L_x\pi^2\hbar^2E_0}\sum_{n=0}^{n_{\max}}\left[T_{413HD}\left(n,\bar{E}_{FBHD},\eta_g\right)+T_{414HD}\left(n,\bar{E}_{FBHD},\eta_g\right)\right] \tag{8.53}$$

where

$$T_{413HD}\left(n,\bar{E}_{FBHD},\eta_g\right)\equiv\left[\frac{\sqrt{2m_3^*\left(\bar{E}_{FBHD},\eta_g\right)}}{\rho^*\left(\bar{E}_{FBHD},\eta_g\right)}\right]$$
$$\left[\left[T_5\left(n,\bar{E}_{FBHD},\eta_g\right)+\frac{E_0}{B}\rho^*\left(\bar{E}_{FBHD},\eta_g\right)\hbar x_{hHD}\left(\bar{E}_{FBHD},\eta_g\right)\rho^*\left(\bar{E}_{FBHD},\eta_g\right)\right]^{\frac{3}{2}}\right.$$
$$\left.-\left[T_5\left(n,\bar{E}_{FBHD},\eta_g\right)+\frac{E_0}{B}\rho^*\left(\bar{E}_{FBHD},\eta_g\right)\hbar x_{hHD1}\left(\bar{E}_{FBHD},\eta_g\right)\rho^*\left(\bar{E}_{FBHD},\eta_g\right)\right]^{\frac{3}{2}}\right],$$

$$T_5(n, \bar{E}_{FBHD}, \eta_g) = \left[G^*(\bar{E}_{FBHD}, \eta_g) - \left(n + \frac{1}{2}\right) \hbar\overline{\omega_i}(\bar{E}_{FBHD}, \eta_g) + \frac{m_1^*(\bar{E}_{FBHD}, \eta_g) E_0^2}{2B^2} [\rho^*(\bar{E}_{FBHD}, \eta_g)]^2 \right]$$

$$x_{lHD}(\bar{E}_{FBHD}, \eta_g) = \frac{-m_1^*(\bar{E}_{FBHD}, \eta_g) E_0 [\rho^*(\bar{E}_{FBHD}, \eta_g)]}{B},$$

$$x_{hHD}(\bar{E}_{FBHD}, \eta_g) = \frac{|e|BL_x}{\hbar} + x_{lHD}(\bar{E}_{FBHD}, \eta_g) \quad \text{and}$$

$$T_{414HD}(n, \bar{E}_{FBHD}, \eta_g) \equiv \sum_{r=1}^{s} L(r) T_{413HD}(n, \bar{E}_{FBHD}, \eta_g)$$

The DOS in this case is given by

$$N(\mathrm{E}_0, \mathrm{B}) = \frac{2B}{3L_x \pi^2 \hbar^2 E_0} \sum_{n=0}^{n_{max}} [T_{413HD}(n, \mathrm{E}, \eta_g)]' \mathrm{H}(\mathrm{E} - \bar{\mathrm{E}}_{n8}) \tag{8.54}$$

8.3 Summary and Conclusion

- The DRs in this case are quantized displaced non-parabolas in the $k_z^2 - k_y$ plane.
- The DRs are concentration dependent which is possible only under band-tailing effect.
- The subband energies are complex for non-removable poles in the finite s-plane.
- The subband energies also depend on concentration due to band tails.
- The DOS functions depend on electric field, concentration and the position of branch-cut changes from denominator to numerator.
- The cross fields introduces energy and quantum number dependent mass anisotropy.
- The effective mass exists in the band gap which is impossible without band tailing effect.

Although in a more rigorous statement the many body effects, the hot electron effects, spin and broadening should be considered along with the self-consistent procedure, the simplified analysis as presented in this chapter exhibits the basic qualitative features of the DR in degenerate HD materials having various band structures in the presence of crossed electric and quantizing magnetic fields with reasonable accuracy. As a collateral understanding, we have studied the EMMs along the directions of the magnetic and the electric fields. The characteristic feature of the cross fields is to introduce index-dependent oscillatory mass anisotropy which is also concentration dependent for any value of electron energy in this case.

8.4 Open Research Problems

R.8.1 Investigate the DR in the presence of an arbitrarily oriented quantizing magnetic and crossed electric fields in HD tetragonal semiconductors by including broadening and the electron spin. Study all the special cases for HD III–V, ternary and quaternary materials in this context.

R.8.2 Investigate the DRs for all models of HD IV–VI, II–VI and stressed Kane type compounds in the presence of an arbitrarily oriented quantizing magnetic and crossed electric fields by including broadening and electron spin.

R.8.3 Investigate the DR for all the materials as stated in R.1.1 of Chap. 1 in the presence of an arbitrarily oriented quantizing magnetic and crossed electric fields by including broadening and electron spin.

References

1. W. Zawadzki, B. Lax, Phys. Rev. Lett. **16**, 1001 (1966)
2. M.J. Harrison, Phys. Rev. A **29**, 2272 (1984); J. Zak, W. Zawadzki, Phys. Rev. **145**, 536 (1966)
3. W. Zawadzki, Q.H. Vrehen, B. Lax, Phys. Rev. **148**, 849 (1966); Q.H. Vrehen, W. Zawadzki, M. Reine, Phys. Rev. **158**, 702 (1967); M.H. Weiler, W. Zawadzki, B. Lax, Phys. Rev. **163**, 733 (1967)
4. W. Zawadzki, J. Kowalski, Phys. Rev. Lett. **27**, 1713 (1971); C. Chu, M.S Chu, T. Ohkawa, Phys. Rev. Lett. **41**, 653 (1978); P. Hu, C.S. Ting, Phys. Rev. **B 36**, 9671 (1987)
5. E.I. Butikov, A.S. Kondratev, A.E. Kuchma, Sov. Phys. Sol. State **13**, 2594 (1972)
6. K.P. Ghatak, J.P. Banerjee, B. Goswami, B. Nag, Non. Optics and Quantum Optics **16**, 241 (1996); M. Mondal, K.P. Ghatak, Phys. Stat. Sol. (b) **133**, K67 (1986)
7. M. Mondal, N. Chattopadhyay, K.P. Ghatak, Jour. Low Temp. Phys. **66**, 131 (1987); K. P. Ghatak, M. Mondal, Zeitschrift fur Physik B **69**, 471 (1988)
8. M. Mondal, K.P. Ghatak, Phys. Lett. A **131A**, 529 (1988); M. Mondal, K.P. Ghatak, in Phys. Stat. Sol. (b) Germany **147**, K179 (1988); B. Mitra, K.P. Ghatak, Phys. Lett. **137A**, 413 (1989)
9. B. Mitra, A. Ghoshal, K.P. Ghatak, Phys. Stat. Soli. (b) **154**, K147 (1989)
10. B. Mitra, K.P. Ghatak, Phys. Stat. Sol. (b) **164,** K13 (1991); K.P. Ghatak, B. Mitra, Int. J. Electron. **70**, 345 (1991); S.M. Adhikari, D. De, J.K. Baruah, S. Chowdhury, K.P. Ghatak Adv. Sci. Focus **1**, 57 (2013); S Pahari, S Bhattacharya, D. De, S.M. Adhikari, A. Niyogi, A. Dey, N. Paitya, S.C. Saha, K.P. Ghatak, P.K. Bose, Physica B: Condens. Matter **405**, 4064 (2010); S. Bhattaacharya, S. Choudhary, S. Ghoshal, S.K. Bishwas, D. De, K.P. Ghatak, J. Comp. Theory Nanosci. **3**, 423 (2006); M. Mondal, K.P. Ghatak, Ann. der Physik. **46**, 502 (1989); K.P. Ghatak, B. De, MRS Proceedings **242**, 377 (1992); K.P. Ghatak, B. Mitra, Int. J. Electron. Theory Exp. **70**, 343 (1991); B. Mitra, K.P. Ghatak, Phys. Lett. A **141**, 81 (1989); B. Mitra, K.P. Ghatak, Phys. Lett. A **137**, 413 (1989); M. Mondal, N. Chattopadhyay, K. P. Ghatak, J. Low Temp. Phys. **73**, 321 (1988); K.P. Ghatak, N. Chattopadhyay, S.N. Biswas, Int. Soc. Optics Photonics, Proc. Soc. Photo-Optic. Instrum. Eng., USA, 203 (1987)
11. K.P. Ghatak, M. Mitra, B. Goswami and B. Nag, Nonlinear Opt. **16**, 167 (1996); K.P. Ghatak, D.K. Basu, B. Nag, J. Phys. Chem. Sol. **58**, 133 (1997)

12. S. Biswas, N. Chattopadhyay, K.P. Ghatak Int. Soc. Optics Photonics, Proc. Soc. Photo-Optic. Instrum. Eng., USA, **836**, 175 (1987); K.P. Ghatak, M. Mondal and S. Bhattacharyya, SPIE **1284**, 113 (1990); K.P. Ghatak, Photonic materials and optical bistability. SPIE **1280**, 53 (1990)
13. K.P. Ghatak, and S.N. Biswas, Growth and characterization of materials for infrared detectors and nonlinear optical switches. SPIE **1484**, 149 (1991)
14. K.P. Ghatak, Fiber optic and laser sensors IX. SPIE **1584**, 435 (1992)

Chapter 9
The DR in Heavily Doped (HD) Non-parabolic Semiconductors Under Magneto-Size Quantization

Failure is a tragedy when seen in close-up but a comedy in long-shot.

9.1 Introduction

In this chapter in Sect. 9.2.1, of the theoretical background, the DR has been investigated in ultra thin films of HD non linear optical semiconductors in the presence of a quantizing magnetic field. The Sect. 9.2.2 contains the results for ultra thin films of HD III–V, ternary and quaternary compounds in accordance with the three and the two band models of Kane. In the same section the DR in accordance with the models of Stillman et al. and Palik et al. have also been studied for the purpose of relative comparison. The Sect. 9.2.3 contains the study of the DR for ultra thin films of HD II–VI semiconductors under magnetic quantization. In Sect. 9.2.4, the DR in ultra thin films of HD IV–VI materials has been discussed in accordance with the models of Cohen, Lax, Dimmock, Bangert and Kastner and Foley and Landenberg respectively. In Sect. 9.2.5, the magneto-DR for the stressed ultra thin films of HD Kane type semiconductors has been investigated. In Sect. 9.2.6, the DR in ultra thin films of HD Te has been studied under magnetic quantization. In Sect. 9.2.7, the magneto-DR in ultra thin films of HD n-GaP has been studied. In Sect. 9.2.8, the DR in ultra thin films of HD $PtSb_2$ has been explored under magnetic quantization. In Sect. 9.2.9, the magneto-DR in ultra thin films of HD Bi_2Te_3 has been studied. In Sect. 9.2.10, the DR in ultra thin films of HD Ge has been studied under magnetic quantization in accordance with the models of Cardona et al. and Wang and Ressler respectively. In Sects. 9.2.11 and 9.2.12, the magneto-DR in ultra thin films of HD n-GaSb and II–V compounds has respectively been studied. In Sect. 9.2.13 the magneto DR in ultra thin films of HD $Pb_{1-x}Ge_xTe$ *has been discussed.* The Sect. 9.3 explores the summary and conclusion and the last Sect. 9.4 contains 19 open research problems for this chapter.

K.P. Ghatak, *Dispersion Relations in Heavily-Doped Nanostructures*,
Springer Tracts in Modern Physics 265, DOI 10.1007/978-3-319-21000-1_9

9.2 Theoretical Background

9.2.1 The DR in HD Nonlinear Optical Semiconductors Under Magneto-Size Quantization

The DR of the conduction electrons in ultra thin films of heavily doped non-linear optical semiconductors in the presence of a quantizing magnetic field B can be written following (7.3) as

$$\frac{\hbar^2\left(\frac{n_z\pi}{d_z}\right)^2}{2m_{\parallel}^*} = U_{1,\pm}(\mathrm{e}_{81},\eta,\eta_g) + U_{2,\pm}(\mathrm{e}_{81},\eta,\eta_g) \tag{9.1}$$

where e_{81} is the totally quantized energy in this case.

The DOS function is given by

$$N_{Bn_z} = \frac{g_v eB}{\pi\hbar}\sum_{n=0}^{n_{\max}}\sum_{n_z=1}^{n_{z\max}}\delta'(E - e_{81}) \tag{9.2}$$

9.2.2 The DR in QWs of HD III–V Semiconductors Under Magneto-Size Quantization

(a) Three Band Model of Kane

In accordance with three band model of Kane, the DR in the present case can be written following (7.11) as

$$\frac{\hbar^2\left(\frac{n_z\pi}{d_z}\right)^2}{2m_{\parallel}^*} = U_{3,\pm}(\mathrm{e}_{82},n,\eta_g) + U_{4,\pm}(\mathrm{e}_{82},n,\eta_g) \tag{9.3}$$

where e_{82} is the totally quantized energy in this case.

The DOS function is given by

$$N_{Bn_z} = \frac{g_v eB}{\pi\hbar}\sum_{n=0}^{n_{\max}}\sum_{n_z=1}^{n_{z\max}}\delta'(E - e_{82}) \tag{9.4}$$

(b) Two Band Model of Kane

The DR in this case is given by

$$\frac{\hbar^2\left(\frac{n_z\pi}{d_z}\right)^2}{2m_c} = \gamma_2(e_{83},\eta_g) - \left(n+\frac{1}{2}\right)\hbar\omega_0 \mp \frac{1}{2}g^*\mu_0 B \tag{9.5}$$

where e_{83} is the totally quantized energy in this case.

The DOS function is given by

$$N_{Bn_z} = \frac{g_v eB}{\pi\hbar}\sum_{n=0}^{n_{\max}}\sum_{n_z=1}^{n_{z\max}}\delta'(E-e_{83}) \tag{9.6}$$

(c) Parabolic Energy Bands

The DR in this case is given by

$$\frac{\hbar^2\left(\frac{n_z\pi}{d_z}\right)^2}{2m_c} = \gamma_3(e_{84},\eta_g) - \left(n+\frac{1}{2}\right)\hbar\omega_0 \mp \frac{1}{2}g^*\mu_0 B \tag{9.7}$$

where e_{84} is the totally quantized energy in this case.

The DOS function is given by

$$N_{Bn_z} = \frac{g_v eB}{\pi\hbar}\sum_{n=0}^{n_{\max}}\sum_{n_z=1}^{n_{z\max}}\delta'(E-e_{84}) \tag{9.8}$$

(d) The Model of Stillman et al.

Following (7.21) the DR in the present case can be written as

$$\left(\frac{n_z\pi}{d_z}\right)^2 = U_7(e_{85},n,\eta_g) \tag{9.9}$$

where e_{85} is the totally quantized energy in this case.

The DOS function is given by

$$N_{Bn_z} = \frac{g_v eB}{\pi\hbar}\sum_{n=0}^{n_{\max}}\sum_{n_z=1}^{n_{z\max}}\delta'(E-e_{85}) \tag{9.10}$$

(e) The Model of Palik et al.

Following (7.25) the DR in the present case can be written as

$$\left(\frac{n_z \pi}{d_z}\right)^2 = A_{35,\pm}(e_{86}, n, \eta_g) \tag{9.11}$$

where e_{86} is the totally quantized energy in this case.

The DOS function is given by

$$N_{Bn_z} = \frac{g_v eB}{2\pi\hbar} \sum_{n=0}^{n_{max}} \sum_{n_z=1}^{n_{zmax}} \delta'(E - e_{86}) \tag{9.12}$$

9.2.3 The DR in HD II–VI Semiconductors Under Magneto-Size Quantization

Following (7.28) the DR in the present case can be written as

$$\left(\frac{n_z \pi}{d_z}\right)^2 = U_{8\pm}(e_{87}, n, \eta_g) \tag{9.13}$$

where e_{87} is the totally quantized energy in this case.

The DOS function is given by

$$N_{Bn_z} = \frac{g_v eB}{\pi\hbar} \sum_{n=0}^{n_{max}} \sum_{n_z=1}^{n_{zmax}} \delta'(E - e_{87}) \tag{9.14}$$

9.2.4 The DR in HD IV–VI Semiconductors Under Magneto Size-Quantization

The electron energy spectrum in IV–VI semiconductors are defined by the models of Cohen, Lax, Dimmock and Bangert and Kastner respectively. The magneto DR in HD IV–VI semiconductors is discussed in accordance with the said model for the purpose of relative comparison.

(a) Cohen Model

The DR in the present case can be written as

$$\left(\frac{n_z\pi}{d_z}\right)^2 = U_{16\pm}(e_{88}, n, \eta_g) \tag{9.15}$$

where e_{88} is the totally quantized energy in this case.

The DOS function is given by

$$N_{Bn_z} = \frac{g_v eB}{\pi\hbar}\sum_{n=0}^{n_{\max}}\sum_{n_z=1}^{n_{z\max}}\delta'(E - e_{88}) \tag{9.16}$$

(b) Lax Model

The DR in the present case can be written as

$$\left(\frac{n_z\pi}{d_z}\right)^2 = U_{17\pm}(e_{89}, n, \eta_g) \tag{9.17}$$

where e_{89} is the totally quantized energy in this case.

The DOS function is given by

$$N_{Bn_z} = \frac{g_v eB}{2\pi\hbar}\sum_{n=0}^{n_{\max}}\sum_{n_z=1}^{n_{z\max}}\delta'(E - e_{89}) \tag{9.18}$$

(c) Dimmock Model

The DR in the present case can be written as

$$\left(\frac{n_z\pi}{d_z}\right)^2 = U_{170}(e_{90}, n, \eta_g) \tag{9.19}$$

where e_{90} is the totally quantized energy in this case.

The DOS function is given by

$$N_{Bn_z} = \frac{g_v eB}{\pi\hbar}\sum_{n=0}^{n_{\max}}\sum_{n_z=1}^{n_{z\max}}\delta'(E - e_{90}) \tag{9.20}$$

(d) Model of Bangert and Kastner

The DR in the present case can be written as

$$\left(\frac{n_z\pi}{d_z}\right)^2 = U_{18}(e_{91}, n, \eta_g)\Big|_{\theta=0} \tag{9.21}$$

where e_{91} is the totally quantized energy in this case.

The DOS function is given by

$$N_{Bn_z} = \frac{g_v eB}{\pi\hbar}\sum_{n=0}^{n_{\max}}\sum_{n_z=1}^{n_{z\max}}\delta'(E - e_{91}) \tag{9.22}$$

(e) Model of Foley and Langenberg

The DR in the present case can be written as

$$\left(\frac{n_z\pi}{d_z}\right)^2 = U_{19}(e_{92}, n, \eta_g) \tag{9.23}$$

where e_{92} is the totally quantized energy in this case.

The DOS function is given by

$$N_{Bn_z} = \frac{g_v eB}{\pi\hbar}\sum_{n=0}^{n_{\max}}\sum_{n_z=1}^{n_{z\max}}\delta'(E - e_{92}) \tag{9.24}$$

9.2.5 The DR in HD Stressed Kane Type Semiconductors Under Magneto-Size Quantization

The DR in the present case can be written as

$$\left(\frac{n_z\pi}{d_z}\right)^2 = U_{41}(e_{93}, n, \eta_g)\Big|_{\theta=0} \tag{9.25}$$

where e_{93} is the totally quantized energy in this case.

The DOS function is given by

$$N_{Bn_z} = \frac{g_v eB}{\pi\hbar}\sum_{n=0}^{n_{\max}}\sum_{n_z=1}^{n_{z\max}}\delta'(E - e_{93}) \tag{9.26}$$

9.2.6 The DR in HD Te Under Magneto Size-Quantization

The DR in the present case can be written as

$$\left(\frac{n_z\pi}{d_z}\right)^2 = U_{42\pm}(e_{94}, n, \eta_g) \tag{9.27}$$

where e_{94} is the totally quantized energy in this case.

The DOS function is given by

$$N_{Bn_z} = \frac{g_v eB}{2\pi\hbar}\sum_{n=0}^{n_{\max}}\sum_{n_z=1}^{n_{z\max}}\delta'(E - e_{94}) \tag{9.28}$$

9.2.7 The DR in HD Gallium Phosphide Under Magneto Size Quantization

The DR in the present case can be written as

$$\left(\frac{n_z\pi}{d_z}\right)^2 = U_{43}(e_{94}, n, \eta_g) \tag{9.29}$$

where e_{94} is the totally quantized energy in this case.

The DOS function is given by

$$N_{Bn_z} = \frac{g_v eB}{\pi\hbar}\sum_{n=0}^{n_{\max}}\sum_{n_z=1}^{n_{z\max}}\delta'(E - e_{94}) \tag{9.30}$$

9.2.8 The DR in HD Platinum Antimonide Under Magneto Size Quantization

The DR in the present case can be written as

$$\left(\frac{n_z\pi}{d_z}\right)^2 = U_{44}(e_{95}, n, \eta_g) \tag{9.31}$$

where e_{95} is the totally quantized energy in this case.

The DOS function is given by

$$N_{Bn_z} = \frac{g_v eB}{\pi\hbar} \sum_{n=0}^{n_{\max}} \sum_{n_z=1}^{n_{z\max}} \delta'(E - e_{95}) \tag{9.32}$$

9.2.9 The DR in HD Bismuth Telluride Under Magneto Size Quantization

The DR in the present case can be written as

$$\left(\frac{n_z \pi}{d_z}\right)^2 = U_{45}(e_{96}, n, \eta_g) \tag{9.33}$$

where e_{96} is the totally quantized energy in this case.

The DOS function is given by

$$N_{Bn_z} = \frac{g_v eB}{\pi\hbar} \sum_{n=0}^{n_{\max}} \sum_{n_z=1}^{n_{z\max}} \delta'(E - e_{96}) \tag{9.34}$$

9.2.10 The DR in HD Germanium Under Magneto Size Quantization

(a) Model of Cardona et al.

The DR in the present case can be written as

$$\left(\frac{n_z \pi}{d_z}\right)^2 = U_{46}(e_{97}, n, \eta_g) \tag{9.35}$$

where e_{97} is the totally quantized energy in this case.

The DOS function is given by

$$N_{Bn_z} = \frac{g_v eB}{\pi\hbar} \sum_{n=0}^{n_{\max}} \sum_{n_z=1}^{n_{z\max}} \delta'(E - e_{97}) \tag{9.36}$$

(b) Model of Wang and Ressler

The DR in the present case can be written as

$$\left(\frac{n_z\pi}{d_z}\right)^2 = U_{47}(e_{98}, n, \eta_g) \tag{9.37}$$

where e_{98} is the totally quantized energy in this case.

The DOS function is given by

$$N_{Bn_z} = \frac{g_v eB}{\pi\hbar}\sum_{n=0}^{n_{\max}}\sum_{n_z=1}^{n_{z\max}}\delta'(E - e_{98}) \tag{9.38}$$

9.2.11 *The DR in HD Gallium Antimonide Under Magneto Size Quantization*

The DR in the present case can be written as

$$\left(\frac{n_z\pi}{d_z}\right)^2 = U_{48}(e_{99}, n, \eta_g) \tag{9.39}$$

where e_{99} is the totally quantized energy in this case.

The DOS function is given by

$$N_{Bn_z} = \frac{g_v eB}{\pi\hbar}\sum_{n=0}^{n_{\max}}\sum_{n_z=1}^{n_{z\max}}\delta'(E - e_{99}) \tag{9.40}$$

9.2.12 *The DR in HD II–V Materials Under Magneto Size Quantization*

The DR in the present case can be written as

$$\left(\frac{n_z\pi}{d_z}\right)^2 = U_{49\pm}(e_{100}, n, \eta_g) \tag{9.41}$$

where e_{100} is the totally quantized energy in this case.

The DOS function is given by

$$N_{Bn_z} = \frac{g_v eB}{2\pi\hbar} \sum_{n=0}^{n_{\max}} \sum_{n_z=1}^{n_{z\max}} \delta'(E - e_{100}) \tag{9.42}$$

9.2.13 *The DR in HD Lead Germanium Telluride Under Magneto Size Quantization*

The DR in the present case can be written as

$$\left(\frac{n_z\pi}{d_z}\right)^2 = U_{50\pm}(e_{101}, n, \eta_g) \tag{9.43}$$

where e_{101} is the totally quantized energy in this case.

The DOS function is given by

$$N_{Bn_z} = \frac{g_v eB}{2\pi\hbar} \sum_{n=0}^{n_{\max}} \sum_{n_z=1}^{n_{z\max}} \delta'(E - e_{101}) \tag{9.44}$$

9.3 Summary and Conclusion

The DRs in magneto-size quantized of HD materials exhibit the fact that the total energy is quantized since the corresponding wave vector space is totally quantized by quantizing magnetic field and size quantization along z-direction.

- The DOS functions for all the materials in this case are series of non-uniformly distributed Dirac's Delta functions at specified quantized points in the respective energy axis. The spacing between the consecutive Delta functions are functions of energy band constants and quantization of the wave vector space of a particular material. The DOS function needs two summations namely one summation over the Landau quantum number and the other one is due to size quantization.
- It may be noted that the energy levels in magneto-size quantized HD materials lead to the discrete energy levels, somewhat like atomic energy levels, which produce very large changes. This follows from the inherent nature of the quantum confinement of the carrier gas dealt with here. In QBs, there remain no free carrier states in between any two allowed sets of size-quantized levels unlike that found for QWs, NWs and QBs where the quantum confinements are 1D, 2D and 3D respectively. Consequently, the crossing of the Fermi level by

the size-quantized levels in this case would have much greater impact on the redistribution of the carriers among the allowed levels, as compared to that found for QWs, NWs and QBs respectively.

- It is the band structure which changes in a fundamental way and consequently all the physical properties of all the electronic materials changes radically leading to new physical concepts.

9.4 Open Research Problems

(R.9.1) Investigate the DR in the presence of an arbitrarily oriented quantizing magnetic field for all the QW HD materials as given in problems in (R.1.1) of Chap. 1.

(R.9.2) Investigate the DR in the presence of an arbitrarily oriented quantizing magnetic field in QW HD nonlinear optical semiconductors by including broadening and the electron spin. Study all the special cases for QW HD III–V, ternary and quaternary materials in this context.

(R.9.3) Investigate the DRs for QW HD IV–VI, II–VI and stressed Kane type compounds in the presence of an arbitrarily oriented quantizing magnetic field by including broadening and electron spin.

(R.9.4) Investigate the DR for all the QW HD materials as stated in (R.1.1) of Chap. 1 in the presence of an arbitrarily oriented quantizing magnetic field by including broadening and electron spin.

(R.9.5) Investigate the DR in the presence of an arbitrarily oriented quantizing magnetic field and crossed electric fields in QW HD nonlinear optical semiconductors by including broadening and the electron spin. Study all the special cases for HD III–V, ternary and quaternary materials in this context.

(R.9.6) Investigate the DRs for QW HD IV–VI, II–VI and stressed Kane type compounds in the presence of an arbitrarily oriented quantizing magnetic field and crossed electric field by including broadening and electron spin.

(R.9.7) Investigate the DR for all the QW HD materials as stated in (R.1.1) of Chap. 1 in the presence of an arbitrarily oriented quantizing magnetic field and crossed electric field by including broadening and electron spin.

(R.9.8) Investigate the 2D DR in QW HD nonlinear optical, III–V, II–VI, IV–VI and stressed Kane type semiconductors.

(R.9.9) Investigate the 2D DR for QW HD for all the materials as considered in problems (R.1.1).

(R.9.10) Investigate the 2D DR in the presence of an arbitrarily oriented non-quantizing magnetic field for the QWs HD nonlinear optical semiconductors by including the electron spin. Study all the special cases for III–V, ternary and quaternary materials in this context.

(R.9.11) Investigate the DRs in QWs HD IV–VI, II–VI and stressed Kane type compounds in the presence of an arbitrarily oriented non-quantizing magnetic field by including the electron spin.

(R.9.12) Investigate the 2D DR for QW HD for all the materials as stated in (R.1.1) of Chap. 1 in the presence of an arbitrarily oriented magnetic field by including electron spin and broadening.

(R.9.13) Investigate the 2D DR for all the problems from RA4.32 to RA4.35 in the presence of an additional arbitrarily oriented non-quantizing electric field.

(R.9.14) Investigate the 2D DR for all the problems from RA4.32 to RA4.35 in the presence of arbitrarily oriented crossed electric and magnetic fields.

(R.9.15) Investigate all the problems from RA4.1 to RA4.37, in the presence of arbitrarily oriented light waves and magnetic quantization.

(R.9.16) Investigate all the problems from RA4.1 up to RA4.37 in the presence of exponential, Kane, Halperin and Lax and Bonch-Bruevich band tails [1–14].

(R.9.17) Investigate all the problems of this chapter by removing all the mathematical approximations and establishing the uniqueness conditions in each case.

(R.9.18) (a) Investigate the DR in all the QW HD semiconductors as considered in this chapter in the presence of defects and magnetic quantization.
(b) Investigate the DR as defined in (R.1.1) in the presence of an arbitrarily oriented quantizing magnetic field including broadening and the electron spin (applicable under magnetic quantization) for all the QW HD semiconductors whose unperturbed carrier energy spectra are defined in Chap. 1.

(R.9.19) Investigate all the problems of this section by removing all the mathematical approximations and establishing the respective appropriate uniqueness conditions.

References

1. N. Miura, *Physics of Semiconductors in High Magnetic Fields, Series on Semiconductor Science and Technology* (Oxford University Press, USA, 2007); K.H.J. Buschow, F.R. de Boer, *Physics of Magnetism and Magnetic Materials* (Springer, New York, 2003); D. Sellmyer (Ed.), R. Skomski (Ed.), *Advanced Magnetic Nanostructures* (Springer, New York, 2005); J.A.C. Bland (Ed.), B. Heinrich (Ed.), *Ultrathin Magnetic Structures III: Fundamentals of Nanomagnetism (Pt. 3)* (Springer, Germany, 2005); B.K. Ridley, *Quantum Processes in Semiconductors*, 4th edn. (Oxford Publications, Oxford, 1999); J.H. Davies, *Physics of Low Dimensional Semiconductors* (Cambridge University Press, UK, 1998); S. Blundell, *Magnetism in Condensed Matter, Oxford Master Series in Condensed Matter Physics* (Oxford University Press, USA, 2001); C. Weisbuch, B. Vinter, *Quantum Semiconductor Structures: Fundamentals and Applications* (Academic Publishers, USA, 1991); D. Ferry, *Semiconductor Transport* (CRC, USA, 2000); M. Reed (Ed.), *Semiconductors and*

Semimetals: Nanostructured Systems (Academic Press, USA, 1992); T. Dittrich, *Quantum Transport and Dissipation* (Wiley-VCH Verlag GmbH, Germany, 1998); A.Y. Shik, *Quantum Wells: Physics & Electronics of Two Dimensional Systems* (World Scientific, USA, 1997)
2. K.P. Ghatak, M. Mondal, Zietschrift fur Naturforschung A **41a**, 881 (1986); K.P. Ghatak, M. Mondal, J. Appl. Phys. **62**, 922 (1987); K.P. Ghatak, S.N. Biswas, Phys. Stat. Sol. (b) **140**, K107 (1987); K.P. Ghatak, M. Mondal, J. Mag. Mag. Mat. **74**, 203 (1988); K.P. Ghatak, M. Mondal, Phys. Stat. Sol. (b) **139**, 195 (1987); K.P. Ghatak, M. Mondal, Phys. Stat. Sol. (b) **148**, 645 (1988); K.P. Ghatak, B. Mitra, A. Ghoshal, Phys. Stat. Sol. (b) **154**, K121 (1989); K.P. Ghatak, S.N. Biswas, J. Low Temp. Phys. **78**, 219 (1990); K.P. Ghatak, M. Mondal, Phys. Stat. Sol. (b) **160**, 673 (1990); K.P. Ghatak, B. Mitra, Phys. Lett. A **156**, 233 (1991); K. P. Ghatak, A. Ghoshal, B. Mitra, Nouvo Cimento D **13D**, 867 (1991); K.P. Ghatak, M. Mondal, Phys. Stat. Sol. (b) **148**, 645 (1989); K.P. Ghatak, B. Mitra, Int. J. Elect. **70**, 345 (1991); K.P. Ghatak, S.N. Biswas, J. Appl. Phys. **70**, 299 (1991); K.P. Ghatak, A. Ghoshal, Phys. Stat. Sol. (b) **170**, K27 (1992); K.P. Ghatak, Nouvo Cimento D **13D**, 1321 (1992); K. P. Ghatak, B. Mitra, Int. J. Elect. **72**, 541 (1992); K.P. Ghatak, S.N. Biswas, Nonlinear Opt. **4**, 347 (1993); K.P. Ghatak, M. Mondal, Phys. Stat. Sol. (b) **175**, 113 (1993); K.P. Ghatak, S.N. Biswas, Nonlinear Opt. **4**, 39 (1993); K.P. Ghatak, B. Mitra, Nouvo Cimento **15D**, 97 (1993); K.P. Ghatak, S.N. Biswas, Nanostruct. Mater. **2**, 91 (1993); K.P. Ghatak, M. Mondal, Phys. Stat. Sol. (b) **185**, K5 (1994); K.P. Ghatak, B. Goswami, M. Mitra, B. Nag, Nonlinear Opt. **16**, 9 (1996); K.P. Ghatak, M. Mitra, B. Goswami, B. Nag, Nonlinear Opt. **16**, 167 (1996); K. P. Ghatak, B. Nag, Nanostruct. Mater. **10**, 923 (1998)
3. D. Roy Choudhury, A.K. Choudhury, K.P. Ghatak, A.N. Chakravarti, Phys. Stat. Sol. (b) **98**, K141 (1980); A.N. Chakravarti, K.P. Ghatak, A. Dhar, S. Ghosh, Phys. Stat. Sol. (b) **105**, K55 (1981); A.N. Chakravarti, A.K. Choudhury, K.P. Ghatak, Phys. Stat. Sol. (a) **63**, K97 (1981); A.N. Chakravarti, A.K. Choudhury, K.P. Ghatak, S. Ghosh, A. Dhar, Appl. Phys. **25**, 105 (1981); A.N. Chakravarti, K.P. Ghatak, G.B. Rao, K.K. Ghosh, Phys. Stat. Sol. (b) **112**, 75 (1982); A.N. Chakravarti, K.P. Ghatak, K.K. Ghosh, H.M. Mukherjee, Phys. Stat. Sol. (b) **116**, 17 (1983); M. Mondal, K.P. Ghatak, Phys. Stat. Sol. (b) **133**, K143 (1984); M. Mondal, K.P. Ghatak, Phys. Stat. Sol. (b) **126**, K47 (1984); M. Mondal, K.P. Ghatak, Phys. Stat. Sol. (b) **126**, K41 (1984); M. Mondal, K.P. Ghatak, Phys. Stat. Sol. (b) **129**, K745 (1985); M. Mondal, K.P. Ghatak, Phys. Scr. **31**, 615 (1985); M. Mondal, K.P. Ghatak, Phys. Stat. Sol. (b) **135**, 239 (1986); M. Mondal, K.P. Ghatak, Phys. Stat. Sol. (b) **93**, 377 (1986); M. Mondal, K.P. Ghatak, Phys. Stat. Sol. (b) **135**, K21 (1986); M. Mondal, S. Bhattacharyya, K. P. Ghatak, Appl. Phys. A **42A**, 331 (1987); S.N. Biswas, N. Chattopadhyay,K.P. Ghatak, Phys. Stat. Sol. (b) **141**, K47 (1987); B. Mitra, K.P. Ghatak, Phys. Stat. Sol. (b) **149**, K117 (1988); B. Mitra, A. Ghoshal, K.P. Ghatak, Phys. Stat. Sol. (b) **150**, K67 (1988); M. Mondal, K.P. Ghatak, Phys. Stat. Sol. (b) **147**, K179 (1988); M. Mondal, K.P. Ghatak, Phys. Stat. Sol. (b) **146**, K97 (1988); B. Mitra, A. Ghoshal, K.P. Ghatak, Phys. Stat. Sol. (b) **153**, K209 (1989); B. Mitra, K.P. Ghatak, Phys. Lett. **142A**, 401 (1989); B. Mitra, A. Ghoshal, K. P. Ghatak, Phys. Stat. Sol. (b) **154**, K147 (1989); B. Mitra, K.P. Ghatak, Sol. State Elect. **32**, 515 (1989); B. Mitra, A. Ghoshal, K.P. Ghatak, Phys. Stat. Sol. (b) **155**, K23 (1989); B. Mitra, K.P. Ghatak, Phys. Lett. **135A**, 397 (1989); B. Mitra, K.P. Ghatak, Phys. Lett. A **146A**, 357 (1990); B. Mitra, K.P. Ghatak, Phys. Stat. Sol. (b) **164**, K13 (1991); S.N. Biswas, K.P. Ghatak, Int. J. Elect. **70**, 125 (1991)
4. P.R. Wallace, Phys. Stat. Sol. (b) **92**, 49 (1979)
5. B.R. Nag, *Electron Transport in Compound Semiconductors, Springer Series in Solid-State Sciences*, vol. 11 (Springer, Germany, 1980)
6. K.P. Ghatak, S. Bhattacharya, D. De, *Einstein Relation in Compound Semiconductors and Their Nanostructures, Springer Series in Materials Science*, vol. 116 (Springer, Germany, 2009)
7. C.C. Wu, C.J. Lin, J. Low Temp. Phys. **57**, 469 (1984); M.H. Chen, C.C. Wu, C.J. Lin, J. Low Temp. Phys. **55**, 127 (1984)
8. E. Bangert, P. Kastner, Phys. Stat. Sol (b) **61**, 503 (1974)
9. G.M.T. Foley, P.N. Langenberg, Phys. Rev. B **15B**, 4850 (1977)

10. M. Singh, P.R. Wallace, S.D. Jog, E. Arushanov, J. Phys. Chem. Solids **45**, 409 (1984); Y. Yamada, Phys. Soc. Japan **35**,1600 (1973),**37**, 606 (1974)
11. L.A. Vassilev, Phys. Stat. Sol. (b) **121**, 203 (1984)
12. R.W. Cunningham, Phys. Rev. **167**, 761 (1968)
13. A.I. Yekimov, A.A. Onushchenko, A.G. Plyukhin, Al, L. Efros, J. Expt. Theor. Phys. **88**, 1490 (1985)
14. B.J. Roman, A.W. Ewald, Phys. Rev. B **5**, 3914 (1972)

Chapter 10
The DR in Heavily Doped Ultra-thin Films (HDUFs) Under Cross-Fields Configuration

I will never find a rainbow, if I am looking down.

10.1 Introduction

In this chapter in Sect. 10.2.1 of theoretical background, the DR in ultra thin films of HD nonlinear optical materials in the presence of crossed electric and quantizing magnetic fields has been investigated by formulating the electron dispersion relation. The Sect. 10.2.2 reflects the study of the DR in ultra thin films of HD III–V, ternary and quaternary compounds as a special case of Sect. 10.2.1. The Sect. 10.2.3 contains the study of the DR for the ultra thin films of HD II–VI semiconductors in the present case. In Sect. 10.2.4, the DR under cross field configuration in ultra thin films of HD IV–VI semiconductors has been investigated in accordance with the models of the Cohen, the Lax non-parabolic ellipsoidal and the parabolic ellipsoidal respectively. In the Sect. 10.2.5, the DR for the ultra thin films of HD stressed Kane type semiconductors has been investigated. The Sect. 10.3 contains summary and conclusion. The last Sect. 10.4 eleven open research problems.

10.2 Theoretical Background

10.2.1 The DR in Heavily Doped Ultra-thin Films (HDUFs) of Nonlinear Optical Semiconductors Under Cross-Fields Configuration

The DR of the conduction electrons in HD ultrathin films of nonlinear optical material in the presence of cross-fields configuration can be written as

K.P. Ghatak, *Dispersion Relations in Heavily-Doped Nanostructures*,
Springer Tracts in Modern Physics 265, DOI 10.1007/978-3-319-21000-1_10

$$T_{22}(E,\eta_g) = \left[\left(\left(n+\frac{1}{2}\right)\hbar\omega_{01}\right) + \left(\frac{[\hbar^2]}{2a(E,n_g)}\right)\left(\frac{n\pi_z}{d_z}\right)^2 - \left(\frac{E_o\hbar k_y\rho(E,\eta_g)}{B}\right) - \left(\frac{M_\perp\rho^2(E,\eta_g)E_o^2}{2B^2}\right)\right] \tag{10.1}$$

The use of (10.1) leads to the expression of EEM along y direction as

$$m_y^*(e_{fA1},\eta_g,n,E_0,n_z) = \text{Real part of } (B/E_0)^2 T_{49}(e_{fA1},\eta_g,n_z)[T_{49}(e_{fA1},\eta_g,n_z)]' \tag{10.2a}$$

where e_{fA1} is the Fermi energy in this case and

$$T_{49}(e_{fA1},\eta_g,n_z) = [T_{22}(e_{fA1},\eta_g)] - \left[\left(\left(n+\frac{1}{2}\right)\hbar\omega_{01}\right) + \left(\frac{[\hbar]^2}{2a(e_{fA1},\eta_g)}\right)\left(\frac{\pi n_z}{d_z}\right)^2 - \left(\frac{M_\perp\rho^2(e_{fA1},\eta_g)E_o^2}{2B^2}\right)\right][\rho(e_{fA1},\eta_g)]^{-1}$$

The sub band energy $E_{9,1}$ in this case assumes the from

$$T_{22}(E_{9,1},\eta_g) = \left[\left(\left(n+\frac{1}{2}\right)\hbar\omega_{01}\right) + \left(\frac{[\hbar^2]}{2a(E_{9,1},n_g)}\right)\left(\frac{\pi n_z}{d_z}\right)^2 - \left(\frac{M_\perp\rho^2(E_{9,1},\eta_g)E_o^2}{2B^2}\right)\right] \tag{10.2b}$$

10.2.2 The DR in Heavily Doped Ultra-thin Films (HDUFs) of Type III–V Semiconductors Under Cross-Fields Configuration

(a) Under the conditions $\delta = 0$, $\Delta_\parallel = \Delta_\perp = \Delta$ and $m_\parallel^* = m_\perp^* = m_c$, (9.1) assumes the form

$$T_{33}(E,\eta_g) = \left(n+\frac{1}{2}\right)\hbar\omega_0 + \frac{\left[\hbar\frac{n\pi_z}{d_z}\right]^2}{2m_c} - \frac{E_0}{B}\hbar k_y\{T_{33}(E,\eta_g)\}' - \frac{m_c E_0^2[\{T_{33}(E,\eta_g)\}']^2}{2B^2} \tag{10.3}$$

The use of (10.3) leads to the expression of EEM along y direction as

$$m_y^*(e_{fA2}, \eta_g, n, E_0, n_z) = \text{Real part of } (B/E_0)^2 T_{50}(e_{fA2}, \eta_g, n_z)[T_{50}(e_{fA2}, \eta_g, n_z)]' \tag{10.4a}$$

where e_{fA2} is the Fermi energy in this case and

$$T_{50}(e_{fA2}, \eta_g, n_z) = \left[T_{33}(e_{fA2}, \eta_g) - \left(n + \frac{1}{2}\right)\hbar\omega_0 . - \frac{\left[\hbar\frac{\pi n_z}{d_z}\right]^2}{2m_c} + \frac{m_c E_0^2 [\{T_{33}(e_{fA2}, \eta_g)\}']^2}{2B^2}\right] [\{T_{33}(e_{fA2}, \eta_g)\}']^{-1}$$

The sub band energy $E_{9,2}$ in this case assumes the from

$$T_{33}(E_{9,2}, \eta_g) = \left(n + \frac{1}{2}\right)\hbar\omega_0 + \frac{\left[\hbar\frac{\pi n_z}{d_z}\right]^2}{2m_c} - \frac{m_c E_0^2 [\{T_{33}(E_{9,2}, \eta_g)\}']^2}{2B^2} \tag{10.4b}$$

(b) HD two band model of Kane

Under the condition $\Delta \gg E_g$, (10.3) assumes the form

$$\gamma_2(E, \eta_g) = \left(n + \frac{1}{2}\right)\hbar\omega_0 - \frac{E_0}{B}\hbar k_y \gamma_2'(E, \eta_g) - \frac{m_c E_0^2}{2B^2}\left(\gamma_2'(E, \eta_g)\right)^2 + \frac{\left[\hbar\frac{n_z \pi}{d_z}\right]^2}{2m_c} \tag{10.5}$$

The use of (10.5) leads to the expression of EEM along y direction as

$$m_y^*(e_{fA3}, \eta_g, n, E_0, n_z) = (B/E_0)^2 T_{51}(e_{fA3}, \eta_g, n_z)[T_{51}(e_{fA3}, \eta_g, n_z)]' \tag{10.6a}$$

where e_{fA3} is the Fermi energy in this case and

$$T_{51}(e_{fA3}, \eta_g, n_z) = \left[\gamma_2(e_{fA3}, \eta_g) - \left(n + \frac{1}{2}\right)\hbar\omega_0 . - \frac{\left[\hbar\frac{\pi n_z}{d_z}\right]^2}{2m_c} + \frac{m_c E_0^2 [\{\gamma_2(e_{fA3}, \eta_g)\}']^2}{2B^2}\right] [\{\gamma_2(e_{fA3}, \eta_g)\}']^{-1}$$

The sub band energy $E_{9,3}$ in this case assumes the from

$$\gamma_2(E_{9,3},\eta_g) = \left(n + \frac{1}{2}\right)\hbar\omega_0 - \frac{m_c E_0^2}{2B^2}\left(\gamma_2'(E_{9,3},\eta_g)\right)^2 + \frac{\left[\hbar\frac{n_z\pi}{d_z}\right]^2}{2m_c} \tag{10.6b}$$

(c) HD Parabolic energy bands

The DR and for this model under this condition $\alpha \to 0$ can be written as

$$\gamma_3(E,\eta_g) = \left(n + \frac{1}{2}\right)\hbar\omega_0 - \frac{E_0}{B}\hbar k_y\gamma_3'(E,\eta_g) - \frac{m_c E_0^2}{2B^2}\left(\gamma_3'(E,\eta_g)\right)^2 + \frac{\left[\hbar\frac{n_z\pi}{d_z}\right]^2}{2m_c} \tag{10.7}$$

The use of (10.7) leads to the expression of EEM along y direction as

$$m_y^*(e_{fA4},\eta_g,n,E_0,n_z) = (B/E_0)^2 T_{52}(e_{fA4},\eta_g,n_z)[T_{52}(e_{fA4},\eta_g,n_z)]' \tag{10.8a}$$

where e_{fA4} is the Fermi energy in this case and

$$T_{52}(e_{fA4},\eta_g,n_z) = \left[\gamma_3(e_{fA4},\eta_g) - \left(n + \frac{1}{2}\right)\hbar\omega_0. - \frac{\left[\hbar\frac{\pi n_z}{d_z}\right]^2}{2m_c} + \frac{m_c E_0^2[\{\gamma_3(e_{fA4},\eta_g)\}']^2}{2B^2}\right][\{\gamma_3(e_{fA4},\eta_g)\}']^{-1}$$

The sub band energy $E_{9,4}$ in this case assumes the from

$$\gamma_2(E_{9,4},\eta_g) = \left(n + \frac{1}{2}\right)\hbar\omega_0 - \frac{m_c E_0^2}{2B^2}\left(\gamma_3'(E_{9,4},\eta_g)\right)^2 + \frac{\left[\hbar\frac{n_z\pi}{d_z}\right]^2}{2m_c} \tag{10.8b}$$

10.2.3 The DR in Heavily Doped Ultra-thin Films (HDUFs) of II–VI Semiconductors Under Cross-Fields Configuration

The DR in this case in ultrathin films of HD II–VI semiconductors can be written as

$$\gamma_3(E,\eta_g) = \beta_1(n,E_0) - \frac{E_0}{B}\hbar k_y\gamma_3'(E,\eta_g) - \frac{m_\parallel^* E_0^2}{2B^2}\left(\gamma_3'(E,\eta_g)\right)^2 + \frac{\left[\hbar\left(\frac{n_z\pi}{d_z}\right)\right]^2}{2m_\parallel^*} \tag{10.9}$$

The use of (10.9) leads to the expression of EEM along y direction as

$$m_y^*(e_{fA5}, \eta_g, n, E_0, n_z) = (B/E_0)^2 T_{53}(e_{fA5}, \eta_g, n_z)[T_{53}(e_{fA5}, \eta_g, n_z)]' \quad (10.10a)$$

where e_{fA5} is the Fermi energy in this case and

$$T_{53}(e_{fA5}, \eta_g, n_z) = \left[\gamma_3(e_{fA5}, \eta_g) - \beta_1(n, E_0). - \frac{\left[\hbar \frac{n\pi_z}{d_z}\right]^2}{2m_{\parallel}^*} + \frac{m_{\parallel}^* E_0^2 [\{\gamma_3(e_{fA5}, \eta_g)\}']^2}{2B^2}\right] [\{\gamma_3(e_{fA5}, \eta_g)\}']^{-1}$$

The sub band energy $E_{9,5}$ in this case assumes the from

$$\gamma_3(E_{9,5}, \eta_g) = \beta_1(n, E_0) - \frac{m_{\parallel}^* E_0^2}{2B^2} (\gamma_3'(E_{9,5}, \eta_g))^2 + \frac{\left[\hbar \left(\frac{n_z \pi}{d_z}\right)\right]^2}{2m_{\parallel}^*} \quad (10.10b)$$

10.2.4 The DR in Heavily Doped Ultra-thin Films (HDUFs) of IV–VI Semiconductors Under Cross-Fields Configuration

The DR in this case is given by

$$g^*(E, \eta_g) = \left(n + \frac{1}{2}\right) \hbar \overline{\omega_{i1}}(E, \eta_g) + \frac{\hbar^2 \left(\frac{n_z \pi}{d_z}\right)^2}{2M_3^*(E, \eta_g)} - \frac{E_0}{B} \rho_1^*(E, \eta_g) \hbar k_y - \frac{E_0^2}{2B^2} [\rho_1^*(E, \eta_g)]^2 M_1^*(E, \eta_g) \quad (10.11)$$

The use of (10.11) leads to the expression of EEM along y direction as

$$m_y^*(e_{fA6}, \eta_g, n, E_0, n_z) = \text{Real part of } (B/E_0)^2 T_{54}(e_{fA6}, \eta_g, n_z)[T_{54}(e_{fA6}, \eta_g, n_z)]' \quad (10.12a)$$

where e_{fA6} is the Fermi energy in this case and

$$T_{49}(e_{fA6},\eta_g,n_z) = \left[g^*(e_{fA6},\eta_g) - \left(n+\frac{1}{2}\right)\hbar\overline{\omega_{i1}}(e_{fA6},\eta_g). - \frac{\hbar^2\left(\frac{n_z\pi}{d_z}\right)^2}{2M_3^*(e_{fA6},\eta_g)} + \frac{E_0^2}{2B^2}[\rho_1^*(e_{fA6},\eta_g)]^2 M_1^*(e_{fA6},\eta_g)\right][\rho_1^*(e_{fA6},\eta_g)]^{-1}$$

The sub band energy $E_{9,6}$ in this case assumes the from

$$g^*(E_{9,6},\eta_g) = \left(n+\frac{1}{2}\right)\hbar\overline{\omega_{i1}}(E_{9,6},\eta_g) + \frac{\hbar^2\left(\frac{n_z\pi}{d_z}\right)^2}{2M_3^*(E_{9,6},\eta_g)} - \frac{E_0^2}{2B^2}[\rho_1^*(E_{9,6},\eta_g)]^2 M_1^*(E_{9,6},\eta_g) \quad (10.12b)$$

10.2.5 The DR in Heavily Doped Ultra-thin Films (HDUFs) of Stressed Semiconductors Under Cross-Fields Configuration

The DR in this case assumes the form

$$G^*(E,\eta_g) = \left(n+\frac{1}{2}\right)\hbar\overline{\omega_i}(E,\eta_g) + \frac{\hbar^2\left(\frac{n_z\pi}{d_z}\right)^2}{2m_3^*(E,\eta_g)} - \frac{E_0}{B}\rho^*(E,\eta_g)\left[\frac{m_1^*(E,\eta_g)}{m_2^*(E,\eta_g)}\right]^{\frac{1}{2}}\hbar k_y - \frac{E_0^2}{2B^2}[\rho^*(E,\eta_g)]^2 m_1^*(E,\eta_g) \quad (10.13)$$

The use of (10.13) leads to the expression of EEM along y direction as

$$m_y^*(e_{fA7},\eta_g,n,E_0,n_z) = \text{Real part of } (B/E_0)^2 T_{55}(e_{fA7},\eta_g,n_z)[T_{55}(e_{fA7},\eta_g,n_z)]' \quad (10.14a)$$

where e_{fA7} is the Fermi energy in this case and

$$T_{55}(e_{fA7},\eta_g,n_z) = \left[G^*(e_{fA7},\eta_g) - \left(n+\frac{1}{2}\right)\hbar\overline{\omega_i}(e_{fA7},\eta_g).\right.$$
$$\left. - \frac{\hbar^2\left(\frac{n_z\pi}{d_z}\right)^2}{2m_3^*(e_{fA7},\eta_g)} + \frac{E_0^2}{2B^2}[\rho^*(e_{fA7},\eta_g)]^2 m_1^*(e_{fA7},\eta_g)\right][m_4^*(e_{fA7},\eta_g)]^{-1}$$

The sub band energy $E_{9,7}$ in this case assumes the from

$$G^*(E_{9,7},\eta_g) = \left(n+\frac{1}{2}\right)\hbar\overline{\omega_i}(E_{9,7},\eta_g) + \frac{\hbar^2\left(\frac{n_z\pi}{d_z}\right)^2}{2m_3^*(E_{9,7},\eta_g)} - \frac{E_0^2}{B}[\rho^*(E_{9,7},\eta_g)]^2 m_1^*(E_{9,7},\eta_g) \tag{10.14b}$$

10.3 Summary and Conclusion

The DRs under cross field configuration in ultra-thin films of HD materials exhibit the fact that the electron motion along k_y direction is free i.e. the materials become less degenerate.

- The DRs in this case are quantized straight lines.
- The DRs are concentration dependent which is possible only under band-tailing effect.
- The subband energies also depend on concentration due to band tails.
- The cross fields introduces energy and quantum number dependent mass anisotropy.
- The effective mass exists in the band gap which is impossible without band tailing effect.

It is the band structure which changes in a fundamental way and consequently all the physical properties of all the electronic materials change radically leading to new physical concepts.

10.4 Open Research Problems

(R.10.1) Investigate the 2D DR in the presence of cross fields for all the HD UFs of the materials whose bulk DRs are given in problems in (R.1.1) of Chap. 1.

(R.10.2) Investigate the 2D DR in the presence of cross fields in ultra-thin films of HD nonlinear optical semiconductors by including broadening and the electron spin. Study all the special cases for HD III–V, ternary and quaternary materials in this context.

(R.10.3) Investigate the 2D DRs for HD UFs of IV–VI, II–VI and stressed Kane type compounds in the presence of cross fields by including broadening and electron spin.

(R.10.4) Investigate the 2D DR in the presence of cross fields in HD UFs of nonlinear optical semiconductors by including broadening and the electron spin. Study all the special cases for HD III–V, ternary and quaternary materials in this context.

(R.10.5) Investigate the 2D DRs for HD UFs of IV–VI, II–VI and stressed Kane type compounds in the presence of cross fields by including broadening and electron spin.

(R.10.6) Investigate the 2D DR for all the UFs of materials whose bulk DRs are stated in (R.1.1) of Chap. 1 in the presence of an arbitrarily oriented quantizing magnetic field and crossed electric field by including broadening and electron spin under the condition of heavy doping.

(R.10.7) Investigate the 2D DR in HD UFs of nonlinear optical, III–V, II–VI, IV–VI and stressed Kane type semiconductors in the presence of non-uniform strain.

(R.10.8) Investigate the 2D DR in the presence of an arbitrarily oriented non-quantizing magnetic field for UFs of HD nonlinear optical semiconductors by including the electron spin. Study all the special cases for III–V, ternary and quaternary materials in this context.

(R.10.9) Investigate the DRs in QWs of HD IV–VI, II–VI and stressed Kane type compounds in the presence of an arbitrarily oriented non-quantizing magnetic field by including the electron spin.

(R.10.10) Investigate the 2D DR for HD UFs of all the materials as stated in (R.1.1) of Chap. 1 in the presence of an arbitrarily oriented magnetic field by including electron spin and broadening.

(R.10.11) Investigate all the problems of this section by removing all the mathematical approximations and establishing the respective appropriate uniqueness conditions.

Chapter 11
The DR in Doping Superlattices of HD Non-parabolic Semiconductors Under Magnetic Quantization

One week means ten thousand and eighty minutes. If by applying the maximum power transfer theorem, I can really use five thousand and forty minutes, then instead of weak I will be rather strong mentally.

11.1 Introduction

In Sect. 11.2.1, of the theoretical background, the DR in doping superlattices of HD non-linear optical semiconductors has been investigated under magnetic quantization. The Sect. 11.2.2 contains the results for doping superlattices under magnetic quantization of HD III–V, ternary and quaternary semiconductors in accordance with the three and the two band models of Kane together with parabolic energy bands and they form the special cases of Sect. 11.2.1. The Sects. 11.2.3–11.2.5 contain the study of the DR under magnetic quantization for doping superlattices of HD II–VI, IV–VI and stressed Kane type semiconductors respectively. The Sects. 11.3 and 11.4 contain the summary and conclusion and the 5 open research problems for this chapter.

11.2 Theoretical Background

11.2.1 The DR in Doping Superlattices of HD Nonlinear Optical Semiconductors Under Magnetic Quantization

The DR of the conduction electrons in doping superlattices under magnetic quantization of HD nonlinear optical materials can be expressed by using (6.1) as

K.P. Ghatak, *Dispersion Relations in Heavily-Doped Nanostructures*,
Springer Tracts in Modern Physics 265, DOI 10.1007/978-3-319-21000-1_11

$$\frac{\left(n_i + \frac{1}{2}\right)}{\hbar T_{21}\left(E_{10,1}, \eta_g\right)} \omega_{8HD}(E_{10,1}, \eta_g) + \frac{\hbar eB\left(n + \frac{1}{2}\right)}{m_{\perp}^* T_{22}(E_{10,1}, \eta_g)} = 1 \tag{11.1}$$

where $E_{10,1}$ is the totally quantized energy in this case.

The DOS function is given by

$$N_{Bn_z} = \frac{g_v eB}{\pi\hbar} \sum_{n=0}^{n_{\max}} \sum_{n_z=1}^{n_{z\max}} \delta'(E - E_{10,1}) \tag{11.2}$$

In the absence of band-tails we can write

$$\psi_1(E_{10,2}) = \psi_2(E_{10,2}) \frac{2eB}{\hbar}\left(n + \frac{1}{2}\right) + \psi_3(E_{10,2})\left(n_i + \frac{1}{2}\right)\frac{2m_{\parallel}^*}{\hbar}\omega_8(E_{10,2}) \tag{11.3}$$

The DOS function is given by

$$N_{Bn_z} = \frac{g_v eB}{\pi\hbar} \sum_{n=0}^{n_{\max}} \sum_{n_z=1}^{n_{z\max}} \delta'(E - E_{10,2}) \tag{11.4}$$

11.2.2 *The DR in Doping Superlattices of HD III–V, Ternary and Quaternary Semiconductors Under Magnetic Quantization*

(a) The electron energy spectrum in doping superlattices under magnetic quantization of HD III–V, ternary and quaternary materials can be expressed from (11.1) under the conditions $\Delta_{\parallel} = \Delta_{\perp} = \Delta$, $\delta = 0$ and $m_{\parallel}^* = m_{\perp}^* = m_c$, as [1–20]

$$\left(n + \frac{1}{2}\right)\hbar\omega_0 = \Big[T_{31}(E_{10,3}, \eta_g) + iT_{32}(E_{10,3}, \eta_g) - \left(n_i + \frac{1}{2}\right)\hbar\omega_{9HD}(E_{10,3}, \eta_g)\Big] \tag{11.5}$$

where $E_{10,3}$ *is the totally quantized energy*

The DOS function is given by

$$N_{Bn_z} = \frac{g_v eB}{\pi\hbar} \sum_{n=0}^{n_{\max}} \sum_{n_z=1}^{n_{z\max}} \delta'(E - E_{10,3}) \tag{11.6}$$

In the absence of band tails, the DR in this case assumes the form

$$I_{11}(E_{10,4}) = \left(n_i + \frac{1}{2}\right)\hbar\omega_{19}(E_{10,4}) + \left(n + \frac{1}{2}\right)\hbar\omega_0 \tag{11.7}$$

The DOS function is given by

$$N_{Bn_z} = \frac{g_v eB}{\pi\hbar}\sum_{n=0}^{n_{\max}}\sum_{n_z=1}^{n_{z\max}}\delta'(E - E_{10,4}) \tag{11.8}$$

(b) The DR in doping superlattics of HD III–V, ternary and quaternary materials under magnetic quantization whose energy band structures in the absence of band tails are described by the two band model of Kane can be expressed from (11.5) under the conditions $\Delta \gg E_g$ or $\Delta \ll E_g$, as

$$\left(n + \frac{1}{2}\right)\hbar\omega_0 = \left[\gamma_2(E_{10,5}, \eta_g) - \left(n_i + \frac{1}{2}\right)\hbar\omega_{10HD}(E_{10,5}, \eta_g)\right] \tag{11.9}$$

The DOS function is given by

$$N_{Bn_z} = \frac{g_v eB}{\pi\hbar}\sum_{n=0}^{n_{\max}}\sum_{n_z=1}^{n_{z\max}}\delta'(E - E_{10,5}) \tag{11.10}$$

In the absence of band tails, the DR in this case assumes the form

$$E_{10,6}(1 + \alpha E_{10,6}) = \left(n_i + \frac{1}{2}\right)\hbar\omega_{20}(E_{10,6}) + \left(n + \frac{1}{2}\right)\hbar\omega_0 \tag{11.11}$$

The DOS function is given by

$$N_{Bn_z} = \frac{g_v eB}{\pi\hbar}\sum_{n=0}^{n_{\max}}\sum_{n_z=1}^{n_{z\max}}\delta'(E - E_{10,6}) \tag{11.12}$$

(c) The electron energy spectrum in nipi structures of HD III–V, ternary and quaternary materials under magnetic quantization whose energy band structures in the absence of band tails are described by the parabolic energy bands can be expressed as

$$\left(n + \frac{1}{2}\right)\hbar\omega_0 = \left[\gamma_3(E_{10,7}, \eta_g) - \left(n + \frac{1}{2}\right)\hbar\omega_{11HD}(E_{10,7}, \eta_g)\right] \tag{11.13}$$

The DOS function is given by

$$N_{Bn_z} = \frac{g_v eB}{\pi\hbar}\sum_{n=0}^{n_{\max}}\sum_{n_z=1}^{n_{z\max}} \delta'(E - E_{10,7}) \tag{11.14}$$

In the absence of band tails, the DR in this case assumes the form

$$E_{10,7} = \left(n_i + \frac{1}{2}\right)\hbar\omega_{21} + \left(n + \frac{1}{2}\right)\hbar\omega_0 \tag{11.15}$$

The DOS function is given by

$$N_{Bn_z} = \frac{g_v eB}{\pi\hbar}\sum_{n=0}^{n_{\max}}\sum_{n_z=1}^{n_{z\max}} \delta'(E - E_{10,6}) \tag{11.16}$$

11.2.3 The DR in Doping Superlattices of HD II–VI Semiconductors Under Magnetic Quantization

The 2D DR in doping superlattices of HD II–VI semiconductors under magnetic quantization can be expressed as

$$\begin{aligned}\gamma_3(E_{10,7}, \eta_g) = a_0'\left[\frac{2eB}{\hbar}\left(n + \frac{1}{2}\right)\right] + \left(n_i + \frac{1}{2}\right)\hbar\omega_{30}(E_{10,7}, \eta_g)\\ \pm \bar{\lambda}_0\sqrt{\left[\frac{2eB}{\hbar}\left(n + \frac{1}{2}\right)\right]}\end{aligned} \tag{11.17}$$

where $E_{10,7}$ is the totally quantized energy in this case.

The DOS function is given by

$$N_{Bn_z} = \frac{g_v eB}{\pi\hbar}\sum_{n=0}^{n_{\max}}\sum_{n_z=1}^{n_{z\max}} \delta'(E - E_{10,7}) \tag{11.18}$$

In the absence of band-tails, the carrier dispersion law in doping superlattices of II–VI compounds can be expressed as

$$\begin{aligned}E_{10,8} = a_0'\left[\frac{2eB}{\hbar}\left(n + \frac{1}{2}\right)\right] + \left(n_i + \frac{1}{2}\right)\hbar\bar{\omega}_{10}\\ \pm \bar{\lambda}_0\left[\frac{2eB}{\hbar}\left(n + \frac{1}{2}\right)\right]^{1/2},\end{aligned} \tag{11.19}$$

The DOS function is given by

$$N_{Bn_z} = \frac{g_v eB}{2\pi\hbar} \sum_{n=0}^{n_{max}} \sum_{n_z=1}^{n_{zmax}} \delta'(E - E_{10,8}) \tag{11.20}$$

11.2.4 The DR in Doping Superlattices of HD IV–VI Semiconductor Under Magnetic Quantization

The 2D magneto DR in this case is given by

$$\frac{2eB}{\hbar}\left(n + \frac{1}{2}\right) = \delta_{15}(E_{10,9}, \eta_g, n_i) \tag{11.21}$$

where

$$\delta_{15}(E, \eta_g, n_i) = [2\delta_{12}(E, \eta_g)]^{-1}\Big[-\delta_{13}(E, \eta_g, n_i) + \sqrt{\delta^2{}_{13}(E, \eta_g, n_i) - 4\delta_{12}(E, \eta_g)\delta_{14}(E, \eta_g, n_i)}\Big],$$

$$\delta_{12}(E, \eta_g) = \frac{\alpha\hbar^4 Z_0(E, \eta_g)}{4m_t^+ m_l^-},$$
$$\delta_{13}(E, \eta_g, n_i) = \hbar^2[\lambda_{71}(E, \eta_g)\delta_{11}(E, \eta_g, n_i) + \lambda_{12}(E, \eta_g)],$$

$$\delta_{14}(E, \eta_g, n_i) = [\lambda_{73}(E, \eta_g)\delta_{11}^2(E, \eta_g, n_i) + \lambda_{74}(E, \eta_g)\delta_{11}^4(E, \eta_g, n_i) - \lambda_{74}(E, \eta_g)],$$

$$\delta_{11}(E, \eta_g, n_i) = \frac{2}{\hbar} m_{HD}^*(0, \eta_g)\left(n_i + \frac{1}{2}\right)\left[\frac{e^2 n_0}{d_0 \varepsilon_{sc} m_{HD}^*(E, \eta_g)}\right]^{1/2},$$

$$m_{HD}^*(E, \eta_g) = \frac{\hbar^2}{4\lambda_{76}^2(E, \eta_g)}\Big[2\lambda_{74}(E, \eta_g)\left\{-\lambda'_{73}(E, \eta_g) + \frac{\lambda_{73}(E, \eta_g)\lambda'_{73}(E, \eta_g) + 2\lambda'_{74}(E, \eta_g)\lambda_{75}(E, \eta_g) + 2\lambda_{74}(E, \eta_g)\lambda'_{75}(E, \eta_g)}{\sqrt{\lambda_{73}^2(E, \eta_g) + 4\lambda_{74}(E, \eta_g)\lambda_{75}(E, \eta_g)}}\right\}$$
$$-2\lambda'_{74}(E, \eta_g)\left\{-\lambda_{73}(E, \eta_g) + \sqrt{\lambda_{73}^2(E, \eta_g) + 4\lambda_{74}(E, \eta_g)\lambda_{75}(E, \eta_g)}\right\}\Big]$$

and $E_{10,9}$ is the totally quantized energy in this case.

The carrier energy spectrum in doping superlattices of IV–VI compounds in the absence of band tails can be written as

$$\frac{2eB}{\hbar}\left(n+\frac{1}{2}\right)=(\hbar^2 S_{19})^{-1}\left[-S_{20}(E_{10,10},n_i)+\sqrt{S_{20}^2(E_{10,10},n_i)+4S_{19}S_{21}(E_{10,10},n_i)}\right] \tag{11.22}$$

The DOS function is given by

$$N_{Bn_z}=\frac{g_v eB}{\pi\hbar}\sum_{n=0}^{n_{\max}}\sum_{n_z=1}^{n_{z\max}}\delta'(E-E_{10,10}) \tag{11.23}$$

11.2.5 The DR in Doping Superlattices of HD Stressed Kane Type Semiconductors Under Magnetic Quantization

The 2D DR in this case is given by

$$\left(n+\frac{1}{2}\right)\hbar\omega_{90}(E_{10,11},\eta_g)+S_{11}(E_{10,11},\eta_g)\delta_{19}(E_{10,11},\eta_g,n_i)=1 \tag{11.24}$$

where $E_{10,11}$ is the totally quantized energy in this case,

$$\omega_{90}(E_{10,11},\eta_g)=\frac{eB}{\sqrt{m_1^*(E_{10,11},\eta_g)m_2^*(E_{10,11},\eta_g)}},$$

$$m_1^*(E_{10,11},\eta_g)=\frac{\hbar^2}{2P_{11}(E_{10,11},\eta_g)}\quad\text{and}$$

$$m_2^*(E_{10,11},\eta_g)=\frac{\hbar^2}{2Q_{11}(E_{10,11},\eta_g)}$$

The DOS function is given by

$$N_{Bn_z}=\frac{g_v eB}{\pi\hbar}\sum_{n=0}^{n_{\max}}\sum_{n_z=1}^{n_{z\max}}\delta'(E-E_{10,11}) \tag{11.25}$$

The electron dispersion law in the doping superlattices of stressed Kane type semiconductors can be written as

$$\left(n+\frac{1}{2}\right)\hbar\omega_{100}(E_{10,12})+\frac{1}{[\bar{c}_0(E_{10,12})]^2}\frac{2m_z^*(0)}{\hbar}\left(n_i+\frac{1}{2}\right)\omega_{12}(E_{10,12})=1 \quad (11.26)$$

where $E_{10,12}$ is the totally quantized energy in this case and $\omega_{100}(E_{10,12})=\frac{eB}{\hbar\bar{a}_0(E_{10,12})\bar{b}_0(E_{10,12})}$

The DOS function is given by

$$N_{Bn_z}=\frac{g_v eB}{\pi\hbar}\sum_{n=0}^{n_{\max}}\sum_{n_z=1}^{n_{z\max}}\delta'(E-E_{10,12}) \quad (11.27)$$

11.3 Summary and Conclusion

From the 2D DR in doping superlattices of HD nonlinear optical and tetragonal materials (11.1) under magnetic quantization, we observe that the electron energy is totally quantized in 3D wave vector surfaces in the complex energy plane which is the consequence of non removable poles in the corresponding DR in the absence of band tails. From (11.5) we have the same inference for doping superlattices of HD III–V materials under magnetic quantization whose un-perturbed conduction electrons obey the three band model of Kane, which contains one non removal pole in energy axis. The 0D electrons in HD doping superlattices of III–V materials under magnetic quantization are also described by two band model of Kane and parabolic energy bands with the DRs as given by (11.19) and (11.13) respectively. Besides the DRs in the case for IV–VI materials is given by (11.21). Since all the said DRs possess no poles in the finite energy planes, the constant quantized energies in this case exist in the real plane. The DR (11.17) in HD doping superlattices of II–VI materials under magnetic quantization reflects the fact that the totally quantized energies exist in the real plane. The DR (11.24) in doping superlattices of HD stressed Kane type semiconductors under magnetic quantization reflects the identical conclusion. Besides in all the cases the DOS functions are non uniformly placed Dirac delta functions in the energy axis.

11.4 Open Research Problems

(R.11.1) Investigate the DR in the presence of an arbitrarily oriented quantizing magnetic field for nipi structures of HD nonlinear optical semiconductors by including the electron spin. Study all the special cases for HD III–V, ternary and quaternary materials in this context.

(R.11.2) Investigate the DRs in nipi structures of HD IV–VI, II–VI and stressed Kane type compounds in the presence of an arbitrarily oriented quantizing magnetic field by including the electron spin.

(R.11.3) Investigate the DR for HD nipi structures of all the materials as stated in (R.1.1) of Chap. 1.

(R.11.4) Investigate the DR for all the problems from (R.11.1) to (R.11.3) in the presence of an additional arbitrarily oriented electric field.

(R.11.5) Investigate the DR for all the problems from (R.11.1) to (R.11.3) in the presence of arbitrarily oriented crossed electric and magnetic fields.

References

1. N.G. Anderson, W.D. Laidig, R.M. Kolbas, Y.C. Lo, J. Appl. Phys. **60**, 2361 (1986)
2. F. Capasso, Semiconductors Semimetals **22**, 2 (1985)
3. F. Capasso, K. Mohammed, A.Y. Cho, R. Hull, A.L. Hutchinson, Appl. Phys. Letts. **47**, 420 (1985)
4. F. Capasso, R.A. Kiehl, J. Appl. Phys. **58**, 1366 (1985)
5. K. Ploog, G.H. Doheler, Adv. Phys. **32**, 285 (1983)
6. F. Capasso, K. Mohammed, A.Y. Cho, Appl. Phys. Lett. **48**, 478 (1986)
7. R. Grill, C. Metzner, G.H. Döhler, Phys. Rev. B **63**, 235316 (2001)
8. A.R. Kost, M.H. Jupina, T.C. Hasenberg, E.M. Garmire, J. Appl. Phys. **99**, 023501 (2006)
9. A.G. Smirnov, D.V. Ushakov, V.K. Kononenko, Proc. SPIE **4706**, 70 (2002)
10. D.V. Ushakov, V.K. Kononenko, I.S. Manak, Proc. SPIE **4358**, 171 (2001)
11. J.Z. Wang, Z.G. Wang, Z.M. Wang, S.L. Feng, Z. Yang, Phys. Rev. B **62**, 6956 (2000)
12. A.R. Kost, L. West, T.C. Hasenberg, J.O. White, M. Matloubian, G.C. Valley, Appl. Phys. Lett. **63**, 3494 (1993)
13. S. Bastola, S.J. Chua, S.J. Xu, J. Appl. Phys. **83**, 1476 (1998)
14. Z.J. Yang, E.M. Garmire, D. Doctor, J. Appl. Phys. **82**, 3874 (1997)
15. G.H. Avetisyan, V.B. Kulikov, I.D. Zalevsky, P.V. Bulaev, Proc. SPIE **2694**, 216 (1996)
16. U. Pfeiffer, M. Kneissl, B. Knüpfer, N. Müller, P. Kiesel, G.H. Döhler, J.S. Smith, Appl. Phys. Lett. **68**, 1838 (1996)
17. H.L. Vaghjiani, E.A. Johnson, M.J. Kane, R. Grey, C.C. Phillips, J. Appl. Phys. **76**, 4407 (1994)
18. P. Kiesel, K.H. Gulden, A. Hoefler, M. Kneissl, B. Knuepfer, S.U. Dankowski, P. Riel, X.X. Wu, J.S. Smith, G.H. Doehler, Proc. SPIE **1985**, 278 (1993)
19. G.H. Doheler, Phys. Script. **24**, 430 (1981)
20. S. Mukherjee, S.N. Mitra, P.K. Bose, A.R. Ghatak, A. Neoigi, J.P. Banerjee, A. Sinha, M. Pal, S. Bhattacharya, K.P. Ghatak, J. Comput. Theor. Nanosci. **4**, 550 (2007); N. Paitya, K. P. Ghatak, J. Adv. Phys. **1**, 161 (2012); N. Paitya, S. Bhattacharya, D. De, K.P. Ghatak Adv. Sci. Eng. Med. **4**, 96 (2012); S. Bhattacharya, D. De, S.M. Adhikari, K.P. Ghatak Superlattices Microstruct. **51**, 203 (2012); D. De, S. Bhattacharya, S.M. Adhikari, A. Kumar, P. K. Bose, K. P. Ghatak, Beilstein J. Nanotech. **2**, 339 (2012); D. De, A. Kumar, S.M. Adhikari, S. Pahari, N. Islam, P. Banerjee, S.K. Biswas, S. Bhattacharya, K.P. Ghatak, Superlattices Microstruct. 47, 377 (2010); S. Pahari, S. Bhattacharya, S. Roy, A. Saha, D. De, K.P. Ghatak, Superlattices Microstruct. **46**, 760 (2009); S. Pahari, S. Bhattacharya, K.P. Ghatak, J. Comput. Theor. Nanosci. **6**, 2088 (2009); S.K. Biswas, A.R. Ghatak, A. Neogi, A. Sharma, S. Bhattacharya, K. P. Ghatak, Phys. E: Low-dimen. Syst. Nanostruct. **36**, 163 (2007); L.J. Singh, S. Choudhury,

D. Baruah, S.K. Biswas, S. Pahari, K.P. Ghatak, Phys. B: Conden. Matter, **368**, 188 (2005); S. Chowdhary, L.J. Singh, K.P. Ghatak, Phys. B: Conden. Matter. **365**, 5 (2005); L. J. Singh, S. Choudhary, A. Mallik, K.P. Ghatak, J. Comput. Theor. Nanosci. **2**, 287 (2005); K.P. Ghatak, J. Mukhopadhyay J.P. Banerjee, SPIE Proce. Ser. **4746**, 1292 (2002); K.P. Ghatak, S. Dutta, D.K. Basu, B. Nag, Il Nuovo Cimento. **D 20**, 227 (1998); K.P. Ghatak, D.K. Basu, B. Nag, J. Phys. Chem. Solids. **58**, 133 (1997); K.P. Ghatak, B. De, Mat. Resc. Soc. Proc. **300**, 513 (1993); K.P. Ghatak, B. Mitra, Il Nuovo Cimento. **D 15**, 97 (1993); K.P. Ghatak, Int. Soci.Opt. Photon. Proc. Soc. Photo Opt. Instru. Eng. **1626**, 115 (1992); K.P. Ghatak, A. Ghoshal, Phys. Stat. Sol. (b) **170**, K27 (1992); K.P. Ghatak, S. Bhattacharya, S.N. Biswas, Proc. Soc. Photo Opt. Instru. Eng. **836**, 72 (1988); K.P. Ghatak, A Ghoshal, S.N. Biswas, M. Mondal, Proc. Soc. Photo Opt. Instru. Eng. **1308**, 356 (1990); K.P. Ghatak, G. Mazumder, Proc. Matt. Res. Soc. **484**, 679 (1998); K.P. Ghatak, B. De, Defect Eng. Semi. Growth, Process. Device Tech. Symp., Mat. Res. Soc. **262**, 911 (1992); S.N. Biswas, K.P. Ghatak, Int. J. Electron. Theor. Exp. **70**, 125 (1991); B Mitra, KP Ghatak, Phys. Lett. A. **146**, 357 (1990); B. Mitra, K. P. Ghatak, Phys. Lett. A. **142**, 401 (1989); K.P. Ghatak, B. Mitra, A. Ghoshal, Phys. Stat. Sol. (b) **154**, K121 (1989); B. Mitra, K.P. Ghatak, Phys. Stat. Sol. (b). **149**, K117 (1988); K. P. Ghatak, S.N. Biswas, Proc. Soc. Photo Optical Instru. Eng. **792**, 239 (1987); S. Bhattacharyya, K.P. Ghatak, S. Biswas, OE/Fibers' 87, Int. Soc. Opt. Photon. 836, 73 (1988); M. Mondal, K.P. Ghatak, Czech. J. Phys. B. **36**, 1389 (1986); K.P. Ghatak, A.N. Chakravarti, Phys. Stat. Sol. (b). **117**, 707 (1983)

Chapter 12
The DR in Accumulation and Inversion Layers of Non-parabolic Semiconductors Under Magnetic Quantization

Blessed are those who can give without remembering and take without forgetting.

12.1 Introduction

In this chapter, in Sect. 12.2.1, of the theoretical background, the DR in accumulation and Inversion layers of nonlinear optical semiconductors under magnetic quantization has been studied under weak electric field limit. The Sect. 12.2.2 contains the results for accumulation and Inversion layers of III–V, ternary and quaternary semiconductors under magnetic quantization for the weak electric field limit whose bulk electrons obey the three and the two band models of Kane together with parabolic energy bands and they form the special cases of Sect. 12.2.1. The Sect. 12.2.3 contains the study of the DR for accumulation and Inversion layers of II–VI semiconductors under magnetic quantization, which is valid for all values of electric field. The Sects. 12.2.4 and 12.2.5 contain the study of the DR in accumulation and Inversion layers of IV–VI and stressed semiconductors under magnetic quantization respectively. The Sect. 12.2.6 contains the study of the DR in accumulation and Inversion layers of Ge under magnetic quantization. The Sect. 12.3 contains the summary and conclusion of this chapter. The last Sect. 12.4 contains 12 open research problems of this chapter.

K.P. Ghatak, *Dispersion Relations in Heavily-Doped Nanostructures*,
Springer Tracts in Modern Physics 265, DOI 10.1007/978-3-319-21000-1_12

12.2 Theoretical Background

12.2.1 The DR in Accumulation and Inversion Layers of Nonlinear Optical Semiconductors Under Magnetic Quantization

Following (6.3) the DR of the 2D electrons in accumulation layers of HD non-linear optical materials under the condition of weak electric field limit under magnetic quantization as

$$\left(n+\frac{1}{2}\right)\frac{\hbar eB}{m_{\parallel}^{*}}=L_{6}(E_{11,1},i,\eta_{g}) \tag{12.1}$$

where $E_{11,1}$ is the totally quantized energy in this case.

The DOS function is given by

$$N_{Bi}=\frac{g_{v}eB}{\pi\hbar}\sum_{n=0}^{n_{\max}}\sum_{i=0}^{n_{i\max}}\delta'(E-E_{11,1}) \tag{12.2}$$

Thus, the 2D electron dispersion law in inversion layers of nonlinear optical materials under the weak electric field limit under magnetic quantization can approximately be written following (6.10) as

$$\psi_{1}(E_{11,2})=P_{7}(E_{11,2},i)\frac{2eB}{\hbar}\left(n+\frac{1}{2}\right)+Q_{7}(E_{11,2},i) \tag{12.3}$$

where $E_{11,2}$ is the totally quantized energy in this case.

The DOS function is given by

$$N_{Bi}=\frac{g_{v}eB}{\pi\hbar}\sum_{n=0}^{n_{\max}}\sum_{i=0}^{n_{i\max}}\delta'(E-E_{11,12}) \tag{12.4}$$

12.2.2 The DR in Accumulation and Inversion Layers of III–V Semiconductors Under Magnetic Quantization

(a) The DR in the present case can be written following (6.15) as

$$T_{90}(E_{11,3},\eta_{g})=\left(n+\frac{1}{2}\right)\frac{\hbar eB}{m_{c}}+S_{i}\left[\frac{\hbar|e|F_{s}[T_{90}(E_{11,3},\eta_{g})]'}{\sqrt{2m_{c}}}\right]^{2/3} \tag{12.5}$$

where $E_{11,3}$ is the totally quantized energy in this case.

The DOS function is given by

$$N_{Bi} = \frac{g_v eB}{\pi\hbar} \sum_{n=0}^{n_{\max}} \sum_{i=0}^{n_{i\max}} \delta'(E - E_{11,3}) \tag{12.6}$$

The (12.5) represents the DR of the 2D electrons in accumulation layers of HD III–V, ternary and quaternary materials under the weak electric field limit under magnetic quantization whose bulk electrons obey the HD three band model of Kane. Since the electron energy spectrum in accordance with the HD three-band model of Kane is complex in nature, the (12.5) will also be complex. The both complexities occur due to the presence of poles in the finite complex plane of the dispersion relation of the materials in the absence of band tails.

Using the substitutions $\delta = 0$, $\Delta_{\|} = \Delta_{\perp} = \Delta$ and $m_{\|}^{*} = m_{\perp}^{*} = m_c$, (6.21) under the condition of weak electric field limit, assumes the form

$$I_{11}(E_{11,4}) = \left(n + \frac{1}{2}\right)\frac{\hbar eB}{m_c} + S_i\left[\frac{\hbar|e|F_s[I_{11}(E_{11,4})]'}{\sqrt{2m_c}}\right]^{2/3} \tag{12.7}$$

where $E_{11,4}$ is the totally quantized energy in this case.

The DOS function is given by

$$N_{Bi} = \frac{g_v eB}{\pi\hbar} \sum_{n=0}^{n_{\max}} \sum_{i=0}^{n_{i\max}} \delta'(E - E_{11,4}) \tag{12.8}$$

(e) Using the constraints $\Delta \gg E_g$ or $\Delta \ll E_g$, the (6.25) under the low electric field limit assumes the form

$$\gamma_2(E_{10,5}, \eta_g) = \left(n + \frac{1}{2}\right)\frac{\hbar eB}{m_c} + S_i\left[\frac{\hbar|e|F_s[\gamma_2(E_{11,5}, \eta_g)]'}{\sqrt{2m_c}}\right]^{2/3} \tag{12.9}$$

where $E_{11,5}$ is the totally quantized energy in this case.

The DOS function is given by

$$N_{Bi} = \frac{g_v eB}{\pi\hbar} \sum_{n=0}^{n_{\max}} \sum_{i=0}^{n_{i\max}} \delta'(E - E_{11,5}) \tag{12.10}$$

Using the constraints $\Delta \gg E_g$ or $\Delta \ll E_g$, the (6.31) under the low electric field limit in the present case assumes the form

$$E_{11,6}(1 + \alpha E_{11,6}) = \left(n + \frac{1}{2}\right)\frac{\hbar eB}{m_c} + S_i\left[\frac{\hbar|e|F_s(1 + 2\alpha E_{11,6})}{\sqrt{2m_c}}\right]^{2/3} \tag{12.11}$$

where $E_{11,6}$ is the totally quantized energy in this case.

The DOS function is given by

$$N_{Bi} = \frac{g_v eB}{\pi\hbar} \sum_{n=0}^{n_{\max}} \sum_{i=0}^{n_{i\max}} \delta'(E - E_{11,6}) \tag{12.12}$$

For large values of i, the (12.11) gets simplified as

$$E_{11,6}(1 + \alpha E_{11,6}) = \left(n + \frac{1}{2}\right)\frac{\hbar eB}{m_c} + S_i\left[\frac{3\pi\hbar|e|F_s}{2}\left(i + \frac{3}{4}\right)\frac{(1 + 2\alpha E_{11,6})}{\sqrt{2m_c}}\right]^{2/3} \tag{12.13}$$

(f) Using the constraints $\alpha \to 0$, following (6.36) the DR under the low electric field limit and in the presence of magnetic quantization assumes the form

$$\gamma_3(E_{10,7}, \eta_g) = \left(n + \frac{1}{2}\right)\frac{\hbar eB}{m_c} + S_i\left[\frac{\hbar|e|F_s[\gamma_3(E_{11,7}, \eta_g)]'}{\sqrt{2m_c}}\right]^{2/3} \tag{12.14}$$

where $E_{11,7}$ is the totally quantized energy in this case.

The DOS function is given by

$$N_{Bi} = \frac{g_v eB}{\pi\hbar} \sum_{n=0}^{n_{\max}} \sum_{i=0}^{n_{i\max}} \delta'(E - E_{11,7}) \tag{12.15}$$

For $\alpha \to 0$, as for inversion layers, whose bulk electrons are defined by the parabolic energy bands, from (12.13), we can write,

$$E_{11,8} = \left(n + \frac{1}{2}\right)\frac{\hbar eB}{m_c} + S_i\left[\frac{\hbar|e|F_s}{\sqrt{2m_c}}\right]^{2/3} \tag{12.16}$$

where $E_{11,8}$ is the totally quantized energy in this case.

The DOS function is given by

$$N_{Bi} = \frac{g_v eB}{\pi\hbar} \sum_{n=0}^{n_{\max}} \sum_{i=0}^{n_{i\max}} \delta'(E - E_{11,8}) \tag{12.17}$$

12.2.3 The DR in Accumulation and Inversion Layers of II–VI Semiconductors Under Magnetic Quantization

The magneto 2D DR for accumulation layers of HD II–VI semiconductors can be expressed following (6.46) as

$$\gamma_3(E_{11,9},\eta_g) = a_0'\left[\frac{2eB}{\hbar}\left(n+\frac{1}{2}\right)\right] \pm \bar{\lambda}_0\left[\frac{2eB}{\hbar}\left(n+\frac{1}{2}\right)\right]^{1/2} + S_i\left(\frac{\hbar|e|F_s\gamma_3'(E_{11,9},\eta_g)}{\sqrt{2m_\parallel^*}}\right)^{2/3} \quad (12.18)$$

where $E_{11,9}$ is the totally quantized energy in this case.

The DOS function is given by

$$N_{Bi} = \frac{g_v eB}{\pi\hbar}\sum_{n=0}^{n_{\max}}\sum_{i=0}^{n_{i\max}}\delta'(E - E_{11,9}) \quad (12.19)$$

Therefore, the magneto 2D DR for inversion layers of II–VI semiconductors can be expressed for all values of F_s following (6.46) as

$$E_{11,10} = a_0'\left[\frac{2eB}{\hbar}\left(n+\frac{1}{2}\right)\right] \pm \bar{\lambda}_0\left[\frac{2eB}{\hbar}\left(n+\frac{1}{2}\right)\right]^{1/2} + S_i\left(\frac{\hbar|e|F_s}{\sqrt{2m_\parallel^*}}\right)^{2/3} \quad (12.20)$$

where $E_{11,10}$ is the totally quantized energy in this case.

The DOS function is given by

$$N_{Bi} = \frac{g_v eB}{\pi\hbar}\sum_{n=0}^{n_{\max}}\sum_{i=0}^{n_{i\max}}\delta'(E - E_{11,10}) \quad (12.21)$$

12.2.4 The DR in Accumulation and Inversion Layers of IV–VI Semiconductors Under Magnetic Quantization

The 2D magneto DR in accumulation layers of IV–VI semiconductors can be written following (6.57) as

$$\left(n+\frac{1}{2}\right)\frac{eB}{\hbar}\sqrt{\theta_1(E_{11,11},i,\eta_g)\theta_2(E_{11,11},i,\eta_g)} = \theta_3(E_{11,11},i,\eta_g) \quad (12.22)$$

where $E_{11,11}$ is the totally quantized energy in this case.

The DOS function is given by

$$N_{Bi} = \frac{g_v eB}{\pi\hbar} \sum_{n=0}^{n_{\max}} \sum_{i=0}^{n_{i\max}} \delta'(E - E_{11,11}) \tag{12.23}$$

The 2D magneto DR of the inversion layers of IV–VI semiconductors in the low electric field limit can be written following (6.62) as

$$\frac{2eB}{\hbar}\left(n + \frac{1}{2}\right) = \beta_3(E_{11,12}, i) \tag{12.24}$$

where $E_{11,12}$ is the totally quantized energy in this case.

The DOS function is given by

$$N_{Bi} = \frac{g_v eB}{\pi\hbar} \sum_{n=0}^{n_{\max}} \sum_{i=0}^{n_{i\max}} \delta'(E - E_{11,12}) \tag{12.25}$$

12.2.5 The DR in Accumulation and Inversion Layers of Stressed Kane Type Semiconductors Under Magnetic Quantization

The 2D magneto DR in accumulation layers of stressed III–V semiconductors can be written following (6.65) as

$$\left(n + \frac{1}{2}\right)\frac{eB}{\hbar}\sqrt{\theta_{13}(E_{11,13}, i, \eta_g)\theta_{23}(E_{11,13}, i, \eta_g)} = \theta_{33}(E_{11,13}, i, \eta_g) \tag{12.26}$$

where $E_{11,13}$ is the totally quantized energy in this case.

The DOS function is given by

$$N_{Bi} = \frac{g_v eB}{\pi\hbar} \sum_{n=0}^{n_{\max}} \sum_{i=0}^{n_{i\max}} \delta'(E - E_{11,13}) \tag{12.27}$$

The expression of the magneto DR of the 2D electrons inversion layers of stressed III–V materials under the low electric field limit can be written following (6.70) as

$$\left(n + \frac{1}{2}\right)\frac{2eB}{\hbar}\sqrt{T_{57}(E_{11,14}, i)T_{67}(E_{11,14}, i)} = T_{77}(E_{11,14}, i) \tag{12.28}$$

where $E_{11,14}$ is the totally quantized energy in this case.

The DOS function is given by

$$N_{Bi} = \frac{g_v eB}{\pi\hbar} \sum_{n=0}^{n_{\max}} \sum_{i=0}^{n_{i\max}} \delta'(E - E_{11,14}) \tag{12.29}$$

12.2.6 The DR in Accumulation and Inversion Layers of Germanium Under Magnetic Quantization

The 2D DR in accumulation layers of Ge can be written as

$$\left(n + \frac{1}{2}\right) \frac{\hbar eB}{\sqrt{m_1^* m_2^*}} = \gamma_{10}(E_{11,15}, i, \eta_g) \tag{12.30}$$

where $E_{11,15}$ is the totally quantized energy in this case.

The DOS function is given by

$$N_{Bi} = \frac{g_v eB}{\pi\hbar} \sum_{n=0}^{n_{\max}} \sum_{i=0}^{n_{i\max}} \delta'(E - E_{11,15}) \tag{12.31}$$

The 2D magneto DR in inversion layers of Ge at low electric field limit can be expressed following (6.80) as

$$\left(n + \frac{1}{2}\right) \frac{\hbar eB}{\sqrt{m_1 m_2}} = [E_{11,16}(1 + \alpha E_{11,16}) + \alpha E_{i20}^2 - E_{i20}(1 + 2\alpha E_{11,16})] \tag{12.32}$$

where $E_{11,16}$ is the totally quantized energy in this case.

The DOS function is given by

$$N_{Bi} = \frac{g_v eB}{\pi\hbar} \sum_{n=0}^{n_{\max}} \sum_{i=0}^{n_{i\max}} \delta'(E - E_{11,16}) \tag{12.33}$$

12.3 Summary and Conclusion

The DRs in inversion and accumulation layers of various materials exhibit the fact that the total energy is quantized since the corresponding wave vector space is totally quantized by quantizing magnetic field and quantization due to formation of such layers along z-direction.

- The DOS functions for all the materials in this case are series of non-uniformly distributed Dirac's Delta functions at specified quantized points in the respective energy axis. The spacing between the consecutive Delta functions are functions of energy band constants and quantization of the wave vector space of a particular material. The DOS function needs two summations namely one summation over the Landau quantum number and the other one is due to formation of such layers.
- It may be noted that the energy levels in inversion and accumulation layers of various materials lead to the discrete energy levels, somewhat like atomic energy levels, which produce very large changes. This follows from the inherent nature of the quantum confinement of the carrier gas dealt with here. In the present case, there remain no free carrier states in between any two allowed sets of totally quantized levels in this case unlike that found for QWs, NWs and QBs where the quantum confinements are 1D, 2D and 3D respectively. Consequently, the crossing of the Fermi level by the totally quantized levels in this case would have much greater impact on the redistribution of the carriers among the allowed levels, as compared to that found for QWs, NWs and QBs respectively.
- It is the band structure which changes in a fundamental way and consequently all the physical properties of all the electronic materials changes radically leading to new physical concepts.

12.4 Open Research Problems

(R.12.1) Investigate the DR in the presence of an arbitrarily oriented electric quantization for accumulation layers of tetragonal semiconductors. Study all the special cases for III–V, ternary and quaternary materials in this context.

(R.12.2) Investigate the DR in accumulation layers of IV–VI, II–VI and stressed Kane type compounds in the presence of an arbitrarily oriented quantizing electric field.

(R.12.3) Investigate the DR in accumulation layers of all the materials as stated in (R.1.1) of Chap. 1 in the presence of an arbitrarily oriented quantizing electric field.

(R.12.4) Investigate the DR in the presence of an arbitrarily oriented non-quantizing magnetic field in accumulation layers of tetragonal semiconductors by including the electron spin. Study all the special cases for III–V, ternary and quaternary materials in this context.

(R.12.5) Investigate the DR in accumulation layers of IV–VI, II–VI and stressed Kane type compounds in the presence of an arbitrarily oriented non-quantizing magnetic field by including the electron spin.

(R.12.6) Investigate the DR in accumulation layers of all the materials as stated in (R.1.1) of Chap. 1 in the presence of an arbitrarily oriented non-quantizing magnetic field by including electron spin.

(R.12.7) Investigate the DR in accumulation layers for all the problems from (R.12.1) to (R.12.5) in the presence of an additional arbitrarily oriented electric field.

(R.12.8) Investigate the DR in accumulation layers for all the problems from (R.12.1) to (R.12.5) in the presence of arbitrarily oriented crossed electric and magnetic fields.

(R.12.9) Investigate the DR in accumulation layers for all the problems of this chapter in the presence of surface states.

(R.12.10) Investigate the DR in accumulation layers for all the problems from (R.12.1) to (R.12.9) in the presence of hot electron effects.

(R.12.11) Investigate the DR in accumulation layers for all the problems from (R.3.1) to (R.3.6) by including the occupancy of the electrons in various electric subbands.

(R.12.12) Investigate the problems from (R.12.1) to (R.12.11) for the appropriate p-channel accumulation layers.

References

1. T. Ando, H. Fowler, F. Stern, Rev. Mod. Phys. **54**, 437 (1982)
2. J.J. Quinn, P.J. Styles (eds.), *Electronic Properties of Quasi Two Dimensional Systems* (North Holland, Amsterdam, 1976)
3. G.A. Antcliffe, R.T. Bate, R.A. Reynolds, in *Proceedings of the International Conference, Physics of Semi-metals and Narrow-Gap semiconductors*, ed. by D.L. Carter, R.T. Bate (Pergamon Press, Oxford, 1971), p. 499
4. Z.A. Weinberg, Sol. Stat. Electron. **20**, 11 (1977)
5. G. Paasch, T. Fiedler, M. Kolar, I. Bartos, Phys. Stat. Sol. (b) **118**, 641 (1983)
6. S. Lamari, Phys. Rev. B **64**, 245340 (2001)
7. T. Matsuyama, R. Kürsten, C. Meißner, U. Merkt, Phys. Rev. B **61**, 15588 (2000)
8. P.V. Santos, M. Cardona, Phys. Rev. Lett. **72**, 432 (1994)
9. L. Bu, Y. Zhang, B.A. Mason, R.E. Doezema, J.A. Slinkman, Phys. Rev. B **45**, 11336 (1992)
10. P.D. Dresselhaus, C.M. Papavassiliou, R.G. Wheeler, R.N. Sacks, Phys. Rev. Lett. **68**, 106 (1992)
11. U. Kunze, Phys. Rev. B **41**, 1707 (1990)
12. E. Yamaguchi, Phys. Rev. B **32**, 5280 (1985)
13. Th. Lindner, G. Paasch, J. Appl. Phys. **102**, 054514 (2007)
14. S. Lamari, J. Appl. Phys. **91**, 1698 (2002)
15. K.P. Ghatak, M. Mondal, J. Appl. Phys. **70**, 299 (1991)
16. K.P. Ghatak, S.N. Biswas, J. Vac. Sci. Tech. **7B**, 104 (1989)
17. B. Mitra, K.P. Ghatak, Sol. State Electron. **32**, 177 (1989)
18. K.P. Ghatak, M. Mondal, J. Appl. Phys. **62**, 922 (1987)
19. M. Mondal, K.P. Ghatak, J. Magnet. Mag. Mat., **62**, 115 (1986); M. Mondal, K.P. Ghatak, Phys. Script. **31**, 613 (1985)

20. K.P. Ghatak, M. Mondal, Z. fur Physik B **64**, 223 (1986); K.P. Ghatak, S.N. Biswas, Sol. State Electron. **37**, 1437 (1994); D.R. Choudhury, A.K. Chowdhury, K.P. Ghatak, A.N. Chakravarti, Phys. Stat. Sol. (b) **98**, (K141) (1980); A.N. Chakravarti, A.K. Chowdhury, K. P. Ghatak, Phys. Stat. Sol. (a) **63**, K97 (1981); M. Mondal, K.P. Ghatak, Acta Phys. Polon. A **67**, 983 (1985); M. Mondal, K.P. Ghatak, Phys. Stat. Sol. (b) **128**, K21 (1985); M. Mondal, K.P. Ghatak, Phys. Stat. Sol. (a) **93**, 377 (1986); K.P. Ghatak, M. Mondal, Phys. Stat. Sol. (b) **135**, 819 (1986); M. Mondal, K.P. Ghatak, Phys. Stat. Sol. (b) **139**, 185 (1987); K. P. Ghatak, N. Chattopadhyay, S.N. Biswas, OE/Fibers' **87**, 203 (1987); K.P. Ghatak, N. Chatterjee, M. Mondal, Phys. Stat. Sol. (b) **139**, K25 (1987); K.P. Ghatak, M. Mondal, Phys. Stat. Sol. **(b) 148**, 645 (1988); K.P. Ghatak, A. Ghosal, Phys. Stat. Sol. (b) **151**, K135 (1989); K.P. Ghatak, N. Chattopadhyay, M. Mondal, Appl. Phys. A **48**, 365 (1989)

Part III
The DR in Heavily Doped (HD) Quantum Confined Superlattices

The best book is life and the best teacher is experience.

Chapter 13
The DR in QWHDSLs

F-E-A-R means face everything and rise.

13.1 Introduction

In recent years, modern fabrication techniques have generated altogether a new dimension in the arena of quantum effect devices through the experimental realization of an important artificial structure known as semiconductor superlattice (SL) by growing two similar but different semiconducting compounds in alternate layers with finite thicknesses [1]. The materials forming the alternate layers have the same kind of band structure but different energy gaps. The concept of SL was developed for the first time by Keldysh [2] and was successfully fabricated by Esaki and Tsu [2]. The SLs are being extensively used in thermal sensors [3], quantum cascade lasers [4], photodetectors [5], light emitting diodes [6], multiplication [7], frequency multiplication [8], photocathodes [9], thin film transistor [10], solar cells [11], infrared imaging [12], thermal imaging [13], infrared sensing [14] and also in other microelectronic devices.

The most extensively studied III–V SL is the one consisting of alternate layers of GaAs and $Ga_{1-x}Al_xAs$ owing to the relative easiness of fabrication. The GaAs and $Ga_{1-x}Al_xAs$ layers form the quantum wells and the potential barriers respectively. The III–V SL's are attractive for the realization of high speed electronic and optoelectronic devices [15]. In addition to SLs with usual structure, other types of SLs such as II–VI [16], IV–VI [17] and HgTe/CdTe [18] SL's have also been investigated in the literature. The IV–VI SLs exhibit quite different properties as compared to the III–V SL due to the specific band structure of the constituent materials [19]. The epitaxial growth of II–VI SL is a relatively recent development and the primary motivation for studying the mentioned SLs made of materials with the large band gap is in their potential for optoelectronic operation in the blue [19]. HgTe/CdTe SL's have raised a great deal of attention since 1979, when as a promising new materials for long wavelength infrared detectors and other electro-optical applications [20]. Interest in Hg-based SL's has been further

K.P. Ghatak, *Dispersion Relations in Heavily-Doped Nanostructures*,
Springer Tracts in Modern Physics 265, DOI 10.1007/978-3-319-21000-1_13

increased as new properties with potential device applications were revealed [20, 21]. These features arise from the unique zero band gap material HgTe [22] and the direct band gap semiconductor CdTe which can be described by the three band mode of Kane [23]. The combination of the aforementioned materials with specified dispersion relation makes HgTe/CdTe SL very attractive, especially because of the tailoring of the material properties for various applications by varying the energy band constants of the SLs.

We note that all the aforementioned SLs have been proposed with the assumption that the interfaces between the layers are sharply defined, of zero thickness, i.e., devoid of any interface effects. The SL potential distribution may be then considered as a one dimensional array of rectangular potential wells. The aforementioned advanced experimental techniques may produce SLs with physical interfaces between the two materials crystallographically abrupt; adjoining their interface will change at least on an atomic scale. As the potential form changes from a well (barrier) to a barrier (well), an intermediate potential region exists for the electrons [24]. The influence of finite thickness of the interfaces on the electron dispersion law is very important, since; the electron energy spectrum governs the electron transport in SLs. In addition to it, for effective mass SLs, the electronic subbands appear continually in real space [25].

In this chapter, the DR in III–V, II–VI, IV–VI, HgTe/CdTe and strained layer quantum well heavily doped superlattices (QWHDSLs) with graded interfaces has been studied in Sects. 13.2.1 to 13.2.5. From Sects. 13.2.6 to 13.2.10, the DR from III–V, II–VI, IV–VI, HgTe/CdTe and strained layer quantum well heavily doped effective mass superlattices respectively has been presented. The Sect. 13.3 contains the summary and conclusion pertinent to this chapter. The last Sect. 13.4 presents 1 multi-dimensional open research problem.

13.2 Theoretical Background

13.2.1 The DR in III–V Quantum Well HD Superlattices with Graded Interfaces

The electron dispersion law in bulk specimens of the heavily doped constituent materials of III–V SLs whose un-doped energy band structures are defined by three band model of Kane can be expressed as

$$\frac{\hbar^2 k^2}{2m_{cj}^*} = T_{1j}(E, \Delta_j, E_{gj}, \eta_{gj}) + iT_{2j}(E, \Delta_j, E_{gj}, \eta_{gj}) \tag{13.1}$$

where

$$\begin{aligned}
&\mathrm{j}=1,2,\ \ T_{ij}(E,\Delta_j,E_{gj},\eta_{gj})\\
&=(2/(1/Erf(E/\eta_{gj})))\left[(\alpha_j b_j/c_j)\cdot\theta_0(E,\eta_{gj})+\left[(\alpha_j c_j+b_j c_j-\alpha_j b_j)/c_j^2\right].\right.\\
&\quad\gamma_0(E,\eta_{gj})+\left[(1/c_j)(1-(\alpha_j/c_j))(1-(b_j/c_j))\frac{1}{2}\left[1+Erf(E/\eta_{gj})\right]\right.\\
&\quad-(1/c_j)(1-(\alpha_j/c_j))(1-(b_j/c_j))\left(2/(c_j\eta_{gj}\sqrt{\pi})\exp\left(-u_j^2\right)\right)\\
&\quad\left.\left.\left[\sum_{p=1}^{\infty}(\exp(-p^2/4)/p)\sinh(pu_j)\right]\right]\right],
\end{aligned}$$

$b_j\equiv(E_{gj}+\Delta_j)^{-1}$, $c_j\equiv\left(E_{gj}+\frac{2}{3}\Delta_j\right)^{-1}$, $u_j\equiv\frac{1+c_jE}{c_j\eta_{gi}}$ and $T_{2j}(E,\Delta_j,E_{gj},\eta_{gj})$

$$\equiv\left(\frac{2}{1+Erf\left(E/\eta_{gj}\right)}\right)\frac{1}{c_j}\left(1-\frac{\alpha_j}{c_j}\right)\left(1-\frac{b_j}{c_j}\right)\frac{\sqrt{\pi}}{c_j\eta_{gj}}\exp\left(-u_j^2\right).$$

Therefore, the dispersion law of the electrons of heavily doped III–V SLs with graded interfaces can be expressed as [25]

$$k_z^2=G_8+iH_8 \tag{13.2}$$

where

$$G_8=\left[\frac{C_7^2-D_7^2}{L_0^2}-k_s^2\right],\ \ C_7=\cos^{-1}(\overline{\omega_7}),\ \overline{\omega_7}=(2)^{\frac{-1}{2}}\left[(1-G_7^2-H_7^2)-\sqrt{(1-G_7^2-H_7^2)^2+4G_7^2}\right]^{\frac{1}{2}}$$

$$a_{20}=\left[\sqrt{\frac{M_{s2}(0,\eta_{g2})}{M_{s1}(0,\eta_{g1})}}+1\right]^2\left[4\left(\frac{M_{s2}(0,\eta_{g2})}{M_{s1}(0,\eta_{g1})}\right)^{1/2}\right]^{-1},$$

$$a_{21}=\left[\sqrt{\frac{M_{s2}(0,\eta_{g2})}{M_{s1}(0,\eta_{g1})}}-1\right]^2\left[4\left(\frac{M_{s2}(0,\eta_{g2})}{M_{s1}(0,\eta_{g1})}\right)^{1/2}\right]^{-1}$$

$$C_{40}(E,k_x,k_y,\eta_{g1})=[1-\overline{P_1}(E,\eta_{g1})k_x^2-\overline{Q_1}(E,\eta_{g1})k_y^2]^{1/2}[\overline{S_1}(E,\eta_{g1})]^{-1/2}$$
$$D_{40}(E,k_x,k_y,\eta_{g2})=[1-\overline{P_2}(E,\eta_{g2})k_x^2-\overline{Q_2}(E,\eta_{g2})k_y^2]\,[\overline{S_2}(E,\eta_{g2})]^{-1/2}$$

$$G_7 = [G_1 + (\rho_5 G_2/2) - (\rho_6 H_2/2) + (\Delta_0/2)\{\rho_6 H_2 - \rho_8 H_3 + \rho_9 H_4 - \rho_{10} H_4 + \rho_{11} H_5 - \rho_{12} H_5 + (1/12)(\rho_{12} G_6 - \rho_{14} H_6)\}],$$

$$G = [(\cos(\mathrm{h}_1))(\cosh(\mathrm{h}_2))(\cosh(g_1))(\cos(g_2)) + (\sin(\mathrm{h}_1))(\sinh(\mathrm{h}_2))(\sinh(g_1))(\sin(g_2))],$$

$$\mathrm{h}_1 = e_1(b_0 - \Delta_0), e_1 = 2^{\frac{-1}{2}}\left(\sqrt{t_1^2 + t_2^2} + t_1\right)^{\frac{1}{2}},$$

$$t_1 = \left[(2m_{c1}^*/\hbar^2)\,.\,T_{11}(E, E_{g1}, \Delta_1, \eta_{g1}) - k_s^2\right],$$
$$t_2 = \left[(2m_{c1}^*/\hbar^2)T_{21}(E, E_{g1}, \Delta_1, \eta_{g1})\right],$$

$$\mathrm{h}_2 = e_2(b_0 - \Delta_0),\ e_2 = 2^{\frac{-1}{2}}\left(\sqrt{t_1^2 + t_2^2} - t_1\right)^{\frac{1}{2}},$$

$$\mathrm{g}_1 = d_1(a_0 - \Delta_0),\ \mathrm{d}_1 = 2^{\frac{-1}{2}}\left(\sqrt{x_1^2 + y_1^2} + x_1\right)^{\frac{1}{2}},$$

$$x_1 = \left[-(2m_{c2}^*/\hbar^2)\,.\,T_{11}(E - V_0, E_{g2}, \Delta_2, \eta_{g2}) + k_s^2\right],$$
$$y_1 = \left[(2m_{c2}^*/\hbar^2)T_{22}(E - V_0, E_{g2}, \Delta_2, \eta_{g2})\right],$$

$$g_2 = d_2(a_0 - \Delta_0), d_2 = 2^{\frac{-1}{2}}\left(\sqrt{x_1^2 + y_1^2} - x_1\right)^{\frac{1}{2}},$$
$$\rho_5 = (\rho_3^2 + \rho_4^2)^{-1}[\rho_1\rho_3 - \rho_2\rho_4],$$

$$\rho_1 = \left[d_1^2 + e_2^2 - d_2^2 - e_1^2\right], \rho_3 = [d_1 e_1 + d_2 e_2],$$
$$\rho_2 = 2[d_1 d_2 + e_1 e_2], \rho_4 = [d_1 e_2 - e_1 d_2],$$

$$G_2 = [(\sin(\mathrm{h}_1))(\cosh(\mathrm{h}_2))(\sinh(g_1))(\cos(g_2)) + (\cos(\mathrm{h}_1))(\sinh(\mathrm{h}_2))(\cosh(g_1))(\sin(g_2))],$$

$$\rho_6 = (\rho_3^2 + \rho_4^2)^{-1}[\rho_1\rho_4 + \rho_2\rho_3],$$

$$H_2 = [(\sin(\mathrm{h}_1))(\cosh(\mathrm{h}_2))(\sin(g_2))(\cosh(g_1)) - (\cos(\mathrm{h}_1))(\sinh(\mathrm{h}_2))(\sinh(g_1))(\cos(g_2))],$$

$$\rho_7 = \left[(e_1^2 + e_2^2)^{-1}\left[e_1(d_1^2 - d_2^2) - 2d_1 d_2 e_2\right] - 3e_1\right],$$

$$G_3 = [(\sin(\mathrm{h}_1))(\cosh(\mathrm{h}_2))(\cosh(g_1))(\cos(g_2)) + (\cos(\mathrm{h}_1))(\sinh(\mathrm{h}_2))(\sinh(g_1))(\sin(g_2))],$$

$$\rho_8 = \left[(e_1^2+e_2^2)^{-1}\left[e_2(d_1^2-d_2^2)-2d_1d_2e_1\right]+3e_2\right],$$

$$H_3 = [(\sin(\mathrm{h}_1))(\cosh(\mathrm{h}_2))(\sin(g_2))(\sinh(g_1)) \\ -(\cos(\mathrm{h}_1))(\sinh(\mathrm{h}_2))(\cosh(g_1))(\cos(g_2))],$$

$$\rho_9 = \left[(d_1^2+d_2^2)^{-1}\left[d_1(e_2^2-e_1^2)+2e_2d_2e_1\right]+3d_1\right],$$

$$G_4 = [(\cos(\mathrm{h}_1))(\cosh(\mathrm{h}_2))(\cos(g_2))(\sinh(g_1)) \\ -(\sin(\mathrm{h}_1))(\sinh(\mathrm{h}_2))(\cosh(g_1))(\sin(g_2))],$$

$$\rho_{10} = \left[-(d_1^2+d_2^2)^{-1}\left[d_2(-e_2^2+e_1^2)+2e_2d_2e_1\right]+3d_2\right],$$

$$H_4 = [(\cos(\mathrm{h}_1))(\cosh(\mathrm{h}_2))(\cosh(g_1))(\sin(g_2)) \\ +(\sin(\mathrm{h}_1))(\sinh(\mathrm{h}_2))(\sinh(g_1))(\cos(g_2))],$$

$$\rho_{11} = 2[d_1^2+e_2^2-d_2^2-e_1^2],$$

$$G_5 = [(\cos(\mathrm{h}_1))(\cosh(\mathrm{h}_2))(\cos(g_2))(\cosh(g_1)) \\ -(\sin(\mathrm{h}_1))(\sinh(\mathrm{h}_2))(\sinh(g_1))(\sin(g_2))]$$

$$\rho_{12} = 4[d_1d_2+e_1e_2],$$

$$H_5 = [(cos(\mathrm{h}_1))(\cosh(\mathrm{h}_2))(\sinh(g_1))(\sin(g_2)) \\ +(\sin(\mathrm{h}_1))(\sinh(\mathrm{h}_2))(\cosh(g_1))(\cos(g_2))],$$

$$\rho_{13} = \Big[\{5(d_1e_1^3-3e_1e_2^2d_1)+5d_2(e_1^3-3e_1^2e_2)\}(d_1^2+d_2^2)^{-1}+(e_1^2+e_2^2)^{-1} \\ \{5(e_1d_1^3-3d_2e_1^2d_1)+5(d_2^3e_2-3d_1^2d_2e_2)\}-34(d_1e_1+d_2e_2)\Big],$$

$$G_6 = [(\sin(\mathrm{h}_1))(\cosh(\mathrm{h}_2))(\sinh(g_1))(\cos(g_2)) \\ +(\cos(\mathrm{h}_1))(\sinh(\mathrm{h}_2))(\cosh(g_1))(\sin(g_2))],$$

$$\rho_{14} = \Big[\{5(d_1e_2^3-3e_2e_1^2d_1)+5d_2(-e_1^3+3e_2^2e_1)\}(d_1^2+d_2^2)^{-1} \\ +(e_1^2+e_2^2)^{-1}\{5(-e_1d_2^3+3d_1^2d_2e_1)+5(-d_1^3e_2+3d_2^2d_1e_2)\Big\}+34(d_1e_2-d_2e_1)\Big],$$

$$H_6 = [(\sin(\mathrm{h}_1))(\cosh(\mathrm{h}_2))(\cosh(g_1))(\sin(g_2)) \\ -(\cos(\mathrm{h}_1))(\sinh(\mathrm{h}_2))(\sinh(g_1))(\cos(g_2))],$$

$$H_7 = [H_1+(\rho_5H_2/2)+(\rho_6G_2/2)+(\Delta_0/2)\{\rho_8G_3+\rho_7H_3+\rho_{10}G_4+\rho_{10}H_4+ \\ \rho_{12}G_5+\rho_{11}H_5+(1/12)(\rho_{14}G_6+\rho_{13}H_6)\}],$$

$$H_1 = [(\sin(\mathrm{h}_1))(\sinh(\mathrm{h}_2))(\cosh(g_1))(\cos(g_2)) + (\cos(\mathrm{h}_1))(\cosh(\mathrm{h}_2))(\sinh(g_1))(\sin(g_2))],$$

$$D_7 = \sinh^{-1}(\overline{\omega_7}),\ \mathrm{H}_8 = (2C_7 D_7 / L_0^2)$$

The simplified DR of heavily doped quantum well III–V super-lattices with graded interfaces can be expressed as

$$\left(\frac{n_z \pi}{d_z}\right)^2 = G_8 + iH_8 \tag{13.3a}$$

The sub-band equation in this case can be expressed as

$$\left(\frac{n_z \pi}{d_z}\right)^2 = [G_8 + iH_8]\Big|_{k_s=0 \text{ and } E=E_{12,1}} \tag{13.3b}$$

where $E_{12,1}$ is the sub-band energy in this case.

13.2.2 The DR in II–VI Quantum Well HD Superlattices with Graded Interfaces

The electron energy spectra of the heavily doped constituent materials of II–VI SLs are given by

$$\gamma_3(E, \eta_1) = \frac{\hbar^2 k_s^2}{2m^*_{\perp,1}} + \frac{\hbar^2 k_z^2}{2m^*_{\|,1}} \pm C_0 k_s \tag{13.4}$$

and

$$\frac{\hbar^2 k^2}{2m^*_{c2}} = T_{12}(E, \Delta_2, E_{g2}, \eta_{g2}) + iT_{22}(E, \Delta_2, E_{g2}, \eta_{g2}) \tag{13.5}$$

where $m^*_{\perp,1}$ and $m^*_{\|,1}$ are the transverse and longitudinal effective electron masses respectively at the edge of the conduction band for the first material. The energy-wave vector dispersion relation of the conduction electrons in heavily doped II–VI SLs with graded interfaces can be expressed as

$$k_z^2 = G_{19} + iH_{19} \tag{13.6}$$

where

$$G_{19} = \left[\frac{C_{18}^2 - D_{18}^2}{L_0^2} - k_s^2\right],$$

$$C_{18} = \cos^{-1}(\omega_{18}), \omega_{18} = (2)^{\frac{-1}{2}}\left[(1 - G_{18}^2 - H_{18}^2) - \sqrt{(1 - G_{18}^2 - H_{18}^2)^2 + 4G_{18}^2}\right]^{\frac{1}{2}},$$

$$\begin{aligned} G_{18} &= \frac{1}{2}[G_{11} + G_{12} + \Delta_0(G_{13} + G_{14}) + \Delta_0(G_{15} + G_{16})], \\ G_{11} &= 2(\cos(g_1))(\cos(g_2))(\cos\gamma_{11}(E, k_s)) \end{aligned}$$

$$\begin{aligned} \gamma_{11}(E, k_s) &= k_{21}(E, k_s)(b_0 - \Delta_0), k_{21}(E, k_s) \\ &= \left\{\left[\gamma_3(E, \eta_{g1}) - \frac{\hbar^2 k_s^2}{2m_{\perp,1}^*} \pm C_0 k_s\right]\frac{2m_{\|,1}^*}{\hbar^2}\right\}^{1/2}, \end{aligned}$$

$$G_{12} = ([\Omega_1(E, k_s)(\sinh g_1)(\cos g_2) - \Omega_2(E, k_s)(\sin g_2)(\cosh g_1)](\sin\gamma_{11}(E, k_s)))$$

$$\Omega_1(E, k_s) = \left[\frac{d_1}{k_{21}(E, k_s)} - \frac{k_{21}(E, k_s)d_1}{d_1^2 + d_2^2}\right] \text{ and } \Omega_2(E, k_s) = \left[\frac{d_2}{k_{21}(E, k_s)} + \frac{k_{21}(E, k_s)d_2}{d_1^2 + d_2^2}\right]$$

$$G_{13} = ([\Omega_3(E, k_s)(\cosh g_1)(\cos g_2) - \Omega_4(E, k_s)(\sin g_1)(\sin g_2)](\sin\gamma_{11}(E, k_s)))$$

$$\Omega_3(E, k_s) = \left[\frac{d_1^2 - d_2^2}{k_{21}(E, k_s)} - 3k_{21}(E, k_s)\right], \Omega_4(E, k_s) = \left[\frac{2d_1 d_2}{k_{21}(E, k_s)}\right]$$

$$G_{14} = ([\Omega_5(E, k_s)(\sinh g_1)(\cos g_2) - \Omega_6(E, k_s)(\sin g_1)(\cosh g_2)](\cos\gamma_{11}(E, k_s))).$$

$$\Omega_5(E, k_s) = \left[3d_1 - \frac{d_1}{d_1^2 + d_2^2}k_{21}^2(E, k_s)\right],$$

$$\Omega_6(E, k_s) = \left[3d_2 + \frac{d_2}{d_1^2 + d_2^2}k_{21}^2(E, k_s)\right]$$

$$G_{15} = ([\Omega_9(E, k_s)(\cosh g_1)(\cos g_2) - \Omega_{10}(E, k_s)(\sinh g_1)(\sin g_2)](\cos\gamma_{11}(E, k_s)))$$

$$\Omega_9(E, k_s) = [2d_1^2 - 2d_2^2 - k_{21}^2(E, k_s)], \Omega_{10}(E, k_s) = [2d_1 d_2]$$

$$G_{16} = ([\Omega_7(E, k_s)(\sinh g_1)(\cos g_2) - \Omega_8(E, k_s)(\sin g_1)(\cosh g_2)](\sin\gamma_{11}(E, k_s)/12)),$$

$$\Omega_7(E, k_s) = \left[\frac{5d_1}{d_1^2 + d_2^2}k_{21}^3(E, k_s) + \frac{5(d_1^3 - 3d_2^2 d_1)}{k_{21}(E, k_s)} - 34k_{21}(E, k_s)d_1\right],$$

$$\Omega_8(E, k_s) = \left[\frac{5d_2}{d_1^2 + d_2^2}k_{21}^3(E, k_s) + \frac{5(d_1^3 - 3d_2^2 d_1)}{k_{21}(E, k_s)} + 34k_{21}(E, k_s)d_2\right]$$

$$H_{18} = \frac{1}{2}[H_{11} + H_{12} + \Delta_0(H_{13} + H_{14}) + \Delta_0(H_{15} + H_{16})],$$

$$H_{11} = 2(\sinh g_1 \sin g_2 \cos \gamma_{11}(E, k_s)),$$
$$H_{12} = ([\Omega_2(E, k_s)(\sinh g_1)(\cos g_2) + \Omega_1(E, k_s)(\sin g_2)(\cosh g_1)](\sin \gamma_{11}(E, k_s))),$$
$$H_{13} = ([\Omega_4(E, k_s)(\cosh g_1)(\cos g_2) + \Omega_3(E, k_s)(\sinh g_1)(\sin g_2)](\sin \gamma_{11}(E, k_s))),$$
$$H_{14} = ([\Omega_6(E, k_s)(\sinh g_1)(\cos g_2) + \Omega_5(E, k_s)(\sin g_1)(\cosh g_2)](\cos \gamma_{11}(E, k_s))),$$
$$H_{15} = ([\Omega_{10}(E, k_s)(\cosh g_1)(\cos g_2) + \Omega_9(E, k_s)(\sinh g_1)(\sin g_2)](\cos \gamma_{11}(E, k_s))),$$
$$H_{16} = ([\Omega_8(E, k_s)(\sinh g_1)(\cos g_2) + \Omega_7(E, k_s)(\sin g_1)(\cosh g_2)](\sin \gamma_{11}(E, k_s)/12)),$$

$$H_{19} = \left[\frac{2C_{18}D_{18}}{L_0^2}\right] \text{ and } D_{18} = \sinh^{-1}(\omega_{18})$$

The simplified DR in heavily doped quantum well II–VI super-lattices with graded interfaces can be expressed as

$$\left(\frac{n_z\pi}{d_z}\right)^2 = G_{19} + iH_{19} \tag{13.7a}$$

The sub-band equation in this case can be expressed as

$$\left(\frac{n_z\pi}{d_z}\right)^2 = [G_{19} + iH_{19}]\Big|_{k_s=0 \text{ and } E=E_{12,2}} \tag{13.7b}$$

where $E_{12,2}$ is the sub-band energy in this case.

13.2.3 The DR in IV–VI Quantum Well HD Superlattices with Graded Interfaces

The E-k DR of the conduction electrons of the heavily doped constituent materials of the IV–VI SLs can be expressed as

$$k_z^2 = [2\overline{p_{9,i}}]^{-1}\left[-\overline{q_{9,i}}(E, k_s, \eta_{gi}) + \left[[\overline{q_{9,i}}(E, k_s, \eta_{gi})]^2 + [4\overline{p_{9,i}}\overline{R_{9,i}}(E, k_s, \eta_{gi})]^{\frac{1}{2}}\right]\right] \tag{13.8}$$

where,

$$\overline{p_{9,i}} = (\alpha_i\hbar^4)/(4m_{l,i}^- m_{li}^+), \quad \mathrm{i} = 1, 2, \quad \overline{\mathrm{q}_{9,i}}(\mathrm{E}, \mathrm{k_s}, \eta_{\mathrm{gi}})$$
$$= \left[(\hbar^2/2)\left((1/m_{li}^*) + (1/m_{li}^-)\right) + \alpha_i(\hbar^4/4)k_s^2\right.$$
$$\left.\left((1/m_{li}^+ m_{ti}^-) + (1/m_{ti}^+ m_{li}^-)\right) - \alpha_i\gamma_3(E, \eta_{gi})\left((1/m_{li}^+) - (1/m_{li}^-)\right)\right] \text{ and}$$
$$\overline{R_{9,i}}(E, k_s, \eta_{gi}) = \left[\gamma_2(E, \eta_{gi}) + \gamma_3(E, \eta_{gi})\left[(\hbar^2/2)\alpha_i k_s^2\left((1/m_{ti}^*)\right.\right.\right.$$
$$\left.\left.-(1/m_{ti}^-)\right)\right] - \left[(\hbar^2/2)k_s^2\left((1/m_{ti}^*) + (1/m_{ti}^-)\right)\right] - \alpha_i(\hbar^6/4)k_s^4\left((1/m_{ti}^+ m_{ti}^-)\right)\Big]$$

The electron dispersion law in heavily doped IV–VI SLs with graded interfaces can be expressed as

$$\cos(L_o k) = \frac{1}{2}\Phi_2(E, k_s) \tag{13.9}$$

where

$$\Phi_2(E,k_s) \equiv \Bigg[2\cosh\{\beta_2(E,k_s)\}\cos\{\gamma_2(E,k_s)\} + \varepsilon_2(E,k_s)\sinh\{\beta_2(E,k_s)\}\sin\{\gamma_{22}(E,k_s)\} + \Delta_0\Bigg[\left(\frac{\{K_{112}(E,k_s)\}^2}{K_{212}(E,k_s)} - 3K_{212}(E,k_s)\right) \cosh\{\beta_2(E,k_s)\}\sin\{\gamma_{22}(E,k_s)\} + \left(3K_{112}(E,k_s)\frac{\{K_{112}(E,k_s)\}^2}{K_{212}(E,k_s)}\right)\sinh\{\beta_2(E,k_s)\} \cosh\{\gamma_2(E,k_s)\}\Big] + \Delta_0[2\left(\{K_{112}(E,k_s)\}^2 - \{K_{212}(E,k_s)\}^2\right)\cosh\{\beta_2(E,k_s)\} \cos\{\gamma_{22}(E,k_s)\} + \frac{1}{12}\left[\frac{5\{K_{112}(E,k_s)\}^3}{K_{212}(E,k_s)} + \frac{5\{K_{212}(E,k_s)\}^3}{K_{112}(E,k_s)} - 34K_{212}(E,k_s)K_{112}(E,k_s)\right]\sinh\{\beta_2(E,k_s)\}\sin\{\gamma_{22}(E,k_s)\}]\Bigg],$$

$$\beta_2(E,k_s) \equiv K_{112}(E,k_s)[a_0 - \Delta_0],$$

$$k_{112}^2(E,k_s) = [2\overline{p_{9,2}}]^{-1}[-\overline{q_{9,2}}(E - V_0, k_s, \eta_{g2}) - [[\overline{q_{9,2}}(E - V_0, k_s, \eta_{g2})]^2 + 4\overline{p_{9,2}}\overline{R_{9,2}}(E - V_0, k_s, \eta_{g2})]^{\frac{1}{2}}],$$

$$\gamma_{22}(E,k_s) = K_{212}(E,k_s)[b_0 - \Delta_0],$$

$$k_{212}^2(E,\, k_s) = [2\overline{p_{9,1}}]^{-1}[-\overline{q_{9,1}}(E, k_s, \eta_{g1}) + [[\overline{q_{9,1}}(E,\, k_s, \eta_{g1})]^2 + 4\overline{p_{9,1}}\overline{R_{9,1}}(E,\, k_s, \eta_{g1})]^{\frac{1}{2}}]$$ and

$$\varepsilon_2(E,k_s) \equiv \left[\frac{K_{112}(E,k_s)}{K_{212}(E,k_s)} - \frac{K_{212}(E,k_s)}{K_{112}(E,k_s)}\right].$$

The simplified DR in heavily doped quantum well IV–VI super-lattices with graded interfaces can be expressed as

$$\left(\frac{n_z\pi}{d_z}\right)^2 = \frac{1}{L_0^2}\left[\cos^{-1}\left\{\frac{1}{2}\Phi_2(E,k_s)\right\}\right]^2 - k_s^2 \tag{13.10a}$$

The sub-band equation in this case can be expressed as

$$\left(\frac{n_z \pi}{d_z}\right)^2 = \frac{1}{L_0^2}\left[\cos^{-1}\left\{\frac{1}{2}\Phi_2(E_{12,3},0)\right\}\right]^2 \tag{13.10b}$$

where $E_{12,3}$ is the sub-band energy in this case.

13.2.4 The DR in HgTe/CdTe Quantum Well HD Superlattices with Graded Interfaces

The electron energy spectra of the constituent materials of HgTe/CdTe SLs are given by

$$k^2 = \left[\frac{B_{01}^2 + 4A_1E - B_{01}\sqrt{B_{01}^2 + 4A_1E}}{2A_1^2}\right] \tag{13.11}$$

and

$$\frac{\hbar^2 k^2}{2m_{c2}^*} = T_{12}(E, \Delta_2, E_{g2}, \eta_{g2}) + iT_{22}(E, \Delta_2, E_{g2}, \eta_{g2}) \tag{13.12}$$

where $B_{01} = \left(3|e|^2 / 128\varepsilon_{sc1}\right)$, $A_1 = (\hbar^2/2m_{c1}^*)$ · ε_{sc1} is the semiconductor permittivity of the first material. The energy-wave vector dispersion relation of the conduction electrons in heavily doped HgTe/CdTe SLs with graded interfaces can be expressed as

$$k_z^2 = G_{192} + iH_{192} \tag{13.13}$$

where

$$G_{192} = [((C_{182}^2 - D_{182}^2)/L_0^2) - k_s^2],$$

$$C_{182} = \cos^{-1}(\omega_{182}),$$

$$\omega_{182} = (2)^{\frac{-1}{2}}[(1 - G_{182}^2 - H_{182}^2) - \sqrt{(1 - G_{182}^2 - H_{182}^2) + 4G_{182}^2}]^{\frac{1}{2}},$$

$$G_{182} = \frac{1}{2}[G_{112} + G_{122} + \Delta_0(G_{132} + G_{142}) + \Delta_0(G_{152} + G_{162})],$$

$$G_{112} = 2(\cos(g_{12}))(\cos(g_{22}))(\cos\gamma_8(E, k_s))$$

$$\gamma_8(E,k_s) = k_8(E,k_s)(b_0 - \Delta_0),$$

$$k_8(E,k_s) = \left[\frac{B_{01}^2 + 4A_1E - B_{01}\sqrt{B_{01}^2 + 4A_1E}}{2A_1^2} - k_s^2\right]^{1/2},$$

$$G_{122} = ([\Omega_{12}(E,k_s)(\sinh g_{12})(\cos g_{22}) - \Omega_{22}(E,k_s)(\sin g_{22})(\cosh g_{12})](\sin \gamma_8(E,k_s)))$$

$$\Omega_{12}(E,k_s) = \left[\frac{d_{12}}{k_8(E,k_s)} - \frac{k_8(E,k_s)d_{12}}{d_{12}^2 + d_{22}^2}\right]$$

$$\Omega_{22}(E,k_s) = \left[\frac{d_{22}}{k_8(E,k_s)} + \frac{k_8(E,k_s)d_{22}}{d_{12}^2 + d_{22}^2}\right],$$

$$G_{132} = ([\Omega_{32}(E,k_s)(\cosh g_{12})(\cos g_{22}) - \Omega_{42}(E,k_s)(\sin g_{12})(\sin g_{22})](\sin \gamma_8(E,k_s))),$$

$$\Omega_{32}(E,k_s) = \left[\frac{d_{12}^2 - d_{22}^2}{k_8(E,k_s)} - 3k_8(E,k_s)\right],$$

$$\Omega_{42}(E,k_s) = \left[\frac{2d_{12}d_{22}}{k_8(E,k_s)}\right],$$

$$G_{142} = ([\Omega_{52}(E,k_s)(\sinh g_{12})(\cos g_{22}) - \Omega_{62}(E,k_s)(\sin g_{12})(\cosh g_{22})](\cos \gamma_8(E,k_s))),$$

$$\Omega_{52}(E,k_s) = \left[3d_{12} - \frac{d_{12}}{d_{12}^2 + d_{22}^2}k_8^2(E,k_s)\right],$$

$$\Omega_{62}(E,k_s) = \left[3d_{22} + \frac{d_{22}}{d_{12}^2 + d_{22}^2}k_8^2(E,k_s)\right],$$

$$G_{152} = ([\Omega_{92}(E,k_s)(\cosh g_{12})(\cos g_{22}) - \Omega_{102}(E,k_s)(_{12})(\sin g_{22})](\cos \gamma_8(E,k_s))),$$

$$\Omega_{92}(E,k_s) = [2d_{12}^2 - 2d_{22}^2 - k_8^2(E,k_s)], \Omega_{102}(E,k_s) = [2d_{12}d_{22}]$$

$$G_{162} = ([\Omega_{72}(E,k_s)(_{12})(\cos g_{22}) - \Omega_{82}(E,k_s)(\sin g_{12})(\cosh g_{22})](\sin \gamma_8(E,k_s)/12)),$$

$$\Omega_{72}(E,k_s) = \left[\frac{5d_{12}}{d_{12}^2 + d_{22}^2}k_8^3(E,k_s) + \frac{5(d_{12}^3 - 3d_{22}^2 d_{12})}{k_8(E,k_s)} - 34k_8(E,k_s)d_{12}\right],$$

$$\Omega_{82}(E,k_s) = \left[\frac{5d_{22}}{d_{12}^2 + d_{22}^2}k_8^3(E,k_s) + \frac{5(d_{22}^3 - 3d_{22}^2 d_{12})}{k_8(E,k_s)} + 34k_8(E,k_s)d_{22}\right]$$

$$H_{182} = \tfrac{1}{2}[H_{112} + H_{122} + \Delta_0(H_{132} + H_{142}) + \Delta_0(H_{152} + H_{162})],$$
$$H_{112} = 2(\sinh g_{12} \sin g_{22} \cos \gamma_8(E, k_s)),$$
$$H_{122} = ([\Omega_{22}(E, k_s)(\sinh g_{12})(\cos g_{22}) + \Omega_{12}(E, k_s)(\sin g_{22})(\cosh g_{12})](\sin \gamma_8(E, k_s))),$$
$$H_{132} = ([\Omega_{42}(E, k_s)(\cosh g_{12})(\cos g_{22}) + \Omega_{32}(E, k_s)(\sinh g_{12})(\sin g_{22})](\sin \gamma_8(E, k_s))),$$
$$H_{142} = ([\Omega_{62}(E, k_s)(\sinh g_{12})(\cos g_{22}) + \Omega_{52}(E, k_s)(\sin g_{12})(\cosh g_{22})](\sin \gamma_8(E, k_s))),$$
$$H_{152} = ([\Omega_{102}(E, k_s)(\cosh g_{12})(\cos g_{22}) + \Omega_{92}(E, k_s)(\sinh g_{12})(\sin g_{22})](\cos \gamma_8(E, k_s))),$$
$$H_{162} = ([\Omega_{82}(E, k_s)(\sinh g_{12})(\cos g_{22}) + \Omega_{72}(E, k_s)(\sin g_{12})(\cosh g_{22})](\sin \gamma_8(E, k_s)/12)),$$
$$H_{192} = \left[((2C_{182}D_{182})/L_0^2)\right] \text{ and } D_{182} = \sinh^{-1}(\omega_{182})$$

The simplified DR in heavily doped quantum well HgTe/CdTe superllatices with graded interfaces can be expressed as

$$\left(\frac{n_z\pi}{d_z}\right)^2 = G_{192} + iH_{192} \tag{13.14a}$$

The sub-band equation in this case can be expressed as

$$\left(\frac{n_z\pi}{d_z}\right)^2 = [G_{192} + iH_{192}]\Big|_{k_s=0 \text{ and } E=E_{12,4}} \tag{13.14b}$$

where $E_{12,4}$ is the sub-band energy in this case.

13.2.5 The DR in Strained Layer Quantum Well HD Superlattices with Graded Interfaces

The DR of the conduction electrons of the constituent materials of the strained layer super lattices can be expressed as

$$[E - T_{1i}]k_x^2 + [E - T_{2i}]k_y^2 + [E - T_{3i}]k_z^2 = q_iE^3 - R_iE^2 + V_iE + \zeta_i \tag{13.15}$$

where

$$T_{1i} = \overline{\theta}_i, \quad \overline{\theta}_i = \left[E_{gi} - C_{1i}^c\varepsilon_i - (a_i + C_{1i}^c)\varepsilon_i + \tfrac{3}{2}b_i\varepsilon_{xxi} - \tfrac{b_i\varepsilon_i}{2} + \tfrac{\sqrt{3}d_i\varepsilon_{xyi}}{2}\right],$$
$$T_{2i} = \omega_i, \quad \omega_i = \left[E_{gi} - C_{1i}^c\varepsilon_i - (a_i + C_{1i}^c)\varepsilon_i + \tfrac{3}{2}b_i\varepsilon_{xxi} - \tfrac{b_i\varepsilon_i}{2} - \tfrac{\sqrt{3}d_i\varepsilon_{xyi}}{2}\right],$$
$$T_{3i} = \delta_i, \quad \delta_i = \left[E_{gi} - C_{1i}^c\varepsilon_i + (a_i + C_{1i}^c)\varepsilon_i + \tfrac{3}{2}b_i\varepsilon_{zzi} - \tfrac{b_i\varepsilon_i}{2}\right]$$
$$R_i = q_i\left[2A_i + C_{1i}^c\varepsilon_i\right], \quad q_i = \tfrac{3}{2B_{2i}^2}, \quad A_i = E_{gi} - C_{1i}^c\varepsilon_i,$$
$$V_i = q_i\left[A_i^2 - \tfrac{2C_{2i}^2\varepsilon_{xyi}}{3} + 2A_iC_{1i}^c\varepsilon_i\right], \quad \zeta_i = q_i\left[\tfrac{2C_{2i}^2\varepsilon_{xyi}}{3} - C_{1i}^c\varepsilon_iA_i^2\right]$$

Therefore the electron energy spectrum in HD stressed materials can be written as

$$\overline{P_i}(E,\eta_{gi})k_x^2+\overline{Q_i}(E,\eta_{gi})k_y^2+\overline{S_i}(E,\eta_{gi})k_z^2=1 \tag{13.16}$$

where

$$\overline{P_i}(E,\eta_{gi})=\frac{[\gamma_0(E,\eta_i-I_0T_{1i})]}{\overline{\Delta}_i(E,\eta_{gi})},$$
$$\overline{\Delta}_i(E,\eta_{gi})=\left[\frac{-q_i\eta_{gi}^3}{2\sqrt{\pi}}\exp\left(\frac{-E^2}{\eta_{gi}^2}\right)\left[1+\frac{E^2}{\eta_{gi}^2}\right]-R_i\theta_0(E,\eta_{gi})+V_i\gamma_0(E,\eta_{gi})+\frac{\zeta_i}{2}\left[1+Erf\left(\frac{E}{\eta_{gi}}\right)\right]\right],$$
$$I_0=\frac{1}{2}\left[1+Erf(E/\eta_{gi})\right],\overline{Q_i}(E,\eta_{gi})=\frac{[\gamma_0(E,\eta_{gi})-I_0T_{2i}]}{\bar{\Delta}_i(E,\eta_{gi})}\text{ and }\bar{S}_i(E,\eta_{gi})=\frac{[\gamma_0(E,\eta_{gi})-I_0T_{3i}]}{\bar{\Delta}_i(E,\eta_{gi})}$$

The energy-wave vector dispersion relation of the conduction electrons in heavily doped strained layer SLs with graded interfaces can be expressed as

$$\cos(L_0k)=\frac{1}{2}\overline{\phi_6}(E,k_s) \tag{13.17}$$

where

$$\begin{aligned}\overline{\phi_6}(E,k_s)\equiv&\Big[2\cosh[T_4(E,\eta_{g2})]\cos[T_5(E,\eta_{g1})]\Big]\\&+[T_6(E,k_s)]\sinh[T_4(E,\eta_{g2})]\sin[T_5(E,\eta_{g1})]\\&+\Delta_0\left[\left(\frac{k_0^2(E,\eta_{g2})}{k'(E,\eta_{g1})}-3k'(E,\eta_{g1})\right)\cosh\Big[T_4(E,\eta_{g2})\Big]\sin\Big[T_5(E,\eta_{g1})\Big]\right.\\&\left.+\left(3k_0(E,\eta_{g2})\frac{k'^2(E,\eta_{g1})}{k_0(E,\eta_{g2})}\right)\sinh[T_4(E,\eta_{g2})]\cos[T_5(E,\eta_{g1})]\right]\\&+\Delta_0\Big[2(k_0^2(E,\eta_{g2})-k'^2(E,\eta_{g1}))\cosh\Big[T_4(E,\eta_{g2})\Big]\cos\Big[T_5(E,\eta_{g1})\Big]\Big]\\&+\frac{1}{12}\left(\frac{5k_0^3(E,\eta_{g2})}{k'(E,\eta_{g1})}+\frac{5k'^3(E,\eta_{g1})}{k_0(E,\eta_{g2})}-34k_0(E,\eta_{g2})k'(E,\eta_{g1})\right)\\&\sinh\Big[T_4(E,\eta_{g2})\Big]\sin\Big[T_4(E,\eta_{g1})\Big]\end{aligned}$$

$$[T_4(E,\eta_{g2})] = k_0(E,\eta_{g2})[a_0 - \Delta_0],$$
$$k_0(E,\eta_{g2}) = [\overline{S_2}(E - V_0,\eta_{g2})]^{-1/2}$$
$$\left[\overline{P_2}(E - V_0,\eta_{g2})k_x^2 + \overline{Q_2}(E - V_0,\eta_{g2})k_y^2 - 1\right]^{-1/2},$$
$$T_5(E,\eta_{g1}) = k'(E,\eta_{g1})[b_0 - \Delta_0],$$
$$k'(E,\eta_{g1}) = [\overline{S_1}(E,\eta_{g1})]^{-1/2}\left[1 - \overline{P_1}(E,\eta_{g1})k_x^2 - \overline{Q_1}(E,\eta_{g1})k_y^2\right]^{1/2} \text{ and}$$
$$T_6(E,k_s) = \left[\frac{k_0(E,\eta_{g2})}{k'(E,\eta_{g1})} - \frac{k'(E,\eta_{g1})}{k_0(E,\eta_{g2})}\right]$$

Therefore the DR of the conduction electrons in heavily doped strained layer quantum well SL with graded interfaces can be expressed as

$$\left(\frac{n_z\pi}{d_z}\right)^2 = \frac{1}{L_0^2}\left[\cos^{-1}\left\{\frac{1}{2}\overline{\Phi_6}(E,k_s)\right\}\right]^2 - k_s^2 \tag{13.18a}$$

The sub-band equation in this case can be expressed as

$$\left(\frac{n_z\pi}{d_z}\right)^2 = \frac{1}{L_0^2}\left[\cos^{-1}\left\{\frac{1}{2}\overline{\Phi_6}(E_{12,5},0)\right\}\right]^2 \tag{13.18b}$$

where $E_{12,5}$ is the sub-band energy in this case.

13.2.6 The DR in III–V Quantum Well HD Effective Mass Super Lattices

Following Sasaki [24], the electron dispersion law in III–V heavily doped effective mass super-lattices (EMSLs) can be written as

$$k_x^2 = \left[\frac{1}{L_0^2}\{\cos^{-1}(f_{21}(E,k_y,k_z))\}^2 - k_\perp^2\right] \tag{13.19}$$

in which $f_{21}(E,k_y,k_z) = a_1\cos[a_0C_{21}(E,k_\perp,\eta_{g1}) + b_0D_{21}(E,k_\perp,\eta_{g2})]$
$-a_2\cos[a_0C_{21}(E,k_\perp,\eta_{g1}) - b_0D_{21}(E,k_\perp,\eta_{g2})], \quad k_\perp^2 = k_y^2 + k_z^2,$

$$a_1 = \left[\sqrt{\frac{M_2(0,\eta_{g2})}{M_1(0,\eta_{g1})}} + 1\right]^2 \left[4\left(\frac{M_2(0,\eta_{g2})}{M_1(0,\eta_{g1})}\right)^{1/2}\right]^{-1},$$

$$a_2 = \left[\sqrt{\frac{M_2(0,\eta_{g2})}{M_1(0,\eta_{g1})}} - 1\right]^2 \left[4\left(\frac{M_2(0,\eta_{g2})}{M_1(0,\eta_{g1})}\right)^{1/2}\right]^{-1},$$

$$M_i(0,\eta_{gi}) = m^*_{ci}\left[\frac{-2}{\sqrt{\pi}}T(0,\eta_{gi}) + 2\left[\frac{\alpha_i b_i}{c_i}\frac{\eta_{gi}}{\sqrt{\pi}} + \frac{1}{2}\left(\frac{\alpha_i c_i + c_i b_i - \alpha_i b_i}{c_i^2}\right)\right.\right.$$
$$+\frac{1}{\sqrt{\pi c_i}}\left(1-\frac{\alpha_i}{c_i}\right)\left(1-\frac{b_i}{c_i}\right) - \frac{1}{c_i}\left(1-\frac{\alpha_i}{c_i}\right)\left(1-\frac{b_i}{c_i}\right)\frac{2}{c_i\eta_{gi}\sqrt{\pi}}$$
$$\left\{\frac{-2}{c_i\eta_{gi}^2}\exp\left(\frac{-1}{c_i^2\eta_{gi}^2}\right)\left(\sum_{p=1}^{\alpha}\left(\exp\left(\frac{-p^2}{4}\right)\right)\frac{1}{p}\sinh\left(\frac{p}{c_i\eta_{gi}}\right)\right)\right.$$
$$\left.\left.\left.+\exp\left(\frac{-1}{c_i^2\eta_{gi}^2}\right)\left(\sum_{p=1}^{\alpha}\left(\exp\left(\frac{-p^2}{4}\right)\right)\frac{1}{\eta_{gi}}\cosh\left(\frac{p}{c_i\eta_{gi}}\right)\right)\right\}\right]\right],$$

$$\mathrm{T}(0,\eta_{gi}) = 2\left[\frac{\alpha_i b_i}{c_i}\frac{\eta_{gi}^2}{\sqrt{4}} + \left(\frac{\alpha_i c_i + b_i c_i - \alpha_i b_i}{c_i^2}\right)\frac{\eta_{gi}}{2\sqrt{\pi}} + \frac{1}{2c_i}\left(1-\frac{\alpha_i}{c_i}\right)\left(1-\frac{b_i}{c_i}\right)\right.$$
$$\left.-\frac{1}{c_i}\left(1-\frac{\alpha_i}{c_i}\right)\left(1-\frac{b_i}{c_i}\right)\frac{2}{c_i\eta_{gi}\sqrt{\pi}}\exp\left(\frac{-1}{c_i^2\eta_{gi}^2}\right)\sum_{p=1}^{\alpha}\frac{\exp(-p^2/4)}{p}\sinh\left(\frac{p}{c_i\eta_{gi}}\right)\right],$$

$$C_{21}(E,k_\perp,\eta_{g1}) = e_1 + ie_2, D_{21}(E,k_\perp,\eta_2) = e_3 + ie_4,$$

$$e_1 = \left[\left(\left(\sqrt{t_1^2+t_2^2}+t_1\right)/2\right)\right]^{\frac{1}{2}}, e_2 = \left[\left(\left(\sqrt{t_1^2+t_2^2}-t_1\right)/2\right)\right]^{\frac{1}{2}},$$

$$t_1 = \left[\frac{2m^*_{c1}}{\hbar^2}T_{11}(E,\Delta_1,\eta_{g1},E_{g1}) - k_\perp^2\right], \mathrm{t}_2 = \frac{2m^*_{c1}}{\hbar^2}T_{21}(E,\Delta_1,\eta_{g1},E_{g1}),$$

$$e_3 = \left[\frac{\sqrt{t_3^2+t_4^2}+t_3}{2}\right]^{1/2}, e_4 = \left[\frac{\sqrt{t_3^2+t_4^2}-t_3}{2}\right]^{1/2}$$

$$t_3 = \left[\frac{2m^*_{c2}}{\hbar^2}T_{12}(E,\Delta_2,\eta_{g2},E_{g2}) - k_\perp^2\right], t_4 = \frac{2m^*_{c2}}{\hbar^2}T_{22}(E,\Delta_2,\eta_{g2},E_{g2}),$$

Therefore (12.19) can be expressed as

$$k_x^2 = \delta_7 + i\delta_8 \tag{13.20}$$

where

$$\delta_7 = \left[\frac{1}{L_0^2}\left(\delta_5^2 - \delta_6^2\right) - k_\perp^2\right], \quad \delta_5 = \cos^{-1} p_5,$$
$$p_5 = \left[\frac{1-\delta_3^2-\delta_4^2-\sqrt{(1-\delta_3^2-\delta_4^2)+4\delta_4^2}}{2}\right]^{1/2},$$

$$\delta_3 = (a_1 \cos\Delta_1 \cosh\Delta_2 - a_2 \cos\Delta_3 \cosh\Delta_4),$$
$$\delta_4 = (a_1 \sin\Delta_1 \sinh\Delta_2 - a_2 \sin\Delta_3 \sinh\Delta_4),$$
$$\Delta_1 = (a_0e_1 + b_0e_3), \Delta_2 = (a_0e_2 + b_0e_4),$$
$$\Delta_3 = (a_0e_1 - b_0e_3), \Delta_4 = (a_0e_2 - b_0e_4),$$
$$\delta_6 = \sinh^{-1} p_5 \text{ and } \delta_8 = \left[2\delta_5\delta_6/L_0^2\right]$$

The DR in III–V heavily doped effective mass quantum well super-lattices can be written as

$$\left(\frac{n_x\pi}{d_x}\right)^2 = \delta_7 + i\delta_8 \tag{13.21a}$$

The sub-band equation in this case can be expressed as

$$\left(\frac{n_x\pi}{d_x}\right)^2 = [\delta_7 + i\delta_8]\Big|_{k_\perp=0 \text{ and } E=E_{12,6}} \tag{13.21b}$$

where $E_{12,6}$ is the sub-band energy in this case.

13.2.7 The DR in II–VI Quantum Well HD Effective Mass Super Lattices

Following Sasaki [24], the electron dispersion law in heavily doped II–VI EMSLs can be written as

$$k_z^2 = \Delta_{13} + i\Delta_{14}, \tag{13.22}$$

where

$$\Delta_{13} = \left[\frac{1}{L_0^2}(\Delta_{11}^2 - \Delta_{12}^2) - k_s^2\right]$$

$$\Delta_{11} = \cos^{-1} p_6, p_6 = \left[\frac{1 - \Delta_9^2 - \Delta_{10}^2 - \sqrt{(1 - \Delta_9^2 - \Delta_{10}^2)^2 + 4\Delta_{10}^2}}{2}\right]^{1/2},$$

$$\Delta_9 = (\overline{a_1} \cos \Delta_6 \cosh \Delta_7 - \overline{a_2} \cos \Delta_8 \cosh \Delta_7),$$
$$\Delta_{10} = (\overline{a_1} \sin \Delta_6 \sinh \Delta_7 + \overline{a_2} \cos \Delta_8 \sinh \Delta_7),$$
$$\Delta_6 = \left[a_0 C_{22}(E, k_s, \eta_{g1}) + b_0 e_3\right],$$
$$\Delta_7 = b_0 e_4,$$
$$\Delta_8 = \left[a_0 C_{22}(E, k_s, \eta_{g1}) - b_0 e_3\right],$$

$$C_{22}(E, k_s, \eta_{g1}) = \left[\frac{2m_{\parallel}^*}{\hbar^2}\left\{\gamma_3(E, \eta_{g1}) - \frac{\hbar^2 k_s^2}{2m_{\perp,1}^*} \mp C_0 k_s\right\}\right]^{1/2},$$

$$\overline{a_1} = \left[\sqrt{\frac{M_2(0, \eta_{g2})}{\overline{M}_1(0, \eta_{g1})}} + 1\right]^2 4\left[\left(\frac{M_2(0, \eta_{g2})}{\overline{M}_1(0, \eta_{g1})}\right)^{1/2}\right]^{-1},$$

$$\overline{M}_1(0, \eta_{g1}) = m_{c1}^*\left(1 - \frac{2}{\pi}\right),$$

$$\overline{a_2} = \left[\sqrt{\frac{M_2(0, \eta_{g2})}{\overline{M}_1(0, \eta_{g1})}} - 1\right]^2 4\left[\left(\frac{M_2(0, \eta_{g2})}{\overline{M}_1(0, \eta_{g1})}\right)^{1/2}\right]^{-1}$$

$$\Delta_{12} = \cos^{-1} p_6, \Delta_{14} = \frac{2\Delta_{11}\Delta_{12}}{L_0^2}$$

The DR in III–V heavily doped effective mass quantum well super-lattices can be written as

$$\left(\frac{n_z \pi}{d_z}\right)^2 = \delta_{13} + i\delta_{14} \tag{13.23a}$$

The sub-band equation in this case can be expressed as

$$\left(\frac{n_z \pi}{d_z}\right)^2 = [\delta_{13} + i\delta_{14}]\bigg|_{k_s=0 \text{ and } E=E_{12,7}} \tag{13.23b}$$

where $E_{12,7}$ is the sub-band energy in this case.

13.2.8 The DR in IV–VI Quantum Well HD Effective Mass Super Lattices

Following Sasaki [24], the electron dispersion law in IV–VI, EMSLs can be written as

$$k_z^2 = \left[\frac{1}{L_0^2}\left\{\cos^{-1}\left(f_{23}(E,k_x,k_y)\right)\right\}^2 - k_s^2\right] \tag{13.24}$$

where

$$\begin{aligned} f_{23}(E,k_x,k_y) = {} & a_3\cos\left[a_0C_{23}(E,k_x,k_y,\eta_{g1})\right. \\ & \left.+ b_0D_{23}(E,k_x,k_y,\eta_{g1})\right] - a_4\cos\left[a_0C_{23}(E,k_x,k_y,\eta_{g2})\right. \\ & \left.- b_0D_{23}(E,k_x,k_y,\eta_{g2})\right], \end{aligned}$$

$$a_3 = \left[\sqrt{\frac{M_3(0,\eta_{g2})}{M_3(0,\eta_{g1})}} + 1\right]^2 \left[4\left(\frac{M_3(0,\eta_{g2})}{M_3(0,\eta_{g1})}\right)^{1/2}\right]^{-1},$$

$$a_4 = \left[\sqrt{\frac{M_3(0,\eta_{g2})}{M_3(0,\eta_{g1})}} - 1\right]^2 \left[4\left(\frac{M_3(0,\eta_{g2})}{M_3(0,\eta_{g1})}\right)^{1/2}\right]^{-1}$$

$$\begin{aligned} M_3(0,\eta_{gi}) = {} & (4\overline{p_{9,i}})^{-1}\left[\left\{\alpha_i\left(1-\frac{2}{\pi}\right)\left(\frac{1}{m_{l,i}^+}-\frac{1}{m_{l,i}^-}\right)\right\} + \left[\left[\overline{q_{9,i}}(0,\eta_{gi})\right]^2\right.\right. \\ & \left.+ (4\overline{p_{9,i}})\overline{R_{9,i}}(0,\eta_{gi})\right]^{-1/2}\left[\alpha_i\left(1-\frac{2}{\pi}\right)\left(\frac{1}{m_{l,i}^+}-\frac{1}{m_{l,i}^-}\right)\overline{q_{9,i}}(0,\eta_{gi})\right. \\ & \left.\left. + 2\overline{p_{9,i}}\left(1-\frac{2}{\pi}+\frac{\alpha_i\eta_{gi}}{\sqrt{\pi}}\right)\right]\right], \overline{p_{9,i}} = \frac{\alpha_i\hbar^4}{4m_{l,i}^+ m_{l,i}^-}, \end{aligned}$$

$$\overline{q_{9,i}}(0,\eta_{gi}) = \left[\frac{\hbar^2}{2}\left(\frac{1}{m_{l,i}^+}+\frac{1}{m_{l,i}^-}\right) - \frac{\alpha_i\eta_{gi}}{\sqrt{\pi}}\left(\frac{1}{m_{l,i}^+}-\frac{1}{m_{l,i}^-}\right)\right],$$

$$\overline{R_{9,i}}(0,\eta_{gi}) = \left[\frac{\eta_{gi}}{\sqrt{\pi}} + \frac{\alpha_i\eta_{gi}^2}{2}\right]$$

$$\begin{aligned} C_{23}(E,k_x,k_y,\eta_{g1}) = {} & \left[\left[2\overline{p_{9,1}}\right]^{-1}\left[-\overline{q_{9,1}}(E,k_x,k_y,\eta_{g1}) + \left[\left\{\overline{q_{9,1}}(E,k_x,k_y,\eta_{g1})\right\}^2\right.\right.\right. \\ & \left.\left.\left. + (4\overline{p_{9,1}})\overline{R_{9,1}}(E,k_x,k_y,\eta_{g1})\right]^{1/2}\right]\right]^{1/2}, \end{aligned}$$

$$D_{23}(E,k_x,k_y,\eta_{g2}) = \left[\left[2\overline{p_{9,2}}\right]^{-1}\left[-\overline{q_{9,2}}(E,k_x,k_y,\eta_{g2}) + \left[\{\overline{q_{9,2}}(E,k_x,k_y,\eta_{g2})\}^2 + \left(4\overline{p_{9,2}}\right)\overline{R_{9,2}}(E,k_x,k_y,\eta_{g2})\right]^{1/2}\right]\right]^{1/2},$$

$$\overline{-q_{9,i}}(E,k_x,k_y,\eta_{gi}) = \left[\frac{\hbar^2}{2}\left(\frac{1}{m^*_{l,i}} + \frac{1}{m^-_{l,i}}\right) + \alpha_i\frac{\hbar^4}{4}k_s^2\left(\frac{1}{m^+_{l,i}m^-_{t,i}} + \frac{1}{m^+_{t,i}m^-_{l,i}}\right) - \alpha_i\gamma_3(E,\eta_{gi})\left(\frac{1}{m^+_{l,i}} - \frac{1}{m^-_{l,i}}\right)\right],$$

$$\overline{R_{9,i}}(E,k_x,k_y,\eta_{gi}) = \left[\gamma_2(E,\eta_{gi}) + \gamma_3(E,\eta_{gi})\alpha_i\frac{\hbar^2}{2}k_s^2\left(\frac{1}{m^+_{t,i}} - \frac{1}{m^-_{t,i}}\right) - \frac{\hbar^2}{2}k_s^2\left(\frac{1}{m^*_{t,i}} - \frac{1}{m^-_{t,i}}\right) - \frac{\alpha\hbar^6}{4}\frac{k_s^4}{m^-_{t,i}m^+_{t,i}}\right], \quad a_5 = \left[\sqrt{\frac{m^*_2}{m^*_1}} + 1\right]^2\left[4\left(\frac{m^*_2}{m^*_1}\right)^{1/2}\right]^{-1}$$

Therefore the DR in heavily doped IV–VI, quantum well EMSLs can be written as

$$\left(\frac{n_z\pi}{d_z}\right)^2 = \left[\frac{1}{L_0^2}\{\cos^{-1}(f_{23}(E,k_x,k_y))\}^2 - k_s^2\right] \tag{13.25a}$$

The sub-band equation in this case can be expressed as

$$\left(\frac{n_z\pi}{d_z}\right)^2 = \left[\frac{1}{L_0^2}\{\cos^{-1}(f_{23}(E_{12,8},0,0))\}^2\right] \tag{13.25b}$$

where $E_{12,8}$ is the sub-band energy in this case.

13.2.9 The DR in HgTe/CdTe Quantum Well HD Effective Mass Super Lattices

Following Sasaki [24], the DR in heavily doped HgTe/CdTe EMSLs can be written as

$$k_z^2 = \Delta_{13H} + i\Delta_{14H} \tag{13.26}$$

where

$$\Delta_{13H} = \left[\frac{1}{L_0^2}\left(\Delta_{11H}^2 - \Delta_{12H}^2\right) - k_s^2\right]$$

$$\Delta_{11H} = \cos^{-1} p_{6H}, p_{6H} = \left[\frac{1 - \Delta_{9H}^2 - \Delta_{10H}^2 - \sqrt{\left(1 - \Delta_{9H}^2 - \Delta_{10H}^2\right)^2 + 4\Delta_{10H}^2}}{2}\right]^{1/2},$$

$$\Delta_{9H} = \left(\overline{a_{1H}} \cos \Delta_{5H} \cosh \Delta_{6H} - \overline{a_{2H}} \cos \Delta_{7H} \cosh \Delta_{6H}\right),$$
$$\Delta_{10H} = \left(\overline{a_{1H}} \sin \Delta_{5H} \sinh \Delta_{6H} + \overline{a_{2H}} \cos \Delta_{7H} \sinh \Delta_{6H}\right),$$
$$\Delta_{5H} = \left[a_0 C_{22H}(E, k_s, \eta_{g1}) + b_0 e_3\right], \Delta_{6H} = b_0 e_4, \Delta_{7H} = \left[a_0 C_{22H}(E, k_s, \eta_{g1}) - b_0 e_3\right],$$

$$C_{22H}(E, k_s, \eta_{g1}) = \left[\frac{B_{01}^2 + 2A_1 E - B_{01}(B_{01}^2 + 4A_1 E)}{2A_1^2} - k_s^2\right]^{1/2},$$

$$\overline{a_{1H}} = \left[\sqrt{\frac{M_2(0, \eta_{g2})}{m_{c1}^*}} + 1\right]^2 \left[4\left[\left(\frac{M_2(0, \eta_{g2})}{m_{c1}^*}\right)^{1/2}\right]\right]^{-1},$$

$$\overline{a_{2H}} = \left[\sqrt{\frac{M_2(0, \eta_{g2})}{m_{c1}^*}} - 1\right]^2 \left[4\left[\left(\frac{M_2(0, \eta_{g2})}{m_{c1}^*}\right)^{1/2}\right]\right]^{-1}$$

$$\Delta_{12H} = \cos^{-1} p_{6H}, \Delta_{14H} = \frac{2\Delta_{11H}\Delta_{12H}}{L_0^2}$$

The DR in heavily doped HgTe/CdTe QWEMSLs can be written as

$$\left(\frac{n_z \pi}{d_z}\right)^2 = \Delta_{13H} + i\Delta_{14H} \tag{13.27a}$$

The sub-band equation in this case can be expressed as

$$\left(\frac{n_z \pi}{d_z}\right)^2 = \left[\Delta_{13H} + i\Delta_{14H}\right]\Big|_{k_s=0 \text{ and } E=E_{12,9}} \tag{13.27b}$$

where $E_{12,9}$ is the sub-band energy in this case.

13.2.10 The DR in Strained Layer Quantum Well HD Effective Mass Super Lattices

The DR of the constituent materials of heavily doped III–V super lattices can be written as

$$\overline{P_i}(\mathrm{E}, \eta_{gi})\mathrm{k}_x^2 + \overline{Q_i}(\mathrm{E}, \eta_{gi})\mathrm{k}_y^2 + \overline{S_i}(\mathrm{E}, \eta_{gi})\mathrm{k}_z^2 = 1 \tag{13.28}$$

where,

$$\overline{P_i}(\mathrm{E}, \eta_{\mathrm{gi}}) = (\gamma_0(\mathrm{E}, \eta_{\mathrm{gi}}) - \mathrm{I_0 T_{1i}})(\overline{\Delta_i}(\mathrm{E}, \eta_{\mathrm{gi}}))^{-1}, \mathrm{I_0} = (1/2)[1 + \mathrm{Erf}(\mathrm{E}/\eta_{\mathrm{gi}})],$$
$$T_{1i} = [E_{gi} - C_{1i}^c \varepsilon_i - (a_i + C_{1i}^c)\varepsilon_i + (3/2) b_i \varepsilon_{xxi} - (b_i \varepsilon_i/2) + (\sqrt{3d_i}\varepsilon_{xyi}/2)],$$
$$\overline{\Delta_i}(E, \eta_{gi}) = [(-q_i \eta_{gi}^3/2\sqrt{\pi})\exp(-(E^2/\eta_{gi}^2))[1 + (E^2/\eta_{gi}^2)] - R_i \theta_0(E, \eta_{gi}) + V_i \gamma_0(E, \eta_{gi})$$
$$+ (\zeta_i/2)[1 + Erf(E/\eta_{gi})]], q_i = (3/2B_{2i}^2), R_i = q_i[2A_i + C_{1i}^c \varepsilon_i], A_i = E_{gi} - C_{1i}^c \varepsilon_i,$$
$$V_i = q_i[A_i^2 - (2C_{2i}^2 \varepsilon_{xyi}/3) + 2A_i C_{1i}^c \varepsilon_i], \zeta_i = q_i[(2C_{2i}^2 \varepsilon_{xyi}/3) - C_{1i}^c \varepsilon_i A_i^2],$$
$$\overline{Q_i}(E, \eta_{gi}) = (\gamma_0(E, \eta_{gi}) - I_0 T_{2i})(\overline{\Delta_i}(E, \eta_{gi}))^{-1},$$
$$T_{2i} = [E_{gi} - C_{1i}^c \varepsilon_i - (a_i + C_{1i}^c)\varepsilon_i + (3/2) b_i \varepsilon_{xxi} - (b_i \varepsilon_i/2)$$
$$- (\sqrt{3d_i}\varepsilon_{xyi}/2)], \overline{S_i}(E, \eta_{gi}) = (\gamma_0(E, \eta_{gi}) - I_0 T_{3i})(\overline{\Delta_i}(E, \eta_{gi}))^{-1},$$
$$T_{3i} = [E_{gi} - C_{1i}^c \varepsilon_i + (a_i + C_{1i}^c)\varepsilon_i + (3/2) b_i \varepsilon_{zzi} - (b_i \varepsilon_i/2)],$$

The electron energy spectrum in heavily doped strained layer effective mass super-lattices can be written as

$$k_z^2 = \left[\frac{1}{L_0^2}\left\{\cos^{-1}\left(f_{40}\left(E, k_x, k_y\right)\right)\right\}^2 - k_s^2\right] \tag{13.29}$$

where

$$f_{40}\left(E, k_x, k_y\right) = a_{20}\cos\left[a_0 C_{40}\left(E, k_x, k_y, \eta_{g1}\right) + b_0 D_{40}\left(E, k_x, k_y, \eta_{g1}\right)\right]$$
$$- a_{21}\cos\left[a_0 C_{40}\left(E, k_x, k_y, \eta_{g2}\right) - b_0 D_{40}\left(E, k_x, k_y, \eta_{g2}\right)\right],$$

$$a_{20} = \left[\sqrt{\frac{M_{s2}(0, \eta_{g2})}{M_{s1}(0, \eta_{g1})}} + 1\right]^2 4\left[\left(\frac{M_{s2}(0, \eta_{g2})}{M_{s1}(0, \eta_{g1})}\right)^{1/2}\right]^{-1},$$

$$M_{si}(0, \eta_{gi}) = (\hbar/2)\rho_i(\eta_{gi})$$
$$\rho_i(\eta_{gi}) = [(\eta_{gi}/2\sqrt{\pi}) - (T_{3i}/2)]^{-2} \times \left[\{(\eta_{gi}/2\sqrt{\pi}) - (T_{3i}/2)\}\right.$$
$$\{(R_i\eta_{gi}/\sqrt{\pi}) + (\zeta_i/\eta_{gi}\sqrt{\pi})\} - ((1/2) - (T_{3i}/\eta_{gi}\sqrt{\pi}))$$
$$\left.\left\{(\zeta_i/2) + (V_i\eta_{gi}/2\sqrt{\pi}) - (R_i\eta_{gi}^2/4) - (q_i\eta_{gi}^3/2\sqrt{\pi})\right\}\right]$$

$$a_{20} = \left[\sqrt{\frac{M_{s2}(0, \eta_{g2})}{M_{s1}(0, \eta_{g1})}} + 1\right]^2 4\left[\left(\frac{M_{s2}(0, \eta_{g2})}{M_{s1}(0, \eta_{g1})}\right)^{1/2}\right]^{-1},$$

$$a_{21} = \left[\sqrt{\frac{M_{s2}(0, \eta_{g2})}{M_{s1}(0, \eta_{g1})}} - 1\right]^2 4\left[\left(\frac{M_{s2}(0, \eta_{g2})}{M_{s1}(0, \eta_{g1})}\right)^{1/2}\right]^{-1}$$

$$C_{40}(E,k_x,k_y,\eta_{g1}) = [1 - \overline{P_1}(E,\eta_{g1})k_x^2 - \overline{Q_1}(E,\eta_{g1})k_y^2]^{1/2}[\overline{S_1}(E,\eta_{g1})]^{-1/2}$$
$$D_{40}(E,k_x,k_y,\eta_{g2}) = [1 - \overline{P_2}(E,\eta_{g2})k_x^2 - \overline{Q_2}(E,\eta_{g2})k_y^2]^{1/2}[\overline{S_2}(E,\eta_{g2})]^{-1/2}$$

Therefore, the DR in heavily doped strained layer effective mass quantum well super-lattices can be expressed as

$$\left(\frac{n_z\pi}{d_z}\right)^2 = \left[\frac{1}{L_0^2}\left\{\cos^{-1}(f_{40}(E,k_x,k_z))\right\}^2 - k_s^2\right] \tag{13.30a}$$

The sub-band equation in this case can be expressed as

$$\left(\frac{n_z\pi}{d_z}\right)^2 = \left[\frac{1}{L_0^2}\left\{\cos^{-1}(f_{40}(E_{12,10},0,0))\right\}^2\right] \tag{13.30b}$$

where $E_{12,10}$ is the sub-band energy in this case.

13.3 Summary and Conclusion

This chapter explore the DR in III–V, II–VI, IV–VI, HgTe/CdTe and strained layer quantum well heavily doped superlattices (QWHDSLs) with graded interfaces has been studied in Sects. 13.2.1 to 13.2.5. From Sects. 13.2.6 to 13.2.10, the DR from III–V, II–VI, IV–VI, HgTe/CdTe and strained layer quantum well heavily doped effective mass superlattices respectively has been presented. The presence of essential poles in the constituent materials of HD III–V and HgTe/CdTe SLs makes the DR in the said SLs complex (for both SLs with graded interfaces and effective mass SLs). The $E - k_s^2$ plots for HD III–V and HgTe/CdTe QWSLs are quantized closed 2D surfaces and they exist in the complex energy plane. The DR for other cases are quantized closed 2D surfaces and the exist in the real energy plane. The DRs in all the cases are concentration dependent for any value of the electron energy. The sub band energies will be real or complex in accordance with said logic. The EEM will be wave vector dependent and analytical expressions of the DOS function are not possible so that the numerical integration method should be used for the determination of the DOS functions in the respective cases.

13.4 Open Research Problem

(R.13.1) Investigate the influence of arbitrarily oriented alternating quantizing magnetic field and strain on the DR for all types of HD super-lattices whose carrier energy spectra are described in this book.

References

1. S. Mukherjee, S.N. Mitra, P.K. Bose, A.R. Ghatak, A. Neoigi, J.P. Banerjee, A. Sinha, M. Pal, S. Bhattacharya, K.P. Ghatak, J. Comput. Theor. Nanosci. **4**, 550 (2007); N.G. Anderson, W. D. Laidig, R.M. Kolbas, Y.C. Lo, J. Appl. Phys. **60**, 2361 (1986); N. Paitya, K.P. Ghatak, J. Adv. Phys. **1**, 161 (2012); N. Paitya, S. Bhattacharya, D. De, K.P. Ghatak Adv. Sci. Eng. Med. **4**, 96 (2012); S. Bhattacharya, D. De, S.M. Adhikari, K.P. Ghatak, Superlattices Microstruct. **51**, 203 (2012); D. De, S. Bhattacharya, S.M. Adhikari, A. Kumar, P.K. Bose, K. P. Ghatak, Beilstein J. Nanotech. **2**, 339 (2012); D. De, A. Kumar, S.M. Adhikari, S. Pahari, N. Islam, P. Banerjee, S.K. Biswas, S. Bhattacharya, K.P. Ghatak, Superlattices Microstruct. 47, (377) (2010); S. Pahari, S. Bhattacharya, S. Roy, A. Saha, D. De, K.P. Ghatak, Superlattices Microstruct. **46**, 760 (2009); S. Pahari, S. Bhattacharya, K.P. Ghatak J. Comput. Theor. Nanosci. **6**, 2088 (2009); S.K. Biswas, A.R. Ghatak, A. Neogi, A. Sharma, S. Bhattacharya, K.P. Ghatak, Phys. E: Low-dimens. Syst. Nanostruct. **36**, 163 (2007); L. J. Singh, S. Choudhury, D. Baruah, S.K. Biswas, S. Pahari, K.P. Ghatak, Phys. B: Conden. Matter, **368**, 188 (2005); S. Chowdhary, L.J. Singh, K.P. Ghatak, Phys. B: Condens. Matter, **365**, 5 (2005); L.J. Singh, S. Choudhary, A. Mallik, K.P. Ghatak, J. Comput. Theor. Nanosci. **2**, 287 (2005); K.P. Ghatak, J. Mukhopadhyay, J.P. Banerjee, SPIE Proc. **4746**, 1292 (2002); K.P. Ghatak, S. Dutta, D.K. Basu, B. Nag, Il Nuovo Cimento. **D 20**, 227 (1998); K.P. Ghatak, D.K. Basu, B. Nag, J. Phys. Chem. Solids **58**, 133 (1997); K.P. Ghatak, B. De, Mater. Res. Soc. Proc. **300**, 513 (1993); K.P. Ghatak, B. Mitra, Il Nuovo Cimento. **D 15**, 97 (1993); K. P. Ghatak, Inter. Soci. Opt. Photon. Proc. Soc. Photo Opt. Instrum. Engg., **1626**, 115 (1992); K.P. Ghatak, A. Ghoshal, Phys. Stat. Sol. (b) **170**, K27 (1992); K.P. Ghatak, S. Bhattacharya, S.N. Biswas, Proc. Soc. Photo Opt. Instrum. Eng, **836**, 72 (1988); K.P. Ghatak, A Ghoshal, S. N. Biswas, M. Mondal, Proc. Soc. Photo Opt. Instrum. Eng. **1308**, (356) (1990); K.P. Ghatak, B. De, Proc. Wide bandgap Semi. Symp. Mat. Res. Soc. 377 (1992); K.P. Ghatak, B. De, Defect Eng. Semi. Growth, Proc. Device Tech. Symp., Mat. Res. Soc. **262**, 911 (1992); S.N. Biswas, K.P. Ghatak, Internat. Jour. Electronics Theo. Exp. **70**, 125 (1991); B. Mitra, K. P. Ghatak, Phys. Lett. A. **146**, 357 (1990); B. Mitra, K.P. Ghatak, Phys. Lett. A. **142**, 401 (1989); K.P. Ghatak, B Mitra, A Ghoshal, Phy. Stat. Sol. (b) **154**, K121 (1989); B. Mitra, K. P. Ghatak, Phys. Stat. Sol. (b). **149**, K117 (1988); K.P. Ghatak, S. N. Biswas, Proc. Soc. Photo Optical Instru. Engg.. **792**, (239) (1987); S. Bhattacharyya, K.P. Ghatak, S. Biswas, OE/Fibers' 87, Inter. Soc. Opt. Photon. 73 (1987); M. Mondal, K.P. Ghatak, Czech. Jour. Phys. B. **36**, 1389 (1986); K.P. Ghatak, A.N. Chakravarti, Phys. Stat. Sol. (b). **117**, 707 (1983)
2. L.V. Keldysh, Sov. Phys. Solid State **4**, 1658 (1962); L. Esaki, R. Tsu, IBM J. Res. Develop. **14**, 61 (1970); G. Bastard, *Wave Mechanics Applied to Heterostructures*, (Editions de Physique, Les Ulis, France, 1990); E.L. Ivchenko, G. Pikus, *Superlattices and Other Heterostructures*, (Springer, Berlin, 1995); R. Tsu, *Superlattices to Nanoelectronics*, (Elsevier, The Netherlands, 2005)
3. P. Fürjes, Cs. Dücs, M. Ádám, J. Zettner, I. Bársony, Superlattices Microstruct. **35**, 455 (2004); T. Borca-Tasciuc, D. Achimov, W. L. Liu, G. Chen, H.W. Ren, C.H. Lin, S.S. Pei, Microscale Thermophys. Eng. **5**, 225 (2001)

4. B.S. Williams, Nat. Photonics **1**, 517 (2007); A. Kosterev, G. Wysocki, Y. Bakhirkin, S. So, R. Lewicki, F. Tittel, R.F. Curl, Appl. Phys. B **90**, 165 (2008); M.A. Belkin, F. Capasso, F. Xie, A. Belyanin, M. Fischer, A. Wittmann. J. Faist Appl. Phys. Lett. **92**, 201101 (2008)
5. G.J. Brown, F. Szmulowicz, R. Linville, A. Saxler, K. Mahalingam, C.-H. Lin, C.H. Kuo, W. Y. Hwang, IEEE Photonics Technol. Lett. **12**, 684 (2000); H.J. Haugan, G.J. Brown, L. Grazulis, K. Mahalingam, D.H. Tomich, Phys. E: Low-dimens. Syst. Nanostruct. **20**, 527 (2004)
6. S.A. Nikishin, V.V. Kuryatkov, A. Chandolu, B.A. Borisov, G.D. Kipshidze, I. Ahmad, M. Holtz, H. Temkin, Jpn. J. Appl. Phys. **42**, L1362 (2003); Y.K. Su, H.C. Wang, C.L. Lin, W.B. Chen, S.M. Chen, Jpn. J. Appl. Phys. **42**, L751 (2003); C.H. Liu, Y.K. Su, L.W. Wu, S. J. Chang, R.W. Chuang, Semicond. Sci. Technol. **18**, 545 (2003); S.B. Che, I. Nomura, A. Kikuchi, K. Shimomura, K. Kishino, Phys. Stat. Sol. (b) **229**, 1001 (2002)
7. C.P. Endres, F. Lewen, T.F. Giesen, S. SchlEEMer, D.G. Paveliev, Y.I. Koschurinov, V.M. Ustinov, A.E. Zhucov, Rev. Sci. Instrum. **78**, 043106 (2007)
8. F. Klappenberger, K.F. Renk, P. Renk, B. Rieder, Y.I. Koshurinov, D.G. Pavelev, V. Ustinov, A. Zhukov, N. Maleev, A. Vasilyev, Appl. Phys. Lett. **84**, 3924 (2004)
9. X. Jin, Y. Maeda, T. Saka, M. Tanioku, S. Fuchi, T. Ujihara, Y. Takeda, N. Yamamoto, Y. Nakagawa, A. Mano, S. Okumi, M. Yamamoto, T. Nakanishi, H. Horinaka, T. Kato, T. Yasue, T. Koshikawa, J. Crystal Growth **310**, 5039 (2008); X. Jin, N. Yamamoto, Y. Nakagawa, A. Mano, T. Kato, M. Tanioku, T. Ujihara, Y. Takeda, S. Okumi, M. Yamamoto, T. Nakanishi, T. Saka, H. Horinaka, T. Kato, T. Yasue, T. Koshikawa, Appl. Phys. Express **1**, 045002 (2008)
10. B.H. Lee, K.H. Lee, S. Im, M.M. Sung, Org. Electron. **9**, 1146 (2008)
11. P.H. Wu, Y.K. Su, I.L. Chen, C.H. Chiou, J.T. Hsu, W.R. Chen, Jpn. J. Appl. Phys. **45**, L647 (2006); A.C. Varonides. Renewable Energy **33**, 273 (2008)
12. M. Walther, G. Weimann, Phys. Stat. Sol. (b) **203**, 3545 (2006)
13. R. Rehm, M. Walther, J. Schmitz, J. Fleiβner, F. Fuchs, J. Ziegler, W. Cabanski, Opto-Electron. Rev. **14**, 19 (2006); R. Rehm, M. Walther, J. Scmitz, J. Fleissner, J. Ziegler, W. Cabanski, R. Breiter, Electron. Lett. **42**, 577 (2006)
14. G.J. Brown, F. Szmulowicz, H. Haugan, K. Mahalingam, S. Houston, Microelectronics Jour. **36**, 256 (2005)
15. K.V. Vaidyanathan, R.A. Jullens, C.L. Anderson, H.L. Dunlap, Solid State Electron. **26**, 717 (1983)
16. B.A. Wilson, IEEE, *J. Quantum Electron.* **24**, 1763 (1988)
17. M. Krichbaum, P. Kocevar, H. Pascher, G. Bauer, IEEE. J. Quantum Electron. **24**, 717 (1988)
18. J.N. Schulman, T.C. McGill, Appl. Phys. Lett. **34**, 663 (1979)
19. H. Kinoshita, T. Sakashita, H. Fajiyasu, J. Appl. Phys. **52**, 2869 (1981)
20. L. Ghenin, R.G. Mani, J.R. Anderson, J.T. Cheung, Phys. Rev. B **39**, 1419 (1989)
21. C.A. Hoffman, J.R. Mayer, F.J. Bartoli, J.W. Han, J.W. Cook, J.F. Schetzina, J.M. Schubman, Phys. Rev. B. **39**, 5208 (1989)
22. V.A. Yakovlev, Sov. Phys. Semicon. **13**, 692 (1979)
23. E.O. Kane., J. Phys. Chem. Solids **1**, 249 (1957)
24. H.X. Jiang, J.Y. Lin, J. Appl. Phys. **61**, 624 (1987)
25. H. Sasaki, Phys. Rev. B **30**, 7016 (1984)

Chapter 14
The DR in Quantum Wire HDSLs

P-A-I-N means positive attitude in negativity.

14.1 Introduction

In this chapter, the DR from III–V, II–VI, IV–VI, HgTe/CdTe and strained layer quantum wire heavily doped superlattices (QWHDSLs) with graded interfaces has been studied in Sects. 14.2.1 to 14.2.5. From Sects. 14.2.6 to 14.2.10, the DR from III–V, II–VI, IV–VI, HgTe/CdTe and strained layer quantum wire heavily doped effective mass superlattices respectively has been presented. The Sect. 14.3 contains the summary and conclusion pertinent to this chapter. The last Sect. 14.4 presents single open research problem.

14.2 Theoretical Background

14.2.1 The DR in III–V Quantum Wire HD Superlattices with Graded Interfaces

The simplified DR of heavily doped quantum wire III–V super-lattices with graded interfaces can be expressed as [1, 2]

$$k_z^2 = [G_8 + iH_8]\Big|_{k_x=\frac{n_x\pi}{d_x} \text{ and } k_y=\frac{n_y\pi}{d_y}} \tag{14.1a}$$

The DOS function can be written as

$$N(E) = \frac{g_v}{\pi}\sum_{n_x=1}^{n_{x\max}}\sum_{n_y=1}^{n_{y\max}} \frac{H(E - E_{13,1})[G_8' + iH_8']}{\sqrt{G_8 + iH_8}} \tag{14.1b}$$

K.P. Ghatak, *Dispersion Relations in Heavily-Doped Nanostructures*,
Springer Tracts in Modern Physics 265, DOI 10.1007/978-3-319-21000-1_14

where $E_{13,1}$ is the sub band energy and the sub-band equation in this case can be expressed as

$$0 = [G_8 + iH_8]|_{k_x = \frac{n_x \pi}{d_x},\ k_y = \frac{n_y \pi}{d_y} \text{ and } E = E_{13,1}} \tag{14.1c}$$

The EEM in this case is given by

$$m^*(E, \eta_g, n_x, n_y) = \frac{\hbar^2}{2} G_8' \tag{14.2}$$

14.2.2 The DR in II–VI Quantum Wire HD Superlattices with Graded Interfaces

The simplified DR of heavily doped quantum wire III–V super-lattices with graded interfaces can be expressed as

$$k_z^2 = [G_{19} + iH_{19}]|_{k_x = \frac{n_x \pi}{d_x} \text{ and } k_y = \frac{n_y \pi}{d_y}} \tag{14.3a}$$

The DOS function can be written as

$$N(E) = \frac{g_v}{\pi} \sum_{n_x=1}^{n_{x\max}} \sum_{n_y=1}^{n_{y\max}} \frac{H(E - E_{13,2})[G_{19}' + iH_{19}']}{\sqrt{G_{19} + iH_{19}}} \tag{14.3b}$$

where $E_{13,2}$ is the sub band energy and the sub-band equation in this case can be expressed as

$$0 = [G_{19} + iH_{19}]|_{k_x = \frac{n_x \pi}{d_x},\ k_y = \frac{n_y \pi}{d_y} \text{ and } E = E_{13,2}} \tag{14.3c}$$

The EEM in this case is given by

$$m^*(E, \eta_g, n_x, n_y) = \frac{\hbar^2}{2} G_{19}' \tag{14.4}$$

14.2.3 The DR in IV–VI Quantum Wire HD Superlattices with Graded Interfaces

The simplified DR in heavily doped quantum wire IV–VI super-lattices with graded interfaces can be expressed as

$$k_z^2 = \left[\frac{1}{L_0^2}\left[\cos^{-1}\left\{\frac{1}{2}\Phi_2(E,k_s)\right\}\right]^2 - k_s^2\right]\Bigg|_{k_x=\frac{n_x\pi}{d_x} \text{ and } k_y=\frac{n_y\pi}{d_y}} \tag{14.5a}$$

The DOS function can be written as

$$N(E) = \frac{g_v}{\pi L_0}\sum_{n_x=1}^{n_{x\max}} \sum_{n_y=1}^{n_{y\max}} \frac{\cos^{-1}\{\frac{1}{2}\Phi_2(E,k_s)\}\Phi_2(E,k_s)\Phi_2'(E,k_s)H(E-E_{13,3})}{\left(\sqrt{\cos^{-1}\{\frac{1}{2}\Phi_2(E,k_s)\}^2 - L_0^2\left\{\left(\frac{n_x\pi}{d_x}\right)^2 + \left(\frac{n_x\pi}{d_x}\right)^2\right\}}\right)\sqrt{1-\frac{1}{4}\Phi_2^2(E,k_s)}} \tag{14.5b}$$

where $E_{13,3}$ is the sub-band energy and the sub-band equation in this case can be expressed as

$$0 = \left[\frac{1}{L_0^2}\left[\cos^{-1}\left\{\frac{1}{2}\Phi_2(E_{13,3},k_s)\right\}\right]^2 - k_s^2\right]\Bigg|_{k_x=\frac{n_x\pi}{d_x} \text{ and } k_y=\frac{n_y\pi}{d_y}} \tag{14.5c}$$

The EEM in this case is given by

$$m^*(E,\eta_g,n_x,n_y) = \frac{\hbar^2}{2L_0^2}\cos^{-1}\left[\frac{1}{2}\Phi_2(E,k_s)\right]\Phi_2'(E,k_s)\left[1-\frac{1}{4}\Phi_2^2(E,k_s)\right]^{-1/2} \tag{14.6}$$

14.2.4 The DR in HgTe/CdTe Quantum Wire HD Superlattices with Graded Interfaces

The simplified DR in heavily doped quantum wire HgTe/CdTe superlattices with graded interfaces can be expressed as

$$k_z^2 = [G_{192} + iH_{192}]\big|_{k_x=\frac{n_x\pi}{d_x} \text{ and } k_y=\frac{n_y\pi}{d_y}} \tag{14.7a}$$

The DOS function can be written as

$$N(E) = \frac{g_v}{\pi}\sum_{n_x=1}^{n_{x\max}}\sum_{n_y=1}^{n_{y\max}} \frac{H(E-E_{13,4})[G_{192}' + iH_{192}']}{\sqrt{G_{192}+iH_{192}}} \tag{14.7b}$$

where $E_{13,4}$ is the sub band energy and the sub-band equation in this case can be expressed as

$$0 = [G_{192} + iH_{192}]|_{k_x=\frac{n_x\pi}{d_x}, k_y=\frac{n_y\pi}{d_y} \text{ and } E=E_{13,4}} \tag{14.7c}$$

The EEM in this case is given by

$$m^*(E, \eta_g, n_x, n_y) = \frac{\hbar^2}{2} G'_{192} \tag{14.8}$$

14.2.5 The DR in Strained Layer Quantum Wire HD Superlattices with Graded Interfaces

Therefore the DR of the conduction electrons in heavily doped strained layer quantum well SL with graded interfaces can be expressed as

$$k_z^2 = \left[\frac{1}{L_0^2}\left[\cos^{-1}\left\{\frac{1}{2}\bar{\Phi}_6(E, k_s)\right\}\right]^2 - k_s^2\right]\Bigg|_{k_x=\frac{n_x\pi}{d_x} \text{ and } k_y=\frac{n_y\pi}{d_y}} \tag{14.9a}$$

The DOS function can be written as

$$N(E) = \frac{g_v}{\pi L_0} \sum_{n_x=1}^{n_{x\max}} \sum_{n_y=1}^{n_{y\max}} \frac{\cos^{-1}\{\frac{1}{2}\bar{\Phi}_6(E, k_s)\}\bar{\Phi}_6(E, k_s)[\bar{\Phi}_6(E, k_s)]' H(E - E_{13,5})}{\left(\sqrt{\cos^{-1}\{\frac{1}{2}\bar{\Phi}_6(E, k_s)\}^2 - L_0^2\left\{\left(\frac{n_x\pi}{d_x}\right)^2 + \left(\frac{n_x\pi}{d_x}\right)^2\right\}}\right)\sqrt{1 - \frac{1}{4}[\bar{\Phi}_6(E, k_s)]^2}} \tag{14.9b}$$

where $E_{13,6}$ is the sub-band energy and the sub-band equation in this case can be expressed as

$$0 = \left[\frac{1}{L_0^2}\left[\cos^{-1}\left\{\frac{1}{2}\bar{\Phi}_6(E_{13,5}, k_s)\right\}\right]^2 - k_s^2\right]\Bigg|_{k_x=\frac{n_x\pi}{d_x} \text{ and } k_y=\frac{n_y\pi}{d_y}} \tag{14.9c}$$

The EEM in this case is given by

$$m^*(E,\eta_g,n_x,n_y)=\frac{\hbar^2}{2L_0^2}\cos^{-1}\left[\frac{1}{2}\bar{\Phi}_6(E,k_s)\right][\bar{\Phi}_6(E,k_s)]'\left[1-\frac{1}{4}[\bar{\Phi}_6(E,k_s)]^2\right]^{-1/2} \tag{14.10}$$

14.2.6 *The DR in III–V Quantum Wire HD Effective Mass Super Lattices*

The DR in III–V heavily doped effective mass quantum wire super-lattices can be written as

$$k_z^2=[\delta_7+i\delta_8]\big|_{k_x=\frac{n_x\pi}{d_x}\text{ and }k_y=\frac{n_y\pi}{d_y}} \tag{14.11a}$$

The DOS function can be written as

$$N(E)=\frac{g_v}{\pi}\sum_{n_x=1}^{n_{x\max}}\sum_{n_y=1}^{n_{y\max}}\frac{H(E-E_{13,6})[\delta_7'+i\delta_8']}{\sqrt{\delta_7+i\delta_8}} \tag{14.11b}$$

where $E_{13,6}$ is the sub band energy and the sub-band equation in this case can be expressed as

$$0=[\delta_7+i\delta_8]\big|_{k_x=\frac{n_x\pi}{d_x},k_y=\frac{n_y\pi}{d_y}\text{ and }E=E_{13,6}} \tag{14.11c}$$

The EEM in this case is given by

$$m^*(E,\eta_g,n_x,n_y)=\frac{\hbar^2}{2}\delta_7' \tag{14.12}$$

14.2.7 *The DR in II–VI Quantum Wire HD Effective Mass Super Lattices*

The DR in III–V heavily doped effective mass quantum wire super-lattices can be written as

$$k_z^2=[\delta_{13}+i\delta_{14}]\big|_{k_x=\frac{n_x\pi}{d_x}\text{ and }k_y=\frac{n_y\pi}{d_y}} \tag{14.13a}$$

The DOS function can be written as

$$N(E) = \frac{g_v}{\pi} \sum_{n_x=1}^{n_{x\max}} \sum_{n_y=1}^{n_{y\max}} \frac{H(E - E_{13,6})[\delta'_{13} + i\delta'_{14}]}{\sqrt{\delta_{13} + i\delta_{14}}} \tag{14.13b}$$

where $E_{13,7}$ is the sub band energy and the sub-band equation in this case can be expressed as

$$0 = [\delta_{13} + i\delta_{14}]|_{k_x = \frac{n_x\pi}{d_x}, k_y = \frac{n_y\pi}{d_y} \text{ and } E = E_{13,7}} \tag{14.13c}$$

The EEM in this case is given by

$$m^*(E, \eta_g, n_x, n_y) = \frac{\hbar^2}{2} \delta'_{13} \tag{14.14}$$

14.2.8 The DR in IV–VI Quantum Wire HD Effective Mass Super Lattices

Therefore the DR in heavily doped IV–VI, quantum wire EMSLs can be written as

$$k_z^2 = \left[\frac{1}{L_0^2}\{\cos^{-1}(f_{23}(E, k_x, k_y))\}^2 - k_s^2\right]\Bigg|_{k_x = \frac{n_x\pi}{d_x} \text{ and } k_y = \frac{n_y\pi}{d_y}} \tag{14.15a}$$

The DOS function can be written as

$$N(E) = \frac{g_v}{\pi L_0} \sum_{n_x=1}^{n_{x\max}} \sum_{n_y=1}^{n_{y\max}} \frac{\cos^{-1}\left\{\frac{1}{2} f_{23}\left(E, \frac{n_x\pi}{d_x}, \frac{n_y\pi}{d_y}\right)\right\} f_{23}\left(E, \frac{n_x\pi}{d_x}, \frac{n_y\pi}{d_y}\right) \left[f_{23}\left(E, \frac{n_x\pi}{d_x}, \frac{n_y\pi}{d_y}\right)\right]' H(E - E_{13,8})}{\left(\sqrt{\cos^{-1}\left\{\frac{1}{2} f_{23}\left(E, \frac{n_x\pi}{d_x}, \frac{n_y\pi}{d_y}\right)\right\}^2 - L_0^2\left\{\left(\frac{n_x\pi}{d_x}\right)^2 + \left(\frac{n_x\pi}{d_x}\right)^2\right\}}\right) \sqrt{1 - \frac{1}{4}\left[f_{23}\left(E, \frac{n_x\pi}{d_x}, \frac{n_y\pi}{d_y}\right)\right]^2}} \tag{14.15b}$$

where $E_{13,8}$ is the sub band energy and the sub-band equation in this case can be expressed as

$$0 = \left[\frac{1}{L_0^2}\{\cos^{-1}(f_{23}(E_{13,8}, k_x, k_y))\}^2 - k_s^2\right]\Bigg|_{k_x = \frac{n_x\pi}{d_x} \text{ and } k_y = \frac{n_y\pi}{d_y}} \tag{14.15c}$$

The EEM in this case is given by

$$m^*(E,\eta_g,n_x,n_y) = \frac{\hbar^2}{2L_0^2}\cos^{-1}\left[\frac{1}{2}f_{23}\left(E,\frac{n_x\pi}{d_x},\frac{n_y\pi}{d_y}\right)\right]$$
$$\left[f_{23}\left(E,\frac{n_x\pi}{d_x},\frac{n_y\pi}{d_y}\right)\right]'\left[1-\frac{1}{4}\left[f_{23}\left(E,\frac{n_x\pi}{d_x},\frac{n_y\pi}{d_y}\right)\right]^2\right]^{-1/2} \tag{14.16}$$

14.2.9 The DR in HgTe/CdTe Quantum Wire HD Effective Mass Super Lattices

The DR in heavily doped HgTe/CdTe QWEMSLs can be written as

$$k_z^2 = [\Delta_{13H} + i\Delta_{14H}]\big|_{k_x=\frac{n_x\pi}{d_x}\text{ and }k_y=\frac{n_y\pi}{d_y}} \tag{14.17a}$$

The DOS function can be written as

$$N(E) = \frac{g_v}{\pi}\sum_{n_x=1}^{n_{x\max}}\sum_{n_y=1}^{n_{y\max}}\frac{H(E-E_{13,9})[\Delta'_{13H}+i\Delta'_{13H}]}{\sqrt{\Delta_{13H}+i\Delta_{14H}}} \tag{14.17b}$$

where $E_{13,9}$ is the sub band energy and the sub-band equation in this case can be expressed as

$$0 = [\Delta_{13H} + i\Delta_{14H}]\big|_{k_x=\frac{n_x\pi}{d_x},k_y=\frac{n_y\pi}{d_y}\text{ and }E=E_{13,9}} \tag{14.17c}$$

The EEM in this case is given by

$$m^*(E,\eta_g,n_x,n_y) = \frac{\hbar^2}{2}\Delta'_{13H} \tag{14.18}$$

14.2.10 The DR in Strained Layer Quantum Wire HD Effective Mass Super Lattices

Therefore the DR in heavily doped IV–VI, quantum wire EMSLs can be written as

$$k_z^2 = \left[\frac{1}{L_0^2}\left\{\cos^{-1}(f_{40}(E,k_x,k_y))\right\}^2 - k_s^2\right]\Bigg|_{k_x=\frac{n_x\pi}{d_x}\text{ and }k_y=\frac{n_y\pi}{d_y}} \tag{14.19a}$$

The DOS function can be written as

$$N(E) = \frac{g_v}{\pi L_0} \sum_{n_x=1}^{n_{x\max}} \sum_{n_y=1}^{n_{y\max}} \frac{\cos^{-1}\left\{\frac{1}{2} f_{40}\left(E, \frac{n_x\pi}{d_x}, \frac{n_y\pi}{d_y}\right)\right\} f_{40}\left(E, \frac{n_x\pi}{d_x}, \frac{n_y\pi}{d_y}\right) \left[f_{40}\left(E, \frac{n_x\pi}{d_x}, \frac{n_y\pi}{d_y}\right)\right]' H(E - E_{13,10})}{\left(\sqrt{\cos^{-1}\left\{\frac{1}{2} f_{40}\left(E, \frac{n_x\pi}{d_x}, \frac{n_y\pi}{d_y}\right)\right\}^2 - L_0^2\left\{\left(\frac{n_x\pi}{d_x}\right)^2 + \left(\frac{n_x\pi}{d_x}\right)^2\right\}}\right) \sqrt{1 - \frac{1}{4}\left[f_{40}\left(E, \frac{n_x\pi}{d_x}, \frac{n_y\pi}{d_y}\right)\right]^2}} \tag{14.19b}$$

where $E_{13,10}$ is the sub band energy and the sub-band equation in this case can be expressed as

$$0 = \left[\frac{1}{L_0^2}\left\{\cos^{-1}(f_{40}(E_{13,10}, k_x, k_y))\right\}^2 - k_s^2\right]\Bigg|_{k_x = \frac{n_x\pi}{d_x} \text{ and } k_y = \frac{n_y\pi}{dy}} \tag{14.19c}$$

The EEM in this case is given by

$$m^*(E, \eta_g, n_x, n_y) = \frac{\hbar^2}{2L_0^2} \cos^{-1}\left[\frac{1}{2} f_{40}\left(E, \frac{n_x\pi}{d_x}, \frac{n_y\pi}{d_y}\right)\right] \left[f_{40}\left(E, \frac{n_x\pi}{d_x}, \frac{n_y\pi}{d_y}\right)\right]' \left[1 - \frac{1}{4}\left[f_{40}\left(E, \frac{n_x\pi}{d_x}, \frac{n_y\pi}{d_y}\right)\right]^2\right]^{-1/2} \tag{14.20}$$

14.3 Summary and Conclusion

In this chapter, the DR from III–V, II–VI, IV–VI, HgTe/CdTe and strained layer quantum wire heavily doped superlattices (QWHDSLs) with graded interfaces has been studied in Sects. 14.2.1 to 14.2.5. From Sects. 14.2.6 to 14.2.10, the DR from III–V, II–VI, IV–VI, HgTe/CdTe and strained layer quantum wire heavily doped effective mass superlattices respectively has been presented. The presence of essential poles in the constituent materials of HD III–V and HgTe/CdTe SLs makes the DR in the said SLs complex (for both SLs with graded interfaces and effective mass SLs). The $E - k_x^2$ plots for HD III–V and HgTe/CdTe QWSLs are quantized closed 2D surfaces and they exist in the complex energy plane. The DR for other cases are quantized closed 2D surfaces and they exist in the real energy plane. The DRs in all the cases are concentration dependent for any value of the electron energy. The sub band energies are either real or complex in accordance with said logic. The DOS functions have been derived in all the cases. The EEM are functions of quantum numbers, screening potential and for any value of the electron energy, the EEM is concentration dependent.

14.4 Open Research Problem

(R.14.1) Investigate the influence of arbitrarily oriented alternating quantizing magnetic field and strain on the DR for all types of HD super-lattices whose carrier energy spectra are described in this book

References

1. S. Chakrabarti, S.K. Sen, S. Chakraborty, L.S. Singh, K.P. Ghatak, Advanced science. Engineering and Medicine **6**, 1042 (2014)
2. K.P. Ghatak, S. Dutta, D.K. Basu, B Nag Il Nuovo Cimento D **20**, 227–240 (1998)
3. K.P. Ghatak, S. Karmakar, D. De, S. Pahari, S.K. Charaborty, S.K. Biswas, S. Chowdhury, Journal of Computational and Theoretical Nanoscience **3**, 153 (2006)

Chapter 15
The DR in Quantum Dot HDSLs

If I realize that my ego pressure is much greater that of my blood pressure, then I am to some extent wise.

15.1 Introduction

In this chapter, the DR from III–V, II–VI, IV–VI, HgTe/CdTe and strained layer heavily doped superlattices (QDHDSLs) with graded interfaces has been studied in Sects. 15.2.1 to 15.2.5. From Sects. 15.2.6 to 15.2.10, the DR from III–V, II–VI, IV–VI, HgTe/CdTe and strained layer quantum dot heavily doped effective mass superlattices respectively has been presented. The Sect. 15.3 contains the summary and conclusion pertinent to this chapter. The last Sect. 15.4 presents single open research problem.

15.2 Theoretical Background

15.2.1 The DR in III–V Quantum Dot HD Superlattices with Graded Interfaces

The simplified DR of heavily doped quantum dot III–V super-lattices with graded interfaces can be expressed as [1–9]

$$\left(\frac{n_z\pi}{d_z}\right)^2 = [G_8 + iH_8]|_{k_x=\frac{n_x\pi}{d_x},k_y=\frac{n_y\pi}{d_y}\text{ and }E=E_{14,1}} \tag{15.1}$$

where $E_{14,1}$ is the totally quantized energy in this case.

The DOS function is given by

K.P. Ghatak, *Dispersion Relations in Heavily-Doped Nanostructures*,
Springer Tracts in Modern Physics 265, DOI 10.1007/978-3-319-21000-1_15

$$N_{QDSL}(E) = \frac{2g_v}{d_x d_y d_z} \sum_{n_x=1}^{n_{x\max}} \sum_{n_{y=1}}^{n_{y\max}} \sum_{n_z=1}^{n_{z\max}} \delta'(E - E_{14,1}) \tag{15.2}$$

15.2.2 The DR in II–VI Quantum Dot HD Superlattices with Graded Interfaces

The simplified DR of heavily doped quantum dot III–V super-lattices with graded interfaces can be expressed as

$$\left(\frac{n_z \pi}{d_z}\right)^2 = [G_{19} + iH_{19}]|_{k_x=\frac{n_x\pi}{d_x}, k_y=\frac{n_y\pi}{d_y} \text{ and } E=E_{14,2}} \tag{15.3}$$

where $E_{14,2}$ is the totally quantized energy in this case.

The DOS function is given by

$$N_{QDSL}(E) = \frac{2g_v}{d_x d_y d_z} \sum_{n_x=1}^{n_{x\max}} \sum_{n_{y=1}}^{n_{y\max}} \sum_{n_z=1}^{n_{z\max}} \delta'(E - E_{14,2}) \tag{15.4}$$

15.2.3 The DR in IV–VI Quantum Dot HD Superlattices with Graded Interfaces

The simplified DR in heavily doped quantum dot IV–VI super-lattices with graded interfaces can be expressed as

$$\left(\frac{n_z \pi}{d_z}\right)^2 = \left[\frac{1}{L_0^2}\left[\cos^{-1}\left\{\frac{1}{2}\Phi_2(E_{14,3}, k_s)\right\}\right]^2 - k_s^2\right]\Bigg|_{k_x=\frac{n_x\pi}{d_x} \text{ and } k_y=\frac{n_y\pi}{d_y}} \tag{15.5}$$

where $E_{14,3}$ is the totally quantized energy in this case.

The DOS function is given by

$$N_{QDSL}(E) = \frac{2g_v}{d_x d_y d_z} \sum_{n_x=1}^{n_{x\max}} \sum_{n_{y=1}}^{n_{y\max}} \sum_{n_z=1}^{n_{z\max}} \delta'(E - E_{14,3}) \tag{15.6}$$

15.2.4 The DR in HgTe/CdTe Quantum Dot HD Superlattices with Graded Interfaces

The simplified DR of heavily doped quantum dot III–V super-lattices with graded interfaces can be expressed as

$$\left(\frac{n_z\pi}{d_z}\right)^2 = [G_{192} + iH_{192}]\big|_{k_x=\frac{n_x\pi}{d_x},k_y=\frac{n_y\pi}{d_y}\text{ and }E=E_{14,4}} \tag{15.7}$$

where $E_{14,4}$ is the totally quantized energy in this case.

The DOS function is given by

$$N_{QDSL}(E) = \frac{2g_v}{d_xd_yd_z}\sum_{n_x=1}^{n_{x\max}}\sum_{n_y=1}^{n_{y\max}}\sum_{n_z=1}^{n_{z\max}}\delta'(E - E_{14,4}) \tag{15.8}$$

15.2.5 The DR in Strained Layer Quantum Dot HD Superlattices with Graded Interfaces

The DR of the conduction electrons in heavily doped strained layer quantum dot SL with graded interfaces can be expressed as

$$\left(\frac{n_z\pi}{d_z}\right)^2 = \left[\frac{1}{L_0^2}\left[\cos^{-1}\left\{\frac{1}{2}\overline{\Phi}_6(E_{14,5}, k_s)\right\}\right]^2 - k_s^2\right]\Bigg|_{k_x=\frac{n_x\pi}{d_x}\text{ and }k_y=\frac{n_y\pi}{d_y}} \tag{15.9}$$

where $E_{14,5}$ is the totally quantized energy in this case.

The DOS function is given by

$$N_{QDSL}(E) = \frac{2g_v}{d_xd_yd_z}\sum_{n_x=1}^{n_{x\max}}\sum_{n_y=1}^{n_{y\max}}\sum_{n_z=1}^{n_{z\max}}\delta'(E - E_{14,5}) \tag{15.10}$$

15.2.6 The DR in III–V Quantum Dot HD Effective Mass Super Lattices

The DR in III–V heavily doped effective mass quantum dot super-lattices can be written as

$$\left(\frac{n_z\pi}{d_z}\right)^2 = [\delta_7 + i\delta_8]|_{k_x=\frac{n_x\pi}{d_x},k_y=\frac{n_y\pi}{d_y}\text{ and }E=E_{14,6}} \tag{15.11}$$

where $E_{14,6}$ is the totally quantized energy in this case.

The DOS function is given by

$$N_{QDSL}(E) = \frac{2g_v}{d_x d_y d_z}\sum_{n_x=1}^{n_{x\max}}\sum_{n_y=1}^{n_{y\max}}\sum_{n_z=1}^{n_{z\max}}\delta'(E - E_{14,6}) \tag{15.12}$$

15.2.7 The DR in II–VI Quantum Dot HD Effective Mass Super Lattices

The DR in III–V heavily doped effective mass quantum dot super-lattices can be written as

$$\left(\frac{n_z\pi}{d_z}\right)^2 = [\delta_{13} + i\delta_{14}]|_{k_x=\frac{n_x\pi}{d_x},k_y=\frac{n_y\pi}{d_y}\text{ and }E=E_{14,7}} \tag{15.13}$$

where $E_{14,7}$ is the totally quantized energy in this case.

The DOS function is given by

$$N_{QDSL}(E) = \frac{2g_v}{d_x d_y d_z}\sum_{n_x=1}^{n_{x\max}}\sum_{n_y=1}^{n_{y\max}}\sum_{n_z=1}^{n_{z\max}}\delta'(E - E_{14,7}) \tag{15.14}$$

15.2.8 The DR in IV–VI Quantum Dot HD Effective Mass Super Lattices

The DR in heavily doped IV–VI, quantum dot EMSLs can be written as

$$\left(\frac{n_z\pi}{d_z}\right)^2 = \left[\frac{1}{L_0^2}\left\{\cos^{-1}(f_{23}(E_{14,8}, k, k_y))\right\}^2 - k_s^2\right]\Bigg|_{k_x=\frac{n_x\pi}{d_x}\text{ and }k_y=\frac{n_y\pi}{d_y}} \tag{15.15}$$

where $E_{14,8}$ is the totally quantized energy in this case.

The DOS function is given by

$$N_{QDSL}(E) = \frac{2g_v}{d_x d_y d_z} \sum_{n_x=1}^{n_{x\max}} \sum_{n_y=1}^{n_{y\max}} \sum_{n_z=1}^{n_{z\max}} \delta'(E - E_{14,8}) \tag{15.16}$$

15.2.9 The DR in HgTe/CdTe Quantum Dot HD Effective Mass Super Lattices

The DR in heavily doped HgTe/CdTe QWEMSLs can be written as

$$\left(\frac{n_z \pi}{d_z}\right)^2 = [\Delta_{13H} + i\Delta_{14H}]|_{k_x = \frac{n_x \pi}{d_x}, k_y = \frac{n_y \pi}{d_y} \text{ and } E = E_{14,9}} \tag{15.17}$$

where $E_{14,9}$ is the totally quantized energy in this case.

The DOS function is given by

$$N_{QDSL}(E) = \frac{2g_v}{d_x d_y d_z} \sum_{n_x=1}^{n_{x\max}} \sum_{n_y=1}^{n_{y\max}} \sum_{n_z=1}^{n_{z\max}} \delta'(E - E_{14,9}) \tag{15.18}$$

15.2.10 The DR in Strained Layer Quantum Dot HD Effective Mass Super Lattices

The DR in heavily doped IV–VI, quantum dot EMSLs can be written as

$$\left(\frac{n_z \pi}{d_z}\right)^2 = \left[\frac{1}{L_0^2}\left\{\cos^{-1}(f_{40}(E_{14,10}, k_x, k_y))\right\}^2 - k_s^2\right]\bigg|_{k_x = \frac{n_x \pi}{d_x} \text{ and } k_y = \frac{n_y \pi}{d_y}} \tag{15.19}$$

where $E_{14,10}$ is the totally quantized energy in this case.

The DOS function is given by

$$N_{QDSL}(E) = \frac{2g_v}{d_x d_y d_z} \sum_{n_x=1}^{n_{x\max}} \sum_{n_y=1}^{n_{y\max}} \sum_{n_z=1}^{n_{z\max}} \delta'(E - E_{14,10}) \tag{15.20}$$

15.3 Summary and Conclusion

In this chapter, the DR from III–V, II–VI, IV–VI, HgTe/CdTe and strained layer quantum dot heavily doped superlattices (QDHDSLs) with graded interfaces has been studied in Sects. 15.2.1 to 15.2.5. From Sects. 15.2.6 to 15.2.10, the DR from III–V, II–VI, IV–VI, HgTe/CdTe and strained layer quantum dot heavily doped effective mass superlattices respectively has been presented

- The DRs for QDHDSLs of HD materials exhibit the fact that the total energy is quantized since the corresponding wave vector space is totally quantized.
- The DOS functions for all the materials in this case are series of non-uniformly distributed Dirac's Delta functions at specified quantized points in the respective energy axis. The spacing between the consecutive Delta functions are functions of energy band constants and quantization of the wave vector space of a particular material.
- It may be noted that the HD QDHDSLs lead to the discrete energy levels, somewhat like atomic energy levels, which produce very large changes. This follows from the inherent nature of the quantum confinement of the carrier gas dealt with here. In QDHDSLs, there remain no free carrier states in between any two allowed sets of size-quantized levels unlike that found for QWs and NWs super-lattices where the quantum confinements are 1D and 2D, respectively. Consequently, the crossing of the Fermi level by the size-quantized levels in HD QDHDSLs would have much greater impact on the redistribution of the carriers among the allowed levels, as compared to that found for QWs and NWs super-lattices respectively. The quantum signature of HD QDHDSLs for the DR is rather prominent as compared to the same from QWs and NWs super-lattices.
- It is the band structure which changes in a fundamental way and consequently all the physical properties of all the electronic materials changes radically leading to new physical concepts.

15.4 Open Research Problem

(R.15.1) Investigate the influence of arbitrarily oriented alternating quantizing magnetic field and strain on the DR for all types of HD super-lattices whose carrier energy spectra are described in this book.

References

1. N. Paitya, K.P. Ghatak, J. Adv. Phys. **1**, 161 (2012)
2. K.P. Ghatak, S. Bhattacharyya, B. Mitra, Fizika **22**, 391 (1990)

3. A. Ghoshal, M. Mondal, K.P. Ghatak, Phys. Status Solidi (b) **158**, K123 (1990)
4. K.P. Ghatak, M. Mondal, Acta Phys. Hungarica **69**, 113 (1991)
5. K.P. Ghatak, *Acta Phys. Hungarica* **70**, 91 (1991)
6. K.P. Ghatak, *Fizika A* **1**, 207(1992)
7. K.P. Ghatak, D. Bhattacharyya, Phys. Status Solidi **181**, 127 (1994)
8. K.P. Ghatak, B. Nag, Nanostruct. Mater. **5**, 769 (1995)
9. K.P. Ghatak, P.K. Bose, G. Majumder, MRS Proc. **484**, 673 (1997)

Chapter 16
The DR in HDSLs Under Magnetic Quantization

A man is measured by not what he says or does, but by what he becomes.

16.1 Introduction

In recent years the influence of magnetic quantization on the various electronic properties of semiconductor supperlattices having various band structures has been investigated in the literature [1–10]. In this chapter, the magneto DR in III–V, II–VI, IV–VI, HgTe/CdTe and strained layer heavily doped superlattices (HDSLs) with graded interfaces has been studied in Sects. 16.2.1 to 16.2.5. From Sects. 16.2.6 to 16.2.10, the magneto DR in III–V, II–VI, IV–VI, HgTe/CdTe and strained layer heavily doped effective mass superlattices respectively has been presented. The Sect. 16.3 contains the summary and conclusion pertinent to this chapter. The last Sect. 16.4 presents 4 open research problems.

16.2 Theoretical Background

16.2.1 The DR in III–V HD Superlattices with Graded Interfaces Under Magnetic Quantization

The simplified DR of heavily doped quantum well III–V super-lattices with graded interfaces under magnetic quantization can be expressed as

$$k_z^2 = G_{8E,n} + iH_{8E,n} \tag{16.1a}$$

K.P. Ghatak, *Dispersion Relations in Heavily-Doped Nanostructures*,
Springer Tracts in Modern Physics 265, DOI 10.1007/978-3-319-21000-1_16

where

$$G_{8E,n} = \left[\frac{C_{7E,n}^2 - D_{7E,n}^2}{L_0^2} - \left\{\frac{2eB}{\hbar}\left(n+\frac{1}{2}\right)\right\}\right], C_{7E,n} = \cos^{-1}\left(\overline{\omega_{7E,n}}\right),$$

$$\left(\overline{\omega_{7E,n}}\right) = (2)^{\frac{-1}{2}}\left[\left(1 - G_{7E,n}^2 - H_{7E,n}^2\right) - \sqrt{\left(1 - G_{7E,n}^2 - H_{7E,n}^2\right)^2 + 4G_{7E,n}^2}\right]^{\frac{1}{2}}$$

$$\begin{aligned} G_{7E,n} = {} & \left[G_{1E,n} + (\rho_{5E,n}G_{2E,n}/2) - (\rho_{6E,n}H_{2E,n}/2) + (\Delta_0/2)\right. \\ & \{\rho_{6E,n}H_{2E,n} - \rho_{8E,n}H_{3E,n} + \rho_{9E,n}H_{4E,n} - \rho_{10E,n}H_{4E,n} \\ & \left. + \rho_{11E,n}H_{5E,n} - \rho_{12E,n}H_{5E,n} + (1/12)(\rho_{12E,n}G_{6E,n} - \rho_{14E,n}H_{6E,n})\}\right], \end{aligned}$$

$$\begin{aligned} G_{1E,n} = {} & [(\cos(\mathrm{h}_{1E,n}))(\cosh(\mathrm{h}_{2E,n}))(\cosh(g_{1E,n}))(\cos(g_{2E,n})) \\ & + (\sin(\mathrm{h}_{1E,n}))(\sinh(\mathrm{h}_{2E,n}))(\sinh(g_{1E,n}))(\sin(g_{2E,n}))], \end{aligned}$$

$$\mathrm{h}_{1E,n} = e_{1E,n}(b_0 - \Delta_0), e_{1E,n} = 2^{\frac{-1}{2}}\left(\sqrt{t_{1E,n}^2 + t_2^2} + t_{1E,n}\right)^{\frac{1}{2}},$$

$$t_{1E,n} = \left[\left(2m_{c1}^*/\hbar^2\right) \cdot T_{11}(E, E_{g1}, \Delta_1, \eta_{g1}) - \left\{\frac{2eB}{\hbar}\left(n+\frac{1}{2}\right)\right\}\right],$$

$$t_2 = \left[(2m_{c1}^*/\hbar^2)T_{21}(E, E_{g1}, \Delta_1, \eta_{g1})\right],$$

$$\mathrm{h}_{2E,n} = e_{2E,n}(b_0 - \Delta_0),\ e_{2E,n} = 2^{\frac{-1}{2}}\left(\sqrt{t_{1E,n}^2 + t_2^2} - t_{1E,n}\right)^{\frac{1}{2}},$$

$$g_{1E,n} = d_{1E,n}(a_0 - \Delta_0),\ \mathrm{d}_{1E,n} = 2^{\frac{-1}{2}}\left(\sqrt{x_{1E,n}^2 + y_1^2} + x_{1E,n}\right)^{\frac{1}{2}},$$

$$x_{1E,n} = \left[-\left(2m_{c2}^*/\hbar^2\right) \cdot T_{11}(E - V_0, E_{g2}, \Delta_2, \eta_{g2}) + \left\{\frac{2eB}{\hbar}\left(n+\frac{1}{2}\right)\right\}\right],$$

$$\mathrm{y}_1 = \left[(2m_{c2}^*/\hbar^2)T_{22}(E - V_0, E_{g2}, \Delta_2, \eta_{g2})\right],$$

$$g_{2E,n} = d_{2E,n}(a_0 - \Delta_0), \mathrm{d}_{2E,n} = 2^{\frac{-1}{2}}\left(\sqrt{x_{1E,n}^2 + y_1^2} - x_{1E,n}\right)^{\frac{1}{2}},$$

$$\rho_{5E,n} = \left(\rho_{3E,n}^2 + \rho_{4E,n}^2\right)^{-1}\left[\rho_{1E,n}\rho_{3E,n} - \rho_{2E,n}\rho_{4E,n}\right],$$

$$\rho_{1E,n} = \left[d_{1E,n}^2 + e_{2E,n}^2 - d_{2E,n}^2 - e_{1E,n}^2\right], \rho_{3E,n} = \left[d_{1E,n}e_{1E,n} + d_{2E,n}e_{2E,n}\right],$$

$$\rho_{2E,n} = 2\left[d_{1E,n}d_{2E,n} + e_{1E,n}e_{2E,n}\right], \rho_{4E,n} = \left[d_{1E,n}e_{2E,n} - e_{1E,n}d_{2E,n}\right],$$

$$\begin{aligned} G_{2E,n} = {} & [(\sin(\mathrm{h}_{1E,n}))(\cosh(\mathrm{h}_{2E,n}))(\sinh(g_{1E,n}))(\cos(g_{2E,n})) \\ & + (\cos(\mathrm{h}_{1E,n}))(\sinh(\mathrm{h}_{2E,n}))(\cosh(g_{1E,n}))(\sin(g_{2E,n}))], \end{aligned}$$

$$\rho_{6E,n} = \left(\rho_{3E,n}^2 + \rho_{4E,n}^2\right)^{-1}\left[\rho_{1E,n}\rho_{4E,n} + \rho_{2E,n}\rho_{3E,n}\right],$$

$$H_{2E,n} = \left[(\sin(\mathrm{h}_{1E,n}))(\cosh(\mathrm{h}_{2E,n}))(\sin(g_{2E,n}))(\cosh(g_{1E,n})) - (\cos(\mathrm{h}_{1E,n}))(\sinh(\mathrm{h}_{2E,n}))(\sinh(g_{1E,n}))(\cos(g_{2E,n}))\right],$$

$$\rho_{7E,n} = \left[\left(e_{1E,n}^2 + e_{2E,n}^2\right)^{-1}\left[e_{1E,n}\left(d_{1E,n}^2 - d_{2E,n}^2\right) - 2d_{1E,n}d_{2E,n}e_{2E,n}\right] - 3e_{1E,n}\right],$$

$$G_{3E,n} = \left[(\sin(\mathrm{h}_{1E,n}))(\cosh(\mathrm{h}_{2E,n}))(\cosh(g_{1E,n}))(\cos(g_{2E,n})) + (\cos(\mathrm{h}_{1E,n}))(\sinh(\mathrm{h}_{2E,n}))(\sinh(g_{1E,n}))(\sin(g_{2E,n}))\right],$$

$$\rho_{8E,n} = \left[\left(e_{1E,n}^2 + e_{2E,n}^2\right)^{-1}\left[e_{2E,n}\left(d_{1E,n}^2 - d_{2E,n}^2\right) + 2d_{1E,n}d_{2E,n}e_{1E,n}\right] + 3e_{2E,n}\right],$$

$$H_{3E,n} = \left[(\sin(\mathrm{h}_{1E,n}))(\cosh(\mathrm{h}_{2E,n}))(\sin(g_{2E,n}))(\sinh(g_{1E,n})) - (\cos(\mathrm{h}_{1E,n}))(\sinh(\mathrm{h}_{2E,n}))(\cosh(g_{1E,n}))(\cos(g_{2E,n}))\right],$$

$$\rho_{9E,n} = \left[\left(d_{1E,n}^2 + d_{2E,n}^2\right)^{-1}\left[d_{1E,n}\left(e_{2E,n}^2 - e_{1E,n}^2\right) + 2e_{2E,n}d_{2E,n}e_{1E,n}\right] + 3d_{1E,n}\right],$$

$$G_{4E,n} = \left[(\cos(\mathrm{h}_{1E,n}))(\cosh(\mathrm{h}_{2E,n}))(\cos(g_{2E,n}))(\sinh(g_{1E,n})) - (\sin(\mathrm{h}_{1E,n}))(\sinh(\mathrm{h}_{2E,n}))(\cosh(g_{1E,n}))(\sin(g_{2E,n}))\right],$$

$$\rho_{10E,n} = \left[-\left(d_{1E,n}^2 + d_{2E,n}^2\right)^{-1}\left[d_{2E,n}\left(-e_{2E,n}^2 + e_{1E,n}^2\right) + 2e_{2E,n}d_{2E,n}e_{1E,n}\right] + 3d_{2E,n}\right],$$

$$H_{4E,n} = \left[(\cos(\mathrm{h}_{1E,n}))(\cosh(\mathrm{h}_{2E,n}))(\cosh(g_{1E,n}))(\sin(g_{2E,n})) + (\sin(\mathrm{h}_{1E,n}))(\sinh(\mathrm{h}_{2E,n}))(\sinh(g_{1E,n}))(\cos(g_{2E,n}))\right],$$

$$\rho_{11E,n} = 2\left[d_{1E,n}^2 + e_{2E,n}^2 - d_{2E,n}^2 - e_{1E,n}^2\right],$$

$$G_{5E,n} = \left[(\cos(\mathrm{h}_{1E,n}))(\cosh(\mathrm{h}_{2E,n}))(\cos(g_{2E,n}))(\cosh(g_{1E,n})) - (\sin(\mathrm{h}_{1E,n}))(\sinh(\mathrm{h}_{2E,n}))(\sinh(g_{1E,n}))(\sin(g_{2E,n}))\right],$$

$$\rho_{12E,n} = 4\left[d_{1E,n}d_{2E,n} + e_{1E,n}e_{2E,n}\right],$$

$$H_{5E,n} = \left[(\cos(\mathrm{h}_{1E,n}))(\cosh(\mathrm{h}_{2E,n}))(\sinh(g_{1E,n}))(\sin(g_{2E,n})) + (\sin(\mathrm{h}_{1E,n}))(\sinh(\mathrm{h}_{2E,n}))(\cosh(g_1))(\cos(g_2))\right],$$

$$\begin{aligned}\rho_{13E,n} = &\left[\left\{5\left(d_{1E,n}e_{1E,n}^3 - 3e_{1E,n}e_{2E,n}^2d_{1E,n}\right) + 5d_{2E,n}\left(e_{1E,n}^3 - 3e_{1E,n}^2e_{2E,n}\right)\right\}\left(d_{1E,n}^2 + d_{2E,n}^2\right)^{-1}\right.\\ &+ \left(e_{1E,n}^2 + e_{2E,n}^2\right)^{-1}\left\{5\left(e_{1E,n}d_{1E,n}^3 - 3d_{2E,n}e_{1E,n}^2d_{1E,n}\right)\right.\\ &\left.\left. + 5\left(d_{2E,n}^3e_{2E,n} - 3d_{1E,n}^2d_{2E,n}e_{2E,n}\right)\right\} - 34\left(d_{1E,n}e_{1E,n} + d_{2E,n}e_{2E,n}\right)\right],\end{aligned}$$

$$G_{6E,n} = \left[(\sin(\mathrm{h}_{1E,n}))(\cosh(\mathrm{h}_{2E,n}))(\sinh(g_{1E,n}))(\cos(g_{2E,n})) + (\cos(\mathrm{h}_{1E,n}))(\sinh(\mathrm{h}_{2E,n}))(\cosh(g_1))(\sin(g_2))\right],$$

$$\rho_{14E,n} = \left[\left\{5\left(d_{1E,n}e_{2E,n}^3 - 3e_{2E,n}e_{1E,n}^2 d_{1E,n}\right) + 5d_{2E,n}\left(-e_{1E,n}^3 + 3e_{2E,n}^2 e_{1E,n}\right)\right\}\left(d_{1E,n}^2 + d_{2E,n}^2\right)^{-1} + \left(e_{1E,n}^2 + e_{2E,n}^2\right)^{-1}\left\{5\left(-e_{1E,n}d_{2E,n}^3 + 3d_{1E,n}^2 d_{2E,n}e_{1E,n}\right) + 5\left(-d_{1E,n}^3 e_{2E,n} + 3d_{2E,n}^2 d_{1E,n}e_{2E,n}\right)\right\} + 34\left(d_{1E,n}e_{2E,n} - d_{2E,n}e_{1E,n}\right)\right],$$

$$H_{6E,n} = \left[(\sin(\mathrm{h}_{1E,n}))(\cosh(\mathrm{h}_{2E,n}))(\cosh(g_{1E,n}))(\sin(g_{2E,n})) - (\cos(\mathrm{h}_{1E,n}))(\sinh(\mathrm{h}_{2E,n}))(\sinh(g_{1E,n}))(\cos(g_{2E,n}))\right],$$

$$H_{7E,n} = \left[H_{1E,n} + \left(\rho_{5E,n}H_{2E,n}/2\right) + \left(\rho_{6E,n}G_{2E,n}/2\right) + (\Delta_0/2)\left\{\rho_{8E,n}G_{3E,n} + \rho_{7E,n}H_{3E,n} + \rho_{10E,n}G_{4E,n} + \rho_{9E,n}H_{4E,n} + \rho_{12E,n}G_{5E,n} + \rho_{11E,n}H_{5E,n} + (1/12)\left(\rho_{14E,n}G_{6E,n} + \rho_{13E,n}H_{6E,n}\right)\right\}\right],$$

$$H_{1E,n} = \left[(\sin(\mathrm{h}_{1E,n}))(\sinh(\mathrm{h}_{2E,n}))(\cosh(g_{1E,n}))(\cos(g_{2E,n})) + (\cos(\mathrm{h}_{1E,n}))(\cosh(\mathrm{h}_{2E,n}))(\sinh(g_{1E,n}))(\sin(g_{2E,n}))\right],$$

$$D_{7E,n} = \sinh^{-1}\left(\overline{\omega_{7E,n}}\right), \quad \mathrm{H}_{8E,n} = \left(2C_{7E,n}D_{7E,n}/L_0^2\right)$$

The DOS function can be written as

$$N(E) = \frac{eB}{2\pi^2\hbar}\sum_{n=0}^{n_{\max}} \frac{\left(G'_{8E,n} + iH'_{8E,n}\right)H(E - E_{15,1})}{\sqrt{G_{8E,n} + iH_{8E,n}}} \tag{16.1b}$$

where $E_{15,1}$ is the sub-band energy in this case and is given by

$$0 = \left[G_{8E,n} + iH_{8E,n}\right]\Big|_{E=E_{15,1}} \tag{16.1c}$$

The EEM can be written as

$$m^*\left(E, \eta_g, n\right) = \frac{\hbar^2}{2} G'_{8E,n} \tag{16.2}$$

16.2.2 The DR in II–VI HD Superlattices with Graded Interfaces Under Magnetic Quantization

The simplified DR in heavily doped II–VI super-lattices with graded interfaces under magnetic quantization can be expressed as

$$k_z^2 = G_{19E,n} + iH_{19E,n} \tag{16.3a}$$

where

$$\mathrm{G}_{19E,n} = \left[\frac{C_{18E,n}^2 - D_{18E,n}^2}{L_0^2} - \left(\frac{2eB}{\hbar}\left(n + \frac{1}{2}\right)\right)\right],$$

$$C_{18E,n} = \cos^{-1}(\omega_{18E,n}),$$

$$\omega_{18E,n} = (2)^{\frac{-1}{2}}\left[\left(1 - G_{18E,n}^2 - H_{18E,n}^2\right) - \sqrt{\left(1 - G_{18E,n}^2 - H_{18E,n}^2\right)^2 + 4G_{180D}^2}\right]^{\frac{1}{2}},$$

$$C_{18E,n} = \frac{1}{2}\left[G_{11E,n} + G_{12E,n} + \Delta_0\left(G_{13E,n} + G_{14E,n}\right) + \Delta_0\left(G_{15E,n} + G_{16E,n}\right)\right],$$

$$G_{11E,n} = 2(\cos(g_{1E,n}))(\cos(g_{2E,n}))(\cos\gamma_{11}(E,n)), \gamma_{11}(E,n) = k_{21}(E,n)(b_0 - \Delta_0),$$

$$k_{21}(E,n) = \left\{\left[\gamma_3(E,\eta_{g1}) - \frac{\hbar^2}{2m_{\perp,1}^*}\left\{\frac{2eB}{\hbar}\left(n + \frac{1}{2}\right)\right\} \pm C_0\left\{\frac{2eB}{\hbar}\left(n + \frac{1}{2}\right)\right\}^{1/2}\right]\frac{2m_{\parallel,1}^*}{\hbar^2}\right\}^{1/2},$$

$$\begin{aligned}G_{12E,n} = {} & ([\Omega_1(E,n)(\sinh\ g_{1E,n})(\cos\ g_{2E,n}) \\ & - \Omega_2(E,n)(\sinh\ g_{2E,n})(\cosh\ g_{1E,n})](\sin\ \gamma_{11}(E,n)))\end{aligned}$$

$$\Omega_1(E,n) = \left[\frac{d_{1E,n}}{k_{21}(E,n)} - \frac{k_{21}(E,n)d_{1E,n}}{d_{1E,n}^2 + d_{2E,n}^2}\right],$$

$$\Omega_2(E,n) = \left[\frac{d_{2E,n}}{k_{21}(E,n)} + \frac{k_{21}(E,n)d_{2E,n}}{d_{1E,n}^2 + d_{2E,n}^2}\right],$$

$$\begin{aligned}G_{13E,n} = {} & ([\Omega_3(E,n)(\cosh\ g_{1E,n})(\cos\ g_{2E,n}) \\ & - \Omega_4(E,n)(\sinh\ g_{1E,n})(\sin\ g_{2E,n})](\sin\ \gamma_{11}(E,n)))\end{aligned}$$

$$\Omega_3(E,n) = \left[\frac{d_{1E,n}^2 - d_{2E,n}^2}{k_{21}(E,n)} - 3k_{21}(E,n)\right],$$

$$\Omega_4(E,n) = \left[\frac{2d_{1E,n}d_{2E,n}}{k_{21}(E,n)}\right]$$

$$\begin{aligned}G_{14E,n} = {} & ([\Omega_5(E,n)(\sinh g_{1E,n})(\cos g_{2E,n}) \\ & - \Omega_6(E,n)(\sin g_{1E,n})(\cosh g_{2E,n})](\cos\gamma_{11}(E,n))).\end{aligned}$$

$$\Omega_5(E,n) = \left[3d_{1E,n} - \frac{d_{1E,n}}{d_{1E,n}^2 + d_{2E,n}^2} k_{21}^2(E,n)\right],$$

$$\Omega_6(E,n) = \left[3d_{2E,n} + \frac{d_{2E,n}}{d_{1E,n}^2 + d_{2E,n}^2} k_{21}^2(E,n)\right]$$

$$G_{15E,n} = \left(\left[\Omega_9(E,n)(\cosh g_{1E,n})(\cos g_{2E,n}) - \Omega_{10}(E,n)(\sinh g_{1E,n})(\sin_{g_{2E,n}})\right](\cos\gamma_{11}(E,n))\right)$$

$$\Omega_9(E,n) = \left[2d_{1E,n}^2 - 2d_{2E,n}^2 - k_{21}^2(E,n)\right],$$

$$\Omega_{10}(E,n) = \left[2d_{1E,n}d_{2E,n}\right]$$

$$G_{16E,n} = \left(\left[\Omega_7(E,n)(\sinh g_{1E,n})(\cos g_{2E,n}) - \Omega_8(E,n)(\sin g_{1E,n})(\cosh_{g_{2E,n}})\right](\sin\gamma_{11}(E,n)/12)\right),$$

$$\Omega_7(E,n) = \left[\frac{5d_{1E,n}}{d_{1E,n}^2 + d_{2E,n}^2} k_{21}^3(E,n) + \frac{5\left(d_{1E,n}^3 - 3d_{2E,n}^2 d_{1E,n}\right)}{k_{21}(E,n)} - 34k_{21}(E,n)d_{1E,n}\right],$$

$$\Omega_8(E,n) = \left[\frac{5d_{2E,n}}{d_{1E,n}^2 + d_{2E,n}^2} k_{21}^3(E,n) + \frac{5\left(d_{2E,n}^3 - 3d_{2E,n}^2 d_{1E,n}\right)}{k_{21}(E,n)} + 34k_{21}(E,n)d_{2E,n}\right]$$

$$H_{18E,n} = \frac{1}{2}\left[H_{11E,n} + H_{12E,n} + \Delta_0\left(H_{13E,n} + H_{14E,n}\right) + \Delta_0\left(H_{15E,n} + H_{16E,n}\right)\right],$$

$$H_{11E,n} = 2(\sinh\ g_{1E,n})(\sin\ g_{2E,n})(\cos\gamma_{11}(E,n))),$$

$$H_{12E,n} = \left(\left[\Omega_2(E,n)(\sinh g_{1E,n})(\cos g_{2E,n}) + \Omega_1(E,n)(\sin g_{2E,n})(\cosh g_{1E,n})\right](\sin\gamma_{11}(E,n))\right),$$

$$H_{13E,n} = \left(\left[\Omega_4(E,n)(\cosh g_{1E,n})(\cos g_{2E,n}) + \Omega_3(E,n)(\sinh g_{1E,n})(\sin g_{2E,n})\right](\sin\gamma_{11}(E,n))\right),$$

$$H_{14E,n} = \left(\left[\Omega_6(E,n)(\sinh g_{1E,n})(\cos g_{2E,n}) + \Omega_5(E,n)(\sin g_{1E,n})(\cosh g_{2E,n})\right](\cos\gamma_{11}(E,n))\right),$$

$$H_{15E,n} = \left(\left[\Omega_{10}(E,n)(\cosh g_{1E,n})(\cos g_{2E,n}) + \Omega_9(E,n)(\sinh g_{1E,n})(\sin g_{2E,n})\right](\cos\gamma_{11}(E,n))\right),$$

$$H_{16E,n} = \left(\left[\Omega_8(E,n)(\sinh g_{1E,n})(\cos g_{2E,n}) + \Omega_7(E,n)(\sin g_{1E,n})(\cosh g_{2E,n})\right](\sin\gamma_{11}(E,n)/2)\right),$$

$$H_{19E,n} = \left[\frac{2C_{18E,n}D_{18E,n}}{L_0^2}\right] \quad \text{and} \quad D_{18E,n} = \sinh^{-1}(\omega_{18E,n})$$

The DOS function can be written as

$$N(E) = \frac{eB}{2\pi^2\hbar}\sum_{n=0}^{n_{\max}} \frac{\left(G'_{19E,n} + iH'_{19E,n}\right)H(E - E_{15,2})}{\sqrt{G_{19E,n} + iH_{19E,n}}} \tag{16.3b}$$

where $E_{15,2}$ is the sub-band energy in this case and is given by

$$0 = \left[G_{19E,n} + iH_{19E,n}\right]\big|_{E=E_{15,2}} \tag{16.3c}$$

The EEM can be written as

$$m^*\left(E, \eta_g, n\right) = \frac{\hbar^2}{2} G'_{19E,n} \tag{16.4}$$

16.2.3 The DR in IV–VI HD Superlattices with Graded Interfaces Under Magnetic Quantization

The simplified DR in heavily doped IV–VI super-lattices with graded interfaces under magnetic quantization can be expressed as

$$k_z^2 = \frac{1}{L_0^2}\left[\cos^{-1}\left\{\frac{1}{2}\Phi_2(E,n)\right\}\right]^2 - \frac{2eB}{\hbar}\left(n + \frac{1}{2}\right) \tag{16.5a}$$

where

$$\begin{aligned}
\Phi_2(E,n) \equiv \Bigg[& 2\cosh\{\beta_2(E,n)\}\cos\{\gamma_2(E,n)\} + \varepsilon_2(E,n)\sinh\{\beta_2(E,n)\}\sin\{\gamma_{22}(E,n)\} \\
& + \Delta_0\Bigg[\left(\left(\{K_{112}(E,n)\}^2/K_{212}(E,n)\right) - 3K_{212}(E,n)\right)\cosh\{\beta_2(E,n)\}\sin\{\gamma_{22}(E,n)\}. \\
& \quad + \left(3K_{112}(E,n) - \frac{\{K_{212}(E,n)\}^2}{K_{112}(E,n)}\right)\sinh\{\beta_2(E,n)\}\cos\{\gamma_{22}(E,n)\}\Bigg] \\
& + \Delta_0\Bigg[2\left(\{K_{112}(E,n)\}^2 - \{K_{212}(E,n)\}^2\right)\cosh\{\beta_2(E,n)\}\cos\{\gamma_{22}(E,n)\}. \\
& \quad + \frac{1}{12}\left[\frac{5\{K_{112}(E,n)\}^3}{K_{212}(E,n)} + \frac{5\{K_{212}(E,n)\}^3}{K_{112}(E,n)} - 34K_{212}(E,n)K_{112}(E,n)\right] \\
& \quad \sinh\{\beta_2(E,n)\}\sin\{\gamma_{22}(E,n)\}\Bigg]\Bigg],
\end{aligned}$$

$$\beta_2(E,n) \equiv K_{112}(E,n)[a_0 - \Delta_0],$$

$$k_{112}^2(E,n) = \left[2\bar{p}_{9,2n}\right]^{-1}\left[-\bar{q}_{9,2n}(E-V_0,\eta_{g2}) - \left[\left[\bar{q}_{9,2n}(E-V_0,\eta_{g2})\right]^2 + 4\bar{p}_{9,2n}\bar{R}_{9,2n}(E-V_0,\eta_{g2})\right]^{\frac{1}{2}}\right],$$

$$\bar{q}_{9,2n}(E-V_0,\eta_{g2}) = \left[(\hbar^2/2)\left((1/m_{l2}^*) + (1/m_{l2}^-)\right) + \alpha_2(\hbar^4/4)\frac{2eB}{\hbar}\left(n+\frac{1}{2}\right)\left((1/m_{l2}^+ m_{t2}^-) + (1/m_{t2}^+ m_{l2}^-)\right) - \alpha_2\gamma_3(E-V_0,\eta_{g2})\left((1/m_{l2}^+) - (1/m_{l2}^-)\right)\right],$$

$$\bar{R}_{9,2n}(E,\eta_{g2}) = \left[\gamma_2(E-V_0,\eta_{g2}) + \gamma_3(E-V_0,\eta_{g2})\left[(\hbar^2/2)\alpha_2\frac{2eB}{\hbar}\left(n+\frac{1}{2}\right)\left((1/m_{t2}^*) - (1/m_{t2}^-)\right)\right] - \left[(\hbar^2/2)k_{s0}^2\left((1/m_{t2}^*) + (1/m_{t2}^-)\right)\right] - \alpha_2(\hbar^6/4)\left[\frac{2eB}{\hbar}\left(n+\frac{1}{2}\right)\right]^2\left((1/m_{t2}^+ m_{t2}^-)\right)\right],$$

$$\gamma_2(E,n) = K_{212}(E,n)[b_0 - \Delta_0],$$

$$k_{212}^2(E,n) = \left[2\bar{p}_{9,1n}\right]^{-1}\left[-\bar{q}_{9,1n}(E,\eta_{g1}) + \left[\left[\bar{q}_{9,1n}(E,\eta_{g1})\right]^2 + 4\bar{p}_{9,1n}\bar{R}_{9,1n}(E,\eta_{g1})\right]^{\frac{1}{2}}\right]$$

$$\bar{q}_{9,1n}(E,\eta_{g1}) = \left[(\hbar^2/2)\left((1/m_{l1}^*) + (1/m_{l1}^-)\right) + \alpha_1(\hbar^4/4)\frac{2eB}{\hbar}\left(n+\frac{1}{2}\right)\left((1/m_{l1}^+ m_{t1}^-) + \left(1/m_{t1}^+ m_{m_{l1}}^-\right)\right) - \alpha_1\gamma_3(E,\eta_{g1})\left((1/m_{l1}^+) - (1/m_{l1}^-)\right)\right],$$

$$\bar{R}_{9,1}(E,\eta_{g1}) = \Bigg[\gamma_2(E,\eta_{g1}) + \gamma_3(E,\eta_{g1})$$

$$\left[(\hbar^2/2)\alpha_1(2eB/\hbar)\left(n+\frac{1}{2}\right)((1/m_{t1}^*) - (1/m_{t1}^-))\right]$$

$$- \left[(\hbar^2/2)k_{s0}^2((1/m_{t1}^*) + (1/m_{t1}^-))\right]$$

$$- \alpha_1(\hbar^6/4)\left((2eB/\hbar)\left(n+\frac{1}{2}\right)\right)^2((1/m_{t1}^+ m_{t1}^-))\Bigg] \text{ and}$$

$$\varepsilon_2(E,n) \equiv \left[\frac{K_{112}(E,n)}{K_{212}(E,n)} - \frac{K_{212}(E,n)}{K_{112}(E,n)}\right].$$

The DOS function can be written as

$$N(E) = \frac{eBg_v}{2\pi^2\hbar L_0}\sum_{n=0}^{n_{\max}} \frac{\cos^{-1}\left[\frac{1}{2}\varphi_2(E,n)\left(1-\frac{1}{4}\varphi_2^2(E,n)\right]^{-1/2}\varphi_2'(E,n)\mathrm{H}(E-E_{15,3})}{\left[\left[\cos^{-1}\left[\frac{1}{2}\varphi_2(E,n)\right]\right]^2 - L_0^2\frac{2eB}{\hbar}\left(1+\frac{1}{2}\right)\right]^{1/2}} \tag{16.5b}$$

where $E_{15,3}$ is the sub-band energy in this case and is given by

$$0 = \frac{1}{L_0^2}\left[\cos^{-1}\left\{\frac{1}{2}\Phi_2(E_{15,3},\, n)\right\}\right]^2 - \frac{2eB}{\hbar}\left(n+\frac{1}{2}\right) \tag{16.5c}$$

The EEM can be written as

$$m^*(E,\eta_g,n) = \frac{\hbar^2}{2L_0^2}\cos^{-1}\left[\frac{1}{2}\varphi_2(E,n)\left(1-\frac{1}{4}\varphi_2^2(E,n)\right)\right]^{-1/2}\varphi_2'(E,n) \tag{16.6}$$

16.2.4 The DR in HgTe/CdTe HD Superlattices with Graded Interfaces Under Magnetic Quantization

The simplified DR in heavily doped HgTe/CdTe superlattices with graded interfaces under magnetic quantization can be expressed as

$$(k_z)^2 = G_{192E,n} + iH_{192E,n} \tag{16.7a}$$

where

$$G_{192E,n} = \left[\frac{C_{182E,n}^2 - D_{182E,n}^2}{L_0^2} - (2eB/\hbar)(n+(1/2))\right],$$

$$C_{1820D} = \cos^{-1}\left(\omega_{182E,n}\right),$$

$$\left(\omega_{182E,n}\right) = (2)^{\frac{-1}{2}}\left[\left(1 - G^2_{182E,n} - H^2_{182E,n}\right) - \sqrt{\left(1 - G^2_{182E,n} - H^2_{182E,n}\right)^2 + 4G^2_{182E,n}}\right]^{\frac{1}{2}},$$

$$G_{182E,n} = \frac{1}{2}\left[G_{112E,n} + G_{122E,n} + \Delta_0\left(G_{132E,n} + G_{142E,n}\right) + \Delta_0\left(G_{152E,n} + G_{162E,n}\right)\right],$$

$$G_{112E,n} = 2(\cos(g_{12}))(\cos(g_{22}))(\cos\gamma_8(E,n)),$$
$$\gamma_8(E,n) = k_8(E,n)(b_0 - \Delta_0),$$

$$k_8(E,n) = \left[\frac{B^2_{01} + 4A_1E - B_{01}\sqrt{B^2_{01} + 4A_1E}}{2A^2_1} - (2eB/\hbar)(n + (1/2))\right]^{1/2},$$

$$G_{120D} = \left(\left[\Omega_{12}(E,n)\left(\sinh\ g_{12E,n}\right)\left(\cos g_{22E,n}\right) - \Omega_{22}(E,n)\left(\sin g_{22E,n}\right)\left(\cosh g_{12E,n}\right)\right](\sin\gamma_8(E,n))\right),$$

$$\Omega_{12}(E,n) = \left[\frac{d_{12E,n}}{k_8(E,n)} - \frac{k_8(E,n)d_{12E,n}}{d^2_{12E,n} + d^2_{22E,n}}\right],$$

$$\Omega_{22}(E,n) = \left[\frac{d_{22E,n}}{k_8(E,n)} + \frac{k_8(E,n)d_{22E,n}}{d^2_{12E,n} + d^2_{22E,n}}\right],$$

$$G_{1320D} = \left(\left[\Omega_{32}(E,n)\left(\cosh\ g_{12E,n}\right)\left(\cos g_{22E,n}\right) - \Omega_{42}(E,n)\left(\sinh g_{12E,n}\right)\left(\sin g_{22E,n}\right)\right](\sin\gamma_8(E,n))\right),$$

$$\Omega_{32}(E,n) = \left[\frac{d^2_{12E,n} - d^2_{2E,n}}{k_8(E,n)} - 3k_8(E,n)\right],$$

$$\Omega_{42}(E,n) = \left[\frac{2d_{12E,n}d_{22E,n}}{k_8(E,n)}\right],$$

$$G_{1420D} = \left(\left[\Omega_{52}(E,n)\left(\sinh\ g_{12E,n}\right)\left(\cos g_{22E,n}\right) - \Omega_{62}(E,n)\left(\sin g_{12E,n}\right)\left(\cosh g_{22E,n}\right)\right](\cos\gamma_8(E,n))\right),$$

$$\Omega_{52}(E,n) = \left[3d_{12E,n} - \frac{d_{12E,n}}{d^2_{12E,n} + d^2_{22E,n}}k^2_8(E,n)\right],$$

$$\Omega_{62}(E,n) = \left[3d_{22E,n} + \frac{d_{22E,n}}{d^2_{12E,n} + d^2_{22E,n}}k^2_8(E,n)\right],$$

$$G_{1520D} = \left(\left[\Omega_{92}(E,n)\left(\cosh\ g_{12E,n}\right)\left(\cos g_{22E,n}\right)\right.\right.$$
$$\left.\left. - \Omega_{102}(E,n)\left(\sinh g_{12E,n}\right)\left(\sin g_{22E,n}\right)\right]\left(\cos\gamma_8(E,n)\right)\right),$$

$$\Omega_{92}(E,n) = \left[2d_{12E,n}^2 - 2d_{22E,n}^2 - k_8^2(E,n)\right],$$
$$\Omega_{102}(E,n) = \left[2d_{12E,n}d_{22E,n}\right],$$

$$G_{162E,n} = \left(\left[\Omega_{72}(E,n)\left(\sinh\ g_{12E,n}\right)\left(\cos g_{22E,n}\right)\right.\right.$$
$$\left.\left. - \Omega_{82}(E,n)\left(\sin g_{12E,n}\right)\left(\cosh g_{22E,n}\right)\right]\left(\sin\gamma_{80D}(E,n)/12\right)\right),$$

$$\Omega_{72}(E,n) = \left[\frac{5d_{12E,n}}{d_{12E,n}^2 + d_{22E,n}^2}k_8^2(E,n) + \frac{5\left(d_{12E,n}^3 - 3d_{22E,n}^2 d_{12E,n}\right)}{k_8(E,n)} - 34k_8(E,n)d_{12E,n}\right],$$

$$\Omega_{82}(E,n) = \left[\frac{5d_{22E,n}}{d_{12E,n}^2 + d_{22E,n}^2}k_8^2(E,n) + \frac{5\left(d_{22E,n}^3 - 3d_{22E,n}^2 d_{12E,n}\right)}{k_8(E,n)} + 34k_8(E,n)d_{22E,n}\right],$$

$$H_{182E,n} = \frac{1}{2}\left[H_{112E,n} + H_{122E,n} + \Delta_0\left(H_{132E,n} + H_{142E,n}\right) + \Delta_0\left(H_{152E,n} + H_{162E,n}\right)\right],$$
$$H_{112E,n} = 2\left(\sinh g_{12E,n}\right)\left(\sin g_{22E,n}\right)\left(\cos\gamma_8(E,n)\right),$$
$$H_{1220D} = \left(\left[\Omega_{22}(E,n)\left(\sinh\ g_{12E,n}\right)\left(\cos g_{22E,n}\right)\right.\right.$$
$$\left.\left. + \Omega_{12}(E,n)\left(\sin g_{22E,n}\right)\left(\cosh g_{12E,n}\right)\right]\left(\sin\gamma_8(E,n)\right)\right),$$
$$H_{132E,n} = \left(\left[\Omega_{42}(E,n)\left(\cosh\ g_{12E,n}\right)\left(\cos g_{22E,n}\right)\right.\right.$$
$$\left.\left. + \Omega_{32}(E,n)\left(\sinh g_{12E,n}\right)\left(\sin g_{22E,n}\right)\right]\left(\sin\gamma_8(E,n)\right)\right),$$
$$H_{142E,n} = \left(\left[\Omega_{62}(E,n)\left(\sinh\ g_{12E,n}\right)\left(\cos g_{22E,n}\right)\right.\right.$$
$$\left.\left. + \Omega_{52}(E,n)\left(\sin g_{12E,n}\right)\left(\cosh g_{22E,n}\right)\right]\left(\cos\gamma_8(E,n)\right)\right),$$
$$H_{1520D} = \left(\left[\Omega_{102}(E,n)\left(\cosh\ g_{12E,n}\right)\left(\cos g_{22E,n}\right)\right.\right.$$
$$\left.\left. + \Omega_{92}(E,n)\left(\sinh g_{12E,n}\right)\left(\sin g_{22E,n}\right)\right]\left(\cos\gamma_8(E,n)\right)\right),$$
$$H_{162E,n} = \left(\left[\Omega_{82}(E,n)\left(\sinh\ g_{12E,n}\right)\left(\cos g_{22E,n}\right)\right.\right.$$
$$\left.\left. + \Omega_{72}(E,n)\left(\sin g_{12E,n}\right)\left(\cosh g_{22E,n}\right)\right]\left(\sin\gamma_8(E,n)/2\right)\right),$$
$$H_{192E,n} = \left[\left(\left(2C_{182E,n}D_{182E,n}\right)/L_0^2\right)\right] \text{ and } D_{182E,n} = \sinh^{-1}\left(\omega_{182E,n}\right)$$

The DOS function can be written as

$$N(E) = \frac{eB}{2\pi^2\hbar}\sum_{n=0}^{n_{\max}}\frac{\left(G'_{192E,n} + iH'_{192E,n}\right)H\left(E - E_{15,4}\right)}{\sqrt{G_{192E,n} + iH_{192E,n}}} \tag{16.7b}$$

where $E_{15,4}$ is the sub-band energy in this case and is given by

$$0 = \left[G_{192E,n} + H_{192E,n}\right]\Big|_{E=E_{15,4}} \tag{16.7c}$$

The EEM can be written as

$$m^*\left(E, \eta_g, n\right) = \frac{\hbar^2}{2} G'_{192E,n} \tag{16.8}$$

16.2.5 The DR in Strained Layer HD Superlattices with Graded Interfaces Under Magnetic Quantization

The DR of the conduction electrons in heavily doped strained layer SL with graded interfaces can be expressed as

$$k_z^2 = \frac{1}{L_0^2}\left[\cos^{-1}\left\{\frac{1}{2}\bar{\varphi}_6(E,n)\right\}\right]^2 - \frac{2|e|B}{\hbar}\left(n + \frac{1}{2}\right) \tag{16.9a}$$

where

$$\begin{aligned}
\bar{\phi}_6(E,n) = {} & \left[2\cosh\left[T_4\left(E,n,\eta_{g2}\right)\right]\cos\left[T_5\left(E,n,\eta_{g1}\right)\right]\right] \\
& + \left[T_6(E,n)\right]\sinh\left[T_4\left(E,n,\eta_{g2}\right)\right]\sin\left[T_5\left(E,n,\eta_{g1}\right)\right] \\
& + \Delta_0\left[\left(\frac{k_0^2\left(E,n,\eta_{g2}\right)}{k_0'\left(E,n,\eta_{g1}\right)} - 3k_0'\left(E,n,\eta_{g1}\right)\right)\cosh\left[T_4\left(E,n,\eta_{g2}\right)\right]\sin\left[T_5\left(E,n,\eta_{g1}\right)\right]\right. \\
& \left. + \left(3k_0\left(E,n,\eta_{g2}\right) - \frac{k_0'^2\left(E,n,\eta_{g1}\right)}{k_0\left(E,n,\eta_{g2}\right)}\right)\sinh\left[T_4\left(E,n,\eta_{g2}\right)\right]\cos\left[T_5\left(E,n,\eta_{g1}\right)\right]\right] \\
& + \Delta_0\left[2\left(k_0^2\left(E,n,\eta_{g2}\right) - k_{0D}'^2\left(E,n,\eta_{g1}\right)\right)\cosh\left[T_4\left(E,n,\eta_{g2}\right)\right]\cos\left[T_5\left(E,n,\eta_{g1}\right)\right]\right] \\
& + \frac{1}{12}\left(\frac{5k_0^3\left(E,n,\eta_{g2}\right)}{k_0'\left(E,n,\eta_{g1}\right)} + \frac{5k_0'^3\left(E,n,\eta_{g1}\right)}{k_0\left(E,n,\eta_{g2}\right)} - 34k_0\left(E,n,\eta_{g2}\right)k_0'\left(E,n,\eta_{g1}\right)\right) \\
& \sinh\left[T_4\left(E,n,\eta_{g2}\right)\right]\sin\left[T_5\left(E,n,\eta_{g1}\right)\right]
\end{aligned}$$

$$\left[T_4\left(E,n,\eta_{g2}\right)\right] = k_0\left(E,n,\eta_{g2}\right)\left[a_0 - \Delta_0\right],$$

$$k_0(E,n,\eta_{g2}) = \left[\bar{S}_2(E-V_0,\eta_{g2})\right]^{-1/2} \cdot \left[\left[(n+1/2)\hbar eB/\left(\sqrt{\rho_1(E)\rho_2(E)}\right)\right]-1\right]^{1/2},$$
$$\rho_1(E) = \hbar^2/(2\bar{P}_2(E-V_0,\eta_{g2})),\ \rho_2(E) = \hbar^2/(2\bar{Q}_2(E-V_0,\eta_{g2}))$$
$$T_5(E,n,\eta_{g1}) = k_0'(E,n,\eta_{g1})[b_0-\Delta_0],$$
$$k_0'(E,n,\eta_{g1}) = \left[\bar{S}_1(E,n,\eta_{g1})\right]^{-1/2}\left[1-\left[(n+1/2)\hbar eB/\left(\sqrt{\rho_3(E)\rho_4(E)}\right)\right]\right]^{1/2}$$
$$\rho_3(E) = \hbar^2/(2\bar{P}_1(E,\eta_{g1})),\ \rho_4(E) = \hbar^2/(2\bar{Q}_1(E,\eta_{g1}))$$
$$T_6(E,n) = \left[\frac{k_0(E,n,\eta_{g2})}{k_0'(E,n,\eta_{g1})} - \frac{k_0'(E,n,\eta_{g1})}{k_0(E,n,\eta_{g2})}\right]$$

The DOS function can be written as

$$N(E) = \frac{eBg_v}{2\pi^2\hbar L_0}\sum_{n=0}^{n_{\max}} \frac{\cos^{-1}\left[\frac{1}{2}\bar{\varphi}_6(E,n)\left(1-\frac{1}{4}[\bar{\varphi}_6(E,n)]^2\right)\right]^{-1/2}[\bar{\varphi}_6(E,n)]'H(E-E_{15,5})}{\left[\left[\cos^{-1}\left[\frac{1}{2}\bar{\varphi}_6(E,n)\right]\right]^2 - L_0^2\frac{2eB}{\hbar}\left(1+\frac{1}{2}\right)\right]^{1/2}} \tag{16.9b}$$

where $E_{15,5}$ is the sub-band energy in this case and is given by

$$0 = \frac{1}{L_0^2}\left[\cos^{-1}\left\{\frac{1}{2}\bar{\varphi}_6(E_{15,5},n)\right\}\right]^2 - \frac{2|e|B}{\hbar}\left(n+\frac{1}{2}\right) \tag{16.9c}$$

The EEM can be written as

$$m^*(E,\eta_g,n) = \frac{\hbar^2}{2L_0^2}\cos^{-1}\left[\frac{1}{2}\bar{\varphi}_6(E,n)\left(1-\frac{1}{4}[\bar{\varphi}_6(E,n)]^2\right)\right]^{-1/2}[\bar{\varphi}_6(E,n)]' \tag{16.10}$$

16.2.6 The DR in III–V HD Effective Mass Superlattices Under Magnetic Quantization

The magneto electron dispersion relation in this case assumes the form

$$(\mathrm{k}_x)^2 = \delta_{7E,n} + i\delta_{8E,n} \tag{16.11a}$$

where

$$\delta_{7E,n} = \left[\frac{1}{L_0^2}\left(\delta_{5E,n}^2 - \delta_{6E,n}^2\right) - \left\{\frac{2eB}{\hbar}\left(n+\frac{1}{2}\right)\right\}\right], \quad \delta_{5E,n} = \cos^{-1} p_{5E,n},$$

$$p_{5E,n} = \left[\frac{1-\delta_{3E,n}^2 - \delta_{4E,n}^2 - \sqrt{\left(1-\delta_{3E,n}^2-\delta_{4E,n}^2\right)^2 + 4\delta_{4E,n}^2}}{2}\right]^{1/2},$$

$$\begin{aligned}
\delta_{3E,n} &= \left(a_1 \cos\Delta_{1E,n}\cosh\Delta_{2E,n} - a_2\cos\Delta_{3E,n}\cosh\Delta_{4E,n}\right),\\
\delta_{4E,n} &= \left(a_1 \sin\Delta_{1E,n}\sinh\Delta_{2E,n} - a_2\sin\Delta_{3E,n}\sinh\Delta_{4E,n}\right)\\
\Delta_{1E,n} &= \left(a_0 e_{1E,n} + b_0 e_{3E,n}\right),\ \Delta_{2E,n} = \left(a_0 e_{2E,n} + b_0 e_{4E,n}\right),\\
\Delta_{3E,n} &= \left(a_0 e_{1E,n} - b_0 e_{3E,n}\right),\ \Delta_{4E,n} = \left(a_0 e_{2E,n} - b_0 e_{4E,n}\right),\\
\delta_{6E,n} &= \sinh^{-1} \mathrm{p}_{5E,n} \text{ and } \delta_{8E,n} = \left[2\delta_{5E,n}\delta_{6E,n}/L_0^2\right],
\end{aligned}$$

$$e_{1E,n} = \left[\left(\left(\sqrt{t_{1E,n}^2 + t_2^2} + t_{1E,n}\right)/2\right)\right]^{\frac{1}{2}}, \quad \mathrm{e}_{2E,n} = \left[\left(\left(\sqrt{t_{1E,n}^2 + t_2^2} - t_{1E,n}\right)/2\right)\right]^{\frac{1}{2}},$$

$$e_{3E,n} = \left[\frac{\sqrt{t_{3E,n}^2 + t_4^2} + t_{3E,n}}{2}\right]^{1/2}, \quad e_{4E,n} = \left[\frac{\sqrt{t_{3E,n}^2 + t_4^2} - t_{3E,n}}{2}\right]^{1/2},$$

$$\begin{aligned}
t_{1E,n} &= \left[\frac{2m_{c1}^*}{\hbar^2} T_{11}(E,\Delta_1,\eta_{g1},E_{g1}) - \frac{2eB}{\hbar}\left(n+\frac{1}{2}\right)\right],\\
t_{3E,n} &= \left[\frac{2m_{c1}^*}{\hbar^2} T_{12}(E,\Delta_2,\eta_{g2},E_{g2}) - \frac{2eB}{\hbar}\left(n+\frac{1}{2}\right)\right]
\end{aligned}$$

The DOS function can be written as

$$N(E) = \frac{eB}{2\pi^2\hbar}\sum_{n=0}^{n_{\max}} \frac{\left(\delta'_{7E,n} + i\delta'_{8E,n}\right) H(E - E_{15,6})}{\sqrt{\delta_{7E,n} + i\delta_{8E,n}}} \tag{16.11b}$$

where $E_{15,6}$ is the sub-band energy in this case and is given by

$$0 = \delta_{7E_{15,6,n}} + i\delta_{8E_{15,6,n}} \tag{16.11c}$$

The EEM can be written as

$$m^*(\mathrm{E},\eta_g,n) = \frac{\hbar^2}{2}\delta'_{7E,n} \tag{16.12}$$

16.2.7 The DR in II–VI HD Effective Mass Superlattices Under Magnetic Quantization

The DR in heavily doped II–VI EMSL can be written as

$$(k_z)^2 = \Delta_{13E,n} + i\Delta_{14E,n}, \tag{16.13a}$$

where

$$\Delta_{13E,n} = \left[\frac{1}{L_0^2}\left(\Delta_{11E,n}^2 - \Delta_{12E,n}^2\right) - \left\{\frac{2eB}{\hbar}\left(n+\frac{1}{2}\right)\right\}\right]$$

$$\Delta_{11E,n} = \cos^{-1} p_{6E,n}, p_{6E,n}$$
$$= \left[\frac{1 - \Delta_{9E,n}^2 - \Delta_{10E,n}^2 - \sqrt{\left(1 - \Delta_{9E,n}^2 - \Delta_{10E,n}^2\right)^2 + 4\Delta_{10E,n}^2}}{2}\right]^{1/2},$$

$$\Delta_{9E,n} = \left(\bar{a}_1 \cos\Delta_{6E,n}\cosh\Delta_{7E,n} - \bar{a}_2\cos\Delta_{8E,n}\cosh\Delta_{7E,n}\right),$$
$$\Delta_{10E,n} = \left(\bar{a}_1 \sin\Delta_{6E,n}\sinh\Delta_{7E,n} + \bar{a}_2\sin\Delta_{8E,n}\sinh\Delta_{7E,n}\right),$$
$$\Delta_{6E,n} = \left[a_0 C_{22E,n}\left(E_{E,n}, \eta_{g1}\right) + b_0 e_{3E,n}\right],$$
$$\Delta_{7E,n} = b_0 e_{4E,n},$$
$$\Delta_{8E,n} = \left[a_0 C_{22E,n}\left(E_{E,n}, \eta_{g1}\right) - b_0 e_{3E,n}\right],$$

$$C_{22E,n}\left(E_{E,n}, \eta_{g1}\right) = \left[\frac{2m_{\parallel,1}^*}{\hbar^2}\left\{\gamma_3\left(E_{E,n}, \eta_{g1}\right) - \frac{\hbar^2}{2m_{\perp,1}^*}\left\{\frac{2eB}{\hbar}\left(n+\frac{1}{2}\right)\right\} \mp C_0\left[\left\{\frac{2eB}{\hbar}\left(n+\frac{1}{2}\right)\right\}\right]^{1/2}\right\}\right]^{1/2},$$

$$\Delta_{12E,n} = \cos^{-1} p_{6E,n},\ \Delta_{14E,n} = \frac{2\Delta_{11E,n}\Delta_{12E,n}}{L_0^2},$$

The DOS function can be written as

$$N(E) = \frac{eB}{2\pi^2\hbar}\sum_{n=0}^{n_{\max}}\frac{\left(\Delta'_{13E,n} + i\Delta'_{14E,n}\right)H(E - E_{15,7})}{\sqrt{\Delta_{13E,n} + i\Delta_{14E,n}}} \tag{16.13b}$$

where $E_{15,7}$ is the sub-band energy in this case and is given by

$$0 = \Delta_{13E_{15,7,n}} + i\Delta_{14E_{15,7,n}} \tag{16.13c}$$

The EEM can be written as

$$m^{*}\left(E,\eta_{g},n\right)=\frac{\hbar^{2}}{2}\Delta'_{13E,n} \tag{16.14}$$

16.2.8 The DR in IV–VI HD Effective Mass Superlattices Under Magnetic Quantization

The DR in heavily doped IV–VI, EMSLs under magnetic quantization can be written as

$$(k_z)^2=\left[\left[1/L_0^2\right]\left\{\cos^{-1}(f_{23}(E,n))\right\}^2-\left(\frac{2eB}{\hbar}\left(n+\frac{1}{2}\right)\right)\right] \tag{16.15a}$$

where

$$f_{23}(E,n)=a_3\cos\left[a_0C_{23E,n}\left(E,n,\eta_{g1}\right)+b_0D_{23E,n}\left(E,n,\eta_{g1}\right)\right]$$
$$-a_4\cos\left[a_0C_{23E,n}\left(E,n,\eta_{g2}\right)-b_0D_{23E,n}\left(E,n,\eta_{g2}\right)\right],$$

$$C_{23}\left(E,n,\eta_{g1}\right)=\left[\left[2\overline{p_{9,1}}\right]^{-1}\left[-\overline{q_{9,1}}\left(E,n,\eta_{g1}\right)+\left[\left\{\overline{q_{9,1}}\left(E,n,\eta_{g1}\right)\right\}^{2}+\left(4\overline{p_{9,1}}\right)\overline{R_{9,1}}\left(E,n,\eta_{g1}\right)\right]^{1/2}\right]^{1/2},$$

$$D_{23}\left(E,n,\eta_{g2}\right)=\left[\left[2\overline{p_{9,2}}\right]^{-1}\left[-\overline{q_{9,2}}\left(E,n,\eta_{g2}\right)+\left[\left\{\overline{q_{9,2}}\left(E,n,\eta_{g2}\right)\right\}^{2}+\left(4\overline{p_{9,2}}\right)\overline{R_{9,2}}\left(E,n,\eta_{g2}\right)^{1/2}\right]\right]\right]^{1/2},$$

$$\overline{q_{9,i}}\left(E,n,\eta_{gi}\right)=\left[\frac{\hbar^{2}}{2}\left(\frac{1}{m_{l,i}^{*}}+\frac{1}{m_{l,i}^{-}}\right)+\alpha_i\frac{\hbar^{4}}{4}\left(\frac{2eB}{\hbar}\left(n+\frac{1}{2}\right)\right)\left(\frac{1}{m_{l,i}^{+}m_{t,i}^{-}}+\frac{1}{m_{t,i}^{+}m_{l,i}^{-}}\right)\right.$$
$$\left.-\alpha_i\gamma_3\left(E,\eta_{gi}\right)\left(\frac{1}{m_{l,i}^{+}}-\frac{1}{m_{l,i}^{-}}\right)\right],$$

$$\overline{R_{9,i}}\left(E,n,\eta_{gi}\right)=\left[\gamma_2\left(E,\eta_{gi}\right)+\gamma_3\left(E,\eta_{gi}\right)\alpha_i\frac{\hbar^{2}}{2}\left(\frac{2eB}{\hbar}\left(n+\frac{1}{2}\right)\right)\left(\frac{1}{m_{t,i}^{+}}-\frac{1}{m_{t,i}^{-}}\right)\right.$$
$$\left.-\frac{\hbar^{2}}{2}\left(\frac{2eB}{\hbar}\left(n+\frac{1}{2}\right)\right)\left(\frac{1}{m_{t,i}^{*}}-\frac{1}{m_{t,i}^{-}}\right)-\frac{\alpha\hbar^{6}}{4}\frac{\left(\frac{2eB}{\hbar}\left(n+\frac{1}{2}\right)\right)^{2}}{m_{t,i}^{-}m_{t,i}^{+}}\right],$$

The DOS function can be written as

$$N(E) = \frac{eBg_v}{2\pi^2\hbar L_0}\sum_{n=0}^{n_{\max}} \frac{\cos^{-1}\left[\frac{1}{2}f_{23}(E,n)\left(1-\frac{1}{4}f_{23}^2(E,n)\right)\right]^{-1/2} f_{23}'(E,n)H(E-E_{15,8})}{\left[\left[\cos^{-1}\left[\frac{1}{2}f_{23}(E,n)\right]\right]^2 - L_0^2\frac{2eB}{\hbar}\left(1+\frac{1}{2}\right)\right]^{1/2}} \tag{16.15b}$$

where $E_{15,8}$ is the sub-band energy in this case and is given by

$$0 = \left[\left[1/L_0^2\right]\left\{\cos^{-1}\left(f_{23}\left(E_{15,8},n\right)\right)\right\}^2 - \left(\frac{2eB}{\hbar}\left(n+\frac{1}{2}\right)\right)\right] \tag{16.15c}$$

The EEM can be written as

$$m^*\left(E,\eta_g,n\right) = \frac{\hbar^2}{2L_0^2}\cos^{-1}\left[\frac{1}{2}f_{23}(E,n)\left(1-\frac{1}{4}f_{23}^2(E,n)\right)\right]^{-1/2} f_{23}'(E,n) \tag{16.16}$$

16.2.9 The DR in HgTe/CdTe HD Effective Mass Superlattices Under Magnetic Quantization

The DR in heavily doped HgTe/CdTe EMSLs under magnetic quantization can be written as

$$(k_z)^2 = \Delta_{13HE,n} + i\Delta_{14HE,n} \tag{16.17a}$$

where

$$\Delta_{13HE,n} = \left[\left(1/L_0^2\right)\left(\Delta_{11HE,n}^2 - \Delta_{12HE,n}^2\right) - \frac{2eB}{\hbar}\left(n+\frac{1}{2}\right)\right]$$

$$\Delta_{11HE,n} = \cos^{-1} p_{6HE,n},$$

$$p_{6HE,n} = \left[\left(\left(1-\Delta_{9HE,n}^2 - \Delta_{10HE,n}^2 - \sqrt{\left(1-\Delta_{9HE,n}^2-\Delta_{10HE,n}^2\right)^2 + 4\Delta_{10HE,n}^2}\right)/2\right)\right]^{\frac{1}{2}}$$

$$\begin{aligned}
\Delta_{9HE,n} &= \left(\overline{a_{1H}}\cos\Delta_{5HE,n}\cosh\Delta_{6HE,n} - \overline{a_{2H}}\cos\Delta_{7HE,n}\cosh\Delta_{6HE,n}\right),\\
\Delta_{10HE,n} &= \left(\overline{a_{1H}}\sin\Delta_{5HE,n}\sinh\Delta_{6HE,n} + \overline{a_{2H}}\sin\Delta_{7HE,n}\sinh\Delta_{6HE,n}\right),\\
\Delta_{5HE,n} &= \left[a_0C_{22HE,n}\left(E_{E,n},\eta_{g1}\right) + b_0e_3\right],\\
\Delta_{6HE,n} &= b_0e_4,\\
\Delta_{7HE,n} &= \left[a_0C_{22HE,n}\left(E_{E,n},\eta_{g1}\right) - b_0e_3\right],
\end{aligned}$$

$$C_{22HE,n}(E_{E,n},\eta_{g1})=\left[\frac{B_{01}^2+2A_1E_{E,n}-B_{01}\left(B_{01}^2+4A_1E_{E,n}\right)}{2A_1^2}-\left[\frac{2eB}{\hbar}\left(n+\frac{1}{2}\right)\right]\right]^{1/2},$$

$$\Delta_{12HE,n}=\cos^{-1}p_{6HE,n},\ \Delta_{14HE,n}=\frac{2\Delta_{11HE,n}\Delta_{12HE,n}}{L_0^2}$$

The DOS function can be written as

$$N(E)=\frac{eB}{2\pi^2\hbar}\sum_{n=0}^{n_{\max}}\frac{\left(\Delta'_{13HE,n}+i\Delta'_{14HE,n}\right)H(E-E_{15,9})}{\sqrt{\Delta_{13HE,n}+i\Delta_{14HE,n}}}\tag{16.17b}$$

where $E_{15,9}$ is the sub-band energy in this case and is given by

$$0=\Delta_{13HE_{15,9},n}+i\Delta_{14HE_{15,9},n}\tag{16.17c}$$

The EEM can be written as

$$m^*(E,\eta_g,n)=\frac{\hbar^2}{2}\Delta'_{13HE,n}\tag{16.18}$$

16.2.10 The DR in Strained Layer HD Effective Mass Superlattices Under Magnetic Quantization

The DR in heavily doped strained layer effective mass super-lattices under magnetic quantization can be expressed as

$$(k_z)^2=\left[\frac{1}{L_0^2}\left\{\cos^{-1}(\mathrm{f}_{40}(E,n))\right\}^2-\left(\frac{2eB}{\hbar}\left(n+\frac{1}{2}\right)\right)\right]\tag{16.19a}$$

where

$$f_{40}(E,n)=a_{20}\cos\left[a_0C_{40}(E,n,\eta_{g1})+b_0D_{40}(E,n,\eta_{g1})\right]\\-a_{21}\cos\left[a_0C_{40}(E,n,\eta_{g2})-b_0D_{40}(E,n,\eta_{g2})\right],$$

$$C_{40}(E,n,\eta_{g1})=\left[1-\frac{\hbar eB}{\phi_{50}(E,\eta_{g1})}\left(n+\frac{1}{2}\right)\right]^{1/2}\left[\bar{S}_1(E,\eta_{g1})\right]^{-1/2},$$

$$\phi_{50}(E,\eta_{g1})=\sqrt{\psi_{50}(E,\eta_{g1})\psi_{51}(E,\eta_{g1})},$$

$$\psi_{50}(E,\eta_{g1}) = \frac{\hbar^2}{2\bar{P}_1(E,\eta_{g1})},\quad \psi_{51}(E,\eta_{g1}) = \frac{\hbar^2}{2\bar{Q}_1(E,\eta_{g1})}$$

$$D_{40}(E,n,\eta_{g2}) = \left[1 - \frac{\hbar eB}{\phi_{501}(E,\eta_{g2})}\left(n+\frac{1}{2}\right)\right]^{1/2}\left[\bar{S}_2(E,\eta_{g2})\right]^{-1/2}$$

$$\phi_{501}(E,\eta_{g2}) = \sqrt{\psi_{501}(E,\eta_{g2})\psi_{511}(E,\eta_{g2})}$$

$$\psi_{501}(E,\eta_{g2}) = \frac{\hbar^2}{2\bar{P}_2(E,\eta_{g2})},\ \psi_{511}(E,\eta_{g2}) = \frac{\hbar^2}{2\bar{Q}_2(E,\eta_{g2})}$$

The DOS function can be written as

$$N(E) = \frac{eBg_v}{2\pi^2\hbar L_0}\sum_{n=0}^{n_{\max}} \frac{\cos^{-1}\left[\frac{1}{2}f_{40}(E,n)\left(1-\frac{1}{4}f_{40}^2(E,n)\right)\right]^{-1/2} f_{40}'(E,n)H(E-E_{15,10})}{\left[\left[\cos^{-1}\left[\frac{1}{2}f_{40}(E,n)\right]\right]^2 - L_0^2\frac{2eB}{\hbar}\left(1+\frac{1}{2}\right)\right]^{1/2}} \tag{16.19b}$$

where $E_{15,10}$ is the sub-band energy in this case and is given by

$$0 = \left[\left[1/L_0^2\right]\left\{\cos^{-1}\left(f_{40}(E_{15,10},n)\right)\right\}^2 - \left(\frac{2eB}{\hbar}\left(n+\frac{1}{2}\right)\right)\right] \tag{16.19c}$$

The EEM can be written as

$$m^*(E,\eta_g,n) = \frac{\hbar^2}{2L_0^2}\cos^{-1}\left[\frac{1}{2}f_{40}(E,n)\left(1-\frac{1}{4}f_{40}^2(E,n)\right)\right]^{-1/2} f_{40}'(E,n) \tag{16.20}$$

16.3 Summary and Conclusion

In this chapter, the magneto DR in III–V, II–VI, IV–VI, HgTe/CdTe and strained layer heavily doped superlattices (HDSLs) with graded interfaces has been studied in Sects. 16.2.1 to 16.2.5. From Sects. 16.2.6 to 16.2.10, the magneto DR in III–V, II–VI, IV–VI, HgTe/CdTe and strained layer heavily doped effective mass superlattices respectively has been presented

- It appears that the magneto DRs is in general quantized surfaces and the consecutive different between two curves are not constant but varies with other physical constants scattering potentials.
- The magneto DR is concentration dependent, a fact only possible for HD materials forming band-tails
- The Landau sub-bands are concentration dependent

16.4 Open Research Problems

(R.16.1) Investigate the magneto DR in the presence of an arbitrarily oriented non-quantizing magnetic field in III–V, II–VI, IV–VI, HgTe/CdTe and strained layer heavily doped superlattices (HDSLs) with graded interfaces by including the electron spin

(R.16.2) Investigate the magneto DR in III–V, II–VI, IV–VI, HgTe/CdTe and strained layer heavily doped effective mass superlattices in the presence of an arbitrarily oriented non-quantizing magnetic field by including the electron spin

(R.16.3) Investigate the DR for all the problems from (R.16.1) to (R.16.2) in the presence of an additional arbitrarily oriented electric field

(R.16.4) Investigate the DR for all the problems from (R.16.1) to (R.16.3) in the presence of arbitrarily oriented crossed electric and magnetic fields

References

1. M. Mondal, K.P. Ghatak, Czech. J. Phys. B **36**, 1389 (1986)
2. S. Bhattacharyya, K.P. Ghatak, S. Biswas, in *OE/Fibers'87*, vol. 73 (International Society for Optics and Photonics, 1987)
3. B. Mitra, K.P. Ghatak, Phys. Status Solidi B **149**, K117 (1988)
4. K.P. Ghatak, S.N. Biswas, in *M.R.S. Proceedings* **161**, 371(1989)(USA)
5. B. Mitra, K.P. Ghatak, Phys. Lett. A **142**, 401 (1989)
6. K.P. Ghatak, SN Biswas in *Properties of II-VI Semiconductors: Bulk Crystals, Epitaxial Films, Quantum Well Structures, and Dilute Magnetic Systems* (USA, 1990)
7. K.P. Ghatak, in *Properties of II-VI Semiconductors: Bulk Crystals, Epitaxial Films, Quantum Well Structures, and Dilute Magnetic Systems*, vol. 161, p. 371 (USA, 1990)
8. K.P. Ghatak, B. Mitra, S.N. Biswas, in *Optics, Electro-Optics, and Laser Applications in Science and Engineering* 137 (International Society for Optics and Photonics, 1991)
9. K.P. Ghatak, Il Nuovo Cimento D **13**, 1321 (1991)
10. K.P. Ghatak, A. Ghoshal, Phys. Status Solidi B **170**, K27 (1992)

Chapter 17
The DR in QWHDSLs Under Magnetic Quantization

Life is too short to waste time hating any one.

17.1 Introduction

In this chapter, the magneto DR in III–V, II–VI, IV–VI, HgTe/CdTe and strained layer quantum well heavily doped superlattices (QWHDSLs) with graded interfaces has been studied in Sects. 17.2.1 to 17.2.5. From Sects. 17.2.6 to 17.2.10, the magneto DR in III–V, II–VI, IV–VI, HgTe/CdTe and strained layer quantum well heavily doped effective masssuperlattices respectively has been presented. The Sect. 17.3 contains the summary and conclusion pertinent to this chapter. The last Sect. 17.4 presents 4 open research problems.

17.2 Theoretical Background

17.2.1 The DR in III–V Quantum Well HD Superlattices with Graded Interfaces Under Magnetic Quantization

The DR in heavily doped quantum well III–V superlattices under magnetic quantization assumes the form

$$\left(\frac{n_z\pi}{d_z}\right)^2 = G_{8E_{41,n}} + iH_{8E_{41,n}} \quad (17.1)$$

where $E_{41,n}$ is the totally quantized energy in this case.

K.P. Ghatak, *Dispersion Relations in Heavily-Doped Nanostructures*,
Springer Tracts in Modern Physics 265, DOI 10.1007/978-3-319-21000-1_17

The DOS function in this case can be expressed as

$$N_{MQWSL}(E) = \frac{g_v eB}{\pi\hbar} \sum_{n_z=1}^{n_{z\max}} \sum_{n=0}^{n_{\max}} \delta'(E - E_{41,n}) \tag{17.2}$$

17.2.2 The DR in II–VI Quantum Well HD Superlattices with Graded Interfaces Under Magnetic Quantization

The DR in quantum well heavily doped II–VI superlattices under magnetic quantization assumes the form

$$\left(\frac{n_z \pi}{d_z}\right)^2 = G_{19E_{42,n}} + iH_{19E_{42,n}} \tag{17.3}$$

where $E_{42,n}$ is the totally quantized energy in this case.

The DOS function in this case can be expressed as

$$N_{MQWSL}(E) = \frac{g_v eB}{2\pi\hbar} \sum_{n_z=1}^{n_{z\max}} \sum_{n=0}^{n_{\max}} \delta'(E - E_{42,n}) \tag{17.4}$$

17.2.3 The DR in IV–VI Quantum Well HD Superlattices with Graded Interfaces Under Magnetic Quantization

The DR in quantum well heavily doped quantum well IV–VI superlattices under magnetic quantization assumes the form

$$\left(\frac{\pi n_z}{d_z}\right)^2 = \frac{1}{L_0^2}\left[\cos^{-1}\left\{\frac{1}{2}\varphi_2(E_{43,n}, n)\right\}\right]^2 - \frac{2|e|B}{\hbar}\left(n + \frac{1}{2}\right) \tag{17.5}$$

where $E_{43,n}$ is the totally quantized energy in this case.

The DOS function in this case can be expressed as

$$N_{MQWSL}(E) = \frac{g_v eB}{2\pi\hbar} \sum_{n_z=1}^{n_{z\max}} \sum_{n=0}^{n_{\max}} \delta'(E - E_{43,n}) \tag{17.6}$$

17.2.4 The DR in HgTe/CdTe Quantum Well HD Superlattices with Graded Interfaces Under Magnetic Quantization

The DR in quantum well heavily doped HgTe/CdTe superlattices under magnetic quantization assumes the form

$$\left(\frac{n_z\pi}{d_z}\right)^2 = G_{192E_{44,n}} + iH_{192E_{44,n}} \tag{17.7}$$

where $E_{44,n}$ is the totally quantized energy in this case.

The DOS function in this case can be expressed as

$$N_{MQWSL}(E) = \frac{g_v eB}{2\pi\hbar} \sum_{n_z=1}^{n_{z\max}} \sum_{n=0}^{n_{\max}} \delta'(E - E_{44,n}) \tag{17.8}$$

17.2.5 The DR in Strained Layer Quantum Well HD Superlattices with Graded Interfaces Under Magnetic Quantization

The DR of the conduction electrons in heavily doped quantum well strained layer SLs with graded interfaces can be expressed as

$$\left(\frac{\pi n_z}{d_z}\right)^2 = \frac{1}{L_0^2}\left[\cos^{-1}\left\{\frac{1}{2}\bar{\varphi}_6\left(E_{47,n}, n\right)\right\}\right]^2 - \frac{2|e|B}{\hbar}\left(n + \frac{1}{2}\right) \tag{17.9}$$

where $E_{47,n}$ is the totally quantized energy in this case.

The DOS function in this case can be expressed as

$$N_{MQWSL}(E) = \frac{g_v eB}{2\pi\hbar} \sum_{n_z=1}^{n_{z\max}} \sum_{n=0}^{n_{\max}} \delta'(E - E_{47,n}) \tag{17.10}$$

17.2.6 The DR in III–V Quantum Well HD Effective Mass Super Lattices Under Magnetic Quantization

The dispersion relation in quantum well heavily doped III–V superlattices under magnetic quantization assumes the form

$$\left(\frac{n_x \pi}{d_x}\right)^2 = \delta_{7A1,n} + i\delta_{8A1,n} \tag{17.11}$$

where $A1$ is the totally quantized energy in this case.

The DOS function in this case can be expressed as

$$N_{MQWSL}(E) = \frac{g_v eB}{2\pi\hbar} \sum_{n_z=1}^{n_{z\max}} \sum_{n=0}^{n_{\max}} \delta'(E - A1) \tag{17.12}$$

17.2.7 The DR in II–VI Quantum Well HD Effective Mass Super Lattices Under Magnetic Quantization

The dispersion relation in quantum well heavily doped III–V superlattices under magnetic quantization assumes the form

$$\left(\frac{n_z \pi}{d_z}\right)^2 = \Delta_{13,A2,n} + i\Delta_{14,A2,n} \tag{17.13}$$

where $A2$ is the totally quantized energy in this case.

The DOS function in this case can be expressed as

$$N_{MQWSL}(E) = \frac{g_v eB}{2\pi\hbar} \sum_{n_z=1}^{n_{z\max}} \sum_{n=0}^{n_{\max}} \delta'(E - A2) \tag{17.14}$$

17.2.8 The DR in IV–VI Quantum Well HD Effective Mass Super Lattices Under Magnetic Quantization

The DR of the conduction electrons in heavily doped quantum well strained layer SLs with graded interfaces can be expressed as

$$\left(\frac{\pi n_z}{d_z}\right)^2 = \frac{1}{L_0^2}\left[\cos^{-1}\left\{\frac{1}{2} f_{23}(A3, n)\right\}\right]^2 - \frac{2|e|B}{\hbar}\left(n + \frac{1}{2}\right) \tag{17.15}$$

where $A3$ is the totally quantized energy in this case.

The DOS function in this case can be expressed as

$$N_{MQWSL}(E) = \frac{g_v eB}{2\pi\hbar} \sum_{n_z=1}^{n_{z\max}} \sum_{n=0}^{n_{\max}} \delta'(E - A3) \tag{17.16}$$

17.2.9 The DR in HgTe/CdTe Quantum Well HD Effective Mass Super Lattices Under Magnetic Quantization

The DR in quantum well heavily doped III–V superlattices under magnetic quantization assumes the form

$$\left(\frac{n_z\pi}{d_z}\right)^2 = \Delta_{13,A4,n} + i\Delta_{14,A4,n} \tag{17.17}$$

where $A4$ is the totally quantized energy in this case.

The DOS function in this case can be expressed as

$$N_{MQWSL}(E) = \frac{g_v eB}{2\pi\hbar} \sum_{n_z=1}^{n_{z\max}} \sum_{n=0}^{n_{\max}} \delta'(E - A4) \tag{17.18}$$

17.2.10 The DR in Strained Layer Quantum Well HD Effective Mass Super Lattices Under Magnetic Quantization

The DR of the conduction electrons in heavily doped quantum well strained layer SLs with graded interfaces can be expressed as

$$\left(\frac{\pi n_z}{d_z}\right)^2 = \frac{1}{L_0^2}\left[\cos^{-1}\left\{\frac{1}{2}f_{40}(A8, n)\right\}\right]^2 - \frac{2|e|B}{\hbar}\left(n + \frac{1}{2}\right) \tag{17.19}$$

where $A8$ is the totally quantized energy in this case.

The DOS function in this case can be expressed as

$$N_{MQWSL}(E) = \frac{g_v eB}{2\pi\hbar} \sum_{n_z=1}^{n_{z\max}} \sum_{n=0}^{n_{\max}} \delta'(E - A8) \tag{17.20}$$

17.3 Summary and Conclusion

In this chapter, the magneto DR in III–V, II–VI, IV–VI, HgTe/CdTe and strained layer quantum well heavily doped superlattices (QWHDSLs) with graded interfaces has been studied in Sects. 17.2.1 to 17.2.5. From Sects. 17.2.6 to 17.2.10, the magneto DR in III–V, II–VI, IV–VI, HgTe/CdTe and strained layer quantum well heavily doped effective mass superlattices respectively has been presented.

The magneto DRs in QWHDSLs exhibit the fact that the total energy is quantized since the corresponding wave vector space is totally quantized by quantizing magnetic field and size quantization along z-direction.

- The DOS functions for all the cases are series of non-uniformly distributed Dirac's Delta functions at specified quantized points in the respective energy axis. The spacing between the consecutive Delta functions are functions of energy band constants and quantization of the wave vector space of a particular material. The DOS function needs two summations namely one summation over the Landau quantum number and the other one is due to size quantization.
- It may be noted that the energy levels in this case lead to the discrete energy levels, somewhat like atomic energy levels, which produce very large changes. This follows from the inherent nature of the quantum confinement of the carrier gas dealt with here. In QWHDSLs, under magnetic quantization there remain no free carrier states in between any two allowed sets of size-quantized levels unlike that found for QWs, NWs and quantum dot superlattices where the quantum confinements are 1D, 2D and 0D respectively. Consequently, the crossing of the Fermi level by the quantized levels in this case would have much greater impact on the redistribution of the carriers among the allowed levels, as compared to that found for QWs, NWs and quantum dot superlattices respectively.
- It is the band structure which changes in a fundamental way and consequently all the physical properties of all the electronic materials changes radically leading to new physical concepts.

17.4 Open Research Problems

(R.17.1) Investigate the magneto DR in the presence of an arbitrarily oriented non-quantizing magnetic field in III–V, II–VI, IV–VI, HgTe/CdTe and strained layer heavily doped quantum well superlattices with graded interfaces by including the electron spin.

(R.17.2) Investigate the magneto DR in III–V, II–VI, IV–VI, HgTe/CdTe and strained layer heavily doped effective mass quantum well superlattices in the presence of an arbitrarily oriented non-quantizing magnetic field by including the electron spin.

(R.17.3) Investigate the DR for all the problems from (R.17.1) to (R.17.2) in the presence of an additional arbitrarily oriented electric field.

(R.17.4) Investigate the DR for all the problems from (R.17.1) to (R.17.3) in the presence of arbitrarily oriented crossed electric and magnetic fields.

Part IV
Dispersion Relations in HD Kane Type Semiconductors in the Presence of Light Waves

If my only desire is to be desire-less, then my consciousness is reversed.

Chapter 18
The DR Under Photo Excitation in HD Kane Type Semiconductors

If instead of locating the faults of others, I can locate a single fault of mine and wipe it out permanently from my mind, then surely I am in the state of wiseness.

18.1 Introduction

With the advent of nano-photonics, there has been a considerable interest in studying the optical processes in semiconductors and their nanostructures [1]. It appears from the literature, that the investigations have been carried out on the assumption that the carrier energy spectra are invariant quantities in the presence of intense light waves, which is not fundamentally true. The physical properties of semiconductors in the presence of light waves which change the basic dispersion relation have relatively less investigated in the literature [2–5]. In this appendix we shall study the DR in HD III–V, ternary and quaternary semiconductors on the basis of newly formulated electron dispersion law under external photo excitation.

In Sect. 18.2.1 of the theoretical background 18.2, we have formulated the DR of the conduction electrons of HD III–V, ternary and quaternary materials in the presence of light waves whose unperturbed electron energy spectrum is described by the three-band model of Kane in the absence of band tailing. In the same section, we have studied the DR for the said HD materials in the presence of external photo-excitation when the unperturbed energy spectra are defined by the two band model of Kane and that of parabolic energy bands in the absence of band tails respectively for the purpose of relative comparison. In Sect. 18.2.2, we have studied the opto DR in the said HD materials under magnetic quantization. In Sect. 18.2.3, we have studied the opto DR in the presence of crossed electric and quantizing magnetic fields. In Sect. 18.2.4, we have studied the DR in QWs in HD Kane type semiconductors. In Sect. 18.2.5, we have investigated the DR in doping superlattices of HD Kane type semiconductors in the presence of light waves. In Sect. 18.2.6, we have studied the DR in QBs of HD Kane type semiconductors in the presence of light waves. In Sect. 18.2.7, we have studied the magneto DR in

K.P. Ghatak, *Dispersion Relations in Heavily-Doped Nanostructures*,
Springer Tracts in Modern Physics 265, DOI 10.1007/978-3-319-21000-1_18

QWs of HD Kane type semiconductors in the presence of light Waves. In Sect. 18.2.8, we have studied the DR in accumulation and inversion layers of Kane type semiconductors in the presence of light waves. In Sect. 18.2.9, we have studied the DR in NWs of HD of Kane type semiconductors in the presence of light waves. In Sect. 18.2.10, we have studied the magneto DR in accumulation and inversion layers of Kane type semiconductors in the presence of light waves. In Sect. 18.2.11, we have investigated the magneto DR in doping superlattices of Kane type semiconductors in the presence of light waves. In Sect. 18.2.12, we have investigated the DR in QWHD effective mass superlattices of Kane type semiconductors. In Sect. 18.2.13, we have studied the DR in NWHD effective mass superlattices of Kane type semiconductors. In Sect. 18.2.14, we have studied the magneto DR in HD effective mass superlattices of Kane type semiconductors in the presence of light waves. In Sect. 18.2.15, we have studied the magneto DR in QWHD effective mass superlattices of Kane type semiconductors in the presence of light waves. In Sect. 18.2.16, we have studied the DR in QWHD superlattices of Kane type semiconductors with graded interfaces in the presence of light waves. In Sect. 18.2.17, we have studied the DR in NWHD superlattices of Kane type semiconductors with graded interfaces in the presence of light waves. In Sect. 18.2.18, we have studied the DR in Quantum dot HD superlattices of Kane type semiconductors with graded interfaces in the presence of light waves. In Sect. 18.2.19, we have studied the magneto DR in HD superlattices of Kane type semiconductors with graded interfaces in the presence of light waves. In Sect. 18.2.20, we have studied the magneto DR in QWHD superlattices of Kane type semiconductors with graded interfaces in the presence of light waves. The Sect. 18.3 contains the summary and conclusion pertinent to this chapter. The last Sect. 18.4 presents 6 open research problems.

18.2 Theoretical Background

18.2.1 The Formulation of the Electron Dispersion Law in the Presence of Light Waves in HD III–V, Ternary and Quaternary Semiconductors

The Hamiltonian $\left(\widehat{H}\right)$ of an electron in the presence of light wave characterized by the vector potential $\vec{A}$ can be written following [3] as

$$\widehat{H} = \left[\left|\left(\widehat{p} + |e|\vec{A}\right)\right|^2 \Big/ 2m\right] + V(\vec{r}) \tag{18.1}$$

in which, $\widehat{p}$ is the momentum operator, $V(\vec{r})$ is the crystal potential and m is the free electron mass. (18.1) can be expressed as

$$\widehat{H} = \widehat{H}_0 + \widehat{H}' \tag{18.2}$$

where,

$$\widehat{H}_0 = \frac{\widehat{p}^2}{2m} + V(\vec{r})$$

and

$$\widehat{H}' = \frac{|e|}{2m}\vec{A} \cdot \widehat{p} \tag{18.3}$$

The perturbed Hamiltonian $\widehat{H}'$ can be written as

$$\widehat{H}' = \left(\frac{-i\hbar|e|}{2m}\right)\left(\vec{A} \cdot \nabla\right) \tag{18.4}$$

where, $i = \sqrt{-1}$ and $\widehat{p} = -i\hbar\nabla$

The vector potential $(\vec{A})$ of the monochromatic light of plane wave can be expressed as

$$\vec{A} = A_0\vec{\varepsilon}_s \cos(\vec{s}_0 \cdot \vec{r} - \omega t) \tag{18.5}$$

where A_0 is the amplitude of the light wave, $\vec{\varepsilon}_s$ is the polarization vector, $\vec{s}_0$ is the momentum vector of the incident photon, $\vec{r}$ is the position vector, ω is the angular frequency of light wave and t is the time scale. The matrix element of $\widehat{H}'_{nl}$ between initial state, $\psi_1(\vec{q}, \vec{r})$ and final state $\psi_n\left(\vec{k}, \vec{r}\right)$ in different bands can be written as

$$\widehat{H}'_{nl} = \frac{|e|}{2m}\left\langle n\vec{k}|\vec{A} \cdot \vec{p}|l\vec{q}\right\rangle \tag{18.6}$$

Using (18.4) and (18.5), we can re-write (18.6) as

$$\widehat{H}'_{nl} = \left(\frac{-i\hbar|e|A_0}{4m}\right)\vec{\varepsilon}_s \cdot \left[\left\{\left\langle n\vec{k}|e^{(i\vec{s}_0\cdot\vec{r})}\nabla|l\vec{q}\right\rangle e^{-i\omega t}\right\} + \left\{\left\langle n\vec{k}|e^{(-i\vec{s}_0\cdot\vec{r})}\nabla|l\vec{q}\right\rangle e^{i\omega t}\right\}\right] \tag{18.7}$$

The first matrix element of (18.7) can be written as

$$\left\langle n\vec{k}|e^{(i\vec{s}_0\cdot\vec{r})}\nabla|l\vec{q}\right\rangle = \int e^{\left(i\left[\vec{q}+\vec{s}_0-\vec{k}\right]\cdot\vec{r}\right)} i\vec{q}u_n^*\left(\vec{k}, \vec{r}\right)u_1(\vec{q}, \vec{r})d^3r + \int e^{\left(i\left[\vec{q}+\vec{s}_0-\vec{k}\right]\cdot\vec{r}\right)} u_n^*\left(\vec{k}, \vec{r}\right)\nabla u_1(\vec{q}, \vec{r})d^3r \tag{18.8}$$

The functions $u_n^* u_1$ and $u_n^* \nabla u_1$ are periodic. The integral over all space can be separated into a sum over unit cells times an integral over a single unit cell. It is assumed that the wave length of the electromagnetic wave is sufficiently large so that if $\vec{k}$ and $\vec{q}$ are within the Brillouin zone, $\left(\vec{q}+\vec{s}_0-\vec{k}\right)$ is not a reciprocal lattice vector.

Therefore, we can write (18.8) as

$$\left\langle n\vec{k}\middle|e^{(i\vec{s}_0\cdot\vec{r})}\nabla\middle|l\vec{q}\right\rangle = \left[\frac{(2\pi)^3}{\Omega}\right]\left\{i\vec{q}\delta\left(\vec{q}+\vec{s}_0-\vec{k}\right)\delta_{nl}+\delta\left(\vec{q}+\vec{s}_0-\vec{k}\right)\int\limits_{\text{cell}} u_n^*\left(\vec{k},\vec{r}\right)\nabla u_1(\vec{q},\vec{r})d^3r\right\}$$
$$= \left[\frac{(2\pi)^3}{\Omega}\right]\left\{\delta\left(\vec{q}+\vec{s}_0-\vec{k}\right)\int\limits_{\text{cell}} u_n^*\left(\vec{k},\vec{r}\right)\nabla u_1(\vec{q},\vec{r})d^3r\right\} \tag{18.9}$$

where, Ω is the volume of the unit cell and $\int u_n^*\left(\vec{k},\vec{r}\right)u_1(\vec{q},\vec{r})d^3r = \delta\left(\vec{q}-\vec{k}\right)\delta_{nl}=0$, since $n \neq l$.

The delta function expresses the conservation of wave vector in the absorption of light wave and $\vec{s}_0$ is small compared to the dimension of a typical Brillouin zone and we set $\vec{q}=\vec{k}$.

From (18.8) and (18.9), we can write,

$$\widehat{H}'_{nl} = \frac{|e|A_0}{2m}\vec{\varepsilon}_s\cdot\widehat{p}_{nl}(\vec{k})\delta(\vec{q}-\vec{k})\cos(\omega t) \tag{18.10}$$

where,

$$\widehat{p}_{nl}(\vec{k}) = -i\hbar\int u_n^*\nabla u_1 d^3r = \int u_n^*(\vec{k},\vec{r})\widehat{p}u_1(\vec{k},\vec{r})d^3r$$

Therefore, we can write

$$\widehat{H}'_{nl} = \frac{|e|A_0}{2m}\vec{\varepsilon}\cdot\widehat{p}_{nl}(\vec{k}) \tag{18.11}$$

where,

$$\vec{\varepsilon} = \vec{\varepsilon}_s\cos\omega t.$$

When a photon interacts with a semiconductor, the carriers (i.e., electrons) are generated in the bands which are followed by the inter-band transitions.

For example, when the carriers are generated in the valence band, the carriers then make inter-band transition to the conduction band. The transition of the electrons within the same band i.e., $\widehat{H}'_{nn} = \left\langle n\vec{k}\middle|\widehat{H}'\middle|n\vec{k}\right\rangle$ is neglected. Because, in such a case, i.e., when the carriers are generated within the same bands by photons, are lost by recombination within the aforementioned band resulting zero carriers.

Therefore,

$$\left\langle n\vec{k}\middle|\widehat{H}'\middle|n\vec{k}\right\rangle = 0 \tag{18.12}$$

With $n = c$ stands for conduction band and $l = v$ stand for valance band, the energy equation for the conduction electron can approximately be written as

$$I_{11}(E) = \left(\frac{\hbar^2 k^2}{2m_c}\right) + \frac{\left(\frac{|e|A_0}{2m}\right)^2 \left\langle \left|\vec{\varepsilon}\cdot\widehat{p}_{cv}(\vec{k})\right|^2\right\rangle_{av}}{E_c(\vec{k}) - E_v(\vec{k})} \tag{18.13}$$

where, $I_{11}(E) = E(aE+1)(bE+1)/(cE+1)$, $a \equiv 1/E_{g_0}$, E_{g_0} is the un-perturbed band-gap, $b \equiv 1/(E_{g_0}+\Delta)$, $c \equiv 1/(E_{g_0}+2\Delta/3)$, and $\left\langle \left|\vec{\varepsilon}\cdot\widehat{p}_{cv}(\vec{k})\right|^2\right\rangle_{av}$ represents the average of the square of the optical matrixelement (OME).

For the three-band model of Kane, we can write,

$$\xi_{1k} = E_c(\vec{k}) - E_v(\vec{k}) = (E_{g_0}^2 + E_{g_0}\hbar^2 k^2/m_r)^{1/2} \tag{18.14}$$

where, m_r is the reduced mass and is given by $m_r^{-1} = (m_c)^{-1} + m_v^{-1}$, and m_v is the effective mass of the heavy hole at the top of the valance band in the absence of any field.

The doubly degenerate wave functions $u_1(\vec{k},\vec{r})$ and $u_2(\vec{k},\vec{r})$ can be expressed as

$$u_1(\vec{k},\vec{r}) = a_{k+}[(is)\downarrow'] + b_{k+}\left[\frac{X'-iY'}{\sqrt{2}}\uparrow'\right] + c_{k+}[Z'\downarrow'] \tag{18.15}$$

and

$$u_2(\vec{k},\vec{r}) = a_{k-}[(is)\uparrow'] - b_{k-}\left[\frac{X'-iY'}{\sqrt{2}}\downarrow'\right] + c_{k-}[Z'\uparrow'] \tag{18.16}$$

s is the s-type atomic orbital in both unprimed and primed coordinates, $\downarrow'$ indicates the spin down function in the primed coordinates,

$$a_{k\pm} \equiv \beta\left[\left[E_{g_0} - (\gamma_{0k\pm})^2 (E_{g_0} - \delta')\right]^{1/2} (E_{g_0} + \delta')^{-1/2}\right],$$
$$\beta \equiv \left[(6(E_{g_0} + 2\Delta/3)(E_{g_0} + 2\Delta))/\chi\right]^{1/2},$$
$$\chi \equiv \left(6E_{g_0}^2 + 9E_{g_0}\Delta + 4\Delta^2\right),\ \gamma_{0k\pm} \equiv \left[\frac{(\xi_{1k} \mp E_{g_0})}{2(\xi_{1k} + \delta')}\right]^{1/2},$$
$$\xi_{1k} \equiv E_c(\vec{k}) - E_v(\vec{k}) = E_{g_0}\left[1 + 2\left(1 + \frac{m_c}{m_v}\right)\frac{I_{11}(E)}{E_{g_0}}\right]^{1/2},$$

$\delta' \equiv \left(E_{g_0}^2\Delta\right)(\chi)^{-1}$, X', Y', and Z' are the p-type atomic orbitals in the primed coordinates, $\uparrow'$ indicates the spin-up function in the primed coordinates, $b_{k\pm} \equiv \rho\gamma_{0k\pm}$, $\rho \equiv (4\Delta^2/3\chi)^{1/2}$, $c_{k\pm} \equiv t\gamma_{0k+}$ and $t \equiv \left[6(E_{g_0} + 2\Delta/3)^2/\chi\right]^{1/2}$.

We can, therefore, write the expression for the optical matrix element (OME) as

$$\text{OME} = \hat{p}_{cv}\left(\vec{k}\right) = \left\langle u_1(\vec{k},\vec{r})|\hat{p}|u_2(\vec{k},\vec{r})\right\rangle \tag{18.17}$$

Since the photon vector has no interaction in the same band for the study of inter-band optical transition, we can therefore write

$$\langle S|\hat{p}|S\rangle = \langle X|\hat{p}|X\rangle = \langle Y|\hat{p}|Y\rangle = \langle Z|\hat{p}|Z\rangle = 0$$
$$\text{and } \langle X|\hat{p}|Y\rangle = \langle Y|\hat{p}|Z\rangle = \langle Z|\hat{p}|X\rangle = 0$$

There are finite interactions between the conduction band (CB) and the valance band (VB) and we can obtain

$$\left\langle S\middle|\hat{P}\middle|X\right\rangle = \hat{i}\cdot\hat{P} = \hat{i}\cdot\hat{P}_x$$
$$\left\langle S\middle|\hat{P}\middle|Y\right\rangle = \hat{j}\cdot\hat{P} = \hat{j}\cdot\hat{P}_y$$
$$\left\langle S\middle|\hat{P}\middle|Z\right\rangle = \hat{k}\cdot\hat{P} = \hat{k}\cdot\hat{P}_z$$

where, $\hat{i}$, $\hat{j}$ and $\hat{k}$ are the unit vectors along x, y and z axes respectively.

It is well known that

$$\begin{bmatrix}\uparrow'\\ \downarrow'\end{bmatrix} = \begin{bmatrix} e^{-i\phi/2}\cos(\theta/2) & e^{i\phi/2}\sin(\theta/2)\\ -e^{-i\phi/2}\sin(\theta/2) & e^{i\phi/2}\cos(\theta/2)\end{bmatrix}\begin{bmatrix}\uparrow\\ \downarrow\end{bmatrix}$$

$$\text{and}\ \begin{bmatrix}X'\\ Y'\\ Z'\end{bmatrix} = \begin{bmatrix}\cos\theta\cos\phi & \cos\theta\sin\phi & -\sin\theta\\ -\sin\phi & \cos\phi & 0\\ \sin\theta\cos\phi & \sin\theta\sin\phi & \cos\theta\end{bmatrix}\begin{bmatrix}X\\ Y\\ Z\end{bmatrix}$$

Besides, the spin vector can be written as

$$\vec{S} = \frac{\hbar}{2}\vec{\sigma}, \text{ where } \sigma_x = \begin{bmatrix} 0 & 1 \\ 1 & 0 \end{bmatrix}, \ \sigma_y = \begin{bmatrix} 0 & -i \\ i & 0 \end{bmatrix} \text{ and } \sigma_z = \begin{bmatrix} 1 & 0 \\ 0 & -1 \end{bmatrix}$$

From above, we can write

$$\begin{aligned}\widehat{p}_{cv}\left(\vec{k}\right) &= \left\langle u_1\left(\vec{k},\vec{r}\right)\middle|\widehat{P}\middle|u_2\left(\vec{k},\vec{r}\right)\right\rangle \\ &= \left\langle\left\{a_{k_+}[(iS)\downarrow'] + b_{k_+}\left[\left(\frac{X'-iY'}{\sqrt{2}}\right)\uparrow'\right] + c_{k_+}[Z'\downarrow']\right\}\middle|\widehat{P}\middle|\{a_{k_-}[(iS)\uparrow'] \right. \\ &\quad \left. - b_{k_-}\left[\left(\frac{X'+iY'}{\sqrt{2}}\right)\downarrow' + c_{k_-}[Z'\uparrow']\right]\right\}\right\rangle.\end{aligned}$$

Using above relations, we get

$$\begin{aligned}\widehat{p}_{cv}\left(\vec{k}\right) &= \left\langle u_1\left(\vec{k},\vec{r}\right)\middle|\widehat{P}\middle|u_2\left(\vec{k},\vec{r}\right)\right\rangle \\ &= \frac{b_{k_+}a_{k_-}}{\sqrt{2}}\left\{\left\langle(X'-iY')\middle|\widehat{P}\middle|iS\right\rangle\langle\uparrow'\mid\uparrow'\rangle\right\} + c_{k_+}a_{k_-}\left\{\left\langle Z'\middle|\widehat{P}\middle|iS\right\rangle\langle\downarrow'\mid\uparrow'\rangle\right\} \\ &\quad - \frac{a_{k_+}b_{k_-}}{\sqrt{2}}\left\{\left\langle iS\middle|\widehat{P}\middle|(X'+iY')\right\rangle\langle\downarrow'\mid\uparrow'\rangle\right\} \\ &\quad + a_{k_+}c_{k_-}\left\{\left\langle iS\middle|\widehat{P}\middle|Z'\right\rangle\langle\downarrow'\mid\uparrow'\rangle\right\}\end{aligned} \tag{18.18}$$

We can also write

$$\begin{aligned}\left\langle(X'-iY')\middle|\widehat{P}\middle|iS\right\rangle &= \left\langle(X')\middle|\widehat{P}\middle|iS\right\rangle - \left\langle(iY')\middle|\widehat{P}\middle|iS\right\rangle \\ &= i\int u^*_{X'}\widehat{P}S - \int -iu^*_{Y'}\widehat{P}iu_X = i\left\langle X'\middle|\widehat{P}\middle|S\right\rangle - \left\langle Y'\middle|\widehat{P}\middle|S\right\rangle\end{aligned}$$

From the above relations, for X', Y' and Z', we get

$$|X'\rangle = \cos\theta\cos\phi|X\rangle + \cos\theta\sin\phi|Y\rangle - \sin\theta|Z\rangle$$

Thus,

$$\left\langle X'\middle|\widehat{P}\middle|S\right\rangle = \cos\theta\cos\phi\left\langle X\middle|\widehat{P}\middle|S\right\rangle + \cos\theta\sin\phi\left\langle Y\middle|\widehat{P}\middle|S\right\rangle - \sin\theta\left\langle Z\middle|\widehat{P}\middle|S\right\rangle = \widehat{P}\widehat{r}_1$$

where,

$$\widehat{r}_1 = \widehat{i}\cos\theta\cos\phi + \widehat{j}\cos\theta\sin\phi - \widehat{k}\sin\theta$$

$$|Y'\rangle = -\sin\phi|X\rangle + \cos\phi|Y\rangle + 0|Z\rangle$$

Thus,

$$\left\langle Y'\middle|\widehat{P}\middle|S\right\rangle = -\sin\phi\left\langle X\middle|\widehat{P}\middle|S\right\rangle + \cos\phi\left\langle Y\middle|\widehat{P}\middle|S\right\rangle + 0\left\langle Z\middle|\widehat{P}\middle|S\right\rangle = \widehat{P}\widehat{r}_2$$

where,

$$\widehat{r}_2 = -\widehat{i}\sin\phi + \widehat{j}\cos\phi$$

so that $\left\langle (X' - iY')\middle|\widehat{P}\middle|S\right\rangle = \widehat{P}(i\widehat{r}_1 - \widehat{r}_2)$.

Thus,

$$\frac{a_{k_-}b_{k_+}}{\sqrt{2}}\left\langle (X' - iY')\middle|\widehat{P}\middle|S\right\rangle\langle\uparrow'\mid\uparrow'\rangle = \frac{a_{k_-}b_{k_+}}{\sqrt{2}}\widehat{P}(i\widehat{r}_1 - \widehat{r}_2)\langle\uparrow'\mid\uparrow'\rangle \tag{18.19}$$

Now since,

$$\left\langle iS\middle|\widehat{P}\middle|(X' + iY')\right\rangle = i\left\langle S\middle|\widehat{P}\middle|X'\right\rangle - \left\langle S\middle|\widehat{P}\middle|Y'\right\rangle = \widehat{P}(i\widehat{r}_1 - \widehat{r}_2)$$

We can write,

$$-\left[\frac{a_{k_+}b_{k_-}}{\sqrt{2}}\left\{\left\langle iS\middle|\widehat{P}\middle|(X' + iY')\right\rangle\langle\downarrow'\mid\downarrow'\rangle\right\}\right] = -\left[\frac{a_{k_+}b_{k_-}}{\sqrt{2}}\widehat{P}(i\widehat{r}_1 - \widehat{r}_2)\langle\downarrow'\mid\downarrow'\rangle\right] \tag{18.20}$$

Similarly, we get

$$|Z'\rangle = \sin\theta\cos\phi|X\rangle + \sin\theta\sin\phi|Y\rangle + \cos\theta|Z\rangle$$

So that, $\left\langle Z'\middle|\widehat{P}\middle|iS\right\rangle = i\left\langle Z'\middle|\widehat{P}\middle|S\right\rangle = i\widehat{P}\left\{\sin\theta\cos\phi\widehat{i} + \sin\theta\sin\phi\widehat{j} + \cos\theta\widehat{k}\right\} =$ $i\widehat{P}\widehat{r}_3$

where

$$\widehat{r}_3 = \widehat{i}\sin\theta\cos\phi + \widehat{j}\sin\theta\sin\phi + \widehat{k}\cos\theta$$

Thus,

$$c_{k_+}a_{k_-}\left\langle Z'\middle|\widehat{P}\middle|iS\right\rangle\langle\downarrow'\mid\uparrow'\rangle = c_{k_+}a_{k_-}i\widehat{P}\widehat{r}_3\langle\downarrow'\mid\uparrow'\rangle \tag{18.21}$$

Similarly, we can write,

$$c_{k_-}a_{k_+}\left\langle iS\middle|\widehat{P}\middle|Z'\right\rangle\langle\downarrow'\mid\uparrow'\rangle = c_{k_-}a_{k_+}i\widehat{P}\widehat{r}_3\langle\downarrow'\mid\uparrow'\rangle \tag{18.22}$$

Therefore, we obtain

$$\begin{aligned}&\frac{a_{k_-}b_{k_+}}{\sqrt{2}}\left\{\left\langle(X'-iY')\middle|\widehat{P}\middle|S\right\rangle\langle\uparrow'\mid\uparrow'\rangle\right\} - \frac{a_{k_+}b_{k_-}}{\sqrt{2}}\left\{\left\langle iS\middle|\widehat{P}\middle|(X'+iY')\right\rangle\langle\downarrow'\mid\downarrow'\rangle\right\}\\ &= \frac{\widehat{P}}{\sqrt{2}}\left(-a_{k_+}b_{k_-}\langle\downarrow'\mid\downarrow'\rangle + a_{k_-}b_{k_+}\langle\uparrow\mid\uparrow'\rangle\right)(i\widehat{r}_1 - \widehat{r}_2)\end{aligned} \tag{18.23}$$

Also, we can write,

$$\begin{aligned}&c_{k_+}a_{k_-}\left\langle Z'\middle|\widehat{P}\middle|iS\right\rangle\langle\downarrow'\mid\uparrow'\rangle + c_{k_-}a_{k_+}\left\langle iS\middle|\widehat{P}\middle|Z'\right\rangle\langle\downarrow'\mid\uparrow'\rangle\\ &= i\widehat{P}\left(c_{k_+}a_{k_-} + c_{k_-}a_{k_+}\right)\widehat{r}_3[\langle\downarrow'\mid\downarrow'\rangle]\end{aligned} \tag{18.24}$$

Combining (18.23) and (18.24), we find

$$\begin{aligned}\widehat{p}_{CV}\left(\vec{k}\right) &= \frac{\widehat{P}}{\sqrt{2}}(i\widehat{r}_1 - \widehat{r}_2)\left\{\left(b_{k_+}a_{k_-}\right)\langle\uparrow\mid\uparrow'\rangle - \left(b_{k_-}a_{k_+}\right)\langle\downarrow'\mid\downarrow'\rangle\right\}\\ &\quad + i\widehat{P}\widehat{r}_3\left(c_{k_+}a_{k_-} - c_{k_-}a_{k_+}\right)\langle\downarrow'\mid\uparrow'\rangle\end{aligned} \tag{18.25}$$

From the above relations, we obtain,

$$\left.\begin{aligned}\uparrow' &= e^{-i\phi/2}\cos(\theta/2)\uparrow + e^{i\phi/2}\sin(\theta/2)\downarrow\\ \downarrow' &= -e^{-i\phi/2}\sin(\theta/2)\uparrow + e^{i\phi/2}\cos(\theta/2)\downarrow\end{aligned}\right\} \tag{18.26}$$

Therefore,

$$\begin{aligned}\langle\downarrow'\mid\uparrow'\rangle_x = &-\sin(\theta/2)\cos(\theta/2)\langle\uparrow\mid\uparrow\rangle_x + e^{-i\phi}\cos^2(\theta/2)\langle\downarrow\mid\uparrow\rangle_x\\ &- e^{-i\phi}\sin^2(\theta/2)\langle\uparrow\mid\downarrow\rangle_x + \sin(\theta/2)\cos(\theta/2)\langle\downarrow\mid\downarrow\rangle_x\end{aligned} \tag{18.27}$$

But we know from above that
$\langle\uparrow\mid\uparrow\rangle_x = 0$, $\langle\downarrow\mid\uparrow\rangle = \frac{1}{2}$, $\langle\downarrow\mid\uparrow\rangle_x = \frac{1}{2}$ and $\langle\downarrow\mid\downarrow\rangle_x = 0$
Thus, we get

$$\langle \downarrow' \mid \uparrow' \rangle_x = \frac{1}{2}\left[e^{-i\phi}\cos^2(\theta/2) - e^{i\phi}\sin^2(\theta/2)\right]$$
$$= \frac{1}{2}\left[(\cos\phi - i\sin\phi)\cos^2(\theta/2) - (\cos\phi + i\sin\phi)\sin^2(\theta/2)\right] = \frac{1}{2}[\cos\phi\cos\theta - i\sin\phi] \quad (18.28)$$

Similarly, we obtain
$\langle \downarrow' \mid \uparrow' \rangle_y = \frac{1}{2}[i\cos\phi + \sin\phi\cos\theta]$ and $\langle \downarrow' \mid \uparrow' \rangle_z = \frac{1}{2}[-\sin\theta]$.
Therefore,

$$\langle \downarrow' \mid \uparrow' \rangle = \hat{i}\langle \downarrow' \mid \uparrow' \rangle_x + \hat{j}\langle \downarrow' \mid \uparrow' \rangle_y + \hat{k}\langle \downarrow' \mid \uparrow' \rangle_z = \frac{1}{2}\left\{(\cos\theta\cos\phi - i\sin\phi)\hat{i} + (i\cos\phi + \sin\phi\cos\theta)\hat{j} - \sin\theta\hat{k}\right\}$$
$$= \frac{1}{2}\left[\left\{(\cos\theta\cos\phi)\hat{i} + (\sin\phi\cos\theta)\hat{j} - \sin\theta\hat{k}\right\} + i\left\{-\hat{i}\sin\phi + \hat{j}\cos\phi\right\}\right]$$
$$= \frac{1}{2}[\hat{r}_1 + i\hat{r}_2] = -\frac{1}{2}i[\hat{r}_1 - i\hat{r}_2]$$

Similarly, we can write

$$\langle \uparrow' \mid \uparrow' \rangle = \frac{1}{2}\left[\hat{i}\sin\theta\cos\phi + \hat{j}\sin\theta\sin\phi + \hat{k}\cos\theta\right] = \frac{1}{2}\hat{r}_3 \text{ and } \langle \downarrow' \mid \downarrow' \rangle = -\frac{1}{2}\hat{r}_3$$

Using the above results, we can write

$$\begin{aligned}\hat{p}_{cv}\left(\vec{k}\right) &= \frac{\hat{P}}{\sqrt{2}}(i\hat{r}_1 - \hat{r}_2)\left\{\left(a_{k\_}b_{k_+}\right)\langle \uparrow' \mid \uparrow' \rangle - \left(b_{k\_}a_{k_+}\right)\langle \downarrow' \mid \downarrow' \rangle\right\} \\ &\quad + i\hat{P}\hat{r}_3\left\{\left(c_{k_+}a_{k_+} - c_{k\_}a_{k_+}\right)\langle \downarrow' \mid \uparrow' \rangle\right\} \\ &= \frac{\hat{P}}{2}\hat{r}_3(i\hat{r}_1 - \hat{r}_2)\left\{\left(\frac{a_{k\_}b_{k_+}}{\sqrt{2}} + \frac{b_{k\_}b_{k_+}}{\sqrt{2}}\right)\right\} \\ &\quad + \frac{\hat{P}}{2}\hat{r}_3(i\hat{r}_1 - \hat{r}_2)\left\{\left(c_{k_+}a_{k\_} + c_{k\_}a_{k_+}\right)\right\}\end{aligned}$$

Thus,

$$\hat{p}_{CV}\left(\vec{k}\right) = \frac{\hat{P}}{2}\hat{r}_3(i\hat{r}_1 - \hat{r}_2)\left\{a_{k_+}\left(\frac{b_{k\_}}{\sqrt{2}} + c_{k\_}\right) + a_{k\_}\left(\frac{b_{k_+}}{\sqrt{2}} + c_{k_+}\right)\right\} \quad (18.29)$$

We can write that,

$$|\hat{r}_1| = |\hat{r}_2| = |\hat{r}_3| = 1, \text{ also, } \hat{P}\hat{r}_3 = \hat{P}_x\sin\theta\cos\phi\hat{i} + \hat{P}_y\sin\theta\sin\phi\hat{j} + \hat{P}_z\cos\theta\hat{k}$$

where, $\widehat{P} = \left\langle S|\widehat{P}|X\right\rangle = \left\langle S|\widehat{P}|Y\right\rangle = \left\langle S|\widehat{P}|Z\right\rangle$, $\left\langle S|\widehat{P}|X\right\rangle = \int u_C^*(0,\vec{r})\widehat{P}u_{VX}(0,\vec{r})d^3r = \widehat{P}_{CVX}(0)$ and $\left\langle S|\widehat{P}|Z\right\rangle = \widehat{P}_{CVZ}(0)$.

Thus,

$$\widehat{P} = \widehat{P}_{CVX}(0) = \widehat{P}_{CVY}(0) = \widehat{P}_{CVZ}(0) = \widehat{P}_{CV}(0)$$

where

$$\widehat{P}_{CV}(0) \equiv \int u_c^*(0,\vec{r})\widehat{P}u_V(0,\vec{r})d^3r \equiv \widehat{P}$$

For a plane polarized light wave, we have the polarization vector $\vec{\varepsilon}_s = \widehat{k}$, when the light wave vector is traveling along the z-axis. Therefore, for a plane polarized light-wave, we have considered $\vec{\varepsilon}_s = \widehat{k}$.

Then, from (18.29) we get

$$\left(\vec{\varepsilon}\cdot\widehat{p}_{CV}\left(\vec{k}\right)\right) = \vec{k}\cdot\frac{\widehat{P}}{2}\widehat{r}_3(i\widehat{r}_1 - \widehat{r}_2)\left[A\left(\vec{k}\right) + B\left(\vec{k}\right)\right]\cos\omega t \tag{18.30}$$

and

$$\left.\begin{aligned} A(\vec{k}) &= a_{k_-}\left(\frac{b_{k_+}}{\sqrt{2}} + c_{k_+}\right) \\ B(\vec{k}) &= a_{k_+}\left(\frac{b_{k_-}}{\sqrt{2}} + c_{k_-}\right) \end{aligned}\right\} \tag{18.31}$$

Thus,

$$\begin{aligned} \left|\vec{\varepsilon}\cdot\widehat{p}_{cv}\left(\vec{k}\right)\right|^2 &= \left|\widehat{k}\cdot\frac{\widehat{P}}{2}\widehat{r}_3\right|^2 |i\widehat{r}_1 - i\widehat{r}_2|^2\left[A(\vec{k}) + B(\vec{k})\right]^2\cos^2\omega t \\ &= \frac{1}{4}\left|\widehat{P}_z\cos\theta\right|^2\left[A(\vec{k}) + B(\vec{k})\right]^2\cos^2\omega t \end{aligned} \tag{18.32}$$

So, the average value of $\left|\vec{\varepsilon}\cdot\widehat{p}_{cv}\left(\vec{k}\right)\right|^2$ for a plane polarized light-wave is given by

$$\begin{aligned} \left\langle\left|\vec{\varepsilon}\cdot\widehat{p}_{cv}\left(\vec{k}\right)\right|^2\right\rangle_{av} &= \frac{2}{4}\left|\widehat{P}_z\right|^2\left[A(\vec{k}) + B(\vec{k})\right]^2\left(\int_0^{2\pi} d\phi\int_0^{\pi}\cos^2\theta\sin\theta d\theta\right)\left(\frac{1}{2}\right) \\ &= \frac{2\pi}{3}\left|\widehat{P}_z\right|^2\left[A(\vec{k}) + B(\vec{k})\right]^2 \end{aligned} \tag{18.33}$$

where, $\left|\widehat{P}_z\right|^2 = \left(\frac{1}{2}\right)\left|\vec{k}\cdot\widehat{p}_{cv}(0)\right|^2$ and

$$\left|\vec{k}\cdot\widehat{p}_{cv}(0)\right|^2 = \frac{m^2}{4m_r}\frac{E_{g_0}\left(E_{g_0}+\Delta\right)}{\left(E_{g_0}+\frac{2}{3}\Delta\right)} \tag{18.34}$$

We shall express $A(\vec{k})$ and $B(\vec{k})$ in terms of constants of the energy spectra in the following way:

Substituting $a_{k_\pm}$, $b_{k_\pm}$, $c_{k_\pm}$, and $\gamma_{0k_\pm}$ in $A(\vec{k})$ and $B(\vec{k})$ in (18.31) we get

$$A\left(\vec{k}\right) = \beta\left(t+\frac{\rho}{\sqrt{2}}\right)\left\{\left(\frac{E_{g0}}{E_{g_0}+\delta'}\right)\gamma_{0k_+}^2 - \gamma_{0k_+}^2\gamma_{0k_-}^2\left(\frac{E_{g0}-\delta'}{E_{g_0}+\delta'}\right)\right\}^{1/2} \tag{18.35}$$

$$B\left(\vec{k}\right) = \beta\left(t+\frac{\rho}{\sqrt{2}}\right)\left\{\left(\frac{E_{g0}}{E_{g_0}+\delta'}\right)\gamma_{0k_-}^2 - \gamma_{0k_+}^2\gamma_{0k_-}^2\left(\frac{E_{g0}-\delta'}{E_{g_0}+\delta'}\right)\right\}^{1/2} \tag{18.36}$$

in which, $\gamma_{0k_+}^2 \equiv \frac{\xi_{1k}-E_{g0}}{2(\xi_{1k}+\delta')} \equiv \frac{1}{2}\left[1-\left(\frac{E_{g0}+\delta'}{\xi_{1k}+\delta'}\right)\right]$ and $\gamma_{0k_-}^2 \equiv \frac{\xi_{1k}+E_{g0}}{2(\xi_{1k}+\delta')} \equiv \frac{1}{2}\left[1+\left(\frac{E_{g0}+\delta'}{\xi_{1k}+\delta'}\right)\right]$

Substituting $x \equiv \xi_{1k}+\delta'$ in $\gamma_{0k\pm}^2$, we can write,

$$A\left(\vec{k}\right) = \beta\left(t+\frac{\rho}{\sqrt{2}}\right)\left\{\left(\frac{E_{g0}}{E_{g_0}+\delta'}\right)\frac{1}{2}\left(1-\frac{E_{g_0}+\delta'}{x}\right) - \frac{1}{4}\left(\frac{E_{g_0}-\delta'}{E_{g_0}+\delta'}\right)\left(1-\frac{E_{g_0}+\delta'}{x}\right)\left(1+\frac{E_{g_0}-\delta'}{x}\right)\right\}^{1/2}$$

Thus, $A\left(\vec{k}\right) = \frac{\beta}{2}\left(t+\frac{\rho}{\sqrt{2}}\right)\left\{1-\frac{2a_0}{x}+\frac{a_1}{x^2}\right\}^{1/2}$

where, $a_0 \equiv \left(E_{g_0}^2+\delta'^2\right)\left(E_{g_0}+\delta'\right)^{-1}$ and $a_1 \equiv \left(E_{g_0}-\delta'\right)^2$.

After tedious algebra, one can show that

$$A\left(\vec{k}\right) = \frac{\beta}{2}\left(t+\frac{\rho}{\sqrt{2}}\right)\left(E_{g_0}-\delta'\right)\left[\frac{1}{\xi_{1k}+\delta'}-\frac{1}{E_{g_0}+\delta'}\right]^{1/2}\left[\frac{1}{\xi_{1k}+\delta'}-\frac{\left(E_{g_0}+\delta'\right)}{\left(E_{g_0}-\delta'\right)^2}\right]^{1/2} \tag{18.37}$$

Similarly, from (18.36), we can write,

$$B\left(\vec{k}\right) = \beta\left(t+\frac{\rho}{\sqrt{2}}\right)\left\{\left(\frac{E_{g0}}{E_{g_0}+\delta'}\right)\frac{1}{2}\left(1+\frac{E_{g_0}-\delta'}{x}\right) - \frac{1}{4}\left(\frac{E_{g_0}-\delta'}{E_{g_0}+\delta'}\right)\left(1-\frac{E_{g_0}+\delta'}{x}\right)\left(1+\frac{E_{g_0}-\delta'}{x}\right)\right\}^{1/2}$$

So that, finally we get,

$$B\left(\vec{k}\right) = \frac{\beta}{2}\left(t + \frac{\rho}{\sqrt{2}}\right)\left(1 + \frac{E_{g0} - \delta'}{\xi_{1k} + \delta'}\right) \tag{18.38}$$

Using (18.33), (18.34), (18.37) and (18.38), we can write

$$\left(\frac{|e|A_0}{2m}\right)^2 \frac{\left\langle \left|\vec{\varepsilon}\cdot\widehat{p}_{cv}(\vec{k})\right|^2\right\rangle_{av}}{E_c(\vec{k}) - E_v(\vec{k})} = \left(\frac{|e|A_0}{2m}\right)^2 \frac{2\pi}{3}\left|\vec{k}\cdot\widehat{p}_{cv}(0)\right|^2 \frac{\beta^2}{4}\left(t + \frac{\rho}{\sqrt{2}}\right)^2$$
$$\frac{1}{\xi_{1k}}\left\{\left(1 + \frac{E_{g0} - \delta'}{\xi_{1k} + \delta'}\right) + (E_{g0} - \delta')\left[\frac{1}{\xi_{1k} + \delta'} - \frac{1}{E_{g0} + \delta'}\right]^{1/2}\left[\frac{1}{\xi_{1k} + \delta'} - \frac{E_{g0} + \delta'}{(E_{g0} - \delta')^2}\right]^{1/2}\right\}^2 \tag{18.39}$$

Following Nag [4], it can be shown that

$$A_0^2 = \frac{I\lambda^2}{2\pi^2 c^3\sqrt{\varepsilon_{sc}\varepsilon_0}} \tag{18.40}$$

where, I is the light intensity of wavelength λ, ε_0 is the permittivity of free space and c is the velocity of light. Thus, the simplified electron energy spectrum in III–V, ternary and quaternary materials in the presence of light waves can approximately be written as

$$\frac{\hbar^2 k^2}{2m_c} = \beta_0(E, \lambda) \tag{18.41}$$

where,

$$\beta_0(E, \lambda) \equiv [I_{11}(E) - \theta_0(E, \lambda)],$$

$$\theta_0(E, \lambda) \equiv \frac{|e|^2}{96 m_r \pi c^3}\frac{I\lambda^2}{\sqrt{\varepsilon_{sc}\varepsilon_0}}\frac{E_{g0}(E_{g0} + \Delta)}{\left(E_{g0} + \frac{2}{3}\Delta\right)}\frac{\beta^2}{4}\left(t + \frac{\rho}{\sqrt{2}}\right)^2 \frac{1}{\phi_0(E)}$$

$$\left\{\left(1 + \frac{E_{g0} - \delta'}{\phi_0(E) + \delta'}\right) + (E_{g0} - \delta')\left[\frac{1}{\phi_0(E) + \delta'} - \frac{1}{E_{g0} + \delta'}\right]^{1/2}\left[\frac{1}{\phi_0(E) + \delta'} - \frac{E_{g0} + \delta'}{(E_{g0} - \delta')}\right]^{1/2}\right\}^2$$

and

$$\phi_0(E) \equiv E_{g0}\left(1 + 2\left(1 + \frac{m_c}{m_v}\right)\frac{I_{11}(E)}{E_{g0}}\right)^{1/2}$$

Thus, under the limiting condition $\vec{k} \to 0$, from (18.41), we observe that $E \neq 0$ and is positive. Therefore, in the presence of external light waves, the energy of the electron does not tend to zero when $\vec{k} \to 0$, where as for the un-perturbed three band model of Kane, $I_{11}(E) = \left[\hbar^2 k^2/(2m_c)\right]$ in which $E \to 0$ for $\vec{k} \to 0$. As the conduction band is taken as the reference level of energy, therefore the lowest positive value of E for $\vec{k} \to 0$ provides the increased band gap (ΔE_g) of the semiconductor due to photon excitation. The values of the increased band gap can be obtained by computer iteration processes for various values of I and λ respectively.

Special Cases:

(1) For the two-band model of Kane, we have $\Delta \to 0$. Under this condition, $I_{11}(E) \to E(1 + aE) = \frac{\hbar^2 k^2}{2m_c}$. Since, $\beta \to 1$, $t \to 1$, $\rho \to 0$, $\delta' \to 0$ for $\Delta \to 0$, from (18.41), we can write the energy spectrum of III–V, ternary and quaternary materials in the presence of external photo-excitation whose unperturbed conduction electrons obey the two band model of Kane as

$$\frac{\hbar^2 k^2}{2m_c} = \tau_0(E, \lambda) \tag{18.42}$$

where,

$$\tau_0(E, \lambda) \equiv E(1 + aE) - B_0(E, \lambda),$$

$$B_0(E, \lambda) \equiv \frac{|e|^2 I \lambda^2 E_{g0}}{384\pi c^3 m_r \sqrt{\varepsilon_{sc}\varepsilon_0}} \frac{1}{\phi_1(E)} \left\{ \left(1 + \frac{E_{g0}}{\phi_1(E)}\right) + E_{g0}\left[\frac{1}{\phi_1(E)} - \frac{1}{E_{g0}}\right] \right\}^2,$$

$$\phi_1(E) \equiv E_{g0}\left\{1 + \frac{2m_c}{m_r} aE(1 + aE)\right\}^{1/2}.$$

(2) For relatively wide band gap semiconductors, one can write, $a \to 0$, $b \to 0$, $c \to 0$ and $I_{11}(E) \to E$.
Thus, from (18.42), we get,

$$\frac{\hbar^2 k^2}{2m_c} = \rho_0(E, \lambda) \tag{18.43}$$

where,

$$\rho_0(E, \lambda) \equiv E - \frac{|e|^2 I \lambda^2}{96\pi c^3 m_r \sqrt{\varepsilon_{sc}\varepsilon_0}} \left[1 + \left(\frac{2m_c}{m_r}\right) aE\right]^{-3/2}$$

The (18.41), (18.42) and (18.43) can approximately be written as

$$\frac{\hbar^2 k^2}{2m_c} = U_\lambda I_{11}(\mathrm{E}) - \mathrm{P}_\lambda \tag{18.44}$$

$$\frac{\hbar^2 k^2}{2m_c} = t_{1\lambda}\mathrm{E} + t_{2\lambda}\mathrm{E}^2 - \delta_\lambda \tag{18.45}$$

and

$$\frac{\hbar^2 k^2}{2m_c} = t_{1\lambda}\mathrm{E} - \delta_\lambda \tag{18.46a}$$

where,

$$U_\lambda = (1+\theta_\lambda),\ \theta_\lambda = \frac{C_0}{A}\left(\mathrm{t}_\lambda + \frac{BJ_\lambda}{A}\right),\ C_0 = \left[\frac{|e|^2 I\lambda^2 E_{g_0}(E_{g0}+\Delta)}{96\pi c^3 m_r\sqrt{\varepsilon_{sc}\varepsilon_0}\left(E_{g_0}+\frac{2}{3}\Delta\right)}\frac{\beta^2}{4}\left(1+\frac{\rho}{\sqrt{2}}\right)^2\right],$$

$$A = E_{g_0},$$

$$B = \left[1+\frac{m^*}{m_V}\right],\ G_\lambda = \left[\frac{2B}{(A+\delta')^3} - \frac{BC_\lambda}{(A+\delta')}\right]$$

$$C_\lambda = [(\mathrm{E}_{g_0}+\delta')^{-1} + (\mathrm{E}_{g_0}+\delta')(\mathrm{E}_{g_0}-\delta')^{-2}](A+\delta')^{-1}$$

$$P_\lambda = \frac{C_0}{A}J_\lambda,\ J_\lambda = (\mathrm{D}_\lambda + 2(\mathrm{E}_{g_0}-\delta')\sqrt{f_\lambda}),$$

$$D_\lambda = \left(1+\frac{2(\mathrm{E}_{g_0}-\delta')}{(A+\delta')}\right),\ f_\lambda = \left[\frac{1}{(A+\delta')^2} + \frac{1}{(\mathrm{E}_{g_0}-\delta')^2} - C_\lambda\right],$$

$$t_{1\lambda} = \left(1+\frac{3m_c}{m_r}\alpha\delta_\lambda\right),\ \alpha = \frac{1}{E_{g0}},\ \delta_\lambda = \frac{|e|^2 I\lambda^2}{96 m_r \pi c^3\sqrt{\varepsilon_{sc}\varepsilon_0}}\ \text{and}\ t_{2\lambda} = \alpha t_{1\lambda}$$

Under the condition of heavy doping, following the methods as developed in Chap. 1, the HD dispersion relations in this case in the presence of light waves can be written as

$$\frac{\hbar^2 k^2}{2m_c} = T_1(\mathrm{E}, \eta_g, \lambda) \tag{18.46b}$$

$$\frac{\hbar^2 k^2}{2m_c} = T_2(\mathrm{E}, \eta_g, \lambda) \tag{18.47}$$

$$\frac{\hbar^2 k^2}{2m_c} = T_3(\mathrm{E}, \eta_g, \lambda) \tag{18.48a}$$

where

$$\begin{aligned} T_1(\mathrm{E}, \eta_g, \lambda) &= [U_\lambda[T_{31}(\mathrm{E}, \eta_g) + iT_{32}(\mathrm{E}, \eta_g)] - \mathrm{P}_\lambda], \\ T_2(\mathrm{E}, \eta_g, \lambda) &= [t_{1\lambda}\gamma_3(\mathrm{E}, \eta_g) + (t_{2\lambda})2\theta_0(\mathrm{E}, \eta_g)[1 + \mathrm{E}rf(\mathrm{E}/\eta_g)]^{-1} - \delta_\lambda] \\ \text{and } T_2(\mathrm{E}, \eta_g, \lambda) &= [t_{1\lambda}\gamma_3(\mathrm{E}, \eta_g) - \delta_\lambda] \end{aligned}$$

The DOS functions for (18.46b), (18.47) and (18.48a) can, respectively, be written as

$$N_L(\mathrm{E}) = 4\pi g_v \left(\frac{2m_c}{\hbar^2}\right)^{\frac{3}{2}} \sqrt{[T_1(\mathrm{E}, \eta_g, \lambda)]}[T_1(\mathrm{E}, \eta_g, \lambda)]' \tag{18.48b}$$

$$N_L(\mathrm{E}) = 4\pi g_v \left(\frac{2m_c}{\hbar^2}\right)^{\frac{3}{2}} \sqrt{[T_2(\mathrm{E}, \eta_g, \lambda)]}[T_2(\mathrm{E}, \eta_g, \lambda)]' \tag{18.48c}$$

$$N_L(\mathrm{E}) = 4\pi g_v \left(\frac{2m_c}{\hbar^2}\right)^{\frac{3}{2}} \sqrt{[T_3(\mathrm{E}, \eta_g, \lambda)]}[T_3(\mathrm{E}, \eta_g, \lambda)]' \tag{18.48d}$$

The EEM can be expressed in this case by using (18.46a), (18.47) and (18.48a).

$$m^*(\mathrm{E}_{FHDL}, \eta_g, \lambda) = \mathrm{m}_c \text{ Real part of } [T_1(\mathrm{E}_{FHDL}, \eta_g, \lambda)]' \tag{18.49}$$

$$m^*(\mathrm{E}_{FHDL}, \eta_g, \lambda) = \mathrm{m}_c\ [T_2(\mathrm{E}_{FHDL}, \eta_g, \lambda)]' \tag{18.50}$$

$$m^*(\mathrm{E}_{FHDL}, \eta_g, \lambda) = \mathrm{m}_c\ [T_3(\mathrm{E}_{FHDL}, \eta_g, \lambda)]' \tag{18.51}$$

18.2.2 The DR Under Magnetic Quantization in HD Kane Type Semiconductors in the Presence of Light Waves

(i) Using (18.46b), the magneto-dispersion law, in the absence of spin, for HD III–V, ternary and quaternary semiconductors, in the presence of photo-excitation, whose unperturbed conduction electrons obey the three band model of Kane, is given by

$$T_1(E, \eta_g) = \left(n + \frac{1}{2}\right)\hbar\omega_0 + \frac{\hbar^2 k_z^2}{2m_c} \tag{18.52}$$

Using (18.52), the DOS function in the present case can be expressed as

$$D_B(E,\eta_g,\lambda) = \frac{g_v|e|\sqrt{2m_c}}{2\pi^2\hbar^2}\sum_{n=0}^{n_{\max}}\left[\{T_1(E,\eta_g,\lambda)\}'\left\{T_1(E,\eta_g,\lambda)-\left(n+\frac{1}{2}\right)\hbar\omega_0\right\}^{-1/2}H(E-E_{n_{l1}})\right] \tag{18.53}$$

where, $E_{n_{l1}}$ is the Landau sub-band energies in this case and is given as

$$T_1(E_{n_{l1}},\eta_g,\lambda) = \left(n+\frac{1}{2}\right)\hbar\omega_0 \tag{18.54}$$

The EEM in this case assumes the form

$$m^*(E_{FHDL},\eta_g,\lambda) = m_c \text{ Real part of } \{T_1(E_{FHDL},\eta_g,\lambda)\}' \tag{18.55}$$

where, E_{FHDLB} is the Fermi energy under quantizing magnetic field in the presence of light waves as measured from the edge of the conduction band in the vertically upward direction in the absence of any quantization.

(ii) Using (18.47), the magneto-dispersion law, in the absence of spin, for HD III–V, ternary and quaternary semiconductors, in the presence of photo-excitation, whose unperturbed conduction electrons obey the two band model of Kane, is given by

$$T_2(E,\eta_g) = \left(n+\frac{1}{2}\right)\hbar\omega_0 + \frac{\hbar^2k_z^2}{2m_c} \tag{18.56}$$

Using (18.56), the DOS function in the present case can be expressed as

$$D_B(E,\eta_g,\lambda) = \frac{g_v|e|\sqrt{2m_c}}{2\pi^2\hbar^2}\sum_{n=0}^{n_{\max}}\Big[\{T_2(E,\eta_g,\lambda)\}'\{T_2(E,\eta_g,\lambda) - \left(n+\frac{1}{2}\right)\hbar\omega_0\Big\}^{-1/2}H(E-E_{n_{l2}})\Big] \tag{18.57}$$

where, $E_{n_{l2}}$ is the Landau sub-band energies in this case and is given as

$$T_2(E_{n_{l2}},\eta_g) = \left(n+\frac{1}{2}\right)\hbar\omega_0 \tag{18.58}$$

The EEM in this case assumes the form

$$m^*(E_{FHDLB},\eta_g,\lambda) = m_c\{T_2(E_{FHDLB},\eta_g,\lambda)\}' \tag{18.59}$$

(iii) Using (18.47), the magneto-dispersion law, in the absence of spin, for HD III–V, ternary and quaternary semiconductors, in the presence of photo-excitation, whose unperturbed conduction electrons obey the parabolic energy bands, is given by

$$T_2(E,\eta_g) = \left(n+\frac{1}{2}\right)\hbar\omega_0 + \frac{\hbar^2 k_z^2}{2m_c} \tag{18.60}$$

Using (18.60), the DOS function in the present case can be expressed as

$$\begin{aligned} D_B(E,\eta_g,\lambda) = & \frac{g_v|e|\sqrt{2m_c}}{2\pi^2\hbar^2} \sum_{n=0}^{n_{\max}} \Big[\{T_3(E,\eta_g,\lambda)\}'\{T_3(E,\eta_g,\lambda) \\ & -\left(n+\frac{1}{2}\right)\hbar\omega_0\Big\}^{-1/2} H(E-E_{n_{l2}})\Big] \end{aligned} \tag{18.61}$$

where, $E_{n_{l2}}$ is the Landau sub-band energies in this case and is given as

$$T_3(E_{n_{l2}},\eta_g,\lambda) = \left(n+\frac{1}{2}\right)\hbar\omega_0 \tag{18.62}$$

The EEM in this case assumes the form

$$m^*(E_{FHDLB},\eta_g,\lambda) = m_c\{T_3(E_{FHDLB},\eta_g,\lambda)\}' \tag{18.63}$$

18.2.3 The DR Under Crossed Electric and Quantizing Magnetic Fields in HD Kane Type Semiconductors in the Presence of Light Waves

(i) The electron dispersion law in the present case is given by

$$\begin{aligned} T_1(E,\eta_g,\lambda) = & \left(n+\frac{1}{2}\right)\hbar\omega_0 + \frac{[\hbar^2 k_z(E)]^2}{2m_c} \\ & -\frac{E_0}{B}\hbar k_y\{T_1(E,\eta_g,\lambda)\}' - \left\{\frac{m_c E_0^2\left[\{T_1(E,\eta_g,\lambda)\}'\right]^2}{2B^2}\right\} \end{aligned} \tag{18.64}$$

The use of (18.64) leads to the expressions of the EEM s' along z and y directions as

$$m_z^*(E_{F_{BLHDC}}, n, E_0, B, \lambda) = \text{Real part of } m_c\Bigg[\{T_1(E_{F_{BLHDC}}, \eta_g, \lambda)\}'' + \frac{m_c E_0^2\{T_1(E_{F_{BLHDC}}, \eta_g, \lambda)\}'\{T_1(E_{F_{BLHDC}}, \eta_g, \lambda)\}''}{B^2}\Bigg] \tag{18.65}$$

$$\begin{aligned} m_y^*(E_{F_{BLHDC}}, n, E_0, B, \lambda) = {} & \text{Real part of } \left(\frac{B}{E_0}\right)^2 \left[\{T_1(E_{F_{BLHDC}}, \eta_g, \lambda)\}'\right]^{-1} \Bigg[T_1(E_{F_{BLHDC}}, \eta_g, \lambda) - \left(n+\frac{1}{2}\right)\hbar\omega_0 \\ & + \frac{m_c E_0^2\left[\{T_1(E_{F_{BLHDC}}, \eta_g, \lambda)\}'\right]^2}{2B^2}\Bigg]\Bigg[\frac{\{T_1(E_{F_{BLHDC}}, \eta_g, \lambda)\}''}{\left[\{T_1(E_{F_{BLHDC}}, \eta_g, \lambda)\}'\right]^2}\Bigg[T_1(E_{F_{BLHDC}}, \eta_g, \lambda) - \left(n+\frac{1}{2}\right)\hbar\omega_0 \\ & + \frac{m_c E_0^2\left[\{T_1(E_{F_{BLHDC}}, \eta_g, \lambda)\}'\right]^2}{2B^2}\Bigg] + 1 + \frac{m_c E_0^2\{T_1(E_{F_{BLHDC}}, \eta_g, \lambda)\}''}{B^2}\Bigg] \end{aligned} \tag{18.66}$$

where, $E_{F_{BLHDC}}$ is the Fermi energy in this case.

The Landau energy $(E_{n_{l41}})$ can be written as

$$T_1(E_{n_{l41}}, \eta_g, \lambda) = \left(n+\frac{1}{2}\right)\hbar\omega_0 - \left\{\frac{m_c E_0^2\left[\{T_1(E_{n_{l41}}, \eta_g, \lambda)\}'\right]^2}{2B^2}\right\} \tag{18.67}$$

(ii) Similarly, the electron dispersion law in this case is given by

$$\begin{aligned} T_2(E, \eta_g, \lambda) = {} & \left(n+\frac{1}{2}\right)\hbar\omega_0 \\ & - \frac{E_0}{B}\hbar k_y\{T_1(E, \eta_g, \lambda)\}' - \frac{m_c E_0^2}{2B^2}\left[\{T_2(E, \eta_g, \lambda)\}'\right]^2 + \frac{[\hbar k_z(E)]^2}{2m_c} \end{aligned} \tag{18.68}$$

The use of (18.78) leads to the expressions of the EEM s' along z and y directions as

$$m_z^*(E_{F_{BLHDC}}, n, E_0, B, \lambda) = m_c\Bigg[\{T_2(E_{F_{BLHDC}}, \eta_g, \lambda)\}'' + \frac{m_c E_0^2\{T_2(E_{F_{BLHDC}}, \eta_g, \lambda)\}'\{T_2(E_{F_{BLHDC}}, \eta_g, \lambda)\}''}{B^2}\Bigg] \tag{18.69}$$

$$m_y^*(E_{F_{BLHDC}}, n, E_0, B, \lambda) = \left(\frac{B}{E_0}\right)^2 \left[\{T_2(E_{F_{BLHDC}}, \eta_g, \lambda)\}'\right]^{-1} \left[T_2(E_{F_{BLHDC}}, \eta_g, \lambda) - \left(n + \frac{1}{2}\right)\hbar\omega_0 + \frac{m_c E_0^2 \left[\{T_1(E_{F_{BLHDC}}, \eta_g, \lambda)\}'\right]^2}{2B^2}\right] \left[\frac{\{T_2(E_{F_{BLHDC}}, \eta_g, \lambda)\}''}{\left[\{T_2(E_{F_{BLHDC}}, \eta_g, \lambda)\}'\right]^2} \left[T_2(E_{F_{BLHDC}}, \eta_g, \lambda) - \left(n + \frac{1}{2}\right)\hbar\omega_0 + \frac{m_c E_0^2 \left[\{T_2(E_{F_{BLHDC}}, \eta_g, \lambda)\}'\right]^2}{2B^2}\right] + 1 + \frac{m_c E_0^2 \{T_2(E_{F_{BLHDC}}, \eta_g, \lambda)\}''}{B^2}\right] \tag{18.70}$$

The Landau energy $(E_{n_{l42}})$ can be written as

$$T_2(E_{n_{l42}}, \eta_g, \lambda) = \left(n + \frac{1}{2}\right)\hbar\omega_0 - \left\{\frac{m_c E_0^2 \left[\{T_2(E_{n_{l42}}, \eta_g, \lambda)\}'\right]^2}{2B^2}\right\} \tag{18.71}$$

(iii) Similarly, the electron dispersion law in this case is given by

$$T_3(E, \eta_g, \lambda) = \left(n + \frac{1}{2}\right)\hbar\omega_0 - \frac{E_0}{B}\hbar k_y \{T_3(E, \eta_g, \lambda)\}' - \frac{m_c E_0^2}{2B^2}\left[\{T_3(E, \eta_g, \lambda)\}'\right]^2 + \frac{[\hbar k_z(E)]^2}{2m_c} \tag{18.72}$$

The use of (18.72) leads to the expressions of the EEM s' along z and y directions as

$$m_z^*(E_{F_{BLHDC}}, n, E_0, B, \lambda) = m_c\left[\{T_3(E_{F_{BLHDC}}, \eta_g, \lambda)\}'' + \frac{m_c E_0^2 \{T_3(E_{F_{BLHDC}}, \eta_g, \lambda)\}'\{T_3(E_{F_{BLHDC}}, \eta_g, \lambda)\}''}{B^2}\right] \tag{18.73}$$

$$m_y^*(E_{F_{BLHDC}}, n, E_0, B, \lambda) = \left(\frac{B}{E_0}\right)^2 \left[\{T_3(E_{F_{BLHDC}}, \eta_g, \lambda)\}'\right]^{-1} \left[T_3(E_{F_{BLHDC}}, \eta_g, \lambda) - \left(n + \frac{1}{2}\right)\hbar\omega_0 + \frac{m_c E_0^2 \left[\{T_3(E_{F_{BLHDC}}, \eta_g, \lambda)\}'\right]^2}{2B^2}\right] \left[\frac{\{T_3(E_{F_{BLHDC}}, \eta_g, \lambda)\}''}{\left[\{T_3(E_{F_{BLHDC}}, \eta_g, \lambda)\}'\right]^2} \left[T_3(E_{F_{BLHDC}}, \eta_g, \lambda) - \left(n + \frac{1}{2}\right)\hbar\omega_0 + \frac{m_c E_0^2 \left[\{T_3(E_{F_{BLHDC}}, \eta_g, \lambda)\}'\right]^2}{2B^2}\right] + 1 + \frac{m_c E_0^2 \{T_3(E_{F_{BLHDC}}, \eta_g, \lambda)\}''}{B^2}\right] \tag{18.74}$$

The Landau energy $(E_{n_{l43}})$ can be written as

$$T_3\left(E_{n_{l42}}, \eta_g, \lambda\right) = \left(n + \frac{1}{2}\right)\hbar\omega_0 - \left\{\frac{m_c E_0^2\left[\left\{T_3\left(E_{n_{l42}}, \eta_g, \lambda\right)\right\}'\right]^2}{2B^2}\right\} \tag{18.75}$$

18.2.4 The DR in QWs of HD Kane Type Semiconductors in the Presence of Light Waves

(i) The 2D DR in QWs of HD III–V, ternary and quaternary materials, whose unperturbed band structure is defined by the three band model of Kane, in the presence of light waves can be expressed following (18.46b) as

$$\frac{\hbar^2 k_s^2}{2m_c} + \frac{\hbar^2}{2m_c}\left(\frac{n_z \pi}{d_z}\right)^2 = T_1\left(E, \eta_g, \lambda\right) \tag{18.76}$$

The sub band energies $(E_{n_{l7HD}})$ can be written as

$$T_1\left(E_{n_{l7HD}}, \eta_g, \lambda\right) = \frac{\hbar^2}{2m_c}\left(n_z \pi / d_z\right)^2 \tag{18.77}$$

The expression of the EEM in this case is given by

$$m^*(E_{F2DLHD}, n_z, \lambda) = m_c \text{ Real part of}\left\{T_1\left(E_{F2DLHD}, \eta_g, \lambda\right)\right\}' \tag{18.78}$$

where, E_{F2DLHD} is the Fermi energy in the present case as measured from the edge of the conduction band in the vertically upward direction in absence of any quantization.

The DOS function can be written as

$$N_{2D}(E, \eta_g, \lambda) = \left(\frac{m_c g_v}{\pi \hbar^2}\right) \sum_{n_z = 1}^{n_{z\max}} \left[T_1\left(E, \eta_g, \lambda\right)\right]' H(E - E_{n_{l7HD}}) \tag{18.79}$$

(ii) The 2D DR in QWs of HD III–V, ternary and quaternary materials, whose unperturbed band structure is defined by the two band model of Kane, in the presence of light waves can be expressed following (18.47) as

$$\frac{\hbar^2 k_s^2}{2m_c} + \frac{\hbar^2}{2m_c}\left(\frac{n_z \pi}{d_z}\right)^2 = T_2\left(E, \eta_g, \lambda\right) \tag{18.80}$$

The sub band energies $(E_{n_{l8HD}})$ can be written as

$$T_2\left(E_{n_{l8HD}}, \eta_g, \lambda\right) = \frac{\hbar^2}{2m_c}(n_z\pi/d_z)^2 \tag{18.81}$$

The expression of the EEM in this case is given by

$$m^*(E_{F2DLHD}, n_z, \lambda) = m_c\left\{T_2\left(E_{F2DLHD}, \eta_g, \lambda\right)\right\}' \tag{18.82}$$

The DOS function can be written as

$$N_{2D}(E, \eta_g, \lambda) = \left(\frac{m_c g_v}{\pi\hbar^2}\right)\sum_{n_z=1}^{n_{z\max}}\left[T_2\left(E, \eta_g, \lambda\right)\right]' H(E - E_{n_{l8HD}}) \tag{18.83}$$

(iii) The 2D DR in QWs of HD III–V, ternary and quaternary materials, whose unperturbed band structure is defined by the parabolic energy bands, in the presence of light waves can be expressed following (18.48a) as

$$\frac{\hbar^2 k_s^2}{2m_c} + \frac{\hbar^2}{2m_c}\left(\frac{n_z\pi}{d_z}\right)^2 = T_3\left(E, \eta_g, \lambda\right) \tag{18.84}$$

The sub band energies $(E_{n_{l9HD}})$ can be written as

$$T_2\left(E_{n_{l9HD}}, \eta_g, \lambda\right) = \frac{\hbar^2}{2m_c}(n_z\pi/d_z)^2 \tag{18.85}$$

The expression of the EEM in this case is given by

$$m^*(E_{F2DLHD}, n_z, \lambda) = m_c\left\{T_3\left(E_{F2DLHD}, \eta_g, \lambda\right)\right\}' \tag{18.86}$$

The DOS function can be written as

$$N_{2D}(E, \eta_g, \lambda) = \left(\frac{m_c g_v}{\pi\hbar^2}\right)\sum_{n_z=1}^{n_{z\max}}\left[T_3\left(E, \eta_g, \lambda\right)\right]' H(E - E_{n_{l9HD}}) \tag{18.87}$$

18.2.5 The DR in Doping Superlattices of HD Kane Type Semiconductors in the Presence of Light Waves

(i) The DR in doping superlattices of HD III–V, ternary and quaternary materials in the presence of external photo-excitation whose unperturbed electrons are defined by the three band model of Kane can be expressed following (18.46b) as

$$T_1(E,\eta_g,\lambda) = \left(n_i + \frac{1}{2}\right)\hbar\omega_{91HD}(E,\eta_g,\lambda) + \frac{\hbar^2 k_s^2}{2m_c} \tag{18.88}$$

where,

$$\omega_{91HD}(E,\eta_g,\lambda) \equiv \left(\frac{n_0|e|^2}{d_0\varepsilon_{sc}T_1'(E,\eta_g,\lambda)m_c}\right)^{1/2}.$$

The sub band energies $(E_{n_{i10HD}})$ can be written as

$$T_1\left(E_{n_{i10HD}},\eta_g,\lambda\right) = \left(n_i + \frac{1}{2}\right)\hbar\omega_{91HD}\left(E_{n_{i10HD}},\eta_g,\lambda\right) \tag{18.89}$$

The expression of the EEM in this case is given by

$$m^*(E_{F2DLHDD},\eta_g,\lambda,n_i) = m_c\{\mathrm{M}_{40HD}(E_{F2DLHDD},\eta_g,\lambda,n_i)\}' \tag{18.90}$$

where,

$$\mathrm{M}_{40HD}(E_{F2DLHDD},\eta_g,\lambda,n_i) = \text{Real part of}\left\{T_1(E_{F2DLHDD},\eta_g,\lambda) - \left(n_i + \frac{1}{2}\right)\hbar\omega_{91HD}(E_{F2DLHDD},\eta_g,\lambda)\right\}$$

and $E_{F2DLHDD}$ is the Fermi energy in the present case as measured from the edge of the conduction band in the vertically upward direction in absence of any quantization.

The DOS function can be written as

$$N_{2D}(E,\eta_g,\lambda) = \left(\frac{m_c g_v}{\pi\hbar^2}\right)\sum_{n_z=1}^{n_{z\max}}\left[\mathrm{M}_{40HD}(E,\eta_g,\lambda)\right]' H(E - E_{n_{i10HD}}) \tag{18.91}$$

(ii) The DR in doping superlattices of HD III–V, ternary and quaternary materials in the presence of external photo-excitation whose unperturbed electrons are defined by the two band model of Kane can be expressed following (18.47) as

$$T_2(E,\eta_g,\lambda) = \left(n_i + \frac{1}{2}\right)\hbar\omega_{92HD}(E,\eta_g,\lambda) + \frac{\hbar^2 k_s^2}{2m_c} \tag{18.92}$$

where

$$\omega_{92HD}(E,\eta_g,\lambda) \equiv \left(\frac{n_0|e|^2}{d_0\varepsilon_{sc}T_2'(E,\eta_g,\lambda)m_c}\right)^{1/2}.$$

The sub band energies $(E_{n_{i11HD}})$ can be written as

$$T_2\left(E_{n_{i11HD}}, \eta_g, \lambda\right) = \left(n_i + \frac{1}{2}\right)\hbar\omega_{92HD}\left(E_{n_{i11HD}}, \eta_g, \lambda\right) \tag{18.93}$$

The expression of the EEM in this case is given by

$$m^*(E_{F2DLHDD}, \eta_g, \lambda, n_i) = m_c\left\{\mathrm{M}_{41HD}\left(E_{F2DLHDD}, \eta_g, \lambda, n_i\right)\right\}' \tag{18.94}$$

where

$\mathrm{M}_{41HD}\left(E_{F2DLHDD}, \eta_g, \lambda, n_i\right) = \left\{T_2\left(E_{F2DLHDD}, \eta_g, \lambda\right) - \left(n_i + \frac{1}{2}\right)\hbar\omega_{92HD}\left(E_{F2DLHDD}, \eta_g, \lambda\right)\right\}$ and

The DOS function can be written as

$$N_{2D}(E, \eta_g, \lambda) = \left(\frac{m_c g_v}{\pi\hbar^2}\right)\sum_{n_z=1}^{n_{z\max}}\left[\mathrm{M}_{41HD}\left(E, \eta_g, \lambda\right)\right]' H(E - E_{n_{i11HD}}) \tag{18.95}$$

(iii) The DR in doping superlattices of HD III–V, ternary and quaternary materials in the presence of external photo-excitation whose unperturbed electrons are defined by the two band model of Kane can be expressed following (18.48a) as

$$T_3\left(E, \eta_g, \lambda\right) = \left(n_i + \frac{1}{2}\right)\hbar\omega_{93HD}\left(E, \eta_g, \lambda\right) + \frac{\hbar^2 k_s^2}{2m_c} \tag{18.96}$$

where

$$\omega_{93HD}\left(E, \eta_g, \lambda\right) \equiv \left(\frac{n_0|e|^2}{d_0\varepsilon_{sc}T_3'\left(E, \eta_g, \lambda\right)m_c}\right)^{1/2}.$$

The sub band energies $(E_{n_{i12HD}})$ can be written as

$$T_3\left(E_{n_{i12HD}}, \eta_g, \lambda\right) = \left(n_i + \frac{1}{2}\right)\hbar\omega_{93HD}\left(E_{n_{i12HD}}, \eta_g, \lambda\right) \tag{18.97}$$

The expression of the EEM in this case is given by

$$m^*(E_{F2DLHDD}, \eta_g, \lambda, n_i) = m_c\left\{\mathrm{M}_{42HD}\left(E_{F2DLHDD}, \eta_g, \lambda, n_i\right)\right\}' \tag{18.98}$$

where

$\mathrm{M}_{42HD}\left(E_{F2DLHDD}, \eta_g, \lambda, n_i\right) = \left\{T_3\left(E_{F2DLHDD}, \eta_g, \lambda\right) - \left(n_i + \frac{1}{2}\right)\hbar\omega_{93HD}\left(E_{F2DLHDD}, \eta_g, \lambda\right)\right\}$ and

The DOS function can be written as

$$N_{2D}(E,\eta_g,\lambda) = \left(\frac{m_c g_v}{\pi\hbar^2}\right)\sum_{n_z=1}^{n_{z\max}}\left[\mathrm{M}_{42HD}(E,\eta_g,\lambda)\right]' H(E - E_{n_{i12HD}}) \tag{18.99}$$

18.2.6 The DR of QDs of HD Kane Type Semiconductors in the Presence of Light Waves

(i) The DR in QDs of HD III–V, ternary and quaternary materials in the presence of external photo-excitation whose unperturbed electrons are defined by the three band model of Kane can be expressed following (18.46b) as

$$\frac{\hbar^2(\mathrm{n}_z\pi/\mathrm{d}_z)^2}{2m_c} + \frac{\hbar^2(\mathrm{n}_y\pi/\mathrm{d}_y)^2}{2m_c} + \frac{\hbar^2(\mathrm{n}_x\pi/\mathrm{d}_x)^2}{2m_c} = T_1(\mathrm{E}_{17,1},\eta_g,\lambda) \tag{18.100}$$

where $\mathrm{E}_{17,1}$ is the totally quantized energy in this case.

The DOS function is given by

$$N_{ODT}(E) = \frac{2g_v}{d_x d_y d_z}\sum_{n_x=1}^{n_{x\max}}\sum_{n_y=1}^{n_{y\max}}\sum_{n_z=1}^{n_{z\max}}\delta'(E - \mathrm{E}_{17,1}) \tag{18.101}$$

In the absence of band-tails, the totally quantized energy $(\mathrm{E}_{17,2})$ in this case is given by

$$\beta_0(\mathrm{E}_{17,2},\lambda) = \frac{\hbar^2\pi^2}{2m_c}\left[\left(\frac{n_x}{d_x}\right)^2 + \left(\frac{n_y}{d_y}\right)^2 + \left(\frac{n_z}{d_z}\right)^2\right] \tag{18.102}$$

The DOS function is given by

$$N_{OL}(E) = \frac{2g_v}{d_x d_y d_z}\sum_{n_x=1}^{n_{x\max}}\sum_{n_y=1}^{n_{y\max}}\sum_{n_z=1}^{n_{z\max}}\delta'(E - \mathrm{E}_{17,2}) \tag{18.103}$$

(ii) The DR in QDs of HD III–V, ternary and quaternary materials in the presence of external photo-excitation whose unperturbed electrons are defined by the two band model of Kane can be expressed following (18.47) as

$$\frac{\hbar^2(\mathrm{n}_z\pi/\mathrm{d}_z)^2}{2m_c} + \frac{\hbar^2(\mathrm{n}_y\pi/\mathrm{d}_y)^2}{2m_c} + \frac{\hbar^2(\mathrm{n}_x\pi/\mathrm{d}_x)^2}{2m_c} = T_2(\mathrm{E}_{17,3},\eta_g,\lambda) \tag{18.104}$$

where $\mathrm{E}_{17,3}$ is the totally quantized energy in this case

The DOS function is given by

$$N_{ODT}(E) = \frac{2g_v}{d_x d_y d_z} \sum_{n_x=1}^{n_{x\max}} \sum_{n_y=1}^{n_{y\max}} \sum_{n_z=1}^{n_{z\max}} \delta'\left(E - \mathrm{E}_{17,3}\right) \tag{18.105}$$

In the absence of band-tails, the totally quantized energy $\left(\mathrm{E}_{17,4}\right)$ in this case is given by

$$\tau_0\left(\mathrm{E}_{17,4}, \lambda\right) = \frac{\hbar^2\pi^2}{2m_c}\left[\left(\frac{n_x}{d_x}\right)^2 + \left(\frac{n_y}{d_y}\right)^2 + \left(\frac{n_z}{d_z}\right)^2\right] \tag{18.106}$$

The DOS function is given by

$$N_{OL}(E) = \frac{2g_v}{d_x d_y d_z} \sum_{n_x=1}^{n_{x\max}} \sum_{n_y=1}^{n_{y\max}} \sum_{n_z=1}^{n_{z\max}} \delta'\left(E - \mathrm{E}_{17,4}\right) \tag{18.107}$$

(iii) The DR in QDs of HD III–V, ternary and quaternary materials in the presence of external photo-excitation whose unperturbed electrons are defined by the parabolic energy bands can be expressed as

$$\frac{\hbar^2(\mathrm{n}_z\pi/\mathrm{d}_z)^2}{2m_c} + \frac{\hbar^2(\mathrm{n}_y\pi/\mathrm{d}_y)^2}{2m_c} + \frac{\hbar^2(\mathrm{n}_x\pi/\mathrm{d}_x)^2}{2m_c} = T_3(\mathrm{E}_{17,5}, \eta_g, \lambda) \tag{18.108}$$

where $E_{17,5}$ is the totally quantized energy in this case.

The DOS function is given by

$$N_{ODT}(E) = \frac{2g_v}{d_x d_y d_z} \sum_{n_x=1}^{n_{x\max}} \sum_{n_y=1}^{n_{y\max}} \sum_{n_z=1}^{n_{z\max}} \delta'\left(E - \mathrm{E}_{17,5}\right) \tag{18.109}$$

In the absence of band-tails, the totally quantized energy $\left(\mathrm{E}_{17,6}\right)$ in this case is given by

$$\rho_0\left(\mathrm{E}_{17,6}, \lambda\right) = \frac{\hbar^2\pi^2}{2m_c}\left[\left(\frac{n_x}{d_x}\right)^2 + \left(\frac{n_y}{d_y}\right)^2 + \left(\frac{n_z}{d_z}\right)^2\right] \tag{18.110}$$

The DOS function is given by

$$N_{OL}(E) = \frac{2g_v}{d_x d_y d_z} \sum_{n_x=1}^{n_{x\max}} \sum_{n_y=1}^{n_{y\max}} \sum_{n_z=1}^{n_{z\max}} \delta'\left(E - \mathrm{E}_{17,6}\right) \tag{18.111}$$

18.2.7 The Magneto DR in QWs of HD Kane Type Semiconductors in the Presence of Light Waves

(i) The magneto DR in QWs of HD III–V, ternary and quaternary materials, whose unperturbed band structure is defined by the three band model of Kane, in the presence of light waves can be expressed following (18.76) as

$$\left(n+\frac{1}{2}\right)\hbar\omega_0+\frac{\hbar^2}{2m_c}\left(\frac{n_z\pi}{d_z}\right)^2=T_1(\mathrm{E}_{17,7},\eta_g,\lambda) \tag{18.112}$$

where $E_{17,7}$ is the totally quantized energy in this case.

The DOS function is given by

$$N_{ODT}(E)=\frac{2g_v}{d_xd_yd_z}\sum_{n_x=1}^{n_{x\max}}\sum_{n_y=1}^{n_{y\max}}\sum_{n_z=1}^{n_{z\max}}\delta'(E-\mathrm{E}_{17,7}) \tag{18.113}$$

(ii) The 2D magneto DR in QWs of HD III–V, ternary and quaternary materials, whose unperturbed band structure is defined by the two band model of Kane, in the presence of light waves can be expressed following (18.47) as

$$\left(n+\frac{1}{2}\right)\hbar\omega_0+\frac{\hbar^2}{2m_c}\left(\frac{n_z\pi}{d_z}\right)^2=T_2(\mathrm{E}_{17,8},\eta_g,\lambda) \tag{18.114}$$

where $\mathrm{E}_{17,8}$ is the totally quantized energy in this case.

The DOS function is given by

$$N_{ODT}(E)=\frac{2g_v}{d_xd_yd_z}\sum_{n_x=1}^{n_{x\max}}\sum_{n_y=1}^{n_{y\max}}\sum_{n_z=1}^{n_{z\max}}\delta'(E-\mathrm{E}_{17,8}) \tag{18.115}$$

(iii) The 2D magneto DR in QWs of HD III–V, ternary and quaternary materials, whose unperturbed band structure is defined by the parabolic energy bands, in the presence of light waves can be expressed following (18.48a) as

$$\left(n+\frac{1}{2}\right)\hbar\omega_0+\frac{\hbar^2}{2m_c}\left(\frac{n_z\pi}{d_z}\right)^2=T_3(\mathrm{E}_{17,9},\eta_g,\lambda) \tag{18.116}$$

where $\mathrm{E}_{17,9}$ is the totally quantized energy in this case

The DOS function is given by

$$N_{ODT}(E)=\frac{2g_v}{d_xd_yd_z}\sum_{n_x=1}^{n_{x\max}}\sum_{n_y=1}^{n_{y\max}}\sum_{n_z=1}^{n_{z\max}}\delta'(E-\mathrm{E}_{17,9}) \tag{18.117}$$

18.2.8 The DR in Accumulation and Inversion Layers of Kane Type Semiconductors in the Presence of Light Waves

(a) The 2D DR in accumulation layers of HD III–V, ternary and quaternary materials, whose unperturbed band structure is defined by the two band model of Kane, in the presence of light waves can be expressed following (18.46b) as

$$T_1(\mathrm{E},\eta_g,\lambda)=\frac{\hbar^2k_s^2}{2m_c}+S_i\left[\frac{\hbar|e|\mathrm{F}_s[T_1(\mathrm{E},\eta_g,\lambda)]'}{\sqrt{2m_c}}\right]^{\frac{2}{3}} \tag{18.118}$$

(18.118) represents the DR of the 2D electrons in accumulation layers of HD III–V, ternary and quaternary materials under the weak electric field limit in the presence of light waves whose bulk electrons obey the HD three band model of Kane. Since the DR in accordance with the HD three-band model of Kane is complex in nature, the (18.118) will also be complex. The both complexities occur due to the presence of poles in the finite complex plane of the dispersion relation of the materials in the absence of band tails.

The EEM can be expressed as

$$m_L^*(\mathrm{E}'_{fl},i,\eta_g)=m_c\ \text{Real part of}\ P'_{3HDL}(\mathrm{E}'_{fL},i,\eta_g,\lambda) \tag{18.119}$$

where

$$P_{3HDL}(\mathrm{E}'_{fl},i,\eta_g,\lambda)=\left[T_1(\mathrm{E},\eta_g,\lambda)-S_i\left[\frac{\hbar|e|\mathrm{F}_s[T_1(\mathrm{E},\eta_g,\lambda)]'}{\sqrt{2m_c}}\right]^{\frac{2}{3}}\right]$$

and E'_{fl} is the Fermi energy in this case.

Thus, one can observe that the EEM is a function of light intensity, scattering potential, the sub-band index, surface electric field, the Fermi energy and the other spectrum constants due to the combined influence of E_g and Δ.

The sub-band energy E_{i1L} is given by

$$0=\text{Real part of}\left[T_1(E_{i1L},\eta_g,\lambda)-S_i\left[\frac{\hbar|e|\mathrm{F}_s[T_1(E_{i1L},\eta_g,\lambda)]'}{\sqrt{2m_c}}\right]^{\frac{2}{3}}\right] \tag{18.120}$$

The DOS function can be written as

$$N_{2DiL}(E)=\frac{m_cg_v}{\pi\hbar^2}\sum_{i=0}^{i_{\max}}\left[P_{3HDL}(E,i,\eta_g,\lambda)H(E-E_{i1L})\right] \tag{18.121}$$

Thus the DOS function is complex in nature.

In the absence of band-tails and under the condition of weak electric field limit, (18.118) assumes the form

$$\beta_0(E,\lambda) = \frac{\hbar^2 k_s^2}{2m_c} + S_i\left[\frac{\hbar|e|F_s[\beta_0(E,\lambda)]'}{\sqrt{2m_c}}\right]^{2/3} \tag{18.122}$$

(18.122) represents the DR of the 2D electrons in inversion layers of III–V, ternary and quaternary materials under the weak electric field limit in the present of light waves waves whose bulk electrons in the absence of any perturbation obey the three band model of Kane.

The EEM can be expressed as

$$m_L^*(E_{Fiwl},i,\lambda) = m_c[P_{3L}(E,i,\lambda)]'|_{E=E_{FiwL}} \tag{18.123}$$

where,

$$P_{3L}(E,i,\lambda) = \left[\beta_0(E,\lambda) - S_i\left[\frac{\hbar|e|F_s[\beta_0(E,\lambda)]'}{\sqrt{2m_c}}\right]^{2/3}\right]$$

Thus, one can observe that the EEM is a function of the subband index, the light intensity, surface electric field, the Fermi energy and the other spectrum constants due to the combined influence of E_g and Δ.

The subband energy $E_{n_{iw2L}}$ in this case can be obtained from the (18.123) as

$$0 = \left[\beta_0(E_{n_{iw2L}},\lambda) - S_i\left[\frac{\hbar|e|F_s[\beta_0(E_{n_{iw2L}},\lambda)]'}{\sqrt{2m_c}}\right]^{2/3}\right] \tag{18.124}$$

Thus the 2D total DOS function in weak electric field limit can be expressed as

$$N_{2D_{iL}}(E) = \frac{m_c g_v}{\pi\hbar^2}\sum_{i=0}^{i_{\max}}[P_{3L}(E,i,\lambda)H(E-E_{n_{iwL}})] \tag{18.125}$$

(b) The 2D DR in accumulation layers of HD III–V, ternary and quaternary materials, whose unperturbed band structure is defined by the two band model of Kane in the absence of any field, in the presence of light waves can be expressed following (18.47) as

$$T_2(\mathrm{E},\eta_g,\lambda) = \frac{\hbar^2 k_s^2}{2m_c} + S_i\left[\frac{\hbar|e|F_s[T_2(\mathrm{E},\eta_g,\lambda)]'}{\sqrt{2m_c}}\right]^{\frac{2}{3}} \tag{18.126}$$

(18.126) represents the DR of the 2D electrons in accumulation layers of HD III–V, ternary and quaternary materials under the weak electric field limit in the presence of light waves whose bulk electrons obey the HD two band model of Kane. Since the electron energy spectrum in accordance with the HD two-band model of Kane is real in nature, the (18.126) will also be real.

The EEM can be expressed as

$$m_L^*(E'_{fl}, i, \eta_g) = m_c P'_{3HDL1}(E'_{fl}, i, \eta_g, \lambda) \tag{18.127}$$

where,

$$P_{3HDL1}(E'_{fl}, i, \eta_g, \lambda) = \left[T_2(E, \eta_g, \lambda) - S_i \left[\frac{\hbar|e|F_s[T_2(E, \eta_g, \lambda)]'}{\sqrt{2m_c}}\right]^{\frac{2}{3}}\right]$$

Thus, one can observe that the EEM is a function of light intensity, scattering potential, the sub-band index, surface electric field, the Fermi energy and the other spectrum constants due to the combined influence of E_g and Δ.

The sub-band energy E_{i1L1} is given by

$$0 = \left[T_2(E_{i1L1}, \eta_g, \lambda) - S_i \left[\frac{\hbar|e|F_s[T_2(E_{i1L1}, \eta_g, \lambda)]'}{\sqrt{2m_c}}\right]^{\frac{2}{3}}\right] \tag{18.128}$$

The DOS function can be written as

$$N_{2DiL}(E) = \frac{m_c g_v}{\pi\hbar^2} \sum_{i=0}^{i_{\max}} \left[P_{3HDL1}(E, i, \eta_g, \lambda) H(E - E_{i1L1})\right] \tag{18.129}$$

In the absence of band-tails and under the condition of weak electric field limit, (18.126) assumes the form

$$\tau_0(E, \lambda) = \frac{\hbar^2 k_s^2}{2m_c} + S_i \left[\frac{\hbar|e|F_s[\tau_0(E, \lambda)]'}{\sqrt{2m_c}}\right]^{2/3} \tag{18.130}$$

(18.130) represents the DR of the 2D electrons in inversion layers of III–V, ternary and quaternary materials under the weak electric field limit in the present of light waves waves whose bulk electrons in the absence of any perturbation obey the two band model of Kane.

The EEM can be expressed as

$$m_L^*(E_{FiwL}, i, \lambda) = m_c[P_{3L2}(E, i, \lambda)]'|_{E=E_{FiwL}} \tag{18.131}$$

where,

$$P_{3L2}(E, i, \lambda) = \left[\tau_0(E, \lambda) - S_i\left[\frac{\hbar|e|F_s[\tau_0(E, \lambda)]'}{\sqrt{2m_c}}\right]^{2/3}\right]$$

Thus, one can observe that the EEM is a function of the subband index, the light intensity, surface electric field, the Fermi energy and the other spectrum constants due to the combined influence of E_g and Δ.

The subband energy $E_{n_{iw2L2}}$ in this case can be obtained from the (18.130) as

$$0 = \left[\tau_0(E_{n_{iw2L2}}, \lambda) - S_i\left[\frac{\hbar|e|F_s[\tau_0(E_{n_{iw2L2}}, \lambda)]'}{\sqrt{2m_c}}\right]^{2/3}\right] \tag{18.132}$$

Thus the 2D total DOS function in weak electric field limit can be expressed as

$$N_{2D_{iL}}(E) = \frac{m_c g_v}{\pi\hbar^2}\sum_{i=0}^{i_{\max}}[P_{3L2}(E, i, \lambda)]H(E - E_{n_{iwL2}}) \tag{18.133}$$

(c) The 2D DR in accumulation layers of HD III–V, ternary and quaternary materials, whose unperturbed band structure is defined by the two band model of Kane in the absence of any field, in the presence of light waves can be expressed following (18.48a) as

$$T_3(\mathrm{E}, \eta_g, \lambda) = \frac{\hbar^2 k_s^2}{2m_c} + S_i\left[\frac{\hbar|\mathrm{e}|\mathrm{F}_s[T_3(\mathrm{E}, \eta_g, \lambda)]'}{\sqrt{2m_c}}\right]^{\frac{2}{3}} \tag{18.134}$$

(18.134) represents the DR of the 2D electrons in accumulation layers of HD III–V, ternary and quaternary materials under the weak electric field limit in the presence of light waves whose bulk electrons obey the HD model of parabolic energy bands. Since the electron energy spectrum in this case is real in nature, the (18.134) will also be real.

The EEM can be expressed as

$$m_L^*(\mathrm{E}_{fl}', i, \eta_g) = \mathrm{m}_c P_{3HDL2}'(\mathrm{E}_{fl}', i, \eta_g, \lambda) \tag{18.135}$$

where,

$$P_{3HDL2}(E'_{fl}, i, \lambda) = \left[T_3(E, \eta_g, \lambda) - S_i \left[\frac{\hbar |e| F_s [T_3(E, \eta_g, \lambda)]'}{\sqrt{2m_c}} \right]^{\frac{2}{3}} \right]$$

Thus, one can observe that the EEM is a function of light intensity, scattering potential, the sub-band index, surface electric field, the Fermi energy and the other spectrum constants due to the combined influence of E_g and Δ.

The sub-band energy E_{i1L2} is given by

$$0 = \left[T_3(E_{i1L2}, \eta_g, \lambda) - S_i \left[\frac{\hbar |e| F_s [T_3(E_{i1L2}, \eta_g, \lambda)]'}{\sqrt{2m_c}} \right]^{\frac{2}{3}} \right] \tag{18.136}$$

The DOS function can be written as

$$N_{2DiL}(E) = \frac{m_c g_v}{\pi \hbar^2} \sum_{i=0}^{i_{max}} \left[P_{3HDL2}(E, i, \eta_g, \lambda) \right] H(E - E_{i1L2}) \tag{18.137}$$

In the absence of band-tails and under the condition of weak electric field limit, (18.134) assumes the form

$$\rho_0(E, \lambda) = \frac{\hbar^2 k_s^2}{2m_c} + S_i \left[\frac{\hbar |e| F_s [\rho_0(E, \lambda)]'}{\sqrt{2m_c}} \right]^{2/3} \tag{18.138}$$

(18.138) represents the DR of the 2D electrons in inversion layers of III–V, ternary and quaternary materials under the weak electric field limit in the present of light waves waves whose bulk electrons in the absence of any perturbation obey the model of isotropic parabolic energy bands.

The EEM can be expressed as

$$m_L^*(E_{FiwL}, i, \lambda) = m_c [P_{3L3}(E, i, \lambda)]' \big|_{E=E_{FiwL}} \tag{18.139}$$

where,

$$P_{3L3}(E, i, \lambda) = \left[\rho_0(E, \lambda) - S_i \left[\frac{\hbar |e| F_s [\rho_0(E, \lambda)]'}{\sqrt{2m_c}} \right]^{2/3} \right]$$

Thus, one can observe that the EEM is a function of the subband index, the light intensity, surface electric field, the Fermi energy and the other spectrum constants due to the combined influence of E_g and Δ.

The subband energy $E_{n_{iw2L3}}$ in this case can be obtained from the (18.138) as

$$0 = \left[\rho_0(E_{n_{iw2L3}}, \lambda) - S_i\left[\frac{\hbar|e|F_s[\rho_0(E_{n_{iw2L3}}, \lambda)]'}{\sqrt{2m_c}}\right]^{2/3}\right] \quad (18.140)$$

Thus the 2D total DOS function in weak electric field limit can be expressed as

$$N_{2D_{iL}}(E) = \frac{m_c g_v}{\pi\hbar^2}\sum_{i=0}^{i_{\max}}[P_{3L3}(E, i, \lambda)]H(E - E_{n_{iwL3}}) \quad (18.141)$$

18.2.9 The DR in NWs of HD Kane Type Semiconductors in the Presence of Light Waves

(a) The 1D DR in NWs of HD III–V, ternary and quaternary materials, whose unperturbed band structure is defined by the three band model of Kane in the absence of any field, in the presence of light waves can be expressed following (18.46b) as

$$\frac{\hbar^2(n_z\pi/d_z)^2}{2m_c} + \frac{\hbar^2(n_y\pi/d_y)^2}{2m_c} + \frac{\hbar^2 k_x^2}{2m_c} = T_1(E, \eta_g, \lambda) \quad (18.142)$$

The DOS function in this case assumes the form

$$N_{1DHD\Gamma L}(E, \eta_g) = \frac{2g_v}{\pi}\sum_{n_y=1}^{n_{y\max}}\sum_{n_z=1}^{n_{z\max}} T'_{3L1}(E, n_y, n_x, \eta_g, \lambda)H(E - E'_{3HDNWL1}) \quad (18.143)$$

where

$$T_{3L1}(E, n_y, n_z, \eta_g, \lambda) = \left[\left[T_1(E, \eta_g, \lambda) - \left[\frac{\hbar^2(n_z\pi/d_z)^2}{2m_c} + \frac{\hbar^2(n_y\pi/d_y)^2}{2m_c}\right]\right]\frac{2m_c}{\hbar^2}\right]^{\frac{1}{2}}$$

In (18.143), $E'_{3HDNWL1}$ is the sub-band energy in this case which can be expressed as

$$\frac{\hbar^2(n_z\pi/d_z)^2}{2m_c} + \frac{\hbar^2(n_y\pi/d_y)^2}{2m_c} = T_1(E'_{3HDNWL1}, \eta_g, \lambda) \quad (18.144)$$

The EEM in this case is given by

$$m^*\left(\mathrm{E}_{F1HDNWL1,\eta_g,\lambda}\right) = m_c \text{ Real part of}\left[T_1'\left(\mathrm{E}_{F1HDNWL1}, \eta_g, \lambda\right)\right] \tag{18.145}$$

The 1D DR, for NWs of III–V materials whose energy band structures are defined by the three-band model of Kane in the absence of band tailing assumes the form

$$\frac{\hbar^2(\mathrm{n}_z\pi/\mathrm{d}_z)^2}{2m_c} + \frac{\hbar^2(\mathrm{n}_y\pi/\mathrm{d}_y)^2}{2m_c} + \frac{\hbar^2 k_x^2}{2m_c} = \beta_0(\mathrm{E}, \lambda) \tag{18.146}$$

The DOS function can be expressed as

$$N_{1D\Gamma L}(\mathrm{E}) = \frac{2g_v}{\pi}\sum_{n_y=1}^{n_{y\max}}\sum_{n_z=1}^{n_{z\max}} f_{12L1}'(\mathrm{E}, \mathrm{n}_y, \mathrm{n}_z, \lambda)\mathrm{H}(\mathrm{E} - \mathrm{E}_{3L1}') \tag{18.147}$$

where

$$f_{12L1}(\mathrm{E}, \mathrm{n}_y, \mathrm{n}_z, \lambda) = \left[\left[\beta_0(\mathrm{E}, \lambda) - \left[\frac{\hbar^2(\mathrm{n}_z\pi/\mathrm{d}_z)^2}{2m_c} + \frac{\hbar^2(\mathrm{n}_y\pi/\mathrm{d}_y)^2}{2m_c}\right]\right]\frac{2m_c}{\hbar^2}\right]^{\frac{1}{2}}$$

and E_{3L1}' is the sub-band energy in this case which can be obtained from the following equation

$$\frac{\hbar^2(\mathrm{n}_z\pi/\mathrm{d}_z)^2}{2m_c} + \frac{\hbar^2(\mathrm{n}_y\pi/\mathrm{d}_y)^2}{2m_c} + \frac{\hbar^2 k_x^2}{2m_c} = \beta_0(E_{3L1}', \lambda) \tag{18.148}$$

The EEM in this case is given by

$$m^*(\mathrm{E}_{F1HDNWL2}, \lambda) = \mathrm{m}_c\left[\beta_0'(\mathrm{E}_{F1HDNWL2}, \lambda)\right] \tag{18.149}$$

where $\mathrm{E}_{F1HDNWL2}$ is the Fermi energy in this case.

(b) The 1D DR in NWs of HD III–V, ternary and quaternary materials, whose unperturbed band structure is defined by the two band model of Kane in the absence of any field, in the presence of light waves can be expressed following (18.47) as

$$\frac{\hbar^2(\mathrm{n}_z\pi/\mathrm{d}_z)^2}{2m_c} + \frac{\hbar^2(\mathrm{n}_y\pi/\mathrm{d}_y)^2}{2m_c} + \frac{\hbar^2 k_x^2}{2m_c} = T_2(\mathrm{E}, \eta_g, \lambda) \tag{18.150}$$

The DOS function in this case assumes the form

$$N_{1DHD\Gamma L}(\mathrm{E},\eta_g) = \frac{2g_v}{\pi}\sum_{n_y=1}^{n_{y\max}}\sum_{n_z=1}^{n_{z\max}} T'_{3L2}(\mathrm{E},\mathrm{n}_y,\mathrm{n}_z,\eta_g,\lambda)\mathrm{H}(\mathrm{E}-\mathrm{E}'_{3HDNWL2}) \quad (18.151)$$

where

$$T_{3L2}(\mathrm{E},\mathrm{n}_y,\mathrm{n}_z,\eta_g,\lambda) = \left[\left[T_2(\mathrm{E},\eta_g,\lambda) - \left[\frac{\hbar^2(\mathrm{n}_z\pi/\mathrm{d}_z)^2}{2m_c} + \frac{\hbar^2(\mathrm{n}_y\pi/\mathrm{d}_y)^2}{2m_c}\right]\right]\frac{2m_c}{\hbar^2}\right]^{\frac{1}{2}}$$

In (18.151), $E'_{3HDNWL2}$ is the sub-band energy in this case which can be expressed as

$$\frac{\hbar^2(\mathrm{n}_z\pi/\mathrm{d}_z)^2}{2m_c} + \frac{\hbar^2(\mathrm{n}_y\pi/\mathrm{d}_y)^2}{2m_c} = T_2(E'_{3HDNWL2},\eta_g,\lambda) \quad (18.152)$$

The EEM in this case is given by

$$m^*(\mathrm{E}_{F1HDNWL2},\eta_g,\lambda) = \mathrm{m}_c\left[T'_2(\mathrm{E}_{F1HDNWL2},\eta_g,\lambda)\right] \quad (18.153)$$

The 1D DR, for NWs of III–V materials whose energy band structures are defined by the two-band model of Kane in the absence of band tailing assumes the form

$$\frac{\hbar^2(\mathrm{n}_z\pi/\mathrm{d}_z)^2}{2m_c} + \frac{\hbar^2(\mathrm{n}_y\pi/\mathrm{d}_y)^2}{2m_c} + \frac{\hbar^2k_x^2}{2m_c} = \tau_0(\mathrm{E},\lambda) \quad (18.154)$$

The DOS function can be expressed as

$$N_{1D\Gamma L}(\mathrm{E}) = \frac{2g_v}{\pi}\sum_{n_y=1}^{n_{y\max}}\sum_{n_z=1}^{n_{z\max}} f'_{12L2}(\mathrm{E},\mathrm{n}_y,\mathrm{n}_z,\lambda)\mathrm{H}(\mathrm{E}-\mathrm{E}'_{3L2}) \quad (18.155)$$

where

$$f_{12L2}(\mathrm{E},\mathrm{n}_y,\mathrm{n}_z,\lambda) = \left[\left[\tau_0(\mathrm{E},\lambda) - \left[\frac{\hbar^2(\mathrm{n}_z\pi/\mathrm{d}_z)^2}{2m_c} + \frac{\hbar^2(\mathrm{n}_y\pi/\mathrm{d}_y)^2}{2m_c}\right]\right]\frac{2m_c}{\hbar^2}\right]^{\frac{1}{2}}$$

and E'_{3L2} is the sub-band energy in this case which can be obtained from the following equation

$$\frac{\hbar^2(\mathrm{n}_z\pi/\mathrm{d}_z)^2}{2m_c}+\frac{\hbar^2(\mathrm{n}_y\pi/\mathrm{d}_y)^2}{2m_c}+\frac{\hbar^2k_x^2}{2m_c}=\tau_0(\mathrm{E}'_{3L2},\lambda) \tag{18.156}$$

The EEM in this case is given by

$$m^*(\mathrm{E}_{F1HDNWL2},\lambda)=\mathrm{m}_c\left[\tau'_2(\mathrm{E}_{F1HDNWL2},\lambda)\right] \tag{18.157}$$

where $\mathrm{E}_{F1HDNWL2}$ is the Fermi energy in this case.

(c) The 1D DR in NWs of HD III–V, ternary and quaternary materials, whose unperturbed band structure is defined by the parabolic energy bands in the absence of any field, in the presence of light waves can be expressed following (18.48a) as

$$\frac{\hbar^2(\mathrm{n}_z\pi/\mathrm{d}_z)^2}{2m_c}+\frac{\hbar^2(\mathrm{n}_y\pi/\mathrm{d}_y)^2}{2m_c}+\frac{\hbar^2k_x^2}{2m_c}=T_3(\mathrm{E},\eta_g,\lambda) \tag{18.158}$$

The DOS function in this case assumes the form

$$N_{1DHD\Gamma L}(\mathrm{E},\eta_g)=\frac{2g_v}{\pi}\sum_{n_y=1}^{n_{y\max}}\sum_{n_z=1}^{n_{z\max}}T'_{3L3}(\mathrm{E},\mathrm{n}_y,\mathrm{n}_z,\eta_g,\lambda)\mathrm{H}(\mathrm{E}-\mathrm{E}'_{3HDNWL3}) \tag{18.159}$$

where

$$T_{3L3}(\mathrm{E},\mathrm{n}_y,\mathrm{n}_z,\lambda)=\left[\left[T_3(\mathrm{E},\eta_g,\lambda)-\left[\frac{\hbar^2(\mathrm{n}_z\pi/\mathrm{d}_z)^2}{2m_c}+\frac{\hbar^2(\mathrm{n}_y\pi/\mathrm{d}_y)^2}{2m_c}\right]\right]\frac{2m_c}{\hbar^2}\right]^{\frac{1}{2}}$$

In (18.159), $E'_{3HDNWL3}$ is the sub-band energy in this case which can be expressed as

$$\frac{\hbar^2(\mathrm{n}_z\pi/\mathrm{d}_z)^2}{2m_c}+\frac{\hbar^2(\mathrm{n}_y\pi/\mathrm{d}_y)^2}{2m_c}=T_3(E'_{3HDNWL3},\eta_g,\lambda) \tag{18.160}$$

The EEM in this case is given by

$$m^*(\mathrm{E}_{F1HDNWL2},\eta_g,\lambda)=\mathrm{m}_c\left[T'_3(\mathrm{E}_{F1HDNWL2},\eta_g,\lambda)\right] \tag{18.161}$$

The 1D DR, for NWs of III–V materials whose energy band structures are defined by the parabolic energy bands in the absence of band tailing assumes the form

$$\frac{\hbar^2(n_z\pi/d_z)^2}{2m_c}+\frac{\hbar^2(n_y\pi/d_y)^2}{2m_c}+\frac{\hbar^2k_x^2}{2m_c}=\rho_0(E,\lambda) \tag{18.162}$$

The DOS function can be expressed as

$$N_{1D\Gamma L}(E)=\frac{2g_v}{\pi}\sum_{n_y=1}^{n_{ymax}}\sum_{n_z=1}^{n_{zmax}}f'_{12L3}(E,n_y,n_z,\lambda)H(E-E'_{3L3}) \tag{18.163}$$

where

$$f_{12L3}(E,n_y,n_z,\lambda)=\left[\left[\rho_0(E,\lambda)-\left[\frac{\hbar^2(n_z\pi/d_z)^2}{2m_c}+\frac{\hbar^2(n_y\pi/d_y)^2}{2m_c}\right]\right]\frac{2m_c}{\hbar^2}\right]^{\frac{1}{2}}$$

and E'_{3L3} is the sub-band energy in this case which can be obtained from the following equation

$$\frac{\hbar^2(n_z\pi/d_z)^2}{2m_c}+\frac{\hbar^2(n_y\pi/d_y)^2}{2m_c}+\frac{\hbar^2k_x^2}{2m_c}=\rho_0(E'_{3L3},\lambda) \tag{18.164}$$

The EEM in this case is given by

$$m^*(E_{F1HDNWL2},\lambda)=m_c\left[\rho'_3(E_{F1HDNWL2},\lambda)\right] \tag{18.165}$$

18.2.10 The Magneto DR in Accumulation and Inversion Layers of Kane Type Semiconductors in the Presence of Light Waves

(a) The 2D DR in accumulation layers of HD III–V, ternary and quaternary materials, whose unperturbed band structure is defined by the three band model of Kane, in the presence of light waves under magnetic quantization can be expressed following (18.118) as

$$T_1(E_{17,30},\eta_g,\lambda)=\left(n+\frac{1}{2}\right)\hbar\omega_0+S_i\left[\frac{\hbar|e|F_s[T_1(E_{17,30},\eta_g,\lambda)]'}{\sqrt{2m_c}}\right]^{\frac{2}{3}} \tag{18.166}$$

where $E_{17,30}$ is the totally quantized energy in this case.

(18.166) represents the magneto DR of the 2D electrons in accumulation layers of HD III–V, ternary and quaternary materials under the weak electric field limit in the presence of light waves whose bulk electrons obey the HD three band model of

Kane. Since the DR in accordance with the HD three-band model of Kane is complex in nature, the (18.166) will also be complex in the energy plane. The tolal energy is quantized since the wave vector space is totally quantized.

The DOS function is given by

$$N_{2DALB}(E) = \frac{g_v eB}{\pi\hbar} \sum_{n=0}^{n_{\max}} \sum_{i=0}^{i_{\max}} \delta'(E - E_{17,30}) \quad (18.167)$$

In the absence of band-tails and under the condition of weak electric field limit, (18.166) assumes the form

$$\beta_0(E_{17,31}, \lambda) = \left(n + \frac{1}{2}\right)\hbar\omega_0 + S_i \left[\frac{\hbar|e|F_s[\beta_0(E_{17,31}, \lambda)]'}{\sqrt{2m_c}}\right]^{2/3} \quad (18.168)$$

where $E_{17,31}$ is the totally quantized energy in this case.

The DOS function is given by

$$N_{2DALB}(E) = \frac{g_v eB}{\pi\hbar} \sum_{n=0}^{n_{\max}} \sum_{i=0}^{i_{\max}} \delta'(E - E_{17,31}) \quad (18.169)$$

(b) The 2D DR in accumulation layers of HD III–V, ternary and quaternary materials, whose unperturbed band structure is defined by the two band model of Kane, in the presence of light waves under magnetic quantization can be expressed as

$$T_2(E_{17,32}, \eta_g, \lambda) = \left(n + \frac{1}{2}\right)\hbar\omega_0 + S_i \left[\frac{\hbar|e|F_s[T_2(E_{17,32}, \eta_g, \lambda)]'}{\sqrt{2m_c}}\right]^{\frac{2}{3}} \quad (18.170)$$

where $E_{17,32}$ is the totally quantized energy in this case.

(18.170) represents the magneto DR of the 2D electrons in accumulation layers of HD III–V, ternary and quaternary materials under the weak electric field limit in the presence of light waves whose bulk electrons obey the HD two band model of Kane. Since the DR in accordance with the HD two-band model of Kane is real in nature, the (18.170) will also be real in the energy plane. The total energy is quantized since the wave vector space is totally quantized.

The DOS function is given by

$$N_{2DALB}(E) = \frac{g_v eB}{\pi\hbar} \sum_{n=0}^{n_{\max}} \sum_{i=0}^{i_{\max}} \delta'(E - E_{17,32}) \quad (18.171)$$

In the absence of band-tails and under the condition of weak electric field limit, (18.170) assumes the form

$$\tau_0(E_{17,33},\lambda) = \left(n+\frac{1}{2}\right)\hbar\omega_0 + S_i\left[\frac{\hbar|e|F_s[\tau_0(E_{17,33},\lambda)]'}{\sqrt{2m_c}}\right]^{2/3} \quad (18.172)$$

where $E_{17,33}$ is the totally quantized energy in this case.

The DOS function is given by

$$N_{2DALB}(E) = \frac{g_v eB}{\pi\hbar}\sum_{n=0}^{n_{\max}}\sum_{i=0}^{i_{\max}}\delta'(E - E_{17,33}) \quad (18.173)$$

(c) The 2D magneto DR in accumulation layers of HD III–V, ternary and quaternary materials, whose unperturbed band structure is defined by the parabolic energy bands, in the presence of light waves under magnetic quantization can be expressed as

$$T_3(\mathrm{E}_{17,34},\eta_g,\lambda) = \left(n+\frac{1}{2}\right)\hbar\omega_0 + S_i\left[\frac{\hbar|\mathrm{e}|\mathrm{F}_s[T_3(\mathrm{E}_{17,34},\eta_g,\lambda)]'}{\sqrt{2m_c}}\right]^{\frac{2}{3}} \quad (18.174)$$

where $E_{17,34}$ is the totally quantized energy in this case.

(18.174) represents the magneto DR of the 2D electrons in accumulation layers of HD III–V, ternary and quaternary materials under the weak electric field limit in the presence of light waves whose bulk electrons obey the isotropic parabolic energy bands. Since the DR in accordance with the HD parabolic energy bands is real in nature, the (18.174) will also be real in the energy plane. The total energy is quantized since the wave vector space is totally quantized.

The DOS function is given by

$$N_{2DALB}(E) = \frac{g_v eB}{\pi\hbar}\sum_{n=0}^{n_{\max}}\sum_{i=0}^{i_{\max}}\delta'(E - E_{17,34}) \quad (18.175)$$

In the absence of band-tails and under the condition of weak electric field limit, (18.174) assumes the form

$$\rho_0(E_{17,35},\lambda) = \left(n+\frac{1}{2}\right)\hbar\omega_0 + S_i\left[\frac{\hbar|e|F_s[\rho_0(E_{17,35},\lambda)]'}{\sqrt{2m_c}}\right]^{2/3} \quad (18.176)$$

where $\mathrm{E}_{17,35}$ is the totally quantized energy in this case.

The DOS function is given by

$$N_{2DALB}(E) = \frac{g_v eB}{\pi\hbar}\sum_{n=0}^{n_{\max}}\sum_{i=0}^{i_{\max}}\delta'(E - E_{17,35}) \quad (18.177)$$

18.2.11 The Magneto DR in Doping Superlattices of HD Kane Type Semiconductors in the Presence of Light Waves

(a) The magneto DR in doping superlattices of HD III–V, ternary and quaternary materials in the presence of external photo-excitation whose unperturbed electrons are defined by the three band model of Kane can be expressed following (18.46b) as

$$T_1(E_{17,40}, \eta_g, \lambda) = \left(n_i + \frac{1}{2}\right)\hbar\omega_{91HD}(E_{17,40}, \eta_g, \lambda) + \left(n + \frac{1}{2}\right)\hbar\omega_0 \qquad (18.178)$$

where $E_{17,40}$ is the total energy in this case.

The DOS function is given by

$$N_{2DALB}(E) = \frac{g_v eB}{\pi\hbar}\sum_{n=0}^{n_{\max}}\sum_{i=0}^{i_{\max}}\delta'(E - E_{17,40}) \qquad (18.179)$$

(b) The magneto DR in doping superlattices of HD III–V, ternary and quaternary materials in the presence of external photo-excitation whose unperturbed electrons are defined by the two band model of Kane can be expressed following (18.47) as

$$T_2(E_{17,41}, \eta_g, \lambda) = \left(n_i + \frac{1}{2}\right)\hbar\omega_{92HD}(E_{17,41}, \eta_g, \lambda) + \left(n + \frac{1}{2}\right)\hbar\omega_0 \qquad (18.180)$$

where $E_{17,41}$ is the total energy in this case.

The DOS function is given by

$$N_{2DALB}(E) = \frac{g_v eB}{\pi\hbar}\sum_{n=0}^{n_{\max}}\sum_{i=0}^{i_{\max}}\delta'(E - E_{17,41}) \qquad (18.181)$$

(c) The magneto DR in doping superlattices of HD III–V, ternary and quaternary materials in the presence of external photo-excitation whose unperturbed electrons are defined by the parabolic energy bands can be expressed following (18.48a) as

$$T_2(E_{17,42}, \eta_g, \lambda) = \left(n_i + \frac{1}{2}\right)\hbar\omega_{93HD}(E_{17,42}, \eta_g, \lambda) + \left(n + \frac{1}{2}\right)\hbar\omega_0 \qquad (18.182)$$

where $E_{17,42}$ is the total energy in this case.

The DOS function is given by

$$N_{2DALB}(E) = \frac{g_v eB}{\pi\hbar}\sum_{n=0}^{n_{\max}}\sum_{i=0}^{i_{\max}}\delta'(E - E_{17,42}) \qquad (18.183)$$

18.2.12 The DR in QWHD Effective Mass Superlattices of Kane Type Semiconductors in the Presence of Light Waves

(a) The electron dispersion law in HD III–V effective mass superlattices (EMSLs) in the presence of light waves, the dispersion relations of whose constituent materials in the absence of any perturbation are defined by the three band model of Kane can be written as

$$k_x^2 = \left[\frac{1}{L_0^2}\left\{\cos^{-1}\left(f_{HD1}(E, k_y, k_z, \lambda)\right)\right\}^2 - k_\perp^2\right] \tag{18.184}$$

in which, $f_{HD1}(E, k_y, k_z, \lambda) = a_{1HD1}\cos[a_0 C_{1HD1}(E, k_\perp) + b_0 D_{1HD1}(E, k_\perp)] - a_{2HD1}\cos[a_0 C_{1HD1}(E, k_\perp) - b_0 D_{1HD1}\ (E, k_\perp)], k_\perp^2 = k_y^2 + k_z^2,\ L_0 = a_0 + b_0,$

$$a_{1HD1} = \left[\sqrt{\frac{m_{c2}\text{ Real part of }[\mathrm{T}_1(0, \eta_{g2}, \lambda)]}{m_{c1}\text{ Real part of }[\mathrm{T}_1(0, \eta_{g1}, \lambda)]}} + 1\right]^2 \left[4\left(\frac{m_{c2}\text{ Real part of }[\mathrm{T}_1(0, \eta_{g2}, \lambda)]}{m_{c1}\text{ Real part of }[\mathrm{T}_1(0, \eta_{g1}, \lambda)]}\right)^{1/2}\right]^{-1},$$

$$a_{2HD1} = \left[-1 + \sqrt{\frac{m_{c2}\text{ Real part of }[\mathrm{T}_1(0, \eta_{g2}, \lambda)]}{m_{c1}\text{ Real part of }[\mathrm{T}_1(0, \eta_{g1}, \lambda)]}}\right]^2 \left[4\left(\frac{m_{c2}\text{ Real part of }[\mathrm{T}_1(0, \eta_{g2}, \lambda)]}{m_{c1}\text{ Real part of }[\mathrm{T}_1(0, \eta_{g1}, \lambda)]}\right)^{1/2}\right]^{-1},$$

$$C_{1HD1}(E, k_\perp, \lambda) \equiv \left[\left(\frac{2m_{c1}}{\hbar^2}\right) T_1(E, \eta_{g1}, \lambda) - k_\perp^2\right]^{1/2}$$

and $D_{1HD1}(E, k_\perp, \lambda) \equiv \left[\left(\frac{2m_{c2}}{\hbar^2}\right) T_1(E, \eta_{g2}, \lambda) - k_\perp^2\right]^{1/2}$.

The DR in QWHD effective mass superlattices of Kane type semiconductors in the presence of light waves, the dispersion relations of whose constituent materials in the absence of any perturbation are defined by the three band model of Kane can be written as

$$\left(\frac{n_x \pi}{d_x}\right)^2 = \left[\frac{1}{L_0^2}\left\{\cos^{-1}\left(f_{HD1}(E, k_y, k_z, \lambda)\right)\right\}^2 - k_\perp^2\right] \tag{18.185}$$

The EEM in this case assumes the form

$$m^*(k_\perp, E, \lambda) = \frac{\hbar^2}{L_0^2}\left|\left[\frac{\cos^{-1}[f_{HD1}(E, k_y, k_z)] f'_{HD1}(E, k_y, k_z)}{\sqrt{1 - f_{HD1}^2(E, k_y, k_z)}}\right]\right| \tag{18.186}$$

The Subband energies $(E_{nSL5HD1})$ can be written as

$$\left(\frac{n_x \pi}{d_x}\right)^2 = \left[\frac{1}{L_0^2}\left\{\cos^{-1}(f_{HD1}(E_{nSL5HD1}, 0, 0, \lambda))\right\}^2\right] \quad (18.187)$$

(b) The electron dispersion law in HD III–V effective mass superlattices (EMSLs) in the presence of light waves, the dispersion relations of whose constituent materials in the absence of any perturbation are defined by the two band model of Kane can be written as.

$$k_x^2 = \left[\frac{1}{L_0^2}\left\{\cos^{-1}(f_{HD2}(E, k_y, k_z, \lambda))\right\}^2 - k_\perp^2\right] \quad (18.188)$$

in which $f_{HD2}(E, k_y, k_z, \lambda) =$
$a_{1HD2}\cos[a_0 C_{1HD2}(E, k_\perp) + b_0 D_{1HD2}(E, k_\perp)] - a_{2HD2}\cos[a_0 C_{1HD2}(E, k_\perp) - b_0 D_{1HD2}(E, k_\perp)]$,

$$a_{1HD2} = \left[\sqrt{\frac{m_{c2}\,[\mathrm{T}_2(0, \eta_{g2}, \lambda)]}{m_{c1}\,[\mathrm{T}_2(0, \eta_{g1}, \lambda)]}} + 1\right]^2 \left[4\left(\frac{m_{c2}\,[\mathrm{T}_2(0, \eta_{g2}, \lambda)]}{m_{c1}\,[\mathrm{T}_2(0, \eta_{g1}, \lambda)]}\right)^{1/2}\right]^{-1},$$

$$a_{2HD2} = \left[-1 + \sqrt{\frac{m_{c2}\,[\mathrm{T}_2(0, \eta_{g2}, \lambda)]}{m_{c1}\,[\mathrm{T}_2(0, \eta_{g1}, \lambda)]}}\right]^2 \left[4\left(\frac{m_{c2}\,[\mathrm{T}_2(0, \eta_{g2}, \lambda)]}{m_{c1}\,[\mathrm{T}_2(0, \eta_{g1}, \lambda)]}\right)^{1/2}\right]^{-1},$$

$C_{1HD2}(E, k_\perp, \lambda) \equiv \left[\left(\frac{2m_{c1}}{\hbar^2}\right) T_2(E, \eta_{g1}, \lambda) - k_\perp^2\right]^{1/2}$ and $D_{1HD2}(E, k_\perp, \lambda) \equiv \left[\left(\frac{2m_{c2}}{\hbar^2}\right) T_2(E, \eta_{g2}, \lambda) - k_\perp^2\right]^{1/2}$.

The DR in QWHD effective mass superlattices of Kane type semiconductors in the presence of light waves, the dispersion relations of whose constituent materials in the absence of any perturbation are defined by the two band model of Kane can be written as

$$\left(\frac{n_x \pi}{d_x}\right)^2 = \left[\frac{1}{L_0^2}\left\{\cos^{-1}(f_{HD2}(E, k_y, k_z, \lambda))\right\}^2 - k_\perp^2\right] \quad (18.189)$$

The EEM in this case assumes the form

$$m^*(k_\perp, E, \lambda) = \frac{\hbar^2}{L_0^2}\left|\left[\frac{\cos^{-1}[f_{HD2}(E, k_y, k_z)] f'_{HD2}(E, k_y, k_z)}{\sqrt{1 - f_{HD2}^2(E, k_y, k_z)}}\right]\right| \quad (18.190)$$

The Subband energies $(E_{nSL5HD2})$ can be written as

$$\left(\frac{n_x\pi}{d_x}\right)^2 = \left[\frac{1}{L_0^2}\left\{\cos^{-1}(f_{HD2}(E_{nSL5HD2}, 0, 0, \lambda))\right\}^2\right] \tag{18.191}$$

(C) The electron dispersion law in HD III–V effective mass superlattices (EMSLs) in the presence of light waves, the dispersion relations of whose constituent materials in the absence of any perturbation are defined by the parabolic energy bands can be written as.

$$k_x^2 = \left[\frac{1}{L_0^2}\left\{\cos^{-1}(f_{HD3}(E, k_y, k_z, \lambda))\right\}^2 - k_\perp^2\right] \tag{18.192}$$

in which $f_{HD3}(E, k_y, k_z, \lambda) =$
$a_{1HD3}\cos[a_0 C_{1HD3}(E, k_\perp) + b_0 D_{1HD3}(E, k_\perp)] - a_{2HD3}\cos[a_0 C_{1HD3}(E, k_\perp) - b_0 D_{1HD3}(E, k_\perp)]$,

$$a_{1HD3} = \left[\sqrt{\frac{m_{c2}\,[T_3(0, \eta_{g2}, \lambda)]}{m_{c1}\,[T_3(0, \eta_{g1}, \lambda)]}} + 1\right]^2 \left[4\left(\frac{m_{c2}\,[T_3(0, \eta_{g2}, \lambda)]}{m_{c1}\,[T_3(0, \eta_{g1}, \lambda)]}\right)^{1/2}\right]^{-1},$$

$$a_{2HD3} = \left[-1 + \sqrt{\frac{m_{c2}\,[T_3(0, \eta_{g2}, \lambda)]}{m_{c1}\,[T_3(0, \eta_{g1}, \lambda)]}}\right]^2 \left[4\left(\frac{m_{c2}\,[T_3(0, \eta_{g2}, \lambda)]}{m_{c1}\,[T_3(0, \eta_{g1}, \lambda)]}\right)^{1/2}\right]^{-1},$$

$$C_{1HD3}(E, k_\perp, \lambda) \equiv \left[\left(\frac{2m_{c1}}{\hbar^2}\right)T_3(E, \eta_{g1}, \lambda) - k_\perp^2\right]^{1/2} \quad \text{and}$$

$$D_{1HD3}(E, k_\perp, \lambda) \equiv \left[\left(\frac{2m_{c1}}{\hbar^2}\right)T_3(E, \eta_{g2}, \lambda) - k_\perp^2\right]^{1/2}.$$

The DR in in QWHD effective mass superlattices of Kane type semiconductors in the presence of light waves, the dispersion relations of whose constituent materials in the absence of any perturbation are defined by the parabolic energy bands can be written as

$$\left(\frac{n_x\pi}{d_x}\right)^2 = \left[\frac{1}{L_0^2}\left\{\cos^{-1}(f_{HD3}(E, k_y, k_z, \lambda))\right\}^2 - k_\perp^2\right] \tag{18.193}$$

The EEM in this case assumes the form

$$m^*(k_\perp, E, \lambda) = \frac{\hbar^2}{L_0^2}\left|\left[\frac{\cos^{-1}[f_{HD3}(E, k_y, k_z)]f'_{HD3}(E, k_y, k_z)}{\sqrt{1 - f_{HD3}^2(E, k_y, k_z)}}\right]\right| \tag{18.194}$$

The Subband energies $(E_{nSL5HD3})$ can be written as

$$\left(\frac{n_x\pi}{d_x}\right)^2 = \left[\frac{1}{L_0^2}\left\{\cos^{-1}(f_{HD2}(E_{nSL5HD3}, 0, 0, \lambda))\right\}^2\right] \quad (18.195)$$

18.2.13 The DR in NWHD Effective Mass Superlattices of Kane Type Semiconductors in the Presence of Light Waves

(a) The DR in NWHD effective mass superlattices of Kane type semiconductors in the presence of light waves, the dispersion relations of whose constituent materials in the absence of any perturbation are defined by the three band model of Kane can be written as

$$k_x^2 = \left[\frac{1}{L_0^2}\left\{\cos^{-1}\left(f_{HD1}\left(E, \frac{n_y\pi}{d_y}, \frac{n_z\pi}{d_z}, \lambda\right)\right)\right\}^2 - \left[\left(\frac{n_y\pi}{d_y}\right)^2 + \left(\frac{n_z\pi}{d_z}\right)^2\right]\right] \quad (18.196)$$

The EEM in this case assumes the form

$$m^*\left(\frac{n_y\pi}{d_y}, \frac{n_z\pi}{d_z}, E, \lambda\right) = \frac{\hbar^2}{L_0^2}\left|\left[\frac{\cos^{-1}\left[f_{HD1}\left(E, \frac{n_y\pi}{d_y}, \frac{n_z\pi}{d_z}\right)\right] f'_{HD1}\left(E, \frac{n_y\pi}{d_y}, \frac{n_z\pi}{d_z}\right)}{\sqrt{1 - f_{HD1}^2\left(E, \frac{n_y\pi}{d_y}, \frac{n_z\pi}{d_z}\right)}}\right]\right| \quad (18.197)$$

The Subband energies $(E_{nSL5HD4})$ can be written as

$$0 = \left[\frac{1}{L_0^2}\left\{\cos^{-1}\left(f_{HD1}\left(E_{nSL5HD4}, \frac{n_y\pi}{d_y}, \frac{n_z\pi}{d_z}, \lambda\right)\right)\right\}^2 - \left[\left(\frac{n_y\pi}{d_y}\right)^2 + \left(\frac{n_z\pi}{d_z}\right)^2\right]\right] \quad (18.198)$$

(b) The DR in NWHD effective mass superlattices of Kane type semiconductors in the presence of light waves, the dispersion relations of whose constituent materials in the absence of any perturbation are defined by the two band model of Kane can be written as

$$k_x^2 = \left[\frac{1}{L_0^2}\left\{\cos^{-1}\left(f_{HD2}\left(E, \frac{n_y\pi}{d_y}, \frac{n_z\pi}{d_z}, \lambda\right)\right)\right\}^2 - \left[\left(\frac{n_y\pi}{d_y}\right)^2 + \left(\frac{n_z\pi}{d_z}\right)^2\right]\right] \quad (18.199)$$

The EEM in this case assumes the form

$$m^{*}\left(\frac{n_y\pi}{d_y},\frac{n_z\pi}{d_z},E,\lambda\right)=\frac{\hbar^2}{L_0^2}\left|\left[\frac{\cos^{-1}\left[f_{HD2}\left(E,\frac{n_y\pi}{d_y},\frac{n_z\pi}{d_z}\right)\right]f'_{HD2}\left(E,\frac{n_y\pi}{d_y},\frac{n_z\pi}{d_z}\right)}{\sqrt{1-f_{HD2}^2\left(E,\frac{n_y\pi}{d_y},\frac{n_z\pi}{d_z}\right)}}\right]\right| \tag{18.200}$$

The Subband energies $(E_{nSL5HD5})$ can be written as

$$0=\left[\frac{1}{L_0^2}\left\{\cos^{-1}\left(f_{HD2}\left(E_{nSL5HD5},\frac{n_y\pi}{d_y},\frac{n_z\pi}{d_z},\lambda\right)\right)\right\}^2-\left[\left(\frac{n_y\pi}{d_y}\right)^2+\left(\frac{n_z\pi}{d_z}\right)^2\right]\right] \tag{18.201}$$

(C) (b) The DR in NWHD effective mass superlattices of Kane type semiconductors in the presence of light waves, the dispersion relations of whose constituent materials in the absence of any perturbation are defined by the parabolic energy bands can be written as

$$k_x^2=\left[\frac{1}{L_0^2}\left\{\cos^{-1}\left(f_{HD3}\left(E,\frac{n_y\pi}{d_y},\frac{n_z\pi}{d_z},\lambda\right)\right)\right\}^2-\left[\left(\frac{n_y\pi}{d_y}\right)^2+\left(\frac{n_z\pi}{d_z}\right)^2\right]\right] \tag{18.202}$$

The EEM in this case assumes the form

$$m^{*}\left(\frac{n_y\pi}{d_y},\frac{n_z\pi}{d_z},E,\lambda\right)=\frac{\hbar^2}{L_0^2}\left|\left[\frac{\cos^{-1}\left[f_{HD3}\left(E,\frac{n_y\pi}{d_y},\frac{n_z\pi}{d_z}\right)\right]f'_{HD3}\left(E,\frac{n_y\pi}{d_y},\frac{n_z\pi}{d_z}\right)}{\sqrt{1-f_{HD3}^2\left(E,\frac{n_y\pi}{d_y},\frac{n_z\pi}{d_z}\right)}}\right]\right| \tag{18.203}$$

The Subband energies $(E_{nSL5HD6})$ can be written as

$$0=\left[\frac{1}{L_0^2}\left\{\cos^{-1}\left(f_{HD3}\left(E_{nSL5HD6},\frac{n_y\pi}{d_y},\frac{n_z\pi}{d_z},\lambda\right)\right)\right\}^2-\left[\left(\frac{n_y\pi}{d_y}\right)^2+\left(\frac{n_z\pi}{d_z}\right)^2\right]\right] \tag{18.204}$$

18.2.14 The Magneto DR in HD Effective Mass Superlattices of Kane Type Semiconductors in the Presence of Light Waves

(a) In the presence of an external magnetic field along x-direction, the simplified magneto dispersion law in this case can be written as

$$k_x^2 = [\rho_{4HD1}(n, E, \lambda)] \tag{18.205}$$

in which, $\rho_{4HD1}(n, E, \lambda) = \frac{1}{L_0^2}[\cos^{-1}(f_{HD1}(n, E, \lambda))]^2 - \left\{\frac{2|e|B}{\hbar}\left(n + \frac{1}{2}\right)\right\}$,

$$f_{HD1}(E, n, \lambda) = a_{1HD1}\cos[a_0 C_{1HD1}(E, n, \lambda) + b_0 D_{1HD1}(E, n, \lambda)] - a_{2HD1}\cos[a_0 C_{1HD1}(E, n, \lambda) - b_0 D_{1HD1}(E, n, \lambda)]$$

$$C_{1HD1}(E, n, \lambda) \equiv \left[\left(\frac{2m_{c1}}{\hbar^2}\right)T_1(E, \eta_{g1}, \lambda) - \left\{\frac{2|e|B}{\hbar}\right\}\left(n + \frac{1}{2}\right)\right]^{1/2} \quad \text{and}$$

$$D_{1HD1}(E, n, \lambda) \equiv \left[\left(\frac{2m_{c2}}{\hbar^2}\right)T_1(E, \eta_{g2}, \lambda) - \left\{\frac{2|e|B}{\hbar}\right\}\left(n + \frac{1}{2}\right)\right]^{1/2}.$$

The EEM in this case assumes the form

$$m^*(n, E_{fSLHDB}, \lambda) = \text{Real part of } \frac{\hbar^2}{2}[\rho_{4HD}(n, E_{fSLHDB}, \lambda)]' \tag{18.206a}$$

where E_{fSLHDB} is the Fermi energy in this case.

The EEM in III–V EMSLs under magnetic quantization depends on both, the Fermi energy, magnetic quantum number and wavelength which is the intrinsic property of such SLs.

The DOS function is given by

$$N(E) = \frac{eB}{\pi\hbar}\sum_{n=0}^{n_{\max}}\frac{\rho'_{4HD1}(n, E, \lambda)H(E - E_{nSL5HD})}{\sqrt{\rho_{4HD1}(n, E, \lambda)}} \tag{18.206b}$$

where the Landau Subband energies (E_{nSL5HD}) can be written as

$$\rho_{4HD1}(n, E_{nSL5HD}, \lambda) = 0 \tag{18.207}$$

(b) In the presence of an external magnetic field along x-direction, the simplified magneto dispersion law in this case can be written as

$$k_x^2 = [\rho_{4HD2}(n, E, \lambda)] \tag{18.208}$$

in which, $\rho_{4HD2}(n,E,\lambda) = \frac{1}{L_0^2}\left[\cos^{-1}(f_{HD2}(n,E,\lambda))\right]^2 - \left\{\frac{2|e|B}{\hbar}\left(n+\frac{1}{2}\right)\right\}$,

$f_{HD2}(E,n,\lambda) = a_{1HD2}\cos[a_0C_{1HD2}(E,n,\lambda)+b_0D_{1HD2}(E,n,\lambda)] - a_{2HD2}\cos[a_0C_{1HD2}(E,n,\lambda)-b_0D_{1HD2}(E,n,\lambda)]$

$$C_{1HD2}(E,n,\lambda) \equiv \left[\left(\frac{2m_{c1}}{\hbar^2}\right)T_2(E,\eta_{g1},\lambda) - \left\{\frac{2|e|B}{\hbar}\left(n+\frac{1}{2}\right)\right\}\right]^{1/2} \quad \text{and}$$

$$D_{1HD2}(E,n,\lambda) \equiv \left[\left(\frac{2m_{c2}}{\hbar^2}\right)T_2(E,\eta_{g2},\lambda) - \left\{\frac{2|e|B}{\hbar}\left(n+\frac{1}{2}\right)\right\}\right]^{1/2}.$$

The EEM in this case assumes the form

$$m^*(n,E_{fSLHDB},\lambda) = \frac{\hbar^2}{2}[\rho_{4HD2}(n,E_{fSLHDB},\lambda)]' \tag{18.209a}$$

where E_{fSLHDB} is the Fermi energy in this case.

The EEM in III–V EMSLs under magnetic quantization depends on both the Fermi energy, magnetic quantum number and wavelength which is the intrinsic property of such SLs.

The DOS function is given by

$$N(E) = \frac{eB}{2\pi^2\hbar}\sum_{n=0}^{n_{\max}}\frac{\rho'_{4HD2}(n,E,\lambda)H(E-E_{nSL5HD2})}{\sqrt{\rho_{4HD2}(n,E,\lambda)}} \tag{18.209b}$$

where the Landau Subband energies $(E_{nSL5HD2})$ can be written as

$$\rho_{4HD2}(n,E_{nSL5HD2},\lambda) = 0 \tag{18.210}$$

(C) In the presence of an external magnetic field along x-direction, the simplified magneto dispersion law in this case can be written as

$$k_x^2 = [\rho_{4HD3}(n,E,\lambda)] \tag{18.211}$$

in which, $\rho_{4HD3}(n,E,\lambda) = \frac{1}{L_0^2}\left[\cos^{-1}(f_{HD3}(n,E,\lambda))\right]^2 - \left\{\frac{2|e|B}{\hbar}\left(n+\frac{1}{2}\right)\right\}$,

$f_{HD3}(E,n,\lambda) = a_{1HD3}\cos[a_0C_{1HD3}(E,n,\lambda)+b_0D_{1HD3}(E,n,\lambda)] - a_{2HD3}\cos[a_0C_{1HD3}(E,n,\lambda)-b_0D_{1HD3}(E,n,\lambda)]$

$$C_{1HD3}(E,n,\lambda) \equiv \left[\left(\frac{2m_{c1}}{\hbar^2}\right)T_3(E,\eta_{g1},\lambda) - \left\{\frac{2|e|B}{\hbar}\left(n+\frac{1}{2}\right)\right\}\right]^{1/2} \quad \text{and}$$

$$D_{1HD3}(E,n,\lambda) \equiv \left[\left(\frac{2m_{c2}}{\hbar^2}\right)T_3(E,\eta_{g2},\lambda) - \left\{\frac{2|e|B}{\hbar}\left(n+\frac{1}{2}\right)\right\}\right]^{1/2}.$$

The EEM in this case assumes the form

$$m^*(n, E_{fSLHDB}, \lambda) = \frac{\hbar^2}{2}[\rho_{4HD3}(n, E_{fSLHDB}, \lambda)]' \tag{18.212a}$$

where E_{fSLHDB} is the Fermi energy in this case.

The EEM in III–V EMSLs under magnetic quantization depends on both the Fermi energy, magnetic quantum number and wavelength which is the intrinsic property of such SLs.

The DOS function is given by

$$N(E) = \frac{eB}{2\pi^2\hbar}\sum_{n=0}^{n_{\max}}\frac{\rho'_{4HD3}(n, E, \lambda)H(E - E_{nSL5HD3})}{\sqrt{\rho_{4HD3}(n, E, \lambda)}} \tag{18.212b}$$

where the Landau Subband energies $(E_{nSL5HD3})$ can be written as

$$\rho_{4HD3}(n, E_{nSL5HD3}, \lambda) = 0 \tag{18.213}$$

18.2.15 The Magneto DR in QWHD Effective Mass Superlattices of Kane Type Semiconductors in the Presence of Light Waves

(a) In the presence of an external magnetic field along x-direction, the simplified magneto dispersion law in QWHD III–V effective mass superlattices (EMSLs) in the presence of light waves, the dispersion relations of whose constituent materials in the absence of any perturbation are defined by the three band model of Kane can be written as

$$\left(\frac{n_x\pi}{d_x}\right)^2 = [\rho_{4HD1}(n, E_{17,50}, \lambda)] \tag{18.214}$$

where $E_{17,50}$ is the totally quantized energy in this case.

The DOS function is given by

$$N_{BQWHDEMSL}(E) = \frac{g_v eB}{\pi\hbar}\sum_{n=0}^{n_{\max}}\sum_{n_x=1}^{n_{x\max}}\delta'(E - E_{17,50}) \tag{18.215}$$

(b) In the presence of an external magnetic field along x-direction, the simplified magneto dispersion law in QWHD III–V effective mass superlattices (EMSLs) in the presence of light waves, the dispersion relations of whose constituent materials in the absence of any perturbation are defined by the two band model of Kane can be written as

$$\left(\frac{n_x \pi}{d_x}\right)^2 = \left[\rho_{4HD1}(n, E_{17,51}, \lambda)\right] \quad (18.216)$$

where $E_{17,51}$ is the totally quantized energy in this case.

The DOS function is given by

$$N_{BQWHDEMSL}(E) = \frac{g_v eB}{\pi\hbar} \sum_{n=0}^{n_{\max}} \sum_{n_x=1}^{n_{x\max}} \delta'(E - E_{17,51}) \quad (18.217)$$

(C) In the presence of an external magnetic field along x-direction, the simplified magneto dispersion law in QWHD III–V effective mass superlattices (EMSLs) in the presence of light waves, the dispersion relations of whose constituent materials in the absence of any perturbation are defined by the isotropic parabolic energy bands can be written as

$$\left(\frac{n_x \pi}{d_x}\right)^2 = \left[\rho_{4HD3}(n, E_{17,52}, \lambda)\right] \quad (18.218)$$

where $E_{17,52}$ is the totally quantized energy in this case.

The DOS function is given by

$$N_{BQWHDEMSL}(E) = \frac{g_v eB}{\pi\hbar} \sum_{n=0}^{n_{\max}} \sum_{n_x=1}^{n_{x\max}} \delta'(E - E_{17,52}) \quad (18.219)$$

18.2.16 The DR in QWHD Superlattices of Kane Type Semiconductors with Graded Interfaces in the Presence of Light Waves

The electron dispersion law in bulk specimens of the heavily doped constituent materials of III–V SLs whose energy band structures are defined by (17.46b) be expressed as

$$\frac{\hbar^2 k^2}{2m_{cj}} = V_{1j}(E, \eta_{gj}, \lambda, \Delta_j, E_{g0j}) + iV_{2j}(E, \eta_{gj}, \lambda, \Delta_j, E_{g0j}) \quad (18.220)$$

where

$$j = 1, 2, \quad V_{1j}(E, \eta_{gj}, \lambda, \Delta_j, E_{g0j}) = \left[U_{\lambda j} T_{1j}(E, \Delta_j, E_{gj}, \eta_{gj}) - P_{\lambda j}\right]$$

$$U_{\lambda j}=(1+\theta_j),\ \ \theta_{\lambda j}=\frac{C_{0j}}{A_j}\left(t_{\lambda j}+\frac{B_j J_{\lambda j}}{A_j}\right),C_{0j}=\left[\frac{|e|^2 I\lambda^2 E_{g0j}(E_{gj}+\Delta_j)\beta_j^2\left(1+\frac{\rho_j}{\sqrt{2}}\right)^2}{384 m_{rj}\pi c^3\sqrt{\varepsilon_{sc}\varepsilon_0}\left(E_{g0j}+\frac{2}{3}\Delta_j\right)}\right],$$

$$\beta_j=\left[6\left(E_{g0j}+\frac{2}{3}\Delta_j\right)\frac{(E_{g0j}+\Delta_j)}{\chi_j}\right]^{1/2},\ \chi_j=(6E_{g0}^2+9E_{g0j}\Delta_j+4\Delta_j^2),\ \ \rho_j=\left[\frac{4\Delta_j^2}{3\chi_j}\right]^{1/2},\ A_j=E_{g0j},$$

$$t_{\lambda j}=\left[E_{\lambda j}-\frac{G_{\lambda j}(E_{g0j}-\delta_j')}{\sqrt{t_{\lambda j}}}\right],\ E_{\lambda j}=\frac{2B_j(E_{g0j}-\delta_j')}{\left(A_j+\delta_j'\right)},\ \ B_j=\left[1+\frac{m_{cj}}{m_{vj}}\right],\ G_{\lambda j}=\left[\frac{2B_j}{\left(A_j+\delta_j'\right)^3}-\frac{B_j C_{\lambda j}}{\left(A_j+\delta_j'\right)}\right],$$

$$C_{\lambda j}=\left[(E_{g0j}+\delta_j')^{-1}+(E_{g0j}+\delta_j')(E_{g0j}-\delta_j')^{-2}\right]\left(A_j+\delta_j'\right)^{-1},\ \ P_{\lambda j}=\frac{C_{0j}}{A_j}J_{\lambda j},\ \ J_{\lambda j}=(D_{\lambda j}+2(E_{g0j}-\delta_j')\sqrt{f_{\lambda j}}),$$

$$\mathrm{D}_{\lambda j}=\left(1+\frac{2(E_{g0j}-\delta_j')}{(A_j+\delta_j')}\right),\ \ f_{\lambda j}=\left[\frac{1}{(A_j+\delta_j')^2}+\frac{1}{(E_{g0j}-\delta_j')^2}-C_{\lambda j}\right],$$

$$T_{1j}(E,\Delta_j,E_{gj},\eta_{gj})=(2/(1+Erf(E/\eta_{gj})))\Big[(\alpha_j b_j/c_j).\theta_0(E,\eta_{gj})+[\alpha_j c_j+b_j c_j-\alpha_j b_j/c_j^2].$$
$$\gamma_0(E,\eta_{gj})+\Big[(1/c_j)(1-(\alpha_j/c_j))(1-b_j/c_j))\frac{1}{2}[1+Erf(E/\eta_{gj})]$$
$$-\ (1/c_j)(1-(\alpha_j/c_j))(1-b_j/c_j))(2/c_j\eta_{gj}\sqrt{\pi})\exp(-u_j^2)\left[\sum_{p=1}^{\infty}(\exp(-p^2/4)/p)\sinh(pu_j)\right]\Big]\Big],$$

$$b_j\equiv(E_{gj}+\Delta_j)^{-1},c_j\equiv\left(E_{gj}+\frac{2}{3}\Delta_j\right)^{-1},\ \ u_j\equiv(1+c_jE)(c_j\eta_{gi})^{-1},$$

$$V_{2j}(E,\eta_{gj},\lambda,\Delta_j,E_{g0j})=[\mathrm{U}_{\lambda j}\mathrm{T}_{2j}(E,\Delta_j,E_{gj},\eta_{gj})]$$
$$T_{2j}(E,\Delta_j,E_{gj},\eta_{gj})\equiv\left(\frac{2}{(1+Erf(E/\eta_{gj}))}\right)\frac{1}{c_j}\left(1-\frac{\alpha_j}{c_j}\right)\left(1-\frac{b_j}{c_j}\right)\frac{\sqrt{\pi}}{c_j\eta_{gj}}\exp(-u_j^2)$$

Therefore, the DR in HD III–V SLs with graded interfaces in the presence of light waves can be expressed as

$$k_z^2=G_8+iH_8 \tag{18.221}$$

where,

$$G_8=\left[\frac{C_7^2-D_7^2}{L_0^2}-k_s^2\right],\ \ C_7=\cos^{-1}(\bar{\omega}_7),$$

$$\bar{\omega}_7=(2)^{\frac{-1}{2}}\left[(1-G_7^2-H_7^2)-\sqrt{(1-G_7^2-H_7^2)^2+4G_7^2}\right]^{\frac{1}{2}},$$

$$G_7 = [G_1 + (\rho_5 G_2/2) - (\rho_6 H_2/2) + (\Delta_0/2)\{\rho_6 H_2 - \rho_8 H_3 + \rho_9 H_4 - \rho_{10} H_4 + \rho_{11} H_5 - \rho_{12} H_5 + (1/12)(\rho_{12} G_6 - \rho_{14} H_6)\}],$$

$$G_1 = [(\cos(h_1))(\cosh(h_2))(\cosh(g_1))(\cos(g_2)) + (\sin(\mathrm{h}_1))(\sinh(\mathrm{h}_2))(\sinh(g_1))(\sin(g_2))],$$

$$\mathrm{h}_1 = e_1(b_0 - \Delta_0), e_1 = 2^{\frac{-1}{2}}\left(\sqrt{t_1^2 + t_2^2} + t_1\right)^{\frac{1}{2}},$$

$$t_1 = \left[(2m_{c1}/\hbar^2).\mathrm{V}_{11}(E, E_{g1}, \lambda, \Delta_1, \eta_{g1}) - k_s^2\right],$$
$$t_2 = \left[(2m_{c1}/\hbar^2)\mathrm{V}_{21}(E, E_{g1}, \lambda, \Delta_1, \eta_{g1})\right],$$

$$h_2 = e_2(b_0 - \Delta_0), e_2 = 2^{\frac{-1}{2}}\left(\sqrt{t_1^2 + t_2^2} - t_1\right)^{\frac{1}{2}}, g_1 = d_1(a_0 - \Delta_0), d_2 = 2^{\frac{-1}{2}}\left(\sqrt{x_1^2 + y_1^2} + x_1\right)^{\frac{1}{2}},$$

$$x_1 = \left[-(2m_{c2}/\hbar^2) \cdot \mathrm{V}_{11}(E - V_0, E_{g2}, \lambda, \Delta_2, \eta_{g2}) + k_s^2\right], y_1 = \left[(2m_{c2}/\hbar^2)\mathrm{V}_{22}(E - V_0, E_{g2}, \lambda, \Delta_2, \eta_{g2})\right],$$

$$g_2 = d_2(a_0 - \Delta_0), d_2 = 2^{\frac{-1}{2}}\left(\sqrt{x_1^2 + y_1^2} - x_1\right)^{\frac{1}{2}}, \rho_5 = (\rho_3^2 + \rho_4^2)^{-1}[\rho_1\rho_3 - \rho_2\rho_4],$$

$$\rho_1 = \left[d_1^2 + e_2^2 - d_2^2 - e_1^2\right], \rho_3 = [d_1 e_1 + d_2 e_2], \rho_2 = 2[d_1 d_2 + e_1 e_2], \rho_4 = [d_1 e_2 - e_1 d_2],$$

$$G_2 = [(\sin(h_1))(\cosh(h_2))(\sinh(g_1))(\cos(g_2)) + (\cos(h_1))(\sinh(h_2))(\cosh(g_1))(\sin(g_2))],$$
$$\rho_6 = (\rho_3^2 + \rho_4^2)^{-1}[\rho_1\rho_4 + \rho_2\rho_3],$$

$$H_2 = [(\sin(h_1))(\cosh(h_2))(\sin(g_2))(\cosh(g_1)) - (\cos(h_1))(\sinh(h_2))(\sinh(g_1))(\cos(g_2))],$$
$$\rho_7 = \left[(e_1^2 + e_2^2)^{-1}\left[e_1(d_1^2 - d_2^2) - 2d_1 d_2 e_2\right] - 3e_1\right],$$

$$G_3 = [(\sin(h_1))(\cosh(h_2))(\cosh(g_1))(\cos(g_2)) + (\cos(h_1))(\sinh(h_2))(\sinh(g_1))(\sin(g_2))],$$
$$\rho_8 = \left[(e_1^2 + e_2^2)^{-1}\left[e_2(d_1^2 - d_2^2) - 2d_1 d_2 e_2\right] + 3e_2\right],$$

$$H_3 = [(\sin(h_1))(\cosh(h_2))(\sin(g_2))(\sinh(g_1)) - (\cos(h_1))(\sinh(h_2))(\cosh(g_1))(\cos(g_2))],$$
$$\rho_9 = \left[(d_1^2 + d_2^2)^{-1}\left[d_1(e_2^2 - e_1^2) + 2e_2 d_2 e_1\right] + 3d_1\right],$$

$$G_4 = [(\cos(h_1))(\cosh(h_2))(\cos(g_2))(\sinh(g_1)) - (\sin(h_1))(\sinh(h_2))(\cosh(g_1))(\sin(g_2))],$$
$$\rho_{10} = \left[-(d_1^2+d_2^2)^{-1}\left[d_2(-e_2^2+e_1^2)+2e_2d_2e_1\right]+3d_2\right],$$

$$H_4 = [(\cos(h_1))(\cosh(h_2))(\cosh(g_1))(\sin(g_2)) + (\sin(h_1))(\sinh(h_2))(\sinh(g_1))(\cos(g_2))],$$
$$G_5 = [(\cos(h_1))(\cosh(h_2))(\cos(g_2))(\cosh(g_1)) - (\sin(h_1))(\sinh(h_2))(\sinh(g_1))(\sin(g_2))],$$
$$\rho_{12} = 4[d_1d_2+e_1e_2],$$

$$H_5 = [(\cos(h_1))(\cosh(h_2))(\sinh(g_1))(\sin(g_2)) + (\sin(h_1))(\sinh(h_2))(\cosh(g_1))(\cos(g_2))],$$
$$\rho_{13} = \Big[\{5(d_1e_1^3-3e_1e_2^2d_1)+5d_2(e_1^3-3e_1^2e_2)\}(d_1^2+d_2^2)^{-1}+(e_1^2+e_2^2)^{-1}\{5(e_1d_1^3-3d_2e_1^2d_1) + 5(d_2^3e_2-3d_1^2d_2e_2)\} - 34(d_1e_1+d_2e_2)\Big],$$

$$G_6 = [(\sin(h_1))(\cosh(h_2))(\sinh(g_1))(\cos(g_2)) + (\cos(h_1))(\sinh(h_2))(\cosh(g_1))(\sin(g_2))],$$
$$\rho_{14} = \Big[\{5(d_1e_2^3-3e_2e_1^2d_1)+5d_2(-e_1^3+3e_2^2e_1)\}(d_1^2+d_2^2)^{-1}+(e_1^2+e_2^2)^{-1}\{5(-e_1d_2^3+3d_1^2d_2e_1) + 5(-d_1^3e_2+3d_2^2d_1e_2)\} + 34(d_1e_2-d_2e_1)\Big],$$

$$H_6 = [(\sin(h_1))(\cosh(h_2))(\cosh(g_1))(\sin(g_2)) - (\cos(h_1))(\sinh(h_2))(\sinh(g_1))(\cos(g_2))],$$
$$H_7 = [H_1+(\rho_5H_2/2)+(\rho_6G_2/2)+(\Delta_0/2)\{\rho_6G_3+\rho_7H_3+\rho_{10}G_4+\rho_{10}H_4 + \rho_{12}G_5+\rho_{11}H_5+(1/12)(\rho_{14}G_6+\rho_{13}H_6)\}],$$

$$H_1 = [(\sin(h_1))(\sinh(h_2))(\cosh(g_1))(\cos(g_2)) + (\cos(h_1))(\cosh(h_2))(\sinh(g_1))(\sin(g_2))],$$
$$D_7 = \sinh^{-1}(\bar{\omega}_7),\ H_8 = (2C_7D_7/L_0^2)$$

The simplified DR of HD QWs of III–V super-lattices with graded interfaces can be expressed as

$$\left(\frac{n_z\pi}{d_z}\right)^2 = G_8 + iH_8 \tag{18.222}$$

The sub-band equation in this case can be expressed as

$$\left(\frac{n_z\pi}{d_z}\right)^2 = [G_8 + iH_8]\Big|_{k_s=0 \text{ and } E=E_{17,100}} \tag{18.223}$$

where $E_{17,100}$ is the sub-band energy in this case.

The EEM and the DOS function should be obtained numerically in this case.

18.2.17 The DR in NWHD Superlattices of Kane Type Semiconductors with Graded Interfaces in the Presence of Light Waves

The DR in NWHD superlattices of Kane type semiconductors with graded interfaces in the presence of light waves can be expressed as

$$k_z^2 = G_{8,17,50} + iH_{8,17,50} \tag{18.224}$$

$$G_{8,17.50} = \left[\frac{C_{7,50}^2 - D_{7,50}^2}{L_0^2} - \left[\left(\frac{n_x\pi}{dx}\right)^2 + \left(\frac{n_y\pi}{dy}\right)^2\right]\right], \mathrm{C}_{7,50} = \cos^{-1}(\bar{\omega}_{7,50}),$$

where

$$\bar{\omega}_{7,50} = (2)^{\frac{-1}{2}}\left[(1 - G_{7,50}^2 - H_{7,50}^2) - \sqrt{(1 - G_{7,50}^2 - H_{7,50}^2)^2 + 4G_{7,50}^2}\right]^{\frac{1}{2}},$$

$$\begin{aligned} G_{7,50} =& \big[G_{1,50} + (\rho_{5,50}G_{2,50}/2) - (\rho_{6,50}H_{2,50}/2) + (\Delta_0/2) \\ & \{\rho_{6,50}H_{2,50} - \rho_{8,50}H_{3,50} + \rho_{9,50}H_{4,50} - \rho_{10,50}H_{4,50} + \\ & \rho_{11,50}H_{5,50} - \rho_{12,50}H_{5,50} + (1/12)(\rho_{12,50}G_{6,50} - \rho_{14,50}H_{6,50})\}\big], \\ G_{1,50} =& \big[(\cos(h_{1,50}))(\cosh(h_{2,50}))(\cosh(g_{1,50}))(\cos(g_{2,50})) \\ & + (\sin(h_{1,50}))(\sinh(h_{2,50}))(\sinh(g_{1,50}))(\sin(g_{2,50}))\big], \end{aligned}$$

$$h_{1,50} = e_{1,50}(b_0 - \Delta_0), e_{1,50} = 2^{\frac{-1}{2}}\left(\sqrt{t_{1,50}^2 + t_{2,50}^2} + t_{1,50}\right)^{\frac{1}{2}},$$

$$\begin{aligned} t_{1,50} =& \left[(2m_{c1}/\hbar^2).\mathrm{V}_{11}(E, E_{g1}, \lambda, \Delta_1, \eta_{g1}) - \left[\left(\frac{n_x\pi}{dx}\right)^2 + \left(\frac{n_y\pi}{dy}\right)^2\right]\right], t_{2,50} \\ =& \left[(2m_{c1}/\hbar^2)\mathrm{V}_{21}(E, E_{g1}, \lambda, \Delta_1, \eta_{g1})\right], \end{aligned}$$

$$\begin{aligned} h_{2,50} =& e_{2,50}(b_0 - \Delta_0), e_{2,50} = 2^{\frac{-1}{2}}\left(\sqrt{t_{1,50}^2 + t_{2,50}^2} - t_{1,50}\right)^{\frac{1}{2}}, g_{1,50} \\ =& d_{1,50}(a_0 - \Delta_0), d_{2,50} = 2^{\frac{-1}{2}}\left(\sqrt{x_{1,50}^2 + y_{1,50}^2} + x_{1,50}\right)^{\frac{1}{2}}, \end{aligned}$$

$$x_{1,50} = \left[-(2m_{c2}/\hbar^2).\mathrm{V}_{11}(E - V_0, E_{g2}, \lambda, \Delta_2, \eta_{g2}) + \left[\left(\frac{n_x\pi}{dx}\right)^2 + \left(\frac{n_y\pi}{dy}\right)^2\right]\right],$$

$$y_{1,50} = \left[(2m_{c2}/\hbar^2)\mathrm{V}_{22}(E - V_0, E_{g2}, \lambda, \Delta_2, \eta_{g2})\right],$$

$$\begin{aligned} g_{2,50} =& d_{2,50}(a_0 - \Delta_0), d_{2,50} = 2^{\frac{-1}{2}}\left(\sqrt{x_{1,50}^2 + y_{1,50}^2} - x_{1,50}\right)^{\frac{1}{2}}, \rho_{5,50} \\ =& (\rho_{3,50}^2 + \rho_{4,50}^2)^{-1}\left[\rho_{1,50}\rho_{3,50} - \rho_{2,50}\rho_{4,50}\right], \end{aligned}$$

$$\rho_{1,50} = \left[d_{1,50}^2 + e_{2,50}^2 - d_{2,50}^2 - e_{1,50}^2\right],$$
$$\rho_{3,50} = [d_{1,50}e_{1,50} + d_{2,50}e_{2,50}],$$
$$\rho_{2,50} = 2\left[d_{1,50}d_{2,50} + e_{1,50}e_{2,50}\right],$$
$$\rho_{4,50} = \left[d_{1,50}e_{2,50} - e_{1,50}d_{2,50}\right],$$

$$G_{2,50} = [(\sin(h_{1,50}))(\cosh(h_{2,50}))(\sinh(g_{1,50}))(\cos(g_{2,50})) + (\cos(h_{1,50}))(\sinh(h_{2,50}))(\cosh(g_{1,50}))(\sin(g_{2,50}))],$$
$$\rho_{6,50} = (\rho_{3,50}^2 + \rho_{4,50}^2)^{-1}[\rho_{1,50}\rho_{4,50} + \rho_{2,50}\rho_{3,50}],$$

$$H_{2,50} = [(\sin(h_{1,50}))(\cosh(h_{2,50}))(\sin(g_{2,50}))(\cosh(g_{1,50})) - (\cos(h_{1,50}))(\sinh(h_{2,50}))(\sinh(g_{1,50}))(\cos(g_{2,50}))],$$
$$\rho_{7,50} = \left[(e_{1,50}^2 + e_{2,50}^2)^{-1}\left[e_{1,50}(d_{1,50}^2 - d_{2,50}^2) - 2d_{1,50}d_{2,50}e_{2,50}\right] - 3e_{1,50}\right],$$

$$G_{3,50} = [(\sin(h_{1,50}))(\cosh(h_{2,50}))(\cosh(g_{1,50}))(\cos(g_{2,50})) + (\cos(h_{1,50}))(\sinh(h_{2,50}))(\sinh(g_{1,50}))(\sin(g_{2,50}))],$$
$$\rho_{8,50} = \left[(e_{1,50}^2 + e_{2,50}^2)^{-1}\left[e_{2,50}(d_{1,50}^2 - d_{2,50}^2) - 2d_{1,50}d_{2,50}e_{2,50}\right] + 3e_{2,50}\right],$$

$$H_{3,50} = [(\sin(h_{1,50}))(\cosh(h_{2,50}))(\sin(g_{2,50}))(\sinh(g_{1,50})) - (\cos(h_{1,50}))(\sinh(h_{2,50}))(\cosh(g_{1,50}))(\cos(g_{2,50}))],$$
$$\rho_{9,50} = \left[(d_{1,50}^2 + d_{2,50}^2)^{-1}\left[d_{1,50}(e_{2,50}^2 - e_{1,50}^2) + 2e_{2,50}d_{2,50}e_{1,50}\right] + 3d_{1,50}\right],$$

$$G_{4,50} = [(\cos(h_{1,50}))(\cosh(h_{2,50}))(\cos(g_{2,50}))(\sinh(g_{1,50})) - (\sin(h_{1,50}))(\sinh(h_{2,50}))(\cosh(g_{1,50}))(\sin(g_{2,50}))],$$
$$\rho_{10,50} = \left[-(d_{1,50}^2 + d_{2,50}^2)^{-1}\left[d_{2,50}(-e_{2,50}^2 + e_{1,50}^2) + 2e_{2,50}d_{2,50}e_{1,50}\right] + 3d_{2,50}\right],$$

$$H_{4,50} = [(\cos(h_{1,50}))(\cosh(h_{2,50}))(\cosh(g_{1,50}))(\sin(g_{2,50})) + (\sin(h_{1,50}))(\sinh(h_{2,50}))(\sinh(g_{1,50}))(\cos(g_{2,50}))],$$
$$\rho_{11,50} = 2\left[d_{1,50}^2 e_{2,50}^2 - d_{2,50}^2 - e_{1,50}^2\right],$$

$$G_{5,50} = [(\cos(h_{1,50}))(\cosh(h_{2,50}))(\cos(g_{2,50}))(\cosh(g_{1,50})) - (\sin(h_{1,50}))(\sinh(h_{2,50}))(\sinh(g_{1,50}))(\sin(g_{2,50}))],$$
$$\rho_{12,50} = 4\left[d_{1,50}d_{2,50} + e_{1,50}e_{2,50}\right],$$

$$H_{5,50} = \left[(\cos(h_{1,50}))(\cosh(h_{2,50}))(\sinh(g_{1,50}))(\sin(g_{2,50})) + (\sin(h_{1,50}))(\sinh(h_{2,50}))(\cosh(g_{1,50}))(\cos(g_{2,50})) \right],$$

$$\rho_{13,50} = \left[\left\{ 5(d_{1,50}e_{1,50}^3 - 3e_{1,50}e_{2,50}^2 d_{1,50}) + 5d_{2,50}(e_{1,50}^3 - 3e_{1,50}^2 e_{2,50}) \right\} (d_{1,50}^2 + d_{2,50}^2)^{-1} + (e_{1,50}^2 + e_{2,50}^2)^{-1} \left\{ 5(e_{1,50}d_{1,50}^3 - 3d_{2,50}e_{1,50}^2 d_{1,50}) + 5(d_{2,50}^3 e_{2,50} - 3d_{1,50}^2 d_{2,50}e_{2,50}) \right\} - 34(d_{1,50}e_{1,50} + d_{2,50}e_{2,50}) \right],$$

$$G_{6,50} = \left[(\sin(h_{1,50}))(\cosh(h_{2,50}))(\sinh(g_{1,50}))(\cos(g_{2,50})) + (\cos(h_{1,50}))(\sinh(h_{2,50}))(\cosh(g_{1,50}))(\sin(g_{2,50})) \right],$$

$$\rho_{14,50} = \left[\left\{ 5(d_{1,50}e_{2,50}^3 - 3e_{2,50}e_{1,50}^2 d_{1,50}) + 5d_{2,50}(-e_{1,50}^3 + 3e_{2,50}^2 e_{1,50}) \right\} (d_{1,50}^2 + d_{2,50}^2)^{-1} + (e_{1,50}^2 + e_{2,50}^2)^{-1} \left\{ 5(-e_{1,50}d_{2,50}^3 + 3d_{1,50}^2 d_{2,50}e_{1,50}) + 5(-d_{1,50}^3 e_{2,50} + 3d_{2,50}^2 d_{1,50}e_{2,50}) \right\} + 34(d_{1,50}e_{2,50} - d_{2,50}e_{1,50}) \right],$$

$$H_{6,50} = \left[(\sin(h_{1,50}))(\cosh(h_{2,50}))(\cosh(g_{1,50}))(\sin(g_{2,50})) - (\cos(h_{1,50}))(\sinh(h_{2,50}))(\sinh(g_{1,50}))(\cos(g_{2,50})) \right],$$

$$H_{7,50} = \left[H_{1,50} + (\rho_{5,50}H_{2,50}/2) + (\rho_{6,50}G_{2,50}/2) + (\Delta_0/2) \left\{ \rho_{6,50}G_{3,50} + \rho_{7,50}H_{3,50} + \rho_{10,50}G_{4,50} + \rho_{10,50}H_{4,50} + \rho_{12,50}G_{5,50} + \rho_{11,50}H_{5,50} + (1/12)(\rho_{14,50}G_{6,50} + \rho_{13,50}H_{6,50}) \right\} \right],$$

$$H_{1,50} = \left[(\sin(h_{1,50}))(\sinh(h_{2,50}))(\cosh(g_{1,50}))(\cos(g_{2,50})) + (\cos(h_{1,50}))(\cosh(h_{2,50}))(\sinh(g_{1,50}))(\sin(g_{2,50})) \right],$$

$$D_{7,50} = \sinh^{-1}(\bar{\omega}_{7,50}),$$

$$H_{8,50} = (2C_{7,50}D_{7,50}/L_0^2)$$

The sub-band equation in this case can be expressed as

$$0 = \left[G_{8,17,50} + iH_{8,17,50} \right]\big|_{E=E_{17,51}} \tag{18.225}$$

where $E_{17,51}$ is the sub-band energy in this case.

At low temperatures where the quantum effects become prominent, the DOS function for the lowest SL mini-band is given by

$$N_{HDSL}(E, \eta_g, \lambda) = \frac{g_v}{\pi} \sum_{n_x=1}^{n_{x\max}} \sum_{n_y=1}^{n_{y\max}} \frac{\left[G'_{8,17,50} + iH'_{8,17,50} \right]}{\sqrt{G_{8,17,50} + iH_{8,17,50}}} H(E - E_{17,51}) \tag{18.226}$$

The EEM can be written as

$$m^*(E, n_x, n_y, \lambda, \eta_g) = \frac{\hbar^2}{2}(G'_{8,17,50}) \tag{18.227}$$

18.2.18 The DR in Quantum Dot HD Superlattices of Kane Type Semiconductors with Graded Interfaces in the Presence of Light Waves

The DR in QDHD superlattices of Kane type semiconductors with graded interfaces in the presence of light waves can be expressed as

$$\left(\frac{n_z \pi}{d_z}\right)^2 = [G_{8,17,50} + iH_{8,17,50}]\Big|_{E=E_{17,52}} \tag{18.228}$$

where $E_{17,52}$ is the totally quantized energy in this case.The DOS function is given by

$$N_{QDHDSL}(E, \eta_g, \lambda) = \frac{2g_v}{d_x d_y d_z} \sum_{n_x=1}^{n_{x\max}} \sum_{n_y=1}^{n_{y\max}} \sum_{n_z=1}^{n_{z\max}} \delta'(E - E_{17,52}) \tag{18.229}$$

18.2.19 The Magneto DR in HD Superlattices of Kane Type Semiconductors with Graded Interfaces in the Presence of Light Waves

The magneto DR in HD superlattices of Kane type semiconductors with graded interfaces in the presence of light waves can be expressed as

$$k_z^2 = G_{8,17,54} + iH_{8,17,54} \tag{18.230}$$

where

$$G_{8,17,54} = \left[\frac{C_{7,54}^2 - D_{7,54}^2}{L_0^2} - \left[\frac{2eB}{\hbar}\left(n + \frac{1}{2}\right)\right]\right], \mathrm{C}_{7,54} = \cos^{-1}(\bar{\omega}_{7,54}),$$

$$\bar{\omega}_{7,54} = (2)^{\frac{-1}{2}}\left[(1 - G_{7,54}^2 - H_{7,54}^2) - \sqrt{(1 - G_{7,54}^2 - H_{7,54}^2)^2 + 4G_{7,54}^2}\right]^{\frac{1}{2}},$$

$$G_{7,54} = \big[G_{1,54} + (\rho_{5,54}G_{2,54}/2) - (\rho_{6,54}H_{2,54}/2) + (\Delta_0/2)$$
$$\{\rho_{6,54}H_{2,54} - \rho_{8,54}H_{3,54} + \rho_{9,54}H_{4,54} - \rho_{10,54}H_{4,54}$$
$$+ \rho_{11,54}H_{5,54} - \rho_{12,54}H_{5,54} + (1/12)(\rho_{12,54}G_{6,54} - \rho_{14,54}H_{6,54})\}\big],$$

$$G_{1,54} = \big[(\cos(h_{1,54}))(\cosh(h_{2,54}))(\cosh(g_{1,54}))(\cos(g_{2,54}))$$
$$+ (\sin(\mathrm{h}_{1,54}))(\sinh(\mathrm{h}_{2,54}))(\sinh(g_{1,54}))(\sin(g_{2,54}))\big],$$

$$\mathrm{h}_{1,54} = e_{1,54}(b_0 - \Delta_0), e_{1,54} = 2^{\frac{-1}{2}}\left(\sqrt{t_{1,54}^2 + t_{2,54}^2} + t_{1,54}\right)^{\frac{1}{2}},$$

$$t_{1,54} = \left[(2m_{c1}/\hbar^2).\mathrm{V}_{11}(E, E_{g1}, \lambda, \Delta_1, \eta_{g1}) - \left[\frac{2eB}{\hbar}\left(n + \frac{1}{2}\right)\right]\right],$$
$$t_{2,54} = \big[(2m_{c1}/\hbar^2)\mathrm{V}_{21}(E, E_{g1}, \lambda, \Delta_1, \eta_{g1})\big],$$

$$h_{2,54} = e_{2,54}(b_0 - \Delta_0),\ e_{2,54} = 2^{\frac{-1}{2}}\left(\sqrt{t_{1,54}^2 + t_{2,54}^2} - t_{1,54}\right)^{\frac{1}{2}}$$

$$g_{1,54} = d_{1,54}(a_0 - \Delta_0),\ d_{2,54} = 2^{\frac{-1}{2}}\left(\sqrt{x_{1,54}^2 + y_{1,54}^2} + x_{1,54}\right)^{\frac{1}{2}},$$

$$x_{1,54} = \left[-(2m_{c2}/\hbar^2).\mathrm{V}_{11}(E - V_0, E_{g2}, \lambda, \Delta_2, \eta_{g2}) + \left[\frac{2eB}{\hbar}\left(\mathrm{n} + \frac{1}{2}\right)\right]\right],$$
$$y_{1,54} = \big[(2m_{c2}/\hbar^2)\mathrm{V}_{22}(E - V_0, E_{g2}, \lambda, \Delta_2, \eta_{g2})\big],$$

$$g_{2,54} = d_{2,54}(a_0 - \Delta_0), d_{2,54} = 2^{\frac{-1}{2}}\left(\sqrt{x_{1,54}^2 + y_{1,54}^2} - x_{1,54}\right)^{\frac{1}{2}}, \rho_{5,54}$$
$$= (\rho_{3,54}^2 + \rho_{4,54}^2)^{-1}\big[\rho_{1,54}\rho_{3,54} - \rho_{2,54}\rho_{4,54}\big],$$

$$\rho_{1,54} = \left[d_{1,54}^2 + e_{2,54}^2 - d_{2,54}^2 - e_{1,54}^2\right],\ \rho_{3,54} = [d_{1,54}e_{1,54} + d_{2,54}e_{2,54}],$$
$$\rho_{2,54} = 2\big[d_{1,54}d_{2,54} + e_{1,54}e_{2,54}\big],\ \rho_{4,54} = \big[d_{1,54}e_{2,54} - e_{1,54}d_{2,54}\big],$$

$$G_{2,54} = \big[(\sin(h_{1,54}))(\cosh(h_{2,54}))(\sinh(g_{1,54}))(\cos(g_{2,54}))$$
$$+ (\cos(h_{1,54}))(\sinh(h_{2,54}))(\cosh(g_{1,54}))(\sin(g_{2,54}))\big],$$
$$\rho_{6,54} = (\rho_{3,54}^2 + \rho_{4,54}^2)^{-1}[\rho_{1,54}\rho_{4,54} + \rho_{2,54}\rho_{3,54}],$$

$$H_{2,54} = \big[(\sin(h_{1,54}))(\cosh(h_{2,54}))(\sin(g_{2,54}))(\cosh(g_{1,54}))$$
$$-(\cos(h_{1,54}))(\sinh(h_{2,54}))(\sinh(g_{1,54}))(\cos(g_{2,54}))\big],$$
$$\rho_{7,54} = \left[(e_{1,54}^2 + e_{2,54}^2)^{-1}\left[e_{1,54}(d_{1,54}^2 - d_{2,54}^2) - 2d_{1,54}d_{2,54}e_{2,54}\right] - 3e_{1,54}\right],$$

$$G_{3,54} = \left[(\sin(h_{1,54}))(\cosh(h_{2,54}))(\cosh(g_{1,54}))(\cos(g_{2,54})) + (\cos(h_{1,54}))(\sinh(h_{2,54}))(\sinh(g_{1,54}))(\sin(g_{2,54}))\right],$$

$$\rho_{8,54} = \left[(e_{1,54}^2 + e_{2,54}^2)^{-1}\left[e_{2,54}(d_{1,54}^2 - d_{2,54}^2) - 2d_{1,54}d_{2,54}e_{2,54}\right] + 3e_{2,54}\right],$$

$$H_{3,54} = \left[(\sin(h_{1,54}))(\cosh(h_{2,54}))(\sin(g_{2,54}))(\sinh(g_{1,54})) - (\cos(h_{1,54}))(\sinh(h_{2,54}))(\cosh(g_{1,54}))(\cos(g_{2,54}))\right],$$

$$\rho_{9,54} = \left[(d_{1,54}^2 + d_{2,54}^2)^{-1}\left[d_{1,54}(e_{2,54}^2 - e_{1,54}^2) + 2e_{2,54}d_{2,54}e_{1,54}\right] + 3d_{1,54}\right],$$

$$G_{4,54} = \left[(\cos(h_{1,54}))(\cosh(h_{2,54}))(\cos(g_{2,54}))(\sinh(g_{1,54})) - (\sin(h_{1,54}))(\sinh(h_{2,54}))(\cosh(g_{1,54}))(\sin(g_{2,54}))\right],$$

$$\rho_{10,54} = \left[-(d_{1,54}^2 + d_{2,54}^2)^{-1}\left[d_{2,54}(-e_{2,54}^2 + e_{1,54}^2) + 2e_{2,54}d_{2,54}e_{1,54}\right] + 3d_{2,54}\right],$$

$$H_{4,54} = \left[(\cos(h_{1,54}))(\cosh(h_{2,54}))(\cosh(g_{1,54}))(\sin(g_{2,54})) + (\sin(h_{1,54}))(\sinh(h_{2,54}))(\sinh(g_{1,54}))(\cos(g_{2,54}))\right],$$

$$\rho_{11,54} = 2\left[d_{1,54}^2 e_{2,54}^2 - d_{2,54}^2 - e_{2,54}^2\right],$$

$$G_{5,54} = \left[(\cos(h_{1,54}))(\cosh(h_{2,54}))(\cos(g_{2,54}))(\cosh(g_{1,54})) - (\sin(h_{1,54}))(\sinh(h_{2,54}))(\sinh(g_{1,54}))(\sin(g_{2,54}))\right],$$

$$\rho_{12,54} = 4\left[d_{1,54}d_{2,54} + e_{1,54}e_{2,54}\right],$$

$$H_{5,54} = \left[(\cos(h_{1,54}))(\cosh(h_{2,54}))(\sinh(g_{1,54}))(\sin(g_{2,54})) + (\sin(h_{1,54}))(\sinh(h_{2,54}))(\cosh(g_{1,54}))(\cos(g_{2,54}))\right],$$

$$\rho_{13,54} = \left[\left\{5(d_{1,54}e_{1,54}^3 - 3e_{1,54}e_{2,54}^2 d_{1,54}) + 5d_{2,54}(e_{1,54}^3 - 3e_{1,54}^2 e_{2,54})\right\} (d_{1,54}^2 + d_{2,54}^2)^{-1} + (e_{1,54}^2 + e_{2,54}^2)^{-1} \left\{5(e_{1,54}d_{1,54}^3 - 3d_{2,54}e_{1,54}^2 d_{1,54}) + 5(d_{2,54}^3 e_{2,54} - 3d_{1,54}^2 d_{2,54}e_{2,54})\right\} - 34(d_{1,54}e_{1,54} + d_{2,54}e_{2,54})\right],$$

$$G_{6,54} = \left[(\sin(h_{1,54}))(\cosh(h_{2,54}))(\sinh(g_{1,54}))(\cos(g_{2,54})) + (\cos(h_{1,54}))(\sinh(h_{2,54}))(\cosh(g_{1,54}))(\sin(g_{2,54}))\right],$$

$$\rho_{14,54} = \left[\left\{5(d_{1,54}e_{2,54}^3 - 3e_{2,54}e_{1,54}^2 d_{1,54}) + 5d_{2,54}(-e_{1,54}^3 + 3e_{2,54}^2 e_{1,54})\right\} (d_{1,54}^2 + d_{2,54}^2)^{-1} + (e_{1,54}^2 + e_{2,54}^2)^{-1} \left\{5(-e_{1,54}d_{2,54}^3 + 3d_{1,54}^2 d_{2,54}e_{1,54}) + 5(-d_{1,54}^3 e_{2,54} + 3d_{2,54}^2 d_{1,54}e_{2,54})\right\} + 34(d_{1,54}e_{2,54} - d_{2,54}e_{1,54})\right],$$

$$
\begin{aligned}
H_{6,54} =& \left[(\sin(h_{1,54}))(\cosh(h_{2,54}))(\cosh(g_{1,54}))(\sin(g_{2,54})) \right.\\
& \left. (\cos(h_{1,54}))(\sinh(h_{2,54}))(\sinh(g_{1,54}))(\cos(g_{2,54})) \right],\\
H_{7,54} =& \left[H_{1,54} + (\rho_{5,54} H_{2,54}/2) + (\rho_{6,54} G_{2,54}/2) + (\Delta_0/2) \right.\\
& \{ H_{1,54} + (\rho_{5,54} H_{2,54}/2) + (\rho_{6,54} G_{2,54}/2) + (\Delta_0/2)\\
& \left. + \rho_{12,54} G_{5,54} + \rho_{11,54} H_{5,54} + (1/12)(\rho_{14,54} G_{6,54} + \rho_{13,54} H_{6,54}) \} \right],
\end{aligned}
$$

$$
\begin{aligned}
H_{1,54} =& \left[(\sin(h_{1,54}))(\sinh(h_{2,54}))(\cosh(g_{1,54}))(\cos(g_{2,54})) \right.\\
& \left. + (\cos(h_{1,54}))(\cosh(h_{2,54}))(\sinh(g_{1,54}))(\sin(g_{2,54})) \right],\\
D_{7,54} =& \sinh^{-1}(\bar{\omega}_{7,54}),\ H_{8,54} = (2C_{7,54} D_{7,54} / L_0^2)
\end{aligned}
$$

The sub-band equation in this case can be expressed as

$$
0 = [G_{8,17,54} + iH_{8,17,54}]\big|_{E=E_{17,54}} \tag{18.231}
$$

where $E_{17,54}$ is the Landau sub-band energy in this case.

At low temperatures where the quantum effects become prominent, the DOS function for the lowest SL mini-band is given by

$$
N_{HDSL}(E, \eta_g, \lambda) = \frac{g_v eB}{2\pi^2 \hbar} \sum_{n=0}^{n_{\max}} \frac{[G'_{8,17,54} + iH'_{8,17,54}]}{\sqrt{G_{8,17,54} + iH_{8,17,54}}} H(E - E_{17,54}) \tag{18.232}
$$

The EEM can be written as

$$
m^*(E, n_x, n_y, \lambda, \eta_g) = \frac{\hbar^2}{2} (G'_{8,17,54}) \tag{18.233}
$$

18.2.20 The Magneto DR in QWHD Superlattices of Kane Type Semiconductors with Graded Interfaces in the Presence of Light Waves

The magneto DR in QWHD superlattices of Kane type semiconductors with graded interfaces in the presence of light waves can be expressed as

$$
\left(\frac{n_z \pi}{d_z} \right)^2 = [G_{8,17,54} + iH_{8,17,54}]\Big|_{E=E_{17,55}} \tag{18.234}
$$

where $E_{17,55}$ is the totally quantized energy in this case.

The DOS function is given by

$$N_{QDHDSL}(E, \eta_g, \lambda) = \frac{g_v}{\pi\hbar} \sum_{n_z=1}^{n_{z\max}} \sum_{n=0}^{n_{\max}} \delta'(E - E_{17,55}) \tag{18.235}$$

18.3 Summary and Conclusion

The (18.46b), (18.47) and (18.48a) represent the DR in HD Kane type semiconductors (i.e. III–V, ternary, quaternary compounds) and in the presence of strong light waves whose unperturbed conduction electrons obey the three, two and one band models of Kane. The DRs depend on the intensity and wave length of the incident waves in addition to the screening potential. The EEMs ((18.49), (18.50), (18.51)) and the DOS functions ((18.48b), (18.48c) and (18.48d)) for (18.46b), (18.47) and (18.48a) have also been formulated which exhibit the signature of the light waves. The (18.52), (18.54) and (18.60) represent the magneto DR in HD Kane type materials in the present case. The EEMs ((18.55), (18.59), (18.63)) and the DOS functions ((18.53), (18.57), (18.61)) have been derived for ((18.52), (18.54) and (18.60)) and they also depend on light wave length. The (18.64), (18.68) and (18.72) represent the DR in HD Kane type materials under cross fields configurations in the present case. The anisotropic EEMs are given by (18.65) and (18.66) for (18.64) and (18.69) for (18.68) and (18.73) and (18.74) for (18.72) respectively in this case. The mass anisotropy depends on the magnetic field, quantum number, electric field and other physical constants together with the fact that the screening potential dependent EEMs exist in the band gap, a fact impossible without heavy doping. The (18.76), (18.80) and (18.84) represent the DR in QW of HD Kane type materials in the presence of external light waves. The EEMs ((18.79), (18.83), (18.87)) and the DOS functions ((18.78), (18.82), (18.86)) have been derived for ((18.76), (18.80) and (18.84)) and they also depend on light wave length. The (18.88), (18.92) and (18.96) represent the DR in doping superlattices of HD Kane type materials in the presence of external light waves. The EEMs ((18.90), (18.94), (18.98)) and the DOS functions ((18.91), (18.95), (18.99)) have been derived for ((18.88), (18.92) and (18.98)) and they also depend on light wave length. The EEMs in doping superlattices will depend on mini band index which is the characteristics property of such quantized structures together with the fact in the present case the same mass will be strongly influenced by the intensity of the incident light waves.

The DR in QBs of HD Kane type materials under strong light excitation is given by (18.100), (18.104) and (18.108) and the corresponding DOS functions are reflected by (18.101), (18.105) and (18.109) respectively. The DRs are totally quantized due to the quantization of the wave vector space and the DOS functions are unevenly distributed in the energy axis in the forms of Dirac's delta function. The magneto DR in QWs of HD Kane type compounds under light waves is given by (18.112), (18.114) and (18.116) and the corresponding DOS functions can be expressed by (18.113), (18.115) and (18.117) respectively like the previous case, the DOS functions are represented by the Dirac delta function. The DRs in accumulation layers of Kane type compounds in the presence of light waves are given by (18.118), (18.126) and (18.134). The EEMs ((18.119), (18.127), (18.135)) and the DOS functions ((18.121), (18.129), (18.137)) have been derived for ((18.118), (18.126) and (18.134)) and they also depend on light wave length. The (18.142), (18.150) and (18.158) represent the DR in NWs of HD Kane type materials in the presence of external light waves. The EEMs ((18.145), (18.153), (18.161)) and the DOS functions ((18.143), (18.151), (18.159)) have been derived for ((18.145), (18.153) and (18.161)) and they also depend on light wave length. The magneto DR in accumulation layers of HD Kane type compounds under light waves is given by (18.166), (18.170) and (18.174) and the corresponding DOS functions can be expressed by (18.167), (18.171) and (18.175) respectively like the previous case, the DOS functions are represented by the Dirac delta function. The magneto DR in doping superlattices of HD Kane type compounds under light waves is given by (18.178), (18.180) and (18.182) and the corresponding DOS functions can be expressed by (18.179), (18.181) and (18.183) respectively like the previous case, the DOS functions are represented by the Dirac delta function. The DR in QWHD effective mass superlattices of Kane type compounds under light waves is given by (18.185), (18.189) and (18.193) and the corresponding EEMs can be expressed by (18.186), (18.190) and (18.194) respectively. The EEMs are quantum number dependent in addition to other physical variables which is the characteristics feature of such superlattices. The DR in NWHD effective mass superlattices are given by (18.196), (18.199) and (18.202) respectively and their use lead to expressions of EEM and sub-band energies as given by (18.197), (18.200) and (18.203) and (18.198), (18.201) and (18.204) respectively. As usual, the effective masses are functions of quantum numbers in addition to other physical variables. The magneto DRs in HD effective mass superlattices of Kane type semiconductors in the presence of light waves are given by (18.205), (18.208) and (18.211) and their use lead to the expressions of EEM and DOS functions as (18.206a), (18.209) and (18.212a) and (18.206b), (18.209) and (18.212b) respectively. The EEMs are as usual quantum number dependent. The magneto DRs in QWHD effective mass superlattices of Kane type semiconductors in the presence of light waves are given by (18.214), (18.216) and (18.218) and their use lead to the expressions of the DOS functions as represented by (18.215), (18.217) and (18.219) respectively. The DR in QWHD superlattices of Kane type semiconductors with graded interfaces in the presence of light waves is given by (18.222) and sub-band energies are expressed through (18.223). The EEM and the DOS function should be obtained numerically in this case. The DR in NWHD superlattices of Kane Type semiconductors with graded interfaces in the presence of light waves is given by (18.224). The EEM and DOS function for this case are given by (18.227) and (18.226) respectively. The DR in Quantum dot HD superlattices of Kane type semiconductors with graded interfaces in the presence of light waves is given by (18.228) and totally quantized DOS function is represented by (18.229). The magneto DR in HD superlattices of Kane type semiconductors with graded interfaces in the presence of light waves is given by (18.230). The EEM and DOS function for this case are given by (18.233) and (18.232) respectively. The last section of theoretical background investigates the magneto DR in QWHD superlattices of Kane type semiconductors with graded interfaces in the presence of light waves which is given by (18.234) and the delta functional DOS function is represented through (18.235).

18.4 Open Research Problems

(R.18.1) Investigate the DR in the presence of intense external non-uniform light waves for all the HD superlattices whose respective dispersion relations of the carriers are given in this chapter

(R.18.2) Investigate the DR for the heavily–doped semiconductors in SLs the presences of Gaussian, exponential, Kane, Halperian, Lax and Bonch-Burevich types of band tails for all SL systems as discussed in this chapter in the presence of external oscillatory and non-uniform light waves

(R.18.3) Investigate the DR in the presence of external light waves for short period, strained layer, random and Fibonacci HD superlattices in the presence of an arbitrarily oriented alternating electric field

(R.18.4) Investigate all the appropriate problems of this chapter for a Dirac electronDirac electron

(R.18.5) Investigate all the appropriate problems of this chapter by including the many body, broadening and hot carrier effects respectively

(R.18.6) Investigate all the appropriate problems of this chapter by removing all the mathematical approximations and establishing the respective appropriate uniqueness conditions.

References

1. P.K. Basu, *Theory of Optical Process in Semiconductors, Bulk and Microstructures* (Oxford University Press, Oxford, 1997)
2. K.P. Ghatak, S. Bhattacharya, S. Bhowmik, R. Benedictus, S. Chowdhury, J. Appl. Phys. **103**, 094314 (2008); K.P.Ghatak, S.Bhattacharya, J.Appl.Phys.**102**, 073704 (2007); K.P. Ghatak, S. Bhattacharya, S.K. Biswas, A. De, A.K. Dasgupta, Phys.Scr.,**75**, 820 (2007); P.K. Bose, N. Paitya, S. Bhattacharya, D. De, S. Saha, K.M. Chatterjee, S. Pahari, K.P. Ghatak, Quantum Matter, **1**, 89 (2012); K. P. Ghatak, S. Bhattacharya, A. Mondal, S. Debbarma, P. Ghorai, and A. Bhattacharjee, Quantum Matter, **2**, 25 (2013); S. Bhattacharya, D. De, S. Ghosh, J. P. Bannerje, M. Mitra, B. Nag, S. Saha, S. K. Bishwas, M. Paul, J. Comp. Theo. Nanosci, **7**, 1066 (2010); K.P. Ghatak, S. Bhattacharya, S. Pahari, S.N. Mitra, P.K. Bose, J. Phys. Chem. Solids, **70**, 122 (2009), S. Bhattacharya, D. De, R. Sarkar, S. Pahari, A. De, A.K. Dasgupta, S.

N. Biswas, K.P. Ghatak, J. Comp. Theo. Nanosci, **5**, 1345 (2008); S. Mukherjee, D. De, D. Mukherjee, S. Bhattacharya, A. Sinha, K.P. Ghatak Physica B, **393**, 347 (2007)
3. K. Seeger, Semiconductor Physics, 7th edn. (Springer-Verlag, 2006)
4. B.R. Nag, *Physics of Quantum Well Devices* (Kluwer Academic Publishers, The Netherlands, 2000)
5. R.K. Pathria, *Statistical Mechanics*, 2nd edn. (Butterworth-Heinmann, Oxford, 1996)

Part V
Dispersion Relations in HD Kane Type Semiconductors in the Presence of Intense Electric Field

Information is just bits of Data. Knowledge is putting them together, but the wisdom is transcending them.

Chapter 19
The DR Under Intense Electric Field in HD Kane Type Semiconductors

Some cause happiness where ever they go; others whenever they go.

19.1 Introduction

With the advent of modern nano devices, there has been considerable interest in studying the electric field induced processes in semiconductors having different band structures. It appears from the literature that the studies have been made on the assumption that the carrier dispersion laws are invariant quantities in the presence of intense electric field, which is not fundamentally true. In this chapter, we shall study the DR in quantum confined optoelectronic semiconductors under strong electric field. In Sect. 19.2.1, and attempt is made to investigate the DR in the presence of intense electric field in HD III–V, ternary and quaternary semiconductors The Sect. 19.2.2 contains the investigation of the DR under magnetic quantization in HD Kane type semiconductors in the presence of intense electric field. In Sect. 19.2.3, the DR in QWs of HD Kane Type Semiconductors in the Presence of intense electric field has been studied. In the Sect. 19.2.4 we have investigated the DR in NWs of HD Kane type semiconductors in the presence of intense electric field. In Sect. 19.2.5 the DR in QBs in HD Kane type semiconductors in the presence of intense electric field has been studied. In Sect. 19.2.6 the magneto DR in QWs in HD Kane type semiconductors in the presence of intense electric field has been studied. In Sect. 19.2.7 the DR in accumulation and inversion layers of Kane type semiconductors in the presence of intense electric field has been studied. In Sect. 19.2.8 the magneto DR in accumulation and inversion layers of Kane type semiconductors in the presence of intense electric field has been studied. In Sect. 19.2.9 the DR in doping superlattices of HD Kane type semiconductors in the presence of intense electric field has been studied. In Sect. 19.2.10 the magneto DR in doping superlattices of HD Kane type semiconductors in the presence of intense electric field has been investigated. In Sect. 19.2.11 the DR in QWHD effective mass superlattices of Kane type semiconductors in the presence of intense

K.P. Ghatak, *Dispersion Relations in Heavily-Doped Nanostructures*,
Springer Tracts in Modern Physics 265, DOI 10.1007/978-3-319-21000-1_19

electric field has been investigated. In Sect. 19.2.12 the DR in NWHD effective mass superlattices of Kane type semiconductors in the presence of intense electric field has been studied. In Sect. 19.2.13 the DR in Quantum dot HD superlattices of Kane type semiconductors in the presence of intense electric field has been studied. In Sect. 19.2.14 the magneto DR in HD effective mass superlattices of Kane type semiconductors in the presence of intense electric field has been studied. In Sect. 19.2.15 the DR in QWHD superlattices of Kane type semiconductors with graded interfaces in the presence of intense electric field has been studied. In Sect. 19.2.16 the DR in NWHD superlattices of Kane type semiconductors with graded interfaces in the presence of intense electric field has been investigated. In Sect. 19.2.17 the DR in Quantum dot HD superlattices of Kane type semiconductors with graded interfaces in the presence of intense electric field has been studied. In Sect. 19.2.18 the magneto DR in HD superlattices of Kane type semiconductors with graded interfaces in the presence of intense electric field has been investigated. In Sect. 19.2.19 the magneto DR in QWHD superlattices of Kane type semiconductors with graded interfaces in the presence of intense electric field has been investigated. The Sect. 19.3 contains the summary and conclusion pertinent to this chapter. The Sect. 19.4 presents 6 open research problems which challenges the first order creativity from the readers of diverse fields.

19.2 Theoretical Background

19.2.1 The Formulation of the Electron Dispersion Law in the Presence of Intense Electric Field in HD III–V, Ternary and Quaternary Semiconductors

The expression of the inter-band transition matrix element (X_{12}) in this case can be written as

$$X_{12} = i \int u_{\bar{k}1}^{*}(\bar{r}).\frac{\partial}{\partial k_x} u_{\bar{k}2}(\bar{r}) d^3 r \tag{19.1}$$

where $u_{\bar{k}1}(\bar{r}) \equiv u_1(\bar{k}, \bar{r})$ and $u_{\bar{k}2}(\bar{r}) \equiv u_2(\bar{k}, \bar{r})$ in which $u_1(\bar{k}, \bar{r})$ and $u_2(\bar{k}, \bar{r})$ are given by (18.15) and (18.16) respectively.

In the case of the presence of an external electric field, F_s along x axis, the inter-band transition matrix-element, X_{12}, has finite interaction band same band, e.g.,

$$\langle S \mid S \rangle = \langle X \mid X \rangle = \langle Y \mid Y \rangle = \langle Z \mid Z \rangle = 1$$
$$\langle X \mid Y \rangle = \langle Y \mid Z \rangle = \langle Z \mid X \rangle = 0$$
$$\langle S \mid X \rangle = \langle X \mid S \rangle = 0; \qquad \langle S \mid Y \rangle = \langle Y \mid S \rangle = 0 \text{ and } \langle S \mid Z \rangle = \langle Z \mid S \rangle = 0.$$

Using the appropriate equations we can write

$$X_{12} = i\int d^3r\left\{a_{k_+}[(iS)\downarrow'] + b_{k_+}\left[\left(\frac{X' - iY'}{\sqrt{2}}\right)\uparrow'\right] + c_{k_+}[Z'\downarrow']\right\}^* \frac{\partial}{\partial k_x}$$
$$\left\{a_{k_-}[(iS)\uparrow'] - b_{k_-}\left[\left(\frac{X' + iY'}{\sqrt{2}}\right)\downarrow'\right] + c_{k_-}[Z'\uparrow']\right\}$$
$$= i\int d^3r\left[\left\{\left(a_{k_+}\frac{\partial}{x}a_{k_-}\right).([(iS)\downarrow']^*.[(iS)\uparrow']) + \left(\frac{b_{k_+}}{\sqrt{2}}\frac{\partial}{\partial k_x}a_{k_-}\right)[(X' - iY')\uparrow']^*.[(iS)\uparrow']\right.\right.$$
$$\left.\left. + \left(c_{k_+}\frac{\partial}{\partial k_x}a_{k_-}\right)[Z'\downarrow']^*.[(iS)\uparrow']\right\}\right]$$
$$+ i\int d^3r\left[\left\{\left(-\frac{a_{k_+}}{\sqrt{2}}\frac{\partial}{\partial k_x}b_{k_-}\right).([(iS)\downarrow']^*.[(X' - iY')\downarrow']) - \left(\frac{b_{k_+}\frac{\partial}{\partial k_x}b_{k_-}}{2}\right)[(X' - iY')\uparrow']^*.[(X' - iY')\downarrow']\right.\right.$$
$$\left.\left. - \left(\frac{c_{k_+}\frac{\partial}{\partial k_x}b_{k_-}}{\sqrt{2}}\right)[(Z'\downarrow')]^*.[(X' + iY')\downarrow']\right\}\right]$$
$$= i\int d^3r\left[\left\{\left(a_{k_+}\frac{\partial}{\partial k_x}c_{k_-}\right).([(iS)\downarrow']^*.[Z'\uparrow']) + \left(\frac{b_{k_+}}{\sqrt{2}}\frac{\partial}{\partial k_x}c_{k_-}\right)[(X' - iY')\uparrow']^*.[Z'\uparrow']\right.\right.$$
$$\left.\left. + \left(c_{k_+}\frac{\partial}{x}c_{k_-}\right)[(Z'\downarrow')]^*.[Z'\uparrow']\right\}\right] \tag{19.2}$$

Therefore,

$$\frac{X_{12}}{i} = \left[\left\{\left[\left(a_{k_+}\frac{\partial}{\partial k_x}a_{k_-}\right).\langle iS\,|\,iS\rangle\langle\downarrow'\,|\,\uparrow'\rangle\right] + \left[\left(\frac{b_{k_+}}{\sqrt{2}}\frac{\partial}{\partial k_x}a_{k_-}\right)\langle(X' - iY')iS\rangle\langle\downarrow'\,|\,\uparrow'\rangle\right]\right.\right.$$
$$\left.\left. + \left[\left(c_{k_+}\frac{\partial}{\partial k_x}a_{k_-}\right)\langle Z'\,|\,iS\rangle\langle\downarrow'\,|\,\uparrow'\rangle\right]\right\}\right]$$
$$- \left[\left\{\left[\left(\frac{a_{k_+}}{\sqrt{2}}\frac{\partial}{\partial k_x}b_{k_-}\right).\langle iS\,|\,(X' + iY')\rangle\langle\downarrow'\,|\,\downarrow'\rangle\right] + \left[\left(\frac{b_{k_+}}{2}\frac{\partial}{\partial k_x}b_{k_+}\right)\langle(X' - iY')|(X' + iY')\rangle\langle\uparrow'\,|\,\downarrow'\rangle\right]\right.\right.$$
$$\left.\left. + \left[\left(\frac{c_k}{\sqrt{2}}\frac{\partial}{\partial k_x}b_{k_-}\right)\langle Z'|(X' + iY')\rangle\langle\downarrow'\,|\,\downarrow'\rangle\right]\right\}\right]$$
$$+ \left\{\left[\left(a_k\frac{\partial}{\partial k_x}c_{k_-}\right).\langle iS\,|\,Z'\rangle\langle\downarrow'\,|\,\uparrow'\rangle\right] + \left[\left(\frac{b_{k_+}}{\sqrt{2}}\frac{\partial}{\partial k_x}c_{k_-}\right).\langle(X' - iY')|Z'\rangle\langle\uparrow'\,|\,\uparrow'\rangle\right]\right.$$
$$\left. + \left[\left(c_k\frac{\partial}{\partial k_x}c_{k_-}\right).\langle Z'|Z'\rangle\langle\downarrow'\,|\,\uparrow'\rangle\right]\right\} \tag{19.3}$$

Therefore we can write,

$$X_{12} = i\left\{-\left(a_{k_+}\frac{\partial}{\partial k_x}a_{k_-}\right)\langle\downarrow'\,|\,\uparrow'\rangle + \left(c_{k_+}\frac{\partial}{\partial k_x}c_{k_-}\right)\langle\downarrow'\,|\,\uparrow'\rangle\right\} \tag{19.4}$$

We can prove that

$$\langle\downarrow'\,|\,\uparrow'\rangle = \frac{1}{2}(\hat{r}_1 + i\hat{r}_2) \tag{19.5}$$

Therefore (19.4) and (19.5) we get

$$\begin{aligned} X_{12} &= i\left\{-\left(a_{k_+}\frac{\partial}{\partial k_x}a_{k-}\right)+\left(c_{k_+}\frac{\partial}{\partial k_x}c_{k_-}\right)\right\}\langle\downarrow'\mid\uparrow'\rangle \\ &= -i\left\{\left(a_{k_+}\frac{\partial}{\partial k_x}a_{k_-}\right)-\left(c_{k_+}\frac{\partial}{\partial k_x}c_{k_-}\right)\right\}.\frac{1}{2}(\hat{r}_1+i\hat{r}_2) \\ &= \frac{-iA(k)}{2}(\hat{r}_1+i\hat{r}_2) \end{aligned} \tag{19.6}$$

where

$$A(k)=\left(a_{k_+}\frac{\partial}{\partial k_x}a_{k_-}\right)-\left(c_{k+}\frac{\partial}{dk_x}c_{k_-}\right) \tag{19.7}$$

From (19.6), we find,

$$|X_{12}|^2=\frac{1}{4}A^2(k)(1+1)=\frac{1}{2}A^2(k)\ \ [\text{since, } |\hat{r}_1|=|\hat{r}_2|=1] \tag{19.8}$$

Considering spin-up and spin-down, we have to multiply by 2

$$|X_{12}|^2=2\times\frac{1}{2}A^2(k)=A^2(k) \tag{19.9}$$

We can evaluate X_{11} and X_{22}. in the following way :

$$\begin{aligned} X_{11} &= i\int u^*_{\bar{k}1}(\bar{r}).\frac{\partial}{\partial k_x}u_{\bar{k}1}(\bar{r}).d^3r \\ &= i\int d^3r\left\{\left(a_{k_+}\frac{\partial}{\partial k_x}a_{k_+}\right)+\left(b_{k_+}\frac{\partial}{\partial k_x}b_{k_+}\right)+\left(c_{k_+}\frac{\partial}{\partial k_x}c_{k_+}\right)\right\} \\ &= \frac{1}{2}i\int d^3r\left\{\frac{\partial}{\partial k_x}(a_k^2+b_k^2+c_k^2)\right\}=\frac{1}{2}i\int d^3r\left\{\frac{\partial}{\partial k_x}(1)\right\}=0,\ \text{ since } a^2_{k_+}+b^2_{k_+}+c^2_{k_+}=1. \end{aligned}$$

Therefore $X_{11}=0$, and similarly we can prove $X_{22}=0$. Thus we conclude that intraband momentum matrix element due to external electric field (X_{CC}) is zero. From the expression of $a_{k\pm}$ we can write

$$a^2{}_{k_+}=r_0^2\left[\frac{E_g-\gamma^2_{k_+}(E_g-\delta')}{E_g+\delta'}\right]^2 \text{ and } a^2_{k-}={r_0}^2\left[\frac{E_g-\gamma^2_{k-}(E_g-\delta')}{E_g+\delta'}\right]^2.$$

Therefore,

$$2a_{k-}\frac{\partial}{\partial k_x}a_{k_-} = {r_0}^2\left[-\left(\frac{E_g-\delta'}{E_g+\delta'}\right)\right]\frac{\partial\gamma_{k_-}^2}{\partial k_x} \text{ and } \frac{\partial a_{k_-}}{\partial k_x} = -\frac{{r_0}^2}{2}\left(\frac{E_g-\delta'}{E_g+\delta'}\right)\frac{1}{a_{k_-}}\frac{\partial\gamma_{k_-}^2}{\partial k_x}$$

Combining we can write

$$a_{k+}\frac{\partial a_{k-}}{\partial k_x} = -\frac{{r_0}^2}{2}\left[\left(\frac{E_g-\delta'}{E_g+\delta'}\right)\right]\frac{a_{k+}}{a_{k-}}\frac{\partial\gamma_{k-}^2}{\partial k_x}$$

Similarly, $c_{k_+} = t\gamma_{k_+}$ and $c_{k_-} = t\gamma_{k_-}$. Therefore, $c_{k+}\frac{\partial}{\partial k_x}c_{k-} = \frac{t^2}{2}\frac{c_{k_+}}{c_{k_-}}\frac{\partial\gamma_{k_-}^2}{\partial k_x}$

$$A(k) = \left\{-\frac{r_0^2}{2}\left[\left(\frac{E_g-\delta'}{E_g+\delta'}\right)\right]\frac{a_{k_+}}{a_{k_-}} - \frac{t^2}{2}\frac{c_{k_+}}{c_{k_-}}\right\}\frac{\partial\gamma_{k_-}^2}{\partial k_x}$$

Now,

$$\left(\frac{a_{k_+}}{a_{k_-}}\right)^2 = \frac{E_g-\gamma_{k_+}^2(E_g-\delta')}{E_g-\gamma_{k_-}^2(E_g-\delta')} = \frac{E_g-\frac{\eta-E_g}{2(\eta+\delta')}(E_g-\delta')}{E_g-\frac{\eta+E_g}{2(\eta+\delta')}(E_g-\delta')}$$

$$= \frac{2E_g-(\eta+\delta')-(\eta-E_g)(E_g-\delta')}{2E_g(\eta+\delta')-(\eta+E_g)(E_g-\delta')} = \frac{\eta(E_g+\delta')+E_g(E_g+\delta')}{\eta(E_g+\delta')-E_g(E_g-3\delta')}$$

Therefore, $\left(\frac{a_{k_+}}{a_{k_-}}\right)^2 = \frac{\eta+E_g}{\eta-E_g\left(\frac{E_g-3\delta'}{E_g+\delta'}\right)} = \frac{\eta+E_g}{\eta-E_g'}$ where, $E_g' = \frac{E_g(E_g-3\delta')}{E_g+\delta'}$. Thus, $\frac{a_{k_+}}{a_{k_-}} = \sqrt{\frac{\eta+E_g}{\eta-E_g'}}$. Similarly, $\frac{c_{k_+}}{c_{k_-}} = \frac{\gamma_{k_+}}{\gamma_{k_-}} = \sqrt{\frac{\eta-E_g}{\eta+E_g}}$ and thus,

$$A(k) = -\left\{P\left(\frac{\eta+E_g}{\eta-E_g'}\right)^{1/2} + Q\left(\frac{\eta-E_g}{\eta+E_g}\right)^{1/2}\right\}\frac{\partial\gamma_{k_-}^2}{\partial k_x} \tag{19.10}$$

where, $P = \frac{r_0^2}{2}\left(\frac{E_g-\delta'}{E_g+\delta'}\right)$ and $Q = t^2/2$.

Now $\gamma_{k_-}^2 = \frac{\eta+E_g}{2(\eta+\delta')}$ so that $\frac{\partial\gamma_{k_-}^2}{\partial k_x} = \frac{1}{2}\left[\frac{\partial\eta/\partial k_x}{(\eta+\delta')} - \frac{\eta+E_g}{(\eta+\delta')^2}\frac{\partial\eta}{\partial k_x}\right]$

Thus,

$$\frac{\partial\gamma_{k_-}^2}{\partial k_x} = \frac{1}{2}\left[\frac{\eta+\delta'-\eta-E_g}{(\eta+\delta')}\right]\frac{\partial\eta}{\partial k_x} = -\frac{1}{2}\frac{(E_g-\delta')}{(\eta+\delta')^2}\cdot\frac{\partial\eta}{\partial k_x} \tag{19.11}$$

From (19.10) and (19.11), we get

$$A(k) = \frac{1}{2}\frac{(E_g - \delta')}{(\eta + \delta')^2} \cdot \frac{\partial \eta}{\partial k_u} \cdot \left\{ P\left(\frac{\eta + E_g}{\eta - E'_g}\right)^{1/2} + Q\left(\frac{\eta - E_g}{\eta + E_g}\right)^{1/2} \right\} \tag{19.12}$$

This implies

$$\frac{\partial \eta}{\partial k_x} = \frac{E_g \hbar^2}{m_r} \cdot \frac{k_x}{\eta} \tag{19.13}$$

From (19.12) and (19.13), one can write

$$A(k) = \frac{E_g \hbar^2}{2m_r} \cdot \frac{k_u}{\eta} \cdot \frac{(E_g - \delta')}{(\eta + \delta')^2} \cdot \left\{ P\left(\frac{\eta + E_g}{\eta - E'_g}\right)^{1/2} + Q\left(\frac{\eta - E_g}{\eta + E_g}\right)^{1/2} \right\} \tag{19.14}$$

Thus,

$$|A(k)|^2 = \frac{E_g^2 (E_g - \delta')\hbar^2}{4m_r} \frac{\hbar^2 k_x^2}{m_r} \frac{1}{\eta^2} \frac{1}{(\eta + \delta')^4} \cdot \left\{ P\left(\frac{\eta + E_g}{\eta - E'_g}\right)^{1/2} + Q\left(\frac{\eta - E_g}{\eta + E_g}\right)^{1/2} \right\}^2 \tag{19.15}$$

and

$$|X_{12}|^2 = |A(\bar{k})|^2. \tag{19.16}$$

From (19.15) and (19.16) we can write the square of the magnitude of the inter-band transition matrix element due to external electric field $\left(|X_{CV}|^2\right)$

It is well-known that the energy Eigen value, $E_n^{(2)}(\bar{k})$ in the presence of a perturbed Hamiltonian, H', is given by [1]

$$E_n^{(2)}(\bar{k}) = E_n(\bar{k}) + \langle n\bar{k} \,|\, H' \,|\, n\bar{k} \rangle + \left\{ |\langle n\bar{k} \,|\, H' \,|\, n\bar{k} \rangle|^2 \big/ [E_n(\bar{k}) - E_m(\bar{k})] \right\} \tag{19.17}$$

where

$$H\psi_n(\bar{k}, \bar{r}) = E\psi(\bar{k}, \bar{r}) \tag{19.18}$$

$$H = H_0 + H' \tag{19.19}$$

$$H_0 u_n(\bar{k}, \bar{r}) = E_n(\bar{k}) u_n(\bar{k}, \bar{r}) \tag{19.20}$$

in which, H is the total Hamiltonian, $\psi(\bar{k}, \bar{r})$ is the wave function, where $u_n(\bar{k}, \bar{r})$ is the periodic function of it, H_0 is the unperturbed Hamiltonian, n is the band index, and $E_n(\bar{k})$ is the energy of an electron in the periodic lattice.

For an external electric field (F_s) applied along the x-axis, the perturbed Hamiltonian (H') can be written as

$$H' = -F \cdot x \tag{19.21}$$

where

$$F(= eF_s)$$

Therefore we get

$$E_n^{(2)}(\bar{k}) = E_n(\bar{k}) - F\langle n\bar{k} \,|\, H' \,|\, n\bar{k}\rangle + F^2\left\{|\langle n\bar{k} \,|\, H' \,|\, n\bar{k}\rangle|^2 \Big/ [E_n(\bar{k}) - E_m(\bar{k})]\right\} \tag{19.22}$$

In (19.22), the second and the third terms are due to the perturbation factor.

$$\text{For } X_{nm}(\bar{k}) = \langle n\bar{k}|x|m\bar{k}\rangle \tag{19.23}$$

we find

$$X_{nm}(\bar{k}) = i \int u_n^*(\bar{k}, \bar{r})(\partial/\partial u)[u_m(\bar{k}, \bar{r})]d^3r \tag{19.24}$$

where k_x is the x component of the $\bar{k}$ and the integration in (19.24) extends over the unit cell. From (19.22), (19.23) and (19.24) with the n corresponds to the conduction band (C) and m corresponds to the valance band (V), we get

$$E_c^{(2)}(\bar{k}) = E_c(\bar{k}) - FX_{cc} + \left\{F^2|X_{CV}|^2 \Big/ [E_c(\bar{k}) - E_V(\bar{k})]\right\} \tag{19.25}$$

Thus combining the appropriate equations, the dispersion relation of the conduction electrons in the presence of electric field along x-axis can be written as

$$\begin{aligned} I_{11}(E) &= \frac{\hbar^2 k_x^2}{2m_c} + \frac{\hbar^2 k_y^2}{2m_c} + \frac{\hbar^2 k_z^2}{2m_c} + \frac{F^2 |X_{12}|^2}{\eta} \\ &= \left[\frac{\hbar^2 k_x^2}{2m_c} + \frac{\hbar^2 k_y^2}{2m_c} + \frac{\hbar^2 k_z^2}{2m_c} + \left\{ \frac{\hbar^2 k_x^2}{2m_c} \cdot \frac{2m_c}{m_r} \cdot \frac{F^2 \hbar^2 E_g^2 (E_g - \delta')^2}{4m_r} \frac{1}{\eta^3} \frac{1}{(\eta + \delta')^4} \left[P\left(\frac{\eta + E_g}{\eta - E_g'} \right)^{1/2} \right.\right.\right. \\ &\quad \left.\left.\left. + Q\left(\frac{\eta - E_g}{\eta + E_g} \right)^{1/2} \right]^2 \right\} \right] \end{aligned} \tag{19.26}$$

When $F \to 0$, we have from (19.26), $k^2 \to \frac{2m_c}{\hbar^2} I_{11}(E)$ and $\eta_1^2 = \left[E_g^2 + E_g \frac{2m_c}{m_r} I_{11}(E) \right]$. Using the method of successive approximation one can write

$$1 = \frac{\hbar^2 k_x^2}{2m_c I_{11}(E)} + \frac{\hbar^2 k_y^2}{2m_c I_{11}(E)} + \frac{\hbar^2 k_z^2}{2m_c I_{11}(E)} + \frac{\hbar^2 k_x^2}{2m_c I_{11}(E)} \cdot \Phi(E, F) \tag{19.27}$$

where,

$$\Phi(E, F) = \frac{2m_c}{m_r} \frac{F^2 \hbar^2 E_g^2 (E_g - \delta')}{4m_r} \frac{1}{\eta_1^3} \frac{1}{(\eta_1 + \delta')^4} \left[P\left(\frac{\eta_1 + E_g}{\eta_1 - E_g'} \right)^{1/2} + Q\left(\frac{\eta_1 - E_g}{\eta_1 + E_g} \right)^{1/2} \right]^2$$

Therefore, the *E*-*k* dispersion relation in the presence of an external electric field for III–V, ternary and quaternary materials whose unperturbed energy band structures are defined by the three band model of Kane can be expressed as

$$\frac{k_x^2}{\frac{2m_c}{\hbar^2} \left[\frac{I_{11}(E)}{1 + \Phi(E,F)} \right]} + \frac{k_y^2}{\frac{2m_c}{\hbar^2} I_{11}(E)} + \frac{k_z^2}{\frac{2m_c}{\hbar^2} I_{11}(E)} = 1 \tag{19.28}$$

In (19.28), the coefficients of k_x, k_y and k_z are not same and for this reason, this basic equation is "anisotropic" in nature together with the fact that the anisotropic dispersion relation is the ellipsoid of revolution in the k-space.

From (19.28) the expressions of the effective electron masses along x, y, and z directions can, respectively, be written as

$$m_x^*(E,F) = \hbar^2 k_x \frac{\partial k_x}{\partial E}\bigg|_{k_y=0,k_z=0} = m_c[1+\Phi(E,F)]^{-2}\left[[1+\Phi(E,F)]I'_{11}(E) - I_{11}(E)\Phi'(E,F)\right] \tag{19.29}$$

$$m_y^*(E,F) = \hbar^2 k_y \frac{\partial k_y}{\partial E}\bigg|_{k_x=0,k_z=0} = m_0 I'_{11}(E) \tag{19.30}$$

$$m_z^*(E,F) = \hbar^2 k_z \frac{\partial k_z}{\partial E}\bigg|_{k_x=0,k_y=0} = m_0 I'_{11}(E) \tag{19.31}$$

where $I'_{11}(E) = \frac{\partial}{\partial E}(I_{11}(E))$ and $\Phi'(E,F) = \frac{\partial}{\partial E}[\Phi(E,F)]$. It may be noted from (19.29) that the effective mass along x-direction is a function of both electron energy and electric field respectively whereas from (19.30) and (19.31) we can infer the expressions of the effective masses along y and z directions are same and they depend on the electron energy only. Thus in the presence of an electric field, the mass anisotropy for Kane type semiconductors depends both on electron energy and electric field respectively.

The use of the usual approximation [2]

$$k_x^2 \approx \frac{1}{3}k^2 \tag{19.32}$$

in (19.28), leads to the simplified expression of the electron energy spectrum in the present case as

$$I_{11}(E) = \frac{\hbar^2 k^2}{2m_c} + \frac{F^2\hbar^2 E_g^2 (E_g - \delta')^2}{12 m_r} \frac{2m_c}{m_r} I_{11}(E) \frac{1}{\eta_1} \frac{1}{(\eta_1 + \delta')^4} \left\{ P\left(\frac{\eta_1 + E_g}{\eta_1 - E'_g}\right)^{1/2} + Q\left(\frac{\eta_1 - E_g}{\eta_1 + E_g}\right)^{1/2} \right\}^2 \tag{19.33}$$

The (19.33) can approximately be written as

$$\frac{\hbar^2 k^2}{2m_c} = \left[e_1 E^4 + e_2 E^3 + e_3 E^2 + e_4 E + e_5 - \frac{e_6}{1+CE} + e_7 (1+CE)^{-2} \right] \tag{19.34}$$

where

$$e_1 = Q_f.\omega_1,\ Q_f = \frac{m_c}{m_r}E_g^{-4}\left[5e_fE_g^{-2} - 6G_f + 7h_fE_g^{-4}\right],$$
$$e_f = A_fP_f,\ A_f = \left[F\hbar E_g(E_g - \delta')\right]^2 m_c\left(6m_r^2(\delta')^4\right)^{-1},$$
$$F = eF_s,\ G_f = e_f(4\delta' + C_f),\ C_f = \left(2E_gQ^2 + PQ\left(E_g - E_g'\right) - 2P^2E_g\right),$$
$$E_g' = \frac{E_g(E_g - 3\delta')}{E_g + \delta'},\ P = \frac{r_0^2}{2}\left(\frac{E_g - \delta'}{E_g + \delta'}\right),\ r_0 = \left[\frac{6}{\chi}(E_g + \Delta)\left(E_g + \frac{2}{3}\Delta\right)\right]^{1/2},\ Q = \frac{t^2}{2},$$
$$t = \left[\frac{6}{\chi}\left(E_g + \frac{2}{3}\Delta\right)\right]^{1/2},\ h_f = (4\delta' e_f C_f)(B_f)^{-1},\ B_f = (P + Q)^2$$
$$P_f = E_g^{-3}\left(e_fE_g^{-2} - G_f + h_fE_g^{-4}\right),\ \omega_1 = a_1^2,\ a_1 = \frac{ab}{c},\ a = \frac{1}{E_g},\ b = \frac{1}{E_g + \Delta},\ c = \left(E_g + \frac{2}{3}\Delta\right)^{-1},$$
$$e_2 = Q_f.\omega_2,\ \omega_2 = 2a_1b_1,\ b_1 = (c)^{-2}(ac + bc - ab),\ e_3 = (1 - P_f)a_1 + Q_f.\omega_3,\ \omega_3 = (b_1^2 + 2a_1c_1),$$
$$c_1 = \left[\frac{1}{c}\left(1 - \frac{a}{c}\right)\left(1 - \frac{b}{c}\right)\right],\ e_4 = \left[(1 - P_f)b_1 + Q_f\omega_4\right],\ \omega_4 = 2b_1c_1,\ e_5 = \left[(1 - P_f)c_1 + \omega_5Q_f\right]$$
$$\omega_5 = (c_1^2 - 2c_1b_1),\ e_7 = Q_f\omega_7,\ \omega_7 = c_1^2,\ e_6 = \left[(1 - P_f)c_1 - Q_f\omega_6\right] \text{ and } \omega_6 = \frac{2c_1b_1}{c}\left(1 - \frac{cc_1}{b_1}\right)$$

Using (1.26b) and (19.34) we get

$$\begin{aligned}\frac{\hbar^2k^2}{2m_c}\int_{-\infty}^{E}F(V)dV =\ & e_1\int_{-\infty}^{E}(E - V)^4F(V)dV + e_2\int_{-\infty}^{E}(E - V)^3F(V)dV \\ & + e_3\int_{-\infty}^{E}(E - V)^2F(V)dV + e_4\int_{-\infty}^{E}(E - V)F(V)dV \\ & + e_5\int_{-\infty}^{E}F(V)dV - e_6\int_{-\infty}^{E}\frac{F(V)dV}{[1 + c(E - V)]} + e_7\int_{-\infty}^{E}\frac{F(V)dV}{[1 + c(E - V)]^2}\end{aligned} \tag{19.35}$$

We can prove that

$$\begin{aligned}\int_{-\infty}^{E}(E - V)^4F(V)dV =\ & \frac{E^4}{2}\left[1 + Erf(E/\eta_g)\right] \\ & + \frac{3\eta_g^4}{8\pi}\left[1 + Erf(E/\eta_g) - \frac{2\eta_g}{3E}\exp\left(\frac{-E^2}{\eta_g^2}\right)\right] \\ & + \frac{3}{2}(E\eta_g)^2\left[1 + Erf(E/\eta_g)\right] + \frac{E^3\eta_g}{\sqrt{\pi}}\exp\left(\frac{-E^2}{\eta_g^2}\right) \\ & + \frac{2}{\sqrt{\pi}}E\eta_g^3\exp\left(\frac{-E^2}{\eta_g^2}\right)\left[1 + \frac{E^2}{\eta_g^2}\right] = \psi_0(E, \eta_g)\end{aligned} \tag{19.36}$$

From Chap. 2, we know that

$$\int_{-\infty}^{E} \frac{F(V)dV}{[1+c(E-V)]} = c_1(c,E,\eta_g) - ic_2(c,E,\eta_g) \tag{19.37}$$

where

$$c_1(c,E,\eta_g) = \left[\frac{2}{c\eta_g\sqrt{\pi}}\exp(-u^2)\left[\sum_{p=1}^{\infty}\exp\left(\frac{-p^2}{4}\right)(p)^{-1}\sinh(pu)\right]\right], \quad c_2(c,E,\eta_g) = \frac{\sqrt{\pi}}{c\eta_g}\exp(-u^2)$$

and

$$u = \frac{1+cE}{c\eta_g}$$

We know that

$$\frac{\partial}{\partial x}\int_{A(x)}^{B(x)} F(x,\zeta)d\zeta = \int_{A(x)}^{B(x)} \frac{\partial}{\partial x}[F(x,\zeta)]d\zeta + F(x,B(x))\frac{\partial B(x)}{\partial x} - F(x,A(x))\frac{\partial A(x)}{\partial x} \tag{19.38}$$

Using (19.37) and (19.38), we get

$$\int_{-\infty}^{E} \frac{F(V)dV}{[1+c(E-V)]^2} = c_3(c,E,\eta_g) - ic_4(c,E,\eta_g) \tag{19.39}$$

where

$$c_3(c,E,\eta_g) = \left[\frac{-4u\exp(-u^2)}{c^2\eta_g^2\sqrt{\pi}}\left[\sum_{p=1}^{\infty}\exp\left(\frac{-p^2}{4}\right)(p)^{-1}\sinh(pu)\right] + \frac{1}{\eta_g}\exp\left(\frac{-E^2}{\eta_g^2}\right) - \frac{2\exp(-u^2)}{c^2\eta_g^2\sqrt{\pi}}\sum_{p=1}^{\infty}\exp\left(\frac{-p^2}{4}\right)\cosh(pu)\right] \text{ and } D_3(c,E,\eta_g) = \frac{2u}{c^2\eta_g^2}\exp(-u^2)$$

Therefore the DR in HD Kane type semiconductors can be written using (19.35), (19.36), (19.37) and (19.39) as

$$\frac{\hbar^2 k^2}{2m_c} = J_1(E, c, \eta_g, F) + iJ_2(E, c, \eta_g, F) \tag{19.40}$$

where,

$$\begin{aligned} J_1(E, c, \eta_g, F) &= 2[1 + Erf(E/\eta_g)]^{-1}\left[e_1\psi_0(E, \eta_g) + e_2\psi_1(E, \eta_g) + e_3\theta_0(E, \eta_g) + e_4\gamma_0(E, \eta_g)\right. \\ &\quad \left. + e_5\frac{1}{2}[1 + Erf(E/\eta_g)] - e_6c_1(E, c, \eta_g) + e_7c_3(E, c, \eta_g)\right], \ \psi_1(E, \eta_g) \\ &= \left[\frac{E}{2}[1 + Erf(E/\eta_g)]\left[E^2 + \frac{3}{2}\eta_g^2\right] + \frac{\eta_g}{2\sqrt{\pi}}\exp\left(\frac{-E2}{\eta_g^2}\right)(4E^2 + \eta_g^2)\right] \end{aligned}$$

and

$$J_2(E, c, \eta_g, F) = 2\left[1 + Erf(E/\eta_g)\right]^{-1}\left[e_6c_2(E, c, \eta_g) + e_7D_3(E, c, \eta_g)\right]$$

The DOS function is given by

$$N_F(E) = 4\pi g_v\left(\frac{2m_c}{\hbar^2}\right)^{\frac{3}{2}}\left[J_1'(E, c, \eta_g, F) + iJ_2'(E, c, \eta_g, F)\right]\sqrt{\left[J_1(E, c, \eta_g, F) + iJ_2(E, c, \eta_g, F)\right]} \tag{19.41}$$

The EEM in this case is given by

$$m^*(E, \eta_g, F) = m_c J_1'(E, c, \eta_g, F) \tag{19.42}$$

19.2.2 The DR Under Magnetic Quantization in HD Kane Type Semiconductors in the Presence of Intense Electric Field

The DR of the conduction electrons in HD optoelectronic materials under electric field can be written in presence of quantizing magnetic field B along x-direction whose unperturbed electron energy spectra are defined by the three of Kane as

$$\left(n + \frac{1}{2}\right)\hbar\omega_0 + \frac{\hbar^2 k_x^2}{2m_c} = \left[J_1(E, c, \eta_g, F) + iJ_2(E, c, \eta_g, F)\right], \quad \omega_0 = \frac{eB}{m_c} \tag{19.43}$$

From (19.43) we get,

$$k_x^2 = \omega_{11}(E, F, n, \eta_g) \tag{19.44}$$

where,

$$\omega_{11}(E, F, n, \eta_g) = \frac{2m_c}{\hbar^2}\left[[J_1(E, c, \eta_g, F) + iJ_2(E, c, \eta_g, F)] - \left(n + \frac{1}{2}\right)\hbar\omega_0\right]$$

The density-of-states function for both the cases can, respectively, be expressed as

$$N(E, c, \eta_g, F) = \frac{g_v eB\sqrt{2m_c}}{2\pi^2\hbar^2}\sum_{n=0}^{n_{\max}}\frac{[J_1'(E, c, \eta_g, F) + iJ_2'(E, c, \eta_g, F)]H(E - E_{n1HD})}{\sqrt{[J_1(E, c, \eta_g, F) + iJ_2(E, c, \eta_g, F)] - (n + \frac{1}{2})\hbar\omega_0}} \tag{19.45}$$

where, E_{n1HD} is the Landau level in this case and can be expressed as

$$\left(n + \frac{1}{2}\right)\hbar\omega_0 = [J_1(E_{n1HD}, c, \eta_g, F) + iJ_2(E_{n1HD}, c, \eta_g, F)] \tag{19.46}$$

The EEM in this case is given by

$$m^*(E, \eta_g, F) = m_c J_1'(E, c, \eta_g, F) \tag{19.47}$$

19.2.3 The DR in QWs in HD Kane Type Semiconductors in the Presence of Intense Electric Field

The DR in this case is given by

$$\frac{\hbar^2}{2m_c}\left(\frac{n_z\pi}{d_z}\right)^2 + \frac{\hbar^2 k_s^2}{2m_c} = [J_1(E, c, \eta_g, F) + iJ_2(E, c, \eta_g, F)] \tag{19.48}$$

The DOS function in this case is given by

$$N_{2D}(E, c, \eta_g, F) = \frac{m_c g_v}{\pi\hbar^2}\sum_{n_z=1}^{n_{z\max}}[J_1'(E, c, \eta_g, F) + iJ_2'(E, c, \eta_g, F)]H(E - E_{n_{z18,1HD}}) \tag{19.49}$$

where, $E_{n_{z18,1HD}}$ is the sub-band energy in this case and can be expressed as

$$\frac{\hbar^2}{2m_c}\left(\frac{n_z\pi}{d_z}\right)^2=\left[J_1\left(E_{n_{z18,1HD}},c,\eta_g,F\right)+iJ_2\left(E_{n_{z18,1HD}},c,\eta_g,F\right)\right] \tag{19.50}$$

The EEM in this case is given by

$$m^*\left(E,\eta_g,F\right)=m_cJ_1'\left(E,c,\eta_g,F\right) \tag{19.51}$$

19.2.4 The DR in NWs in HD Kane Type Semiconductors in the Presence of Intense Electric Field

The DR in this case is given by

$$\frac{\hbar^2}{2m_c}\left(\frac{n_z\pi}{d_z}\right)^2+\frac{\hbar^2}{2m_c}\left(\frac{n_y\pi}{d_y}\right)^2+\frac{\hbar^2k_x^2}{2m_c}=\left[J_1\left(E,c,\eta_g,F\right)+iJ_2\left(E,c,\eta_g,F\right)\right] \tag{19.52}$$

The DOS function in this case is given by

$$N_{1D}(E,c,\eta_g,F)=\frac{g_v\sqrt{2m_c}}{\pi\hbar}\sum_{n_y=1}^{n_{y\max}}\sum_{n_z=1}^{n_{z\max}}\frac{\left[J_1'\left(E,c,\eta_g,F\right)+iJ_2'\left(E,c,\eta_g,F\right)\right]H\left(E-E_{n_{z18,2HD}}\right)}{\sqrt{\left[J_1\left(E,c,\eta_g,F\right)+iJ_2\left(E,c,\eta_g,F\right)\right]-G_2\left(n_y,n_z\right)}} \tag{19.53}$$

where, $E_{n_{z18,2HD}}$ is the sub-band energy in this case and can be expressed as

$$\frac{\hbar^2}{2m_c}\left(\frac{n_z\pi}{d_z}\right)^2+\frac{\hbar^2}{2m_c}\left(\frac{n_y\pi}{d_y}\right)^2=\left[J_1\left(E_{n_{z18,2HD}},c,\eta_g,F\right)+iJ_2\left(E_{n_{z18,2HD}},c,\eta_g,F\right)\right] \tag{19.54}$$

and $G_2(n_y,n_z)$, is defined in (3.13) of Chap. 3.

The EEM in this case is given by

$$m^*\left(E,\eta_g,F\right)=m_cJ_1'\left(E,c,\eta_g,F\right) \tag{19.55}$$

19.2.5 The DR in QDs in HD Kane Type Semiconductors in the Presence of Intense Electric Field

The DR in this case is given by

$$\frac{\hbar^2}{2m_c}\left(\frac{n_z\pi}{d_z}\right)^2+\frac{\hbar^2}{2m_c}\left(\frac{n_y\pi}{d_y}\right)^2+\frac{\hbar^2}{2m_c}\left(\frac{n_x\pi}{d_x}\right)^2 = \left[J_1\left(E_{n_{z18,3HD}},c,\eta_g,F\right)+iJ_2\left(E_{n_{z18,3HD}},c,\eta_g,F\right)\right] \tag{19.56}$$

where, $E_{n_{z18,3HD}}$ is the totally quantized energy in this case.

The DOS function in this case is given by

$$N_{QBHDL}(E,\eta_g,F)=\frac{2g_v}{d_xd_yd_z}\sum_{n_x=1}^{n_{x\max}}\sum_{n_y=1}^{n_{y\max}}\sum_{n_z=1}^{n_{z\max}}\delta'\left(E-E_{n_{z18,3HD}}\right) \tag{19.57}$$

19.2.6 The Magneto DR in QWs of HD Kane Type Semiconductors in the Presence of Intense Electric Field

The DR in this case is given by

$$\frac{\hbar^2}{2m_c}\left(\frac{n_x\pi}{d_x}\right)^2+\left(n+\frac{1}{2}\right)\hbar\omega_0=\left[J_1\left(E_{n_{z18,4HD}},c,\eta_g,F\right)+iJ_2\left(E_{n_{z18,4HD}},c,\eta_g,F\right)\right] \tag{19.58}$$

where, $E_{n_{z18,4HD}}$ is the totally quantized energy in this case

The DOS function in this case is given by

$$N_{QWBHD}\left(E,\eta_g,F\right)=\frac{g_veB}{\pi\hbar}\sum_{n_x=1}^{n_{x\max}}\sum_{n=0}^{n_{\max}}\delta'\left(E-E_{n_{z18,4HD}}\right) \tag{19.59}$$

19.2.7 The DR in Accumulation and Inversion Layers of Kane Type Semiconductors in the Presence of Intense Electric Field

(a) The 2D DR in accumulation layers of HD III–V, ternary and quaternary materials, in this case can be expressed as

$$\left[J_1\left(E,c,\eta_g,F\right)+iJ_2\left(E,c,\eta_g,F\right)\right]=\frac{\hbar^2k_s^2}{2m_c}+\frac{\hbar|e|F_s}{\sqrt{2m_c}}\left(\frac{2\sqrt{2}S_i^{3/2}}{3}\right)\left[J_1'\left(E,c,\eta_g,F\right)+iJ_2'\left(E,c,\eta_g,F\right)\right]' \tag{19.60}$$

Since the DR in accordance with the HD three-band model of Kane is complex in nature, the (19.60) will also be complex. The both complexities occur due to the presence of poles in the finite complex plane of the dispersion relation of the materials in the absence of band tails.

The EEM can be expressed as

$$m^*\left(E,c,\eta_g,F,i\right)=m_c \text{ Real part of } \left[P_{3HDL1}\left(E,c,\eta_g,F,i\right)\right]' \tag{19.61}$$

where,

$$P_{3HDL1}\left(E,c,\eta_g,F,i\right)=\left[\left[J_1\left(E,c,\eta_g,F\right)+iJ_2\left(E,c,\eta_g,F\right)\right]-\frac{\hbar|e|F_s}{\sqrt{2m_c}}\left(\frac{2\sqrt{2}S_i^{3/2}}{3}\right)\left[J_1'\left(E,c,\eta_g,F\right)+iJ_2'\left(E,c,\eta_g,F\right)\right]'\right]$$

Thus, one can observe that the EEMis a function of electric field, scattering potential, the sub-band index, surface electric field, the Fermi energy and the other spectrum constants due to the combined influence of E_g and Δ.

The sub-band energy E_{400HD} is given by

$$\left[J_1\left(E_{400HD},c,\eta_g,F\right)+iJ_2\left(E_{400HD},c,\eta_g,F\right)\right]=\frac{\hbar|e|F_s}{\sqrt{2m_c}}\left(\frac{2\sqrt{2}S_i^{3/2}}{3}\right)\left[J_1'\left(E_{400HD},c,\eta_g,F\right)+iJ_2'\left(E_{400HD},c,\eta_g,F\right)\right]' \tag{19.62}$$

The DOS function can be written as

$$N_{2D}\left(E,c,\eta_g,F\right)=\frac{m_cg_v}{\pi\hbar^2}\sum_{i=0}^{i_{\max}}\left[P_{3HDL1}\left(E,c,\eta_g,F,i\right)\right]'H(E-E_{400HD}) \tag{19.63}$$

Thus the DOS function is complex in nature.

In the absence of band-tails, the DR in this case assumes the form

$$[J_{11}(E,F)]=\frac{\hbar^2k_s^2}{2m_c}+\frac{\hbar|e|F_s}{\sqrt{2m_c}}\left(\frac{2\sqrt{2}S_i^{3/2}}{3}\right)[J_{11}'(E,F)]' \tag{19.64}$$

where

$$J_{11}(E,F) = \left[e_1E^4 + e_2E^3 + e_3E^3 + e_4E + e_5 - \frac{e_6}{1+CE} + e_7(1+CE)^{-2}\right]$$

(19.63) represents the DR of the 2D electrons in inversion layers of III–V, ternary and quaternary materials under the intense electric field limit whose bulk electrons in the absence of any perturbation obey the three band model of Kane.

The EEM can be expressed as

$$m^*(E,F,i) = m_c[P_{31}(E,F,i)]'$$

where,

$$P_{31}(E,F,i) = \left[[J_{11}(E,F)] - \frac{\hbar|e|F_s}{\sqrt{2m_c}}\left(\frac{2\sqrt{2}S_i^{3/2}}{3}\right)[J'_{11}(E,F)]'\right] \tag{19.65}$$

Thus, one can observe that the EEM is a function of the sub-band index, the light intensity, surface electric field, the Fermi energy and the other spectrum constants due to the combined influence of E_g and Δ.

The sub-band energy E_{401} in this case can be obtained from the (19.64) as

$$0 = \left[[J_{11}(E_{401},F)] - \frac{\hbar|e|F_s}{\sqrt{2m_c}}\left(\frac{2\sqrt{2}S_i^{3/2}}{3}\right)[J'_{11}(E_{401},F)]'\right] \tag{19.66}$$

Thus the 2D total DOS function in weak electric field limit can be expressed as

$$N_{2D}(E,F) = \frac{m_c g_v}{\pi\hbar^2}\sum_{i=0}^{i_{max}}[P_{31}(E,F,i)]'H(E-E_{401}) \tag{19.67}$$

19.2.8 The Magneto DR in Accumulation and Inversion Layers of Kane Type Semiconductors in the Presence of Intense Electric Field

(a) The magneto DR in accumulation layers of HD III–V, ternary and quaternary materials, in this case can be expressed as

$$[J_1(E_{450},c,\eta_g,F) + iJ_2(E_{450},c,\eta_g,F)] = \left(n+\frac{1}{2}\right)\hbar\omega_0 + \frac{\hbar|e|F_s}{\sqrt{2m_c}}\left(\frac{2\sqrt{2}S_i^{3/2}}{3}\right)[J'_1(E_{450},c,\eta_g,F) + iJ'_2(E_{450},c,\eta_g,F)]' \tag{19.68}$$

where E_{450}, is the total energy in this case.

The DOS function is given by

$$N_{HDA}(E,\eta_g) = \frac{g_v eB}{\pi\hbar}\sum_{i=0}^{i_{\max}}\sum_{n=0}^{n_{\max}}\delta'(E - E_{450}) \tag{19.69}$$

In the absence of band-tails, the DR in this case assumes the form

$$[J_{11}(E_{451},F)] = \left(\mathrm{n} + \frac{1}{2}\right)\hbar\omega_0 + \frac{\hbar|e|F_s}{\sqrt{2m_c}}\left(\frac{2\sqrt{2}S_i^{3/2}}{3}\right)[J'_{11}(E_{451},F)]' \tag{19.70}$$

where E_{451}, is the total energy in this case.

The DOS function is given by

$$N_{HDA}(E,\eta_g) = \frac{g_v eB}{\pi\hbar}\sum_{i=0}^{i_{\max}}\sum_{n=0}^{n_{\max}}\delta'(E - E_{451}) \tag{19.71}$$

19.2.9 The DR in Doping Superlattices of HD Kane Type Semiconductors in the Presence of Intense Electric Field

The DR in doping superlattices of HD III–V, ternary and quaternary materials in the presence of intense electric field whose unperturbed electrons are defined by the three band model of Kane can be expressed as

$$[J_1(E,c,\eta_g,F) + iJ_2(E,c,\eta_g,F)] = \left(\mathrm{n}_i + \frac{1}{2}\right)\hbar\omega_{91HD1}(E,c,\eta_g,F) + \frac{\hbar^2 k_s^2}{2m_c} \tag{19.72}$$

where,

$$\omega_{91HD1}(E,c,\eta_g,F) = \left(\frac{n_s|e|^2}{d_0\varepsilon_{sc}[J_1(E,c,\eta_g,F) + iJ_2(E,c,\eta_g,F)]' m_c}\right)^{\frac{1}{2}}$$

The sub band energies E_{452} can be written as

$$\left[J_1\left(E_{452},c,\eta_g,F\right)+iJ_2\left(E_{452},c,\eta_g,F\right)\right]=\left(n_i+\frac{1}{2}\right)\hbar\omega_{91HD1}\left(E_{452},c,\eta_g,F\right) \quad (19.73)$$

The EEM in this case is given by

$$m^*\left(E,c,\eta_g,F,\mathrm{n}_i\right)=m_c \text{ Real part of } \left[P_{3HDL3}\left(E,c,\eta_g,F,\mathrm{n}_i\right)\right]' \quad (19.74)$$

where,

$$\left[P_{3HDL3}\left(E,c,\eta_g,F,\mathrm{n}_i\right)\right]=\left[\left[J_1\left(E,c,\eta_g,F\right)+iJ_2\left(E,c,\eta_g,F\right)\right]-\left(\mathrm{n}_i+\frac{1}{2}\right)\hbar\omega_{91HD1}\left(E,c,\eta_g,F\right)\right]$$

The DOS function in this case is given by

$$N_{2DDSL}\left(E,c,\eta_g,F\right)=\frac{m_c g_v}{\pi\hbar^2}\sum_{n_i=0}^{n_{i\max}}\frac{\left[J_1'\left(E,c,\eta_g,F\right)+iJ_2'\left(E,c,\eta_g,F\right)-\left(\mathrm{n}_i+\frac{1}{2}\right)\hbar\omega_{91HD1}'\left(E,c,\eta_g,F\right)\right]H\left(E-E_{452}\right)}{\sqrt{\left[J_1\left(E,c,\eta_g,F\right)+iJ_2\left(E,c,\eta_g,F\right)\right]-\left(\mathrm{n}_i+\frac{1}{2}\right)\hbar\omega_{91HD1}\left(E,c,\eta_g,F\right)}} \quad (19.75)$$

19.2.10 The Magneto DR in Inversion Layers of Kane Type Semiconductors in the Presence of Intense Electric Field

The magneto DR in inversion layers of HD III–V, ternary and quaternary materials in the presence of intense electric field whose unperturbed electrons are defined by the three band model of Kane can be expressed as

$$\left[J_1\left(E_{555},c,\eta_g,F\right)+iJ_2\left(E_{555},c,\eta_g,F\right)\right]=\left(\mathrm{n}_i+\frac{1}{2}\right)\hbar\omega_{91HD1}\left(E_{555},c,\eta_g,F\right)+\left(n+\frac{1}{2}\right)\hbar\omega_0 \quad (19.76)$$

where, E_{555} is the totally quantized energy in this case.

The DOS function is given by

$$N_{HDA}\left(\mathrm{E},\eta_g\right)=\frac{g_v eB}{\pi\hbar}\sum_{i=0}^{i_{\max}}\sum_{n=0}^{n_{\max}}\delta'\left(E-E_{555}\right) \quad (19.77)$$

19.2.11 The DR in QWHD Effective Mass Superlattices of Kane Type Semiconductors in the Presence of Intense Electric Field

Following Sasaki [3], the electron dispersion law in HD III–V effective mass superlattices (EMSLs) in the presence of light waves, the dispersion relations of whose constituent materials in the absence of any perturbation are defined by the three band model of Kane can be written as

$$k_x^2 = \left[\frac{1}{L_0^2}\left\{\cos^{-1}\left(\mathrm{f}_{18HD1}\left(\mathrm{E},\mathrm{k}_y,\mathrm{k}_z,\mathrm{F}\right)\right)\right\}^2 - \mathrm{k}_\perp^2\right] \tag{19.78}$$

In which,

$$\begin{aligned}\mathrm{f}_{18HD1}\left(\mathrm{E},\mathrm{k}_y,\mathrm{k}_z,\mathrm{F}\right) = {} & a_{1HD1,18}\cos\left[\mathrm{a}_0\mathrm{C}_{1HD1,18}(\mathrm{E},\mathrm{k}_\perp,\mathrm{F}) + b_0 D_{1HD1,18}(\mathrm{E},\mathrm{k}_\perp,\mathrm{F})\right] \\ & - a_{2HD1,18}\cos\left[\mathrm{a}_0\mathrm{C}_{1HD1,18}(\mathrm{E},\mathrm{k}_\perp,\mathrm{F}) + b_0 D_{1HD1,18}(\mathrm{E},\mathrm{k}_\perp,\mathrm{F})\right]\end{aligned}$$

$$a_{1HD1,18} = \left[\sqrt{\frac{m_{c2}\text{Real part of }\left[\mathrm{J}_3(0,c,\eta_{g2},F)\right]}{m_{c1}\text{Real part of }\left[\mathrm{J}_3(0,c,\eta_{g1},F)\right]}}+1\right]^2\left[4\left(\frac{m_{c2}\text{Real part of }\left[\mathrm{J}_3(0,c,\eta_{g2},F)\right]}{m_{c1}\text{Real part of }\left[\mathrm{J}_3(0,c,\eta_{g1},F)\right]}\right)^{\frac{1}{2}}\right]^{-1},$$

$$a_{2HD1,18} = \left[\sqrt{\frac{m_{c2}\text{Real part of }\left[\mathrm{J}_3(0,c,\eta_{g2},F)\right]}{m_{c1}\text{Real part of }\left[\mathrm{J}_3(0,c,\eta_{g1},F)\right]}}-1\right]^2\left[4\left(\frac{m_{c2}\text{Real part of }\left[\mathrm{J}_3(0,c,\eta_{g2},F)\right]}{m_{c1}\text{Real part of }\left[\mathrm{J}_3(0,c,\eta_{g1},F)\right]}\right)^{\frac{1}{2}}\right]^{-1}$$

$$J_1(E,c,\eta_g,F) + iJ_2(E,c,\eta_g,F) = J_3(E,c,\eta_g,F),$$

$$C_{1HD1,18}(E,\mathrm{k}_\perp,F) \equiv \left[\left(\frac{2m_{c1}}{\hbar^2}\right)J_3(E,c,\eta_{g1},F) - \mathrm{k}_\perp^2\right]^{\frac{1}{2}}$$

$$\text{and } D_{1HD1,18}(E,\mathrm{k}_\perp,F) \equiv \left[\left(\frac{2m_{c1}}{\hbar^2}\right)J_3(E,c,\eta_{g2},F) - \mathrm{k}_\perp^2\right]^{\frac{1}{2}}.$$

The DR in QWHD effective mass superlattices of Kane type semiconductors in the presence of intense electric field, the dispersion relations of whose constituent materials in the absence of any perturbation are defined by the three band model of Kane can be written as

$$\left(\frac{n_x\pi}{d_x}\right)^2 = \left[\frac{1}{L_0^2}\left\{\cos^{-1}\left(\mathrm{f}_{18HD1}\left(\mathrm{E},\mathrm{k}_y,\mathrm{k}_z,\mathrm{F}\right)\right)\right\}^2 - \mathrm{k}_\perp^2\right] \tag{19.79}$$

The EEM in this case assumes the form

$$m^*(k_\perp, E, F) = \frac{\hbar^2}{L_0^2}\left|\left[\frac{\cos^{-1}[f_{18HD1}(E, k_y, k_z, F)]f'_{18HD1}(E, k_y, k_z, F)}{\sqrt{1 - f^2_{18HD1}(E, k_y, k_z, F)}}\right]\right| \tag{19.80}$$

The subband energies E_{600} can be written as

$$\left(\frac{n_x \pi}{d_x}\right) = \left[\frac{1}{L}\left\{\cos^{-1}\left(f_{18HD1}\left(E_{600}, k_y, k_z, F\right)\right)\right\}\right] \tag{19.81}$$

19.2.12 The DR in NWHD Effective Mass Superlattices of Kane Type Semiconductors in the Presence of Intense Electric Field

Following Sasaki [3], the magneto DR in HD III–V effective mass superlattices (EMSLs) in the presence of intense electric field, the dispersion relations of whose constituent materials in the absence of any perturbation are defined by the three band model of Kane can be written as

$$k_x^2 = \left[\frac{1}{L_0^2}\left\{\cos^{-1}(f_{18HD2}(E, n, F))\right\}^2 - \frac{2eB}{\hbar}\left(n + \frac{1}{2}\right)\right] \tag{19.82}$$

In which,

$$f_{18HD2}(E, n, F) = a_{1HD1,18}\cos\left[a_0 C_{1HD2,18}(E, n, F) + b_0 D_{1HD2,18}(E, n, F)\right] - a_{2HD1,18}\cos\left[a_0 C_{1HD2,18}(E, n, F) + b_0 D_{1HD2,18}(E, n, F)\right]$$

$$a_{1HD1,18} = \left[\sqrt{\frac{m_{c2}\text{Real part of }\left[J_3(0, c, \eta_{g2}, F)\right]}{m_{c1}\text{Real part of }\left[J_3(0, c, \eta_{g1}, F)\right]}} + 1\right]^2\left[4\left(\frac{m_{c2}\text{Real part of }\left[J_3(0, c, \eta_{g2}, F)\right]}{m_{c1}\text{Real part of }\left[J_3(0, c, \eta_{g1}, F)\right]}\right)^{\frac{1}{2}}\right]^{-1},$$

$$a_{2HD1,18} = \left[\sqrt{\frac{m_{c2}\text{Real part of }\left[J_3(0, c, \eta_{g2}, F)\right]}{m_{c1}\text{Real part of }\left[J_3(0, c, \eta_{g1}, F)\right]}} - 1\right]^2\left[4\left(\frac{m_{c2}\text{Real part of }\left[J_3(0, c, \eta_{g2}, F)\right]}{m_{c1}\text{Real part of }\left[J_3(0, c, \eta_{g1}, F)\right]}\right)^{\frac{1}{2}}\right]^{-1}$$

$$J_1(E, c, \eta_g, F) + iJ_2(E, c, \eta_g, F) = J_3(E, c, \eta_g, F),$$

$$C_{1HD2,18}(E, n, F) \equiv \left[\left(\frac{2m_{c1}}{\hbar^2}\right)J_3(E, c, \eta_{g1}, F) - \frac{2eB}{\hbar}\left(n + \frac{1}{2}\right)\right]^{\frac{1}{2}}$$

and $D_{1HD2,18}(E, n, F) \equiv \left[\left(\frac{2m_{c1}}{\hbar^2}\right)J_3(E, c, \eta_{g2}, F) - \frac{2eB}{\hbar}\left(n + \frac{1}{2}\right)\right]^{\frac{1}{2}}$

The DOS function in this case is given by

$$N_{100}(\mathrm{E},\mathrm{F}) = \frac{g_v eB}{\pi^2\hbar}\sum_{n=0}^{n_{\max}}[\omega_{100}(\mathrm{E},n,\mathrm{F})]' \qquad (19.83)$$

where

$$\omega_{100}(\mathrm{E},n,\mathrm{F}) = \left[\left[\frac{1}{L_0^2}\{\cos^{-1}(\mathrm{f}_{18HD2}(\mathrm{E},n,\mathrm{F}))\}^2 - \frac{2eB}{\hbar}\left(\mathrm{n}+\frac{1}{2}\right)\right]\right]^{\frac{1}{2}}$$

The EEM in this case assumes the form

$$m^*(n,\mathrm{E},F) = \frac{\hbar^2}{L_0^2}\left|\left[\frac{\cos^{-1}[\mathrm{f}_{18HD2}(\mathrm{E},n,\mathrm{F})]\mathrm{f}'_{18HD2}(\mathrm{E},n,\mathrm{F})}{\sqrt{1-\mathrm{f}^2_{18HD2}(\mathrm{E},n,\mathrm{F})}}\right]\right| \qquad (19.84)$$

The Landau sub-band energies E_{602} can be written as

$$0 = \left[\frac{1}{L_0^2}\{\cos^{-1}(\mathrm{f}_{18HD2}(\mathrm{E}_{602},n,\mathrm{F}))\}^2 - \frac{2eB}{\hbar}\left(\mathrm{n}+\frac{1}{2}\right)\right] \qquad (19.85)$$

19.2.13 The DR in Quantum Dot HD Superlattices of Kane Type Semiconductors in the Presence of Intense Electric Field

Following Sasaki [3], the DR in QDHD III–V effective mass superlattices (EMSLs) in the presence of intense electric field, the dispersion relations of whose constituent materials in the absence of any perturbation are defined by the three band model of Kane can be written as

$$\left(\frac{n_x\pi}{d_x}\right)^2 = \left[\frac{1}{L_0^2}\{\cos^{-1}(\mathrm{f}_{18HD4}(\mathrm{E}_{600},n_y,n_z,\mathrm{F}))\}^2 - G_2(n_y,n_z)\right] \qquad (19.86)$$

In which,

$$\begin{aligned}\mathrm{f}_{18HD4}(\mathrm{E}_{600},n_y,\mathrm{n}_z,\mathrm{F}) = {} & a_{1HD1,18}\cos\left[\mathrm{a}_0\mathrm{C}_{1HD4,18}(\mathrm{E}_{600},n_y,\mathrm{n}_z,\mathrm{F}) + b_0 D_{1HD4,18}(\mathrm{E}_{600},n_y,\mathrm{n}_z,\mathrm{F})\right] \\ & - a_{2HD1,18}\cos\left[\mathrm{a}_0\mathrm{C}_{1HD4,18}(\mathrm{E}_{600},n_y,\mathrm{n}_z,\mathrm{F}) + b_0 D_{1HD4,18}(\mathrm{E}_{600},n_y,\mathrm{n}_z,\mathrm{F})\right]\end{aligned}$$

E_{600} is the totally quantized energy in this case,

$$a_{1HD1,18} = \left[\sqrt{\frac{m_{c2}\text{Real part of }[J_3(0,c,\eta_{g2},F)]}{m_{c1}\text{Real part of }[J_3(0,c,\eta_{g1},F)]}}+1\right]^2\left[4\left(\frac{m_{c2}\text{Real part of }[J_3(0,c,\eta_{g2},F)]}{m_{c1}\text{Real part of }[J_3(0,c,\eta_{g1},F)]}\right)^{\frac{1}{2}}\right]^{-1},$$

$$a_{2HD1,18} = \left[\sqrt{\frac{m_{c2}\text{Real part of }[J_3(0,c,\eta_{g2},F)]}{m_{c1}\text{Real part of }[J_3(0,c,\eta_{g1},F)]}}-1\right]^2\left[4\left(\frac{m_{c2}\text{Real part of }[J_3(0,c,\eta_{g2},F)]}{m_{c1}\text{Real part of }[J_3(0,c,\eta_{g1},F)]}\right)^{\frac{1}{2}}\right]^{-1}$$

$$J_1(E,c,\eta_g,F) + iJ_2(E,c,\eta_g,F) = J_3(E,c,\eta_g,F),$$

$$C_{1HD4,18}(E_{600},\mathrm{n}_y,\mathrm{n}_z,F) \equiv \left[\left(\frac{2m_{c1}}{\hbar^2}\right)J_3(E_{600},c,\eta_{g1},F) - G_2(\mathrm{n}_y,\mathrm{n}_z)\right]^{\frac{1}{2}}$$

$$\text{and } D_{1HD4,18}(E_{600},\mathrm{n}_y,\mathrm{n}_z,F) \equiv \left[\left(\frac{2m_{c1}}{\hbar^2}\right)J_3(E_{600},c,\eta_{g2},F) - G_2(\mathrm{n}_y,\mathrm{n}_z)\right]^{\frac{1}{2}}.$$

The DOS function in this case can be written as

$$N_{600}(\mathrm{E,c,F}) = \frac{2g_v}{d_x d_y d_z}\sum_{n_x=1}^{n_{\max}}\sum_{n_y=1}^{n_{\max}}\sum_{n_z=1}^{n_{\max}}\delta'(\mathrm{E}-\mathrm{E}_{600}) \tag{19.87}$$

19.2.14 The Magneto DR in QWHD Effective Mass Superlattices of Kane Type Semiconductors in the Presence of Intense Electric Field

Following Sasaki [3], the electron dispersion law in HD III–V effective mass superlattices (EMSLs) in the presence of light waves, the dispersion relations of whose constituent materials in the absence of any perturbation are defined by the three band model of Kane can be written as

$$\left(\frac{n_x\pi}{d_x}\right)^2 = \left[\frac{1}{L_0^2}\left\{\cos^{-1}(\mathrm{f}_{18HD2}(\mathrm{E}_{601},n,\mathrm{F}))\right\}^2 - \frac{2eB}{\hbar}\left(\mathrm{n}+\frac{1}{2}\right)\right] \tag{19.88}$$

where E_{601} is the totally quantized energy in this case.

The DOS function is given by

$$N_{601}(\mathrm{E,F},\eta_g) = \frac{g_v eB}{\pi\hbar}\sum_{n_x=1}^{n_{x\max}}\sum_{n=0}^{n_{\max}}\delta'(\mathrm{E}-\mathrm{E}_{601}) \tag{19.89}$$

19.2.15 The DR in QWHD Superlattices of Kane Type Semiconductors with Graded Interfaces in the Presence of Intense Electric Field

The DR in bulk specimens of the heavily doped constituent materials of III–V SLs in the presence of intense electric field can be expressed as

$$\frac{\hbar^2 k^2}{2m_{cj}} = J_{1j}(\mathrm{E}, \Delta_j, \mathrm{E}_{g0j}, \eta_j, \mathrm{F}) + iJ_{2j}(\mathrm{E}, \Delta_j, \mathrm{E}_{g0j}, \eta_j, \mathrm{F}) \tag{19.90}$$

where $j = 1, 2$.

$$J_{1j}(\mathrm{E}, \Delta_j, \mathrm{E}_{g0j}, \eta_j, \mathrm{F}) = 2\left[1 + Erf\left(E/\eta_{g_j}\right)\right]^{-1}\left[e_{1j}\psi_{0j}\left(E, \eta_{g_j}\right) + e_{2j}\psi_{1j}\left(E, \eta_{g_j}\right) + e_{3j}\theta_{0j}\left(E, \eta_{g_j}\right)\right.$$
$$\left. + e_{4j}\gamma_0\left(E, \eta_{g_j}\right) + e_{5j}\frac{1}{2}\left[1 + Erf\left(E/\eta_{g_j}\right)\right] - e_{6j}c_{1j}\left(E, c, \eta_{g_j}\right) + e_{7j}c_{3j}\left(E, c, \eta_{g_j}\right)\right],$$

$$\psi_{1j}(E, \eta_{g_j}) = \left[\frac{E}{2}\left[1 + Erf\left(E/\eta_{g_j}\right)\right]\left[E^2 + \frac{3}{2}\eta_{g_j}^2\right] + \frac{\eta_{g_j}}{2\sqrt{\pi}}\exp\left(\frac{-E^2}{\eta_{g_j}^2}\right)\left(4E^2 + \eta_{g_j}^2\right)\right],$$

$$J_{2j}(\mathrm{E}, \Delta_j, \mathrm{E}_{g0j}, \eta_j, \mathrm{F}) = 2\left[1 + Erf\left(E/\eta_{g_j}\right)\right]^{-1}\left[e_{6j}c_{2j}\left(E, c, \eta_{g_j}\right) + e_{7j}D_{3j}\left(E, c, \eta_{g_j}\right)\right],$$

$$e_{1j} = Q_{fj}.\omega_{1j},\quad Q_{fj} = \frac{m_{cj}}{m_{rj}}E_{g_j}^{-4}\left[5e_{fj}E_{g_j}^{-2} - 6G_{fj} + 7h_{fj}E_{g_j}^{-4}\right],$$

$$e_{fj} = A_{fj}P_{fj},\quad A_{fj} = \left[F\hbar E_{g_j}\left(E_{g_j} - \delta_j'\right)\right]^2 m_{cj}\left(6m_{rj}^2\left(\delta_j'\right)^4\right)^{-1},$$

$$F = eF_s,\quad G_{fj} = e_{fj}\left(4\delta_j' + C_{fj}\right),\quad C_{fj} = \left(2E_{g_j}Q_j^2 + P_jQ_j\left(E_{g_j} - E_{g_j}'\right) - 2P_j^2E_{g_j}\right),$$

$$E_{g_j}' = \frac{E_{g_j}\left(E_{g_j} - 3\delta_j'\right)}{E_{g_j} + \delta_j'},\quad P_j = \frac{r_{0j}^2}{2}\left(\frac{E_{g_j} - \delta_j'}{E_{g_j} + \delta_j'}\right),\quad r_{0j} = \left[\frac{6}{\chi_j}\left(E_{g_j} + \Delta_j\right)\left(E_{g_j} + \frac{2}{3}\Delta_j\right)\right]^{1/2},\quad Q_j = \frac{t_j^2}{2},$$

$$t_j = \left[\frac{6}{\chi_j}\left(E_{g_j} + \frac{2}{3}\Delta_j\right)\right]^{1/2},\quad h_{fj} = \left(4\delta_j' e_{fj}C_{fj}\right)\left(B_{fj}\right)^{-1},\quad B_{fj} = \left(P_j + Q_j\right)^2$$

$$P_{fj} = E_{g_j}^{-3}(e_{fj}E_{g_j}^{-2} - G_{fj} + h_{fj}E_{g_j}^{-4}),\quad \omega_{1j} = a_{1j}^2,\quad a_{1j} = \frac{a_jb_j}{c_j},$$

$$a_j = \frac{1}{E_{g_j}},\quad b_j = \frac{1}{E_{g_j} + \Delta_j},\quad c_j = \left(E_{g_j} + \frac{2}{3}\Delta_j\right)^{-1},$$

$$e_{2j} = Q_{fj}.\omega_{2j},\quad \omega_{2j} = 2a_{1j}b_{1j},\quad b_{1j} = (c_j)^{-2}(a_jc_j + b_jc_j - a_jb_j),$$
$$e_{3j} = (1 - P_{fj})a_{1j} + Q_{fj}\cdot\omega_{3j},\quad \omega_{3j} = (b_{1j}^2 + 2a_{1j}c_{1j}),$$

$$c_{1j} = \left[\frac{1}{c_j}\left(1 - \frac{a_j}{c_j}\right)\left(1 - \frac{b_j}{c_j}\right)\right],\quad e_{4j} = \left[(1 - P_{fj})b_{1j} + Q_{fj}\omega_{4j}\right],$$
$$\omega_{4j} = 2b_{1j}c_{1j},\quad e_{5j} = \left[(1 - P_{fj})c_{1j} + \omega_{5j}Q_{fj}\right]$$

$$\omega_{5j} = \left(c_{1j}^2 - 2c_{1j}b_{1j}\right),\ e_{7j} = Q_{fj}\omega_{7j},\ \omega_{7j} = c_{1j}^2,\ e_{6j} = \left[(1 - P_{fj})c_{1j} - Q_{fj}\omega_{6j}\right]$$

and $\omega_{6j} = \dfrac{2c_{1j}b_{1j}}{c_j}\left(1 - \dfrac{c_jc_{1j}}{b_{1j}}\right)$

Therefore, the DR in HD III–V SLs with graded interfaces in the presence of intense electric field can be expressed as [4]

$$k_z^2 = G_{8,19} + iH_{8,19} \tag{19.91}$$

where

$$G_{8,19} = \left[\frac{C_{7,19}^2 - D_{7,19}^2}{L_0^2} - k_s^2\right], C_{7,19} = \cos^{-1}(\bar{\omega}_{7,19}),$$

$$\bar{\omega}_{7,19} = (2)^{\frac{-1}{2}}\left[\left(1 - G_{7,19}^2 - H_{7,19}^2\right) - \sqrt{\left(1 - G_{7,19}^2 - H_{7,19}^2\right)^2 + 4G_{7,19}^2}\right]^{\frac{1}{2}},$$

$$\begin{aligned}G_{7,19} = \big[&G_{1,19} + (\rho_{5,19}G_{2,19}/2) - (\rho_{6,19}H_{2,19}/2)\\ &+ (\Delta_0/2)\{\rho_{6,19}H_{2,19} - \rho_{8,19}H_{3,19} + \rho_{9,19}H_{4,19} - \rho_{10,19}H_{4,19} +\\ &\rho_{11,19}H_{5,19} - \rho_{12,19}H_{5,19} + (1/12)(\rho_{12,19}G_{6,19} - \rho_{14,19}H_{6,19})\}\big],\end{aligned}$$

$$\begin{aligned}G_{1,19} = \big[&(\cos(h_{1,19}))(\cosh(h_{2,19}))(\cosh(g_{1,19}))(\cos(g_{2,19})) +\\ &(\sin(\mathrm{h}_{1,19}))(\sinh(\mathrm{h}_{2,19}))(\sinh(\mathrm{g}_{1,19}))(\sin(\mathrm{g}_{2,19}))\big],\end{aligned}$$

$$\mathrm{h}_{1,19} = e_{1,19}(b_0 - \Delta_0), e_{1,19} = 2^{\frac{-1}{2}}\left(\sqrt{t_{1,19}^2 + t_{2,19}^2} + t_{1,19}\right)^{\frac{1}{2}},$$

$$\begin{aligned}t_{1,19} &= \left[(2m_{c1}/\hbar^2).\mathrm{J}_{11}(\mathrm{E}, \Delta_1, \mathrm{E}_{g01}, \mathrm{F}) - k_s^2\right],\\ t_{2,19} &= \left[(2m_{c1}/\hbar^2)J_{21}(\mathrm{E}, \Delta_1, \mathrm{E}_{g01}, \mathrm{F})\right],\\ h_{2,19} &= e_{2,19}(b_0 - \Delta_0),\\ e_{2,19} &= 2^{\frac{-1}{2}}\left(\sqrt{t_{1,19}^2 + t_{2,19}^2} - t_{1,19}\right)^{\frac{1}{2}},\\ g_{1,19} &= d_{1,19}(a_0 - \Delta_0),\\ d_{1,19} &= 2^{\frac{-1}{2}}\left(\sqrt{x_{1,19}^2 + y_{1,19}^2} + x_{1,19}\right)^{\frac{1}{2}},\end{aligned}$$

$$x_{1,19} = \left[-\left(2m_{c2}/\hbar^2\right) \cdot \mathrm{J}_{11}\left(E - V_0, E_{g2}, \Delta_2, \eta_{g2}, \mathrm{F}\right) + k_s^2\right],$$
$$y_1 = \left[\left(2m_{c2}/\hbar^2\right) \cdot \mathrm{J}_{22}\left(E - V_0, E_{g2}, \Delta_2, \eta_{g2}, \mathrm{F}\right)\right],$$
$$g_{2,19} = d_{2,19}(a_0 - \Delta_0),$$
$$d_{2,19} = 2^{\frac{-1}{2}}\left(\sqrt{x_{1,19}^2 + y_{1,19}^2} - x_{1,19}\right)^{\frac{1}{2}},$$
$$\rho_{5,19} = \left(\rho_{3,19}^2 + \rho_{4,19}^2\right)^{-1}\left[\rho_{1,19}\rho_{3,19} - \rho_{2,19}\rho_{4,19}\right],$$
$$\rho_{1,19} = \left[d_{1,19}^2 + e_{2,19}^2 - d_{2,19}^2 - e_{1,19}^2\right],$$
$$\rho_{3,19} = \left[d_{1,19}e_{1,19} + d_{2,19}e_{2,19}\right],$$
$$\rho_{2,19} = 2\left[d_{1,19}d_{2,19} + e_{1,19}e_{2,19}\right],$$
$$\rho_{4,19} = \left[d_{1,19}e_{2,19} - e_{1,19}d_{2,19}\right],$$

$$G_{2,19} = \left[(\sin(h_{1,19}))(\cosh(h_{2,19}))(\sinh(g_{1,19}))(\cos(g_{2,19})) + (\cos(h_{1,19}))(\sinh(h_{2,19}))(\cosh(g_{1,19}))(\sin(g_{2,19}))\right],$$
$$\rho_{6,19} = \left(\rho_{3,19}^2 + \rho_{4,19}^2\right)^{-1}\left[\rho_{1,19}\rho_{4,19} + \rho_{2,19}\rho_{3,19}\right],$$

$$H_{2,19} = \left[(\sin(h_{1,19}))(\cos(h_{2,19}))(\sin(g_{2,19}))(\cosh(g_{1,19})) - (\cos(h_{1,19}))(\sinh(h_{2,19}))(\sinh(g_{1,19}))(\cos(g_{2,19}))\right],$$
$$\rho_{7,19} = \left[\left(\mathrm{e}_{1,19}^2 + e_{2,19}^2\right)^{-1}\left[\mathrm{e}_{1,19}\left(\mathrm{d}_{1,19}^2 - d_{2,19}^2\right) - 2d_{1,19}d_{2,19}e_{2,19}\right] - 3e_{1,19}\right],$$

$$G_{3,19} = \left[(\sin(h_{1,19}))(\cosh(h_{2,19}))(\cosh(g_{1,19}))(\cos(g_{2,19})) + (\cos(h_{1,19}))(\sinh(h_{2,19}))(\sinh(g_{1,19}))(\sin(g_{2,19}))\right],$$
$$\rho_{8,19} = \left[\left(\mathrm{e}_{1,19}^2 + e_{2,19}^2\right)^{-1}\left[\mathrm{e}_{2,19}\left(\mathrm{d}_{1,19}^2 - d_{2,19}^2\right) + 2d_{1,19}d_{2,19}e_{1,19}\right] + 3e_{2,19}\right],$$

$$H_{3,19} = \left[(\sin(h_{1,19}))(\cosh(h_{2,19}))(\sin(g_{2,19}))(\sinh(g_{1,19})) - (\cos(h_{1,19}))(\sinh(h_{2,19}))(\cosh(g_{1,19}))(\cos(g_{2,19}))\right],$$
$$\rho_{9,19} = \left[\left(\mathrm{d}_{1,19}^2 + d_{2,19}^2\right)^{-1}\left[\mathrm{d}_{1,19}\left(\mathrm{e}_{2,19}^2 - e_{2,19}^2\right) + 2e_{1,19}d_{2,19}e_{1,19}\right] + 3d_{1,19}\right],$$

$$G_{4,19} = \left[(\cos(h_{1,19}))(\cosh(h_{2,19}))(\cos(g_{2,19}))(\sinh(g_{1,19})) - (\sin(h_{1,19}))(\sinh(h_{2,19}))(\cosh(g_{1,19}))(\sin(g_{2,19}))\right],$$
$$\rho_{10,19} = \left[-\left(\mathrm{d}_{1,19}^2 + d_{2,19}^2\right)^{-1}\left[\mathrm{d}_{2,19}\left(-\mathrm{e}_{2,19}^2 + e_{1,19}^2\right) + 2e_{2,19}d_{2,19}e_{1,19}\right] + 3d_{2,19}\right],$$

$$H_{4,19} = \left[\left(\cos\left(h_{1,19}\right)\right)\left(\cosh\left(h_{2,19}\right)\right)\left(\cos\left(g_{1,19}\right)\right)\left(\sin\left(g_{2,19}\right)\right) + \left(\sin\left(h_{1,19}\right)\right)\left(\sinh\left(h_{2,19}\right)\right)\left(\sinh\left(g_{1,19}\right)\right)\left(\cos\left(g_{2,19}\right)\right)\right],$$

$$\rho_{11,19} = 2\left[d_{1,19}^2 + e_{2,19}^2 - d_{2,19}^2 - e_{1,19}^2\right],$$

$$G_{5,19} = \left[\left(\cos\left(h_{1,19}\right)\right)\left(\cosh\left(h_{2,19}\right)\right)\left(\cos\left(g_{1,19}\right)\right)\left(\cos\left(g_{1,19}\right)\right) - \left(\sin\left(h_{1,19}\right)\right)\left(\sinh\left(h_{2,19}\right)\right)\left(\sinh\left(g_{1,19}\right)\right)\left(\sin\left(g_{2,19}\right)\right)\right],$$

$$\rho_{12,19} = 4\left[d_{1,19}d_{2,19} + e_{1,19}e_{2,19}\right],$$

$$H_{5,19} = \left[\left(\cos\left(h_{1,19}\right)\right)\left(\cosh\left(h_{2,19}\right)\right)\left(\sinh\left(g_{1,19}\right)\right)\left(\sin\left(g_{2,19}\right)\right) + \left(\sin\left(h_{1,19}\right)\right)\left(\sinh\left(h_{2,19}\right)\right)\left(\cosh\left(g_{1,19}\right)\right)\left(\cos\left(g_{2,19}\right)\right)\right],$$

$$\rho_{13,19} = \left[\left\{5\left(d_{1,19}e_{1,19}^3 - 3e_{1,19}e_{2,19}^2d_{1,19}\right) + 5d_{2,19}\left(e_{1,19}^3 - 3e_{1,19}^2e_{2,19}\right)\right\}\left(d_{1,19}^2 + d_{2,19}^2\right)^{-1} + \left(e_{1,19}^2 + e_{2,19}^2\right)^{-1} \left\{5\left(e_{1,19}d_{1,19}^3 - 3d_{2,19}e_{1,19}^2d_{1,19}\right) + 5\left(d_{2,19}^3e_{2,19} - 3d_{1,19}^2d_{2,19}e_{2,19}\right)\right\} - 34\left(d_{1,19}e_{1,19} + d_{2,19}e_{2,19}\right)\right]$$

$$G_{6,19} = \left[\left(\sin\left(h_{1,19}\right)\right)\left(\cosh\left(h_{2,19}\right)\right)\left(\sinh\left(g_{1,19}\right)\right)\left(\cos\left(g_{2,19}\right)\right) + \left(\cos\left(h_{1,19}\right)\right)\left(\sinh\left(h_{2,19}\right)\right)\left(\cosh\left(g_{1,19}\right)\right)\left(\sin\left(g_{2,19}\right)\right)\right],$$

$$\rho_{14,19} = \left[\left\{5\left(d_{1,19}e_{2,19}^3 - 3e_{2,19}e_{1,19}^2d_{1,19}\right) + 5d_{2,19}\left(-e_{1,19}^3 + 3e_{2,19}^2e_{1,19}\right)\right\}\left(d_{1,19}^2 + d_{2,19}^2\right)^{-1} + \left(e_{1,19}^2 + e_{2,19}^2\right)^{-1} \left\{5\left(-e_{1,19}d_{1,19}^3 + 3d_{2,19}^2d_{2,19}e_{1,19}\right) + 5\left(-d_{1,19}^3e_{2,19} + 3d_{2,19}^2d_{1,19}e_{2,19}\right)\right\} + 34\left(d_{1,19}e_{2,19} - d_{2,19}e_{1,19}\right)\right],$$

$$H_{6,19} = \left[\left(\sin\left(h_{1,19}\right)\right)\left(\cosh\left(h_{2,19}\right)\right)\left(\cosh\left(g_{1,19}\right)\right)\left(\sin\left(g_{2,19}\right)\right) - \left(\cos\left(h_{1,19}\right)\right)\left(\sinh\left(h_{2,19}\right)\right)\left(\sinh\left(g_{1,19}\right)\right)\left(\cos\left(g_{2,19}\right)\right)\right],$$

$$H_{7,19} = \left[H_{1,19} + \left(\rho_{5,19}H_{2,19}/2\right) + \left(\rho_{6,19}G_{2,19}/2\right) + \left(\Delta_0/2\right) \left\{\rho_{8,19}G_{3,19} + \rho_{7,19}H_{3,19} + \rho_{10,19}G_{4,19} + \rho_{9,19}H_{4,19} + \rho_{12,19}G_{5,19} + \rho_{11,19}H_{5,19} + (1/12)\left(\rho_{14,19}G_{6,19} + \rho_{13,19}H_{6,19}\right)\right\}\right],$$

$$H_{1,19} = \left[\left(\sin\left(h_{1,19}\right)\right)\left(\sinh\left(h_{2,19}\right)\right)\left(\cosh\left(g_{1,19}\right)\right)\left(\cos\left(g_{2,19}\right)\right) + \left(\cos\left(h_{1,19}\right)\right)\left(\cosh\left(h_{2,19}\right)\right)\left(\sinh\left(g_{1,19}\right)\right)\left(\sin\left(g_{2,19}\right)\right)\right],$$

$$D_{7,19} = \sinh^{-1}\left(\bar{\omega}_{7,19}\right), H_{8,19} = \left(2C_{7,19}D_{7,19}/L_0^2\right)$$

The simplified DR of HD QWs of III–V super-lattices with graded interfaces can be expressed as

$$\left(\frac{n_z\pi}{d_z}\right)^2 = G_{8,19} + iH_{8,19} \tag{19.92}$$

The sub-band equation in this case can be expressed as

$$\left(\frac{n_z \pi}{d_z}\right)^2 = \left|G_{8,19} + iH_{8,19}\right|_{k_s=0 \text{ and } E=E_{700}} \quad (19.93)$$

where E_{700} is the sub-band energy in this case.

The EEM and the DOS function should be obtained numerically in this case.

19.2.16 The DR in NWHD Superlattices of Kane Type Semiconductors with Graded Interfaces in the Presence of Intense Electric Field

The DR in NWHD III–V SLs with graded interfaces in the presence of intense electric fieldcan be expressed as [4]

$$k_z^2 = G_{8,20} + iH_{8,20} \quad (19.94)$$

where

$$G_{8,20} = \left[\frac{C_{7,20}^2 - D_{7,20}^2}{L_0^2} - k_s^2\right], C_{7,20} = \cos^{-1}\left(\bar{\omega}_{7,20}\right),$$

$$\bar{\omega}_{7,20} = (2)^{\frac{-1}{2}}\left[\left(1 - G_{7,20}^2 - H_{7,20}^2\right) - \sqrt{\left(1 - G_{7,20}^2 - H_{7,20}^2\right)^2 + 4G_{7,20}^2}\right]^{\frac{1}{2}},$$

$$\begin{aligned} G_{7,20} = {} & [G_{1,20} + (\rho_{5,20}G_{2,20}/2) - (\rho_{6,20}H_{2,20}/2) + (\Delta_0/2) \\ & \{\rho_{6,20}H_{2,20} - \rho_{8,20}H_{3,20} + \rho_{9,20}H_{4,20} - \rho_{10,20}H_{4,20} + \\ & \rho_{11,20}H_{5,20} - \rho_{12,20}H_{5,20} + (1/12)(\rho_{12,20}G_{6,20} - \rho_{14,20}H_{6,20})\}], \end{aligned}$$

$$\begin{aligned} G_{1,20} = {} & [(\cos(h_{1,20}))(\cosh(h_{2,20}))(\cosh(g_{1,20}))(\cos(g_{2,20})) + \\ & (\sin(\mathrm{h}_{1,20}))(\sinh(\mathrm{h}_{2,20}))(\sinh(\mathrm{g}_{1,20}))(\sin(\mathrm{g}_{2,20}))], \end{aligned}$$

$$\mathrm{h}_{1,20} = e_{1,20}(b_0 - \Delta_0), e_{1,20} = 2^{\frac{-1}{2}}\left(\sqrt{t_{1,20}^2 + t_{2,20}^2} + t_{1,20}\right)^{\frac{1}{2}},$$

$$t_{1,20} = \left[(2m_{c1}/\hbar^2)\cdot \mathrm{J}_{11}(\mathrm{E}, \Delta_1, \mathrm{E}_{g01}), \mathrm{F} - G_2(\mathrm{n_y}, \mathrm{n_z})\right],$$

$$t_{2,20} = \left[(2m_{c1}/\hbar^2)\mathrm{J}_{21}(\mathrm{E}, \Delta_1, \mathrm{E}_{g01}, \mathrm{F})\right],$$

$$h_{2,20} = e_{2,20}(b_0 - \Delta_0),$$

$$e_{2,20} = 2^{\frac{-1}{2}}\left(\sqrt{t_{1,20}^2 + t_{2,20}^2} - t_{1,20}\right)^{\frac{1}{2}},$$

$$g_{1,20} = d_{1,20}(a_0 - \Delta_0),$$

$$d_{1,20} = 2^{\frac{-1}{2}}\left(\sqrt{x_{1,20}^2 + y_{1,20}^2} + x_{1,20}\right)^{\frac{1}{2}},$$

$$x_{1,20} = \left[-\left(2m_{c2}/\hbar^2\right) \cdot \mathrm{J}_{11}\left(E - V_0, E_{g2}, \Delta_2, \eta_{g2}, \mathrm{F}\right) + G_2\left(\mathrm{n}_y, \mathrm{n}_z\right)\right],$$

$$y_1 = \left[\left(2m_{c2}/\hbar^2\right) \cdot \mathrm{J}_{22}\left(E - V_0, E_{g2}, \Delta_2, \eta_{g2}, \mathrm{F}\right)\right],$$

$$g_{2,20} = d_{2,20}(a_0 - \Delta_0),$$

$$d_{2,20} = 2^{\frac{-1}{2}}\left(\sqrt{x_{1,20}^2 + y_{1,20}^2} - x_{1,20}\right)^{\frac{1}{2}},$$

$$\rho_{5,20} = \left(\rho_{3,20}^2 + \rho_{4,20}^2\right)^{-1}\left[\rho_{1,20}\rho_{3,20} - \rho_{2,20}\rho_{4,20}\right],$$

$$\rho_{1,20} = \left[d_{1,20}^2 + e_{2,20}^2 - d_{2,20}^2 - e_{1,20}^2\right], \rho_{3,20} = \left[d_{1,20}e_{1,20} + d_{2,20}e_{2,20}\right],$$

$$\rho_{2,20} = 2\left[d_{1,20}d_{2,20} + e_{1,20}e_{2,20}\right], \rho_{4,20} = \left[d_{1,20}e_{2,20} - e_{1,20}d_{2,20}\right],$$

$$G_{2,20} = \left[\left(\sin\left(h_{1,20}\right)\right)\left(\cosh\left(h_{2,20}\right)\right)\left(\sinh\left(\mathrm{g}_{1,20}\right)\right)\left(\cos\left(\mathrm{g}_{2,20}\right)\right) + \left(\cos\left(h_{1,20}\right)\right)\left(\sinh\left(h_{2,20}\right)\right)\left(\cosh\left(\mathrm{g}_{1,20}\right)\right)\left(\sin\left(\mathrm{g}_{2,20}\right)\right)\right],$$

$$\rho_{6,20} = \left(\rho_{3,20}^2 + \rho_{4,20}^2\right)^{-1}\left[\rho_{1,20}\rho_{4,20} + \rho_{2,20}\rho_{3,20}\right],$$

$$H_{2,20} = \left[\left(\sin\left(h_{1,20}\right)\right)\left(\cosh\left(h_{2,20}\right)\right)\left(\sinh\left(\mathrm{g}_{2,20}\right)\right)\left(\cosh\left(\mathrm{g}_{1,20}\right)\right) - \left(\cos\left(h_{1,20}\right)\right)\left(\sinh\left(h_{2,20}\right)\right)\left(\sinh\left(\mathrm{g}_{1,20}\right)\right)\left(\cos\left(\mathrm{g}_{2,20}\right)\right)\right],$$

$$\rho_{7,20} = \left[\left(\mathrm{e}_{1,20}^2 + e_{2,20}^2\right)^{-1}\left[\mathrm{e}_{1,20}\left(\mathrm{d}_{1,20}^2 - d_{2,20}^2\right) - 2d_{1,20}d_{2,20}e_{2,20}\right] - 3e_{1,20}\right],$$

$$G_{3,20} = \left[\left(\sin\left(h_{1,20}\right)\right)\left(\cosh\left(h_{2,20}\right)\right)\left(\cosh\left(\mathrm{g}_{1,20}\right)\right)\left(\cos\left(\mathrm{g}_{2,20}\right)\right) + \left(\cos\left(h_{1,20}\right)\right)\left(\sinh\left(h_{2,20}\right)\right)\left(\sinh\left(\mathrm{g}_{1,20}\right)\right)\left(\sin\left(\mathrm{g}_{2,20}\right)\right)\right],$$

$$\rho_{8,20} = \left[\left(e_{1,20}^2 + e_{2,20}^2\right)^{-1}\left[e_{2,20}\left(d_{1,20}^2 - d_{2,20}^2\right) + 2d_{1,20}d_{2,20}e_{1,20}\right] + 3e_{2,20}\right],$$

$$H_{3,20} = \left[\left(\sin\left(h_{1,20}\right)\right)\left(\cosh\left(h_{2,20}\right)\right)\left(\sin\left(\mathrm{g}_{2,20}\right)\right)\left(\sinh\left(\mathrm{g}_{1,20}\right)\right) - \left(\cos\left(h_{1,20}\right)\right)\left(\sinh\left(h_{2,20}\right)\right)\left(\cosh\left(\mathrm{g}_{1,20}\right)\right)\left(\cos\left(\mathrm{g}_{2,20}\right)\right)\right],$$

$$\rho_{9,20} = \left[\left(\mathrm{d}_{1,20}^2 + d_{2,20}^2\right)^{-1}\left[\mathrm{d}_{1,20}\left(\mathrm{e}_{2,20}^2 - e_{1,20}^2\right) + 2e_{2,20}d_{2,20}e_{1,20}\right] + 3d_{1,20}\right],$$

$$G_{4,20} = [(\cos(h_{1,20}))(\cosh(h_{2,20}))(\cos(g_{2,20}))(\sinh(g_{1,20})) - (\sin(h_{1,20}))(\sinh(h_{2,20}))(\cosh(g_{1,20}))(\sin(g_{2,20}))],$$

$$\rho_{10,20} = \left[-\left(d_{1,20}^2 + d_{2,20}^2\right)^{-1}\left[d_{2,20}\left(-e_{2,20}^2 + e_{1,20}^2\right) + 2e_{2,20}d_{2,20}e_{1,20}\right] + 3d_{2,20}\right],$$

$$H_{4,20} = [(\cos(h_{1,20}))(\cosh(h_{2,20}))(\cosh(g_{1,20}))(\sin(g_{2,20})) + (\sin(h_{1,20}))(\sinh(h_{2,20}))(\sinh(g_{1,20}))(\cos(g_{2,20}))],$$

$$\rho_{11,20} = 2\left[d_{1,20}^2 + e_{2,20}^2 - d_{2,20}^2 - e_{1,20}^2\right],$$

$$G_{5,20} = [(\cos(h_{1,20}))(\cosh(h_{2,20}))(\cos(g_{2,20}))(\cosh(g_{1,20})) - (\sin(h_{1,20}))(\sinh(h_{2,20}))(\sinh(g_{1,20}))(\sin(g_{2,20}))],$$

$$\rho_{12,20} = 4\left[d_{1,20}d_{2,20} + e_{1,20}e_{2,20}\right],$$

$$H_{5,20} = [(\cos(h_{1,20}))(\cosh(h_{2,20}))(\sinh(g_{1,20}))(\sin(g_{2,20})) + (\sin(h_{1,20}))(\sinh(h_{2,20}))(\cosh(g_{1,20}))(\cos(g_{2,20}))],$$

$$\rho_{13,20} = \left[\left\{5\left(d_{1,20}e_{1,20}^3 - 3e_{1,20}e_{2,20}^2d_{1,20}\right) + 5d_{2,20}\left(e_{1,20}^3 - 3e_{1,20}^2e_{2,20}\right)\right\}\left(d_{1,20}^2 + d_{2,20}^2\right)^{-1} + \left(e_{1,20}^2 + e_{2,20}^2\right)^{-1} \left\{5(e_{1,20}d_{1,20}^3 - 3d_{2,20}e_{1,20}^2d_{1,20}) + 5(d_{2,20}^3e_{2,20} - 3d_{1,20}^2d_{2,20}e_{2,20})\right\} - 34(d_{1,20}e_{1,20} + d_{2,20}e_{2,20})\right],$$

$$G_{6,20} = [(\sin(h_{1,20}))(\cosh(h_{2,20}))(\sinh(g_{1,20}))(\cos(g_{2,20})) + (\cos(h_{1,20}))(\sinh(h_{2,20}))(\cosh(g_{1,20}))(\sin(g_{2,20}))],$$

$$\rho_{14,20} = \left[\left\{5\left(d_{2,20}e_{1,20}^3 - 3e_{2,20}e_{1,20}^2d_{1,20}\right) + 5d_{2,20}\left(-e_{1,20}^3 + 3e_{2,20}^2e_{1,20}\right)\right\}\left(d_{1,20}^2 + d_{2,20}^2\right)^{-1} + \left(e_{1,20}^2 + e_{2,20}^2\right)^{-1} \left\{5\left(-e_{1,20}d_{2,20}^3 + 3d_{1,20}^2d_{2,20}e_{1,20}\right) + 5\left(-d_{1,20}^3e_{2,20} + 3d_{2,20}^2d_{1,20}e_{2,20}\right)\right\} + 34(d_{1,20}e_{2,20} - d_{2,20}e_{1,20})\right],$$

$$H_{6,20} = [(\sin(h_{1,20}))(\cosh(h_{2,20}))(\cosh(g_{1,20}))(\sin(g_{2,20})) - (\cos(h_{1,20}))(\sinh(h_{2,20}))(\sinh(g_{1,20}))(\cos(g_{2,20}))],$$

$$H_{7,20} = [H_{1,20} + (\rho_{5,20}H_{2,20}/2) + (\rho_{6,20}G_{2,20}/2) + (\Delta_0/2)\{\rho_{8,20}G_{3,20} + \rho_{7,20}H_{3,20} + \rho_{10,20}G_{4,20} + \rho_{9,20}H_{4,20} + \rho_{12,20}G_{5,20} + \rho_{11,20}H_{5,20} + (1/12)(\rho_{14,20}G_{6,20} + \rho_{13,20}H_{6,20})\}],$$

$$H_{1,20} = [(\sin(h_{1,20}))(\sinh(h_{2,20}))(\cosh(g_{1,20}))(\cos(g_{2,20})) + (\cos(h_{1,20}))(\cosh(h_{2,20}))(\sinh(g_{1,20}))(\sin(g_{2,20})),$$

$$D_{7,20} = \sinh^{-1}(\bar{\omega}_{7,20}), H_{8,20} = (2C_{7,19}D_{7,20}/L_0^2)$$

The sub-band equation in this case can be expressed as

$$0 = [G_{8,20} + iH_{8,20}]\big|_{E=E_{610}} \tag{19.95}$$

where E_{610} is the sub-band energy in this case.

At low temperatures where the quantum effects become prominent, the DOS function for the lowest SL mini-band is given by

$$N_{HDSL}(E, \eta_g, F) = \frac{g_v}{\pi}\sum_{n_x=1}^{n_{x\max}}\sum_{n_y=1}^{n_{y\max}} \frac{[G'_{8,20} + iH'_{8,20}]}{\sqrt{G_{8,20} + iH_{8,20}}} H(E - E_{610}) \tag{19.96a}$$

The EEM can be written as

$$m^*(\mathrm{E}, \mathrm{n_y}, \mathrm{n_z}, \eta_g, \mathrm{F}) = \frac{\hbar^2}{2}(\mathrm{G}'_{8,20}) \tag{19.96b}$$

19.2.17 The DR in Quantum Dot HD Superlattices of Kane Type Semiconductors with Graded Interfaces in the Presence of Intense Electric Field

The DR in QDHD superlattices of Kane type semiconductors with graded interfaces in the presence of intense electric field can be expressed as

$$\left(\frac{n_z \pi}{d_z}\right)^2 = \left|G_{8,20} + iH_{8,20}\right|_{E=E_{620}} \tag{19.97}$$

where E_{620} is the totally quantized energy in this case.

The DOS function is given by

$$N_{QDHDSL}(\mathrm{E}, \eta_g, F) = \frac{2g_v}{d_x d_y d_z} \sum_{n_x=1}^{n_{x\max}} \sum_{n_y=1}^{n_{y\max}} \sum_{n_z=1}^{n_{z\max}} \delta'(\mathrm{E} - \mathrm{E}_{620}) \tag{19.98}$$

19.2.18 The Magneto DR in HD Superlattices of Kane Type Semiconductors with Graded Interfaces in the Presence of Intense Electric Field

The magneto DR in HD III–V SLs with graded interfaces in the presence of intense electric field can be expressed as

$$k_z^2 = G_{8,19n} + iH_{8,19n} \tag{19.99}$$

where

$$G_{8,19n} = \left[\frac{C_{7,19n}^2 - D_{7,19n}^2}{L_0^2} - \frac{2eB}{\hbar}\left(n + \frac{1}{2}\right)\right], \mathrm{C}_{7,19n} = \cos^{-1}\left(\bar{\omega}_{7,19n}\right)$$

$$\bar{\omega}_{7,19n} = (2)^{\frac{-1}{2}}\left[\left(1 - G_{7,19n}^2 - H_{7,19n}^2\right) - \sqrt{\left(1 - G_{7,19n}^2 - H_{7,19n}^2\right)^2 + 4G_{7,19n}^2}\right]^{\frac{1}{2}},$$

$$G_{7,19n} = \left[G_{1,19n} + \left(\rho_{5,19n} G_{2,19n}/2\right) - \left(\rho_{6,19n} H_{2,19n}/2\right) + (\Delta_0/2)\right.$$
$$\left\{\rho_{6,19n} H_{2,19n} - \rho_{8,19n} H_{3,19n} + \rho_{9,19n} H_{4,19n} - \rho_{10,19n} H_{4,19n} +\right.$$
$$\left.\left.\rho_{11,19n} H_{5,19n} - \rho_{12,19n} H_{5,19n} + (1/12)\left(\rho_{12,19n} G_{6,19n} - \rho_{14,19n} H_{6,19n}\right)\right\}\right],$$

$$G_{1,19n} = \left[\left(\cos\left(h_{1,19n}\right)\right)\left(\cosh\left(h_{2,19n}\right)\right)\left(\cosh\left(g_{1,19n}\right)\right)\left(\cos\left(g_{2,19n}\right)\right) + \left(\sin\left(h_{1,19n}\right)\right)\left(\sinh\left(h_{2,19n}\right)\right)\left(\sinh\left(g_{1,19n}\right)\right)\left(\sin\left(g_{2,19n}\right)\right)\right],$$

$$h_{1,19n} = e_{1,19n}(b_0 - \Delta_0), e_{1,19n} = 2^{\frac{-1}{2}}\left(\sqrt{t_{1,19n}^2 + t_{2,19n}^2} + t_{1,19n}\right)^{\frac{1}{2}},$$

$$t_{1,19n} = \left[\left(2m_{c1}/\hbar^2\right) \cdot J_{11}(E, \Delta_1, E_{g01}, F) - \frac{2eB}{\hbar}\left(n + \frac{1}{2}\right)\right],$$

$$t_{2,19n} = \left[\left(2m_{c1}/\hbar^2\right) J_{21}(E, \Delta_1, E_{g01}, F)\right],$$

$$h_{2,19n} = e_{2,19n}(b_0 - \Delta_0),$$

$$e_{2,19n} = 2^{\frac{-1}{2}}\left(\sqrt{t_{1,19n}^2 + t_{2,19n}^2} - t_{1,19n}\right)^{\frac{1}{2}},$$

$$g_{1,19n} = d_{1,19n}(a_0 - \Delta_0), d_{1,19n} = 2^{\frac{-1}{2}}\left(\sqrt{x_{1,19n}^2 + y_{1,19n}^2} + x_{1,19n}\right)^{\frac{1}{2}},$$

$$x_{1,19n} = \left[-\left(2m_{c2}/\hbar^2\right).J_{11}(E - V_0, E_{g2}, \Delta_2, \eta_{g2}, F) + \frac{2eB}{\hbar}\left(n + \frac{1}{2}\right)\right],$$

$$y_1 = \left[\left(2m_{c2}/\hbar^2\right) \cdot J_{22}(E - V_0, E_{g2}, \Delta_2, \eta_{g2}, F)\right],$$

$$g_{2,19n} = d_{2,19n}(a_0 - \Delta_0),$$

$$d_{2,19n} = 2^{\frac{-1}{2}}\left(\sqrt{x_{1,19n}^2 + y_{1,19n}^2} - x_{1,19n}\right)^{\frac{1}{2}},$$

$$\rho_{5,19n} = \left(\rho_{3,19n}^2 + \rho_{4,19n}^2\right)^{-1}\left[\rho_{1,19n}\rho_{3,19n} - \rho_{2,19n}\rho_{4,19n}\right],$$

$$\rho_{1,19n} = \left[d_{1,19n}^2 + e_{2,19n}^2 - d_{2,19n}^2 - e_{1,19n}^2\right],$$

$$\rho_{3,19n} = \left[d_{1,19n}e_{2,19n} + d_{2,19n}e_{2,19n}\right],$$

$$\rho_{2,19n} = 2\left[d_{1,19n}d_{2,19n} + e_{1,19n}e_{2,19n}\right],$$

$$\rho_{4,19n} = \left[d_{1,19n}e_{2,19n} - e_{1,19n}d_{2,19n}\right],$$

$$G_{2,19n} = \left[\left(\sin\left(h_{1,19n}\right)\right)\left(\cosh\left(h_{2,19n}\right)\right)\left(\sinh\left(g_{1,19n}\right)\right)\left(\cos\left(g_{2,19n}\right)\right) + \left(\cos\left(h_{1,19n}\right)\right)\left(\sinh\left(h_{2,19n}\right)\right)\left(\cosh\left(g_{1,19n}\right)\right)\left(\sin\left(g_{2,19n}\right)\right)\right],$$

$$\rho_{6,19n} = \left(\rho_{3,19n}^2 + \rho_{4,19n}^2\right)^{-1}\left[\rho_{1,19n}\rho_{4,19n} + \rho_{2,19n}\rho_{3,19n}\right],$$

$$H_{2,19n} = \left[\left(\sin\left(h_{1,19n}\right)\right)\left(\cos\left(h_{2,19n}\right)\right)\left(\sin\left(g_{2,19n}\right)\right)\left(\cosh\left(g_{1,19n}\right)\right) \right.$$
$$\left. -\left(\cos\left(h_{1,19n}\right)\right)\left(\sinh\left(h_{2,19n}\right)\right)\left(\sinh\left(g_{1,19n}\right)\right)\left(\cos\left(g_{2,19n}\right)\right)\right],$$

$$\rho_{7,19n} = \left[\left(e_{1,19n}^2 + e_{2,19n}^2\right)^{-1}\left[e_{1,19n}\left(d_{1,19n}^2 - d_{2,19n}^2\right) - 2d_{1,19n}d_{2,19n}e_{2,19n}\right] - 3e_{1,19n}\right],$$

$$G_{3,19n} = \left[\left(\sin\left(h_{1,19n}\right)\right)\left(\cosh\left(h_{2,19n}\right)\right)\left(\cosh\left(g_{1,19n}\right)\right)\left(\cos\left(g_{2,19n}\right)\right) \right.$$
$$\left. + \left(\cos\left(h_{1,19n}\right)\right)\left(\sinh\left(h_{2,19n}\right)\right)\left(\sinh\left(g_{1,19n}\right)\right)\left(\sin\left(g_{2,19n}\right)\right)\right],$$

$$\rho_{8,19n} = \left[\left(e_{1,19}^2 + e_{2,19}^2\right)^{-1}\left[e_{2,19}\left(d_{1,19n}^2 - d_{2,19n}^2\right) + 2d_{1,19n}d_{2,19n}e_{1,19n}\right] + 3e_{2,19n}\right],$$

$$H_{3,19n} = \left[\left(\sin\left(h_{1,19n}\right)\right)\left(\cosh\left(h_{2,19n}\right)\right)\left(\sin\left(g_{2,19n}\right)\right)\left(\sinh\left(g_{1,19n}\right)\right) \right.$$
$$\left. -\left(\cos\left(h_{1,19n}\right)\right)\left(\sinh\left(h_{2,19n}\right)\right)\left(\cosh\left(g_{1,19n}\right)\right)\left(\cos\left(g_{2,19n}\right)\right)\right],$$

$$\rho_{9,19n} = \left[\left(d_{1,19n}^2 + d_{2,19n}^2\right)^{-1}\left[d_{1,19n}\left(e_{2,19n}^2 - e_{1,19n}^2\right) + 2e_{2,19n}d_{2,19n}e_{1,19n}\right] + 3d_{1,19n}\right],$$

$$G_{4,19n} = \left[\left(\cos\left(h_{1,19n}\right)\right)\left(\cosh\left(h_{2,19n}\right)\right)\left(\cos\left(g_{2,19n}\right)\right)\left(\sinh\left(g_{1,19n}\right)\right) \right.$$
$$\left. -\left(\sin\left(h_{1,19n}\right)\right)\left(\sinh\left(h_{2,19n}\right)\right)\left(\cosh\left(g_{1,19n}\right)\right)\left(\sin g_{2,19n}\right)\right],$$

$$\rho_{10,19n} = \left[-\left(d_{1,19n}^2 + d_{2,19n}^2\right)^{-1}\left[d_{2,19n}\left(-e_{2,19n}^2 + e_{1,19n}^2\right) + 2e_{2,19n}d_{2,19n}e_{1,19n}\right] + 3d_{2,19n}\right],$$

$$H_{4,19n} = \left[\left(\cos\left(h_{1,19n}\right)\right)\left(\cosh\left(h_{2,19n}\right)\right)\left(\cosh\left(g_{1,19n}\right)\right)\left(\sin\left(g_{2,19n}\right)\right) \right.$$
$$\left. + \left(\sin\left(h_{1,19n}\right)\right)\left(\sinh\left(h_{2,19n}\right)\right)\left(\sinh\left(g_{1,19n}\right)\right)\left(\cos\left(g_{2,19n}\right)\right)\right],$$

$$\rho_{11,19n} = 2\left[d_{1,19n}^2 + e_{2,19n}^2 - d_{2,19n}^2 - e_{1,19n}^2\right],$$

$$G_{5,19n} = \left[\left(\cos\left(h_{1,19n}\right)\right)\left(\cosh\left(h_{2,19n}\right)\right)\left(\cos\left(g_{1,19n}\right)\right)\left(\cos\left(g_{1,19n}\right)\right) \right.$$
$$\left. -\left(\sin\left(h_{1,19n}\right)\right)\left(\sinh\left(h_{2,19n}\right)\right)\left(\sinh\left(g_{1,19n}\right)\right)\left(\sin\left(g_{2,19n}\right)\right)\right],$$

$$\rho_{12,19n} = 4\left[d_{1,19n}d_{2,19n} + e_{1,19n}e_{2,19n}\right],$$

$$H_{5,19n} = \left[\left(\cos\left(h_{1,19n}\right)\right)\left(\cosh\left(h_{2,19n}\right)\right)\left(\sinh\left(g_{1,19n}\right)\right)\left(\sin\left(g_{2,19n}\right)\right) \right.$$
$$\left. + \left(\sin\left(h_{1,19n}\right)\right)\left(\sinh\left(h_{2,19n}\right)\right)\left(\cosh\left(g_{1,19n}\right)\right)\left(\cos\left(g_{2,19n}\right)\right)\right],$$

$$\rho_{13,19n} = \left[\left\{5\left(d_{1,19n}e_{1,19n}^3 - 3e_{1,19n}e_{2,19n}^2d_{1,19n}\right) + 5d_{2,19n}\left(e_{1,19n}^3 - 3e_{1,19n}^2e_{2,19n}\right)\right\}\left(d_{1,19n}^2 + d_{2,19n}^2\right)^{-1} + \left(e_{1,19n}^2 + e_{2,19n}^2\right)^{-1} \right.$$
$$\left. \left\{5\left(e_{1,19n}d_{1,19n}^3 - 3d_{2,19n}e_{1,19n}^2d_{1,19n}\right) + 5(d_{2,19n}^3e_{2,19n} - 3d_{1,19n}^2d_{2,19n}e_{2,19n})\right\} - 34\left(d_{1,19n}e_{1,19n} + d_{2,19n}e_{2,19n}\right)\right],$$

$$
\begin{aligned}
G_{6,19n} &= [(\sin(h_{1,19n}))(\cosh(h_{2,19n}))(\sinh(g_{1,19n}))(\cos(g_{2,19n})) \\
&\quad + (\cos(h_{1,19n}))(\sinh(h_{2,19n}))(\cosh(g_{1,19n}))(\sin(g_{2,19n}))], \\
\rho_{14,19n} &= \Big[\Big\{5\Big(d_{2,19n}e_{1,19n}^3 - 3e_{2,19n}e_{1,19n}^2 d_{1,19n}\Big) + 5d_{2,19n}\Big(-e_{1,19n}^3 + 3e_{2,19n}^2 e_{1,19n}\Big)\Big\}\Big(d_{1,19n}^2 + d_{2,19n}^2\Big)^{-1} + \Big(e_{1,19n}^2 + e_{2,19n}^2\Big)^{-1} \\
&\quad \Big\{5\Big(-e_{1,19n}d_{2,19n}^3 + 3d_{1,19n}^2 d_{2,19n}e_{1,19n}\Big) + 5\Big(-d_{1,19n}^3 e_{2,19n} + 3d_{2,19n}^2 d_{1,19n}e_{2,19n}\Big)\Big\} + 34(d_{1,19n}e_{2,19n} - d_{2,19n}e_{1,19n})\Big],
\end{aligned}
$$

$$
\begin{aligned}
H_{6,19n} &= [(\sin(h_{1,19n}))(\cosh(h_{2,19n}))(\cosh(g_{1,19n}))(\sin(g_{2,19n})) \\
&\quad - (\cos(h_{1,19n}))(\sinh(h_{2,19n}))(\sinh(g_{1,19n}))(\cos(g_{2,19n}))], \\
H_{7,19n} &= [H_{1,19n} + (\rho_{5,19n}H_{2,19n}/2) + (\rho_{6,19n}G_{2,19n}/2) \\
&\quad + (\Delta_0/2)\{\rho_{8,19n}G_{3,19n} + \rho_{7,19n}H_{3,19n} + \rho_{10,19n}G_{4,19n} \\
&\quad + \rho_{9,19n}H_{4,19n} + \rho_{12,19n}G_{5,19n} + \rho_{11,19n}H_{5,19n} + (1/12)(\rho_{14,19n}G_{6,19n} + \rho_{13,19n}H_{6,19n})\}], \\
H_{1,19n} &= [(\sin(h_{1,19n}))(\sinh(h_{2,19n}))(\cosh(g_{1,19n}))(\cos(g_{2,19n})) \\
&\quad + (\cos(h_{1,19n}))(\cosh(h_{2,19n}))(\sinh(g_{1,19n}))(\sin(g_{2,19n}))], \\
D_{7,19n} &= \sinh^{-1}(\bar{\omega}_{7,19n}), H_{8,19n} = (2C_{7,19n}D_{7,19n}/L_0^2)
\end{aligned}
$$

The sub-band equation in this case can be expressed as

$$
0 = [G_{8,19n} + iH_{8,19n}]\big|_{E=E_{630}} \tag{19.100}
$$

where E_{630} is the Landau sub-band energy in this case.

At low temperatures where the quantum effects become prominent, the DOS function for the lowest SL mini-band is given by

$$
N_{HDSL}(\mathrm{E}, \eta_g, \mathrm{F}) = \frac{g_v eB}{2\pi^2\hbar}\sum_{n=0}^{n_{\max}} \frac{[G'_{8,19n} + iH'_{8,19n}]}{\sqrt{G_{8,19n} + iH_{8,19n}}} H(\mathrm{E} - \mathrm{E}_{630}) \tag{19.101}
$$

The EEM can be written as

$$
m^*(\mathrm{E}, \mathrm{n}, \eta_g) = \frac{\hbar^2}{2}(\mathrm{G}'_{8,19n}) \tag{19.102}
$$

19.2.19 The Magneto DR in QWHD Superlattices of Kane Type Semiconductors with Graded Interfaces in the Presence of Intense Electric Field

The magneto DR in QWHD superlattices of Kane type semiconductors with graded interfaces in the presence of intense electric field can be expressed as

$$\left(\frac{n_z \pi}{d_z}\right)^2 = \left[G_{8,19n} + iH_{8,19n}\right]\big|_{E=E_{650}} \tag{19.103}$$

where E_{650} is the totally quantized energy in this case.

The DOS function is given by

$$N_{QWHDSLB}(\mathrm{E}, \eta_g, \mathrm{F}) = \frac{g_v eB}{\pi\hbar} \sum_{n_z=1}^{n_{z\max}} \sum_{n=0}^{n_{\max}} \delta'(\mathrm{E} - \mathrm{E}_{650}) \tag{19.104}$$

19.3 Summary and Conclusion

In this chapter in, (19.28), the coefficients of k_x,k_y and k_y are not same and for this reason, this basic equation is "anisotropic" in nature together with the fact that the anisotropic dispersion relation is the ellipsoid of revolution in the k-space although the basic DR in the absence of any field for Kane type semiconductors is isotropic in accordance with the three band model of Kane. The (19.29), (19.30) and (19.31) exhibit the EEMs along x, y and z directions and they show electric field and energy dependent mass anisotropy due to the presence of electric field only. The (19.40), (19.41) and (19.42) represent the DR, DOS and the EEM in HD Kane type semiconductors under intense electric field. The aforesaid equations depend on the electric field, screening potential and the electron concentration at any value of the electron energy.

The (19.48), (19.49) and (19.51) represent the DR in QW of HD Kane type materials, the DOS function and the EEM in the presence of intense electric field. The sub-band energies as given by (19.50) are complex and electric field dependent. The (19.52), (19.53) and (19.55) represent the DR in NW of HD Kane type materials, the DOS function and the EEM in the presence of intense electric field. The sub-band energies as given by (19.54) are complex and electric field dependent. The DR in QBs of HD Kane type materials under strong electric field is given by (19.56) and the corresponding DOS function is reflected through (19.57). The DR is totally quantized due to the quantization of the wave vector space and the DOS functions are unevenly distributed in the energy axis in the forms of Dirac's delta function. The magneto DR in QWs of HD Kane Type Semiconductors in the Presence of intense electric field is given by (19.58) and the corresponding DOS function is expressed through (19.59). The total energy is being quantized due to the simultaneous combination of magnetic and size quantization's respectively. The DRs in accumulation and inversion layers of Kane Type Semiconductors in the Presence of intense electric field are given by (19.60) and (19.64) respectively. For accumulation layers the DRs are quantized circles for constant complex energies in the $k_x k_y$ plane. For inversion layers the DRs are quantized circles for constant real energies in the same plane. The sub-band index dependent EEMs are given by (19.61) and (19.65) for accumulation and inversions layers respectively. The DOS functions in the respective cases can be expressed through (19.63) and (19.67) respectively. The magneto DRs in accumulation and inversion layers of Kane Type Semiconductors in the Presence of intense electric field are given by (19.68) and (19.70) respectively. It appears that the energy is totally quantized in both the cases due to the application of quantizing magnetic field in inversion layers forming quantized sub-bands. The DOS functions in both the cases can be expressed through (19.69) and (19.71) respectively. The (19.69) and (19.71) exhibit the fact that they are delta functions respectively and summations over the Landau quantum numbers and the quantized mini-band index are needed to generate the total DOS functions in each case. The DR, the sub-band energy, the EEM and the DOS function in electric field aided doping superlattices of HD Kane Type Semiconductors are given by (19.72), (19.73), (19.74) and (19.75) respectively. The DRs are quantized circles in the $k_x k_y$ plane for constant complex energies and the sub-band energies are also complex. The EEM is mini-band index dependent which is the characteristics features of doping superlattices in general. The magneto DR and the DOS function in doping superlattices of HD Kane Type semiconductors in the presence of intense electric field are given by (19.76) and (19.77) respectively. The DR in QWHD effective mass superlattices of Kane Type Semiconductors in the presence of intense electric field is given by (19.88) and (19.89) explores the corresponding DOS function. The DR in QWHD superlattices of Kane Type Semiconductors with graded interfaces in the Presence of intense electric field has been formulated in (19.92) and the sub-band equation for this system is given by (19.93). The derivations of the analytical explicit expressions of the DOS function and EEM are not possible and we have to rely on the numerical computations in this context. The DR, the EEM and the DOS function in NWHD superlattices of Kane Type Semiconductors with graded interfaces in the presence of intense electric field are given by (19.94), (19.96a) and (19.96b) respectively. The EEM is quantum number dependent which is the characteristics property of such superlattices. The DR in Quantum dot HD superlattices of Kane type semiconductors with graded interfaces in the presence of intense electric field is given by (19.97) and the DOS function has been investigated in (19.98). The magneto DR, the DOS function and the EEM in HD superlattices of Kane Type Semiconductors with graded interfaces in the Presence of intense electric field are given by (19.99), (19.101) and (19.102) respectively. The magneto DR and the DOS function in QWHD superlattices of Kane Type Semiconductors with graded interfaces in the Presence of intense electric field are given by (19.103) and (19.104) respectively.

19.4 Open Research Problems

(R.19.1) Investigate the DR in the presence of intense external non-uniform electric field for all the HD superlattices whose respective dispersion relations of the carriers are given in this chapter.

(R.19.2) Investigate the DR for the heavily–doped semiconductors in SLs the presences of Gaussian, exponential, Kane, Halperian, Lax and Bonch-Burevich types of band tails for all SL systems as discussed in this chapter in the presence of external oscillatory and non-uniform electric field.

(R.19.3) Investigate the DR in the presence of external non-uniform electric field for short period, strained layer, random and Fibonacci HD superlattices in the presence of an arbitrarily oriented alternating electric field.

(R.19.4) Investigate all the appropriate problems of this chapter for a Dirac electronDirac electron.

(R.19.5) Investigate all the appropriate problems of this chapter by including the many body, broadening and hot carrier effects respectively.

(R.19.6) Investigate all the appropriate problems of this chapter by removing all the mathematical approximations and establishing the respective appropriate uniqueness conditionsUniqueness conditions.

References

1. L. Schiff, *Quantum Mechanics* (McGraw-Hill, USA, 1968)
2. B.R. Nag, *Electron Transport in compound Semiconductors* (Springer, Heidelberg, 1980)
3. H. Sasaki, Phys. Rev. B **30**, 7016 (1984)
4. H.X. Jiang, J.Y. Lin, J. Appl. Phys. **61**, 624 (1987)

Chapter 20
Few Related Applications

Morning is not only sunrise, but a beautiful miracle of GOD that defeats the DARKNESS and spreads light.

20.1 Introduction

In this monograph, we have investigated different DRs of HD semiconductor nanostructures of different technologically important materials and its implications in materials science in general. The concept of DR is one of the main keys for investigating the carrier transport phenomena of HD quantum effect devices. In this chapter, we shall discuss **twenty eight different applications** in this context. The Sect. 20.3 contains the single gigantic open research problem.

20.2 Different Related Applications

The investigations as presented in this monograph find **twenty eight different applications** in the realm of modern quantum effect devices.

20.2.1 Carrier Statistics

In the whole realm of the electronic properties of the semiconducting materials, in accordance with importance the investigations of carrier statistics occupy the third position in the merit list where the other two's are DR and the DOS function chronologically. Incidentally the first book in this line is Semiconductor Statistics written by famous physicist Prof. J.S. Blakemore as early as in 1962 [1] many electronic properties are connected with carrier statistics and the carrier statistics

K.P. Ghatak, *Dispersion Relations in Heavily-Doped Nanostructures*,
Springer Tracts in Modern Physics 265, DOI 10.1007/978-3-319-21000-1_20

itself has been investigated in the literature extensively [2–10]. The electron statistics can in general be written as

$$n_0 = \int_{E'}^{\infty} N(E) f_0 dE \tag{20.1}$$

where E' is the lower limit of integration. In this book we have formulated many DOS functions for different HD semiconductors and their quantized counter parts. Thus for the expressions of $N(E)$ as formulated in different chapters we can studied the n_0 for each condition and thus we can study the influence of different physical variables on the electron degeneracy.

20.2.2 Thermoelectric Power

In recent years, with the advent of Quantum Hall Effect (QHE) [11], there has been considerable interest in studying the thermoelectric power under strong magnetic field (TPSM) in various types of nanostructured materials having quantum confinement of their charge carriers in one, two and three dimensions of the respective wave vector space leading to different carrier energy spectra [12–41]. The classical TPSM equation $G = \pi^2 k_B / 3e$ (which is a function of three fundamental constants of nature) is valid only under the condition of carrier non-degeneracy, being independent of carrier concentration and reflects the fact that the signature of the band structure of any material is totally absent in the same.

Zawadzki [17] demonstrated that the TPSM for electronic materials having degenerate electron concentration is essentially determined by their respective energy band structures. It has, therefore, different values in different materials and changes with the doping, magnitude of the reciprocal quantizing magnetic field under magnetic quantization, quantizing electric field as in inversion layers, nanothickness as in quantum wells, wires and dots, with superlattice period as in quantum confined semiconductor superlattices with graded interfaces having various carrier energy spectra and also in other types of field assisted nanostructured materials.

The magnitude of the thermoelectric power G can be written as [18]

$$G = \frac{1}{|e| T n_0} \int_{-\infty}^{\infty} (E - E_F) R(E) \left[-\frac{\partial f_0}{\partial E} \right] dE \tag{20.2}$$

where $R(E)$ is the total number of states. The (20.2) can be written under the condition of carrier degeneracy [13] as

$$G = \left(\frac{\pi^2 k_B^2 T}{3|e| n_0} \right) \left(\frac{\partial n_0}{\partial E_F} \right) \tag{20.3}$$

For inversion layers and NIPI structures, under the condition of electric quantum limit, (20.3) assumes the form

$$G = \frac{\pi^2 k_B^2 T}{3en_{02D}}\left[\frac{\partial n_{02D}}{\partial(E_{F2D} - E_{02D})}\right] \tag{20.4}$$

where n_{02D}, E_{F2D} and E_{02D} are the surface electron concentration, the Fermi energy and the sub band energy for the said 2D systems at the electric quantum limit.

For heavily doped semiconductors (20.2) assumes the form

$$G = \frac{\pi^2 k_B^2 T}{3en_{0HD}}\left[\frac{\partial n_{0HD}}{\partial(E_{FHD} - E_{0HD})}\right] \tag{20.5}$$

where n_{0HD}, E_{FHD} and E_{0HD} are the electron concentration, the Fermi energy and the band tail energy for the heavily doped semiconductors where E_{0HD} should be obtained from the heavily doped dispersion relation of the semiconductor under the conditions $E = E_{0HD}$ and $k = 0$.

Thus, we can use the carrier statistics for different low dimensional HD materials to investigate the TPSM in such compounds.

20.2.3 Debye Screening Length

It is well known that the Debye screening Length (DSL) of the carriers in semiconductors is a very important quantity characterizing the screening of the Coulomb field of the ionized impurity centers by the free carriers [42]. It affects many of the special features of modern nano-devices, the carrier motilities under different mechanisms of scattering, and the carrier plasmas in semiconductors [43]. The DSL is a very good approximation to the accurate self-consistent screening in presence of band tails and is also used to illustrate the interaction between the colliding carriers in Auger effect in solids [42]. The classical value of the DSL is equal to $\left[\varepsilon_{sc} k_B T/(e^2 n_0)\right]^{1/2}$ (ε_{sc}, k_B, T, e, and n_0 are the semiconductor permittivity, the Boltzmann's constant, the temperature, the magnitude of the carrier charge, and the electron concentration, respectively) which is valid for both the carriers. In this conventional form, the DSL decreases with increasing carrier concentration at a constant temperature and this relation holds only under the condition of carrier non-degeneracy. It is interesting to note that the under the condition of extreme degeneracy, the expression of DSL for materials having parabolic energy bands can be written as $L_D = \left(\pi^{2/3}\hbar\sqrt{\varepsilon_{sc}}\right)\left(e g_v^{1/3} 3^{1/6} n_0^{1/6}\sqrt{m_c}\right)^{-1}$ ($\hbar$, m_c and g_v are Dirac constant, effective electron mass at the edge of the conduction band and valley degeneracy respectively). Thus we observed that in this case the result is independent of temperature, but depends on n_0, g_v and m_c. Besides, the indices of inverse electron variation changes from half in the former case to one-sixth in the latter case. Since, the performance of the electron devices at the device terminals and the speed of operation of modern switching transistors are significantly influence by the degree of carrier degeneracy present in these devices, the simplest way of analyzing such devices taking into account of the degeneracy of the band is to use the appropriate DSL to express the performance at the device terminal and switching speed in terms of the carrier concentration [44].

The DSL depends on the density-of-states function which, in turn, is significantly affected by the different carrier energy spectra of different semiconductors having various band structures [42, 44]. In recent years, various energy wave vector dispersion relations of the carriers of different materials have been proposed [45] which have created the interest in studying the DSL in such quantized structures under external conditions. It is well known from the fundamental study of Landsberg [42], that the DSL for electronic materials having degenerate electron concentration is essentially determined by their respective energy band structures. It has, therefore, different values in different materials and varies with the electron concentration, with the magnitude of the reciprocal quantizing magnetic field under magnetic quantization, with the quantizing electric field as in inversion layers, with the nano-thickness as in quantum wells, with super-lattice period as in the quantum confined super-lattices of small gap compounds with graded interfaces having various carrier energy spectra. The nature of these variations has been investigated by in the literature [46–54].

The 3D DSL can, in general, be expressed as [44]

$$L_{3D} = \left[\frac{e^2}{\varepsilon_{sc}} \cdot \frac{\partial n_0}{\partial E_F}\right]^{\frac{-1}{2}} \tag{20.6}$$

Using (20.3) and (20.6) we can write

$$L_{3D} = \left[\frac{\pi^2 k_B^2 T \varepsilon_{sc}}{3e^3 n_0 G}\right]^{\frac{1}{2}} \tag{20.7}$$

The 3D DSL for the heavily doped systems is given by

$$L_{3D} = \left[\frac{e^2}{\varepsilon_{sc}} \cdot \frac{\partial n_{0HD}}{\partial (E_{FHD} - E_{0HD})}\right]^{\frac{-1}{2}} \tag{20.8}$$

Using (20.8) and (20.5) we obtain the same equation as given by (9.7). Thus, the 3D DSL for degenerate materials can be determined by knowing the experimental values of G.

For inversion layers and NIPI structures, the 2D DSL is given by

$$L_{2D} = \left[\frac{e^2}{2\varepsilon_{sc}} \frac{\partial n_{02D}}{\partial (E_{F2D} - E_{02D})} \right]^{-1} \qquad (20.9)$$

Using (20.4) and (20.9) we get,

$$L_{2D} = \frac{2\pi^2 k_B^2 T \varepsilon_{sc}}{3e^3 n_{02D} G} \qquad (20.10)$$

For QWs, the G assumes the form

$$G = \frac{\pi^2 k_B^2 T}{3en_{2D}} \cdot \frac{\partial n_{2D}}{\partial E_{Fs}} \qquad (20.11)$$

where n_{2D} and E_{Fs} are 2D electron concentration and Fermi energy in this case. Using (20.10) and (20.11) we get

$$L_{2D} = \frac{2\pi^2 k_B^2 T \varepsilon_{sc}}{3e^3 n_{2D} G} \qquad (20.12)$$

From the suggestion for the experimental determination of the 3D DSL for degenerate materials having arbitrary dispersion laws as given by (20.7), we observe that for a constant T, the DSL varies inversely with the square root of Gn_0. Only the experimental values of G for any material as a function of electron concentration will generate the experimental values of the 3D DSL for that range of n_0 for that material. Since $(Gn_0)^{-1/2}$ decreases with increasing n_0 for constant T, from (20.7) we can conclude that the 3D DSL will decrease with increasing n_0. For 2D DSL L_D at a constant temperature varies inversely with $n_{02D}G$ or $n_{2D}G$ as appear from (20.10) and (20.12) respectively. Since $n_{02D}G$ and $n_{2D}G$ increase with decreasing surface concentration, from (20.10) and (20.12) we can infer that 2D DSL will increase with decreasing n_{02D} or n_{2D} for the appropriate 2D systems. This statement provides a compatibility test of our theoretical analysis. Thus (20.7), (20.10) and (20.12) provide experimental checks of both 3D and 2D DSLs and also a technique for probing the band structures of the degenerate materials having arbitrary band structures.

20.2.4 Carrier Contribution to the Elastic Constants

The knowledge of the carrier contribution to the elastic constants is important in studying the mechanical properties of the materials and has been investigated in the literature [55–58]. The electronic contribution to the second- and third-order elastic constants for HD materials can be written as [55–58]

$$\Delta C_{44} = \frac{-G_0^2}{9}\left[\frac{\partial n_{0HD}}{\partial(E_{FHD} - E_{0HD})}\right] \tag{20.13}$$

and

$$\Delta C_{456} = \frac{G_0^3}{27}\left[\frac{\partial^2 n_{0HD}}{\partial(E_{FHD} - E_{0HD})^2}\right] \tag{20.14}$$

where G_0 is the deformation potential constant. Thus, using (20.5), (20.13) and (20.14), we can write

$$\Delta C_{44} = \left[-n_0(\bar{G}_0)^2|e|G/(3\pi^2 k_B^2 T)\right] \tag{20.15}$$

and

$$\Delta C_{456} = \left(n_0|e|(\overline{G}_0)^3\right)G^2/(3\pi^4 k_B^3 T)\left(1 + \frac{n_0}{G}\frac{\partial G}{\partial n_0}\right) \tag{20.16}$$

Therefore, by using (20.13) and (20.14) we can investigate ΔC_{44} and ΔC_{456} for all the cases of this monograph. Besides, the experimental graph of G versus n_0 allows us to determine the electronic contribution to the elastic constants for materials having arbitrary spectra.

20.2.5 Diffusivity-Mobility Ratio

It is well known that the diffusivity-mobility ratio (DMR) occupies a central position in the whole field of solid state device electronics and the related sciences since the diffusion constant (a quantity very useful for device analysis where exact experimental determination is rather difficult) can be obtained from this ratio by knowing the experimental values of the mobility. The classical value of the DMR is equal to $(k_B T/|e|)$, (k_B, T, and $|e|$ are Boltzmann's constant, temperature and the magnitude of the carrier charge respectively). This relation in this form was first introduced by Einstein to study of the diffusion of gas particles and is known as the Einstein relation [59, 60]. It appears that the DMR increases linearly with increasing T and is independent of electron concentration. This relation is applicable for both types of charge carriers only under non-degenerate carrier concentration although its validity has been suggested erroneously for degenerate materials [61]. Landsberg first pointed out that the DMR for degenerate semiconductors is essentially determined by their energy band structures [62–65]. This relation is useful for semiconductor homo-structures [62, 63], semiconductor-semiconductor hetero-structures [66, 67], metals-semiconductor hetero-structures [68–76] and insulator-semiconductor hetero-structures [77–80]. The nature of the variations of the DMR under different physical conditions has been studied in the literature [59–63, 81–104]. Incidentally, A.N. Chakravarti (a recognized leading expert of DMR in general) and his research group are still contributing significantly under his able leadership regarding this pin pointed research topic on DMR from 1972 [59, 81–85, 91, 96–104] and some of the significant features, which have emerged from these studies, are:

(a) The DMR increases monotonically with increasing carrier concentration in bulk semiconductors and the nature of these variations are significantly influenced by the band structures of different materials.
(b) The DMR increases with the increasing quantizing electric field as in inversion layers.
(c) The DMR oscillates with the inverse quantizing magnetic field under magnetic quantization due to the Shubnikov-de Haas effect.
(d) The DMR shows composite oscillations with the various controlled quantities of semiconductor super lattices.
(e) In ultrathin films, quantum wires and other field assisted low-dimensional systems, the value of the DMR changes appreciably with the external variables depending on the nature of quantum confinements of different materials.

The DMR depends on the density-of-states (DOS) function, which, in turn, is significantly affected by the different carrier energy spectra of different semiconductors having various band structures. It can, in general, be proved that for bulk specimens the DMR is given by [81]

$$\frac{D}{\mu} = \frac{n_{0HD}}{e}\left[\frac{\partial n_{0HD}}{\partial(E_{FHD} - E_{0HD})}\right]^{-1} \tag{20.17}$$

The electric quantum limit as in inversion layers and NIPI structures refers to the lowest electric sub-band and (20.17) assumes the form [81]

$$\frac{D}{\mu} = \frac{n_{02D}}{e}\left[\frac{\partial n_{02D}}{\partial(E_{F2D} - E_{02D})}\right]^{-1} \tag{20.18}$$

Using (20.4), (20.5), (20.17) and (20.18) one obtains

$$\frac{D}{\mu} = \left[\frac{\pi^2 k_B^2 T}{3|e|^2 G}\right] \tag{20.19}$$

Thus, the DMR for degenerate materials can be determined by knowing the experimental values of G.

The suggestion for the experimental determination of the DMR for degenerate semiconductors having arbitrary dispersion laws as given by (20.19) does not contain any energy band constants. For a fixed temperature, the DSL varies inversely as G. Only the experimental values of G for any material as a function of electron concentration will generate the experimental values of the DMR for that range of n_0 for that system. Since G decreases with increasing n_0, from (20.19) one can infer that the DMR will increase with increase in n_0. This statement is the compatibility test so far as the suggestion for the experimental determination of DMR for degenerate materials is concerned.

20.2.6 Measurement of Band-Gap in the Presence of Light Waves

Using the appropriate equations, the normalized incremental band gap (ΔE_g) has been plotted as a function of normalized I_0 (for a given wavelength and considering red light for which $\lambda = 660\,\text{nm}$) at $T = 4.2$ K in Figs. 20.1 and 20.2 for n-$Hg_{1-x}Cd_xTe$ and n-$In_{1-x}Ga_xAs_yP_{1-y}$ lattice matched to InP in accordance with the perturbed three and two band models of Kane and that of perturbed parabolic energy bands respectively. In Figs. 20.3 and 20.4, the normalized incremental band gap has been plotted for the aforementioned optoelectronic compounds as a function of λ. It is worth remarking that the influence of an external photo-excitation is to change radically the original band structure of the material. Because of this

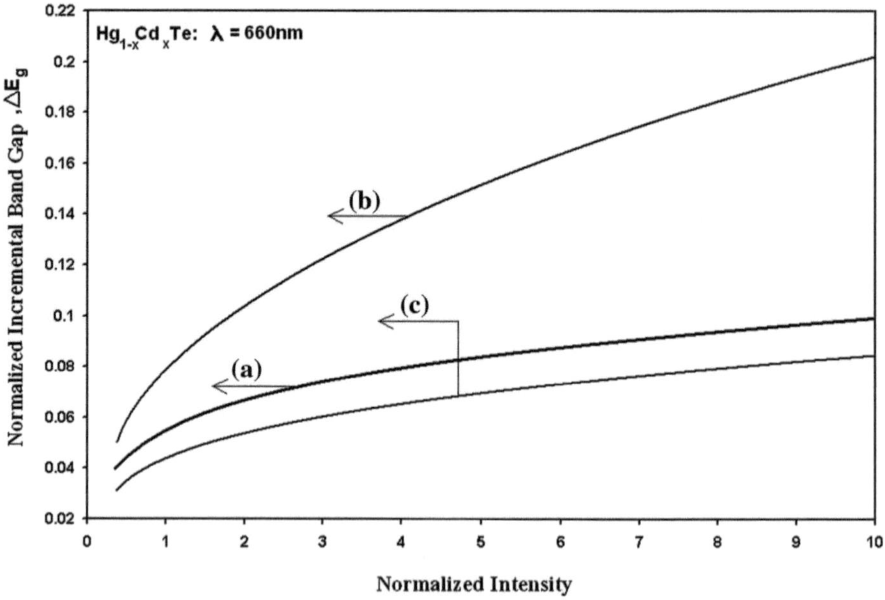

Fig. 20.1 Plots of the normalized incremental band gap (ΔE_g) for n-$Hg_{1-x}Cd_xTe$ as a function of normalized light intensity in which the *curves* (*a*) and (*b*) represent the perturbed three and two band models of Kane respectively. The *curve* (*c*) represents the same variation in n-$Hg_{1-x}Cd_xTe$ in accordance with the perturbed parabolic energy bands

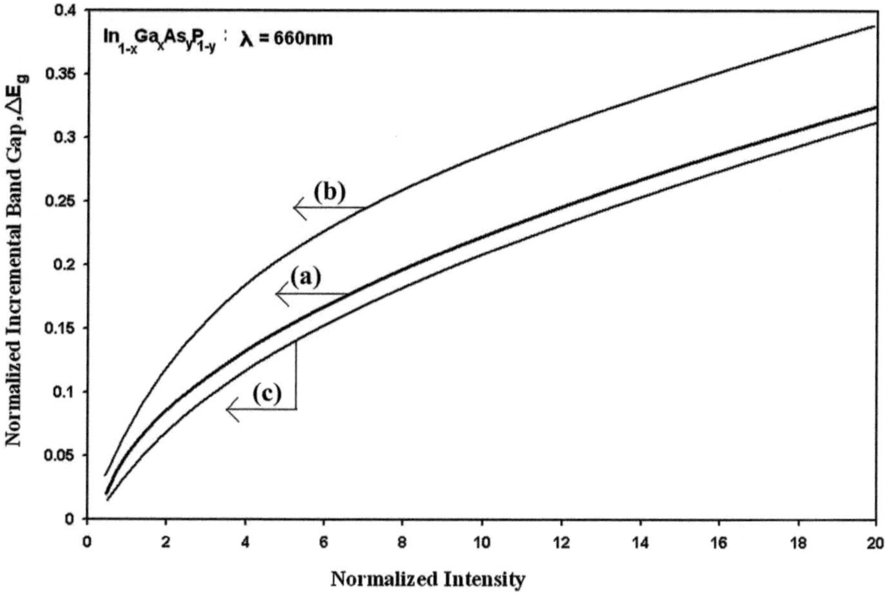

Fig. 20.2 Plots of the normalized incremental band gap (ΔE_g) for $In_{1-x}Ga_xAs_yP_{1-y}$ lattice matched to InP as a function of normalized light intensity for all cases of Fig. 20.1

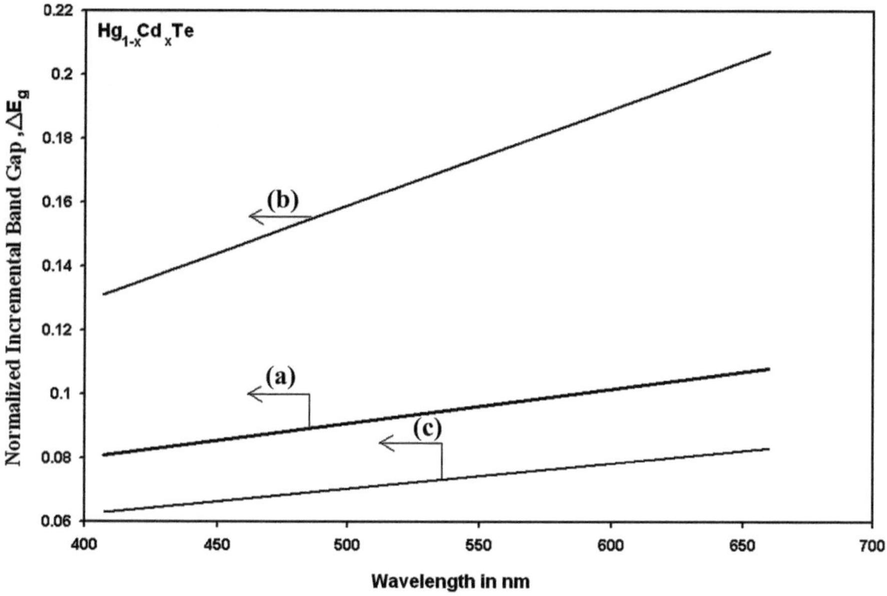

Fig. 20.3 Plots of the normalized incremental band gap (ΔE_g) for $Hg_{1-x}Cd_xTe$ as a function of wavelength for all cases of Fig. 20.2

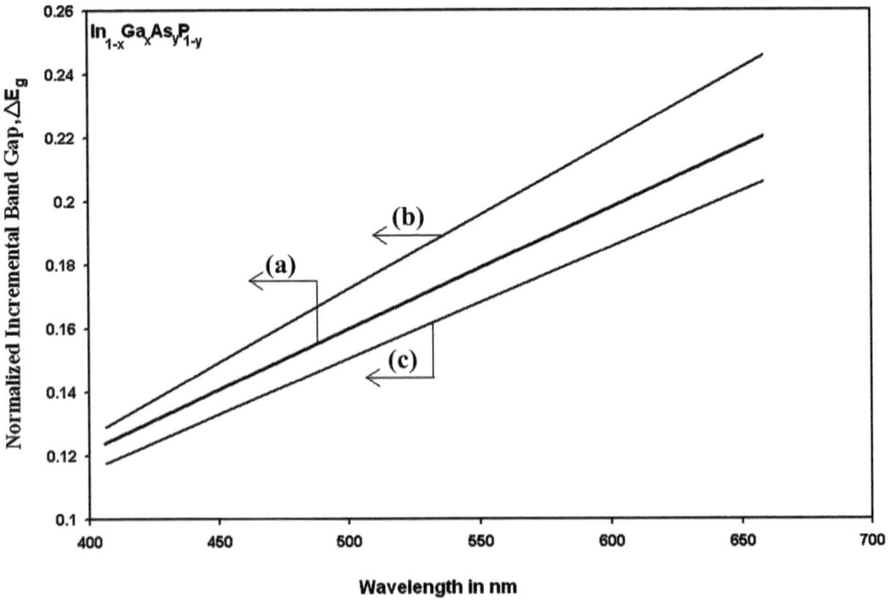

Fig. 20.4 Plots of the normalized incremental band gap (ΔE_g) for $In_{1-x}Gs_xAs_yP_{1-y}$ lattice matched to InP as a function of wavelength for all cases of Fig. 20.1

change, the photon field causes to increase the band gap of semiconductors. We propose the following two experiments for the measurement of band gap of semiconductors under photo-excitation.

(A) A white light with colour filter is allowed to fall on a semiconductor and the optical absorption coefficient $(\bar{\alpha}_0)$ is being measured experimentally. For different colours of light, $\bar{\alpha}_0$ is measured and $\bar{\alpha}_0$ versus $\hbar\omega$ (the incident photon energy) is plotted and we extrapolate the curve such that $\bar{\alpha}_0 \to 0$ at a particular value $\hbar\omega_1$. This $\hbar\omega_1$ is the unperturbed band gap of the semiconductor. During this process, we vary the wavelength with fixed I_0. From our present study, we have observed that the band gap of the semiconductor increases for various values of λ when I_0 is fixed (from Figs. 20.3 and 20.4). This implies that the band gap of the semiconductor measured (i.e. $\hbar\omega_1 = E_g$) is not the unperturbed band gap E_{g_0} but the perturbed band gap E_g; where $E_g = E_{g_0} + \Delta E_g$, ΔE_g is the increased band gap at $\hbar\omega_1$. Conventionally, we consider this E_g as the unperturbed band gap of the semiconductor and this particular concept needs modification. Furthermore, if we vary I_0 for a monochromatic light (when λ is fixed) the band gap of the semiconductor will also change consequently (Figs. 20.1 and 20.2). Consequently, the absorption coefficient will change with the intensity of light [105]. For the overall understanding, the detailed theoretical and experimental investigations are needed in this context for various materials having different band structures.

(B) The conventional idea for the measurement of the band gap of the semiconductors is the fact that the minimum photon energy $h\nu$ (ν is the frequency of the monochromatic light) should be equal to the band gap E_{g_0} (unperturbed) of the semiconductor, i.e.,

$$h\nu = E_{g_0} \tag{20.20}$$

In this case, λ is fixed for a given monochromatic light and the semiconductor is exposed to a light of wavelength λ. Also the intensity of the light is fixed. From Figs. 20.3 and 20.4, we observe that the band gap of the semiconductor is not E_{g_0} (for a minimum value of $h\nu$) but E_g, the perturbed band gap. Thus, we can rewrite the above equality as

$$h\nu = E_g \tag{20.21}$$

Furthermore, if we vary the intensity of light (Figs. 20.1 and 20.2) for the study of photoemission, the minimum photon energy should be

$$h\nu_1 = E_{g_1} \tag{20.22}$$

where E_{g_1} is the perturbed band gap of the semiconductor due to various intensity of light when ν and ν_1 are different.

Thus, we arrive at the following conclusions:

(a) Under different intensity of light, keeping λ fixed, the condition of band gap measurement is given by

$$h\nu_1 = E_{g_1} = E_{g_0} + \Delta E_{g_1} \tag{20.23}$$

(b) Under different colour of light, keeping the intensity fixed, the condition of band gap measurement assumes the form

$$h\nu = E_g = E_{g_0} + \Delta E_g \tag{20.24}$$

and not the conventional result as given by (20.20).

determined by the appropriate carrier statistics. Thus, our present study plays an important role in determining the diffusion coefficients of the minority carriers of quantum-confined lasers with materials having arbitrary band structures. Therefore in the investigation of the optical excitation of the optoelectronic materials which lead to the study of the ambipolar diffusion coefficients the present results contribute significantly.

20.2.7 Diffusion Coefficient of the Minority Carriers

This particular coefficient in quantum confined lasers can be expressed as

$$D_i/D_0 = dE_{Fi}/dE_F \tag{20.25}$$

where D_i and D_0 are the diffusion coefficients of the minority carriers both in the presence and absence of quantum confinements and E_{Fi} and E_F are the Fermi energies in the respective cases. It appears then that, the formulation of the above ratio requires a relation between E_{Fi} and E_F, which, in turn, can be formed through the respective carrier statistics.

20.2.8 Nonlinear Optical Response

The nonlinear response from the optical excitation of the free carriers is given by [106]

$$Z_0 = \frac{-e^2}{\omega^2\hbar^2}\int\limits_0^{\infty}\left(k_x\frac{\partial k_x}{\partial E}\right)^{-1} f_0 N(E)dE \tag{20.26}$$

where ω is the optical angular frequency, $N(E)$ is the density-of-states function. From the various E-k relations of different materials under different physical conditions, we can formulate the expression of $N(E)$ and from band structure we can derive the term $\left(k_x \frac{\partial k_x}{\partial E}\right)$ and thus by using the density-of-states function as formulated, we can study the Z_0 for all types of materials as considered in this monograph.

20.2.9 Third Order Nonlinear Optical Susceptibility

This particular susceptibility can be written as [107]

$$\chi_{NP}(\omega_1, \omega_2, \omega_3) = \frac{n_0 e^4\langle\varepsilon^4\rangle}{24\omega_1\omega_2\omega_3(\omega_1+\omega_2+\omega_3)\hbar^4} \tag{20.27}$$

where $n_0\langle\varepsilon^4\rangle = \int_0^{\infty} \frac{\partial^4 E}{\partial k_z^4} N(E) f_0\, dE$ and the other notations are defined in [107]. The term $\left(\frac{\partial^4 E}{\partial k_z^4}\right)$ can be formulated by using the dispersion relations of different materials as given in appropriate sections of this monograph. Thus one can investigate the $\chi_{NP}(\omega_1, \omega_2, \omega_3)$ for all materials as considered in this monograph.

20.2.10 Generalized Raman Gain

The generalized Raman gain in optoelectronic materials can be expressed as [108]

$$R_G = \bar{I}\left(\frac{16\pi^2 c^2}{\hbar\omega\rho g\omega_s^2 n_s n_p}\right)\left(\frac{\Gamma_\rho}{\Gamma}\right)\left(\left(\frac{e^2}{mc^2}\right)m^2 R^2\right) \quad (20.28)$$

where, $\bar{I} = \sum_{n.t_z} [f_0(n, k_z \uparrow) - f_0(n, k_z \downarrow)]$, $f_0(n, k_z \uparrow)$ is the Fermi factor for spin-up Landau levels, $f_0(n, k_z \downarrow)$ is the Fermi factor for spin down Landau levels, n is the Landau quantum number and the other notations are defined in [108]. It appears then the formulation of R_G is determined by the appropriate derivation requires the magneto-dispersion relations. By using the different appropriate formulas as formulated in various chapters of this monograph R_G can, in general, be investigated.

20.2.11 The Plasma Frequency

The plasma frequency ω_p can be expressed as [109]

$$\omega_p^2 = \frac{e^2}{\varepsilon_{sc}} \int_{E'}^{\infty} N(E) v_x^2 \frac{\partial f_0(E)}{\partial E} dE \quad (20.29)$$

where E' can be obtained from $N(E') = 0$.

20.2.12 The Activity Coefficient

Dilute solution transport equations introduced by Shockley [110] have been very effective in describing the fluxes of electrons and holes in semiconductors. These expressions, however, are inaccurate at high carrier concentrations due to nonidealities associated with carrier degeneracy and bandgap narrowing. Within the framework of dilute solution theory, these nonidealities can be included by introduction of concentration-dependent activity coefficients. Expressions for activity coefficients should also be useful in the area of chemical kinetics where mass-action principles are applied. Reviews of the literature on activity coefficients are available. Of special interest is a paper by Hwan and Brews [110], where majority carrier activity coefficients were presented that included the effects of Fermi statistics and bandgap narrowing due to exchange interactions and impurity effects. Explicit expressions and plots of electron activity coefficients as a function of concentration for GaAs were included. Activity coefficients were derived under the assumption of complete ionization. No mention was made, however, of minority carrierscreening by majority carriers, which are shown here to have a strong influence upon the majority carrier activity coefficients.

The activity coefficient in semiconductors can, in general be expressed as [110]

$$f_e^* = \frac{k_B T}{n_0} \int_0^{\infty} \frac{N(E)}{(1 - f_0)^2} \left[-\frac{\partial f_0}{\partial E} \right] dE \tag{20.30}$$

20.2.13 *Magneto-Thermal Effect in Quantized Structures*

The magneto-thermal effect can be expressed by the dimensionless quantity as [111]

$$\phi = \frac{B}{T}\frac{\partial T}{\partial B} = -\frac{B}{T}\left[\left(\frac{\partial S}{\partial \beta}\right)_{(T,n)}\right]\left[\left(\frac{\partial S}{\partial T}\right)_{(\beta,n)}\right]^{-1} \tag{20.31}$$

where S is the entropy which can be written as [112]

$$S = k_B \int_0^{\infty} -\left(\frac{\partial f_0}{\partial E}\right)\left(\frac{E - E_F}{k_B T}\right)\left[\int_0^{E} N(E) dE\right] dE \tag{20.32}$$

Under magnetic quantization we know, the DOS function assumes the form

$$N(E) = \left(\frac{eBg_v}{2\pi^2 \hbar}\right) \sum_{n=0}^{n_{\max}} \frac{\partial k_z'(E)}{\partial E} H(E - E') \tag{20.33}$$

where E' is the Landau sub-band energy or Landau level.

Therefore,

$$S = k_B \sum_{n=0}^{n_{\max}} \int_{E'}^{\alpha} \left(\frac{-\partial f_0}{\partial E}\right)\left(\frac{E - E_F}{k_B T}\right) \frac{eBg_v}{2\pi^2 \hbar} k_z dE \tag{20.34}$$

Since

$$\int N(E) dE = \frac{eBg_v}{2\pi^2 \hbar} \sum_{n=0}^{n_{\max}} k_z(E) \tag{20.35}$$

$$S = c_0 \sum_{n=0}^{n_{\max}} \left[\int_{E'}^{\alpha} \left\{ k_z E \left(\frac{-\partial f_0}{\partial E} \right) dE \right\} - E_f \int_{E'}^{\alpha} \left\{ k_z E \left(\frac{-\partial f_0}{\partial E} \right) dE \right\} \right] \tag{20.36}$$

where $c_0 = (eBg_v/2\pi^2\hbar T)$.

Using the concept of 3D Quantization (magneto-size quantization, Magneto NIPI, Magneto Inversion layers, Magneto accumulation Layers, magneto quantum dot, Quantum dot superlattices, magneto-size quantized superlattices and the respective heavily doped cases together with influence of light waves and electric fields on the 3D quantized structures), the DOS functions can respectively be written as

$$N(E) = \frac{g_v eB}{2\pi\hbar} \sum_{n_z=1}^{n_{z\max}} \sum_{n=0}^{n_{\max}} \delta'(E - E'') \tag{20.37}$$

$$N(E) = \frac{g_v eB}{2\pi\hbar} \sum_{i=0}^{i_{\max}} \sum_{n=0}^{n_{\max}} \delta'(E - E'') \tag{20.38}$$

and

$$N(E) = \frac{2g_v}{(d_x d_y d_z)} \sum_{n_x=1}^{n_{x\max}} \sum_{n_y=1}^{n_{y\max}} \sum_{n_z=1}^{n_{z\max}} \delta'(E - E''') \tag{20.39}$$

where E'' is the totally quantized energy in 2D systems in the presence of applied quantizing magnetic field along z-direction and E''' is the totally quantized energy in 0D systems in the absence of quantizing magnetic field along z-direction, (20.37) represents all the cases corresponding to E'', (20.38) represents all the cases corresponding to Magneto Inversion layers and Magneto accumulation Layers and (20.39) represents the rest cases.

In the three cases we can show

$$S = c_0'' \sum \sum E''[F_0(\eta'') - (\eta'')F_{-1}(\eta'')] \tag{20.40}$$

$$S = c_0''' \sum \sum E'''[F_0(\eta''') - (\eta''')F_{-1}(\eta''')] \tag{20.41}$$

where $c_0'' = \frac{g_v eBk_B}{2\pi\hbar}$, $c_0''' = \frac{2g_v}{d_x d_y d_z}$, $\eta'' = \frac{E_F - E''}{k_B T}$, $\eta''' = \frac{E_F - E'''}{k_B T}$ and E_F in this case is the Fermi energy for all the previous respective cases of 3D quantization of wave vector space.

20.2.14 Normalized Hall Coefficient

The normalized Hall coefficient in a semiconductor can be expressed as [113]

$$\frac{R(T)}{R(0)} = 1 + \frac{(\pi k_B T)^2}{3}\left[\frac{1}{N(E)}\frac{\partial N(E)}{\partial E}\right]^2\Bigg|_{E=E_{FT}} \tag{20.42}$$

where $R(T)$ is the Hall coefficient at T, $R(0)$ is the same at $T \to 0$ and E_{FT} is the Fermi energy at T.

20.2.15 Reflection Coefficient

In the presence of light waves, the reflection coefficient (R) is given by [114]

$$R = \frac{(x-1)^2}{(x+1)^2} \tag{20.43}$$

where $x = \sqrt{\varepsilon_{sc} - \frac{n_0 e^2 \lambda^2}{m^* \pi c^2}}$, m^* is the effective electron mass and c is the velocity of light.

20.2.16 Heat Capacity

The heat capacity of the carriers in semiconductors is a very important quantity.

The total heat capacity c_t is given by

$$c_t = c_e + c_l \tag{20.44}$$

where c_e is the electronic heat capacity and is given by [115]

$$c_e = \int_0^\infty -\frac{\partial f_0}{\partial E}\left[\frac{E-E_F}{T} + \frac{\partial E_F}{\partial T}\right](E-E_F)N(E)dE \tag{20.45}$$

Thus (20.44) can be written as

$$c_t = \frac{\pi^2 k_B^2 T N(E_F)}{3}\left[1 + T\left(\frac{\partial E_F}{\partial T}\right)\left(\frac{N'(E_F)}{N(E_F)}\right)\right] + c_l \tag{20.46}$$

20.2.17 Magnetic Susceptibilities

The Landau dia-magnetic susceptibility (χ_d) can be written as

$$\chi_d = -\mu_0 \frac{\partial^2 F(B)}{\partial B^2} \tag{20.47}$$

where μ_0 is the free space permeability and $F(B)$ is the free energy which can be written as

$$F(B) = n_0 E_{FB} + \int_0^{\infty} D_2(E) \frac{\partial f_0}{\partial E} dE \tag{20.48}$$

where $D_2(E)$ is twice integration of the DOS function with respect to E.

The Pauli's paramagnetic susceptibility (χ_p) can be expressed as [116]

$$\chi_p = -2\mu_B^2 \int_{E_0}^{\infty} N(E) \frac{\partial f_0}{\partial E} dE \tag{20.49}$$

where μ_B is the Bohr magnetic field.

20.2.18 Faraday Rotation

The Faraday rotation of the electrons in semiconductors can in general, be expressed under the condition $\omega\tau \gg 1$, as [117]

$$\theta = AI \tag{20.50}$$

where ω is the angular frequency of the incident light, τ is the momentum relaxation time, $A \equiv \frac{e^3 l}{8\pi^3 \hbar^4 \omega^2}\left(\frac{1}{2n_r \varepsilon_0 c}\right)$, l is the thickness of the semiconductor specimen, n_r is the real part of the refractive index at that frequency and

$$I = \frac{(2\pi)^3}{2g_v} \int_0^{\infty} \frac{\partial E}{\partial k_x} \left[\frac{\partial E}{\partial k_y} \cdot \frac{\partial^2 E}{\partial k_x \partial k_y} - \frac{\partial E}{\partial k_x} \cdot \frac{\partial^2 E}{\partial k_y^2} \right] \frac{\partial f_0}{\partial E} N(E) dE \tag{20.51}$$

Thus from different DRs as formulated in this book, we can formulate the partial differentials and from the expressions of $N(E)$ as derived in different chapters we can study the Faraday rotations for various HD materials.

20.2.19 Fowler-Nordheim Fied Emission

The Fowler-Nordheim field emission (FNFE) is a well-known quantum-mechanical phenomenon that involves tunneling of electrons through a surface barrier due to the application of an intense external electric field. Normally, at field strengths of the order of 10^8 V/m (below the electrical breakdown), the potential barriers at the surfaces of metals and semiconductors usually become very thin and result in field emission of electrons due to the tunnel effect [118]. This has been well-investigated with reference to three dimensional electron gases in metals and semiconductors and the FNFE from quantum confined structures has also been studied in this context [119–125].

The current (I) due to Fowler-Nordheim {FN}field emission can be written as

$$I = \frac{1}{2}\sum_{n_x=1}^{n_{x\max}}\sum_{n_u=1}^{n_{y\max}}\int_{E11}^{\infty}[e \cdot vn_1t_{11}] \tag{20.52}$$

where e is the magnitude of the electron charge, v is the velocity of the electron and given by $v = \frac{1}{\hbar}\frac{\partial E}{\partial k_z}$ $n_1 = N_{1D}(E) \cdot f_0 dE$, $N_{1D}(E)$ is the 1D DOS function per sub-bands and can be written as $N_{1D}(E) = \frac{2g_v}{\pi}\frac{\partial k_z}{\partial E}$, t_{11} is the transmission coefficient which can be expressed as

$$t_{11} = \exp\left[-2\int_i |k_i| d_i\right]$$

in which $i = x, i, z$.

20.2.20 Optical Effective Mass

The investigation of the energy band structure of the noble metals is one of the standard problems in solid state physics. The topology and geometry of the Fermi surfaces of the noble metals are very well known in detail from the de Hass-van Alphen measurements of Shoenberg. Very accurate values for the cyclotron masses of the noble metals were experimentally obtained from de Hass-van Alphen investigations [126]. The optical effective mass being used to study the different optical properties of semiconductors can be expressed following Madelung [126] as

$$m_{opt} = 3\hbar^2 n_0\left[\int_0^{\infty} N(E)\frac{\partial E}{\partial k} f_0 dE\right]^{-1} \tag{20.53}$$

20.2.21 Einstein's Photoemission

Einstein's Photoemission (EP) is a physical phenomenon and occupies a singular position in the whole arena of materials science and related disciplines in general. The EPin recent years finds extensive applications in modern optoelectronics, characterization and investigation of condensed matter systems, photoemission spectroscopy and related aspects in connection with the investigations of the optical properties of nanostructures [127–131]. Interest in low dimensional silicon nanostructures also grew up and gained momentum, after the discovery of room temperature photoluminescence and electroluminescence of silicon nano-wires in porous silicon [127]. Work on ultrathin layers of $SiSiO_2$ superlattices resulting into visible light emission at room temperature clearly exhibited low dimensional quantum confinement effect [128] and one of the most popular technique for analyzing the low dimensional structures is to employ photoemission techniques. Recent observation of room temperature photoluminescence and electro luminescence in porous silicon has stimulated vigorous research activities in silicon nanostructures [129].

It is well-known that the classical equation of the photo-emitted current density is [133] $J = \left[4\pi\alpha_0 e m_c g_v (k_B T)^2 / h^3\right] \exp[(hv - \phi)/(k_B T)]$, where α_0, e, m^*, g_v, k_B, T, h, hv and ϕ are the probability of photoemission, electron charge, effective electron mass at the edge of the conduction band, valley degeneracy, the Boltzmann constant, temperature, the Planck constant, incident photon energy along z-axis and work function respectively. The afore-mentioned equation is valid for both the charge carriers and in this conventional form it appears that, the photoemission changes with the effective mass, temperature, work function and the incident photon energy respectively. This relation holds only under the condition of carrier non-degeneracy. The EP has different values for different materials and varies with doping and with external fields which creates quantization of the wave-vector space of the carriers leading to various types of quantized structures. The nature of these variations has been studied in [127–158].

20.2.22 Righi-Leduc Coefficient

It is well known that some transport effects under strong magnetic field are independent of relaxation mechanism and are determined by the DRs only. Among them one is thermoelectric power and the other one is Righi-Leduc coefficient, which can in turn, be written as [159]

$$R = \frac{cn_0}{eB^2 \chi_{ph}} \left[\left\langle (E - E_F)^2 \right\rangle - \langle E - E_F \rangle^2 \right] \tag{20.54}$$

where, χ_{ph} is the phonon thermal conductivity, $\left\langle (E - E_F)^2 \right\rangle = \frac{1}{n}\int (E - E_F)^2 G(E)$ $\left[-\frac{\partial f_0}{\partial E}\right] dE$, $G(E)$ is the total number of states and is given by $G(E) = \int_{E_{30}}^{E} \mathrm{N}(\zeta)\mathrm{d}\zeta$ in which ζ is the variable of integration, E_{30} is the lower limit of integration and $\langle E - E_F \rangle = \frac{1}{n}\int (E - E_F) G(E) \left[-\frac{\partial f_0}{\partial E}\right] dE$.

20.2.23 Electric Susceptibility

In semiconductors the effect of free carriers on the optical properties becomes important at wavelengths longer than the intrinsic absorption edge. Free carriers produce absorption and affect also the dispersion at sufficiently long wavelengths. Electric susceptibility can be written as [160]

$$\chi_c = \frac{-e^2}{3\omega^2 \varepsilon_0 \hbar^2} \int_0^{\infty} N(E) \frac{\partial E}{\partial k} f_0 dE \tag{20.55}$$

20.2.24 Electric Susceptibility Mass

In recent years there has been considerable interest in studying the electric susceptibility mass of the carriers in narrow gap materials because of its direct influence in the study of carrier scattering mechanisms in semiconductor. Electric susceptibility mass can be determined from measurements in the infrared region of the frequency dependence of the spectral reflectivity at normal incidence and provides useful information regarding band structures. Electric Susceptibility Mass can be written as [161]

$$m_s^* = 3A_{20} \left[\int_{E_{20}}^{\infty} N(E) \frac{\partial E}{\partial k} f_0 dE \right]^{-1} \tag{20.56}$$

where $A_{20} = n_0 \varepsilon_0 \hbar^2$ and E_{20} is the lower limit of integration.

20.2.25 Electron Diffusion Thermo-Power

The electron diffusion thermopower (S_e) is an important property of AlGaN/GaN QLD structures since it yields useful information about the carrier transport mechanism. (S_e) can be written as [162]

$$S_e = -\frac{1}{n_0\, e}\frac{df}{dT} \tag{20.57}$$

where $f = n_0 E_F - \sum \int_{E_{21}}^{\infty} G(E) f_0 dE$ in which $G(E) = \int_{E_{30}}^{E} \mathrm{N}(\zeta)\mathrm{d}\zeta$, E_{21} and E_{30} are the lower limits of integrations of f and $G(E)$ respectively.

20.2.26 Hydrostatic Piezo-resistance Coefficient

The importance of Hydrostatic Piezo-resistance Coefficient (H_c) is well known in the sensor technology and H_c can be expressed as [163]

$$H_c = \frac{-\partial n_0}{n_0 \partial P} \tag{20.58}$$

where P is the pressure in kilobar.

20.2.27 Relaxation Time for Acoustic Mode Scattering

The importance of momentum relaxation time for acoustic mode scattering in semiconductors is well known for the evaluation of the electron mobility and in the simplest scattering theory, the relaxation time for acoustic mode scattering $(\tau_{am}(E))$ [164] is directly proportional to the inverse of $N(E)$ and can, in general be expressed as

$$\tau_{am}(E) = \frac{\tau_0}{N(E)} \tag{20.59}$$

where τ_0 is a constant. Thus using various expressions of $N(E)$ in different cases we can study not only $\tau_{am}(E)$ but also acoustic mode limited mobility in HD semiconductors and their nanostructures.

20.2.28 Gate Capacitance

In recent years there has been considerable interest in studying the surface capacitance of inversion layers in Si MOS structures. The dependence of the MOS capacitance on gate voltage and on a quantizing magnetic field has been studied both theoretically and experimentally. The fact that the MOS capacitance can be very easily controlled by varying the gate voltage is of much importance from the point of view of technical applications. It may also be stated that, as far as the determination of the effective mass under the degenerate electron distribution at the surface is concerned, measurement of magneto-capacitance as compared to that of conductivity or cyclotron resonance would not be more advantageous regarding the experimental facilities required or accuracies achieved. Nevertheless, it is felt that the theoretical investigation presented here would be of much significance as the interest on gate capacitance has been growing very much in recent years from the point of view of exploration of other fundamental aspects of semiconductor surfaces of MOS structures [165]. The gate capacitance (C_g) can in general be expressed as

$$C_g = \frac{e\partial n_s}{\partial V_g} \tag{20.60}$$

where n_s is the surface electron concentration per unit area and V_g is the gate voltage. It may be noted that although C_g for inversion layers have been studied in literature but investigations of C_g in accumulation layers are rather small. Thus by using the expressions of the DOS functions of inversion and accumulation layers as given in this monograph we can study the gate capacitance for different important cases.

20.3 Open Research Problems

(R.20.1) Investigate Carrier Statistics, Thermoelectric Power, Debye Screening Length, Carrier contribution to the elastic constants, Diffusivity-mobility ratio, Measurement of Band-gap in the presence of Light Waves, Diffusion coefficient of the minority carriers, Nonlinear optical response, Third order nonlinear optical susceptibility, Generalized Raman gain, The plasma frequency, The activity coefficient, Magneto-Thermal effect in Quantized Structures, Normalized Hall coefficient, Reflection coefficient, Heat Capacity, Magnetic Susceptibilities, Faraday rotation, Fowler-NordheimFied Emission, Optical Effective Mass, Einstein's Photoemission, Righi-Leduc coefficient, Electric Susceptibility, Electric Susceptibility Mass, Electron Diffusion Thermo-power, Hydrostatic Piezo-resistance Coefficient, Relaxation time for Acoustic Mode Scattering and Gate Capacitance for the appropriate problems of this monograph and perform related experiments for experimental investigations.

References

1. J.S. Blakemore, *Semiconductor Statistics* (Pergamon Press, London, 1962)
2. A.N. Chakravarti, K.P. Ghatak, G.B. Rao, K.K. Ghosh, Phys. Stat. Sol(b) **112**, 75 (1982)
3. K.P. Ghatak, M. Mondal, Zeitschriftfür Naturforschung. Teil A: Physik, physikalische Chemie, Kosmophysik **41,** 881 (1986)
4. M. Mondal, N. Chattopadhyay, K.P. Ghatak, Czechoslovak J. Phys. B **38**, 885 (1988); S. Pahari, S. Bhattacharya, K.P. Ghatak, J. Comput. Theor. Nanosci. **6**, 2088 (2009); K. P. Ghatak, S. Bhattacharya, K.M. Singh, S. Choudhury, S. Pahari, Physica B: Condensed Matter **403**, 2116 (2008)
5. L.J. Singh, S. Choudhary, A. Mallik, K.P. Ghatak, J. Comput. Theor. Nanosci. **2**, 287 (2005)
6. B. Nag, K.P. Ghatak, Physica Scripta **54**, 657 (1996)
7. B. Mitra, K.P. Ghatak, Solid-State Electronics **32**, 515 (1989); K.P. Ghatak, M. Mondal, Physica Status Solidi (b) **148**(2), 645-652 (1988); M. Mondal, K.P. Ghatak, Physica Status Solidi (b) **147**(2), K179–K183 (1988); M. Mondal, K.P. Ghatak, Physica Scripta **31**(6), 613 (1985)
8. M. Mondal, K.P. Ghatak, Physica Status Solidi (b) **126**, K41 (1984); A.N. Chakravarti, K. P. Ghatak, A. Dhar, K.K. Ghosh, Czechoslovak J. Phys. **33**(1), 65–69 (1983); K.P. Ghatak, M. Mondal, Physica Status Solidi (b) **160**(2), 673–681 (1990)
9. B. Mitra, K.P. Ghatak, Physica Scripta, **42**, (103) (1990); K.P. Ghatak, M. Mondal, Zeitschriftfür Physik B Condensed Matter **64**, 223 (1986); A.N. Chakravarti, K.P. Ghatak, K. K. Ghosh, H.M. Mukherjee. Physica Status Solidi (b) **116**, 17 (1983)
10. K.P. Ghatak, A. Ghoshal, B. Mitra, Il NuovoCimento D **14**, 903 (1992)
11. K.v. Klitzing, G. Dorda, M. Pepper, Phys. Rev. Lett. **45**, 494 (1980); K.v. Klitzing, Rev. Mod. Phys. **58**, 519 (1986)
12. J. Hajdu, G. Landwehr, in *Strong and Ultrastrong Magnetic Fields and Their Applications*, ed. by F. Herlach (Springer, 1985), p. 17
13. I.M. Tsidilkovskii, *Band Structure of Semiconductors* (PergamonPress, Oxford, 1982)
14. E.A. Arushanov, A.F. Knyazev, A.N. Natepov, S.I. Radautsan, Sov. Phys. Semi. **15**, 828 (1981)
15. S.P. Zelenim, A.S. Kondratev, A.E. Kuchma, Sov. Phys. Semi. **16**, 355 (1982)
16. F.M. Peeters, P. Vasilopoulos, Phys. Rev. B **46**, 4667 (1992)
17. W. Zawadzki, in *Two-Dimensional Systems, Heterostructures and Superlattice*. Springer Series in Solid State Sciences, Vol. 53, ed. by G. Bauer, F. Kuchar, H. Heinrich (Springer, Berlin, Heidelberg, 1984)
18. B.M. Askerov, N.F. Gashimzcede, M.M. Panakhov, Sov. Phys. Sol. State **29**, 465 (1987)
19. G.P. Chuiko, Sov. Phys. Semi. **19**, 1279 (1985)
20. S. Pahari, S. Bhattacharya, K.P. Ghatak, J. Comput. Theor. Nanosci. Invited review article, 2009 [In the press]
21. K.P. Ghatak, S. Bhattacharya, S. Bhowmik, R. Benedictus, S. Choudhury, J. Appl. Phys. **103**, 034303 (2008)
22. K.P. Ghatak, S. Bhattacharya, S. Pahari, D. De, S. Ghosh, M. Mitra, Annalen der Physik **17**, 195 (2008)
23. K.P. Ghatak, S.N. Biswas, J. Appl. Phys. **70**, 299 (1991)
24. K.P. Ghatak, S.N. Biswas, J. Low Temp. Phys. **78**, 219 (1990)
25. K.P. Ghatak, M. Mondal, J. Appl. Phys. **65**, 3480 (1989)
26. K.P. Ghatak, B. Nag, Nanostructured Materials **5**, 769 (1995)
27. K.P. Ghatak, M. Mondal, Phys. Stat. Sol. (b) **185**, K5 (1994)
28. K.P. Ghatak, B. Mitra, IlNuovoCimento **D15**, 97 (1993)
29. K.P. Ghatak, S.N. Biswas, Phys. Stat. Solidi (b) **140**, K107 (1987)
30. K.P. Ghatak, Il Nuovo Cimento D **13**, 1321 (1991)
31. K.P. Ghatak, A. Ghoshal, Phys. Stat. Sol. (b) **170**, K27 (1992)

32. K.P. Ghatak, B. De, B. Nag, P.K. Chakraborty, Molecular Crystals and Liquid Crystals Science and Technology Section B: Nonlinear Optics **16**, 221 (1996)
33. K.P. Ghatak, M. Mitra, B. Goswami, B. Nag, Mol. Cryst. Liquid Cryst. Sci. Technol. Section B: Nonlinear Optics **16**, 167 (1996)
34. K.P. Ghatak, D.K. Basu, D. Basu, B. Nag, Nuovo Cimentodella SocietaItaliana di Fisica D—Condensed Matter, Atomic, Molecular and Chemical Physics, Biophysics **18**, 947 (1996)
35. B. Mitra, K.P. Ghatak, Phys. Letts. A **141**, 81 (1989); S.K. Biswas, A. R. Ghatak, A. Neogi, A. Sharma, S. Bhattacharya, K.P. Ghatak, Physica E:Low-Dimensional Systems and Nanostructures **36**, 163 (2007)
36. M. Mondal, A. Ghoshal, K.P. Ghatak, Il NuovoCimento D **14**, 63 (1992); K.P. Ghatak, M. Mondal Phys. Stat. Sol. (b), **135**, 819 (1986); L.J. Singh, S. Choudhury, D. Baruah, S.K. Biswas, S. Pahari, K.P. Ghatak, Phys. B **368**, 188 (2005)
37. K.P. Ghatak, B. De, M. Mondal, S.N. Biswas, *Materials Research Society Symposium—Proceedings* **198**, 327 (1990); K.P. Ghatak, *Proceedings of SPIE —The International Society for Optical Engineering* **1584**, 435 (1992)
38. K.P. Ghatak, B. De, *Materials Research Society Symposium—Proceedings* **234**, 55 and 59 (1991)
39. K.P. Ghatak, B. De, M. Mondal, S.N. Biswas, *Materials Research Society Symposium—Proceedings* **184**, 261 (1990)
40. M. Mondal, K.P. Ghatak, PhysicaScripta **31**, 613 (1985)
41. K.P. Ghatak, S.N. Biswas, *Materials Research Society Symposium—Proceedings* **216**, 465 (1990)
42. P.T. Landsberg, Eur. J. Phys. **2**, 213 (1981)
43. R.B. Dingle, Philos. Mag. **46**, 813 (1955); D. Redfield, M. A. Afromowitz, ibid. **19**, 831 (1969); H.C. Casey, F. Stern, J. Appl. Phys. **47**, 631 (1976)
44. S.N. Mohammad, J. Phys. C **13**, 2685 (1980)
45. K.P. Ghatak, S. Bhattacharya, S. Pahari, S.N. Mitra, P.K. Bose, D. De, J. Phys. Chem. Sol. **70**, 122 (2009); S. Chowdhary, L.J. Singh, K.P. Ghatak, Physica B: Condensed Matter **365**, 5 (2005); B. Nag, K.P. Ghatak, Molecular Crystals and Liquid Crystals Science and Technology Section B: Nonlinear Optics **19**, 1 (1998); K.P. Ghatak, B. Mitra, Phys. Scr. **46**, 182 (1992)
46. K.P. Ghatak, S. Bhattacharya, J. Appl. Phys. **102,** 073704 (2007); M. Mondal, K.P. Ghatak, Phys. Lett. **102A**, 54 (1984); S. Bhattacharya, N.C. Paul, K.P. Ghatak, Physica B, **403,** 4139 (2008); K.P. Ghatak, S. Bhattacharya, H. Saikia, D. Baruah, A. Saikia, K.M. Singh, A. Ali, S.N. Mitra, P.K. Bose, A. Sinha, J. Comp. Theor. Nanosci **3**, 727 (2006); P.K. Chakraborty, G.C. Datta, K.P. Ghatak, Phys. Scr. **68**, 368 (2003); E.O. Kane, Solid-State Electron. **28**, 3 (1985); W. Zawadzki, in *Handbook on Semiconductors*, ed. by W. Paul, (North Holland, New York, 1982), **1**, p. 715
47. A.N. Chakravarti, D. Mukherjee, Phys. Lett. **53A**, 403 (1975)
48. B. Mitra, D.K. Basu, B. Nag, K.P. Ghatak, Nonlinear Opt. **17**, 171 (1997)
49. M. Mondal, K.P. Ghatak, Phys. Status Solidi B **135**, 239 (1986)
50. A.N. Chakravarti, S. Swaminathan, Phys. Status Solidi A **23**, K191 (1974); A.N. Chakravarti, ibid. **25**, K105 (1974)
51. A.N. Chakravarti, K.P. Ghatak, K.K. Ghosh, A. Dhar, Phys. Status Solidi B **103**, K55 (1981)
52. T. Ando, A.H. Fowler, F. Stern, Rev. Mod. Phys. **54**, 437 (1982); P.K. Basu, *Optical Processes in Semiconductors* (Oxford University Press, New York, 2001)
53. N. Kampf, D. Ben-Yaakov, D. Andelman, S.A. Safran, J. Klein, Phys. Rev. Lett. **103**, 118304 (2009); P. Arnold, L.G. Yaf Fe, Phys. Rev. D **52**, 7208 (1995); G.S. Kulkarni, Z. Zhong, Nano Lett. **12**, 719 (2012); E. Stern, R. Wagner, F.J. Sigworth, R. Breaker, T.M. Fahmy, M.A. Reed, Nano Lett. **7**, 3405(2007)
54. A.N. Chakravarti, D. Mukherji, Phys. Lett A, **53**, 57 (1975); W. Zawadzki, Adv. Phys. **23**, 435 (1974)
55. A.K. Sreedhar, S.C. Gupta, Phys. Rev. B **5**, 3160 (1972); R.W. Keyes, IBM. J. Res. Devlopm. **5**, 266 (1961); Solid State Phys. **20**, 37 (1967)

56. S. Bhattacharya, S. Chowdhury, K.P. Ghatak, J. Comput. Theor. Nanosci. **3**, 423 (2006); K. P. Ghatak, J.Y. Siddiqui, B. Nag, Phys. Letts. A, **282**, 428 (2001); K.P. Ghatak, J. P. Banerjee, B. Nag, J. Appl. Phys. **83**, 1420 (1998); K.P. Ghatakand, B. Nag, Nanostruct. Mater. **10**, 923 (1998); B. Nag, K.P. Ghatak, J. Phys. Chem. Sol. **58**, 427 (1997)
57. K.P. Ghatak, J.P. Banerjee, B. Goswami, B. Nag, Nonlin. Opt. Quant. Opt. **16**, 241 (1996); K.P. Ghatak, J.P. Banerjee, D. Bhattacharyya, B. Nag, Nanotechnology, **7**, 110 (1996); K. P. Ghatak, J.P. Banerjee, M. Mitra, B. Nag, Nonlin. Opt. **17**, 193 (1996); K.P. Ghatak, Int. J. Electron. **71**, 239 (1991)
58. K.P. Ghatak, B. De, S.N. Biswas, M. Mondal, Mechanical behavior of materials and structures in microelectronics. MRS Symposium Proceedings, Spring Meeting, **2216**,191, (1991); K.P. Ghatak, B. De, MRS Symposium Proceedings, **226**, 191 (1991); K.P. Ghatak, B. Nag, G. Majumdar, MRS, **379**, 109, (1995); D. Baruah, S. Choudhury, K.M. Singh, K. P. Ghatak, J. Phys., Conf. Series, **61**, 80 (2007)
59. K.P. Ghatak, S. Bhattacharya, D. De, Einstein Relation in Compound Semiconductors and Their Nanostructures. Springer Series in Materials Science (Springer, Heidelberg, 2009), p. 116
60. A. Einstein, Ann. der Physik, **17**, 549 (1905); H. Kroemer, IEEE Trans. Electron Devices **25**, 850 (1978); W. Nernst, Z. Phys. Chem. **2**, 613 (1888); J.S. Townsend, Trans. Roy. Soc. **193A**, 129 (1900); C. Wagner, Z. Physik. Chem. **B21**, 24 (1933); C. Herring, M.H. Nichols, Rev. Mod. Phys. **21**, 185 (1949); P.T. Landsberg, *Thermodynamics and Statistical Mechanics*, (Oxford University Press, Oxford 1978); in *Recombination in Semiconductors* (Cambridge University Press, U.K. 1991)
61. R.W. Lade, Proc. IEEE **52**, (743) (1965)
62. P.T. Landsberg, Proc R. Soc. A **213**, (226) (1952); Proc. Phys. Soc. A **62**, (806) (1949)
63. C.H. Wang, A. Neugroschel, IEEE Electron. Dev. Lett. **ED-11**, (576) (1990)
64. I.Y. Leu, A. Neugroschel, IEEE Trans. Electron. Dev. **ED-40**, (1872) (1993)
65. F. Stengel, S.N. Mohammad, H. Morkoç, J. Appl. Phys. **80**, (3031) (1996)
66. H.J. Pan, W.C. Wang, K.B. Thai, C.C. Cheng, K.H. Yu, K.W. Lin, C.Z. Wu, W.C. Liu, Semicond. Sci. Technol. **15**, (1101) (2000)
67. S.N. Mohammad, J. Appl. Phys. **95**, (4856) (2004)
68. V.K. Arora, Appl. Phys. Lett. **80**, (3763) (2002)
69. S.N. Mohammad, J. Appl. Phys. **95**, (7940) (2004)
70. S.N. Mohammad, Philos. Mag. **84**, (2559) (2004)
71. S.N. Mohammad, J. Appl. Phys. **97**, (063703) (2005)
72. K. Suzue, S.N. Mohammad, Z.F. Fan, W. Kim, O. Aktas, A.E. Botchkarev, H. Morkoç, J. Appl. Phys. **80**, (4467) (1996)
73. S.N. Mohammad, Z.F. Fan, W. Kim, O. Aktas, A.E. Botchkarev, A. Salvador, H. Morkoç, Electron. Lett. **32**, (598) (1996)
74. Z. Fan, S.N. Mohammad,W. Kim, O. Aktas, A.E. Botchkarev, K. Suzue, H. Morkoç, J. Electron. Mater. **25**, (1703) (1996)
75. C. Lu, H. Chen, X. Lv, X. Xia, S. N. Mohammad, J. Appl. Phys. **91**, (9216) (2002)
76. S.G. Dmitriev, Yu V. Markin, Semiconductors **34**, (931) (2000)
77. M. Tao, D. Park, S.N. Mohammad, D. Li, A.E. Botchkerav, H. Morkoç, Philos. Mag. B **73**, (723) (1996)
78. D.G. Park, M. Tao, D. Li, A.E. Botchkarev, Z. Fan, S.N. Mohammad, H. Morkoç, J. Vac. Sci. Technol. B **14**, (2674) (1996)
79. Z. Chen, D.G. Park, S.N. Mohammad, H. Morkoç, Appl. Phys. Lett. **69**, (230) (1996)
80. A.N. Chakravarti, D.P. Parui, Phys. Letts **40A**, (113) (1972)
81. A.N. Chakravarti, D.P. Parui, Phys. Letts **43A** (60), (1973); B.R. Nag, A.N. Chakravarti, Solid State Electron. **18**, (109) (1975), Phys. Stat Sol. (a) **22**, (K153) (1974)
82. B.R. Nag, A.N. Chakravarti, P.K. Basu, Phys. Stat. Sol. (a) **68**, (K75) (1981)
83. B.R. Nag, A.N. Chakravarti, Phys. Stal Sol. (a) **67**, (K113) (1981)
84. A.N. Chakravarti, B.R. Nag, Phys. Stat. Sol. (a) **14**, (K55) (1972); Int. J. Elect. **37**, (281) (1974); Phys. Stat. Sol. (a) **14**, (K23) (1972); A.N. Chakravarti, D.P. Parui, Canad.

J. Phys. **51**, (451) (1973). D. Mukherjee, A.N. Chakravarti, B.R. Nag, Phys. Stat. Sol (a) **26**, (K27) (1974); S. Ghosh, A.N. Chakravarti, Phys. Stat. Sol. (b), **147**, (355) (1988); A.N. Chakravarti, K.K. Ghosh, K.P. Ghatak, H.M. Mukherjee, Phys. Stat. Sol (b) **118**, (843) (1983); A.N. Chakravarti, K.P. Ghatak, K.K. Ghosh, G.B. Rao, Phys. Stat. Sol (b) **111**, (K61) (1982); A.N. Chakravarti, K.P. Ghatak, K.K. Ghosh, H.M. Mukherjee, S. Ghosh, Phys. Stat. Sol (b) **108**, (609) (1981); A.N. Chakravarti, K.P. Ghatak, S. Ghosh, A. Dhar, Phys. Stat. Sol. (b) **105**, (K55) (1981); A.N. Chakravarti, A.K. Choudhury, K. P. Ghatak, D. Roy Choudhury, Phys. Stat. Sol (b) **59**, (K211) (1980); A.N. Chakravarti, A.K. Chowdhury, K.P. Ghatak, D.R. Choudhury, Acta Phys. Polon. A **58**, (251) (1980); K. P. Ghatak, A.K. Chowdhury, S. Ghosh, A.N. Chakravarti, Phys. Stat. Sol (b) **99**, (K55) (1980); A.N. Chakravarti, A.K. Chowdhury, K.P. Ghatak, D.R. Choudhury, Czech. J. Phys. B **30**, (1161) (1980); A.N. Chakravarti, K.P. Ghatak, A. Dhar, K.K. Ghosh, S. Ghosh, Acta Phys. Polon A **60**, (151) (1981); A.N. Chakravarti, A.K. Chowdhury, K.P. Ghatak, A. Dhar, D.R. Choudhury, Czech. J. Phys. B **31**, (905) (1981); A.N. Chakravarti, K.P. Ghatak, K.K. Ghosh, S. Ghosh, H.M. Mukherjee, Czech. J. Phys. B **31,** (1138) (1981)

85. P.T. Landsberg, J. Appl. Phys. **56**, (1696) (1984); P.T. Landsberg, Phys. Rev. B **33**, (8321) (1986); Y. Roichman, N. Tessler, Appl. Phys. Lett. **80**, (1948) (2002); J.M.H. Peters, Eur. J. Phys. **3**, (19) (1982); H. Van Cong, S. Brunet, S. Charar, Phys. Stat. Solidi B **109**, (K1) (1982); H. Van Cong, Phys. Stat. Solidi A **56**, (395) (1979); H. Van Cong, Solid State Electron. **24**, (495) (1981)
86. S.N. Mohammad, S.T.H. Abidi, J. Appl. Phys. **61**, (4909) (1987); Solid State Electron. **27**, (1153) (1985); J. Appl. Phys. **56**, (3341) (1984); M.A. Sobhan, S.N. Mohammad, J. Appl. Phys. **58**, (2634) (1985); S.N. Mohammad, A.V. Bemis, IEEE Trans. Electron. Dev. ED-**39**, (282) (1992); S.N. Mohammad, R.L. Carter, Philos. Mag. B **72**, (13) (1995); S.N. Mohammad, Solid State Electron. **46**, (203) (2002); S.N. Mohammad, J. Chen, J.I. Chyi, H. Morkoç, Appl. Phys. Lett. **56**, (937) (1990)
87. P.T. Landsberg, S.A. Hope Solid State Electronics **20**, (421) (1977); S.A. Hope, G. Feat, P.T. Landsberg, J. Phys. A. Math. Gen. **14**, (2377) (1981)
88. W. Elsäber, E.O. Göbel, Electron. Lett. **19**, (335) (1983); R. Hilfer, A. Blumen, Phys. Rev. A **37**, (578) (1988); T.G. Castner, Phys. Rev. B **55**, (4003) (1997); E. Barkai, V.N. Fleurov, Phys. Rev. E., **58**, (1296) (1998); T.H. Nguyen, S.K. O'Leary, Appl. Phys. Lett. **83**, (1998) (2003); T.H. Nguyen, S.K. O'Leary, J. Appl. Phys. **98**, (076102) (2005); C.G. Rodrigues, Á. R. Vasconcellos, R. Luzzi, J. Appl. Phys. **99**, (073701) (2006)
89. R.K. Jain, Phys. Stat. Sol. (a) **42**, (K221) (1977); B.A. Aronzon, E.Z. Meilikhov, Phys. Stat. Sol. (a) **19**, (313) (1973)
90. S. Choudhury, D. De, S. Mukherjee, A. Neogi, A. Sinha, M. Pal, S.K. Biswas, S. Pahari, S. Bhattacharya, K.P. Ghatak, J. Comp. Theor. Nanosci. **5**, (375) (2008); S.M. Adhikari, K. P. Ghatak, Quantum Matter **2**, (296) (2013); S. Pahari, S. Bhattacharya, D. De, S.M. Adhikari, A. Niyogi, A. Dey, N. Paitya, K.P. Ghatak, Physica B: Condensed Matter **405**, (4064) (2010); K.P. Ghatak, S. Bhattacharya, S. Pahari, D. De, R Benedictus, Superlatt. Microst. **46**, (387) (2009)
91. J.P. Bouchaud, A. Georges, Phys. Rep. **195**, (127) (1996); V. Blickle, T. Speck, C. Lutz, U. Seifert, C. Bechinger, Phys. Rev. Lett. **98**, (210601) (2007); Y. Kang, E. Jean, C.M. Fortmonn, Appl. Phys. Lett., **88**, (112110) (2006); F. Neumann, Y.A. Genenko, H.V. Seggern, J. Appl. Phys. **99**, (013704) (2006); J. van de Lagemaat, Phys. Rev. B **73**, (235319) (2005); Q. Gu, E.A. Schiff, S. Grneber, F. Wang, R. Schwarz, Phys. Rev. Lett. **76**, (3196) (1996); M.Y. Azbel, Phys. Rev. B., **46**, (15004) (1992)
92. A.H. Marshak, Solid State Electron. **30**, (1089) (1987); A.H. Marshak, C.M.V. Vliet, Proc. IEEE **72**, (148) (1984); C.M.V. Vliet, A. van der Zeil, Solid State Electron., **20**, (931) (1977)
93. A. Khan, A. Das, Appl. Phys. A, **89**, (695) (2007)
94. O. Madelung, Semiconductors: Data Handbook, 3rd edn (Springer, 2004); M. Krieehbaum, P. Kocevar, H. Pascher, G. Bauer, IEEE QE, **24** (1727) (1988)
95. K.P. Ghatak, S.N. Biswas, J. Appl. Phys. **70**, (4309) (1991); K.P. Ghatak, B. Mitra, M. Mondal Ann. der Physik. **48**, (283) (1991); B. Mitra, K.P. Ghatak, Phys. Letts. **135A**,

(397) (1989); K.P. Ghatak, N. Chattopadhyay, M. Mondal, J. Appl. Phys. **63**, (4536) (1988); J. Low Temp. Phys. **73**, (321) (1988)
96. K.P. Ghatak, D. Bhattacharyya, Phys. Letts. **A184**, (366) (1994)
97. K.P. Ghatak, D. Bhattacharyya, Physica Scripta **52**, (343) (1995); M. Mondal, K.P. Ghatak, J. Mag. Mag. Materials, **62**, (115) (1986)
98. K.P. Ghatak, B. Nag, D. Bhattacharyya, J. Low Temp. Phys. **14**, (1) (1995)
99. K.P. Ghatak, M. Mondal, Thin Solid Films **148**, (219) (1987)
100. K.P. Ghatak, S.N. Biswas, Nanostruct. Mater. **2**, (91) (1993)
101. K.P. Ghatak, Influence of Band Structure on Some Quantum Processes in Nonlinear optical Semiconductors. D. Eng. Thesis, (Jadavpur University, Kolkata, India, 1991)
102. K.P. Ghatak, N. Chattropadhyay, M. Mondal, Appl. Phys. A, **44**, (305) (1987)
103. S.N. Biswas, K.P. Ghatak, *Proceedings of the Society of Photo-optical and Instrumentation Engineers (SPIE), Quantum Well and Superlattice Physics,* **792**, (239) (1987); K.P. Ghatak, M. Mondal, S. Bhattacharyya, SPIE, **1284**, (113) (1990); K.P. Ghatak, S. Bhattacharyya, M. Mondal, SPIE, **1307**, (588) (1990); K.P. Ghatak, B. De, *Defect Engineering in Semiconductor Growth, Processing and Device technology Materials Research Society Proceedings*, (MRS) Spring meeting, **262**, (911) (1992); S. Bhattacharya, K.P. Ghatak, S.N. Biswas, *Optoelectronic materials, Devices, Packaging Interconnects*. SPIE, **836**, (72) (1988)
104. M. Mondal, K.P. Ghatak, J. Phys. C (Sol. State.), **20**, (1671) (1987); M. Mondal. S.N. Banik, K.P. Ghatak, Canad. J. Phys. **67**, (72) (1989); K.P. Ghatak, M. Mondal, J. Appl. Phys. **70**, (1277) (1992)
105. P.K. Chakraborty, S. Bhattacharya, K.P. Ghatak, J. Appl. Phys. **98**, 053517 (2005)
106. A.S. Filipchenko, I.G. Lang, D.N. Nasledov, S.T. Pavlov, L.N. Radaikine, Phys. Stat. Sol. (b) **66**, 417 (1974)
107. M. Wegener, *Extreme Nonlinear Optics* (Springer-Verlag, Germany, 2005)
108. B.S. Wherreff, W. Wolland, C.R. Pidgeon, R.B. Dennis, S.D. Smith, in *Proceedings of the 12th International Conference of the Physics of the Semiconductors*, ed. by M.H. Pilkahn, R. G. Tenbner (Staffgard, 1978), p. 793
109. C.H. Peschke, Phys. Stat. Sol. (b) **191**, 325 (1995)
110. P.T. Landsberg, H.C. Cheng, Phys. Rev. **32B**, 8021 (1985); Semiconductors (D. van Nostrand Co., Inc., New York, 1950); H. Reiss, C.S. Fuller, F.J. Morin, Bell Syst. Tech. J., **35**, 535 (1956); P.T. Landsberg, A.G. Guy, Phys. Rev. B, **28**, 1187(1983); Handbook on Semiconductors, ed. by W. Paul. (North-Holland Publishing Co., 1982), pp. 359–449
111. A. Isihara, Y. Shiwa, J. Phys. C **18**, 4703 (1985)
112. V.G.I. Ravish, B.A. Efimova, I.A. Smirnov Semiconducting Lead Chalocogenides. (Plenum Press, New York, 1970), p163
113. F.J. Blatt, J. Phys. Chem. Solids **17**, 177(1961)
114. D.K. Roy, Quantum Mechanical Tunnelling and its applications. (World Scientific Publication Co. Ptc. Ltd, 1986), p. 309
115. W. Zawadzki, Physica **127**, 388 (1980)
116. E.V. Rozenfeld, YuP Irkhin, P.G. Guletskii, Sov. Phys. Solid State **29**, 1134 (1987)
117. H. Piller, Semiconductors and Semimetals, ed. by R.K. Willardson, A.C. Beer, vol. 8, p. 124 (1972)
118. R.H. Fowler, L. Nordheim, Proc. Royal Society of London. Series-A, **119**, 173 (1928); A. Van Der Ziel, Solid State Physical Electronics (Prentice-Hall, Inc., Englewood Cliffs U.S. A.1957), p. 176
119. B. Mitra, K.P. Ghatak, Phys. Lett. A, **357**,146 (1990); K.P. Ghatak, M. Mondal, J. Mag. Mag. Mat., **74**, 203(1988); K.P. Ghatak, B. Mitra, Phys. Lett. **156A**, 233, (1991); K. P. Ghatak, A. Ghosal, S.N. Biswas, M. Mondal, Proc SPIE, USA, **1308**, 356 (1990)
120. V.T. Binh, Ch. Adessi, Phys. Rev. Lett. **85**, 864 (2000); R.G. Forbes, Ultramicroscopy, **79**, 11 (1999); J.W. Gadzuk, E.W. Plummer, Rev. Mod. Phys, **45**, 487 (1973); J.M. Beebe, B. Kim, J.W. Gadzuk, C.D. Frisbie,J.G. Kushmerick, Phys. Rev. Lett, **97**, 026801 (1999); Y. Feng, J.P. Verboncoeur, Phys.Plasmas, **12**, 103301 (2005); W.S. Koh, L.K. Ang,

Nanotechnology, **19**, 235402 (2008); M. Razavy, *Quantum Theory of Tunneling* (World Scientific Publishing Co. Pte. Ltd, Singapore, 2003)

121. S.I. Baranchuk, N.V. Mileshkina, Soviet Phys. Solid State, **23**, 1715 (1981); P.G. Borzyak, A.A. Dadykin, Soviet Phys. Docklady **27**, 335 (1982); S. Bono, R.H. Good, Jr., Surface Sci. **134**, 272 (1983); S.M. Lyth, S.R.P. Silva, Appl. Phys. Letts. **90**, 173124 (2007); C. Xu, X. Sun, Int. J. Nanotech. **1**, 452 (2004);S.D. Liang, L. Chen, Phys. Rev. Letts. **101**, 027602 (2008)
122. E.C. Heeres, E.P.A.M. Bakkers, A.L. Roest, M. Kaiser, T. H. Oosterkamp, N. de Jonge, Nano Lett. **7**, 536 (2007); L. Dong, J. Jiao, D.W. Tuggle, J.M. Petty, S.A. Elliff, M. Coulter, Appl. Phys. Lett. **82**, 1096 (2003); S.Y. Li, P. Lin, C.Y. Lee, T.Y. Tseng, J. Appl. Phys. **95**, 3711 (2004); N.N. Kulkarni, J. Bae, C.K. Shih, S.K. Stanley, S.S. Coffee, J.G. Ekerdt, Appl. Phys. Lett. **87**, 213115 (2005)
123. K. Senthil, K. Yong, Mat. Chem. Phys. **112**, 88 (2008); R. Zhou, H.C. Chang, V. Protasenko, M. Kuno, A.K. Singh, D. Jena, H. Xing, J. Appl. Phys. **101**, 073704 (2007); K.S. Yeong, J.T. L. Thong, J. Appl. Phys. **100**, 114325 (2006); C.H. Oon, S.H. Khong, C.B. Boothroyd, J.T. L. Thong, J. Appl. Phys. **99**, 064309 (2006)
124. B.H. Kim, M.S. Kim, K.T. Park, J.K. Lee, D.H. Park, J. Joo, S.G. Yu, S.H. Lee, Appl. Phys. Lett. **83**, 539 (2003); Z.S. Wu, S.Z. Deng, N.S. Xu, J. Chen, J. Zhou, J. Chen, Appl. Phys. Lett. **80**, 3829 (2002); Y.W. Zhu, T. Yu, F.C. Cheong, X.J. Xu, C.T. Lim, V.B.C. Tan, J.T.L. Thong, C.H. Sow, Nanotechnology, **16,** 88 (2005); Y.W. Zhu, H.Z. Zhang, X.C. Sun, S.Q. Feng, J. Xu, Q. Zhao, B. Xiang, R.M. Wang, D.P. Yu, Appl. Phys. Lett. **83**, 144 (2003); S. Bhattacharjee, T. Chowdhury, Appl. Phys. Lett. **95**, 061501 (2009); S. Kher, A. Dixit, D.N. Rawat, M.S. Sodha, Appl. Phys. Lett. **96**, 044101 (2010)
125. I. Shigeo, W. Teruo, O. Kazuyoshi, T. Masateru, U. Satoshi, N. Norio, J. Vac. Sci. Tech. B **13**, 487 (2009); C.A. Spindt, I. Brodie, L. Humphrey, E.R. Westerberg, J. Appl. Phys. **47**, 5248 (2009); Q. Fu, A.V. Nurmikko, L.A. Kolodziejski, R.L. Gunshor, J.W. Wu, Appl. Phys. Lett. **51**, 578 (2009)
126. I. Mertig, E. Mrosan, V.N. Antonov, V.I. Antonov, P. Ziesche, Phys. Stat. Sol. (b), 135, K13 (1986); O. Madelung, *Physics of III-V Compounds.* (John Wiley and Sons, Inc. New York, 1966), p. 80
127. L.T. Canham, Appl. Phys. Letts. **57**, 1046 (1990)
128. Z.H. Lu, D.J. Lockwood, J.M. Baribeam, Nature **378**, 258 (1995)
129. A.G. Cullis, L.T. Canham, P.D.O. Calocott, J. Appl. Phys. **82**, 909 (1997)
130. M. Cardona, L. Ley, *Photoemission in Solids 1 and 2, Topics in Applied Physics*, vols. 26, 27, (Springer-Verlag, Germany, 1978); S. Hüfner, *Photoelectron Spectroscopy,* (Springer, Germany, 2003); S. Hüfner, (Ed.), *Very High Resolution Photoelectron Spectroscopy,* Lecture Notes in Physics, vol. 715 (Springer-Verlag, Germany, 2007); D.W. Lynch, C.G. Olson, *Photoemission Studies of High-Temperature Superconductors*, (Cambridge University Press, UK, 1999); *Photoemission and the Electronic Properties of Surfaces* ed. B. Feuerbacher, B. Fitton, R.F. Willis (Wiley, New York, 1978); W. Schattke, M.A.V. Hove, *Solid-State Photoemission and Related Methods: Theory and Experiment,* (Wiley, USA, 2003); V.V. Afanas'ev, *Internal Photoemission Spectroscopy: Principles and Applications* (Elsevier, North Holland, 2010); D.J. Lockwood, *Light Emission in Silicon in Silicon Based Materials and Devices*, ed. H.S. Nalwa (Academic Press, San Diego, USA, 2001), vol. 2
131. K.P. Ghatak, D. De, S. Bhattacharya, *Photoemission from optoelectronic materials and their nanostructures* (Springer, Springer Series in Nanostructure Science and Technology, 2009)
132. B. Mitra, K.P. Ghatak, Solid-State Electronics **32**, 810 (1989); S. Choudhury, L.J. Singh, K. P. Ghatak, Nanotechnology **15**, 180 (2004); K.P. Ghatak, A.K. Chowdhury, S. Ghosh, A.N. Chakravarti, Appl. Phys. **23**, 241 (1980); K.P. Ghatak, D.K. Basu, B. Nag, J. Phys. Chem. Solids **58**, 133 (1997); P.K. Chakraborty, G.C. Datta, K.P. Ghatak, Physica B: Condensed

Matter **339**, 198 (2003); K.P. Ghatak, A. Ghoshal, B. Mitra, Il NuovoCimento D **14**, 903 (1992); B. Mitra, A. Ghoshal, K.P. Ghatak, Phys. Stat. Sol. (b), **154**, K147 (1989); K. P. Ghatak, S. Bhattacharya, S. Bhowmik, R. Benedictus, S. Choudhury J. Appl. Phys. **103**, 094314 (2008); M. Mondal, S. Bhattacharya, K.P. Ghatak Appl. Phys. **A 42**, 331 (1987); A. N. Chakravarti, K.P. Ghatak, A. Dhar, K.K. Ghosh, S. Ghosh, Appl. Phys. A **26**, 165 (1981)

133. R.K. Pathria, *Statistical Mechanics*, 2nd edn. (Butterworth-Heinemann, Oxford, 1996)
134. D. De, S. Bhattacharya, S.M. Adhikari, A. Kumar, P.K. Bose, K.P. Ghatak, Beilstein J. Nanotech. **2**, 339 (2012); D. De, Kumar, S.M. Adhikari, S. Pahari, N. Islam, P. Banerjee, S.K. Biswas, S. Bhattacharya, K.P. Ghatak, Superlatti. Microstruct. **47**,377 (2010); K.P. Ghatak, M. Mondal, S.N. Biswas, J. Appl. Phys. **68**, 3032 (1990); M. Mondal, S. Banik, K.P. Ghatak, J. Low Temp. Phys. **74,** 423 (1989); B. Mitra, A. Ghoshal, K. P. Ghatak, Phys. Stat. Sol. (b)**150**, K67 (1988); K.P. Ghatak, D. Bhattacharyya, B. Nag, S.N. Biswas, J. Nonlin. Optics and Quant. Optics **13**, 267 (1995); L. Torres, L. Lopez-Diaz, J. Iniguez, Appl. Phys. Lett. **73**, 3766, (1996); A.Y. Toporov, R.M. Langford, A.K. Petford-Long, Appl. Phys. Lett. **77**, 3063 (2000); L. Torres, L. Lopez-Diaz, O. Alejos, J. Iniguez, J. Appl. Phys. **85**, 6208 (1999); S. Tiwari, S. Tiwari, Cryst. Res. Technol. **41**, 78 (2006); R. Houdré, C. Hermann, G. Lampel, P.M. Frijlink, A.C. Gossard, Phys. Rev. Lett. **55**, 734 (1985)
135. S.M. Adhikari, K.P. Ghatak, J. Adv. Phys. **2**, 130 (2013); D. De, S. Bhattacharaya, S. Ghosh, K.P. Ghatak, Adv. Sci., Eng. Med. **4**, 211 (2012); A. Kumar, S. Chowdhury, S.M. Adhikari, S. Ghosh, M. Mitra, D. De, A. Sharma, S. Bhattacharya, A. Dey, K.P. Ghatak, J. Comput. Theor. Nanosci. **7**, 115 (2010); R. Houdré, C. Hermann, G. Lampel, P.M. Frijlink, Surface Sci. **168**, 538 (1986); T.C. Chiang, R. Ludeke, D.E. Eastman, Phys. Rev. **B. 25,** 6518 (1982); S.P. Svensson, J. Kanski, T.G. Andersson, P.O. Nilsson, J. Vacuum Sci. Technol. B **2,** 235 (1984); S.F. Alvarado, F. Ciccacci, M. Campagna, Appl. Phys. Letts. **39**, 615 (1981); L. Fleming, M.D. Ulrich, K. Efimenko, J. Genzer, A.S.Y. Chan, T.E. Madey, S.J. Oh, O. Zhou, J.E. Rowe, Vac. Sci. Technol. B **22**, 2000 (2004); H. Shimoda, B. Gao, X.-P. Tang, A. Kleinhammes, L. Fleming, Y. Wu, O. Zhou, Phys. Rev. Lett. **88**, 015502 (2002); M. Maillard, P. Monchicourt, M.P. Pileni, Chem. Phys. Lett. **380**, 704 (2003)
136. B. Mitra, K.P. Ghatak, Phys. Scrip. **40**, 776 (1989); D. De, S. Bhattacharya, K.P. Ghatak, International work shop in Physics of Semiconductor Devices, IEEE 897 (2007); S. Bhattacharya, D. De, S. Chowdhury, S. Karmakar, D.K. Basu, S. Pahari, K.P. Ghatak, J. Comput. Theor. Nanosci **3**, 280 (2006)
137. I. Matsuda, S. Hasegawa, A. Konchenko, Y. Nakayama, Y. Nakamura, M. Ichikawa, Phys. Rev. B, **73**, 113311 (2006); V.L. Colvin, A.P. Alivisatos, J.G. Tobin, Phys. Rev. Lett. **66**, 2786 (1991); B. Schroeter, K. Komlev, W. Richter, Mat. Sci. Eng. **B88**, 259 (2002); G.F. Bertsch, N. Van Giai, N. Vinh Mau, Phys. Rev. A **61**, 033202 (2000)
138. K.P. Ghatak, S.N. Biswas, Nonlinear Opt. **4**, 39 (1993)
139. K.P. Ghatak, S.N. Biswas, *SPIE, Growth and Characterization of Materials for Infrared Detectors and Nonlinear Optical Switches,* vol. 1484 (USA, 1991) p. 136
140. K.P. Ghatak, B. De, *Polymeric materials for Integrated Optics and Information Storage, Materials Research Society (MRS) Symposium Proceedings*, MRS Spring Meeting, 1991, vol. 228, p. 237; L.J. Heyderman, F. Nolting, Quitmann, Appl. Phys. Letts. **83**, 1797 (2003); R.P. Cowburn, A. O. Adeyeye, J.A.C. Bland, Appl. Phys. Lett. **70**, 2309 (1997)
141. K.P. Ghatak, B. Nag, G. Majumdar, *Strained Layer Expitaxy—Materials, Processing, and Device Application*s. MRS Symposium Proceedings, MRS Spring Meeting, 1995, vol. 379, p.85
142. K.P. Ghatak, SPIE, High Speed Phenomena in Photonic Materials and Optical Bistability (USA, 1990), vol. 1280, p. 53
143. K.P. Ghatak, *Long Wave Length Semiconductor devices.* Materials and Processes Symposium Proceedings, MRS Symposium Proceedings (MRS Spring Meeting, 1990), vol. 216, p. 469
144. K.P. Ghatak, A. Ghoshal, S. Bhattacharyya, SPIE, Nonlinear Optical Materials and Devices for Photonic Switching (USA, 1990), vol. 1216, p. 282

145. K.P. Ghatak, SPIE, Nonlinear Optics III (USA, 1992), vol. 1626, p. 115
146. K.P. Ghatak, A. Ghoshal, B. De, SPIE, Optoelectronic Devices and Applications (USA, 1990), vol. 1338, p. 111
147. K.P. Ghatak, S.N. Biswas, in *Proceedings of the Society of Photo-optical and Instrumentation Engineers (SPIE),* Nonlinear Optics II, (USA, 1991), vol. 1409, p. 28; K. P. Ghatak, SPIE, Process Module Metrology (USA, 1992), vol. 1594, p. 110; K.P. Ghatak, SPIE, International Conference on the Application and Theory of Periodic Structures, 1991, vol. 1545, p. 282
148. K.P. Ghatak, M. Mondal, Solid State Electron. **31**, 1561 (1988)
149. K.P. Ghatak, M. Mondal, J. Appl. Phys. **69**, 1666 (1991)
150. C. Majumdar, A.B. Maity, A.N. Chakravarti, Phys. Stat. Sol. (b) **140**, K7 (1987)
151. C. Majumdar, A.B. Maity, A.N. Chakravarti, Phys. Stat. Sol. (b) **141**, K35 (1987)
152. N.R. Das, K.K. Ghosh, D. Ghoshal, Phys. Stat. Sol. (b) **197**, 97 (1996)
153. C. Majumdar, A.B. Maity, A.N. Chakravarti, Phys. Stat. Sol. (b) **144**, K13 (1987)
154. N.R. Das, A.N. Chakravarti, Phys. Stat. Sol. (b) **176**, 335 (1993)
155. S. Sen, N.R. Das, A.N. Chakravarti, J. Phys: Conden. Mat. **19**, 186205 (2007); N.R. Das, S. Ghosh, A.N. Chakravarti, Phys. Stat. Sol. (b) **174**, 45 (1992)
156. A.B. Maity, C. Majumdar, A.N. Chakravarti, Phys. Stat. Sol. (b) **144**, K93 (1987)
157. A.B. Maity, C. Majumdar, A.N. Chakravarti, Phys. Stat. Sol. (b) **149**, 565 (1988)
158. A.V.D. Ziel, *Solid State Physical Electronics* (Prentice Hall, Inc. Eaglewood Cliffs, 1957); A. Modinos, *Field, Thermionic and Secondary Electron Emission Spectroscopy* (Plenum Press, New York, 1984)
159. L.J. Singh, S. Choudhary, A. Mallik, K.P. Ghatak, J. Comput. Theor. Nanosci. **2**(2), 287 (2005)
160. O. Madelung, Physics of III-V compounds (John Wiley and Sons, Inc. New York, 1966)
161. B. Mitra, A. Ghoshal, K.P. Ghatak, Physica Status Solidi (b), **155**, K23 (1989)
162. J.A. Woollam, Phys. Rev. **3**, 1148 (1971)
163. V.V. Kaminoskii, N.N. Stepanav, I.M. Smirnov, Sov. Phys. Solid State **27**, 1295 (1985)
164. P.I. Baranskii, V.V. Kolomoets, S.S. Korolyuk, Phys. Stat Sol. (b) **116**, K109 (1983)
165. K.P. Ghatak, S. Biswas, J. Vac. Sci. Technol. **B7** (1), 104 (1989); D.R. Choudhury, A.K. Chowdhury, A.N. Chakravarti, Physica Scripta, **22**, 656 (1981); D.R. Choudhury, A.K. Chowdhury, A.N. Chakravarti, Czech. J. Phys. B **30** (1980); D.R. Choudhury, A.K. Chowdhury, A.N. Chakravarti, B.R. Nag, Phys. Stst. Sol. (a), **58**, K51 (1980); B. Mitra, A. Ghoshal, K.P. Ghatak, Phys. Stat. Sol. (b), **153**, K209 (1989); M. Mondal, K.P. Ghatak, Phys. Stat. Sol. (a), 93, 377 (1986); D.R. Choudhury, A.K. Chowdhury, A.N. Chakravarti, Appl. Phys. 22, 145 (1980)

Chapter 21
Conclusion and Scope for Future Research

The most wasted of all days is one without surrendering to GOD.

This monograph deals with the DRs in various types of HD materials and their quantized counter parts. The external photo excitation, quantization and strong electric field alter profoundly the basic band structures, which, in turn, generate pinpointed knowledge regarding DR in various HDS and their nanostructures. The in-depth experimental investigations covering the whole spectrum of solid state and allied science in general, are extremely important to uncover the underlying physics and the related mathematics in this particular aspect. We have formulated the simplified expressions of DR for few HD quantized structures together with the fact that our investigations are based on the simplified $\boldsymbol{k} \cdot \boldsymbol{p}$ formalism of solid-state science without incorporating the advanced field theoretic techniques. In spite of such constraints, the role of band structure, which generates, in turn, new concepts are truly amazing and discussed throughout the text.

We present the last bouquet of open research problem in this pin pointed topic of research of modern physics.

(R.21.1) Investigate the DR in the presence of a quantizing magnetic field under exponential, Kane, Halperin, Lax and Bonch-Bruevich band tails [1] for all the problems of this monograph of all the HD materials whose unperturbed carrier energy spectra are defined in Chap. 1 by including spin and broadening effects.

(R.21.2) Investigate all the appropriate problems after proper modifications introducing new theoretical formalisms for the problems as defined in (R21.1) for HD negative refractive index, macro molecular, nitride and organic materials.

(R.21.3) Investigate all the appropriate problems of this monograph for all types of HD quantum confined p-InSb, p-CuCl and semiconductors having diamond structure valence bands whose dispersion relations of the carriers in bulk materials are given by Cunningham [2], Yekimov et al. [3] and Roman et al. [4] respectively.

K.P. Ghatak, *Dispersion Relations in Heavily-Doped Nanostructures*,
Springer Tracts in Modern Physics 265, DOI 10.1007/978-3-319-21000-1_21

(R.21.4) Investigate the influence of defect traps and surface states separately on the DR of the HD materials for all the appropriate problems of all the chapters after proper modifications.

(R.21.5) Investigate the DR of the HD materials under the condition of non-equilibrium of the carrier states for all the appropriate problems of this monograph.

(R.21.6) Investigate the DR for all the appropriate problems of this monograph for the corresponding HD p-type semiconductors and their nanostructures.

(R.21.7) Investigate the DR for all the appropriate problems of this monograph for all types of HD semiconductors and their nanostructures under mixed conduction in the presence of strain.

(R.21.8) Investigate the DR for all the appropriate problems of this monograph for all types of HD semiconductors and their nanostructures in the presence of hot electron effects.

(R.21.9) Investigate the DR for all the appropriate problems of this monograph for all types of HD semiconductors and their nanostructures for nonlinear charge transport.

(R.21.10) Investigate the DR for all the appropriate problems of this monograph for all types of HD semiconductors and their nanostructures in the presence of strain in an arbitrary direction.

(R.21.11) Investigate all the appropriate problems of this monograph for strongly correlated electronic. Strongly correlated electronic HD systems in the presence of strain.

(R.21.12) Investigate all the appropriate problems of this chapter in the presence of arbitrarily oriented photon field and strain.

(R.21.13) Investigate all the appropriate problems of this monograph for all types of HD nanotubes in the presence of strain.

(R.21.14) Investigate all the appropriate problems of this monograph for HD Bi_2Te_3-Sb_2Te_3 super-lattices in the presence of strain.

(R.21.15) Investigate the influence of the localization of carriers on the DR in HDS for all the appropriate problems of this monograph in the presence of crossed fields.

(R.21.16) Investigate DR for HD p-type SiGe under different appropriate physical conditions as discussed in this monograph in the presence of strain and crossed fields.

(R.21.17) Investigate DR for HD GaN under different appropriate physical conditions as discussed in this monograph in the presence of strain and crossed fields.

(R.21.18) Investigate DR for different disordered HD conductors under different appropriate physical conditions as discussed in this monograph in the presence of strain and crossed fields.

(R.21.19) Investigate all the appropriate problems of this monograph for HD $Bi_2Te_{3-x}Se_x$ and $Bi_{2-x}Sb_xTe_3$ respectively in the presence of strain and crossed fields.

(R.21.20) Investigate all the appropriate problems of this monograph in the presence of crossed electric and alternating quantizing magnetic fields.

(R.21.21) Investigate all the appropriate problems of this monograph in the presence of crossed alternating electric and quantizing magnetic fields.

(R.21.22) Investigate all the appropriate problems of this monograph in the presence of crossed alternating non uniform electric and alternating quantizing magnetic fields.

(R21.23) Investigate all the appropriate problems of this monograph in the presence of alternating crossed electric and alternating quantizing magnetic fields.

(R.21.24) Investigate all the appropriate problems of this monograph in the presence of arbitrarily oriented pulsed electric and quantizing magnetic fields.

(R.21.25) Investigate all the appropriate problems of this monograph in the presence of arbitrarily oriented alternating electric and quantizing magnetic fields.

(R.21.26) Investigate all the appropriate problems of this monograph in the presence of crossed in homogeneous electric and alternating quantizing magnetic fields.

(R.21.27) Investigate all the appropriate problems of this monograph in the presence of arbitrarily oriented electric and alternating quantizing magnetic fields under strain.

(R.21.28) Investigate all the appropriate problems of this monograph in the presence of arbitrarily oriented electric and alternating quantizing magnetic fields under light waves.

(R.21.29)
(a) Investigate the DR for all types of HD materials of this monograph in the presence of many body effects, strain and arbitrarily oriented alternating light waves respectively.
(b) Investigate all the appropriate problems of this chapter for the Dirac electron.
(c) Investigate all the problems of this monograph by removing all the physical and mathematical approximations and establishing the respective appropriate uniqueness conditions.

The formulation of DR for all types of HD materials and their quantum confined counter parts considering the influence of all the bands created due to all types of quantizations after removing all the assumptions and establishing the respective appropriate uniqueness conditions is, in general, an extremely difficult problem. 200 open research problems have been presented in this monograph and we hope that the readers will not only solve them but also will generate new concepts, both theoretical and experimental. Incidentally, we can easily infer how little is presented and how much more is yet to be investigated in this exciting topic which is the signature of coexistence of new physics, advanced mathematics combined with the inner fire for performing creative researches in this context from the young scientists since like Kikoin [5] we firmly believe that "A young scientist is no good if his teacher learns nothing from him and gives his teacher nothing to be proud of". In the mean time our research interest has been shifted and we are leaving this particular beautiful topic with the hope that (R.21.29) alone is sufficient to draw the attention of the researchers from diverse fields and our readers are surely in tune with the fact that "Exposition, criticism, appreciation is the work for second-rate minds" [6].

References

1. B.R. Nag, *Electron Transport in Compound Semiconductors. Springer Series in Solid State Sciences*, vol. 11 (Springer-Verlag, Germany, 1980)
2. R.W. Cunningham, Phys. Rev. **167**, 761 (1968)
3. A.I. Yekimov, A.A. Onushchenko, A.G. Plyukhin, Al, L. Efros, J. Expt. Theor. Phys. **88**, 1490 (1985)
4. B.J. Roman, A.W. Ewald, Phys. Rev. B **5**, 3914 (1972)
5. I.K. Kikoin, *Science for Everyone: Encounters with Physicists and Physics* (Mir Publishers, Russia, 1989), p. 154
6. G.H. Hardy, *A mathematician's Apology* (Cambridge University Press, 1990), p. 61

Index

K.P. Ghatak, *Dispersion Relations in Heavily-Doped Nanostructures*,
Springer Tracts in Modern Physics 265, DOI 10.1007/978-3-319-21000-1

MIX
Papier aus verantwortungsvollen Quellen
Paper from responsible sources
FSC® C105338

If you have any concerns about our products,
you can contact us on
ProductSafety@springernature.com

In case Publisher is established outside the EU,
the EU authorized representative is:
Springer Nature Customer Service Center GmbH
Europaplatz 3, 69115 Heidelberg, Germany

Printed by Libri Plureos GmbH
in Hamburg, Germany